Magnetospheric MHD Oscillations

Magnetospheric MHD Oscillations

A Linear Theory

Anatoly Leonovich
Vitalii Mazur
Dmitri Klimushkin

WILEY-VCH

Authors

Dr. Anatoly Leonovich
Inst. of Solar-Terrestrial Physics
Lermontova st. 126a
Irkutsk
Russia, 664033

Prof. Vitalii Mazur
Inst. of Solar-Terrestrial Physics
Lermontova st. 126a
Irkutsk
Russia, 664033

Dr. Dmitri Klimushkin
Inst. of Solar-Terrestrial Physics
Lermontova st. 126a
Irkutsk
Russia, 664033

Library of Congress Card No.: applied for

British Library Cataloguing-in-Publication Data
A catalogue record for this book is available from the British Library.

Bibliographic information published by the Deutsche Nationalbibliothek
The Deutsche Nationalbibliothek lists this publication in the Deutsche Nationalbibliografie; detailed bibliographic data are available on the Internet at <http://dnb.d-nb.de>.

Print ISBN: 978-3-527-41430-7
ePDF ISBN: 978-3-527-84574-3
ePub ISBN: 978-3-527-84575-0
oBook ISBN: 978-3-527-84573-6

Cover Image: © MARK GARLICK/SCIENCE PHOTO LIBRARY/Getty Images

Typesetting Straive, Chennai, India
Printing and Binding CPI Group (UK) Ltd, Croydon, CR0 4YY

C9783527414307_150324

Contents

List of Figures *xi*
List of Tables *xxv*
Author Biography *xxvii*
Preface *xxix*
Acknowledgements *xxxi*
Acronyms *xxxiii*
Symbols *xxxv*
Introduction *xxxix*

1 Hydromagnetic Oscillations in Homogeneous Plasma *1*

2 MHD Oscillations in 1D-Inhomogeneous Model Magnetosphere *10*
2.1 A Qualitative Picture of MHD Wave Propagation in a 1D-Inhomogeneous Plasma *12*
2.2 Model of a Smooth Transition Layer and Basic Equations for MHD Oscillations *15*
2.3 FMS Wave Reflected from the Transition Layer in a Cold Plasma. Alfvén Resonance *16*
2.4 Alfvén Resonance Excited by a Wave Impulse *20*
2.5 Energy Balance in the Problem of an Incident FMS Wave Reflected from the Transition Layer Containing an Alfvén Resonance Point *25*
2.6 FMS Wave Reflected from the Transition Layer in a 'warm' Plasma. Alfvén and Magnetosonic Resonances *30*
2.7 Alfvén Resonance in Non-ideal Plasma. Kinetic Alfvén Waves *37*
2.8 FMS Waveguide *43*
2.9 Waveguide for Quasilongitudinal Alfvén Waves *47*
2.10 Waveguides for Kinetic Alfvén Waves in a 'cold' Plasma. Waveguide Mode Attenuation *49*
2.11 Waveguide for Kinetic Alfvén and FMS Waves in a 'warm' Plasma. Waveguide Mode Resonance *51*
2.12 Waveguides in Plasma Filaments *54*
2.12.1 Axisymmetric Plasma Waveguide for Quasi-Longitudinal Alfvén Waves *56*
2.12.2 Axisymmetric Plasma Waveguide for Kinetic Alfvén and FMS Waves *58*
2.12.3 Waveguide Propagation of Geomagnetic Pulsations in the Outer Magnetosphere *60*

2.13 FMS Wave Passing Through a Tangential Discontinuity *62*
2.13.1 Model Medium and Matching MHD Equation Solutions *63*
2.13.2 Energy Flux Transferred by FMS Waves Through the Magnetopause *67*
2.14 Unstable MHD Shear Flows in the Presence/Absence of Boundary Walls *70*
2.14.1 Model of Medium and Basic Equations *71*
2.14.2 Types of Boundary Conditions *73*
2.14.3 Unstable Shear Flows in a Boundless Medium *74*
2.14.3.1 The $k_t \parallel B_0$ Case *75*
2.14.3.2 The $k_t \perp B_0$ Case *77*
2.14.4 Instability of the Shear Flow Bounded by One Rigid Wall *77*
2.14.4.1 The $k_t \parallel B_0$ Case *77*
2.14.4.2 The $k_t \perp B_0$ Case *79*
2.14.5 Shear Flow Instability Between Two Boundary Walls *81*
2.14.5.1 The $k_t \parallel B_0$ Case *81*
2.14.5.2 The $k_t \perp B_0$ Case *82*
2.15 Geotail Instability Due to Shear Flow at the Magnetopause *83*
2.15.1 Model Medium and Basic Equations *84*
2.15.2 Calculating the Magnetopause MHD Instability Growth Rate in the Tangential Discontinuity Model *88*
2.15.3 Geotail Instability in a Smooth-Boundary Model *94*
2.15.4 K-H Instability of Global Modes in the Geotail *95*
2.16 Kelvin–Helmholtz Instability in the Geotail Low-Latitude Boundary Layer *98*
2.16.1 Cylindrical Model of the Geotail Near LLBL *99*
2.16.2 Basic Equation and Boundary Conditions for MHD Oscillations in the Cylindrical Coordinate System *103*
2.16.3 Numerical Solution of the Basic Equation and Discussion *105*
2.17 Cherenkov Radiation of the Fast Magnetoacoustic Waves *111*
2.17.1 The Single Fourier-Harmonic *112*
2.17.2 Summation of the Fourier-Harmonics *113*
2.18 MHD Oscillation Field Penetrating from the Magnetosphere to Ground *115*
2.18.1 Boundary Conditions for MHD Waves at the Upper Ionospheric Boundary in a 'Thin Layer' Model with a Vertical Magnetic Field *117*
2.18.2 Alfvén Waves Penetrating to Ground from the Magnetosphere in a Model Geospace with an Inclined Magnetic Field *120*
2.18.2.1 Low-Frequency Electromagnetic Oscillation Field in the Ground and Atmosphere *122*
2.18.2.2 Low-Frequency Electromagnetic Oscillation Field in the Ionosphere *122*
2.18.2.3 Boundary Conditions for Alfvén Waves at the Upper Boundary of the Ionosphere *128*
2.18.2.4 Electromagnetic Oscillations Induced on the Earth Surface by Magnetospheric Alfvén Waves *130*

3 MHD Oscillations in 2D-Inhomogeneous Models *133*
3.1 Resonance Between FMS and Kinetic Alfvén Waves in a Dipole-Like Magnetosphere *137*
3.1.1 Longitudinal Structure of Toroidal Alfvén Waves *141*

3.1.2 Structure of Resonant Kinetic Alfvén Waves Across Magnetic Shells *143*
3.1.3 Feedback from Resonant Alfvén Oscillations to FMS Wave Field *146*
3.2 Alfvén Resonance in a Dipole-Like Magnetosphere *146*
3.2.1 Model of the Medium and Basic Equations *147*
3.2.2 Resonant Alfvén Wave Field Structure *149*
3.2.3 Field Structure of Monochromatic FMS Oscillations in a Dipole-Like Magnetosphere *151*
3.2.4 MHD Oscillation Magnetic Field Amplitude Distribution in the Meridional Plane *154*
3.3 Resonant Alfvén Waves Excited in a Dipole-Like Magnetosphere by Broadband Sources *158*
3.3.1 Monochromatic Source of FMS Waves *159*
3.3.2 Pulse Source of FMS Waves *160*
3.3.3 FMS Wave Source in the Form of a Wave Packet (Substorm Pi2 Model) *161*
3.3.4 Stochastic Source of FMS Waves (Dayside Pc3 Model) *162*
3.4 Magnetosonic Resonance in a Dipole-Like Magnetosphere *164*
3.4.1 Self-Consistent Model of a Dipole Magnetosphere with Rotating Plasma *164*
3.4.2 Basic Equations for Magnetosonic Waves *169*
3.4.3 Structure of Standing SMS Waves Along Magnetic Field Lines *171*
3.4.4 Structure of Resonant SMS Oscillations Across Magnetic Shells *172*
3.4.5 The Field Component Structure of Resonant SMS Oscillations Near the Resonance Surface *174*
3.4.6 Numerical Solutions of Equations for Resonant SMS Waves *175*
3.5 FMS Oscillations in a Dipole-Like Magnetosphere *178*
3.5.1 Longitudinal Structure of FMS Oscillations *178*
3.5.2 FMS Oscillation Structure Across Magnetic Shells *184*
3.6 FMS Resonators in Earth's Magnetosphere *185*
3.6.1 Qualitative Proof that FMS Resonators Exist in the Magnetosphere *185*
3.6.2 FMS Resonators in the Dayside Magnetosphere *190*
3.6.3 FMS Resonator in the Near-Earth Plasma Sheet *193*
3.6.3.1 Model of the Medium *194*
3.6.3.2 Coordinate System and Basic Equations *195*
3.7 Monochromatic Transverse-Small-Scale Alfvén Waves with $m \gg 1$ in a Dipole-Like Magnetosphere *200*
3.7.1 Formulating the Problem of the Alfvén Oscillation Structure in the WKB Approximation *201*
3.7.2 Qualitative Investigation of the Eigenvalue Problem *203*
3.7.3 Structure of High-m Alfvén Waves Along Magnetic Field Lines *208*
3.7.4 Dissipation of Standing Alfvén Waves in the Ionosphere *213*
3.7.5 Amplitude Distribution of High-m Alfvén Oscillations Across Magnetic Shells *215*
3.7.6 Solution Near the Poloidal Resonance Surface *218*
3.7.7 Solution Near the Toroidal Resonance Surface *220*
3.7.8 Global Structure of High-m Alfvén Wave (Matching the Solutions for Different Regions) *222*

3.8 Electromagnetic Oscillations Induced at Earth Surface by Magnetospheric Standing High-m Alfvén Waves *226*
3.9 Linear Transformation of Standing High-m Alfvén Waves Near the Toroidal Resonance Surface *232*
3.10 Magnetospheric Resonator for Standing High-m Alfvén Waves *238*
3.11 High-m Alfvén Waves Generated in the Magnetosphere by Stochastic Sources *241*
3.11.1 Expressions for Physical Components of Alfvén Oscillation Magnetic Field *241*
3.11.2 Statistical Properties of the Oscillation Source *244*
3.11.3 Spectral and Polarisation Properties of Alfvén Noise *246*
3.12 Broadband Standing High-m Alfvén Waves Generated by Correlated Sources *251*
3.12.1 Response of Magnetospheric Alfvén Oscillations to Instantaneous Pulse *253*
3.12.2 Conclusions vs. Observations *258*
3.13 Model Equation to Determine the Transverse Structure of Standing Alfvén Waves in the Magnetosphere *260*
3.13.1 Deriving the Homogeneous Model Equation (in the Absence of an Oscillation Source) *261*
3.13.2 Inhomogeneous Model Equation *264*
3.13.3 Analytical Solution of the Model Equation *267*
3.13.4 Numerical Investigation of the Model Equation Solutions *272*
3.14 Spatial Structure of Alfvén Oscillations Excited in the Magnetosphere by Localised Monochromatic Source *275*
3.14.1 Structure of Monochromatic Alfvén Oscillations from a Source Localised Across Magnetic Field Lines *275*
3.14.2 Transverse Structure of Standing Alfvén Waves from a Source Strongly Localised in One of the Transverse Coordinates *276*
3.14.3 Transverse Structure of Standing Alfvén Waves from a Source Localised in Two Transverse Coordinates *278*
3.14.4 On the Methods of Measuring the Polarisation Splitting of the Alfvén Oscillations *280*
3.15 High-m Alfvén Oscillations Generated in the Magnetosphere by Localised Pulse Sources *282*
3.15.1 From Monochromatic to Broadband Oscillations *283*
3.15.2 Initial Oscillation Regime ($\tau_N \ll 1$) *285*
3.15.3 Asymptotic Regime of Oscillations ($\tau_N \gg 1$) *286*
3.15.4 Model Plasmasphere and Equations for the Field Components of Standing Alfvén Waves *292*
3.15.5 Calculating Alfvén Oscillation Field in the MASSA Experiment *294*
3.16 Ballooning Instability of Alfvén and SMS Oscillations on Field Lines Crossing the Current Sheet *298*
3.16.1 Equation for Ballooning Modes *299*
3.16.2 Model of the Medium *302*
3.16.3 The Ballooning Instability of MHD Oscillations as Studied in the Local Approximation *306*

3.16.4 Calculating the Structure and Spectrum of Standing Alfvén and SMS Waves on Elongated Field Lines, in the WKB Approximation *309*
3.17 Coupled Alfvén and SMS Oscillation Modes in the Geotail *315*
3.17.1 Coupled Mode Structure Along Magnetic Field Lines *315*
3.17.2 Linear Transformation of Alfvén and SMS Waves in the Current Sheet *320*
3.17.3 Coupled MHD Mode Structure Across Magnetic Shells *324*

4 MHD Oscillations in 3D-Inhomogeneous Models of the Magnetosphere *329*
4.1 MHD Oscillation Properties in Non-homogeneous Models of the Magnetosphere of Different Dimension *329*
4.2 Coordinate System *330*
4.3 Basic Equations *331*
4.4 Qualitative Investigation of the Equation for Characteristics *334*
4.5 Wave Singularity in the 3D-Inhomogeneous Magnetosphere *337*

5 Conclusion *341*

Appendixes *349*
A Transverse Dispersion of MHD Waves in a 'Cold' and 'Hot' Plasma *349*
B Deriving an Equation for MHD Oscillations in a 1D-Inhomogeneous Moving Plasma *349*
C A Model of the Spectral Function of the Solar Wind FMS Oscillations *351*
D Stability of MHD Oscillations with $k_t \parallel B_0$, in the Shear Layer, for $\beta^* < 1$ in a Boundless Medium *353*
E Deriving Equations for Potentials φ and ψ for MHD Waves in a 'Warm' Plasma, in a Curvilinear Orthogonal System of Coordinates (x^1, x^2, x^3) *353*
F WKB Solution of the Longitudinal Problem for FMS Waves Having Two Turning Points on the Field Line *355*
G Integrals of Functions $G(z)$ and $g(z)$ Describing the Transverse Structure of Standing Alfvén Waves *357*
H Parameters of the Polarisation Ellipse of Stochastic Oscillations *357*
I Deriving Coefficients of the Differential Equation Based on the Given WKB Solution *358*
J Strictly Deriving a Transverse Model Equation for Standing Alfvén Waves, for the $\kappa \ll 1$ Case *359*
K Calculating Characteristics η_0 *362*
L Calculating the Integral (3.390) Near the Characteristics $\pm\eta_1, \pm\eta_4$ and η_0 *362*
M Calculating the Integral (3.388) Near the Characteristics $\pm\eta_2, \pm\eta_3, \pm\eta_5$ and $\pm\eta_6$ *363*
N Determining the Shape of a Field Line from Given Components of Background Magnetic Field *364*
O Defining Tri-orthogonal System of Coordinates Related to Magnetic-Field Lines *365*
P Determining Metric Tensor Components in a Curvilinear Orthogonal System of Coordinates *366*

Q Coefficients of the Equation for the Coupled Modes of MHD Oscillations 367
R Equation for MHD Oscillations in a Cylindrical Coordinate System 368
S Equality of the Alfvén Oscillation Specific Power Absorbed Near the Resonance Surface and the Density of Energy Carried Away by KAWs 369

References 372
Index 401

List of Figures

Figure 1.1 Friedrichs diagrams for the Alfvén (black), slow (light gray) and fast (dark gray) MHD modes for case when the Alfvén speed v_A is greater than the speed of sound v_s: (a) for the phase velocity v_{ph} and (b) for the group velocity v_g. Here indices || and $\perp$ mean projections on the direction parallel and perpendicular to the ambient magnetic field, respectively *3*

Figure 1.2 Schematic of magnetic field oscillations (field lines), plasma pressure (shades of gray) and group velocity directions for Alfvén (a), fast magnetosonic (b) and slow magnetosonic (c) waves propagating in a homogeneous plasma *5*

Figure 1.3 Relative SMS decrement $\bar{\varepsilon}_s = \gamma_s/\text{Re}\omega$ vs. non-isothermality, lg T_e/T_i, of homogeneous plasma *6*

Figure 1.4 Qualitative behaviour of the function $\Lambda^2(s_e/\rho_s)$ in the complex Λ^2 plane as the argument varies from $s_e/\rho_s \gg 1$ to $s_e/\rho_s \ll 1$ *8*

Figure 2.1 Model 1D-inhomogeneous plasma (with density $\rho_0 \equiv \rho_0(x)$) in a homogeneous magnetic field $\mathbf{B}_0 = const$ *12*

Figure 2.2 Parallel structure of electric E (black lines) and magnetic B (gray lines) field of the MHD wave. Solid and dashed lines depict the fundamental ($N = 1$) and second ($N = 2$) harmonics, respectively *13*

Figure 2.3 (a) Alfvén speed distribution along the x axis in a 1D-inhomogeneous model of the medium. (b) Dependence of the squared WKB component of the wave vector, $k_x^2(x)$, of MHD oscillations in a 'cold' plasma (x_A is the resonance point for Alfvén wave, x_F is the turning point for FMS wave), (c) The structure of wave field components of MHD oscillations in a 1D-inhomogeneous plasma *14*

Figure 2.4 Spatial distribution of MHD oscillation B_x-component (left-hand axis) in the FMS wave which is incident to/reflected from the transition layer with an Alfvén resonance point. The gray line is the full oscillation field. WKB approximation: line 1 is FMS wave incident onto the transition layer, line 2 is FMS wave reflected from the transition layer. Oscillation hodographs for various points x are shown above. The right-hand axis line is Alfvén speed distribution $v_A(x)$ *16*

Figure 2.5 The $k_y\Delta_A$ dependence of the absorption coefficient for FMS waves incident onto the transition layer that are partially absorbed at the Alfvén resonance point. Distributions for various $k_z\Delta_A = 0.1; 1; 10; \infty$ are shown. The black marginal curve, for $k_z\Delta_A \to \infty$, corresponds to an infinite layer with a linear $v_A(x)$ profile. The dark gray dash lines show analytical distributions for two limiting cases: (1) $k_y\Delta_A \gg 1$ and (2) $k_y\Delta_A \ll 1$, described by (2.21) and (2.23) *20*

Figure 2.6 Evolution of an FMS wave packet incident onto a smooth transition layer containing an Alfvén resonance point, $x = x_A$. The distribution of the Alfvén speed $v_A(x)$ (right coordinate axis) and the B_y component of the wave field (left coordinate axis) are shown. (a) Initial state of the unit-wide wave packet. (b) Field structure at the moment the wave packet is reflected from the transition layer. (c) Wave packet structure upon reflection from the transition layer *25*

Figure 2.7 Distributions of Alfvén speed $v_A(x)$, SMS wave speed $c_S(x)$ (black lines, left-hand axis) and the square of the WKB component of the wave vector, $k_x^2(x)$ (gray lines, right-hand axis), across the transition layer, in two limiting cases: (1) (solid gray line for k_x^2) the opaque region (x_{01}, x_{02}) is present for FMS waves, (2) (gray dashed line for k_x^2) the transparent region for SMS waves spreads from the resonance surface for SMS waves, $x = x_S$, to infinity *31*

Figure 2.8 Derivative $\partial\zeta/\partial x$ distribution in the problem of FMS wave incident on/reflected from the transition layer with resonance surfaces for Alfvén ($x = x_A$) and SMS waves ($x = x_S$). The gray line is the (numerically calculated) full oscillation field, (1) FMS wave incident on the transition layer, (2) FMS wave reflected from the transition layer (WKB approximation) *34*

Figure 2.9 Distribution of MHD oscillation magnetic field components (Re(B_x, B_y, B_z)) in the problem of FMS wave incident on/reflected from the transition layer with resonance surfaces for Alfvén ($x = x_A$) and SMS waves ($x = x_S$) *35*

Figure 2.10 Hodograph behaviour for resonant MHD oscillations in the neighbourhood of resonance surfaces, x_A and x_S, for various decrements of Alfvén ($\gamma/\omega = 0.1$) and SMS oscillations. The curves and circles with hodograph rotation directions labelled 1,2 and 3, correspond to three values of SMS oscillation decrement: $\gamma_S/\omega = 0.01$, $\gamma_S/\omega = 0.1$ and $\gamma_S/\omega = 1$ *36*

Figure 2.11 Absorption coefficient dependence for FMS waves incident on the transition layer with two resonance surfaces: for Alfvén and SMS waves. The distributions $D(k_y\tilde{\Delta})$ are shown for the $\beta^* = 0.3$ plasma layer (for waves with $k_z\Delta = 0.1;\ 1;\ 3.5$: curves 1, 2, 3 for $\overline{\gamma}_s = 0.01\omega$ and curves $1', 2', 3'$ for $\overline{\gamma}_s = \omega$) and the $\beta^* = 1$ plasma layer (for $k_z\Delta = 0.1;\ 1;\ 3.5$: curves 4, 5, 6 for $\gamma_s = 0.01\omega$ and curves $4', 5', 6'$ for $\overline{\gamma}_s = \omega$) *36*

Figure 2.12 Distribution of the real and imaginary parts of $G(x)$, a function of a real argument *42*

Figure 2.13 Structure of MHD oscillations across magnetic shells when FMS waves are in resonance with kinetic Alfvén waves. The FMS wave amplitude decreases exponentially into the opaque region. In a 'cold' plasma ($\beta \ll m_e/m_i$), a kinetic Alfvén wave is excited on the resonance magnetic shell $x = x_A$ and travels away leftwards, while in a 'warm' plasma ($\beta \gg m_e/m_i$), an Alfvén wave travels away rightwards from the resonance shell *43*

Figure 2.14 Potential in (2.79) for quasi-parallel MHD oscillations: (a) the form of potential $V(x)$ with Alfvén resonance points (x_A) for waveguide FMS oscillations ($\delta > 0$); (b) potential $V(x)$ for waveguide-travelling quasi-parallel Alfvén waves ($\delta < 0$) *45*

Figure 2.15 The structure of waveguide modes in the potential $V(x)$ decaying as the related MHD waves escape. Top: waveguide FMS mode and related escaping kinetic Alfvén wave (A2) (see Figure A.1, (a) in Appendix A); bottom: waveguide mode of kinetic Alfvén waves that decays due to the FMS wave escaping from the waveguide *50*

Figure 2.16 Graphic solution of dispersion Eqs. (2.122), (2.123). Curves I correspond to the right side of (2.122), and curves II to the right side of (2.123) *53*

Figure 2.17 Alfvén speed distribution in the dayside magnetosphere (gray scale). v_A distribution is shown in the meridional section inside the plasmapause and in the plasma filament (duct for MHD waves) in the outer magnetosphere. Smaller v_A are shown by light gray. The white arrows indicate the direction of the FMS group velocity in the meridional section *55*

Figure 2.18 Schematic representation of a magnetospheric duct model in a cylindric system of coordinates (ρ, ϕ, z): (a) direction of outer magnetic field B_0 and plasma density distribution $\rho_0(\rho)$ in a plasma filament; (b) the form of the potential $V(\rho)$ in (2.128) *56*

Figure 2.19 Model medium and coordinate system. Roman numbers indicate the following regions: *I* – solar wind, *II* – magnetosphere. FMS waves are labelled as: *1* – incident on the magnetosphere, *2* – reflected from the magnetopause, *3* – penetrating into the magnetosphere *63*

Figure 2.20 The k_t dependence of function k_x^2 for fixed ω and θ. (a) – $k_x^2(k_t)$ in the magnetosphere, the points labelled: 1 – $\tilde{k}_t^{(1)}$, 2 – $\tilde{k}_t^{(2)}$, 3 – $k_t = |\omega/c_s \cos\theta|$. (b) – $k_x^2(k_t)$ in the solar wind, the points labelled: 1 – $k_t^{(1)}$, 2 – $k_t^{(3)}$, 3 – $\overline{k}_t^{(1)}$, 4 – $\overline{k}_t^{(2)}$, 5 – $k_t^{(4)}$, 6 – $k_t^{(2)}$ *65*

Figure 2.21 The characteristic form of the k_t dependencies of the relative densities of monochromatic FMS energy fluxes in the solar wind and in the magnetosphere, in the $0 \leq \theta \leq \theta^*$ sector. Energy flux densities are labelled as: 1 – FMS wave incident to the magnetopause; 2 – wave reflected from the magnetopause; 3 – wave penetrating into the magnetosphere *67*

Figure 2.22 Total energy W transferred into the magnetosphere by the 'geoeffective' FMS flux over time $\Delta t \approx 3 \cdot 10^3$s vs. the geotail growth rate index γ in the model Eq. (2.168) *70*

Figure 2.23 Model medium and coordinate system: a is the characteristic scale of the shear layer, $\pm\Delta$ is the location of possible boundaries in the form of rigid walls, $\mathbf{v}_0$, $\mathbf{B}_0$ are the unperturbed speed and background magnetic field vectors, $\mathbf{k}_t$ is the tangential wave vector of the oscillations *72*

Figure 2.24 Growth rate ($c_i = \text{Im}c$) isoline distribution for MHD oscillations with $\mathbf{k}_t \parallel \mathbf{B}_0$ generated by a shear flow in the form of a smooth transition layer in a boundless medium, for two values of β^*: (a) $\beta^* = 10$, (b) $\beta^* = 1.1$ *76*

Figure 2.25 Growth rate ($c_i = \text{Im}c$) isoline distribution for MHD oscillations with $\mathbf{k}_t \perp \mathbf{B}_0$ generated by a shear flow in the form of a smooth transition layer in a boundless medium *77*

Figure 2.26 Growth rate $c_i(M)$ distribution of MHD oscillations with $\mathbf{k}_t \parallel \mathbf{B}_0$ for a shear flow in the form of a tangential discontinuity bounded by a rigid wall on one side, for different values of parameters β^* and κ: (a) $\beta^* = 1$, plots 1–5 refer to $\kappa = 0.1, 0.5, 1, 5, 10$; (b) $\kappa = 1$, plots 1–5 refer to $\beta^* = \infty, 10, 1, 0.5, 0.2$ *78*

Figure 2.27 Growth rate c_i isoline distribution for MHD oscillations with $\mathbf{k}_t \parallel \mathbf{B}_0$ for a shear flow with a smooth transition layer bounded by a rigid wall ($\Delta = 20a$) on one side, for two different values of parameter β^*: (a) – $\beta^* = 10$, (b) – $\beta^* = 1$ *79*

Figure 2.28 Growth rate c_i isoline distribution for MHD oscillations with $\mathbf{k}_t \perp \mathbf{B}_0$ for a shear flow with a smooth transition layer bounded by a rigid wall on one side ($\Delta = -15a$). Thick lines correspond to the surface and radiative oscillation modes, thin lines to the oscillation mode reflected from the wall *80*

Figure 2.29 Dependence of growth rate $c_i(M)$ for MHD oscillations with $\mathbf{k}_t \parallel \mathbf{B}_0$ for a shear flow in the form of a tangential discontinuity between two rigid walls, for different values of parameters β^* and κ: (a) $\beta^* = 1$, plots *1–3* correspond to values $\kappa = 0.1, 1, 5$; (b) $\kappa = 1$, plots *1–4* correspond to values $\beta^* = \infty, 10, 1, 0.1$ *80*

Figure 2.30 Growth rate c_i isoline distribution for MHD oscillations with $\mathbf{k}_t \parallel \mathbf{B}_0$ for a shear flow in the form of a smooth transition layer between two rigid walls ($\Delta = \pm 10a$) for two different values of parameter β^*: (a) $\beta^* = 10$, (b) $\beta^* = 1$ *82*

Figure 2.31 Growth rate c_i isoline distribution for MHD oscillations with $\mathbf{k}_t \perp \mathbf{B}_0$ for a shear flow in the form of a smooth transition layer bounded by rigid walls on both sides ($\Delta = \pm 20a$). The thick lines correspond to the surface and radiative oscillation modes, and the thin lines to oscillation modes reflected from the walls *83*

Figure 2.32 (a) cylindrical model of the geotail wrapped around by the solar wind plasma flow and a schematic structure of the unstable “global mode” of the geotail. (b) Alfvén speed $v_A(\rho)$ and SMS speed $c_s(\rho)$ distributions in the geotail and the solar wind. On the resonance surfaces $\rho = \rho_s$ (points *2* and 3) and $\rho = \rho_A$ (points *1* and *4*), the parallel phase velocity $\overline{\omega}/k_z$ of the monochromatic wave coincides with, respectively, the local SMS speed c_s and the Alfvén speed v_A 85

Figure 2.33 Spatial structure of monochromatic MHD waves with $m = 1$ for two different values of the parallel phase speed $\overline{\omega}/k_z$: (a) oscillations with resonant surfaces for SMS waves in the geotail, $\overline{\omega}_A(\rho_s) = \overline{\omega}_s(\rho_s)$, (b) oscillations with no resonance surfaces in the geotail 88

Figure 2.34 Alfvén $v_A(\rho)$ and SMS wave speed distribution $c_s(\rho)$ (light and dark gray lines; the vertical axis is to the right) inside and outside the geotail in the plasma cylinder model. Squared wave-vector WKB component distribution over radius, $k_\rho^2(\rho)$ (black lines; the vertical axis is to the left, the thick dashed lines correspond to $m = 0$). Coordinates ρ_A and ρ_s correspond to resonance surfaces for Alfvén and SMS oscillations, $\rho_{00}, \rho_{01}, \rho_{02}$ are the turning points for magnetosonic waves. Numbers and shades of grey denote the transparent regions: *1* – for SMS waves, *2* – for FMS waves with $m \neq 0$, and *3* – for FMS waves with $m = 0$ 89

Figure 2.35 The Mach number M_A dependence of the frequency (Re (c), dark gray line) and growth rate (Im (c) light gray lines) for unstable oscillations driven at the geotail boundary. (a) WKB solution for the model with a boundary in the form of a tangential discontinuity, where $c_{01,02,03,04}$ are the roots of the dispersion equation $\tan(\Psi(c_{0n}) + \pi/4) = 0$ determining, in the WKB approximation, the FMS-waveguide eigenfrequencies (c_{0n}) in the geotail lobes. (b) solution for the model with a boundary in the form of a smooth transition layer of characteristic thickness $\Delta \equiv \Delta_\rho/\rho_m = 0,066$ for the same parameters as in panel (a) 93

Figure 2.36 Radial structure of unstable oscillations in the geotail for the azimuthal harmonic $m = 1$, normalised to the maximum value of $|d\zeta/d\xi|_{\max}$: (a) oscillations close to the second harmonic, $n = 2$, of the eigenmodes propagating in the FMS waveguide in the geotail lobes $(k_z\rho_m = 2)$, (b) 'global mode' oscillations for small values of $k_z\rho_m \to 0$ 95

Figure 2.37 The Mach number, M_A, dependence of the growth rate of the 'global' modes in the geotail for the first azimuthal harmonics $m = 0, 1, 2, 3$ and $k_z\rho_m = 0.1$ 96

Figure 2.38 The dependence of the growth rate $\gamma \equiv Im\omega$ of unstable azimuthal harmonics $m = 0$ and $m = 1$ on the global mode frequency $f = \mathrm{Re}\,(\omega)/2\pi$ for various speeds of the solar wind flow around the magnetosphere: *1* – $v_0 = 200$ km/s, *2* – $v_0 = 400$ km/s, *3* – $v_0 = 600$ km/s, *4* – $v_0 = 800$ km/s 98

Figure 2.39 The low-latitude boundary layer (LLBL): (a) schematic of the magnetic field lines and the electric current configuration in the geotail LLBL, (b) cylindrical model of the geotail enwrapped by a helical plasma flow. Here: I is magnetosphere, II is solar wind 99

Figure 2.40 Radial distribution of Alfvén speed v_A, sound speed v_s, SMS speed c_s (left vertical axis) and the β_*^{-1} parameter (right vertical axis) in the equilibrium geotail model for $v_{0z} = 400$ km/s and $S_m = 0.2$. Here (see Fig. 2.39(b)): **I** – magnetosphere, **II** – solar wind *102*

Figure 2.41 Radial structure of unstable MHD oscillations generated by the magnetopause shear flow: (a) – surface mode structure (1), radiative mode structure (2) ; (b) – structure of the first (1) and second (2) eigen-mode harmonics of the FMS waveguide in the geomagnetic tail. On both panels: solid lines are $\mathrm{Re}\,\zeta/(\mathrm{Re}\,\zeta)_{\max}$, dotted lines are $\mathrm{Im}\,\zeta/(\mathrm{Re}\,\zeta)_{\max}$ (in panel (a) for the surface mode only); black curves (3) are the squared wave number, $\mathrm{Re}\,k_\rho^2/(\mathrm{Re}\,k_\rho^2)_{\max}$ *106*

Figure 2.42 $k_z\Delta_m$ dependence of the oscillation growth rate $\mathrm{Im}\,(c)$ for the given azimuthal wave number $m = 1$ and the Mach number $M = 1.4$: (0) – for the surface mode, (1,2,3) – for three harmonics of the FMS waveguide in the geotail *108*

Figure 2.43 Growth rate ($\mathrm{Im}\,c$) isoline maps for the basic $m = 0$ and first ($m = 1$) azimuthal harmonics of MHD oscillations in the $(M, k_z\Delta_m)$ plane in the longitudinal solar wind flows (left-hand panels, helicity index $S_m = 0$) and in the strongly twisted flows (right-hand panels, $S_m = 1$). The bold line depicts the boundary $\mathrm{Im}\,(c) = 0$ separating the areas of unstable ($\mathrm{Im}\,(c) > 0$) and stable ($\mathrm{Im}\,(c) < 0$) oscillations *109*

Figure 2.44 Dependence of the growth rate of unstable MHD oscillations $\mathrm{Im}\,(\omega)$ on their frequency $f = \mathrm{Re}\omega/2\pi$ (for azimuthal harmonics $m = 0, 1, 2, 3$) for different velocities of the solar wind flowing around the magnetosphere: $v_{0m} = 200, 400, 800$ km/s (curves 1,2,3). Solid lines depict the oscillations generated by a longitudinal flow (the helicity index $S_m = 0$), dashed lines stand for the oscillations generated by a strongly twisted flow ($S_m = 1$) *110*

Figure 2.45 The box model of the magnetosphere *112*

Figure 2.46 The spatial structure of the fast mode wave field generated by the Cherenkov mechanism in the source reference frame (normalised amplitude). The parameters chosen in Eq. (2.264) are $l = 1, x_M = 0$. The Mach number on the magnetopause is chosen $M = 5$. The sum of the first 20 harmonics is depicted. The source is denoted by the circle on the top left, the grey line denotes the general reflection surface. The numbers denote different wave branches *114*

Figure 2.47 The section of the wave field along the azimuthal coordinate (exact solution at $x = -0.5$). The numbers denote wave branches, as in Figure 2.46 *114*

Figure 2.48 (a) typical height profiles of the conductivity tensor components $\hat{\sigma}$ and the Alfvén speed v_A. Roman numerals indicate the following layers: I – ground with isotropic conductivity σ_g, II – atmosphere with conductivity σ_a, III – lower ionosphere with transverse Pedersen σ_P and Hall conductivities σ_H and parallel conductivity $\sigma_\parallel$, IV – upper ionosphere, where $\sigma_P, \sigma_H \to 0$, V – magnetosphere. (b) model of near-Earth medium with a vertical magnetic field; Alfvén wave penetrating to Earth is shown schematically: 1 – incident wave from the magnetosphere, 2 – wave reflected from the ionosphere, 3 – field of the wave penetrating to ground *117*

Figure 2.49 Mutual positions of three systems of coordinates used in the problem of MHD wave field penetrating from the magnetosphere to Earth, for a geospace model with inclined geomagnetic field: (x, y, z), (τ, b, z) and (n, y, l). Roman numerals denote the following layers: I – ground with conductivity σ_g, II – atmosphere with conductivity σ_a, III – lower ionosphere (E-layer) and IV – upper ionosphere with anisotropic conductivities, V – magnetosphere *121*

Figure 3.1 Toroidal and poloidal oscillations of field lines. The fundamental ($N = 1$) and second ($N = 2$) harmonics are shown *134*

Figure 3.2 Coordinate systems related to geomagnetic field lines in an axisymmetric model magnetosphere: curvilinear orthogonal system of coordinates (x^1, x^2, x^3), curvilinear non-orthogonal system of coordinates (a, ϕ, θ) *138*

Figure 3.3 Equatorial dependence of Alfvén speed $v_A(L, 0)$ in the model under study and the corresponding dependence of the basic period $t_A(L)$ of magnetospheric Alfvén eigen oscillations on the magnetic shell parameter L *148*

Figure 3.4 Eigenfrequency distribution for the first seven harmonics of standing toroidal Alfvén waves across magnetic shells in a dipole model magnetosphere. Horizontal dashed lines indicate the frequencies of magnetosonic waves incident on the magnetosphere from the solar wind. Vertical dashed lines are conditional boundaries of the transition layer of the magnetopause *150*

Figure 3.5 The structure of monochromatic ($f = \omega/2\pi = 0.01$ Hz) FMS waves across magnetic shells for two first longitudinal harmonics ($n = 1,\ 2$), with (dashed lines) or without (solid lines) feedback from resonant Alfvén waves. The vertical dashed lines denote FMS turning points ($L = 6.1$ and $L = 9.2$) and the magnetopause ($L = 10$) *154*

Figure 3.6 Amplitude distributions of magnetic field components, B_{rf}, $B_{\phi f}$ and B_{lf}, across magnetic shells for monochromatic ($f = 0.01$ Hz) magnetosonic oscillations: (a) inside the magnetosphere, (b) in the solar wind region. Numbers (1) and (2) in the panels correspond to the two first longitudinal harmonics of FMS oscillations ($n = 1, 2$) *155*

Figure 3.7 Mean amplitude distribution for the field component $B_{\phi A}$ of the two first harmonics ($N = 1, 2$) of resonant Alfvén waves across magnetic shells. These harmonics are excited by a monochromatic ($f = 0.01$ Hz, $m = 1$) magnetosonic waves with $n = 1$ (lines 1,2) and $n = 2$ (lines 3,4). The amplitude distribution of the first harmonic ($N = 1$) is shown (line 1(10)) for comparison, as excited by magnetosonic wave with $m = 10$, $n = 1$ which practically does not penetrate into the magnetosphere *156*

Figure 3.8 Distribution of the $B_{\phi A}$ component (thick lines) of the field of resonant Alfvén oscillations excited by monochromatic ($f = 0.01$ Hz) FMS waves incident on the magnetosphere. The numbers indicate the first four harmonics of resonant Alfvén waves ($N = 1, 2, 3, 4$). The distribution of the parallel B_{lf} component (thick line) of the field of the second parallel harmonic ($n = 2$) of the FMS wave is given for comparison *157*

Figure 3.9 Amplitude distribution in the meridional plane for the full field of resonant Alfvén oscillations excited by monochromatic (frequency 0.01 Hz) magnetosonic wave in a dipole magnetosphere. Numbers 1, 2 denote the first ($N = 1$) and the second ($N = 2$) harmonics of standing Alfvén waves *158*

Figure 3.10 (a) FMS wave packet; (b–d) behaviour of the Nth harmonic of standing Alfvén waves excited by FMS wave packet: (b) exciting FMS wave frequency is higher than the frequency of Alfvén eigen oscillations, $\omega_0 \gg \Omega_{TN}$; (c) $|\omega_0 - \Omega_{TN}| \ll \Gamma$ – standing Alfvén waves and FMS oscillations are in resonance; (d) $\omega_0 \ll \Omega_{TN}$ *161*

Figure 3.11 Standing Alfvén waves excited in the magnetosphere by a stochastic source of FMS oscillations. Top left – the spectrum of stochastic FMS oscillations, $F_f(\omega)$. Top right – eigenfrequency Ω_{TN} distribution for three first harmonics ($N = 1, 2, 3$) of resonant Alfvén waves across magnetic shells ($L = a/R_E$ – McIlwain parameter). Bottom – distribution, across magnetic shells, of the oscillation amplitude m_N for the first three harmonics of the excited standing Alfvén waves and their envelope (dash-dotted line) *163*

Figure 3.12 Model axisymmetric magnetosphere and systems of coordinates: (x^1, x^2, x^3) is an orthogonal system of curvilinear coordinates related to magnetic field lines, (a, ϕ, θ) is a non-orthogonal system of curvilinear coordinates. Dashed lines indicate the plasmapause ($a = a_p$) and magnetopause ($a = a_m$) *165*

Figure 3.13 Distribution across magnetic shells of plasma equatorial velocity v_ϕ(curves 1 and 2), Alfvén speed $v_A(a, \theta)$ (curve 3 in the equatorial plane $\theta = 0$, curve 4 along radius $\mathbf{r}$ at angle $\theta = 30°$ to the equator) and the basic harmonic period t_A of magnetospheric standing Alfvén waves (curve 5) *168*

Figure 3.14 Parameter $\beta = 8\pi P_0(a, \theta)/B_0^2(a, \theta)$ isolines in the meridional plane (x, z). Coordinates $x = a\cos^3\theta$ and $z = a\cos^2\theta \sin\theta$ are in Earth radius units *169*

Figure 3.15 (a) Distribution across magnetic shells of the eigenfrequencies of the first three harmonics of standing SMS waves ($\Omega_{sN}/2\pi$) and the SMS transit time t_s along the field line. (b) Frequencies of the first three harmonics of standing toroidal Alfvén waves ($\Omega_{TN}/2\pi$) and Alfvén speed transit time between magneto-conjugated ionospheres, t_A *176*

Figure 3.16 (a) Structure along magnetic field lines of the first three harmonics of standing Alfvén waves ($T_N(\theta)$, $N = 1, 2, 3$) and SMS waves ($S_N(\theta)$, $N = 1, 2, 3$) at magnetic shell $L = 6.6$. (b) Structure of the first three harmonics of standing SMS waves ($S_N(\theta)$, $N = 1, 2, 3$) at magnetic shell $L = 1.3$ *177*

Figure 3.17 Transverse structure of the magnetic field components for the first harmonic ($N = 1$) of resonant SMS oscillations near the resonant shell $L = 6.6$: (a) standing SMS wave amplitude $|B_i|$ and phase α_i distribution

($i = x, y, z$) near the equatorial surface; (b) oscillation amplitude and phase distributions near the ionosphere (longitudinal component B_z turns zero) *177*

Figure 3.18 Dependence of function $F(L, \theta)$ on geomagnetic latitude θ and azimuthal wave number m at magnetic shell $L = 6.6$. Curve 1 refers to $m = 1$, curve 2 to $m = 5$, curve 3 to $m = 7$, curve 4 to $m = 10$. Horizontal dashed lines are possible eigenvalues of parameter k_{1n}^2. Areas where $F(a, \theta) - k_{1n}^2 > 0$ labelled as I, II, III, IV (shown in grey) correspond to the four types of longitudinal structure of FMS eigen oscillations *180*

Figure 3.19 Longitudinal structure of first four harmonics ($n = 0, 1, 2, 3$) of the FMS eigenmodes in the dipole-like magnetosphere, at magnetic shell $L = 6.6$ *181*

Figure 3.20 Squared quasi-classical wave-vector component k_{1n}^2 for the first five harmonics ($n = 0, 1, 2.3, 4$) of the FMS eigenmodes vs. the magnetic shell parameter L: (a) $k_{1n}^2(L)$ distribution inside the magnetosphere and (b) $k_{1n}^2(L)$ distribution in the magnetic shell range $1.5 < L < 15$, including the solar wind region *182*

Figure 3.21 The boundaries of transparent regions (in the meridional plane) for the first four parallel harmonics of FMS oscillations ($n = 0, 1, 2, 3$), in a dipole-like model magnetosphere, for frequency $\omega = 2\pi f_2$ and azimuthal wavenumber $m = 1$ *183*

Figure 3.22 The boundaries of the transparent regions for FMS oscillation harmonic ($n = 0, m = 1$), for different frequencies: (1)$f = f_1$, (2)$f = f_2$, (3)$f = f_3$ *184*

Figure 3.23 Model FMS resonator in the form of a rectangular box with ideally reflecting walls ('box model') *186*

Figure 3.24 Global Alfvén speed distribution in Earth's magnetosphere *188*

Figure 3.25 Axisymmetric model magnetosphere with a plasma sheet *189*

Figure 3.26 Orthogonal system of dimensionless parabolic coordinates (ξ, η), in the meridional section, including the z axis. The focus, $z = 0$, is at Earth's centre. The surface $\xi = 1$ coincides with the magnetopause. The semiaxes $z = (0, \infty)$ and $z = (0, -\infty)$ correspond to the coordinate surfaces $\xi = 0$ and $\eta = 0$, respectively *196*

Figure 3.27 Alfvén speed $v_A(10^3$ km/s) distribution isolines in the meridional plane, in the parabolic model magnetosphere *198*

Figure 3.28 Distribution of the square of the wave-vector WKB component k_{1N}^2 for Alfvén waves with $m \gg 1$, in the transverse coordinate x^1 *206*

Figure 3.29 Lines of constant phase $\vartheta(x^1, x^3) =$ const (characteristic lines) in the transverse section of an axisymmetric magnetosphere. Curves 1 correspond to $k_2 > 0$, and curves 2 to $k_2 < 0$. The circles are the transverse sections of the resonance surfaces: the inner circle is the poloidal ($x^1 = x_{PN}^1$), the outer the toroidal surface ($x^1 = x_{TN}^1$) *207*

Figure 3.30 Structure of the first two harmonics ($N = 1, 2$) of standing toroidal (dark lines) and poloidal (light lines) Alfvén waves, along field-lines, at magnetic shell $L = 6.6$ *211*

Figure 3.31 Poloidal Ω_{PN} and toroidal Ω_{TN} eigenfrequencies ($N = 1, 2$) vs. magnetic shell (the McIlwain) parameter $L = a/R_E$ *211*

Figure 3.32 The polarisation splitting $\Delta\Omega_1$ of the spectrum, equatorial splitting Δx_1^1 of resonance surfaces and the characteristic scale a_1 of the transverse inhomogeneity of the Alfvén speed for the basic harmonic of Alfvén oscillations ($N = 1$) vs. magnetic shell parameter $L = a/R_E$ *212*

Figure 3.33 The group velocity components v_1^1 and v_1^2 for the main harmonic of standing Alfvén waves vs. the radial coordinate in the equatorial plane ($\theta = 0$), the toroidal resonance surface $x_{T1}^1 = x_{P1}^1 + \Delta_1$ for the waves is at magnetic shell $L = 6.6$ *212*

Figure 3.34 Schematic representation of the structure of standing Alfvén wave with $m \gg 1$ in a dipole-like magnetosphere. Function $R_N(x^1, x^3)$ describes the wave structure along magnetic field lines, and $U_N(x^1)$ across magnetic shells *225*

Figure 3.35 Penetration of the high-m Alfvén wave field from the magnetosphere to ground: (a) in the $\lambda_{(P,T)N} > H$ case, (b) in the $\lambda_{(P,T)N} \lesssim H$ case. Shown are the hodographs of oscillations in the plane (B_x, B_y), at various points inside the transparent region (x_{PN}, x_{TN}) at the upper ionospheric boundary and at Earth surface *232*

Figure 3.36 Distribution in x^1 of the quasiclassical wave vector squared k_1^2. Slanting dashed lines show the asymptotics $k_1^2 = (x^1 - x_{TN}^1)/\sigma_N a_N$. The (a) case corresponds to $\sigma_N > 0$, and the (b) case to $\sigma_N < 0$ *235*

Figure 3.37 Possible integration contours in integrals (3.242). Sectors with exponentially growing asymptotics are in grey *236*

Figure 3.38 Spatial structure of Alfvén waves with $m \gg 1$ across magnetic shells. The thick (black) line is 'large-scale' Alfvén wave, the thin line is kinetic Alfvén wave: (a) $\sigma_N > 0$ and (b) $\sigma_N < 0$ *238*

Figure 3.39 Schematic plots of functions $\Omega_{(P,T)N}(x^1)$ in the dayside magnetosphere of Earth. The McIlwain parameter $L = a/R_E$ is used as the x^1 coordinate. *1* – the transparent region is located between the poloidal and toroidal resonant surfaces; *2* – Alfvén resonator; *3* – the presence of two toroidal turning points makes Alfvén oscillations impossible *239*

Figure 3.40 Schematic plots of spectral density of the components $\langle |B_i|^2 \rangle$ ($i = 1, 2, 3$) of perturbed magnetic field of standing Alfvén waves for strong decay of oscillations ($\nu_N \gg 1$) *249*

Figure 3.41 Schematic plots of spectral density of the components $\langle |B_i|^2 \rangle$ ($i = 1, 2, 3$) of perturbed magnetic field of standing Alfvén waves for weak decay of oscillations ($\nu_N \ll 1$) *251*

Figure 3.42 Hodographs of monochromatic Alfvén oscillations with $m \gg 1$, at different points in the transparent region between the resonance surfaces $x_{PN}^1 < x^1 < x_{TN}^1$ *259*

Figure 3.43 Hodographs of non-stationary Alfvén oscillations with $m \gg 1$ excited at the observation point by a source of the 'sudden pulse' type, at different moments of time: 1 – $t \ll m/\omega$ – oscillations of the poloidal type ($B_1 \gg B_2$); 2 – $t \sim m/\omega$ – oscillations of the intermediate type ($B_1 \sim B_2$); 3 – $t \gg m/\omega$ – oscillations of the toroidal type ($B_1 \ll B_2$) *260*

Figure 3.44 Field structure across magnetic shells of standing Alfvén waves with $m \gg 1$, for various values of parameter $\overline{\kappa}$: (a) $\overline{\kappa} = 0.1$ – a typical structure of resonant oscillations; (b) $\overline{\kappa} = 3$ – structure of intermediate type; (c) $\overline{\kappa} = 20$ – 'travelling wave'-type structure *272*

Figure 3.45 Spatial distribution over the transverse coordinate ξ for the poloidal (W_P) and toroidal (W_T) components of the full energy of standing Alfvén wave (in relative units) for 'travelling wave' type oscillations ($\overline{\kappa} \gg 1$, see Figure 3.44c) for various values of the dimensionless dissipation index $\overline{\epsilon} = 2k_y a_N(\gamma_N/\omega)$: (a) $\overline{\epsilon} = 0.4$; (b) $\overline{\epsilon} = 0.8$; (c) $\overline{\epsilon} = 4$ *273*

Figure 3.46 Same as Figure 3.45, for wave with resonant type transverse structure ($\overline{\kappa} \ll 1$, see Figure 3.44a): (a) $\overline{\epsilon} = 0.4$ – weak dissipation – toroidal type wave; (b) $\overline{\epsilon} = 1.2$ – moderate dissipation – intermediate type wave; (c) $\overline{\epsilon} = 6$ – strong dissipation – poloidal type wave *274*

Figure 3.47 Distribution of the amplitude of standing Alfvén waves excited by strongly localised monochromatic sources over dimensionless transverse coordinates (ξ and η). Source location: (a) in the opaque region behind the toroidal surface, (b) at the toroidal surface, (c) at the poloidal surface, and (d) in the opaque region behind the poloidal surface *280*

Figure 3.48 Calculated dependence of the splitting, $\Delta_N^{(i)}$ between the toroidal and poloidal resonance magnetic shells as mapped onto the ionosphere on the magnetic shell parameter $L = a/R_E$, for the first ($N = 1$, bold line, right-hand axis) and two next harmonics ($N = 2, 3$, left-hand axis) of standing Alfvén waves *282*

Figure 3.49 Calculated dependence of the eigenfrequencies $f_{TN} = \Omega_{TN}/2\pi$ of toroidal Alfvén oscillations on the magnetic shell parameter $L = a/R_E$, for the first five eigen harmonics ($N = 1, 2, 3, 4, 5$) of standing Alfvén waves *282*

Figure 3.50 Graphical solution of (3.398): (a) in the tolerance range (3.396), (b) in the tolerance range (3.397). The dark gray curves are the left-hand sides (panel (b) the upper curve is for the '–' sign, the lower for '+'), and straight horizontal black lines are the right-hand sides of (3.398) *288*

Figure 3.51 Region occupied by Alfvén wave oscillations in the asymptotic regime ('butterfly wings'). Solid lines (1–6) are characteristic lines $\eta_1, \eta_2, ..., \eta_6$, dashed lines are characteristic lines η_0. The Roman numerals I, II, III and IV indicate the domains of existence for the roots of (3.398) – $\overline{\kappa}_1, \overline{\kappa}_2, \overline{\kappa}_3$ and $\overline{\kappa}_4$ *289*

Figure 3.52 Amplitude distribution for a separate harmonic of standing Alfvén wave, in the initial regime of oscillations, in the plane of dimensionless transverse coordinates (ξ, η). Sectors with smallest amplitude are in shades of grey *294*

Figure 3.53 Amplitude distribution for a separate harmonic of standing Alfvén wave, in the asymptotic regime of oscillations, in the plane of dimensionless transverse coordinates (ξ, η). Sectors with smallest amplitude are in shades of grey *295*

Figure 3.54 Amplitude distribution of the B_y-component of the full field of standing Alfvén waves excited by a pulsed source near the ionosphere, in the horizontal plane (x, y). Possible view of the Aureole-3 satellite trajectory (dark gray line) in the MASSA experiment relative to the region occupied by the Alfvén oscillations *296*

Figure 3.55 Behaviour of the B_y-component of the full field of Alfvén oscillations onboard the satellite crossing the equatorial boundary of the 'butterfly wing', in the strong decay case. Variants (1–6) correspond to the satellite crossing the boundary at time moments (3.413) *297*

Figure 3.56 Model axisymmetric magnetic field with elongated field lines formed by the vector sum of the dipole magnetic field and the field of the axisymmetric current sheet. The systems of coordinates used in the calculations: (x^1, x^2, x^3) – orthogonal and (a, ϕ, θ) non-orthogonal curvilinear systems of coordinates, (ρ, ϕ, z) – cylindric system of coordinates *299*

Figure 3.57 Shape of magnetic field lines calculated from (3.421), in a geotail model with a thin current sheet *303*

Figure 3.58 Distribution isolines for: (a) Alfvén speed v_A (km/s); (b) sound speed v_s (km/s), in the meridional plane, in the geotail model with a thick current sheet. Lines 1 in panel (b) indicate the meridional sections of the magnetic field-line inflexion surfaces *305*

Figure 3.59 Graphical solution of dispersion equation (3.428). The solution is determined by the points where the parabola crosses the straight lines corresponding to the right-hand side of (3.428), for different ratios between the parameters. Straight line 1 corresponds to solutions for neutral poloidal Alfvén and azimuthally small-scale SMS waves; straight line 2 to the neutral Alfvén and aperiodically unstable SMS wave; straight line 3 to neutral Alfvén and SMS waves, in a force-free magnetic field *307*

Figure 3.60 Eigen frequency distribution for first odd harmonics $N = 1, 3, 5$ of azimuthally small-scale standing Alfvén waves ($f_{AN} = \Omega_{PN}/2\pi$ – solid thick lines) and Eigen frequency ($f_{sN} = \Omega_{sN}/2\pi$ – solid thin lines) and growth rate (δ_{sN} – dashed lines) distribution for standing SMS waves, calculated in the local approximation from the dispersion equation (3.428): (a) for the geotail model with a thin current sheet and (b) for the model with a thick current sheet *308*

Figure 3.61 Distribution of parameter $\overline{\varkappa}_{1B}(\overline{\varkappa}_{1B} + \overline{\varkappa}_{1P})$ along the field line located on magnetic shell $L = 20$, in models with a thick (thick line) and a thin (thin line) current sheet. $\theta = 0$ in the equatorial plane *311*

Figure 3.62 Distribution of the wavenumber squared, along different field lines ($\mathrm{Re}\, k_\parallel^2$ – solid lines, $\mathrm{Im}\, k_\parallel^2$ – dashed line), for the main harmonic ($N = 1$) of standing poloidal Alfvén waves, in the thin current sheet model. Neutrally stable oscillations on magnetic shells: (1) – $L = 6$, (2) – $L = 8$; (3) – $L = 10$; (4) – unstable ($\delta_{P1} > 0$) periodic oscillations on magnetic shell $L = 13$ *312*

Figure 3.63 Eigen frequency distribution for the first five harmonics $N = 1, 2, 3, 4, 5$ of standing poloidal ($f_N = \text{Re}\,(\Omega_{PN})/2\pi \equiv \overline{\Omega}_{PN}/2\pi$ – thin solid lines, $\delta_{Pn} = \text{Im}\,\Omega_{Pn}$ – dashed lines) and toroidal ($f_N = \Omega_{TN}/2\pi$ – thick solid lines) Alfvén waves across magnetic shells: (a) for the geotail model with a thin current sheet and (b) for the model with a thick current sheet *312*

Figure 3.64 Eigen frequency distribution for the first five harmonics $N = 1, 2, 3, 4, 5$ of standing azimuthally small-scale SMS waves ($f_N = \text{Re}\,(\Omega_{sN})/2\pi$ – solid lines, $\delta_{sN} = \text{Im}\,\Omega_{sn}$ – dashed lines), across magnetic shells, in the model geotail with a thin current sheet *313*

Figure 3.65 Longitudinal structure of the main harmonic ($N = 1$) of unstable poloidal Alfvén waves (A) on magnetic shell $L = 13$ and, for comparison, the structure of the SMS wave (SMS) of the same frequency (which fails to satisfy the boundary conditions on the ionosphere). Gaps in the Alfvén wave structure correspond to the neighbourhoods of turning points (where $\text{Re}\,(k_\parallel^2) = 0$) *314*

Figure 3.66 Distribution along the field line of the scalar potential of the electric field of the main ($N = 1$) and first ($N = 2$) harmonics of the coupled modes: (a, b) in the inner magnetosphere, at shell $L = 6$; (c, d) in the current sheet region, at magnetic shell $L = 15$ *317*

Figure 3.67 Distribution along the field line of the electromagnetic field components (E_y, B_x, B_z) of the coupled modes main harmonic ($N = 1$): (a–c) in the inner magnetosphere, at shell $L = 6$; (d–f) in the current sheet region, at magnetic shell $L = 15$ *318*

Figure 3.68 Distribution across magnetic shells of the eigenfrequencies of the main ($N = 1$) and first ($N = 2$) harmonics of toroidal Alfvén (thick dashed lines) and coupled (thick solid lines) modes of MHD oscillations, as well as their growth rates (thin solid lines) in the current sheet region. The transition layer region is shown in grey *319*

Figure 3.69 Parameter κ_2 distribution along the field line, for the main harmonic of coupled modes ($N = 1$), on magnetic shells $L = 6$ (line 1) and $L = 15$ (line 2) *321*

Figure 3.70 Azimuth variations for saddle points S_i ($i = 1, 2, 3, 4$) in (3.452), from $\text{Re}\,z \to \infty$ to $\text{Re}\,z \to -\infty$ and integration contours $\tilde{C}_{1,2}$ for the solutions of (3.447) for $\text{Re}\,\alpha_0 > 0$ *322*

Figure 3.71 Coupled Alfvén and SMS mode structure across magnetic shells, near the resonance surface for poloidal Alfvén waves: (a) oscillations with a growth rate exceeding their decrement due to dissipation in the ionosphere ($\varepsilon_N > 0$); (b) oscillations with a lower instability growth rate than the decrement ($\varepsilon_N < 0$) *326*

Figure 4.1 (a) Characteristics (energy flux lines) of transverse small-scale Alfvén waves for a fixed sign of κ, in an axisymmetrical model magnetosphere.

PT-type characteristics only are present. (b) Characteristics in a 3D-inhomogeneous model magnetosphere. The types of characteristics, *1–6*, are described in the text. Asymmetry can be seen in the behaviour of the characteristics in sectors 0°–180° and 180°–360°. Both PT- (numbered *1–4*) and TT-type characteristics (numbered *5* and *6*) are present *335*

Figure 4.2 The inclination of the channel with respect to the coordinate surfaces *338*

Figure 4.3 The structure transverses the channel of E_x and B_y electromagnetic components of the wave in vicinity of the resonant point x_S (corresponding $x/\Delta = 0.75$) for $k_c > 0$ (a) and $k_c < 0$ (b). The structure was calculated for value $|k_c\Delta| = 5$ *339*

List of Tables

Table 1.1 Amplitudes of wave field components for various MHD modes *4*

Table 2.1 Averaged plasma and magnetic field parameters in the solar wind (I) and the magnetosphere (II) satisfying the full pressure conservation condition (2.59) *68*

Table 2.2 Main parameters of the model medium at the geotail boundary *86*

Table 3.1 Eigen frequencies $\mathrm{Re}\omega_{mnj}$ (rad/s) $\times 10^2$ and decrements $\mathrm{Im}\,\omega_{mnj}(\mathrm{s}^{-1}) \times 10^2$ for several first harmonics of a FMS resonator in the outer part of the dayside magnetosphere *192*

Table 3.2 Eigen frequencies $\mathrm{Re}\,\omega_{mnj}$ (rad/s), of several first harmonics of the FMS resonator below the plasmapause *194*

Table 3.3 Eigen frequencies $f_{mnl}(\mathrm{mHz}) = \omega_{mnl}/2\pi$ of several first harmonics of a FMS resonator in the near-Earth plasma sheet *199*

Author Biography

Anatoly Sergeevich Leonovich was born on July 18, 1957, in Irkutsk, Russia. He graduated from the Irkutsk State University, Department of Physics, in 1979. For most of his life, he worked at the Institute of Solar-Terrestrial physics in Irkutsk. Anatoly Leonovich was an expert in the physics of the magnetohydrodynamic (MHD) waves and oscillations in Earth's magnetosphere. He made an important contribution to the development of the theory of MHD waves in realistic models of the magnetosphere, taking into account the inhomogeneity of the plasma and magnetic field in all three coordinates, including the inhomogeneous curvature of the field lines. His work on the study of the structure of transverse small-scale Alfvén waves in magnetospheric plasma, carried out jointly with his teacher Vitalii Mazur, is rightfully considered classic. Anatoly Leonovich obtained important results in the physics of MHD instabilities of magnetospheric plasma, which play an important role in the transfer of solar wind energy to Earth's magnetosphere and the development of eruptive processes in the near-Earth plasma. A number of papers of Anatoly Leonovich are devoted to the study of the interaction of MHD waves with Earth's ionosphere.

Anatoly Sergeevich Leonovich passed away on April 22, 2023.

Vitalii Aizikovich Mazur (28 December 1946–27 January 2015).

Vitalii Mazur was born on December 28, 1946, in Barnaul, Russia. In his teenager years, he was a student of famous Novosibirsk school of physics and mathematics. Vitalii graduated from Novosibirsk State University; for most of his life he worked in Irkutsk at the Institute of Solar-Terrestrial Physics. At the same time, he was a professor of the Irkutsk State University, Physical Department.

Vitalii's main scientific achievements include the theory of magnetohydrodynamic (MHD) oscillations of Earth's magnetosphere. He was one of the first to consider ultra-frequency waves in realistic models of magnetosphere thus turned from simple, little-resembling reality one-dimensional inhomogeneous box models to the study of MHD oscillations in complex dipole-like models of the magnetosphere.

Among the most important achievements and scientific contributions of Vitalii is the creation of the Irkutsk school of theoretical research of magnetospheric oscillations. Currently, this school is recognised throughout the global scientific community. It includes the third generation of theorists, and it successfully develops promising areas of research initiated by V. Mazur. Among his students are the co-authors of this book: Anatoly Leonovich and Dmitri Klimushkin.

Dmitri Yurievich Klimushkin was born on January 5, 1968, in Irkutsk, Russia. He graduated from the Irkutsk State University, Department of Physics, in 1992. He works at the Institute of Solar-Terrestrial physics in Irkutsk and at the same time teaches physics in the Irkutsk State University. He studies the ultra-low-frequency (ULF) waves and instabilities in Earth's magnetosphere in realistic models of the magnetosphere, taking into account the inhomogeneity of the plasma and magnetic field in all three coordinates, and finite plasma pressure, in both kinetic and MHD theory. An accent is made on the mode coupling, an important phenomenon on the ULF waves' theory in inhomogeneous plasmas.

Preface

The concept of this book evolved gradually as we dug deeper and deeper into the problematic of our research area. Magnetospheric magnetohydrodymanic (MHD) oscillations have proved to be so diverse that, over time, as ever newer problems are solved, they begin to resemble a kaleidoscope picture featuring a rich set of constituent elements. Individual elements change positions, alternately brightening and dimming now dimmer, while the pattern constantly changes. It would be desirable to fix it and explore in more detail, without haste. This monograph is such an effort, attempting to provide basic theoretical notions of magnetospheric MHD oscillations we have acquired to date.

The chief incentive for writing this book was the fact that the currently available literature appears to lack a systematic outline of MHD oscillation theory in inhomogeneous magnetospheric plasma. There are a number of excellent monographs addressing various types of geomagnetic pulsations and invoking various wave processes to interpret them. However, generation and propagation mechanisms for MHD waves (including geomagnetic pulsations) in inhomogeneous magnetosphere have never been consistently expounded.

Vitalii Mazur and Anatoly Leonovich conceived the idea to write a systematic treatise on magnetospheric MHD oscillation theory ca. 2006. Direct work on its text, however, did not start until late 2014. Most regretfully, Vitalii Mazur passed away soon after work on the monograph started. It was a very heavy loss for all his disciples, including me. Together, Vitalii and Anatoly managed to draw up a general plan for the monograph, and it was Vitalii who penned the Introduction. The major part of the book was written by Anatoly. I joined the work only in 2022, when Anatoly felt a sharp deterioration in his health. In April 2023, Anatoly passed away.

I will be much obliged if you inform me of the mistakes or typographical errors you notice by e-mail: klimush@iszf.irk.ru.

Irkutsk, Russia
August, 2023

Dmitri Klimushkin

Acknowledgements

The authors are very grateful to our colleagues from the Institute of Solar-Terrestrial Physics, who took it on themselves to proofread the final text of the monograph: Daniil Kozlov, Irina Dmitrienko and Pavel Mager. All of them are descendants of the school founded by Vitalii Mazur in Irkutsk. A special thank to Dmitri Prokofiev who helped us in preparation of the English version of the manuscript.

Anatoly Leonovich
Dmitri Klimushkin

Acronyms

DPS	distant plasma sheet
FMS	fast magnetosonic (wave)
K-HK	elvin–Helmholtz (instability)
LLBL	low latitude boundary layer
MHD	magnetic hydrodynamics
NEPS	near Earth part of the plasma sheet
SMS	slow magnetosonic (wave)
ULF	ultra low frequency (oscillations)
WKBW	entzel–Kramers–Brillouin (approximation)

Symbols

$\bar{\mathbf{B}} = \mathbf{B}_0 + \mathbf{B}$	Magnetic field vector
$\mathbf{B}_0$	Background magnetic field vector
$\mathbf{B}$	Perturbed magnetic field vector
$\bar{c}$	Speed of light in vacuum
$c_s = \dfrac{v_A v_s}{\sqrt{v_A^2 + v_s^2}}$	Velocity of slow magnetosonic waves
$\mathbf{E}$	Perturbed electric field vector
e	Electron charge
g_1, g_2, g_3	Metric tensor components of (x^1, x^2, x^3) curvilinear coordinate system
$\mathbf{j}$	The perturbed current density vector
$k_{\parallel}$	Longitudinal (magnetic field-aligned) wave vector component
$\mathbf{k}_{\perp}$	Transversal (across magnetic field lines) wave vector component
$k_{\perp}$	Transversal wave number
$\mathbf{k}_t$	Tangential wave vector
k_t	Tangential wave number
$\hat{L}_P$	Longitudinal operator for poloidal Alfvén waves
$\hat{L}_T$	Longitudinal operator for toroidal Alfvén waves
$M = \dfrac{v_0}{v_s} \cos\phi$	Mach number as defined by the flow velocity projection: $(\mathbf{v}_0 \mathbf{k}_t)/k_t = v_0 \cos\phi$
$\bar{M} = M\sqrt{\dfrac{v_s^2}{v_s^2 + v_A^2}}$	Modified Mach number
$M_A = M\dfrac{v_s}{v_A}$	Alfvén Mach number (as defined by the Alfvén speed)
m_e	Electron mass
m_i	Ion mass
$\bar{n} = n_0 + n$	Plasma concentration
n_0	Background plasma concentration
n	Perturbed plasma concentration
$\bar{P} = P_0 + P$	Plasma pressure

P_0	Background plasma pressure
P	Perturbed plasma pressure
q_i	Ion charge
$s_e = \dfrac{\bar{c}}{\omega_{pe}} = \dfrac{\bar{c}}{e}\sqrt{\dfrac{m_e}{4\pi n_0}}$	Electron skin depth
T	Background plasma temperature
T_i	Ion temperature of background plasma
T_e	Electron temperature of background plasma
$\bar{\mathbf{v}} = \mathbf{v}_0 + \mathbf{v}$	Plasma velocity vector
$\mathbf{v}_0$	Background plasma velocity vector
$\mathbf{v}$	Perturbed plasma velocity vector
$v_A = \dfrac{B_0}{\sqrt{4\pi\rho_0}}$	Alfvén speed
$v_f = \sqrt{v_A^2 + v_s^2}$	Velocity of fast magnetosonic waves
$\mathbf{v}_g$	Group velocity vector of a wave packet
$v_s = \sqrt{\bar{\gamma}\dfrac{T}{m_i}}$	Sound speed of plasma
$v_e = \sqrt{\dfrac{T_e}{m_e}}$	Electron thermal velocity
$v_i = \sqrt{\dfrac{T_i}{m_i}}$	Ion thermal velocity
$v_{es} = \sqrt{\dfrac{T_e}{m_i}}$	Ion thermal velocity as defined by the electron temperature
$v_p = \dfrac{\bar{c}^2}{4\pi\Sigma_P}$	Characteristic speed of a low-frequency whistler in the ionosphere
x^1_{PN}	Coordinate of the poloidal resonance surface for Alfvén waves
x^1_{TN}	Coordinate of the toroidal resonance surface for Alfvén waves
x^1_{sN}	Coordinate of the resonance surface for slow magnetosonic waves
$\beta = \dfrac{8\pi P_0}{B_0^2}$	Plasma beta parameter
$\beta^* = v_s^2/v_A^2 = \dfrac{4\bar{\gamma}\,\pi P_0}{B_0^2}$	Modified plasma beta parameter
$\bar{\gamma}$	Adiabatic index
$\hat{\varepsilon}$	Dielectric permittivity tensor of plasma
ν_e	Electron collision frequency
$\bar{\rho} = \rho_0 + \rho$	Plasma density
ρ_0	Background plasma density
ρ	Perturbed plasma density
$\rho_i = \dfrac{v_i}{\omega_i} = \dfrac{\bar{c}\sqrt{T_i m_i}}{q_i B_0}$	Larmor radius of ions
$\rho_s = \dfrac{v_{es}}{\omega_i} = \dfrac{\bar{c}\sqrt{T_e m_i}}{q_i B_0}$	Larmor radius of ions as defined by electron temperature
$\hat{\sigma}$	Plasma conductivity tensor

σ_H	Ionospheric Hall conductivity
σ_P	Ionospheric Pedersen conductivity
Σ_H	Height-integrated ionospheric Hall conductivity
Σ_P	Height-integrated ionospheric Pedersen conductivity
χ	Inclination angle between the geomagnetic field line and a vertical to the ionosphere
ω	Wave frequency
$\omega_e = \dfrac{eB_0}{m_e\bar{c}}$	Cyclotron frequency of electrons
$\omega_i = \dfrac{q_iB_0}{m_i\bar{c}}$	Cyclotron frequency of ions
$\omega_{pe} = \sqrt{\dfrac{4\pi n_0 e^2}{m_e}}$	Electron plasma (Langmuir) frequency
Ω_{PN}	Poloidal Alfvén wave eigen frequency in a dipole-like magnetosphere
Ω_{SN}	Slow magnetosonic wave eigen frequency in a dipole-like magnetosphere
Ω_{TN}	Toroidal Alfvén wave eigen frequency in a dipole-like magnetosphere
Ω_N	Eigen frequency of coupled Alfvén and SMS waves in a dipole-like magnetosphere
ω_{mnj}	Eigen frequencies of resonators for fast magnetosonic waves in magnetosphere models

Introduction

Magnetohydrodynamic (MHD, hydromagnetic) oscillations in Earth's magnetosphere have attracted researchers' close attention for many years. In observations, they are detected as the high frequency component in geomagnetic field variations, its range covering 1 mHz to 5 Hz. Unlike slow variations in the geomagnetic field, these oscillations are characterised by very conspicuous periodicity, because of which they were named 'geomagnetic pulsations'. Many are registered on Earth's surface, which makes them usable as a diagnostic tool for the magnetospheric plasma [1, 2].

Geomagnetic field pulsations were discovered over a century ago (see [3]; also [4] and [5], citing earlier studies by Kristian Birkeland in 1901[1]) and have been examined for many years by means of ground-based methods. Their physical nature, however, was not clear until Hannes Alfvén initiated magnetic hydrodynamics [6], and the magnetosphere and its basic structural elements were discovered at the dawn of the space research era [7–10]. It then became evident that geomagnetic pulsations are magnetospheric MHD oscillations. This perception is supported by an estimated characteristic frequency of the basic tone in magnetospheric oscillations, $f \sim v_A/L$, where v_A is the characteristic value of the Alfvén speed in the magnetosphere, and L is the characteristic scale of the oscillation region. Assuming the characteristic values of $v_A \sim 10^3$ km/s, $L \sim 10^5$ km in the magnetosphere will result in $f \sim 10$ mHz. Higher frequencies can be interpreted as oscillations in limited regions of the magnetosphere with smaller v_A (and, possibly, larger L), or as harmonics of the basic tone. A certain problem arises when trying to explain the existence of the lowest-frequency oscillations, $f \sim 1$ mHz. Even they, however, are given quite a satisfactory theoretical interpretation (see Section 3.6.3 of this monograph).

The classification of geomagnetic pulsations adopted at the XIIIth IQSY General Assembly in 1963 is based on their simple morphological features. Primarily, they are subdivided into two large classes: continuous, Pc (continuous pulsations), quasi-sinusoidal oscillations lasting tens to hundreds of periods, and irregular, Pi (irregular pulsations), lasting a few periods. Either class is subdivided into several frequency ranges. The Pc class encompasses Pc 1, with periods 0.2–5 seconds; Pc 2, 5–10 seconds; Pc 3, 10–45 seconds; Pc 4, 45–150 seconds; Pc 5, 150–600 seconds and Pc 6, with periods of over 600 seconds. The Pi class contains the Pi 1 range, with periods below 40 seconds; Pi 2, 40–150 seconds and

1 Birkeland, Kr., Expédition Norvégienne de 1899–1900 pour l'étude des aurores boréales. Resultats des recherches magnétiques, Kristiania, Skr. Vid. selsk. 1, No. 1, 80 pp., 12 pls. (1901).

Pi 3, longer than 150 seconds. Various types of pulsations are distinguished within certain frequency ranges based on other morphological features. A detailed description of the various types of magnetic pulsations can be found in monographs [11–14] and in review papers [15–18]. It gradually became clear that this subdivision of geomagnetic pulsations into different types does not only reflect their morphological properties but also their physical nature.

The first five decades of research into geomagnetic pulsations accumulated vast amounts of observation data which required theoretical interpretation. First in-depth theoretical papers appeared immediately after the emergence of MHDs, even before the magnetosphere was discovered in experiments [19, 20]. They hypothesised a plasma shell around Earth. The onset of the space-flight era started an ever-increasing flow of experimental data from various spacecraft. This was paralleled by improvements in old methods as well as by the development of new, more informative, methods of ground-based investigations. At the same time, the theory evolved at accelerating rates. At present, theory greatly influences the choice of directions in experimental investigations. Theoretical arguments define a scientific problem, to solve which, a space-based experiment (involving multiple satellites) is carried out, in close collaboration with ground-based observations. The resulting experimental data in turn trigger new theoretical research.

The spatial structure of magnetospheric hydromagnetic oscillations is mainly determined by the global distribution of two medium parameters: Alfvén speed v_A and sound speed v_S. It is these two parameters that are included in equations describing hydromagnetic oscillations in plasma. Hydromagnetic wave properties are, to a large degree, determined by the value of $\beta = 8\pi P_0/B_0^2 \approx v_S^2/v_A^2$, representing the ratio between gas-kinetic pressure in plasma (P_0) and background magnetic field pressure ($B_0^2/8\pi$). In accordance with the conventional terminology employed in plasma physics, plasma is called 'cold' when $\beta \ll 1$, 'hot' when $\beta \gg 1$, and is sometimes called 'warm' in the intermediate case, $\beta \sim 1$. According to this classification, solar wind plasma can be considered as 'hot', and magnetospheric plasma, 'cold' to a certain extent (except for the geotail plasma sheet, where it is 'hot'). Note that the above definition is not contradicted by a situation when a 'hot' plasma in one region is of lower temperature than a 'cold' plasma in another region. Thus, e.g. a 'cold' plasma in the outer magnetosphere is hotter than the 'hot' plasma in the solar wind.

A theoretical investigation of MHD oscillations in the magnetosphere presumes a certain model of the medium. However, a model of the magnetosphere as a whole that could aspire to adequately describe it, at least to a certain degree, is extremely complex (see Figure I.1). In particular, it must be strictly three-dimensional. These circumstances render any rigorous analytical examination of oscillations almost unrealistic, within the confines of such a model. Therefore, one has to resort to simplified models describing individual regions in the magnetosphere and oscillations in these regions.

The selected model must satisfy two, contradictory, in a certain sense, requirements. On the one hand, it must reflect the major properties (for the oscillations in question) of the magnetospheric region under study. On the other hand, it must possibly yield to a simple theoretical examination. The latter requirement is not only related to the fact that it simplifies the theorist's work but also to the fact that results obtained with a simple model are more illustrative. Obviously, different models satisfy the two above requirements to various degrees. Simpler but rougher, or more adequate but more complex, models can be

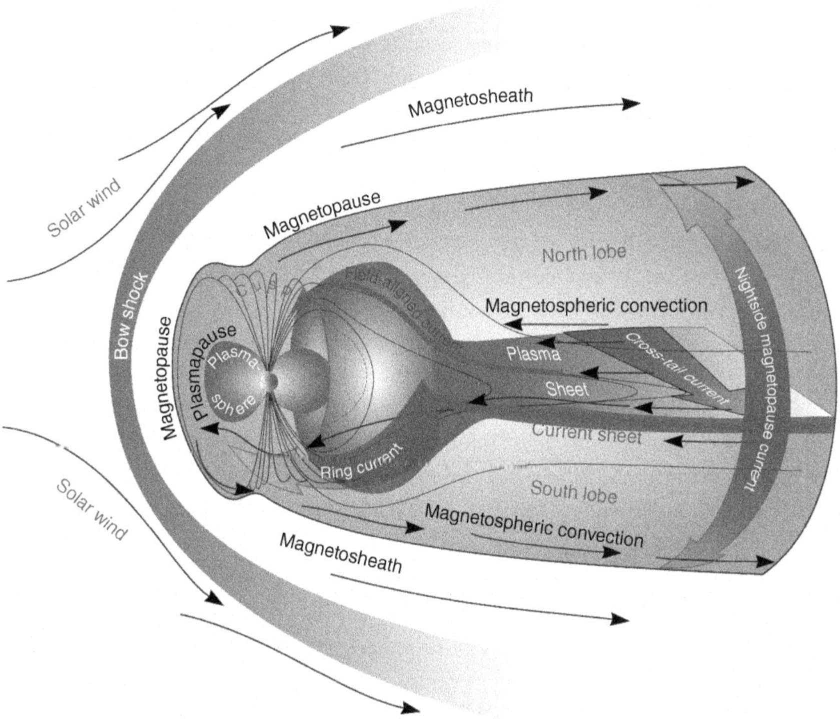

Figure I.1 Basic structural elements of Earth's magnetosphere.

used. Both have a right to exist but each can make its own contribution into understanding the phenomenon.

If we restrict ourselves to linear theory, the main distinction between various models is determined – in terms of a theoretical investigation – by the dimension of the medium inhomogeneity in question. From this standpoint, the simplest model is that of a homogeneous plasma. This model is a far cry from the real plasma medium of the magnetosphere, but is important methodologically. The concepts and language developed within this model (e.g. the concept of three modes in hydromagnetic oscillations) are also used, in a large measure, for describing oscillations in an inhomogeneous plasma. The same concerns dispersion and polarisation properties of the oscillations. Therefore, we provide, in the next Chapter, a short description of hydromagnetic oscillations in a homogeneous plasma. Such a model, however, cannot be used to describe real oscillations of the magnetosphere, one of its major properties being inhomogeneity. Such a model is too remote from the reality.

Next most simple are models with a 1D-inhomogeneous medium. Such models have already been widely applied for interpreting oscillations observed in magnetospheric plasma. The fact is that, in many magnetospheric regions, the medium is much more inhomogeneous in one direction (usually, across magnetic shells) than in the other two. Applying 1D models for such situations seems justifiable. They describe medium oscillations via ordinary differential equations, with highly developed theory and solution techniques. A theoretical investigation therefore can usually be completed. 2D and especially 3D models are much more complex. They describe oscillations using partial differential equations. The theory of such equations is immensely complicated, while methods to obtain a constructible solution have only been developed for a few particular cases. Whereas 2D models sometimes allow much enough progress (making use of some of their particular properties, as a rule), problems solved using 3D models are few and far between.

As is evident from the title of this monograph, we restrict ourselves to linear oscillation theory. This means that our discourse does not cover all those phenomena where nonlinearity plays an important role. At first sight, this appears to seriously restrict the applicability of the theory propounded here. When a nonlinearity is present, the presumption is that the exponential growth of the oscillation amplitude must inevitably end up at a nonlinear level. It turns out, however, that the inhomogeneous magnetosphere substantially broadens the applicability of the linear theory even when an instability is present. Let us explain the effect the inhomogeneity exerts on the instability evolution in terms of wave packets. These terms are not adequate enough for strictly describing magnetospheric oscillations, but are appropriate for a qualitative study. The wave packet motion in a inhomogeneous medium is described by Hamilton-type equations:

$$\frac{d\mathbf{x}}{dt} = \frac{d\omega}{d\mathbf{k}}, \qquad \frac{d\mathbf{k}}{dt} = -\frac{d\omega}{d\mathbf{x}}. \tag{I.1}$$

Here, $\mathbf{x} = \mathbf{x}(t)$, $\mathbf{k} = \mathbf{k}(t)$ are the coordinates of the wave-packet centre in, respectively, the $\mathbf{x}$ space (ordinary space) and the $\mathbf{k}$ space (wave vector space), whereas $\omega(\mathbf{x}, \mathbf{k})$ is the so-called local frequency. The latter generally coincides with the wave frequency, for a set value of $\mathbf{k}$, as found in the homogeneous plasma approximation, i.e. the medium parameters determining this frequency are functions of the $\mathbf{x}$ coordinate.

In accordance with the first equation (I.1), the wave packet centre travels in the $\mathbf{x}$ space at a group velocity $v_g = d\omega/d\mathbf{k}$. The result is that wave-packet amplitude enhancement by an instability is rather limited. Thanks to an inhomogeneous magnetosphere, the instability region (i.e. the region where the growth rate is positive) is of finite dimensions. It takes the packet a finite time to cross it and, consequently, the packet is subject to certain finite amplification. Outside this region, the growth rate is usually negative (i.e. it effectively becomes a decrement), and the packet travelling there is damped.

It is fairly easy to estimate the characteristic value of the full MHD wave amplification coefficient in the magnetosphere: $\Gamma \sim \gamma t$, where γ is the characteristic value of the growth rate, and t is the instability region transit time. The growth rate is practically always much smaller than the wave frequency, which could acceptably be estimated as $\omega \sim kv_A$, where $v_A = B_0/\sqrt{4\pi\rho_0}$ is the Alfvén speed characteristic for MHD waves travelling in the magnetosphere (here, B_0 is magnetic field strength, ρ_0 is plasma density), and k is the characteristic value of the wave vector. The transit time $t \sim L/v_A$, where L is the instability region size. Hence, we have the upper estimate $\Gamma \lesssim kL$. For unstable waves, $k \sim 1/L$, as a rule, that is $\Gamma \lesssim 1$. This means that the wave amplitude increases by a factor of $\exp(\Gamma) \lesssim 10$. In fact, $\exp(\Gamma) \sim 10$–100, in the most unstable cases. Note that a 100-fold increase in amplitude implies a very strong instability indeed.

A phenomenon similar to the one described above also occurs in the wave vector space. In accordance with the second equation in (I.1), the central wave vector in the packet, $\mathbf{k} = \mathbf{k}(t)$, thanks to the inhomogeneous medium, travels in the $\mathbf{k}$ space. As a rule, the growth rate, being a function of $\mathbf{k}$, is positive in a limited region of the $\mathbf{k}$ space too, and the packet, passing through this region within a finite time, is also subjected to a finite amplification only. An estimate of the full amplification coefficient produces the same order-of-magnitude value as above. Very often, however, thanks to certain particular details of how the instability growth rate depends on the wave vector, this effect turns out to be more important than the one mentioned above. In any case, it does limit the wave packet amplification even more. Thus, an instability in an inhomogeneous medium fails to cause a constant exponential increase of the perturbation. It only increases up to a certain finite amplitude. As a result, the perturbation may not – and in many cases, does not – reach a nonlinear level.

The above-described picture of perturbation evolution provides an answer to one ('fatal') question in perturbation theory. Standard instability theory presumes that a certain balanced state exists, which is superimposed by a small perturbation, growing exponentially provided there is an instability. The original state is, however, essentially modified or completely destroyed in the process. The above-mentioned 'fatal' question arises: how did the original balanced state come about and how it persisted up to the moment when it was superimposed by the small perturbation? There are always small perturbations, be it on the thermal noise level. Standard perturbation theory fails to give a satisfactory answer to this question. In the above-described picture of instability evolution, however, the answer to this question is quite clear. The perturbation always remains small in the process of instability evolution and it is therefore quite justified to consider it to be superimposed upon a certain balanced state.

The above does not mean that we completely deny any nonlinear effects for MHD oscillations in the magnetosphere. They, without doubt, do exist and can be observed experimentally (see [21] and the references therein). Thus, for example, the Kelvin–Helmholtz instability on the magnetospheric flanks results in vortices – an explicitly nonlinear effect [22]. To solve the question of what role is played by nonlinearity, however, we must have an adequate nonlinear theory of the phenomenon, or, at least, reliable order-of-magnitude estimates. It should be mentioned that, despite a large number of papers on nonlinear phenomena in the magnetosphere, this theory is still at early stages of development. Most papers on nonlinear MHD oscillations in the magnetosphere either completely ignore its inhomogeneity or fail to properly take it into account. Such papers are of a certain methodological value only. In this monograph, we will not address nonlinear phenomena at all, because we believe that, given the current state of nonlinear theory of MHD oscillations in the magnetosphere, this would only lead us too far away from the main topic of our discussion. As a result, we unfortunately cannot answer the question as to whether the unstable oscillations we discuss here do evolve to a nonlinear level. We are fully aware that this is a considerable drawback of our work.

In conclusion, let us say a few words on the trade-off between analytical and numerical methods in theoretical studies of magnetospheric oscillations. In this monograph, we accentuate analytical methods of solving wave equations. An analytical solution yields much more complete (exhaustive, in fact) and more illustrative information on oscillation properties. Describing it as 'illustrative' may seem paradoxical. On the face of it, what could be more illustrative than numerical calculation results presented as plots? This sort of clearness, however, is in effect illusory. In a problem concerning oscillations, the most important question is how the properties of these oscillations depend on the medium parameters, including on the distribution pattern of these parameters in space. To elucidate this issue using numerical methods, one has to calculate a huge number of variants. The picture expands catastrophically, while all its illustrative merits are lost. A formula obtained as a result of an analytical solution of specially derived equations, in contrast, contains all the information on how the properties of the solution depend on the medium parameters.

Numerical methods possess yet another drawback. The process of an analytical solution of a problem is, as a rule, laid out by the author to a complete enough degree and any sufficiently advanced reader can reproduce it at will, while many important details of numerical calculations usually remain 'behind the scenes'. Therefore, it is quite impossible, as a rule, to reproduce the process of a numerical study. This, to a great degree, discredits the results. Especially, this concerns numerical simulation, when the results are obtained directly from the original system of MHD equations by numerically calculating the evolution of a given initial perturbation. In the thus obtained picture of perturbations, all branches of MHD oscillations are mixed up, and it is impossible to determine which processes are responsible for the final state of the wave field.

This research also uses numerical methods for solving the differential equations describing the wave field structure and dynamics. However, we do not confine ourselves to these methods only. As a rule, a numerical research is used for finding the precise values of the calculated parameters of the oscillations, when approximate analytical solutions have already been found for the derived equations. For example, MHD oscillation structure in magnetospheric resonators is obtained using analytical investigations relying on simplified models of the medium allowing for simple analytical solutions, whereas more complex models employ numerical methods for solving similar wave equations. Comparing the solutions resulting from the two different approaches ensures calculation accuracy for the derived spatial structures and eigenmode spectra of these resonators.

This monograph is constructed along the following principle. Based on the above-mentioned criteria, we select a model of the relevant magnetospheric region. Next, the properties of MHD oscillations are theoretically studied in the framework of this model with all possible mathematical rigour and maximum completeness. The material is arranged according to the complexity of the models of the medium: from simpler to more complex. First, MHD oscillation properties are briefly examined in homogeneous models of the medium; next, 1D and 2D models; and finally, a 3D model (in the last case, a single solved problem is given). In a number of cases, the same magnetospheric region is examined in both 1D and 2D models. This enables oscillation properties in this region to be inspected from different aspects.

The results from each model have their own merit, but can also be regarded as part, or a 'puzzle piece', of the whole picture. Based on this, most of this monograph can be regarded as an accumulation of such puzzle pieces. Eventually, in the Conclusion section, we attempt to compose a general picture of magnetospheric oscillations from them. The resulting picture is of course incomplete. To compose it, we only use pieces that we ourselves examined. This alone leaves large enough gaps in the picture, which we point out whenever possible. Moreover, many pieces fail to fit together properly. We, however, believe that the picture presented, whatever its drawbacks, has a certain merit. It allows the magnetosphere and the MHD oscillation phenomenon in it to be perceived as a whole.

1

Hydromagnetic Oscillations in Homogeneous Plasma

In this section, we will address the basic properties of MHD oscillations in a homogeneous plasma, without going into a detailed research regarding various plasma states. Such examinations of MHD oscillations in a homogeneous plasma can be found in monographs [23–25].

We will generally use the ideal magneto-hydrodynamics approximation to describe hydromagnetic oscillations. According to this approximation, the system of MHD equations has the following form:

$$\bar{\rho}\frac{d\bar{\mathbf{v}}}{dt} = -\nabla\bar{P} + \frac{1}{4\pi}[\text{rot}\,\bar{\mathbf{B}} \times \bar{\mathbf{B}}], \tag{1.1}$$

$$\frac{\partial\bar{\mathbf{B}}}{\partial t} = \text{rot}\,[\bar{\mathbf{v}} \times \bar{\mathbf{B}}], \tag{1.2}$$

$$\frac{\partial\bar{\rho}}{\partial t} + \nabla(\bar{\rho}\bar{\mathbf{v}}) = 0, \tag{1.3}$$

$$\frac{d}{dt}\frac{\bar{P}}{\bar{\rho}^{\bar{\gamma}}} = 0, \tag{1.4}$$

where $\bar{\mathbf{v}}$ and $\bar{\mathbf{B}}$ are the plasma motion velocity and magnetic field vectors, $\bar{\rho}$ and $\bar{P}$ are plasma density and pressure and $\bar{\gamma}$ is the adiabatic index, $d/dt = \partial/\partial t + \bar{\mathbf{v}}\nabla$.

Let us linearise this system with respect to small perturbations. We will subscribe the parameters of an unperturbed background plasma with $_0$, while leaving the perturbation parameters with no subscript. Let us examine small-amplitude oscillations. In the linear approximation, the plasma and magnetic field parameters can then be written as $\bar{\mathbf{B}} = \mathbf{B}_0 + \mathbf{B}$, $\bar{\mathbf{v}} = \mathbf{v}_0 + \mathbf{v}$, $\bar{\rho} = \rho_0 + \rho$, $\bar{P} = P_0 + P$. The system of Eqs. (1.1)–(1.4) linearised with respect to small perturbations reduces to the following equations:

$$\rho_0\left(\frac{\partial\mathbf{v}}{\partial t} + \mathbf{v}_0\nabla\mathbf{v} + \mathbf{v}\nabla\mathbf{v}_0\right) = -\nabla P + \frac{1}{4\pi}\left\{\left[\text{rot}\,\mathbf{B} \times \mathbf{B}_0\right] + \left[\text{rot}\,\mathbf{B}_0 \times \mathbf{B}\right]\right\}, \tag{1.5}$$

$$\frac{\partial\mathbf{B}}{\partial t} = \text{rot}\,[\mathbf{v} \times \mathbf{B}_0] + \text{rot}\,[\mathbf{v}_0 \times \mathbf{B}], \tag{1.6}$$

$$\frac{\partial P}{\partial t} = -\bar{\gamma}P_0\,\text{div}\mathbf{v}. \tag{1.7}$$

Magnetospheric MHD Oscillations: A Linear Theory, First Edition. Anatoly Leonovich, Vitalii Mazur, and Dmitri Klimushkin.

Here, $\mathbf{B}, \mathbf{v}$ are the vectors of perturbed magnetic field and plasma velocity and P is the perturbed pressure. The perturbed electric field is determined by the frozen-in condition:

$$\mathbf{E} = -\frac{1}{\bar{c}}\left([\mathbf{v}\times\mathbf{B}_0] + [\mathbf{v}_0\times\mathbf{B}]\right), \tag{1.8}$$

where $\bar{c}$ is the speed of light in vacuum, and the perturbed electric current is given by

$$\mathbf{j} = \frac{\bar{c}}{4\pi}\,\mathrm{rot}\,\mathbf{B}. \tag{1.9}$$

For a homogeneous immobile plasma ($\mathbf{v}_0 = 0$) inside a homogeneous magnetic field, an arbitrary oscillation can be represented as a superposition of Fourier harmonics of the form $\exp(i\mathbf{kx} - i\omega t)$, where $\mathbf{k}$ is the wave vector, $\mathbf{x}$ is the coordinate vector, ω is the wave frequency and t is the time. Substituting the expressions for the wave field components into (1.5)–(1.7), for each such harmonic, yields

$$\omega^2 = k_\parallel^2 v_A^2, \quad \omega^4 - \omega^2 k^2(v_A^2 + v_s^2) + k_\parallel^2 k^2 v_A^2 v_s^2 = 0, \tag{1.10}$$

– dispersion equations for Alfvén and magnetosonic waves. Here, $v_A = B_0/\sqrt{4\pi\rho_0}$ is the Alfvén speed, $v_s = \sqrt{\bar{\gamma}p_0/\rho_0}$ is the sound speed in plasma, $k = \sqrt{k_\perp^2 + k_\parallel^2}$ is the modulus of the wave vector, and $k_\parallel, k_\perp$ are its components along and across the magnetic field.

The solutions of the second equation (1.10)

$$\omega^2 = \frac{k^2}{2}(v_A^2 + v_s^2) \pm \sqrt{\frac{k^4}{4}(v_A^2 + v_s^2)^2 - k^2 k_\parallel^2 v_A^2 v_s^2}$$

describe two branches of magnetosonic waves: fast magnetosonic (FMS, '+' before the radical) and slow magnetosonic (SMS, '–' before the radical). These equations have an especially simple form if one of the following conditions holds: $v_s \ll v_A$ (equivalent to $\beta \ll 1$, where $\beta = 8\pi p_0/B_0^2$ is the plasma gas-kinetic to magnetic pressure ratio), $v_s \gg v_A$ ($\beta \gg 1$), or $|k_\parallel| \ll |k_\perp|$, found in most real natural plasma formations. In this case, the approximate dispersion equation for FMS waves can be written as

$$\omega^2 \approx k^2 v_f^2, \tag{1.11}$$

where $v_f^2 = v_A^2 + v_s^2$, while the equation for SMS waves can be written as

$$\omega^2 \approx k_\parallel^2 c_s^2, \tag{1.12}$$

where $c_s = v_A v_s/\sqrt{v_A^2 + v_s^2}$. We will hereafter restrict ourselves to these approximations.

It can be seen from the first equation in (1.10) that the Alfvén wave group velocity $\mathbf{v}_{gA} = \partial\omega/\partial\mathbf{k} = v_A(\mathbf{B}_0/B_0)$ is along the magnetic field lines. The FMS group velocity $\mathbf{v}_{gf} = v_f(\mathbf{k}/k)$ is along the wave vector, and SMS group velocity $\mathbf{v}_{gs} = c_s(\mathbf{B}_0/B_0)$ is along the magnetic field, same as for Alfvén waves. As follows from (1.10), the latter statement is only approximate (even in the ideal MHD approximation) and no longer valid when small corrections are taken into account in order to satisfy the above conditions. Given these properties of the MHD-wave group velocity, the FMS is called the 'isotropic' mode and the Alfvén and SMS waves 'guided' modes of MHD oscillations. Useful representation of the phase and group velocities of MHD modes is given by the Friedrichs (polar) diagrams, which show the relation between the parallel ($v_{ph\parallel}$ and $v_{g\parallel}$) and transverse ($v_{ph\perp}$ and $v_{g\perp}$) components of these velocities for each of the three modes (Figure 1.1).

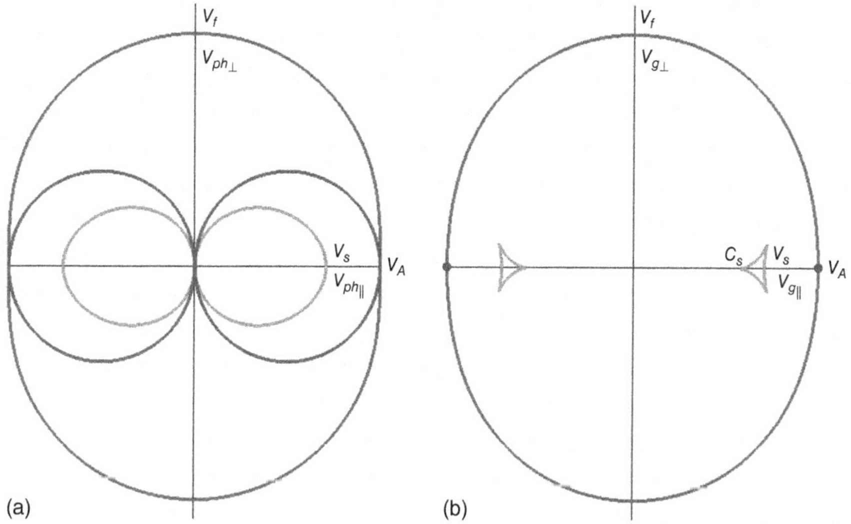

Figure 1.1 Friedrichs diagrams for the Alfvén (black), slow (light gray) and fast (dark gray) MHD modes for case when the Alfvén speed v_A is greater than the speed of sound v_s: (a) for the phase velocity v_{ph} and (b) for the group velocity v_g. Here indices $\|$ and $\perp$ mean projections on the direction parallel and perpendicular to the ambient magnetic field, respectively.

The character of perturbed field oscillations in various MHD modes is presented in Table 1.1. The system of coordinates used here is given by unit vectors $\mathbf{e}_\| = \mathbf{B}_0/B_0$, $\mathbf{e}_\perp = \mathbf{k}_\perp/k_\perp$, while the third unit vector $\mathbf{e}_b$ is chosen such that the three unit vectors $(\mathbf{e}_\perp, \mathbf{e}_b, \mathbf{e}_\|)$ are righthanded. The table lists the (phase-shift corrected) amplitudes of various components of the wave field as expressed through the full amplitudes of the field components $\tilde{B}, \tilde{E}, \tilde{j}, \tilde{v}, \tilde{P}$ ($\tilde{B} \equiv |\mathbf{B}| \ldots$). The relations between the amplitudes of these components are different for different modes. For the Alfvén (A) wave,

$$\tilde{E} = \frac{v_A}{\bar{c}}\tilde{B}, \quad \tilde{v} = v_A\frac{\tilde{B}}{B_0}, \quad \tilde{j} = \frac{\bar{c}}{4\pi}k\tilde{B}. \tag{1.13}$$

For the FMS, the relations between $\tilde{B}, \tilde{E}, \tilde{j}$ and $\tilde{v}$ have the same form, while for the perturbed pressure we have

$$\tilde{P} = \bar{\gamma}P_0\frac{k_\perp}{k}\frac{\tilde{B}}{B_0} = \bar{\gamma}P_0\frac{k_\perp}{k}\frac{\tilde{v}}{v_A}. \tag{1.14}$$

While the energy is distributed equally between magnetic field and plasma oscillations for the Alfvén and FMS waves, the SMS wave is dominated by plasma oscillation energy. In the latter case, it is expedient to express all the values in terms of $\tilde{v}$:

$$\tilde{B} = B_0\frac{k_\perp}{k}\frac{v_s}{v_A}\frac{\tilde{v}}{v_A}, \quad \tilde{E} = B_0\frac{k_\perp k_\|}{k^2}\frac{v_s^2}{v_A^2}\frac{\tilde{v}}{\bar{c}}, \quad \tilde{j} = \frac{\bar{c}}{4\pi}k_\perp B_0\frac{v_s}{v_A}\frac{\tilde{v}}{v_A}, \quad \tilde{P} = \gamma P_0\frac{\tilde{v}}{v_s}. \tag{1.15}$$

One particular consequence of this is the fact that electric and magnetic fields in Alfvén waves are in-phase, while plasma motion velocity is antiphase (phase shift is π), and the current oscillations are $\pi/2$ phase-shifted relative to them. The longitudinal and transverse components of the current are either in-phase or antiphase to each other, depending on the sign of $k_\|$. In magnetosonic waves, plasma motion velocity and pressure oscillate in-phase.

Table 1.1 Amplitudes of wave field components for various MHD modes.

Modes\components	$B_\perp$	B_b	$B_\parallel$	$E_\perp$	E_b	$E_\parallel$
A	0	$\tilde{B}$	0	$\tilde{E}$	0	0
FMS	$-\frac{k_\parallel}{k}\tilde{B}$	0	$\frac{k_\perp}{k}\tilde{B}$	0	$\tilde{E}$	0
SMS	$\frac{k_\parallel}{k}\tilde{B}$	0	$-\frac{k_\perp}{k}\tilde{B}$	0	$-\tilde{E}$	0

Modes\components	$j_\perp$	j_b	$j_\parallel$	$v_\perp$	v_b	$v_\parallel$	P
A	$-i\frac{k_\parallel}{k}\tilde{j}$	0	$i\frac{k_\perp}{k}\tilde{j}$	0	$-\tilde{v}$	0	0
FMS	0	$-i\tilde{j}$	0	$\tilde{v}$	0	0	$\tilde{P}$
SMS	0	$i\tilde{j}$	0	0	0	$\tilde{v}$	$\tilde{P}$

In FMS waves, the electric field and the parallel component of the magnetic field are also in-phase to plasma motion velocity and pressure, while $B_\perp$ is either in-phase or antiphase, depending on the sign of $k_\parallel$. In SMS waves, the electric field and the parallel component of the magnetic field oscillate antiphase to plasma motion velocity and pressure, while $B_\perp$ is either in-phase or antiphase, depending on the sign of $k_\parallel$. Current oscillations in magnetosonic waves are $\pi/2$ phase-shifted relative to the oscillations of the other components.

In Alfvén and FMS waves, the plasma oscillations are directed across the magnetic field lines, while being directed along them in SMS waves. The pressure in Alfvén waves is not perturbed. This means that the field lines move at the same velocity as the plasma is displaced. A certain analogy may be drawn between Alfvén wave propagation and oscillations propagating along a 1D string. In an FMS wave, transverse displacement of the plasma is accompanied by magnetic and plasma pressure perturbations. The propagation of these oscillations is similar to how common sound waves propagate in gas. In SMS waves, the plasma displacement is along the magnetic field. Such oscillations resemble the propagation of sound waves in a straight 1D channel. The magnetic field lines serve as walls determining the propagation direction. A qualitative picture of magnetic field and plasma oscillations in various MHD modes is shown in Figure 1.2.

Let us examine some effects that are beyond the ideal MHD framework. The first is hydromagnetic wave decay. Two effects cause this decay: plasma particle collisions resulting in a viscous plasma with finite conductivity and collisionless – Cherenkov and cyclotron – interaction between the waves and particles. The collision-induced decay of hydromagnetic waves in the magnetosphere is negligibly small. The decay decrement can be represented in the same form for the three modes:

$$\frac{\gamma}{\omega} \sim \frac{\omega \nu_m}{v_A^2} = \frac{\bar{c}^2}{v_A^2}\frac{\omega \nu_e}{\omega_{pe}^2}, \tag{1.16}$$

where γ is the oscillation decrement, $\nu_m = \bar{c}^2/4\pi\sigma = \bar{c}^2\nu_e/\omega_{pe}^2$ is the magnetic viscosity, σ is the longitudinal conductivity of plasma, ν_e is the electron collision frequency, $\omega_{pe} = \sqrt{4\pi n_e e^2/m_e}$ is the electron Langmuir frequency. The parameters in (1.16)

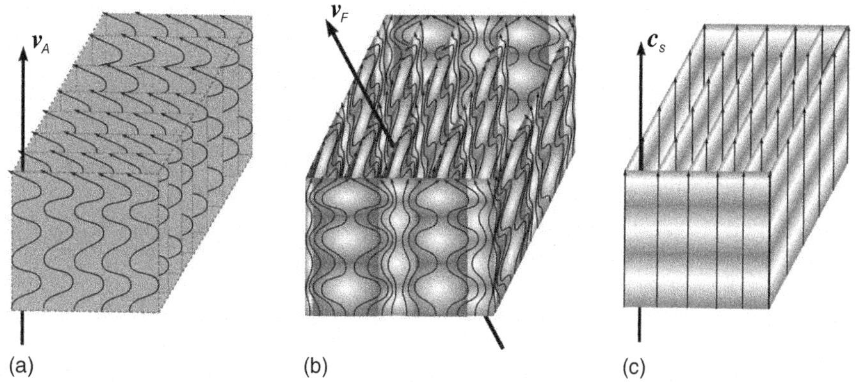

Figure 1.2 Schematic of magnetic field oscillations (field lines), plasma pressure (shades of gray) and group velocity directions for Alfvén (a), fast magnetosonic (b) and slow magnetosonic (c) waves propagating in a homogeneous plasma.

vary very widely in the magnetosphere – from $v_A \sim 3 \cdot 10^3$ km/s, $\omega_{pe} \sim 10^6$ s^{-1}, $\nu_e \sim 10^{-2}$ s^{-1} on inner magnetic shells to $v_A \sim 3 \cdot 10^2$ km/s, $\omega_{pe} \sim 3 \cdot 10^4$ s^{-1}, $\nu_e \sim 10^{-5}$ s^{-1} in the outer magnetosphere. Given the characteristic frequencies of the observable magnetospheric MHD oscillations $\omega = (10^{-2} \div 10)$ s^{-1}, we arrive at an estimated $\gamma/\omega \sim 10^{-7} \div 10^{-14}$. The estimates are somewhat larger if we take into account particle collisions with small-scale wave turbulence of the magnetospheric plasma. However, nearer to the Earth ionosphere, where the collision-induced decay becomes significant, the collision frequency rises dramatically.

The collisionless decay of Alfvén and FMS waves is also small. Their phase velocities ($\sim v_A$) are much higher than the thermal velocities of ions ($v_i \sim v_s$). Their decrement due to ions is therefore proportional to exponentially small factor $\exp(-v_A^2/v_i^2)$ [26]. The thermal velocity of electrons, v_e, can be above, below or of order v_A, in the magnetosphere. The decay due to electrons being small results from the big difference between ion and electron masses. Half of the Alfvén wave and FMS wave energy is contained in the time-averaged kinetic energy of ions and the light electrons cannot slow them down effectively. Therefore, the decrements of the collisionless decay of these waves due to electrons contain a small factor, m_e/m_i (see [24]):

$$\frac{\gamma_{eA}}{\omega} \sim \frac{m_e}{m_i}\frac{v_e}{v_A}\frac{k^2 v_A^2}{\omega_i^2}, \quad \frac{\gamma_{ef}}{\omega} \sim \frac{m_e}{m_i}\frac{v_e}{v_A}.$$

If $v_A \gg v_e$ these formulae also contain the exponentially small factor $\exp(-v_A^2/v_e^2)$.

The picture is completely different for a collisionless Landau decay in SMS waves. The dispersion equation for low-frequency oscillations of plasma (with Maxwell distribution of ions and electrons over velocities) has the following form (see [24]):

$$1 + \sum_{\alpha=i,e} \frac{\omega_{p\alpha}^2}{k^2 v_\alpha^2}\left[1 + i\sqrt{\pi} z_0^\alpha e^{x_\alpha} \sum_{n=-\infty}^{\infty} I_n(x_\alpha) w(z_n^\alpha)\right] = 0, \tag{1.17}$$

where the summing is according to particle types (the α index denotes plasma ions, $\alpha = i$ or electrons, $\alpha = e$) and cyclotron harmonics (the n index). The notations are $k = \sqrt{k_\parallel^2 + k_\perp^2}$ is the wave vector module, $x_\alpha = k_\perp^2 \rho_\alpha^2$, where $\rho_\alpha = v_\alpha/\omega_\alpha$ is the

Larmor radius, $\omega_\alpha = eB_0/m_\alpha \bar{c}$ is the cyclotron frequency, $\omega_{p\alpha} = \sqrt{4\pi n_\alpha e^2/m_\alpha}$ is the plasma frequency and $v_\alpha = \sqrt{T_\alpha/m_\alpha}$ is the thermal velocity of the α type particles, $z_n^\alpha = (\omega - n\omega_\alpha)/\sqrt{2}k_\parallel v_\alpha$. For small values of the argument (the condition $|k_\perp \rho_\alpha| \ll 1$ is assumed to hold), the modified Bessel function $I_n(x_\alpha)$ is approximately represented as $I_n(x_\alpha) \approx (x_\alpha/2)^n/n!$. The function $w(z)$ is the probability integral having the following asymptotic representations (see [27]):

$$w(z) = e^{-z^2}\left(1 + \frac{2i}{\sqrt{\pi}}\int_0^z e^{t^2/2}dt\right) \approx \begin{cases} 1 - z^2 + 2iz/\sqrt{\pi}, & |z| \ll 1, \\ \exp(-z^2) + i/\sqrt{\pi}z, & |z| \gg 1. \end{cases}$$

In the sum over n in (1.17), in the known limiting case $v_i \ll |\omega/k_\parallel| \ll v_e$, we can restrict ourselves to the zero harmonics and write the dispersion equation approximately as

$$\frac{\omega_{pi}^2}{k^2}\left(\frac{1}{v_{es}^2}(1 + i\sqrt{\pi}z_e^0) - \frac{k_\parallel^2}{\omega^2}\right) \approx 0,$$

where $\omega_{pi}^2/v_{es}^2 = \omega_{pe}^2/v_e^2$, $v_{es} = v_i\sqrt{T_e/T_i} = \sqrt{\bar{\gamma}T_e/m_i}$ is the ion thermal velocity determined from the electron temperature. In the zero order of perturbation theory, the solution of this equation gives the dispersion equation for SMS waves in the limit $T_e \gg T_i$: $\omega^2 = k_\parallel^2 v_{es}^2$. Given the next order of perturbation theory, we obtain the dispersion equation taking into account oscillation energy absorption:

$$\omega^2 = k_\parallel^2 v_{es}^2\left(1 - i\sqrt{\frac{\pi m_e}{2m_i}}\right).$$

In this limiting case, the value of the relative SMS decrement we are interested in $\bar{\varepsilon}_s = \gamma_s/\text{Re}(\omega) \equiv \bar{\varepsilon}_{s\infty} = -\sqrt{\pi m_e/2m_i}/2 \approx -0{,}015$, where $\gamma_s \equiv \text{Im}\omega$. The full numerical solution of (1.17) for $\bar{\varepsilon}_s$ as a function of (T_e/T_i) is shown in Figure 1.3. The calculated curve $\bar{\varepsilon}_s(T_e/T_i)$ has a universal form for a large enough variation range of the plasma parameters including the parameter variation range in the Earth magnetosphere (1 nT $\leq B_0 \leq$ 10 T; 1 km/s$\leq v_A \leq 10^4$ km/s; $10^{-2} \leq \beta \leq 1$).

Note that, unlike Alfvén and FMS waves, the SMS decrement is rather large ($|\bar{\varepsilon}_s| \sim 1$) when $T_e/T_i \lesssim 1$ (and $\beta \lesssim 1$). This results from the SMS phase velocity being close to the plasma ion thermal velocity, in this case,

$$c_s \sim v_s = \sqrt{\bar{\gamma}\frac{P_0}{\rho_0}} = \sqrt{\bar{\gamma}\frac{T_e + T_i}{m_i}} \sim v_i,$$

making the decay due to ions very large ($\gamma \sim \omega$), thus preventing SMS waves from propagating.

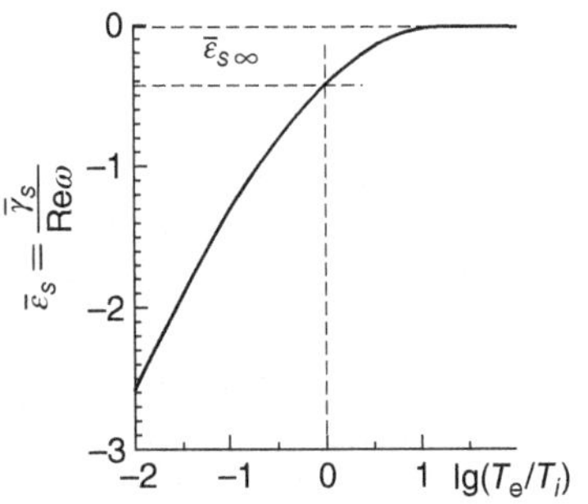

Figure 1.3 Relative SMS decrement $\bar{\varepsilon}_s = \gamma_s/\text{Re}\omega$ vs. non-isothermality, lg T_e/T_i, of homogeneous plasma.

Let us examine another effect unaccounted for within the ideal MHD framework, namely the transverse dispersion of Alfvén waves, absent from (1.10). We will use the following equations for the perturbed electric and magnetic fields:

$$\mathrm{rot}\,\mathbf{E} = i\frac{\omega}{\bar{c}}\,\mathbf{B}, \quad \mathrm{rot}\,\mathbf{B} = -i\frac{\omega}{\bar{c}}\hat{\varepsilon}\,\mathbf{E}, \tag{1.18}$$

where $\hat{\varepsilon}$ is the dielectric permeability tensor of plasma. Let us employ the following representation of the tensor $\hat{\varepsilon}$ for MHD waves (see [24]):

$$\hat{\varepsilon} = \frac{\bar{c}^2}{v_A^2}\begin{pmatrix} 1-\frac{3}{4}k_\perp^2\rho_i^2 & iu & 0 \\ -iu & 1-2k_\perp^2\rho_i^2 & 0 \\ 0 & 0 & \tilde{G}(\omega/k_\parallel v_e)/k_\parallel^2\rho_s^2 \end{pmatrix}, \tag{1.19}$$

where the notations are $u = \omega/\omega_i$, $\rho_s = v_{es}/\omega_i$ (hereafter we assume u, $|k_\perp\rho_i|$, $|k_\perp\rho_s/\tilde{G}| \ll 1$), and the function $\tilde{G}(z)$ is expressed in terms of the familiar Kramp function $\tilde{W}$,

$$\tilde{G}(z) = 1 + i\sqrt{\frac{\pi}{2}}z\tilde{W}\left(\frac{z}{\sqrt{2}}\right) \equiv 1 - ze^{-z^2/2}\int_0^z e^{t^2}dt + i\sqrt{\frac{\pi}{2}}ze^{-z^2/2},$$

and has the following asymptotic representations:

$$\tilde{G}(z) \approx \begin{cases} 1 - z^2 + \cdots + i\sqrt{\pi/2}\,z, & |z| \ll 1, \\ -z^{-2} - \frac{3}{4}z^{-4} + \cdots + \sqrt{\pi/2}\,ze^{-z^2/2}, & |z| \gg 1. \end{cases}$$

For Alfvén waves – even if their transverse dispersion is taken into account – the approximate equality $\omega \approx k_\parallel v_A$ holds true. Hence, $\omega/k_\parallel v_e \approx s_e/\rho_s$, where $s_e = \bar{c}/\omega_{pe}$ is a characteristic electron skin depth in plasma, and we have, in the limiting cases,

$$\frac{\rho_s^2}{\tilde{G}(s_e/\rho_s)} \approx \begin{cases} \rho_s^2, & s_e \ll \rho_s, \\ -s_e^2, & s_e \gg \rho_s. \end{cases} \tag{1.20}$$

Based on (1.18), (1.19), we obtain the following dispersion equation

$$vq^4 - (\alpha - 1)(1 + v)q^2 + (\alpha - 1)^2 - u^2\alpha = 0, \tag{1.21}$$

with these notations: $q = k_\perp/k_\parallel$ is dimensionless transverse wave number, $\alpha = \omega^2/k_\parallel^2 v_A^2$, $v = k_\parallel^2\Lambda^2$,

$$\Lambda^2 \equiv \frac{\rho_s^2}{\tilde{G}(s_e/\rho_s)} + \frac{3}{4}\rho_i^2 = \begin{cases} -s_e^2, & s_e \gg \rho_s, \quad (\beta \ll m_e/m_i), \\ \rho_{si}^2 = \rho_s^2 + \frac{3}{4}\rho_i^2, & s_e \ll \rho_s, \quad (\beta \gg m_e/m_i). \end{cases} \tag{1.22}$$

Accordingly, for these limiting cases,

$$v = \begin{cases} -\mu^4 \equiv -(m_e/m_i)u^2, & (\beta \ll m_e/m_i), \\ \varkappa^4 \equiv (k_\parallel^2/m_i\omega_i^2)(T_i + 3T_e/4), & (\beta \gg m_e/m_i). \end{cases}$$

When $s_e \sim \rho_s$ ($\beta \sim m_e/m_i$), the value of Λ^2 is complex. Its variation in the complex Λ^2 plane for $s_e \gg \rho_s$ to $s_e \ll \rho_s$ is shown in Figure 1.4. Equation (1.21) describes two MHD oscillation branches – the Alfvén and FMS waves. The behaviour of these branches in the plane (q^2, α) for the limiting cases of 'cold' ($\beta \ll m_e/m_i$) and 'hot' ($\beta \gg m_e/m_i$) dispersion of Alfvén waves is examined in Appendix A.

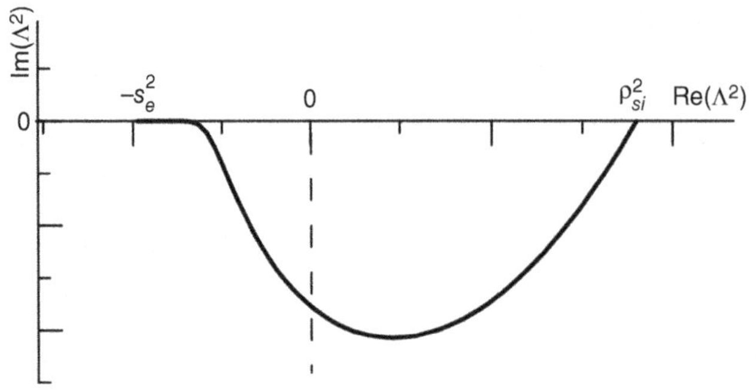

Figure 1.4 Qualitative behaviour of the function $\Lambda^2(s_e/\rho_s)$ in the complex Λ^2 plane as the argument varies from $s_e/\rho_s \gg 1$ to $s_e/\rho_s \ll 1$.

Small corrections for effects unaccounted for within the ideal MHD framework practically do not affect FMS dispersion, so that the dispersion Eq. (1.11) can be used for these waves. Conversely, it is these small corrections that determine the transverse dispersion (dependence of frequency ω on the wave-vector transverse component $k_\perp$) of Alfvén waves. Presence of this dispersion means that Alfvén waves can propagate, not only along magnetic field lines, but also in a transverse direction to them. For quasi-parallel Alfvén waves, $k_\perp \lesssim \sqrt{u}k_\parallel$, the transverse dispersion is determined by ion inertia, and the solution of the dispersion equation has the form:

$$\omega = k_\parallel v_A \left[1 - \frac{u}{2}\left(\frac{k_\perp^2}{k_0^2} + \sqrt{1 + \frac{k_\perp^4}{k_0^4}}\right)^{-1}\right], \tag{1.23}$$

where $k_0^2 = 2uk_\parallel^2$. The characteristic group velocity of the transverse propagation of quasi-parallel Alfvén waves is

$$v_{g\perp} = \frac{\partial\omega}{\partial k_\perp} = v_A \frac{k_\perp}{2k_\parallel} \ll v_A.$$

For large $k_\perp \gg \sqrt{u}k_\parallel$ the Alfvén wave dispersion is called 'kinetic', and the solution of their dispersion equation has the form

$$\omega = k_\parallel v_A \left[1 + k_\perp^2 \Lambda^2\right], \tag{1.24}$$

where Λ^2 is determined from (1.22). When $\beta \sim m_e/m_i$ the value of Λ^2 is complex ($|\Lambda^2| \sim s_e^2 \sim \rho_{si}^2$), and the imaginary part of (1.24) is the decrement of the Cherenkov decay due to background plasma electrons, their thermal velocity being close to the Alfvén wave propagation velocity $v_e \sim v_A$.

The values of s_e and ρ_i characteristic of the magnetosphere lie within the 0.1–10 km range. This means that dispersion (1.24) is only significant for waves that are extremely small scale in the transverse direction. The characteristic transverse group velocity of propagating kinetic Alfvén waves is

$$v_{g\perp} = \mathrm{Re}\,\frac{\partial\omega}{\partial k_\perp} = 2v_A k_\parallel k_\perp \mathrm{Re}\,(\Lambda^2) \ll v_A.$$

Another important feature of Alfvén waves is the fact that, unlike magnetosonic waves, they have a non-zero longitudinal component of the perturbed electric field. As follows from (1.18), (1.19),

$$E_{\parallel} = \frac{k_{\parallel} k_{\perp}}{k_{\perp}^2 - \tilde{G}(s_e/\rho_s)/\rho_s^2} E_{\perp} \approx -k_{\parallel} k_{\perp} \frac{\rho_s^2}{\tilde{G}(s_e/\rho_s)} E_{\perp}. \tag{1.25}$$

The above can be summarised as follows. Taking into account effects determined by the thermal movement of particles results in a strong decay of SMS waves in a $T_i \gtrsim T_e$ plasma. Under typical conditions, $T_i \gg T_e$ in nearly all magnetospheric regions. This makes it impossible for the SMS oscillation eigenmodes to exist, in nearly the entire magnetosphere. The only exception is the inner plasmasphere, where $T_e > T_i$ and the SMS decay decrement is small (see [28]). For Alfvén and FMS waves, the decay effects related to their interaction with plasma particles are small. As for the transverse dispersion of magnetosonic waves, it is large enough even in the ideal MHD approximation so that taking into account small corrections unaccounted for within that approximation fails to result in any significant effects. Situations are possible for Alfvén waves when it is necessary to take into account their transverse dispersion (1.23) or (1.24) determining their slow movement across magnetic field lines or the presence of a parallel component of the electric field (1.25) in them.

2

MHD Oscillations in 1D-Inhomogeneous Model Magnetosphere

As has been stated in Chapter 1, MHD oscillations in an inhomogeneous plasma do not subdivide into separate linearly independent modes but comprise a single wave field. Its dispersion characteristics make this wave field look like one particular branch of MHD oscillations in a homogeneous plasma in some spatial regions, while making it look like a different branch in other regions. They can be clearly distinguished in only a few special cases. In theoretical studies of MHD oscillations in an inhomogeneous plasma, the transition from spatial regions with oscillations of one type of dispersion to regions where oscillations are of another type of dispersion are conventionally described in terms of interaction between the different MHD oscillation branches in a homogeneous plasma. Such interaction is resonant in character and occurs at a small enough spatial scale.

A most fruitful concept in magnetospheric MHD oscillation studies has proven to be one of resonance-generated Alfvén waves (Alfvén resonance, or field line resonance – FLR). This process involves a monochromatic fast magnetosonic (FMS) wave propagating in a plasma that is inhomogeneous across the magnetic field lines and driving Alfvén waves on the magnetic shells where its frequency coincides with the local frequency of the Alfvén oscillations. The Alfvén resonance is often invoked when trying to explain certain types of global magnetospheric oscillations [29], coronal plasma heating [30, 31], as well as being suggested as a plasma heating mechanism in controlled fusion reactors [32, 33].

That Alfvén oscillations in the magnetosphere could be resonance-driven by monochromatic FMS wave was first suggested in [34]. That idea was later elaborated in many papers, of special note being [35–37]. They were the first to examine the characteristic features of resonant Alfvén oscillations, which made these oscillations identifiable in ground-based magnetometer network observations [38, 39].

Observations of such resonant Alfvén oscillations were used to plot the eigenfrequency distributions for resonant Alfvén oscillations in the magnetosphere over latitude and compare them to the calculated theoretical values [40, 41]. This in turn made it possible to model the magnetospheric plasma density distribution in the magnetic meridian plane [42–44].

Moreover, there are numerous evidential proofs attesting to a close link between resonant Alfvén oscillations in the magnetosphere and various types of aurora [45–47]. Since kinetic Alfvén waves possess a longitudinal (to the magnetic field) component of their electric field (see (1.25)), they can precipitate charged particles of the magnetospheric plasma into the ionosphere, thus causing its neutral component to glow [48–50].

Less known is another type of resonant interaction between MHD oscillations – magnetosonic resonance, where FMS wave drives SMS wave on a resonance shell [51–53].

Magnetospheric MHD Oscillations: A Linear Theory, First Edition. Anatoly Leonovich, Vitalii Mazur, and Dmitri Klimushkin.

SMS waves as well as Alfvén waves are easily guided along magnetic field lines. This enables their resonant interaction with FMS, same as in the Alfvén resonance. The reason why this type of MHD-oscillation interaction attracts less attention than the Alfvén resonance is that the SMS propagation velocity in balanced plasma configurations with $\beta < 1$ is close to plasma sound speed so that SMS waves decay strongly because of collisionless Landau decay (see Chapter 1). The only known example of their weak decay is a strongly non-isothermal plasma, $T_e \gg T_i$, where T_e, T_i are electron/ion temperatures (see [54]). The above-mentioned types of resonant MHD interaction are, as a rule, studied separately [55, 56]. A major part of plasma configurations in reality, however, feature both types of resonance, which requires that they are examined in combination in order to understand correctly the processes involved.

In this chapter, we will address a number of problems where an important role is played by plasma inhomogeneity across magnetic field lines. Section 2.1 examines, on a qualitative level, the properties of MHD wave propagation in a 1D-inhomogeneous plasma using a model magnetosphere in the form of a rectangular box with ideally reflecting walls ('box model'). The plasma transition layer is simulated in Section 2.2, and the model is used to solve the problem of fast magnetosound incident and reflected from plasma inhomogeneity. Section 2.3 examines the Alfvén resonance – the process when Alfvén oscillations are driven by a monochromatic FMS wave incident on the 'cold' plasma transition layer. A similar process is studied in Section 2.4 for resonant Alfvén waves excited by an FMS impulse. A problem is solved in Section 2.5 concerning the energy balance of MHD oscillations in Alfvén resonance. Section 2.6 addresses the problem of an incident FMS wave on the 'warm' plasma transition layer when there are resonance surfaces, not only for Alfvén waves, but for slow magnetosonic waves as well. The Alfvén resonance problem is solved in Section 2.7 using a model with a non-ideal plasma, when the transverse dispersion of Alfvén waves is determined by small kinetic effects.

Another important effect related to the transverse inhomogeneity of plasma is the possible waveguide propagation of various types of MHD waves in a plasma with a non-monotonic Alfvén speed profile across magnetic field lines. A waveguide for FMS waves is examined in Section 2.8 formed by a local minimum of the Alfvén speed extended along the magnetic field lines. A similar waveguide is studied in Section 2.9 for quasi-parallel Alfvén waves, the transverse dispersion of which is related to plasma gyrotropy (to the finite value $\omega/\omega_i \ll 1$). A problem is solved in Section 2.10 concerning the waveguide propagation of quasi-transverse Alfvén waves, the transverse dispersion of which is determined by small kinetic effects, when the finite Larmor radius of ions and electron skin depth are taken into account. Section 2.11 analyses the interaction between waveguide modes of various branches of MHD oscillations in a 'warm' plasma. Analogous analysis is done in Section 2.12 for oscillations propagating along cylindric plasma filaments.

Earth's magnetosphere features a sharp enough boundary, the magnetopause, separating the magnetospheric plasma from the solar wind plasma. Of interest, therefore, is the problem of FMS waves crossing the magnetopause while they penetrate from the solar wind to the magnetosphere, where they can excite various types of geomagnetic pulsations. Section 2.13 deals with the problem of FMS waves passing through the magnetopause, which is simulated by a tangential discontinuity in the parameters of the medium. The solar wind shear flow at the magnetopause makes it unstable to FMS oscillations. Plasma flow instability is studied in Section 2.14 for shear flow models in the form of a smooth transition layer separating two homogeneous half-spaces. The inhomogeneity in either half-space is simulated by the presence of reflecting walls in one or both half-spaces. Sections 2.15 and 2.16 solve the stability problem of the geotail boundary, where the geotail is simulated by a plasma cylinder (which is inhomogeneous over radius) enwrapped by the solar wind flow.

Another region of the magnetospheric plasma where its inhomogeneity is important in principle is the near-Earth space including the ionosphere and atmosphere. Alfvén waves penetrating from the magnetosphere to the ground through such a multilayered medium are an important problem in magnetospheric physics. In Section 2.18 we calculate the field of electromagnetic oscillations induced at Earth surface by Alfvén waves from the magnetosphere that are incident on the ionosphere. Geomagnetic field line inclination to the Earth surface is taken into account.

2.1 A Qualitative Picture of MHD Wave Propagation in a 1D-Inhomogeneous Plasma

Both the statement of the Alfvén resonance problem to be discussed below and the model of the medium (box model) correspond to those in [35, 37, 57]. Let us consider a model where the magnetic field $\mathbf{B}_0$ is homogeneous, rectilinear and directed along the z axis. The background plasma is 'cold' ($v_s = 0$) and immobile ($\mathbf{v}_0 = 0$), its density ρ_0 depending on the x coordinate only (see Figure 2.1). To simulate the presence of an ionosphere, let us assume that the plasma is confined between two conducting planes that are normal to the z axis, located at $z_- = 0$ and $z_+ = L$, where L is field-line length. The conductivity of the ionosphere is high enough, and we will assume these planes to be ideally conducting, in the main order. The boundary conditions in the z axis then are

$$E_{x,y}|_{z=z_\pm} = 0. \tag{2.1}$$

In the model in question, the solutions to Eq. (1.5) have the form $\Phi = \tilde{\Phi}(x)\, e^{(ik_y y + ik_z z - i\omega t)}$, where Φ is any of the wave-field components, k_y and k_z are the wave-vector components along the y and z axes. Consequently, the z dependence of $E_{x,y}$, given the boundary conditions (2.1), has the form $\sim \sin k_z(z - z_-)$, where $k_z = \pi N/L$ ($N = 1, 2, 3, \dots$ is the longitudinal wave number). This means that, along the z axis, the field structure of the MHD oscillations is the sum of waves standing along the magnetic field lines. Figure 2.2 shows the behaviour of the electric E as well as magnetic B field of the wave for the fundamental ($N = 1$) and second ($N = 2$) harmonics. Note that the electric field has the antinode at the equator ($z = L/2$), while the magnetic field has the node. For the second harmonic, the situation is the opposite: the electric field at the equator has the node, while the magnetic field has the antinode.

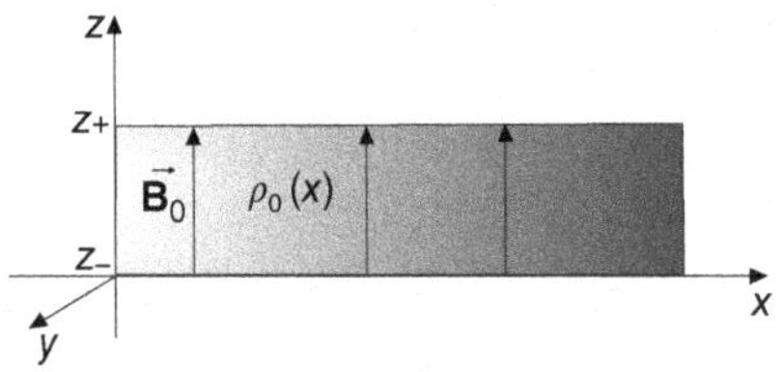

Figure 2.1 Model 1D-inhomogeneous plasma (with density $\rho_0 \equiv \rho_0(x)$) in a homogeneous magnetic field $\mathbf{B}_0 = const$.

Figure 2.2 Parallel structure of electric E (black lines) and magnetic B (gray lines) field of the MHD wave. Solid and dashed lines depict the fundamental ($N = 1$) and second ($N = 2$) harmonics, respectively.

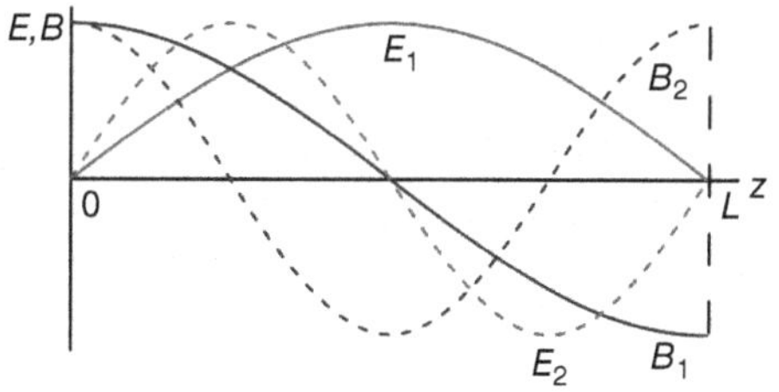

Let us examine, qualitatively, the medium inhomogeneity-induced effects. The most important of these are the reflection of FMS waves from the plasma inhomogeneity, coupling between the Alfvén and FMS modes, and the Alfvén resonance. FMS reflected from the inhomogeneity can be described using the WKB approximation, applicable when the wavelength along the x coordinate is much smaller than the inhomogeneity scale. In the main order of the WKB approximation, the x dependence of the disturbed values is given by the factor $\exp\left(i \int k_x\, dx\right)$, where $k_x = k_x(x)$ is the local value of the wave vector, determined from the local dispersion equation for FMS waves:

$$k_x^2(x) = \omega^2/v_A^2(x) - k_y^2 - k_z^2. \tag{2.2}$$

In the transparent region, where $k_x^2 > 0$, the disturbance has the form of travelling wave, or, more precisely, is, in the general case, a superposition of opposite-travelling waves: $C_1 \exp\left(i \int k_x dx\right) + C_2 \exp\left(-i \int k_x dx\right)$. Not only the wave vector $k_x(x)$, but also the wave amplitudes $C_1(x)$, $C_2(x)$ change as these waves propagate. In the opaque region, where $k_x^2 < 0$, the oscillation does not propagate, but is a superposition of the decreasing and increasing exponential functions. If the opaque region extends to infinity, only the decreasing exponential function is present. These two regions are divided by the turning point $x = x_F$, where $k_x^2(x_F) = 0$ (see Figure 2.3(b)). It is evident from (2.2) that $v_A(x)$ is smaller in the transparent region than in the opaque region (Figure 2.3(a)).

The inhomogeneity affects the Alfvén wave even more radically. In the ideal MHD approximation, this wave has no transverse dispersion (i.e. its dispersion equation $\omega^2 = k_z^2 v_A^2(x)$ contains neither k_x nor k_y). For any given values of ω and k_z, the dispersion equation for Alfvén waves can only be valid at certain points $x = x_A$, called the points of Alfvén resonance. The Alfvén wave is 'tied' to these points and cannot travel along the x coordinate. This is especially obvious in the $k_y = 0$ case, when, in the ideal MHD approximation, the Alfvén and FMS waves are independent modes of inhomogeneous plasma oscillations. The manner in which the FMS propagates was described earlier. For the Alfvén wave we obtain this equation from (1.5), (1.6) and (1.8):

$$[\omega^2 - k_z^2 v_A^2(x)] E_x = 0,$$

which is solved as $E_x = C\delta(x - x_A)$. The Alfvén disturbance is concentrated on magnetic surface $x = x_A$, the frequency of the disturbance coincides with the local Alfvén frequency $k_z v_A(x)$.

When $k_y \neq 0$, the situation is more complex. The spatial structure of an MHD oscillation is such that it is close to FMS wave in one region of space, while being close to Alfvén wave in another. The question arises: what is Alfvén vs. FMS wave, in an inhomogeneous plasma? The disturbance in an inhomogeneous plasma can only be mathematically correctly subdivided into various oscillation modes associated with modes in a homogeneous

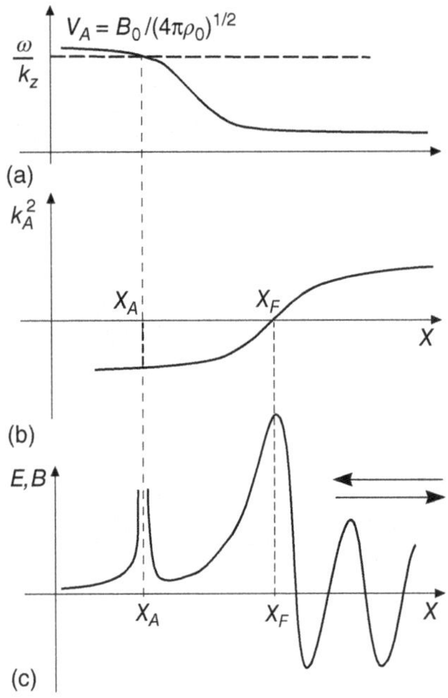

Figure 2.3 (a) Alfvén speed distribution along the x axis in a 1D-inhomogeneous model of the medium. (b) Dependence of the squared WKB component of the wave vector, $k_x^2(x)$, of MHD oscillations in a 'cold' plasma (x_A is the resonance point for Alfvén wave, x_F is the turning point for FMS wave), (c) The structure of wave field components of MHD oscillations in a 1D-inhomogeneous plasma.

plasma if the WKB approximation is applicable to each mode. The concept of dispersion equation retains its sense for this approximation, which allows the different modes to be identified. These modes are linearly transformed at points dividing the domains of existence of the different modes. In the immediate neighbourhood of these points, where the WKB approximation is invalid, the oscillation cannot be treated as any of the above modes.

The WKB approximation cannot be applied when describing the Alfvén wave, because this wave lacks transverse dispersion. Nevertheless, in a close neighbourhood of the Alfvén resonance point, the MHD oscillation can be treated as Alfvén wave, because the dispersion equation for Alfvén waves, $\omega^2 = k_z^2 v_A^2(x)$ is valid at $x = x_A$. Even though we cannot ascribe a particular value to k_x, it should be regarded as very large near $x = x_A$ because the solution at this point is singular (see below). In any case, $|k_x| \gg |k_y|$, yielding $|E_x| \gg |E_y|$, i.e. $\mathbf{E}_\perp \parallel \mathbf{k}_\perp$ (see Table 1.1), which is an Alfvén wave signature. The phenomenon near $x = x_A$ is called 'Alfvén resonance' and treated as an Alfvén wave excited by fast magnetic sound propagating in an inhomogeneous plasma. Note that x_A is in an opaque region for magnetic sound, because it follows from (2.2) that $k_x^2(x_A) = -k_y^2 < 0$ for $\omega^2/v_A^2(x) - k_z^2 = 0$.

We will see from the calculations below that the amplitude of disturbed fields, in the ideal MHD approximation at $x = x_A$, becomes infinite ($E_x, B_y \sim (x - x_A)^{-1}$; $E_y, B_x \sim \ln|x - x_A|$). This singularity is regularised by taking into account the various effects beyond the scope of the ideal MHD, e.g. dissipation. The presence of dissipation results in wave energy absorption. A remarkable property of the Alfvén resonance is that this absorption remains finite as the dissipation coefficients (collision frequency, resistance, etc.) tend to zero. As the dissipation coefficients decrease, the disturbance amplitude increases at $x = x_A$, while the characteristic width of the resonance peak decreases in such a manner that the dissipated power remains finite.

The singularity of the solution is also regularised by taking into account the kinetic dispersion of the Alfvén wave acquiring the opportunity to travel along the x coordinate.

In the region of Alfvén resonance, FMS is transformed into an Alfvén wave travelling across the magnetic field. The energy carried by the Alfvén wave remains finite as the dispersion tends to zero, retaining the same value as the energy absorbed when the oscillations are dissipated [58] (see Appendix S). Thus the energy the FMS loses in the region of Alfvén resonance does not depend on the physical mechanism of its dissipation (provided its effects are small).

Formally, in terms of mathematics, the above-described property of the Alfvén resonance can be explained as follows. Under ideal MHD, the solution is completely determined by the rules of matching both sides of the singular point $x = x_A$. Two possible solutions exist depending on whether this point is bypassed from above or below in complex plane x. Either solution has its characteristic ratio between the energy of the FMS wave falling onto and reflected from the region of Alfvén resonance. In one solution, the energy flux of the incident wave is larger than that of the reflected wave, i.e. the oscillation energy is absorbed in this region; in the other solution, the energy flux of the reflected wave is larger, i.e. energy is generated. The latter solution seems unacceptable in physical terms.

Choosing the rule for bypassing a singular point is dictated by the solution of a problem with initial conditions corresponding to the resonance oscillation amplitude growing with time. This approach was first used in the well-known paper [59], where the collisionless damping of Langmuir waves was discovered. This approach leads to the following rule for bypassing a singular point. The correct type of bypassing is determined by an infinitely small positive imaginary addition to the oscillation frequency, $\omega \to \omega + i\gamma$, where $\gamma > 0$. In an analytic continuation, this addition shifts the singular point $x = x_A$ away from the real axis, thus determining the bypass route along the real x axis. All the above mechanisms for regularising the singularity comply with this rule.

2.2 Model of a Smooth Transition Layer and Basic Equations for MHD Oscillations

The qualitative picture presented in Section 2.1 was based on results of a rigorous study of the problem of MHD waves propagating in a 1D-inhomogeneous 'cold' plasma. A typical problem of this sort is the problem of incident/reflected FMS wave on a plasma inhomogeneity in the form of a smooth transition layer. Let us assume that plasma density $\rho_0(x)$ monotonically increases $x = -\infty$ to $x = +\infty$. The Alfvén speed will then monotonically decrease in this direction. Let us specify the following model of the Alfvén speed distribution

$$v_A(x) = \frac{1}{2}\left[v_A^+ + v_A^- + (v_A^+ - v_A^-)\tanh\frac{x}{\Delta}\right], \tag{2.3}$$

where $v_A^\pm$ are the Alfvén speed values for $x \to \pm\infty$, Δ is the characteristic width of the transition layer. The distribution $v_A(x)$ is shown in Figure 2.3(a).

In the general case, the equation describing the MHD oscillation structure along the x coordinate in a moving plasma (with velocity $\mathbf{v}_0$ and finite pressure P_0), has the following form (see Appendix B):

$$\frac{\partial}{\partial x}\left(\frac{\rho_0\Omega^2}{k_x^2}\frac{\partial\zeta}{\partial x}\right) + \rho_0\Omega^2\zeta = 0. \tag{2.4}$$

Here, $v_x = d\zeta/dt = \partial\zeta/\partial t + (\mathbf{v}_0\nabla)\zeta$ is the x component of the plasma oscillation velocity, ζ is the shift of the plasma element along the x axis, $\Omega^2 = \overline{\omega}^2 - k_z^2 v_A^2$, $\overline{\omega} = \omega - k_z v_0$ is the Doppler-shifted oscillation frequency,

$$k_x^2 = \frac{\overline{\omega}^4}{\overline{\omega}^2(v_A^2 + v_s^2) - k_z^2 v_A^2 v_s^2} - k_z^2 - k_y^2 \tag{2.5}$$

is the squared x component of the wave vector in the WKB approximation, where $v_s = \sqrt{\overline{\gamma} P_0/\rho_0}$ is plasma sound speed, $\overline{\gamma}$ is adiabatic index.

Expressions for the other components of the wave field are obtained in Appendix B.

2.3 FMS Wave Reflected from the Transition Layer in a Cold Plasma. Alfvén Resonance

Let us consider the problem of an incident FMS wave reflected from a plasma inhomogeneity in the form of a smooth transition layer. Let us assume the plasma to be ideal ($T_e = T_i = 0$) and immobile ($\mathbf{v}_0 = 0$). Let the FMS wave of given amplitude fall from $x = +\infty$ onto the transition layer of the form (2.3), the parameters of the oscillations ω, k_y, k_z being such that, in the medium under study, there are both an Alfvén resonance point $x = x_A$, and an FMS wave reflection point $x = x_F$ (see Figure 2.4). Note that this problem simulates an important magnetospheric phenomenon – FMS wave from the solar wind falling onto the magnetosphere and standing Alfvén waves excited inside the magnetosphere.

In the equation for monochromatic oscillations of plasma, (2.4), $k_x^2 = \omega^2/v_A^2 - k_y^2 - k_z^2$ in this case. It is also possible to present (2.4) in the form

$$\eta'' - \left[\ln(\rho_0\Omega^2)\right]'\eta' + k_x^2\eta = 0, \tag{2.6}$$

$$\eta = \frac{\rho_0\Omega^2}{k_x^2}\frac{\partial\zeta}{\partial x},$$

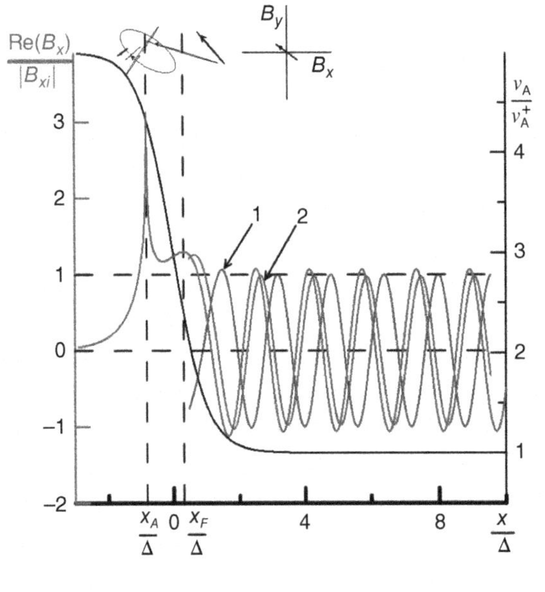

Figure 2.4 Spatial distribution of MHD oscillation B_x-component (left-hand axis) in the FMS wave which is incident to/reflected from the transition layer with an Alfvén resonance point. The gray line is the full oscillation field. WKB approximation: line 1 is FMS wave incident onto the transition layer, line 2 is FMS wave reflected from the transition layer. Oscillation hodographs for various points x are shown above. The right-hand axis line is Alfvén speed distribution $v_A(x)$.

where the prime denotes differentiation over x. In the ideal MHD approximation, (2.4), (2.6) determine the spatial structure of the oscillation field, but outside this approximation they play the role of a zero approximation.

An important feature in (2.4), (2.6) is the presence of a singular point, $x = x_A$, where $\Omega^2 = 0$. This point is the Alfvén resonance point. Another peculiarity of these equations, in the model in question, is the presence of an FMS turning point, $x = x_F$, where the WKB component of the wave vector vanishes, $k_x(x_F) = 0$. The qualitative distribution of function $k_x^2(x)$ is given in Figure 2.3(b). Right of x_F on the asymptotics, in the transparent region, the solution of the problem is the sum of the incident FMS wave and the wave reflected from the transition layer. On the left asymptotic, deep in the opaque region, the solution of the problem is an FMS-type oscillation with decreasing amplitude. These are the boundary conditions of the problem in question.

The full analytical solution of (2.4) can be constructed if the WKB approximation is applicable between the turning point x_F and the Alfvén resonance point x_A. In the neighbourhood of these points, the conditions for applicability of the WKB approximation are broken, therefore, other approximate methods should be used to solve the equations. Let us construct such a solution.

In the opaque region $x < x_A$, the WKB solution of (2.4) has the form

$$\zeta = C\sqrt{\frac{|k_x|}{\rho_0\Omega^2}}\exp\left(\int_{x_A}^{x}|k_x|dx'\right). \tag{2.7}$$

where C is a constant.

As has been noted above, the WKB approximation is inapplicable near the resonance surface $x = x_A$. As $x \to x_A$, we have $k_x^2 \approx -k_y^2$. In this case, (2.4) has a singularity. Let us regularise this singularity by replacing $\omega \to \omega + i\gamma$ as $\gamma \to 0$. In this case, γ acts as a decrement of Alfvén waves at the Alfvén resonance point. An interesting feature of this resonance is that the energy dissipation rate does not depend on γ as $\gamma \to 0$. Near $x = x_A$, let us employ a linear decomposition

$$\Omega^2 \approx -\omega^2[(x - x_A)/a_A + 2i\gamma/\omega] \tag{2.8}$$

in the coefficients of (2.4), where $a_A = (\partial \ln(v_A^2)/\partial x)^{-1}_{x=x_A}$ is the typical scale of the $v_A(x)$ variation at the Alfvén resonance point. Substituting this decomposition into (2.4) yields

$$\frac{\partial}{\partial x}(x - x_A + i\varepsilon_A)\frac{\partial\zeta}{\partial x} - k_y^2(x - x_A + i\varepsilon_A)\zeta = 0, \tag{2.9}$$

where $\varepsilon_A = 2a_A\gamma/\omega$. The solution of (2.9) that matches (2.7)[1]

$$\zeta = C_A K_0[-k_y(x - x_A + i\varepsilon_A)], \tag{2.10}$$

where $K_0(z)$ is a modified Bessel function, C_A is constant. As $x \to x_A$, we have

$$\zeta = -C_A \ln[-k_y(x - x_A + i\varepsilon_A)]. \tag{2.11}$$

1 Solutions (2.10) and (2.7) are matched by means of matching the inner asymptotics of the solution (2.7) as $x \to x_A$ ($\zeta \approx (C/\omega)\sqrt{k_y a_A/\rho_0|x - x_A|}\exp(-k_y|x - x_A|)$), with the outer asymptotics of (2.10) as $k_y(x - x_A) \to -\infty$ ($\zeta \approx C_A\sqrt{\pi/2k_y|x - x_A|}\exp(-k_y|x - x_A|)$).

When $\gamma = 0$, this solution has a well-known logarithmic singularity at the Alfvén resonance point $x = x_A$. It follows from (B.14) that the oscillation magnetic field components have the form

$$B_x = ib_0 \ln[-k_y(x - x_A + i\varepsilon_A)], \quad B_y = \frac{b_0}{k_y(x - x_A + i\varepsilon_A)},$$
$$B_z = -\frac{k_z}{k_y^2 a_A} b_0, \tag{2.12}$$

where

$$b_0 = C_A k_z B_0. \tag{2.13}$$

A characteristic feature of this solution is that the oscillations reverse their polarisation as they pass through the resonance surface, which is due to the B_y component sign reversal. A typical behaviour of the oscillation hodograph in the (B_x, B_y) plane, at various points near the resonance surface, is shown in Figure 2.4. Left and right of the resonance surface $x = x_A$, the hodograph rotates in opposite directions, while the main axis of the polarisation ellipsis changes its inclination. This feature of the resonant Alfvén waves is widely used for identifying them in the MHD oscillations observed in the magnetosphere [38–40]. Matching (2.10) to solution (2.7) connects the constants: $C_A = C(k_y/\omega)\sqrt{2a_A/\pi\rho_{0A}}$, where the subscript $_A$ denotes parameters at the Alfvén resonance point $x = x_A$.

The WKB solution in the opaque region $x_A < x < x_F$ has the form

$$\zeta = \sqrt{\frac{|k_x|}{\rho_0\Omega^2}}\left[C_1 \exp\left(\int_{x_F}^{x} |k_x| dx'\right) + C_2 \exp\left(-\int_{x_F}^{x} |k_x| dx'\right)\right]. \tag{2.14}$$

Matching it to (2.10) produces $C_1 = C_A(k_z B_{0A}/2k_y\sqrt{2a_A})e^{\Gamma}$, where $\Gamma = \int_{x_A}^{x_F} |k_x| dx$. The WKB approximation used between x_A and x_F does not allow the decreasing exponential function to be taken into account on the background of the increasing exponential function, because it would exceed the accuracy of the asymptotic decomposition.

A solution in the neighbourhood of the turning point $x = x_F$ can be constructed for function η (see (2.6). Linearising $k_x^2 \approx (x - x_F)/a_F^3$ near $x = x_F$ (where $a_F^{-3} = -(\partial k_x^2/\partial x)_{x=x_F}$) yields an Airy equation for function η

$$\frac{\partial^2\eta}{\partial x^2} - a_F^{-3}(x - x_F)\eta = 0. \tag{2.15}$$

Its solution matchable to (2.14) is

$$\eta = C_F Ai[-(x - x_F)/a_F], \tag{2.16}$$

where $Ai(z)$ is the Airy function, $C_F = C_1 k_y B_{0F}\sqrt{a_F}$, and the $_F$ subscript denotes values at point $x = x_F$.

Let us represent the WKB solution in the transparent region $x > x_F$ as

$$\zeta = \sqrt{\frac{k_x}{\rho_0\Omega^2}}\left[C_i \exp\left(-i\int_{x_F}^{x} k_x dx'\right) + C_r \exp\left(i\int_{x_F}^{x} k_x dx'\right)\right], \tag{2.17}$$

where C_i is the amplitude of the incident FMS wave, and C_r is the amplitude of the reflected FMS wave. Matching (2.17) to (2.16) at the asymptotic produces $C_r = iC_i = -C_F e^{i\pi/4}/k_y B_{0F}\sqrt{a_F}$ and $C_1 = -C_i e^{i\pi/4}$. The decreasing exponential is lost in the opaque region $x_A < x < x_F$ resulting in $|C_i| = |C_r|$ in the solution (2.17), i.e. the minor effect of incident wave absorption is lost at the Alfvén resonance point. If the characteristic value of the wave vector is such that $k_y\Delta$, $k_z\Delta \sim 1$, the approximation becomes too rough and the MHD oscillation field can only be calculated correctly using numerical techniques.

Note a few essential points. As can be seen from expressions (2.10), (2.11), the singular point x_A is bypassed from above. The disturbance amplitude grows exponentially from x_A to x_F, this growth being determined by factor $\exp(\Gamma)$, where

$$\Gamma = \int_{x_A}^{x_F} |k_x(x)|\, dx. \tag{2.18}$$

If $k_y\Delta_A \gg 1$, where $\Delta_A = x_F - x_A$, this value is large: $\Gamma \gg 1$. If the decomposition (2.8) is applicable within the entire interval (x_A, x_F), the integral can be found easily:

$$\Gamma = (2/3)(k_y\Delta_A)^3. \tag{2.19}$$

This expression can be used for order of magnitude estimates.

In the general case, the problem of an FMS wave incident onto an arbitrary transition layer can only be solved numerically. A numerical solution of such a problem is given in Figure 2.4 for a transition layer of the form (2.3). Shown are the B_x-component distribution of the full field of MHD oscillations and its decomposition, on the asymptotic right of x_F, where a WKB approximation of the form (2.17) applies, into incident FMS wave and FMS wave reflected from the transition layer. The singularity of the solution at the Alfvén resonance point is regularised by the presence of small dissipation $\gamma/\omega = 10^{-3}$. Note that, in this case, the amplitude of the incident wave is larger than that of the reflected wave. This is due to the energy being partially absorbed at the Alfvén resonance point. The absorption coefficient is defined as

$$D = 1 - R, \tag{2.20}$$

where

$$R = \frac{|C_r|^2}{|C_i|^2}$$

is the coefficient of the FMS wave reflection from the transition layer.

In the two limiting cases, expressions for the oscillation absorption coefficient D can be found analytically, by analysing the trade-off between the energy of the incident and reflected FMS waves and the energy absorbed in the Alfvén resonance region. In the above-discussed example, where the WKB approximation is applicable between points x_A and x_F (for $k_y\Delta_A \gg 1$), we have

$$D = 2\exp(-2\Gamma). \tag{2.21}$$

The Γ is a function of k_y, but, in the general case, may also depend on other parameters determining the properties of the wave and the background plasma. $\Gamma(k_y)$ has a universal form in the case when expression (2.19) can be applied, resulting in

$$D = 2\exp[-(4/3)(k_y\Delta_A)^3]. \tag{2.22}$$

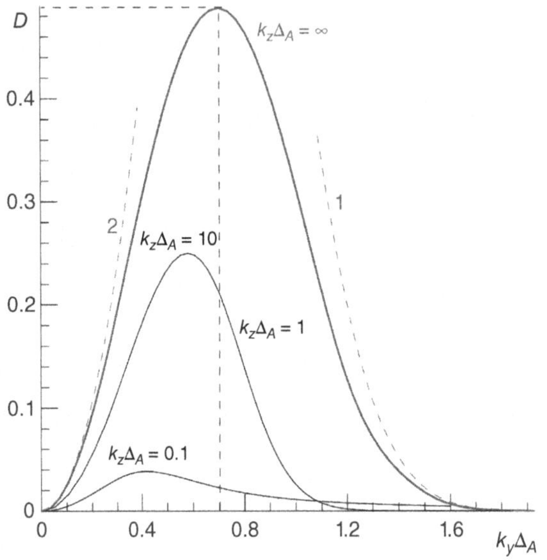

Figure 2.5 The $k_y\Delta_A$ dependence of the absorption coefficient for FMS waves incident onto the transition layer that are partially absorbed at the Alfvén resonance point. Distributions for various $k_z\Delta_A = 0.1; 1; 10; \infty$ are shown. The black marginal curve, for $k_z\Delta_A \to \infty$, corresponds to an infinite layer with a linear $v_A(x)$ profile. The dark gray dash lines show analytical distributions for two limiting cases: (1) $k_y\Delta_A \gg 1$ and (2) $k_y\Delta_A \ll 1$, described by (2.21) and (2.23).

In the other limiting case (for $k_y\Delta_A \ll 1$), we have

$$D = \lambda k_y^2 \Delta_A^2, \tag{2.23}$$

where $\lambda = [2\pi Ai'(0)]^2 \approx 2.6$.

Figure 2.5 shows numerically calculated (for the Alfvén speed profile (2.3) absorption coefficient distributions D as a function of dimensionless parameter $k_y\Delta_A$, for various values of $k_z\Delta_A$. Extremely small values of the decrement were chosen, $\gamma/\omega \to 0$. Also shown are the dependencies $D(k_y\Delta_A)$, given by formulae (2.21) and (2.23) for the two above limiting cases. Parameter $k_y\Delta_A$ determines the characteristic distance from the turning point where the WKB approximation begins to apply. For $k_z\Delta_A \to \infty$, the problem under study becomes similar to that of the linear profile of an inhomogeneity, which has been solved many times (see [52]). Note that the parameters of

2.4 Alfvén Resonance Excited by a Wave Impulse

We have so far considered an individual harmonic of oscillations with given values of ω, k_z, and k_y. Oscillations observed under real circumstances have a more complicated structure – they are a superposition of a certain set of such harmonics.[2] An essential distinction should be noted between a superposition over various values of k_z and k_y and a superposition over frequencies ω. The values of wave vector component $k_z = \pi N/L$ (where L is the field line length, $N = 1, 2, \ldots$ is the standing Alfvén wave number) form a discrete set. Generation mechanisms (i.e. excitation mechanisms) for waves with different N usually differ so much that the main contribution is by a few first harmonics with different N – often by one harmonic only. Given that the y coordinate in the model under consideration corresponds to the azimuthal coordinate in axisymmetric models of the

2 Note that the model magnetosphere discussed here, in the form of a 1D-inhomogeneous transition layer, is also a rough approximation of the reality.

magnetosphere, it is conditionally possible to assume $k_y = m/a$ (where $m = 0, 1, 2, ...$ is the azimuthal wave number, a is the radial coordinate). It is therefore quite justified to examine oscillations with fixed k_z and k_y. Quite different is the case of a superposition over frequencies ω. The frequency can take a continuous series of values, and the properties of the sum of the frequency harmonics can differ radically from the properties of each term. Hereafter we will examine a superposition of such harmonics, while assuming k_z and k_y to be fixed.

Two – opposite in a certain sense – classes can be distinguished among the entire diversity of non-stationary oscillations. One comprises oscillations with strictly correlated phases, exemplified by a wave packet. The other class includes oscillations of random phases. A limiting example of such a phenomenon is white noise, where the phases of individual harmonics are completely uncorrelated. The first class simulates a wave impulse falling from the solar wind onto the frontal part of the magnetosphere; the second class simulates resonant oscillations in the magnetosphere that are pumped by stochastic noises from the solar wind.

Let $\zeta(x, \omega)$ be the solution with given ω, k_z, and k_y, normalised such that, in the transparent region right of the transition layer (see (2.3) and Figure 2.6), where the WKB approximation applies, the incident wave has a unit amplitude ($C_i = 1$ in (2.17). The amplitude of the reflected wave (let us denote it as $S(\omega) \equiv C_r$) is then related to the reflection coefficient as $R = |S(\omega)|^2$. We denote the spatial phase along the x coordinate as

$$\Psi_x(x, \omega) = \int_{x_F(\omega)}^{x} k_x(x', \omega) dx'. \tag{2.24}$$

Let us use the thus normalised solutions (2.17) to construct a wave packet

$$\zeta(x, t) = \int_{-\infty}^{\infty} \tilde{C}(\omega) \zeta(x, \omega) \exp(-i\omega t) d\omega. \tag{2.25}$$

For the amplitude function $\tilde{C}(\omega)$, we will accept the simplest model of the Gauss distribution

$$\tilde{C}(\omega) = \overline{C} \frac{T}{\sqrt{2\pi}} \exp\left(i\omega t_0 - \frac{(\omega - \omega_0)^2 T^2}{2} \right), \tag{2.26}$$

and will assume the characteristic width of the packet, $1/T$, to be much smaller than the carrier frequency ω_0:

$$\omega_0 T \gg 1. \tag{2.27}$$

The sense of the time moment t_0 will be evident from the calculations below. We will also assume that, for FMS waves with carrier frequency ω_0, there is both a turning point, $x = x_F(\omega_0)$, and an Alfvén resonance point, $x_0 = x_A(\omega_0)$, in the region of the transition layer.

Let us examine the solution (2.25) in a transparent region where the WKB approximation (2.17) is applicable. Given the small width of the packet over ω, let us decompose $\Psi_x(x, \omega)$ into a series at $\omega = \omega_0$:

$$\Psi_x(x, \omega) = \Psi_x(x, \omega_0) + \frac{\partial \Psi_x(x, \omega_0)}{\partial \omega}(\omega - \omega_0) + \frac{1}{2}\frac{\partial^2 \Psi_x(x, \omega_0)}{\partial \omega^2}(\omega - \omega_0)^2 + \cdots \tag{2.28}$$

We have $\partial\Psi_x(x,\omega)/\partial\omega = t_g$, where

$$t_g(x,\omega) = \int\limits_{x_F(\omega)}^{x} \frac{dx'}{v_g(x',\omega)}, \tag{2.29}$$

is the time it takes the wave packet, with central frequency ω, to travel from $x_F(\omega)$ to x with local group velocity

$$v_g(x,\omega) = \left[\frac{\partial k_x(x,\omega)}{\partial\omega}\right]^{-1}.$$

(2.29) takes into account the fact that the derivative of the lower limit in (2.24) is zero ($k_x(x_F(\omega)) = 0$).

Let us denote

$$\tilde{t}^{\,2} = \frac{\partial^2\Psi_x(x,\omega)}{\partial\omega^2} = \frac{\partial t_g(x,\omega)}{\partial\omega}.$$

The $\tilde{t}$ value has a dimension of time. On the order of magnitude,

$$\Psi_x(x,\omega) \sim \frac{\omega_0(x-x_0)}{v_A}, \quad t_g \sim \frac{x-x_0}{v_A}, \quad \tilde{t}^{\,2} \sim \frac{x-x_0}{\omega_0 v_A}, \tag{2.30}$$

where v_A is the characteristic value of the Alfvén speed. For function (2.26), integration is effectively over the interval $|\omega-\omega_0| \lessapprox 1/T$. Given (2.30), the sequential terms decrease in this interval in the decomposition (2.28), i.e. the decomposition is justified. Substituting (2.17) and (2.26) into (2.25) and given the decomposition (2.28), the integral over ω is easy to compute:

$$\begin{aligned}\zeta(x,t) = \overline{C}\left[\frac{k_x(x,\omega_0)}{\rho_0(x)\Omega^2(x,\omega_0)}\right]^{1/2} &\exp[-i\omega_0(t-t_0)] \\ \times\left\{\frac{T}{\sqrt{T^2+i\tilde{t}^{\,2}}}\right. &\exp\left[-i\Psi(x,\omega_0) - \frac{(t-t_0+t_g)^2}{T^2+i\tilde{t}^{\,2}}\right] \\ + S(\omega_0)\frac{T}{\sqrt{T^2-i\tilde{t}^{\,2}}} &\left.\exp\left[i\Psi(x,\omega_0) - \frac{(t-t_0-t_g)^2}{T^2-i\tilde{t}^{\,2}}\right]\right\}.\end{aligned} \tag{2.31}$$

Here, the slow-varying functions of frequency, $\Omega^2(x,\omega)$, $k_x(x,\omega)$ and $S(\omega)$, are taken out of the integral sign at $\omega=\omega_0$. Functions $t_g(x,\omega)$ and $\tilde{t}^{\,2}(x,\omega)$ are also taken at $\omega=\omega_0$, in the above sense.

Both terms in braces (2.31) are Gauss functions over t, their maxima being at points determined by this equation

$$t_g(x,\omega_0) = \pm(t-t_0), \tag{2.32}$$

the minus sign corresponding to the first term, the plus sign to the second term. By definition, function $t_g(x,\omega)$ is positive. This means that the first term in (2.31) describes a wave packet from infinity, its maximum crossing x at $t=t_0-t_g(x,\omega_0)$, while the second term describes the reflected packet, crossing the same point at $t=t_0+t_g(x,\omega_0)$. The characteristic duration of the packet $\Delta t = (T^4+\tilde{t}^{\,4})^{1/4}$. At time moment $t=t_0$, it is reflected from the neighbourhood of the turning point, $x_F(\omega_0)$.

For convenience, let us transform (2.31) as follows. Let us introduce function $x = x_g(\tau, \omega_0)$, inverse to $\tau = t_g(x, \omega_0)$. The $x_g(\tau, \omega_0)$ value is the x coordinate reached by the wave packet starting from $x = x_F(\omega_0)$, at time moment $\tau = 0$. The $t_g(x, \omega_0)$ function is positive. This means that, for the minus sign, (2.32) is solved as $x_g(t_0 - t, \omega_0)$ if $t < t_0$, while for the plus sign, the solution will be $x_g(t - t_0, \omega_0)$ if $t > t_0$. If (2.31) is regarded as a function of x for given t, the first term in the braces will have a maximum at $x = x_g(t_0 - t, \omega_0)$ for $t < t_0$; the second, at $x = x_g(t - t_0, \omega_0)$ for $t > t_0$. Examining x values that are not very far from these maxima, (2.31) can be reduced to

$$\zeta(x,t) = \overline{C}\left[\frac{k_x(x,\omega_0)}{\rho_0(x)\Omega^2(x,\omega_0)}\right]^{1/2} \tag{2.33}$$

$$\times\left\{\theta(t_0 - t)\left(\frac{T^2}{T^2 + i\tilde{t}^{\,2}}\right)^{1/2}\exp\left[-i\Psi(x,\omega_0) - \frac{(x - x_g(t_0 - t, \omega_0))^2}{2v_g^2(T^2 + i\tilde{t}^{\,2})}\right]\right.$$

$$\left. + \theta(t - t_0)S(\omega_0)\left(\frac{T^2}{T^2 - i\tilde{t}^{\,2}}\right)^{1/2}\exp\left[i\Psi(x,\omega_0) - \frac{(x - x_g(t - t_0, \omega_0))^2}{2v_g^2(T^2 - i\tilde{t}^{\,2})}\right]\right\}.$$

The first term in the braces in (2.33) describes a wave packet travelling, at $t < t_0$, from $x = +\infty$, at group velocity v_g. The second term describes a packet reflected from the transition layer and travelling at $t > t_0$, at the same group velocity. Its amplitude is multiplied by $S(\omega_0)$. The packet is reflected at approximately $t = t_0$, but this process is not described by (2.33), because the WKB approximation is inapplicable near a turning point $x = x_F(\omega_0)$.

The wave packet width has a minimum value near $x = x_F(\omega_0)$, where $\tilde{t}^{\,2} = 0$, being $v_A T$, on the order of magnitude. Let us assume that this width is much smaller than the characteristic scale of the inhomogeneity, Δ. Combined with (2.27) this results in the following conditions for the T parameter:

$$\frac{1}{\omega_0} \ll T \ll \frac{\Delta}{v_A} = \frac{k_z\Delta}{\omega_0}. \tag{2.34}$$

These inequalities are compatible, because the WKB applicability condition we assume to hold implies $k_z\Delta \gg 1$. Away from the reflection point the characteristic width of the packet is

$$\Delta x = v_g\left(T^4 + \tilde{t}^{\,4}\right)^{1/4} \sim v_A\left(T^4 + \tilde{t}^{\,4}\right)^{1/4}.$$

If $\tilde{t} \gg T$, which according to (2.34) and (2.30) means $k_z(x - x_0) \gg 1$, then

$$\Delta x \sim \left(\frac{x - x_0}{k_z}\right)^{1/2} \ll x - x_0. \tag{2.35}$$

Thus, the packet can always be regarded as narrow enough in x – near the reflection point its width is much smaller than Δ, while away from that point, it is much smaller than the distance to the point.

Let us note a significant circumstance. The reflected wave packet expands, while the incident wave packet constricts. This behaviour does not appear physical. As a rule, a wave packet formed at a certain moment of time will expand. The phases of individual harmonics should be selected specifically, so that the packet should begin to constrict after formation. If we examine a packet arriving from $x = +\infty$, however, we have to content ourselves with the disadvantage of the solution (2.33). If an expanding packet is formed at $x = +\infty$, its width will be infinite upon reaching the reflection point, making it impossible to regard it as a packet. A realistic solution is in the form of an incident packet formed at a finite distance, x_0. Such a solution will be demonstrated below by numerical integration of the wave equation.

The packet contains a small-scale component, described by phases

$$\pm\Psi_x(x, \omega_0) \approx \pm[\Psi_x(x_g, \omega_0) + k_x(x_g, \omega_0)(x - x_g)].$$

The phase incursion inside the packet near the reflection point is

$$\Delta\Psi_x \sim k_x \Delta x \sim k_x v_g T \sim \omega_0 T \gg 1,$$

while away from it,

$$\Delta\Psi_x \sim k_x v_g \tilde{t} \sim [k_z(x - x_0)]^{1/2} \gg 1.$$

In both cases, $\Delta\Psi_x \gg 1$. This means that multiple wavelengths fit within the packet.

The structure of the wave field can be examined in the same manner in the Alfvén resonance region. This analytical investigation is rather cumbersome though. To understand the behaviour of oscillations reflected from the transition layer, one can numerically integrate the equation describing the evolution in time of the original wave packet. To do so, frequency ω in Eq. (2.4) must be replaced with a time derivative $\omega = i\partial/\partial t$, acting on the plasma shift component ζ. Assuming that the plasma inhomogeneity is only related to the inhomogeneity of its density, we can then obtain the following equation

$$\frac{1}{v_A^2}\frac{\partial^2 \zeta}{\partial t^2} - \frac{\partial}{\partial x}\frac{\Omega^2}{\Omega_F^2}\frac{\partial \zeta}{\partial x} + k_z^2 \zeta = 0, \tag{2.36}$$

describing the propagation of an MHD wave packet in a 1D-inhomogeneous plasma. Remember that $v_A(x)$ is the Alfvén speed here, which we simulate with a unit-wide ($\Delta = 1$) transition layer (2.3), $\Omega^2 = (\omega + i\gamma)^2 - k_z^2 v_A^2$, $\Omega_F^2 = \omega^2 - (k_y^2 + k_z^2)v_A^2$. The decrement γ is introduced here for regularising a singularity at the Alfvén resonance point, x_A.

Setting, at time moment t_0, the original wave packet as

$$\zeta(x, t_0) = \exp\left(-\frac{(x - x_0)^2}{2\Delta_x^2} - ik_x x - (i\omega_0 + \gamma)t_0\right),$$

where $\Delta_x = 0.9\Delta$ is the original width of the wave packet, $\omega_0 = \sqrt{k_y^2 + k_z^2} v_A(x_F)$ is its central frequency for which the FMS reflection point $x = x_F$ is at the exact centre of the transition layer, $k_x = \Omega_F(x_0, \omega_0)/v_A(x_0)$ is the asymptotical value of the x component of the wave vector, in the WKB approximation, at the point of the original location of the packet $x_0 = 2\pi\Delta$. Setting also $k_y = k_z = 4/\Delta_x$, $\gamma = 0.1\omega_0$ and $t_0 = 0$, let us integrate (2.36) over two variables, x and t. The result is shown in Figure 2.6 for the oscillation magnetic field component

$$B_y = B_0 \frac{k_y k_z v_A^2}{\Omega_F^2}\frac{\partial \zeta}{\partial x},$$

having the highest amplitude at the Alfvén resonance point.

At the initial time moment, the FMS wave packet travels towards the transition layer keeping its structure intact practically up to the moment of its reflection. As it is reflected, the packet stops ($v_g(x_F) = 0$) and the group velocity reverses its sign. A sharp increase is

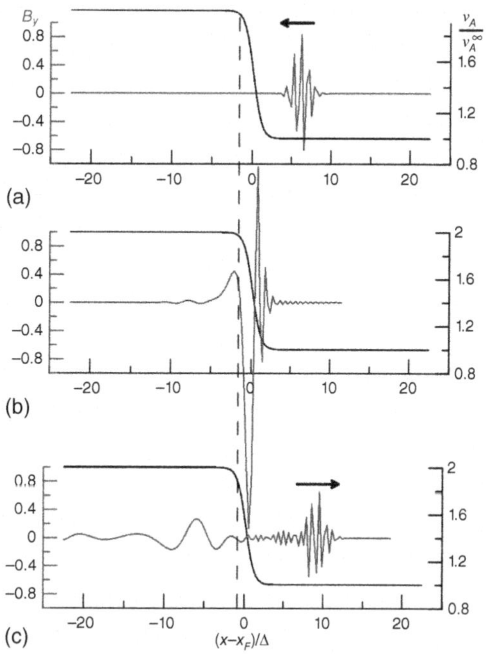

Figure 2.6 Evolution of an FMS wave packet incident onto a smooth transition layer containing an Alfvén resonance point, $x = x_A$. The distribution of the Alfvén speed $v_A(x)$ (right coordinate axis) and the B_y component of the wave field (left coordinate axis) are shown. (a) Initial state of the unit-wide wave packet. (b) Field structure at the moment the wave packet is reflected from the transition layer. (c) Wave packet structure upon reflection from the transition layer.

visible in the oscillation amplitude thanks to resonant Alfvén waves being driven at the same frequency as the central frequency of the wave packet. Upon reflection, the packet amplitude decreases due to part of the packet energy being absorbed at the Alfvén resonance point, while the packet itself expands because its harmonics experience phase mixing.

2.5 Energy Balance in the Problem of an Incident FMS Wave Reflected from the Transition Layer Containing an Alfvén Resonance Point

Let us examine the energy balance in the problem of incident monochromatic FMS wave reflected from the transition layer addressed in Section 2.3. We will determine the difference between the energy fluxes of the FMS wave falling onto and reflected from the transition layer and compare it to the power absorbed by the plasma in the Alfvén resonance region. We will also find the reflection coefficient of the FMS wave falling onto the transition layer, as well as its absorption coefficient in the resonance region, for two limiting cases: $k_y\Delta_A \gg 1$ and $k_y\Delta_A \ll 1$, where $\Delta_A = x_F - x_A$ is the distance between the FMS turning point and the Alfvén resonance point (see Figure 2.4).

Let us represent the dielectric permeability tensor as the sum of the Hermitian and anti-Hermitian parts

$$\hat{\varepsilon} \equiv \varepsilon_{\alpha\beta} = \varepsilon^{(1)}_{\alpha\beta} + i\varepsilon^{(2)}_{\alpha\beta},$$

where both tensors, $\varepsilon^{(1)}_{\alpha\beta}$ and $\varepsilon^{(2)}_{\alpha\beta}$, are Hermitian ($\varepsilon^{(1)*}_{\alpha\beta} = \varepsilon^{(1)}_{\beta\alpha}$, $\varepsilon^{(2)*}_{\alpha\beta} = \varepsilon^{(2)}_{\beta\alpha}$, where ε^* denotes a complex conjugate), while the subscripts $_{\alpha\beta}$ can take the values of x, y, z in the Euclidean system of coordinates (x, y, z). In an ideal plasma, only diagonal Hermitian components are non-zero for MHD waves in the tensor $\hat{\varepsilon}$: $\varepsilon_{xx} = \varepsilon_{yy} = \bar{c}^2/v_A^2, \varepsilon_{zz} = \infty$. Within this approximation, the Alfvén waves lack transverse dispersion, and their group velocity is strictly parallel to the magnetic field lines.

In a non-ideal plasma, the tensor $\hat{\varepsilon}$ components (even non-diagonal components containing an anti-Hermitian part) include terms determining the transverse dispersion of Alfvén waves. The explicit form of the tensor $\hat{\varepsilon}$ for Alfvén oscillations in a non-ideal plasma is given in Section 2.7. For weakly decaying oscillations, $|\varepsilon^{(2)}_{\alpha\beta}| \ll |\varepsilon^{(1)}_{\alpha\beta}|$. The conductivity tensor is represented similarly:

$$\hat{\sigma} = -i\frac{\omega}{4\pi}\hat{\varepsilon}, \tag{2.37}$$

implying

$$\sigma^{(1)}_{\alpha\beta} = (\omega/4\pi)\varepsilon^{(2)}_{\alpha\beta}, \quad \sigma^{(2)}_{\alpha\beta} = -(\omega/4\pi)\varepsilon^{(1)}_{\alpha\beta}.$$

The energy balance of the wave process in question is described by (see [60])

$$\text{div}\mathbf{S} + Q + \partial H/\partial t = 0. \tag{2.38}$$

Here[3]

$$\mathbf{S} = \frac{\bar{c}}{16\pi}\left([\mathbf{E}\times\mathbf{B}^*] + [\mathbf{E}^*\times\mathbf{B}]\right) - \omega\frac{\partial\varepsilon^{(1)}_{\alpha\beta}}{\partial\mathbf{k}}\frac{E^*_\alpha E_\beta}{16\pi}$$

is the time-averaged wave energy flux including not only the electromagnetic energy flux (Poynting vector), but also the mechanical energy flux of the wave. Summing over identical indices $_{\alpha\beta}$ is assumed here. The last term in this expression differs from zero only for MHD oscillations in a non-ideal plasma. This expression, in particular, describes the energy flux of kinetic Alfvén waves across magnetic shells due to their transverse dispersion. The time-averaged density of the dissipated power of the oscillations has the form

$$Q = (1/2)\left(\mathbf{j}\mathbf{E}^* + \mathbf{j}^*\mathbf{E}\right) = (1/2)\sigma^{(1)}_{\alpha\beta}E^*_\alpha E_\beta,$$

while the time-averaged density of the electromagnetic energy

$$H = \frac{\bar{c}}{16\pi}\left(|\mathbf{E}|^2 + |\mathbf{B}|^2]\right)$$

describes an increase in the energy of the oscillation electromagnetic field thanks to increased oscillation amplitude.

Let us examine oscillations in a model of the medium in the form of a plasma layer bounded, in the z coordinate, by two ideally conducting boundaries (see Figure 2.1). We will assume the plasma to be inhomogeneous in x only. The structure of the components of the oscillation electromagnetic field in this model can be represented as

$$\begin{aligned} E_{x,y} &= e_{x,y}(x)V_N(z)e^{ik_y y}, \; E_z = e_z(x)U_N(z)e^{ik_y y}, \\ B_{x,y} &= b_{x,y}(x)U_N(z)e^{ik_y y}, \; B_z = b_z(x)V_N(z)e^{ik_y y}, \end{aligned}$$

where $V_N(z) = \sin k_{zN}z$, $U_N(z) = \cos k_{zN}z$ are functions describing the wave field structure as standing (along magnetic field lines) waves confined between reflecting boundaries

3 The fact that a complex representation is used for perturbed values $F = f\exp(-i\omega t)$, where $f = f_1 + if_2$ is the coordinate function, implies that $\text{Re}F = \text{Re}f\exp(-i\omega t) = f_1\cos(\omega t) + f_2\sin(\omega t)$. To average the bilinear function therefore, we have $\overline{FH} \to \overline{\text{Re}F\cdot\text{Re}H} = (1/2)(f_1h_1 + f_2h_2) = (1/4)(fh^* + f^*h)$.

z_+ and z_-, $k_{zN} = \pi N/L$, $L = z_+ - z_-$ is field line length, $N = 1, 2, 3, ...$ is the number of standing wave harmonic.

Since terms in (2.38) depend on z as $U_N^2(z)$ or as $V_N^2(z)$, averaging them over z produces a factor of 1/2. In our model, $S_z = 0$ and $\partial S_y/\partial y = 0$. We will indicate the averaging over z by an overbar, reducing (2.38) to

$$\frac{\partial \overline{S}_x}{\partial x} + \overline{Q} + \partial \overline{H}/\partial t = 0, \tag{2.39}$$

where

$$\overline{S}_x = \frac{\overline{c}}{32\pi}(e_y b_z^* - e_z b_y^* + e_y^* b_z - e_z^* b_y) - \omega \frac{\partial \varepsilon_{\alpha\beta}^{(1)}}{\partial k_x} \frac{e_\alpha^* e_\beta}{32\pi}, \tag{2.40}$$

$$\overline{Q} = \frac{1}{4}\sigma_{\alpha\beta}^{(1)} e_\alpha^* e_\beta, \tag{2.41}$$

$$\overline{H} = \frac{1}{32\pi}\left(|\vec{e}|^2 + |\vec{b}|^2\right). \tag{2.42}$$

Let us integrate (2.39) over x, from $-\infty$ to $+\infty$, writing the result as

$$\overline{S}_x(\infty) - \overline{S}_x(-\infty) + W = 0, \tag{2.43}$$

where

$$W = \int_{-\infty}^{\infty} \left(\overline{Q} + \frac{\partial \overline{H}}{\partial t}\right) dx. \tag{2.44}$$

In the integral conservation law (2.43), the first two terms determine energy 'credit' coming from infinity, while the last term determines its 'debit' spent on dissipation and increased wave amplitude.

Let us examine the conservation law (2.43), in a cold ideal plasma. In this case, $e_z = 0$, $\partial \varepsilon_{\alpha\beta}/\partial k_x = 0$ and

$$\overline{S}_x = \frac{\overline{c}}{32\pi}(e_y b_z^* + e_y^* b_z).$$

Expressing oscillation field components e_y and b_z through plasma shift ζ along the x coordinate (see Appendix B), as we did in Section 2.3,

$$e_y = -i\frac{\omega}{c} B_0 \zeta, \quad b_z = -B_0 \frac{\Omega_A^2}{\Omega_F^2}\frac{\partial \zeta}{\partial x},$$

produces

$$\overline{S}_x = i b_0^2 \frac{\omega}{32\pi} \frac{\Omega_A^2}{\Omega_F^2}\left(\zeta \frac{\partial \zeta^*}{\partial x} - \zeta^* \frac{\partial \zeta}{\partial x}\right),$$

where $\Omega_F^2 = \Omega_A^2 - k_y^2 v_A^2 = \omega^2 - (k_y^2 + k_z^2) v_A^2$. In the right-hand asymptotic, for $x \to \infty$, where (2.17) is applicable to ζ and $k_x^2 = \Omega_F^2/v_A^2$, we have

$$\overline{S}_x = -\frac{\omega}{4}(|C_i|^2 - |C_r|^2) \tag{2.45}$$

the difference of energy fluxes for the incident FMS wave and the wave reflected from the transition layer. For $x \to -\infty$, function $\zeta(x) \to 0$, therefore $\overline{S}_x(-\infty) = 0$ and

$$\overline{S}_x(+\infty) - \overline{S}_x(-\infty) = -\frac{\omega}{4}(|C_i|^2 - |C_r|^2). \tag{2.46}$$

We assume here that the energy flux transferred by the incident wave is negative, while the flux transferred by the wave reflected from the transition layer is positive. While deriving (2.46), we neglected the small imaginary contribution to the frequency because it is insignificant in the asymptotic region. It is exactly this contribution, however, that determines the entire effect in computations of $\overline{Q}$ and $\partial\overline{H}/\partial t$. As can be seen from (2.37), it results in the presence of a real part in the conductivity tensor

$$\sigma_{\perp} = \frac{(-i\omega+\gamma)\overline{c}^2}{4\pi v_A^2}$$

and thus determines the dissipated power:

$$\overline{Q} = \frac{1}{4}\sigma_{\perp}^{(1)}|\vec{e}|^2 = \frac{\gamma}{16\pi}\frac{\overline{c}^2}{v_A^2}|\vec{e}|^2. \tag{2.47}$$

For the growth rate of the electromagnetic oscillation field energy, we have

$$\frac{\partial\overline{H}}{\partial t} = 2\gamma\overline{H} = \frac{\gamma}{16\pi}(|b|^2+|e|^2).$$

The presence of an imaginary contribution in the frequency means that the disturbance in question grows, at rate γ, and given $|\vec{E}| \approx (v_A/\overline{c})|\vec{B}| \ll |\vec{B}|$ in hydromagnetic waves, we have

$$\frac{\partial\overline{H}}{\partial t} \approx \frac{\gamma}{16\pi}|b|^2 = \frac{\gamma}{16\pi}\frac{\overline{c}^2}{v_A^2}|\vec{e}|^2. \tag{2.48}$$

It follows from (2.47) and (2.48) that increasing oscillation energy and energy dissipation in the Alfvén resonance region give identical contributions to the energy balance of the wave process. Representing the oscillation magnetic field components through shift ζ (see Appendix B) yields

$$|b|^2 = B_0^2\left(k_z^2|\zeta|^2 + \frac{k_y^2k_z^2v_A^4}{|\Omega_F|^4}\left|\frac{\partial\zeta}{\partial x}\right|^2 + \left|\frac{\Omega_A}{\Omega_F}\right|^4\left|\frac{\partial\zeta}{\partial x}\right|^2\right).$$

In the $\gamma\to 0$ limit, the main contribution to this expression, in the Alfvén resonance region (where $\Omega_A = 0$), is by the second term, possessing a singularity of the form x^{-2}. Therefore we have

$$W \approx \frac{\gamma}{8\pi}\int_{-\infty}^{\infty}|b|^2dx \approx \frac{\gamma}{8\pi}k_y^2k_z^2\int_{-\infty}^{\infty}\frac{B_0^2v_A^4}{|\Omega_F|^4}\left|\frac{\partial\zeta}{\partial x}\right|^2dx. \tag{2.49}$$

Let us determine this value for the two limiting cases, $k_y\Delta_A \gg 1$ and $k_y\Delta_A \ll 1$. In the first limiting case, where the WKB approximation is applicable in calculating the wave field between turning point x_F and Alfvén resonance point x_A, we can use expression (2.11) for displacement ζ in the Alfvén resonance region. In this case, we have

$$W = \frac{1}{16\pi}\frac{\omega}{a_A}\frac{(C_Ak_zB_0)^2}{k_y^2}\int_{-\infty}^{\infty}\frac{\varepsilon_A dx}{(x-x_A)^2+\varepsilon_A^2}. \tag{2.50}$$

where $\varepsilon_A = 2a_A\gamma/\omega$. Here, we take into account the fact that $\Omega_F^2 \approx -k_y^2v_A^2$ in the neighbourhood of the Alfvén resonance point, $x = x_A$.

Given the fact that, in the $\gamma\to 0$ limit,

$$\lim_{\gamma\to 0}\frac{\varepsilon_A}{(x-x_A)^2+\varepsilon_A^2} = \pi\delta(x-x_A), \tag{2.51}$$

we obtain

$$W = \frac{\omega}{16a_A}\frac{(C_A k_z B_0)^2}{k_y^2}. \tag{2.52}$$

Note a remarkable property of this expression: it does not depend on the value of γ if the latter is small enough. This may be explained as follows. As γ decreases, the resonance oscillation amplitude increases at the Alfvén resonance point, and, simultaneously, the localisation region of the oscillation decreases in size, so that the full energy of the oscillations remains unchanged in the Alfvén resonance region. Substituting (2.52) and (2.46) into the energy balance equation, (2.43), yields

$$|C_i|^2 - |C_r|^2 = \frac{1}{4a_A}\frac{(k_z B_0)^2}{k_y^2} C_A^2.$$

The matched solutions in Section 2.3 imply that $C_A = -C_i(2k_y\sqrt{2a_A}/k_z B_0)\exp(-\Gamma + i\pi/4)$. This enables us to obtain the expressions for the reflection coefficient

$$R = \frac{|C_r|^2}{|C_i|^2} = 1 - 2\exp(-2\Gamma) \tag{2.53}$$

and absorption coefficient

$$D = 1 - R = 2\exp(-2\Gamma). \tag{2.54}$$

In the second limiting case, $k_y \Delta_A \ll 1$, we will seek the solution to (2.4) using the perturbation theory method. In the main order, let us assume $k_y = 0$. Equation (2.4) has the form

$$\frac{\partial}{\partial x}\rho_0 v_A^2 \frac{\partial \zeta}{\partial x} + \rho_0 \Omega_A^2 \zeta = 0. \tag{2.55}$$

Note that, in this case, the Alfvén resonance is absent, and point x_A coincides with the turning point of the FMS wave, $x_A = x_F$. In the asymptotically remote region right of the turning point, x_F, the WKB solution of (2.55) has the form

$$\zeta = \frac{1}{\sqrt{\rho_0 v_A \Omega_A}}\left[C_i \exp\left(-i\int_{x_F}^{x}\frac{\Omega_A}{v_A}dx'\right) + C_r \exp\left(i\int_{x_F}^{x}\frac{\Omega_A}{v_A}dx'\right)\right]. \tag{2.56}$$

Near the turning point, where the linear expansion is applicable, $\Omega_A^2 = \omega^2 - k_z^2 v_A^2 \approx k_z^2 \overline{v}_A^2 (x - x_F)/a_A$ (here $\overline{v}_A = v_A(x_F)$, $a_A = -(\nabla_x \ln v_A^2)_{x_F}^{-1} > 0$), the solution of (2.55) satisfying the boundary conditions can be represented as

$$\zeta = C_F Ai[-(x - x_F)/a_F], \tag{2.57}$$

where, in this particular case, $a_F = (a_A/k_z^2)^{1/3}$, and $A_i(z)$ is the Airy function. Matching solutions (2.56) and (2.57) in their overlap region produces a relation between the coefficients, $C_F = -C_i 4\pi\sqrt{a_F}e^{i\pi/4}/B_0$. In the $k_y\Delta_A \ll 1$ case, let us use (2.57), an expression obtained in the zero order of the perturbation theory, for solving (2.4) in the Alfvén resonance region. In this case, however, function $|\Omega_F|^4 \approx |\Omega_A|^4 \approx k_z^4 \overline{v}_A^4[(x - x_F)^2 + \epsilon_A^2]/a_A^2$ in (2.49) has a singularity of the form x^{-2} as $\gamma \to 0$. Substituting (2.57) into (2.49) produces

$$W = |C_F|^2 \frac{B_0^2}{16\pi}\frac{k_y^2 a_A \overline{v}_A}{k_z}|Ai'(0)|^2$$
$$\times \int_{-\infty}^{\infty}\frac{\epsilon_A dx}{(x - x_F)^2 + \epsilon_A^2} = |C_F|^2 \frac{B_0^2}{16}\frac{k_y^2 a_A \overline{v}_A}{k_z}|Ai'(0)|^2,$$

where $\varepsilon_A = 2\gamma a_A / k_z \bar{v}_A$. Substituting this expression and (2.46) into the energy balance Eq. (2.43) and making use of the relationship between constants C_F and C_i yields the following expression for the absorption coefficient

$$D = \lambda k_y^2 \Delta_A^2, \tag{2.58}$$

where

$$\lambda = [2\pi Ai'(0)]^2 \approx 2.6.$$

These two limiting cases allow us to obtain a qualitative picture of function $D(k_y\Delta_A)$ as a whole. Evidently, it has a maximum for $k_y\Delta_A \sim 1$, with a height of order unity. In the general case, for an arbitrary profile of the transition layer, the behaviour of $D(k_y\Delta_A)$ can only be found numerically (see Figure 2.5).

2.6 FMS Wave Reflected from the Transition Layer in a 'warm' Plasma. Alfvén and Magnetosonic Resonances

In Section 2.3, we examined the process of Alfvén oscillations driven by an FMS wave travelling in an inhomogeneous 'cold' plasma – the Alfvén resonance. Let us now address a similar problem in a 'warm' plasma, where resonance surfaces, for both Alfvén and SMS waves, are present in the transition layer.

A problem about incident magnetosonic waves on a linear plasma transition layer connecting two homogeneous halfspaces was studied in [52]. It was shown that the presence of a resonance surface for SMS waves in the plasma configuration significantly increases the absorption coefficient for FMS waves incident on the transition layer, as compared to a configuration with only one resonance surface, for Alfvén waves. However, a linear transition layer model does not allow us to obtain certain results that are important in principle.

In [54, 61], a similar problem was solved for a model with a strongly non-isothermic plasma, where the ion and electron temperatures may differ considerably. This model contained a smooth transition layer of the form (2.3). A qualitative explanation was given to findings in [52]. In this section, we will discuss an incident monochromatic FMS wave reflected from such a transition layer. Part of the energy of the incident wave is absorbed in the neighbourhood of the resonance surfaces resulting in plasma heating. As we will see, cases are possible when the incident magnetosonic wave is fully absorbed in the neighbourhood of the resonance surface for SMS waves.

Let us assume an immobile background plasma, ($\mathbf{v}_0 = 0$). In the zero approximation, the x component in (1.1) yields, in a stationary state ($\partial/\partial t = 0$), a condition for the plasma configuration to be balanced

$$P_0 + \frac{B_0^2}{8\pi} = const. \tag{2.59}$$

The basic equation for MHD oscillation field has the form (2.4), where $\Omega_A^2 = \omega^2 - k_z^2 v_A^2$,

$$k_x^2 = -k_y^2 - k_z^2 + \frac{\omega^4}{\omega^2(v_A^2 + v_S^2) - k_z^2 v_A^2 v_S^2}. \tag{2.60}$$

The function $k_x^2(x)$ distribution is important for a correct formulation of the problem. Our task is to examine the process by which incident FMS waves are reflected from a smooth

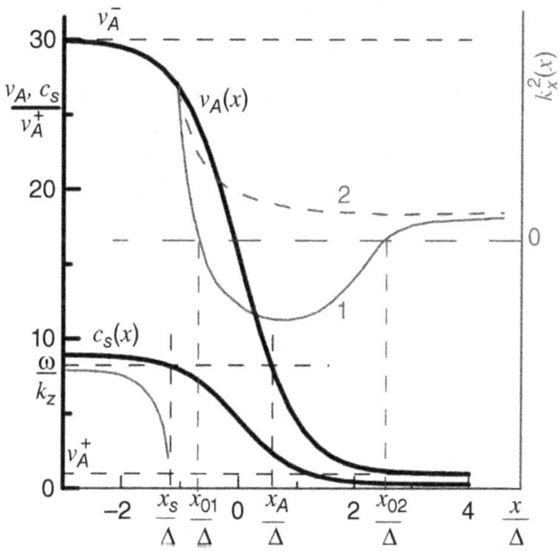

Figure 2.7 Distributions of Alfvén speed $v_A(x)$, SMS wave speed $c_S(x)$ (black lines, left-hand axis) and the square of the WKB component of the wave vector, $k_x^2(x)$ (gray lines, right-hand axis), across the transition layer, in two limiting cases: (1) (solid gray line for k_x^2) the opaque region (x_{01}, x_{02}) is present for FMS waves, (2) (gray dashed line for k_x^2) the transparent region for SMS waves spreads from the resonance surface for SMS waves, $x = x_S$, to infinity.

transition layer containing two resonance surfaces (for Alfvén and SMS waves), where the Alfvén speed distribution $v_A(x)$ is given by (2.3). The sound speed distribution $v_S(x)$ determined by the equilibrium condition (2.59) and the SMS speed distribution $c_S(x)$ are shown in Figure 2.7.

The problem in question must be a solution to (2.4) which is the sum of the incident and reflected wave for $x \to \infty$ and has a finite amplitude when $x \to -\infty$. The value of $v_A^+ \equiv v_A(\infty)$ should be chosen such that $k_x^2(\infty) > 0$ for a chosen value of frequency ω. In this case, the asymptotic $x \to \infty$ is a transparent region for fast magnetosonic waves. Two possible variants of the function $k_x^2(x)$ distribution are shown in Figure 2.7 by curves 1 and 2.

Analysis of (2.60) reveals that, as the x coordinate changes from $+\infty$ to $-\infty$, while $v_A(x)$ increases monotonically, function $k_x^2(x)$ can pass through zero twice, at points we will denote as x_{01} and x_{02}, the opaque region (where $k_x^2(x) < 0$) being between them. Such a behaviour of $k_x^2(x)$ is represented by curve 1 in Figure 2.7 and corresponds to the case of FMS wave incident on the transition layer.

Equation (2.4) also contains two singular points where the coefficient of the higher derivative turns to zero. One is the Alfvén resonance point x_A, determined by $\Omega_A^2(x_A) = 0$, located in the opaque region, within (x_{01}, x_{02}). The other is the magnetosonic resonance point x_S, determined by the denominator turning to zero in (2.60), which yields a local dispersion equation for SMS waves as $|k_x^2| \to \infty$: $\omega^2 = k_z^2 c_S^2(x_S)$, where $c_S^2 = v_A^2 v_S^2/(v_A^2 + v_S^2)$. Point x_S is to the left of the turning point, x_{01}, the transparent region for SMS waves being in between. Left of x_S is the opaque region expanding to $-\infty$ along x.

The value of $v_A^- \equiv v_A(-\infty)$ must be such that a magnetosonic resonance point x_S could exist. Part of the incident wave energy is absorbed in the neighbourhood of two resonance surfaces, $x = x_A$ and $x = x_S$, resulting in the amplitude of the FMS wave reflected from the transition layer being smaller than that of the incident wave. The difference between the incident and reflected wave energy is spent on heating the plasma and increasing the oscillation amplitude in the neighbourhood of the resonance surfaces. The absorption coefficient determined as the ratio of this difference to the incident wave energy will depend on plasma parameters at the resonance surfaces. Below we will examine the field structure of the MHD oscillation in question and the dependence of the FMS wave absorption coefficient on the values of the wave vector components, k_y, k_z, the plasma electron to ion temperature ratio T_e/T_i and parameter $\beta^* = (v_S/v_A)^2$ (differing from the earlier-introduced parameter $\beta = 8\pi P_0/B_0^2$ in the factor only, which is close to unity).

Curve 2 in Figure 2.7 corresponds to the case when the transparent region for SMS waves extends to ∞, and the resonance surface for the Alfvén wave is absent. When analysing (2.60), two ranges of wave field and plasma parameters can be identified on the asymptotic $x \to \infty$, corresponding to curve 1 and 2 in Figure 2.7, with $k_x^2(\infty) > 0$:

$$\frac{\omega^2}{k_z^2 v_A^{+2}} > \omega_{A1}^2$$

and

$$\frac{\beta^*}{1+\beta^*} < \frac{\omega^2}{k_z^2 v_A^{+2}} < \omega_{A2}^2,$$

where $\omega_{A1}^2, \omega_{A2}^2$ are the two roots of a biquadratic equation

$$\omega_A^4 - (1 + k_y^2/k_z^2)[\omega_A^2(1+\beta^*) - \beta^*] = 0,$$

corresponding to the plus or minus sign before the radical in the solution of this equation. Here, $\omega_A^2 = \omega^2/k_z^2 v_A^{+2}$, β^* is determined on the asymptotic $x \to \infty$, while the equation itself is a dispersion equation for magnetosonic waves. The first of the ranges corresponds to an FMS wave incident on the transition layer, the second range refers to an incident SMS wave. Note that the second range is rather narrow. For $\beta^* \to 0$, or $k_y^2/k_z^2 \to \infty$, we have $\omega_{A2}^2 \approx \left[\beta^*/(1+\beta^*) + k_z \beta^{*2}/(1+\beta^*)^3 \sqrt{k_y^2 + k_z^2}\right]$. For $k_y^2/k_z^2 \to 0$, we obtain $\omega_{A2}^2 \approx \beta^*$.

There is yet another condition for the existence of a solution (determined by the shape of curve 2 in Figure 2.7) describing SMS wave incident on the transition layer. For this solution, the condition $\nabla_x(k_x^2) < 0$ holds in the $x > x_S$ region, yielding:

$$\frac{\omega^2}{k_z^2 v_A^{+2}} < \frac{2\beta^*}{1+\beta^*}.$$

Another variant of the $k_x^2(x)$ distribution is possible, of course, when the resonance surface is present for the Alfvén wave, but is absent for SMS waves.

Let us examine a general case, when both resonance surfaces are present in the transition layer, corresponding to the distribution $k_x^2(x)$, described by curve 1 in Figure 2.7. Let us employ the WKB approximation as we did in Section 2.3. In the opaque region left of x_S, the WKB solution of (2.4) satisfying the boundary conditions as $x \to -\infty$ has the form

$$\zeta = C_{S1}\sqrt{\frac{|k_x|}{\rho_0 \Omega_A^2}} \exp\left(\int_{x_S}^{x} |k_x| dx'\right), \tag{2.61}$$

where C_{S1} is an arbitrary constant.

In the neighbourhood of the resonance surface for SMS waves, $x = x_S$, let us linearise the coefficient of the higher derivative in (2.4), representing $k_x^{-2} \approx a_s(x - x_S)$, where $a_s = (\partial k_x^{-2}/\partial x)_{x=x_S}$ is the characteristic scale of the k_x^{-2} variation near $x = x_S$. Equation (2.4) near $x = x_S$ can then be represented as

$$a_s \frac{\partial}{\partial x}(x - x_S)\frac{\partial \zeta}{\partial x} + \zeta = 0. \tag{2.62}$$

Its solution matching the WKB solution (2.61) is

$$\zeta = C_S K_0 \left(2\sqrt{-(x-x_S)/a_s}\right), \tag{2.63}$$

where $K_0(z)$ is a modified Bessel function. Making use of its asymptotic representation for $(x-x_S) \to -\infty$ and matching it to (2.61), we will find the relationship between the constants: $C_S = 2C_{S1}/(\Omega_s\sqrt{\pi\rho_{0s}a_s})$, where the subscript $_S$ indicates that the parameter values are taken at $x = x_S$. To adjust this for the $x > x_S$ region we need to regularise the singularity of solution (2.62). For this purpose, we will formally introduce an increment, γ_s, near the resonance surface $x = x_S$, substituting in the denominator of (2.60): $\omega \to \omega + i\gamma_S$. γ_s can also be regarded as the decrement of magnetosonic waves incident on the transition layer. The specific value for γ_s can be obtained from the solution of the dispersion Eq. (1.17) for SMS waves in a non-isothermic homogeneous plasma. The solution of (2.62) can then be subjected to a formal replacement, $x - x_S \to x - x_S + i\varepsilon_s$, where $\varepsilon_s = a_s\gamma_s/\omega$, and if $x > x_S$ the solution (2.63) has the form

$$\zeta = -i\frac{C_S\pi}{2}H_0^{(2)}\left(2\sqrt{(x-x_S+i\varepsilon_s)/a_s}\right), \tag{2.64}$$

where $H_0^{(2)}(z)$ is the second-type Hankel function, having, for $x - x_S \to \infty$, an asymptotic representation, $H_0^{(2)}(2\sqrt{x-x_S}) \approx \pi^{-1/2}(x-x_S)^{-1/4}\exp(-i2\sqrt{x-x_S}+i\pi/4)$. As $x \to x_S$, the solution (2.63) – (2.64) has the form

$$\zeta \approx -C_S\left(\frac{1}{2}\ln(x-x_S+i\varepsilon_s)+\ln 2\right).$$

As $\gamma_s \to 0$, this solution has the same logarithmic singularity as found on the resonance surface for Alfvén waves. The solution (2.64) describes an SMS wave incident on the resonance surface, $x = x_S$. There is no reflected wave in this case. This means that the wave is completely absorbed in the neighbourhood of the resonance surface. Note that, same as in the Alfvén resonance case, the particular mechanism of SMS wave absorption is not important in this case.

In the transparent region $x_S < x < x_{01}$, the WKB solution matching (2.64) can be represented as

$$\zeta = C_{S2}\sqrt{\frac{k_x}{\rho_0\Omega_A^2}}\exp\left(-i\int_{x_{01}}^{x}k_x dx'\right), \tag{2.65}$$

where the constant $C_{S2} = C_S\Omega_s\sqrt{\pi\rho_{0s}\,a_s}\,\exp(-i\int_{x_{01}}^{x_S}k_x dx - i\pi/4)$ is found by matching with solution (2.64).

What is interesting is that, if the transparent region for SMS waves extends to ∞ (which corresponds to curve 2 for $k_x^2(x)$ in Figure 2.7), the SMS wave incident on the transition layer is fully absorbed in the neighbourhood of the resonance shell. A numerical solution of the equations describing this process in a plasma with a linear transition layer was used in [52] to demonstrate that the waves incident on the transition layer exhibit the highest absorption coefficient ($\sim 90\%$), in this case. However, it is impossible to reach 100% energy absorption for a wave incident on the transition layer based on the model in that paper, consisting of 3 regions with different plasma parameter distributions in the x coordinate. When matching the solutions obtained for the different regions in that model, a reflected wave is inevitable.

To obtain a solution in the neighbourhood of the turning point $x = x_{01}$, let us make use of (2.6), an equation for the function η, same as in Section 2.3. Linearising near $x = x_{01}$

the coefficient $k_x^2 \approx -(x - x_{01})/a_1^3$, where $a_1^{-3} = -(\partial k_x^2/\partial x)_{x=x_{01}}$, yields a homogeneous Airy equation of the form

$$\frac{\partial^2 \eta}{\partial x^2} - a_1^{-3}(x - x_{01})\eta = 0. \tag{2.66}$$

Its solution that matches (2.65) has the form

$$\eta = C_{S3}\left(Ai\left(\frac{x - x_{01}}{a_1}\right) + iBi\left(\frac{x - x_{01}}{a_1}\right)\right), \tag{2.67}$$

where $Ai(z)$, $Bi(z)$ are Airy functions. Matching this solution to (2.65) produces a relation between the constants $C_{S3} = C_{S2}\sqrt{\pi a_1/\rho_{01}\Omega_{A1}^2}$.

The WKB solution in the opaque region $x_{01} < x < x_A$ has precisely the same form, (2.7), as in Section 2.3. Matching it to solution (2.67) relates the constants $C = iC_{S2}$ in the WKB solutions left and right of the magnetosonic resonance point x_S. The exponentially decreasing solution against the background of the exponentially increasing solution is also ignored here. The entire remaining structure of the solution replicates the solution in Section 2.5, when the transition layer possessed only one resonance surface for Alfvén waves $x = x_A$.

To visualise the full structure of the wave field, let us solve the above-stated problem numerically. Figure 2.8 illustrates the structure of the derivative $\partial\zeta/\partial x$ of the wave field in plasma for asymptotic value $\beta^* = 0.5$, for small decrements $\gamma/\omega = 0.01$ and $\gamma_s/\omega = 0.01$ on resonance surfaces. The resonant structure of oscillations near the resonance surfaces $x = x_A$ and $x = x_S$ is clearly evident. Also shown in the $x > x_{02}$ area is the decomposition of the oscillation field into the incident and reflected waves, according to solution (2.17), obtained in the WKB approximation. Note that the incident and reflected wave noticeably differ in amplitude in this calculation, due to oscillation energy being absorbed near the two resonance surfaces. The oscillation field is normalised to the incident wave amplitude as $x \to \infty$.

Figure 2.9 shows the distribution of the components of the oscillation magnetic field of unit amplitude as $x \to \infty$. The same values of the calculation parameters are used here as in Figure 2.8. The dominant field component is B_y, on the resonance surface for Alfvén waves, $x = x_A$, and B_z, on the resonance surface for SMS waves, $x = x_S$. Similar relationships are valid for the velocity field components. For that reason, resonant SMS oscillations

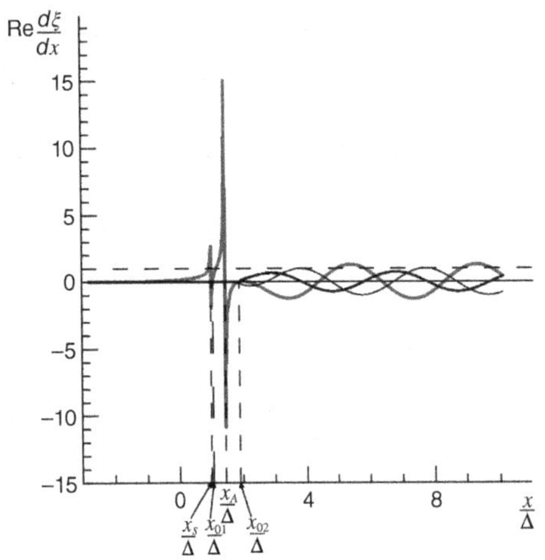

Figure 2.8 Derivative $\partial\zeta/\partial x$ distribution in the problem of FMS wave incident on/reflected from the transition layer with resonance surfaces for Alfvén ($x = x_A$) and SMS waves ($x = x_S$). The gray line is the (numerically calculated) full oscillation field, (1) FMS wave incident on the transition layer, (2) FMS wave reflected from the transition layer (WKB approximation).

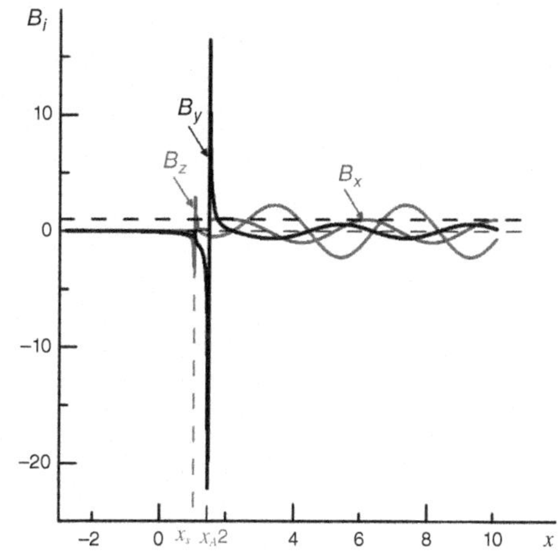

Figure 2.9 Distribution of MHD oscillation magnetic field components ($\text{Re}(B_x, B_y, B_z)$) in the problem of FMS wave incident on/reflected from the transition layer with resonance surfaces for Alfvén ($x = x_A$) and SMS waves ($x = x_S$).

are often called 'longitudinal' (to the magnetic field direction $\mathbf{B}_0$), while resonant Alfvén oscillations are called 'azimuthal' (or toroidal). It is conventional terminology in studies of oscillations in axisymmetric models, where the azimuthal coordinate (e.g. azimuthal angle) corresponds to the coordinate y.

As was shown in Section 2.3, a signature of resonant Alfvén oscillations is hodograph rotation reversal for the transverse wave vector $\mathbf{B}_\perp = (B_x, B_y)$ when the resonance surface is crossed. This follows from the derivative $\partial \zeta / \partial x$ sign reversal on different sides of the resonance surface. In the problem of two resonance surfaces, this rule, for $\gamma, \gamma_s \ll \omega$, is valid when each resonance surface, $x = x_A$ and $x = x_S$, is crossed.

Let us now see, however, what happens when the decrements γ and γ_s are not too small. Figure 2.10 shows a derivative $\partial \zeta / \partial x$ distribution calculated for $\gamma/\omega = 0.1$ and three values $\gamma_s/\omega = 0.01; 0.1; 1$. Here, the circles with rotation-indicating arrows illustrate the behaviour of the hodograph in the (B_x, B_y) plane. When the value is small, $\gamma_s/\omega = 0.01$ ($T_e/T_i \gg 1$), the hodograph behaviour is consistent with what is expected (curve 1). When γ_s/ω is increased to 0.1 ($T_e \sim T_i$), the hodograph rotation direction reversal points are shifted away from the resonance surfaces by a value of the order of the distance between them (curve 2). As γ_s/ω is increased further to 1 ($T_e/T_i \approx 0.1$), no direction reversal occurs in the hodograph rotation (curve 3). This example shows that strongly decaying resonant SMS oscillations that are present in a system can considerably alter the behaviour of the field components, even in the neighbourhood of the resonance surface for Alfvén waves.

Let us now examine the behaviour of the absorption coefficient for FMS waves incident on the transition layer, as defined in (2.20). We will inspect its dependence on the characteristic values of the wave vector component, k_y, k_z, as well as on the parameter β^* and decrement γ_s related to SMS wave dissipation near the resonance surface. Let us choose the decrement related to Alfvén wave dissipation to be extremely small, $\gamma = 10^{-6}\omega$, so that the absorption coefficient, D, be unaffected by its value.

Notably, the decrement ($\gamma_s \sim \omega$) is rather large for $T_e/T_i \gtrsim 1$, therefore it should be localised near $x = x_S$ at such a scale that would allow the MHD oscillations to be regarded as SMS wave. Obviously, this scale is determined by the size of the transparent region for SMS waves, $\Delta_s = x_S - x_{01}$. If a linear approximation for Alfvén speed of the form $v_A^2(x) \approx v_{As}^2[1 - (x - x_S)/a_s]$ is applicable near points x_S, x_{01}, we have $\Delta_s \approx k_z^2 a_s \beta^*/$

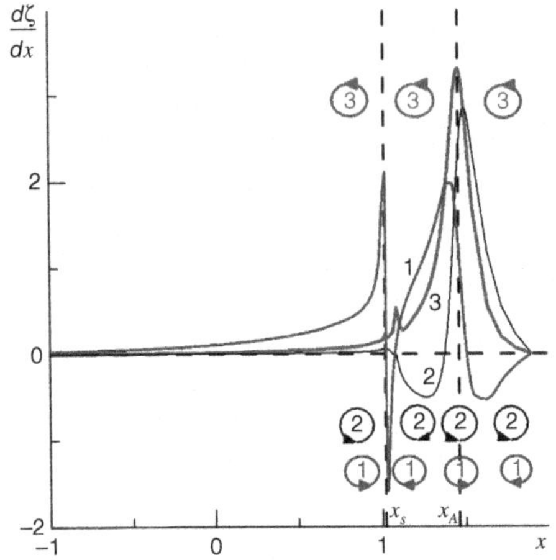

Figure 2.10 Hodograph behaviour for resonant MHD oscillations in the neighbourhood of resonance surfaces, x_A and x_S, for various decrements of Alfvén ($\gamma/\omega = 0.1$) and SMS oscillations. The curves and circles with hodograph rotation directions labelled 1,2 and 3, correspond to three values of SMS oscillation decrement: $\gamma_S/\omega = 0.01$, $\gamma_S/\omega = 0.1$ and $\gamma_S/\omega = 1$.

$[(k_z^2 + k_y^2)(1 + \beta^*)^2 - 2k_z^2\beta^*]$. For small $\beta^* \ll 1$, this scale $\Delta_s \approx a_s\beta^* k_z^2/(k_z^2 + k_y^2)$ is much smaller than the scale $a_s = (\partial \ln(v_A^2(x))/\partial x)^{-1}_{x=x_S}$, while being comparable for $\beta^* \sim 1$. To localise the decrement γ_s near $x = x_S$, let us make use of the model expression

$$\gamma_s = -\overline{\gamma}_s \exp\left(-(x - x_S)^2/\Delta_s^2\right), \tag{2.68}$$

where $\overline{\gamma}_s$ is the decrement value corresponding to plasma parameters at the magnetosonic resonance point (see Figure 1.3). Obviously, this approach cannot be used for the case of SMS waves incident on the transition layer. These waves will be strongly decaying throughout their domain of existence. Solving this problem requires a special formulation, where the SMS wave source must be located at a finite distance from the transition layer.

Figure 2.11 shows the distribution $D(k_y\tilde{\Delta})$, where $\tilde{\Delta} = [\Delta_2/(k_z^2 + k_y^2)]^{1/3}$, $\Delta_2 = |\partial \ln(v_A^2)/\partial x|^{-1}_{x=x_{02}}$ is the characteristic scale of the $v_A^2(x)$ variation at the turning point x_{02}.

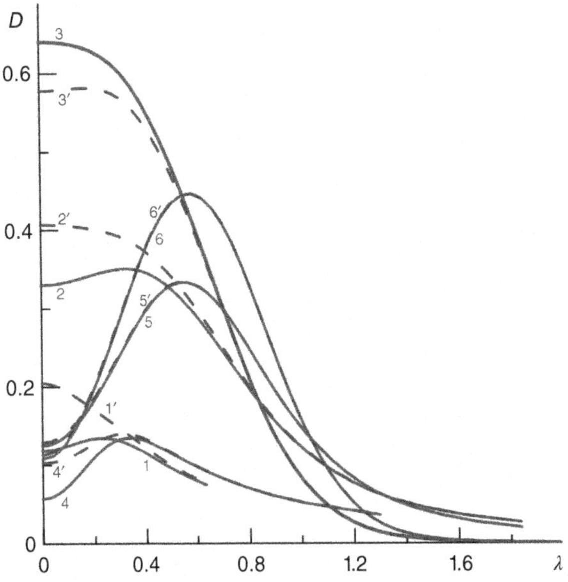

Figure 2.11 Absorption coefficient dependence for FMS waves incident on the transition layer with two resonance surfaces: for Alfvén and SMS waves. The distributions $D(k_y\tilde{\Delta})$ are shown for the $\beta^* = 0.3$ plasma layer (for waves with $k_z\Delta = 0.1;\ 1;\ 3.5$: curves 1, 2, 3 for $\overline{\gamma}_s = 0.01\omega$ and curves 1′, 2′, 3′ for $\overline{\gamma}_s = \omega$) and the $\beta^* = 1$ plasma layer (for $k_z\Delta = 0.1;\ 1;\ 3.5$: curves 4, 5, 6 for $\gamma_s = 0.01\omega$ and curves 4′, 5′, 6′ for $\overline{\gamma}_s = \omega$).

Models of the plasma transition layer with $\beta^* = 0.3; 1$ and oscillations with various values of the decrement $\overline{\gamma}_s = 0.01\omega; \omega$ and the parameter $k_z\Delta = 0.1; 1; 3.5$ are examined. The upper and lower value of the last parameter are close to the limits warranting the presence in the system at hand of both the resonance surfaces x_A and x_S. When $k_z\Delta$ increases, one can see that the value of $D(0)$ increases dramatically, while the maximum $D(k_y\tilde{\Delta})$ shifts to $k_y\tilde{\Delta} = 0$. The maximum value of D can considerably exceed the limiting value in a cold plasma (see Figure 2.5). This conclusion may be essential for problems concerning plasma heating in the solar corona and nuclear fusion reactors.

The presence of a resonance surface for SMS waves results in a full absorption of the energy of oscillations reaching it, with the maximum absorption shifting towards small values of the azimuthal component of the wave vector k_y, and vice versa, towards large values of the parallel component k_z. It is interesting to note that, with $\beta^* = 1$ and $k_z\Delta = 3.5$, the smaller the value of the decrement ($\overline{\gamma}_s = 0.01\omega$, not $\overline{\gamma}_s = \omega$) the larger the value of D, in contrast to the other above-discussed cases. This may be explained by the localisation we imposed on the decrement γ_s (see (2.68), which limits the integral increase of absorbed energy resulting from resonance region expansion.

The chief difference between energy absorption in a plasma with finite β^* and the cold plasma case is a non-zero value of $D(0)$ in the former. This is explained by the fact that the dissipation mechanism of resonant SMS oscillations comes into action and, unlike resonant Alfvén waves, these oscillations will not disappear at $k_y = 0$. This dissipation becomes more effective as β^* grows, which can be explained as follows. The larger the value of β^*, the farther from the resonance surface x_S and the closer to the Alfvén resonance surface x_A is the turning point x_{01}. As a result, the amplitude and, correspondingly, the energy of oscillations that have crossed the opaque region $x_{01} < x < x_A$, increase. All this energy is absorbed in the neighbourhood of the resonance surface x_S, resulting in increased absorption coefficient D.

2.7 Alfvén Resonance in Non-ideal Plasma. Kinetic Alfvén Waves

In Sections 2.1–2.6 dealt with resonance between Alfvén and SMS waves, on the one hand, and monochromatic FMS waves, on the other, in a 1D-inhomogeneous plasma, using the ideal MHD approximation. Let us now address effects arising when Alfvén and FMS waves are coupled beyond the ideal MHD framework. These effects are most important for Alfvén waves because they cause their transverse dispersion. As a result, the Alfvén waves acquire a transverse component of the group velocity and, correspondingly, are able to propagate across magnetic shells. Unlike Alfvén waves, magnetosonic waves exhibit transverse dispersion even in the ideal MHD approximation. The effects discussed below therefore contribute little to their structure or dynamics.

The simplest way to examine the effects resulting in small transverse dispersion in Alfvén waves is to employ the two-liquid MHD approximation. This approximation regards plasma as a medium consisting of two liquids, electrons and ions. To describe the MHD oscillation field, let us employ a system of Eq. (1.18) readily reducible to the vector equation

$$\text{rot rot}\mathbf{E} = \frac{\omega^2}{c^2} \hat{\varepsilon}\, \mathbf{E}. \tag{2.69}$$

The dielectric permeability tensor $\hat{\varepsilon}$ for MHD waves, in the (x, y, z) system of coordinates used in Section 2.1 has the form (see [24] and (1.19) in Chapter 1):

$$\hat{\varepsilon} = \frac{c^2}{v_A^2}\begin{pmatrix} 1 - \frac{3}{4}k_x^2\rho_i^2 & -\frac{3}{4}k_x k_y \rho_i^2 + iu & 0 \\ -\frac{3}{4}k_x k_y \rho_i^2 - iu & 1 - \frac{3}{4}k_y^2\rho_i^2 & 0 \\ 0 & 0 & \frac{\tilde{G}(s_e/\rho_s)}{k_z^2\rho_s^2} \end{pmatrix}, \tag{2.70}$$

the notations being: $u = \omega/\omega_i$, $\rho_i = v_i/\omega_i$ is the Larmor ion radius, $\rho_s = v_{es}/\omega_i$ (where $v_s = v_i\sqrt{T_e/T_i}$), $s_e = \bar{c}/\omega_{pe}$ is the characteristic electron skin depth in plasma, while the properties of the function $\tilde{G}(z)$ are described in Chapter 1. Note that (2.70) involves only small parameters leading to transverse dispersion of Alfvén waves $(u;\ k_\perp^2\rho_i^2;\ k_\perp^2\rho_s^2;\ k_\perp^2 s_e^2 \ll 1)$. It also lacks terms related to slow magnetosonic waves.

Let us consider an Alfvén resonance problem using a model in the form of a transition layer as described in Sections 2.2 and 2.5. We will assume the magnetic field lines to be straight and directed along the z axis, and the plasma density to be inhomogeneous along the x coordinate. When substituting the tensor (2.70) into (2.69) in such a model, the wave vector component k_x should be regarded as a derivative, $k_x = -i\nabla_x$, affecting the oscillation electric field $\mathbf{E}$ components. We will restrict ourselves to exploring the wave field structure in the Alfvén resonance region, where the characteristic scale of oscillations along the x coordinate is much smaller than the background plasma inhomogeneity scale.

From the z component in (2.69) we obtain, in the main order of perturbation theory, the following expression for the parallel component of the oscillation electric field:

$$E_z \approx ik_z \frac{\rho_s^2}{\tilde{G}(s_e/\rho_s)}(\nabla_x E_x + ik_y E_y). \tag{2.71}$$

Recall that (2.71) takes into account the transverse dispersion of Alfvén waves, where

$$\frac{\rho_s^2}{\tilde{G}(s_e/\rho_s)} \approx \begin{cases} \rho_s^2, & s_e \ll \rho_s,\ (\beta \gg m_e/m_i), \\ -s_e^2, & s_e \gg \rho_s,\ (\beta \ll m_e/m_i), \end{cases}$$

and $|(k_\perp\rho_s)^2/\tilde{G}(s_e/\rho_s)| \ll 1$. For magnetosonic waves, $E_z = 0$. Substituting (2.71) into the two other equations of (2.69) produces, for the transverse components of the oscillation electric field $\mathbf{E}_\perp = (E_x, E_y)$, the following system of equations:

$$\frac{\omega^2}{v_A^2}\Lambda^2\nabla_x^2 E_x + \left(\frac{\omega^2}{v_A^2} - k_y^2 - k_z^2\right)E_x = ik_y\left(1 - \frac{\omega^2}{v_A^2}\Lambda^2\right)\nabla_x E_y - iu\frac{\omega^2}{v_A^2}E_y,$$

$$\nabla_x^2 E_y + \left[\frac{\omega^2}{v_A^2}(1 - k_y^2\Lambda^2) - k_z^2\right]E_y = ik_y\left(1 - \frac{\omega^2}{v_A^2}\Lambda^2\right)\nabla_x E_x + iu\frac{\omega^2}{v_A^2}E_x,$$

where $\Lambda^2 = (3/4)\rho_i^2 + \rho_s^2/\tilde{G}(s_e/\rho_s)$.

In further calculations, it is more convenient to use other variables. As follows from the Helmholtz decomposition theorem, an arbitrary vector field (the divergence and curl of which have been determined for each point in space) can be represented as the sum of the vortex-free and solenoidal fields [62]. In particular, the two-dimensional vector field $\mathbf{E}_\perp$ can be represented as

$$\mathbf{E}_\perp = -\nabla_\perp\varphi + [\nabla_\perp \times \Psi], \tag{2.72}$$

where $\varphi(x, y, z)$ is the scalar potential, and Ψ is the vector potential. This decomposition contains derivatives of only the parallel component of the vector potential. Therefore, without losing generality, we can assume $\Psi = (0, 0, \psi(x, y, z))$. Substituting the decomposition (2.72) into the equation for transverse components of the electric field produces

$$\frac{\omega^2}{v_A^2}\Lambda^2\nabla_x^3\varphi + \left[\frac{\omega^2}{v_A^2}\left(1 - k_y^2\Lambda^2\right) - k_z^2\right]\nabla_x\varphi - uk_y\frac{\omega^2}{v_A^2}\varphi =$$
$$ik_y\left[\nabla_x^2\psi + \left(\frac{\omega^2}{v_A^2} - k_y^2 - k_z^2\right)\psi\right] - iu\frac{\omega^2}{v_A^2}\nabla_x\psi, \tag{2.73}$$

$$\nabla_x^3\psi + \left(\frac{\omega^2}{v_A^2} - k_y^2 - k_z^2\right)\nabla_x\psi - uk_y\frac{\omega^2}{v_A^2}\psi =$$
$$-ik_y\left[\frac{\omega^2}{v_A^2}\Lambda^2\nabla_x^2\varphi + \frac{\omega^2}{v_A^2}\left(1 - k_y^2\Lambda^2\right)\varphi - k_z^2\varphi\right] + iu\frac{\omega^2}{v_A^2}\nabla_x\varphi. \tag{2.74}$$

Solving the system of Eqs. (2.73)–(2.74) in the general form is a complex enough problem. For simplicity, we will therefore restrict ourselves to two limiting cases. In quasiparallel Alfvén waves ($k_\perp \ll \sqrt{u}\,k_\parallel$), transverse dispersion is determined by plasma gyrotropy, i.e. ion rotation in the magnetic field. In MHD equations this effect is described by terms proportional to the small parameter $u = \omega/\omega_i \ll 1$. In this approximation, terms proportional to parameter Λ^2 in (2.73)–(2.74) can be ignored. Moreover, we will restrict ourselves to considering oscillations with $k_y = 0$, where gyrotropic effects are dominant. For such oscillations, the system of Eqs. (2.73)–(2.74) is reduced to

$$\left(\frac{\omega^2}{v_A^2} - k_z^2\right)\varphi' = -iu\frac{\omega^2}{v_A^2}\psi', \tag{2.75}$$

$$\nabla_x^2\psi' + \left(\frac{\omega^2}{v_A^2} - k_z^2\right)\psi' = iu\frac{\omega^2}{v_A^2}\varphi', \tag{2.76}$$

where the notations are: $\varphi' \equiv \nabla_x\varphi$, $\psi' \equiv \nabla_x\psi$. It is easily verifiable that for the homogeneous plasma model, the system of Eqs. (2.75)–(2.76) yields a dispersion equation for MHD waves in a cold gyrotropic plasma, one of its roots being the dispersion relation for FMS waves, $\omega^2 = k^2v_f^2$, the other being the dispersion Eq. (1.24) for quasiparallel Alfvén waves.

In the opposite limiting case ($k_\perp \gg \sqrt{u}k_\parallel$) terms proportional to u can be neglected in (2.73)–(2.74). The result is

$$\frac{\omega^2}{v_A^2}\Lambda^2\nabla_x^2\varphi' + \left[\frac{\omega^2}{v_A^2}\left(1 - k_y^2\Lambda^2\right) - k_z^2\right]\varphi'$$
$$= ik_y\left[\nabla_x^2\psi + \left(\frac{\omega^2}{v_A^2} - k_y^2 - k_z^2\right)\psi\right], \tag{2.77}$$

$$\nabla_x^2\psi' + \left(\frac{\omega^2}{v_A^2} - k_y^2 - k_z^2\right)\psi'$$
$$= -ik_y\left[\frac{\omega^2}{v_A^2}\Lambda^2\nabla_x^2\varphi + \frac{\omega^2}{v_A^2}\left(1 - k_y^2\Lambda^2\right)\varphi - k_z^2\varphi\right]. \tag{2.78}$$

As concerns the homogeneous plasma model, (2.77) – (2.78) yields the dispersion equation for FMS waves, $\omega^2 = k^2 v_f^2$, and the dispersion Eq. (1.24) for kinetic Alfvén waves.

The systems of Eqs. (2.75) – (2.76) and (2.77) – (2.78) have a form analogous to equations for two coupled oscillators (see [63]). The operators in the left sides of (2.75) and (2.76) act upon the scalar potential φ producing, in homogeneous plasma, dispersion equations for Alfvén waves; while operators in Eqs. (2.76) and (2.78) act upon the vector potential component ψ producing a dispersion equation for FMS waves [64, 65]. The terms in the right sides of these equations determine the relationship between these MHD oscillation modes. It could be assumed therefore that the scalar potential φ of the full MHD oscillation field describes Alfvén waves, while the parallel component ψ of the vector potential refers to magnetosonic waves.

Let us find a solution to (2.75) – (2.76) for quasiparallel MHD oscillations with $k_y = 0$ near the Alfvén resonance point x_A (see [66]). Expressing φ' from (2.75) and substituting it into (2.76) yields

$$\nabla_x^2 \psi' - V(x)\psi' = 0, \tag{2.79}$$

which has the form of a Schrödinger equation with potential

$$V(x) = -k_x^2 = -\frac{\omega^2}{v_A^2} + k_z^2 + \frac{u^2 k_z^4}{(\omega^2/v_A^2) - k_z^2}, \tag{2.80}$$

where k_x is the x component of the wave vector in the WKB approximation. Near the Alfvén resonance point, we will employ a linear decomposition for the Alfvén speed

$$v_A^2(x) \approx \overline{v}_A^2 \left[1 - \frac{x - x_A}{a_A}\right] \tag{2.81}$$

and, to regularise the singularities in (2.79), we will take into account the small oscillation increment, replacing $\omega \to \omega + i\gamma$. The result, in the main order of perturbation theory, is

$$\nabla_x^2 \psi' + \frac{u^2 k_z^2 a_A}{x - x_A + i\varepsilon_A}\psi' = 0, \tag{2.82}$$

where $\varepsilon_A = 2a_A\gamma/\omega$. Note that in the case at hand, the Alfvén resonance point is also located in the FMS opaque region, at distance $\Delta_A = x_f - x_A = ua_A$ from the turning point x_f, as found from the condition $V(x_f) = 0$. The general solution of (2.82) has the form

$$\psi' = \sqrt{\frac{x - x_A + i\varepsilon_A}{\tilde{a}}}\left[C_1 I_1\left(2\tilde{U}\sqrt{\frac{x - x_A + i\varepsilon_A}{\tilde{a}}}\right) + C_2 K_1\left(2\tilde{U}\sqrt{\frac{x - x_A + i\varepsilon_A}{\tilde{a}}}\right)\right],$$

where the notations are: $\tilde{a} = (a_A/k_z^2)^{1/3}$, $\tilde{U} = u(k_z a_A)^{2/3}$, $I_1(z), K_1(z)$ are modified Bessel functions, $C_{1,2}$ are arbitrary constants. The only role the term $i\varepsilon_A$ plays in this expression is to set the bypass rule for the singular point $x = x_A$:

$$\lim_{\varepsilon_A \to 0} \sqrt{x - x_A + i\varepsilon_A} = \begin{cases} \sqrt{x - x_A}, & x > x_A, \\ i\sqrt{x_A - x}, & x < x_A, \end{cases} \tag{2.83}$$

assuming which, the parameter ε_A can be omitted entirely, because the solution for ψ' has no singularities for $x \to x_A$.

The scalar potential φ does have a singularity, however, and, for $x \to x_A$, it can be expressed from (2.75) as

$$\varphi' = -i\frac{ua_A\psi'}{x - x_A + i\varepsilon_A},$$

whence

$$\varphi = -iua_A\psi' \ln \frac{x - x_A + i\varepsilon_A}{a_A}. \tag{2.84}$$

Thus, the Alfvén oscillation field has the same singularities on the resonance surfaces as do MHD oscillations with $k_y \neq 0$ in an ideal plasma. This means that Alfvén resonance exists in a cold gyrotropic plasma even for $k_y = 0$ (see [66, 67]). The E_y and B_x components of the Alfvén oscillation field have a logarithmic singularity on the resonance surface, while the E_x and B_y components have a singularity of the form x^{-1}. If $k_y = 0$, the parameter u enables Alfvén and FMS waves to be coupled on the resonance surface, but fails to regularise the wave field singularity.

Let us now examine the Alfvén resonance for quasitransverse Alfvén waves as described by the system of Eqs. (2.77)–(2.78). Such waves are called 'kinetic' Alfvén waves (KAWs). In the neighbourhood of the Alfvén resonance point, where $v_A^2(x) \approx \overline{v}_A^2[1 - (x - x_A)/a_A]$, we have

$$\Lambda^2\nabla_x^2\varphi' + \frac{x - x_A}{a_A}\varphi' = i\frac{k_y}{k_z^2}\Delta_\perp\psi, \tag{2.85}$$

$$\nabla_x^2\psi' + \left(\frac{\omega^2}{v_A^2} - k_y^2 - k_z^2\right)\psi' = -ik_y k_z^2\Lambda^2\Delta_\perp\varphi, \tag{2.86}$$

where $\Delta_\perp = \nabla_x^2 - k_y^2$. Let us consider the problem of resonant Alfvén oscillations excited by an FMS wave in an inhomogeneous plasma. The right side in (2.85) is determined by the field of large-scale FMS wave serving as the source of the resonant Alfvén waves. The right-hand side in (2.86) describes the reverse effect of the Alfvén wave on the FMS oscillation field. Since this side is proportional to the small parameter $|k_\perp^2\Lambda^2| \ll 1$, we will assume it to be zero, in the zero approximation. The FMS wave field is then described by a homogeneous equation, (2.86), which has no singularities on the resonance surface. The characteristic variation scale in the right side of (2.85) is determined by the field of large-scale FMS oscillations and can be assumed to be constant in the neighbourhood of the resonance surface.

In this case, the solution of (2.85) satisfying the boundary conditions (finite amplitude for asymptotics $x - x_A \to \pm\infty$) has the form

$$\varphi' = \widetilde{\varphi}' G\left(\frac{x - x_A}{\tilde{a}}\right), \tag{2.87}$$

where $\widetilde{\varphi}' = ik_y(\Delta_\perp\psi)_{x_A}/k_z^2$, $\tilde{a} = \Lambda^{2/3}a_A^{1/3}$, and the function $G(z)$ is the solution of the inhomogeneous Airy equations

$$\nabla_z^2 G + zG = 1 \tag{2.88}$$

and has this integral representation:

$$G(z) = -i\int_0^\infty \exp\left(-i\frac{v^3}{3} + ivz\right)dv. \tag{2.89}$$

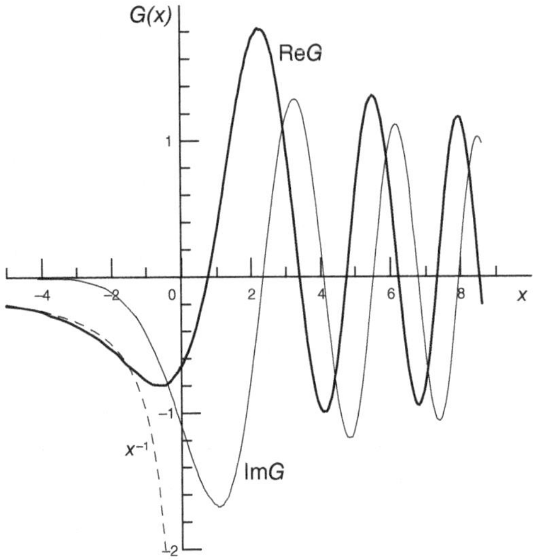

Figure 2.12 Distribution of the real and imaginary parts of $G(x)$, a function of a real argument.

Its asymptotics on the real axis of the complex variable z have the form

$$G(z) = \begin{cases} -\sqrt{\pi}z^{-1/4}\exp(\frac{2}{3}iz^{3/2} + i\pi/4), & z \to \infty, \\ z^{-1}, & z \to -\infty. \end{cases} \tag{2.90}$$

This asymptotic representation is also applicable to the complex values of z in the $|\arg z| < 2\pi/3$ sectors. The distribution of the real and imaginary parts of the function $G(x)$ of a real argument $z = x$ is shown in Figure 2.12.

For negative values of $\Lambda^2 = -s_e^2$ (where $s_e = \overline{c}/\omega_{pe}$ is the electron skin depth) the solution of (2.85) has the form

$$\varphi' = -\widetilde{\varphi}'G\left(-\frac{x - x_A}{\tilde{a}}\right), \tag{2.91}$$

where $\tilde{a} = s_e^{2/3}a_A^{1/3}$. As can be seen from the asymptotic expressions (2.90), for the real positive values of parameter $\Lambda^2 = (3/4)\rho_i^2 + \rho_s^2$, the solution (2.87) describes an Alfvén wave which, for $x > x_A$, has the form of a wave travelling away from the resonance surface $x = x_A$ across the magnetic shells and decreases as x^{-1} into the opaque region ($x < x_A$). As follows from (2.91), the opaque region for Alfvén waves is $x > x_A$ for real negative values of $\Lambda^2 = -s_e^2$, while the wave travels away across magnetic shells in the transparent region $x < x_A$. The wave field structure for these two limiting cases is shown in Figure 2.13.

Note that taking into account the small parameter Λ^2 in (2.85) is not only crucial for the relationship between Alfvén and FMS waves, but also regularises the wave field singularity at the Alfvén resonance point. The singularity is regularised because the oscillation energy is transferred from the resonance region by kinetic Alfvén waves travelling away from the resonance surface. Such regularisation takes place if the decrement γ of Alfvén waves is small: $|\gamma/\omega| \ll |\Lambda/a_A|^{2/3}$. If this decrement is large enough ($|\gamma/\omega| \gg |\Lambda/a_A|^{2/3}$), then the singularity is regularised thanks to the absorption of the energy of resonant Alfvén waves near the resonance surface (see Section 2.3). It is interesting to note that the density of energy carried away by KAW from the resonance surface is equal to the specific power of the resonant Alfvén wave absorbed near the resonance surface (see Appendix S).

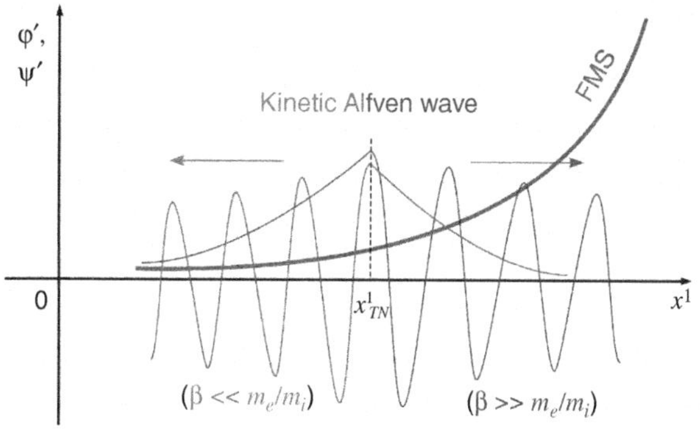

Figure 2.13 Structure of MHD oscillations across magnetic shells when FMS waves are in resonance with kinetic Alfvén waves. The FMS wave amplitude decreases exponentially into the opaque region. In a 'cold' plasma ($\beta \ll m_e/m_i$), a kinetic Alfvén wave is excited on the resonance magnetic shell $x = x_A$ and travels away leftwards, while in a 'warm' plasma ($\beta \gg m_e/m_i$), an Alfvén wave travels away rightwards from the resonance shell.

When the parameter $\Lambda = \lambda e^{-i\alpha}$ (where $0 < \alpha < \pi$) is a complex value, the amplitude of the Alfvén waves travelling away from the resonance surface decreases as $\exp[-(2/3)((x - x_A)/\tilde{a})^{3/2} \sin(\alpha/2)]$, where $\tilde{a} = \lambda^{2/3} a_A^{1/3}$. This is due to wave energy being absorbed by resonant electrons in the background plasma as the waves travel across magnetic shells [58]. Note also that it is in quasitransverse kinetic Alfvén waves that the parallel component of the oscillation electric field has the highest value ($E_z \sim E_\perp k_\perp \omega \rho_s^2 / v_A$, see (2.71).

2.8 FMS Waveguide

In Sections 2.1–2.7 we examined MHD wave propagation in a plasma with a monotonous inhomogeneity along one of the coordinates that are transverse to the background magnetic field. Let us now see what the presence of plasma inhomogeneities with a non-monotonous profile entails. One of the most conspicuous features of wave propagation in such media is that the wave can be captured in waveguides formed by the medium inhomogeneity.

Waveguide propagation for various types of waves is a well-known phenomenon in geophysics. Suffice it to mention acoustic waveguides in the atmosphere and the ocean and the Earth-ionosphere waveguide for radio waves. The interest in waveguide propagation of oscillations is understandable. A waveguide, to a significant degree, restricts wave scattering in space and enables wave propagation over large distances. Of doubtless interest is the waveguide propagation of MHD waves in the near-Earth plasma [68].

Waveguide propagation is due to limits imposed on wave propagation in certain directions. In artificial waveguides (used in, e.g. engineering), this is realised by means of reflecting walls. As a rule, there are no such walls (distinct boundaries) in natural waveguides, the 'waveguide-forming' factor consisting of a smoothly inhomogeneous medium combined with wave dispersion. This thesis is best explained in terms of wave packets. Wave-packet motion equations, in the ray approximation, have the form (I.1). Let us assume the packet central frequency $\omega(\mathbf{x}, \mathbf{k})$ to depend on the transverse (relative to the waveguide) coordinate x and the respective wave vector k_x to be quadratic:

$$\omega = \omega_0 + \frac{ax^2}{2} + \frac{bk_x^2}{2}. \tag{2.92}$$

Here, the second term in the right-hand side is determined by the inhomogeneity of the medium, and the third term, by wave dispersion. A (2.92)-type ratio is generally applicable near the waveguide axis. Equation (I.1) are then reduced to one equation,

$$\frac{d^2x}{dt^2} = -abx,$$

which, for $ab > 0$, describes harmonic oscillations of the packet, which fact, combined with free propagation along the waveguide, results in a wave-like motion of the packet near its axis.

The existence of the waveguide can also be demonstrated in terms of perturbed fields. Let us use the dispersion Eq. (2.92) to reconstruct a differential equation describing the spatial structure of the field. For this purpose, we will multiply the left- and right-hand sides by the field value Φ (it can be any component of the perturbed electric or magnetic field) and replace $k_x^2 \to -d^2/dx^2$. The result will be an equation of the Schrödinger stationary equation type

$$b\frac{d^2\Phi}{dx^2} + [2(\omega - \omega_0) - ax^2]\Phi = 0, \tag{2.93}$$

which, for $ab > 0$, describes a quantum oscillator (see [69]). This produces the waveguide eigenfrequency spectrum

$$\omega = \omega_n \equiv \omega_0 + \sqrt{ab}\left(n + \frac{1}{2}\right),$$

where $n = 0, 1, 2, 3, \ldots$ is the quantum number of the waveguide eigenmodes. The respective eigen-functions have a localisation scale of order $(ab)^{1/4}$ along the x axis. The difference between Eqs. (2.92) and (2.93) is quite similar to the difference between classic and quantum mechanics.

Based on these arguments, it is not difficult to prove that waveguide propagation is possible for FMS waves. The dispersion equation for them has the form $\omega = kv_A$, where $k = \sqrt{k_x^2 + k_y^2 + k_z^2}$ is the modulus of the full wave vector. Let us assume that the magnetic field B_0 is homogeneous and the plasma density depends on the coordinate x and has a maximum at $x = x_0$. The following decomposition can be used near this maximum:

$$v_A^2 \approx \overline{v}_A^2 \left[1 + \frac{(x - x_0)^2}{a_0^2}\right], \tag{2.94}$$

where $a_0 = (\sqrt{v_A^2(x)/|\nabla_x^2 v_A^2(x)|_{x=x_0}})$ is the characteristic scale of the inhomogeneity. Assuming k_x sufficiently small, we have

$$\omega \approx \overline{v}_A\sqrt{k_y^2 + k_z^2}\left(1 + \frac{(x - x_0)^2}{2a_0^2} + \frac{k_x^2}{2(k_y^2 + k_z^2)}\right).$$

This expression has the same form as (2.92), where $ab = \overline{v}_A^2/a_0^2 > 0$. This means that FMS waveguide propagation (in the directions of the y and z axes) is possible in a 1D-inhomogeneous plasma, near its density maximum.

Let us explore this possibility in more detail for quasi-parallel MHD waves described by (2.79), (2.80), where the Alfvén speed near the density maximum is roughly described as (2.94). Recall that, for such oscillations, $k_y = 0$ and the following relation holds: $k_\perp \lessapprox \sqrt{u}k_z$, where $k_\perp$ should be understood, in this case, as a reciprocal of the characteristic wavelength of the oscillations, along the x coordinate. As will be seen later, the characteristic eigenfrequency of the main harmonics of FMS waves travelling down the waveguide is close to the local Alfvén frequency at the plasma density maximum point and can be represented as

$$\omega = k_z \bar{v}_A (1 + \delta),$$

where $|\delta| \ll 1$. For oscillations travelling near the plasma density maximum, the potential in (2.79) can thus be written as

$$V(x) = k_z^2 \left[\frac{(x - x_0)^2}{a_0^2} - \delta - \frac{u^2}{(x - x_0)^2 / a_0^2 - \delta} \right]. \tag{2.95}$$

The form of the potential $V(x)$ for values $\delta > 0$ and $\delta < 0$ is shown in Figure 2.14.

For $\delta > 0$, Eq. (2.79) describes quasi-parallel FMS waves localised between the turning points $x_f = x_0 \pm a_0 \sqrt{\delta \pm u}$. There are also two Alfvén resonance points $x_A = x_0 \pm \sqrt{\delta} a_0$ inside the waveguide (see Figure 2.14(a)). As we know from Section 2.5, these points experience a resonant excitation of Alfvén waves and partial energy absorption of FMS waves travelling in the waveguide. For $u \ll 1$, this absorption is small. As was the case with Alfvén resonance, we well seek a solution to (2.79) using the method of successive approximations. In the main order of perturbation theory, let us set $u = 0$ in the potential (2.95). Equation (2.79) can then be represented as

$$\nabla_x^2 \psi' - k_z^2 \left[\frac{(x - x_0)^2}{a_0^2} - \delta \right] \psi' = 0. \tag{2.96}$$

The solution of (2.96) satisfying the boundary conditions (limited amplitude of oscillations on the asymptotics $x \to \pm\infty$) has the form (see [69])

$$\psi_n'(x) = C_n y_n \left(\frac{x - x_0}{\tilde{a}} \right), \qquad \delta_n = \frac{2n + 1}{k_z a_0}, \tag{2.97}$$

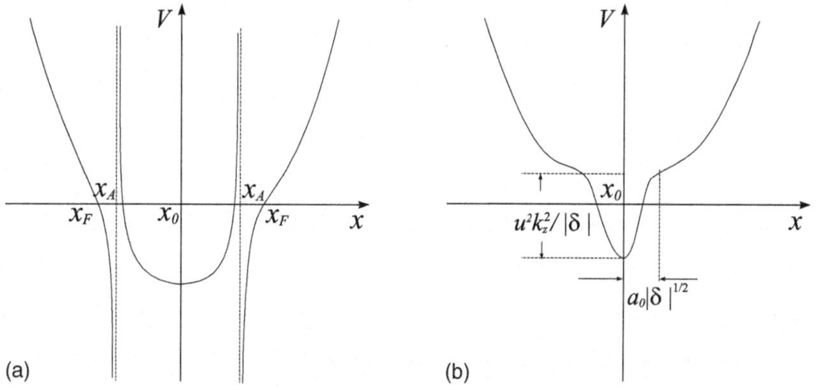

Figure 2.14 Potential in (2.79) for quasi-parallel MHD oscillations: (a) the form of potential $V(x)$ with Alfvén resonance points (x_A) for waveguide FMS oscillations ($\delta > 0$); (b) potential $V(x)$ for waveguide-travelling quasi-parallel Alfvén waves ($\delta < 0$).

where $n = 0, 1, 2, 3, \ldots$ is the waveguide mode harmonic number, C_n is an arbitrary constant, $\tilde{a} = \sqrt{a_0/k_z}$,

$$y_n(\xi) = e^{-\xi^2/2} H_n(\xi), \tag{2.98}$$

$H_n(\xi)$ are Hermitian polynomials ($H_0 = 1,\ H_1 = 2\xi, \ldots$). The eigenfunctions $\psi'_n(x)$ in (2.97) with eigenvalues $\delta = \delta_n$ describe the harmonics of waveguide FMS modes with n nodes in the x coordinate, inside the waveguide. In the WKB approximation along the x coordinate, the waveguide modes are standing waves localised between turning points $x_f = x_0 \pm a_0\sqrt{\delta + u}$.

Let us now determine eigenmode decay at Alfvén resonance points. To determine the decrement for the n-th harmonic γ_n, let us use the equation

$$\gamma_n = \frac{1}{W_n} \int_{-\infty}^{\infty} \frac{\partial F_n}{\partial t} dx, \tag{2.99}$$

where

$$W_n = \frac{1}{8\pi} \int_{-\infty}^{\infty} |B|^2 dx \tag{2.100}$$

is the integral density of the waveguide mode energy in a given area of the plane (y, z), $B^2 = B_x^2 + B_y^2 + B_z^2$ is the squared strength of the FMS wave magnetic field,

$$\frac{\partial F_n}{\partial t} = 4\gamma F_n = 4\gamma \frac{|B|^2}{8\pi} \tag{2.101}$$

is the power dissipated at Alfvén resonance points, $\gamma \to 0$ is the Alfvén wave decrement in the neighbourhood of resonance surfaces. This expression assumes the presence of two Alfvén resonance points inside the waveguide. Integration limits in (2.99) are extended to $\pm\infty$, because the Alfvén wave localisation region determined by a small increment, γ, is much smaller than the waveguide mode localisation region.

From the decomposition (2.72) and the Maxwell equation

$$\text{rot}\mathbf{E} = -\frac{1}{\bar{c}} \frac{\partial \mathbf{B}}{\partial t} \tag{2.102}$$

we have the following representations for the MHD wave magnetic field components:

$$B_x = \frac{k_z \bar{c}}{\omega}(ik_y\varphi + \nabla_x\psi), \quad B_y = -\frac{k_z \bar{c}}{\omega}(\nabla_x\varphi - ik_y\psi),$$
$$B_z = i\frac{\bar{c}}{\omega}(\nabla_x^2\psi - k_y^2\psi). \tag{2.103}$$

Representing the decomposition for the Alfvén speed in the neighbourhood of the Alfvén resonance points, $v_A^2(x) \approx \bar{v}_A^2[1 - (x - x_A)/a_A]$, and adding a small increment, $\omega \to \omega + i\gamma$ (for $\gamma \to 0$) to the eigen-oscillation frequency, we obtain, for the scalar potential φ, an expression of the form (2.84). Let us restrict ourselves to the basic mode of FMS oscillations travelling in the waveguide ($n = 0$). Thus, for the energy density of the basic harmonic, we have

$$\begin{aligned} W_0 &= \frac{\bar{c}^2}{8\pi\omega_0^2} \int_{-\infty}^{\infty} \left(k_z^2\psi_0'^2 + (\nabla_x\psi_0')^2\right) dx \\ &= C_0^2 \frac{\bar{c}^2}{8\pi\omega_0^2} \int_{-\infty}^{\infty} \left(k_z^2 + \frac{1}{\tilde{a}^2} + \frac{(x - x_0)^2}{\tilde{a}^4}\right) \exp\left(-\frac{(x - x_0)^2}{\tilde{a}^2}\right) dx \\ &\approx C_0^2 \frac{k_z^2\bar{c}^2\tilde{a}}{8\sqrt{\pi}\omega_0^2}, \end{aligned} \tag{2.104}$$

where $\omega_0 = \omega + \delta_0$, while, for the power dissipated in the neighbourhood of the Alfvén resonance points:

$$\frac{\gamma}{2\pi}\int_{-\infty}^{\infty}|B|^2dx \approx \frac{\gamma}{2\pi}\int_{-\infty}^{\infty}|B_y(x)|^2dx = \frac{\gamma k_z^2\bar{c}^2}{2\pi\omega_0^2}\int_{-\infty}^{\infty}|\varphi_0'(x)|^2dx =$$

$$\frac{(C_0k_z\bar{c}u)^2a_A}{4\pi\omega_0}\lim_{\varepsilon_A\to 0}\int_{-\infty}^{\infty}\frac{\varepsilon_A\exp\left(-\frac{(x-x_0)^2}{\tilde{a}^2}\right)}{(x-x_A)^2+\varepsilon_A^2}dx \approx$$

$$C_0^2\frac{k_z^2\bar{c}^2u^2a_A}{4\omega_0}e^{-1}. \tag{2.105}$$

These expressions take account of the fact that, for the quasilongitudinal MHD waves, $k_z a_0 \gg 1$ and $|x_A - x_0| = \tilde{a} = \sqrt{a_0/k_z}$ (this follows from the representation (2.94) and the Alfvén resonance condition for the basic harmonic $\omega_0 = k_z v_A(x_A)$). Equating the derivatives $\nabla_x v_A^2(x)$, obtained from the decompositions (2.81) and (2.94), produces the following expression relating to the characteristic scales:

$$a_A = a_0^2/\tilde{a} = a_0\sqrt{k_z a_0/2}.$$

Substituting (2.104) and (2.105) into (2.99) yields the following expression for the waveguide FMS oscillation eigenmode decrement:

$$\gamma_0 = 4\sqrt{\pi}e^{-1}u^2k_za_0\omega_0. \tag{2.106}$$

For small enough values of u (for $uk_za_0 \ll 1$, in any case) we have $\gamma_0 \ll \omega_0$, i.e. FMS waves travelling in the waveguide decay weakly due to interaction with Alfvén waves on resonance surfaces.

2.9 Waveguide for Quasilongitudinal Alfvén Waves

Let us examine the problem of quasilongitudinal Alfvén waves travelling down a waveguide formed by a plasma density inhomogeneity (see [70]). Let us employ the model medium analogous to the one in Section 2.8. As we saw earlier, transverse dispersion of Alfvén waves is determined by small effects. For quasilongitudinal Alfvén waves, it is related to finite plasma gyrotropy described by the small parameter $u = \omega/\omega_i \ll 1$. Similar to the waveguide FMS modes, Alfvén waves travelling down a waveguide are localised near the maximum plasma density. The wave-field structure of such modes is described by (2.79) with potential (2.95), where, in contrast to waveguide FMS modes, $\delta < 0$. In this case, the potential (2.95) has no singularities (see Figure 2.14(b)).

Waveguide modes can only exist in such a potential if the bottom of the 'potential well' is below zero ($V(x_0) < 0$), which is possible for $-\delta < u$. Let us find a solution to (2.79) for two limiting cases. The parameter characterising the potential well is the product of its depth and the square of its width. For a narrow well at the centre of the potential (see Figure 2.14(b)) this parameter equals $(uk_za_0)^2$. If $uk_za_0 \gg 1$, the potential well is 'deep' and oscillations travelling down the waveguide contain numerous eigenmodes. The number of eigenmodes in the discreet spectrum of a 'deep' well is of order $n \sim uk_za_0 \gg 1$. For $uk_za_0 \ll 1$, the potential well is 'shallow' and contains one eigenmode only.

In the case of a 'deep' well, the characteristic scale of localisation Δ for the basic harmonics of oscillations localised near the well bottom is much smaller than the well width $\sqrt{|\delta|}a_0$. For these harmonics, the potential can be assumed to be quadratic and the solutions of (2.79) have the same form as those for FMS modes (2.97) with eigenvalues

$$\delta = \delta_n = -u + \frac{2n+1}{k_z a_0}, \qquad \omega_n = k_z v_A(x_0)\left(1 - \frac{u}{2} + \frac{n+1/2}{k_z a_0}\right).$$

The characteristic localisation scale of the basic modes $\Delta = \sqrt{a_0/k_z}$, and, therefore, the condition that the modes should be localised near the potential well bottom ($\Delta^2 \ll |\delta|a_0^2$) has the form $uk_z a_0 \gg 1$. The quadratic representation of the potential (2.95) is applicable for $\Delta \ll a_0$, which means $k_z a_0 \gg 1$. We will further assume that this condition is satisfied.

In the case of a 'shallow' well ($uk_z a_0 \ll 1$), the eigenmode localisation size is much larger than the well width. The solution of (2.79) can be obtained by matching the solutions within and outside the narrow central well. When describing the wave field structure outside the central well, the term with small parameter u can be omitted in the potential of (2.79). The resulting equation takes the form of (2.96), where $\delta < 0$. Its solution, with the amplitude decreasing as $x \to +\infty$, is:

$$\psi'_+ = D_p[\sqrt{2}(x - x_0)/\tilde{a}], \qquad p = -(1 - \delta k_z a_0)/2, \tag{2.107}$$

where $D_p(z)$ is the parabolic cylinder function (see e.g. [27]), $\tilde{a} = \sqrt{a_0/k_z}$. For the solution inside the potential well, as $x \to x_0$, the main contribution to the potential is made, on the contrary, by the term with u^2. Equation (2.79) in this case has the form

$$\nabla_x^2 \psi' + \frac{u^2 k_z^2}{(x - x_0)^2/a_0^2 - \delta}\psi' = 0.$$

Let us integrate this equation between two points $x_\pm$, the distance between them significantly exceeding the characteristic width of the potential well $\sqrt{|\delta|}a_0$, while being much smaller than the characteristic localisation scale of the function: $\sqrt{|\delta|}a_0 \ll |x_\pm - x_0| \ll \sqrt{a_0/k_z}$. Given that the eigen function, on this scale, varies little, we obtain the following expression for the jump in the logarithmic derivative of the solution in question:

$$\nabla_x \ln \psi' \,|_{x_-}^{x_+} \approx -\int_{-\infty}^{\infty} \frac{u^2 k_z^2}{((x - x_0)/a_0)^2 - \delta} dx. \tag{2.108}$$

Here, integration in the right side is extended to $\pm\infty$, because the integral in the right side is convergent, while the integrand is localised at much smaller scales than the integration limits.

Equating this logarithmic derivative jump to the one of the external solution (2.107), and employing the asymptotical expressions for the parabolic cylinder functions, as $x \to x_0$, produces the following eigenvalue for the only solution:

$$\delta = -\frac{\pi^2}{4}\frac{\Gamma^2(1/4)}{\Gamma^2(3/4)} u^4 (k_z a_0)^3, \tag{2.109}$$

where $\Gamma(x)$ is the gamma function. The solution inside the narrow central well is in fact reduced to sharply changing the derivative and can be interpreted as a kink in the solution (2.107) at point $x = x_0$. The characteristic scale of this solution, $\Delta = \sqrt{a_0/k_z}$. Of interest is the polarisation of the Alfvén waveguide modes found in the two above-discussed limiting cases. From (2.103) and (2.75) we have

$$B_y = -i\frac{u}{(x-x_0)^2/a_0^2 - \delta}B_x.$$

Hence, the polarisation is nearly circular in the 'deep well' case ($uk_z a_0 \gg 1$): $B_y(x)/B_x(x) \approx -i$. This is consistent with the fact that, in a deep well, the characteristic wave vector $k_x \sim \sqrt{k_z/a_0}$ satisfies the quasilongitudinal propagation condition

$$k_x/k_z \sim (k_z a_0)^{-1/2} \ll \sqrt{u}.$$

In the 'shallow well' case, the polarisation is elliptical, with the ratio between the semiaxes of the polarisation ellipsis strongly depending on the x coordinate. At the centre of the waveguide, $x = x_0$, we have

$$|B_x/B_y| \sim (uk_z a_0)^{-3} \gg 1,$$

but even for $|x - x_0| = \sqrt{u}a_0 \ll \Delta$, the semiaxis ratio is reversed: $|B_x/B_y| < 1$.

2.10 Waveguides for Kinetic Alfvén Waves in a 'cold' Plasma. Waveguide Mode Attenuation

Transverse dispersion is small in kinetic Alfvén waves, which is why they can slowly (at much lower speeds than the Alfvén speed) propagate across magnetic shells. The extrema in the Alfvén speed distribution in the transverse x coordinate can also serve as waveguides for these waves (see [67, 71]). To describe waveguide propagation of kinetic Alfvén waves with $k_y = 0$ in a 'cold' plasma ($\beta \ll m_e/m_i$) near such extrema (where $v_A^2(x) \approx \overline{v}_A^2[1 \pm (x-x_0)^2/a_0^2]$), we will follow the procedure in Sections 2.8 and 2.9 and use (2.73), (2.74) to obtain this system of equations

$$\mu^4 \nabla_\xi^2 \varphi' - (\lambda - \sigma\xi^2)\varphi' = -i\epsilon\psi', \tag{2.110}$$

$$\nabla_\xi^2 \psi' + (\lambda - \sigma\xi^2)\psi' = -i\epsilon\varphi', \tag{2.111}$$

where $\mu^4 = k_z^2 s_e^2 \ll 1$ ($s_e = \overline{c}/\omega_{pe}$ is the electron skin depth), $\xi = (x - x_0)/\tilde{a}$ ($\tilde{a} = \sqrt{a_0/k_z}$), $\lambda = \delta k_z a_0$, $\epsilon = uk_z a_0$, $\sigma = \pm 1$ (the plus sign is for the minimum, the minus, for the maximum in the $v_A^2(x)$ distribution).

Let us examine the $\epsilon \ll 1$ case. The solution of the system of Eqs. (2.110), (2.111) can then be sought by successive approximations. In the zero approximation, $\epsilon = 0$, we obtain two sets of eigenmodes. Near the Alfvén speed minimum, the solution is a waveguide FMS mode (2.97). Near the maximum, there is a set of solutions for waveguide kinetic Alfvén waves:

$$\varphi'_m(x) = C_m y_m(\xi/\mu), \quad \lambda_m = -(2m+1)\mu^2, \tag{2.112}$$

where $m = 0, 1, 2, 3, \ldots$ is the waveguide mode number, and the function $y_m(x)$, describing the waveguide mode structure in the x coordinate, is defined by (2.98).

In the next order of perturbation theory, the right sides in (2.110), (2.111) are taken into account resulting in waveguide modes (2.97) and (2.112) transforming into MHD waves

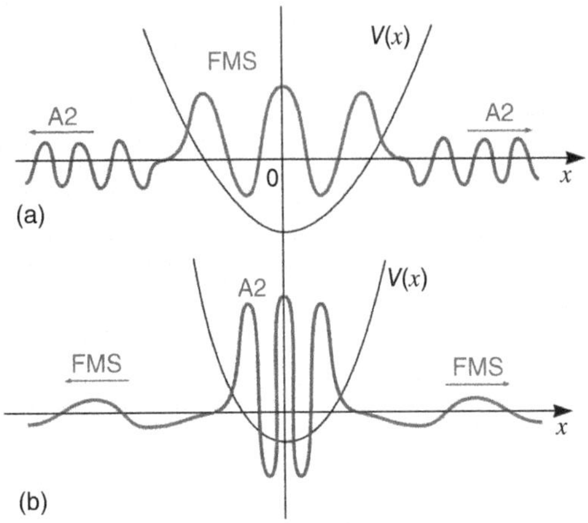

Figure 2.15 The structure of waveguide modes in the potential $V(x)$ decaying as the related MHD waves escape. Top: waveguide FMS mode and related escaping kinetic Alfvén wave (A2) (see Figure A.1, (a) in Appendix A); bottom: waveguide mode of kinetic Alfvén waves that decays due to the FMS wave escaping from the waveguide.

escaping from the waveguide. These waves carry away part of the energy of the eigenmodes travelling in the waveguide. The energy is carried away from the FMS waveguide (same as from the waveguide for quasilongitudinal Alfvén waves) by the kinetic Alfvén wave; from the waveguide for kinetic Alfvén waves, by the large-scale FMS wave (Figure 2.15). This results in waveguide waves decaying with decrement $\gamma = S_x/W$, where S_x is the Poynting vector of the escaping waves, and W is the waveguide mode energy density. Since the terms on the right side of (2.110), (2.111) are assumed to be small, the eigenmode decrements are also small.

A solution for waves escaping from the waveguide can be found using the Green's function method. A Green's function is constructed from solutions of a homogeneous equation that satisfy the boundary conditions (no incident waves on the waveguide). For example, let us determine the structure of an FMS wave carrying away the energy from the waveguide for kinetic Alfvén waves by substituting the zero approximation solution (2.112) into the right-hand side of the equation for FMS waves (2.111). The solution of (2.111) with a given right side can be written as

$$\psi' = -i\epsilon \int_{-\infty}^{\infty} \overline{G}(\xi, \xi')\varphi_m(\xi')d\xi', \tag{2.113}$$

where

$$\overline{G} = \frac{1}{\mu^4 \overline{W}} \begin{cases} \psi'_+(\xi)\psi'_-(\xi'), & \xi > \xi', \\ \psi'_-(\xi)\psi'_+(\xi'), & \xi < \xi', \end{cases} \tag{2.114}$$

is Green's function for (2.111), $\psi'_\pm(\xi)$ are the solutions of a homogeneous equation that satisfy the boundary conditions on the asymptotics $x \to \pm\infty$ (waves carrying energy away from the waveguide are present), $\overline{W}$ is their Wronskian. In the quadratic potential we are dealing with, the solutions of a homogeneous equation that describe FMS waves carrying away the energy of the mth harmonic of waveguide kinetic Alfvén waves have the form

$$\psi_\pm(\xi) = CD_{-(1-i\lambda_m)/2}(\pm\sqrt{2}e^{i\pi/4}\xi),$$

where C is an arbitrary constant, $D_p(z)$ is the parabolic cylinder function.

Substituting (2.114) into (2.113) yields, for $\xi \to +\infty$,

$$\psi' = -i\frac{\epsilon \overline{N}}{\mu^4 \overline{W}}\psi'_+(\xi)$$

– FMS wave carrying away the energy of the m-th harmonic of waveguide kinetic Alfvén waves, where

$$\overline{N} = \int_{-\infty}^{\infty} \psi'_{-}(\xi')\varphi_m(\xi')d\xi'.$$

Without going into too much detail about the calculations that are analogous to the ones in Sections 2.5 and 2.8 (see also [67]), we will present the expressions for the decrements of eigenmodes in different waveguides, determined by their partial transformation into MHD waves escaping from the waveguide. In the above case, the decrement of kinetic Alfvén waves due to their leaking from the waveguide in the form of a large-scale FMS wave has the form

$$\gamma_m = \omega_m \frac{\mu\epsilon^2}{k_z a_0}.$$

This expression holds for even modes $m = 0, 2, 4, \ldots$. The odd modes transform into FMS waves with a node at point $x = x_0$, the waveguide kinetic Alfvén wave being localised in its narrow neighbourhood. As a result, the decrement for the odd modes is μ^2 times smaller.

For waveguide FMS modes, the decrement of which is determined by their transforming into escaping kinetic Alfvén waves, we have

$$\gamma_n = \omega_n \frac{\epsilon^2}{k_z a_0}.$$

Note that the FMS wave decrement in the waveguide does not depend on the small parameter μ and practically coincides with the decrement due to the Joule dissipation at resonance points (2.106). This means that the decrement does not depend on the particular mechanism involved in energy loss in the FMS waveguide. For the only waveguide mode of a large-scale quasilongitudinal Alfvén wave, we obtain, for $\epsilon^2 \gg \mu$,

$$\gamma = \omega \frac{\epsilon^4}{k_z a_0} \exp\left(-\frac{\pi\lambda}{4\mu^2}\right).$$

where $\lambda = \delta k_z a_0$, and δ is determined by (2.109). In the opposite limiting case $\epsilon^2 \ll \mu$, we have

$$\gamma = \omega \frac{\epsilon^6}{\mu k_z a_0}.$$

Note that the decrements are small $\gamma \ll \omega$ in all the above cases. Under this condition only is it at all possible to consider waveguide propagation of MHD waves.

2.11 Waveguide for Kinetic Alfvén and FMS Waves in a 'warm' Plasma. Waveguide Mode Resonance

In Section 2.10, we examined the possibility of waveguide propagation for kinetic Alfvén waves in a 'cold' plasma with $\beta \ll m_e/m_i$. Another type of transverse dispersion exists, however, for kinetic Alfvén waves, causing them to propagate across magnetic shells. They are Alfvén waves in a 'warm' plasma with $\beta \gg m_e/m_i$. Waveguides for such Alfvén, as well as FMS, waves are the regions where the Alfvén speed has a local minimum across magnetic shells. The waveguide modes neither escape from such waveguides or decay. Notably, as will be seen further, these two types (Alfvén and FMS) of waveguide modes being localised in the same region in space may result in their possible internal resonance.

Let us address a possible channelling of kinetic Alfvén and FMS waves in a waveguide formed by a local minimum in the Alfvén speed distribution $v_A(x)$ (near $x = x_0$), extended along magnetic field lines. Same as in Section 2.10, we will use Eqs. (2.73), (2.74) to describe the structure of the oscillations. For the basic modes with $k_y = 0$ localised near the 'potential well' minimum, where a quadratic decomposition (2.94) is applicable for $v_A^2(x)$, we obtain a system of coupled equations

$$\kappa^4 \nabla_\xi^2 \varphi' + (\lambda - \sigma\xi^2)\varphi' = i\epsilon\psi', \tag{2.115}$$

$$\nabla_\xi^2 \psi' + (\lambda - \sigma\xi^2)\psi' = -i\epsilon\varphi', \tag{2.116}$$

where $\kappa^4 = \omega^2[(3/4)\rho_i^2 + \rho_s^2]/v_A^2 = u^2\beta^{*2}$ ($\beta^{*2} = v_s^2/v_A^2$, v_s is the sound speed in plasma), $\xi = (x - x_0)/\tilde{a}$ is a dimensionless coordinate ($\tilde{a} = \sqrt{a_0/k_z}$), $\epsilon = uk_z a_0$. The parameter λ is introduced by the relation

$$\omega = k_z v_A(x_0)(1 + \lambda/k_z a_0),$$

determining the frequency of MHD oscillations propagating in the waveguide. For the parameters that are characteristic of the Earth magnetosphere, we have $\kappa, \epsilon, u \ll 1$. There are two limiting cases for the system of Eqs. (2.115), (2.116). When $\kappa = 0$, we again arrive at the system of Eqs. (2.75), (2.76) describing a large-scale Alfvén waveguide. When $\epsilon = 0$, one of the solutions of the system of Eqs. (2.115), (2.116) describes a waveguide magnetosonic mode (2.97). The other solution, in this case, describes a waveguide large-scale Alfvén mode:

$$\varphi' = Cy_n\left(\frac{\xi}{\kappa}\right), \quad \lambda = \kappa^2\lambda_n = (2n+1)\kappa^2. \tag{2.117}$$

Here, C is an arbitrary constant, $n = 0, 1, 2, \ldots$ is the number of the waveguide harmonic, $y_n(z) = a_n \exp(-z^2/2)H_n(z)$ are unit-normalised eigenfunctions of the form (2.98), $a_n = \pi^{1/4}2^{-n/2}(n!)^{-1/2}$ is the normalisation coefficient, $H_n(z)$ are Hermite polynomials. To find the solution for the system of Eqs. (2.115), (2.116) we will employ the Green's function method. For an infinite discrete set of the solutions of the homogeneous Eq. (2.116) satisfying the given boundary conditions, Green's function has this form

$$G(\xi, \xi') = \sum_{i=0}^{\infty} \frac{y_i(\xi)y_i(\xi')}{\lambda - \lambda_i}.$$

If such solutions of the system of Eqs. (2.115), (2.116) exist that $|\lambda - \lambda_n| \ll 1$ in them, the sum is dominated by the term with $i = n$, and the solution of (2.116) has the form

$$\psi'(\xi) = \frac{c_n y_n(\xi)}{\lambda - \lambda_n}, \tag{2.118}$$

where

$$c_n = -i\epsilon \int_{-\infty}^{\infty} \varphi'(\xi)y_n(\xi)d\xi. \tag{2.119}$$

The solution of (2.115) is obtained similarly:

$$\varphi'(\xi) = \sum_{i=0}^{\infty} \frac{c_i y_i(\xi/\kappa)}{\lambda - \kappa^2\lambda_i}, \tag{2.120}$$

where

$$c_i = i\frac{\epsilon}{\kappa}\int_{-\infty}^{\infty}\psi'(\xi)y_i(\xi/\kappa)d\xi = i\frac{\epsilon}{\kappa}\frac{c_n\alpha_{ni}}{\lambda-\lambda_n}, \qquad \alpha_{ni} = \int_{-\infty}^{\infty}y_n(\xi)y_i(\xi/\kappa)d\xi. \tag{2.121}$$

Substituting (2.120), (2.121) into (2.119) produces the following dispersion equation:

$$\lambda - \lambda_n = \frac{\epsilon^2}{\kappa}\sum_{i=0}^{\infty}\frac{\alpha_{ni}^2}{\lambda-\kappa^2\lambda_i}. \tag{2.122}$$

As we move away from point $\lambda = \lambda_n$, the eigenvalues, on the contrary, approach $\lambda = \kappa^2\lambda_i$, where the dispersion equation can be represented as

$$\lambda - \kappa^2\lambda_i = \frac{\epsilon^2}{\kappa}\sum_{n=0}^{\infty}\frac{\alpha_{ni}^2}{\lambda-\lambda_i}. \tag{2.123}$$

Using this expression for the dispersion equation, one can represent its solutions in the region $|\lambda - \lambda_n| \sim 1$. The full graphic solution of the dispersion equation is shown in Figure 2.16. The large-scale curves in this figure correspond to the right side of (2.123), which near $\lambda = \lambda_n$ is described by the left side of (2.122), and the 'comb' of cotangent-like curves, by its right side. The eigenvalues of parameter λ are determined by the intersection points of these curves.

Of special interest are the solutions for which both $|\lambda - \kappa^2\lambda_m| \ll \kappa^2$ and $|\lambda - \lambda_n| \ll 1$ are true. In this case, the dispersion Eqs. (2.122), (2.123) can be represented in the form of a quadratic equation, its solutions having the form

$$\lambda \approx \frac{\lambda_n + \kappa^2\lambda_m}{2} \pm \sqrt{\frac{(\lambda_n - \kappa^2\lambda_m)^2}{4} + \frac{\epsilon^2}{\kappa}\alpha_{mn}^2} = \kappa^2\lambda_m + \frac{\delta}{2} \pm \sqrt{\frac{\delta^2}{4} + \frac{\epsilon^2}{\kappa}\alpha_{mn}^2},$$

where $\delta = \lambda_n - \kappa^2\lambda_m$. It is easily demonstrable by direct calculation that in this case it is possible to represent $\alpha_{mn}^2 = \kappa^3\sigma_{mn}\tau_{mn}$, where $\sigma_{mn} = 1$ if m and n are of the same parity, and $\sigma_{mn} = 0$ if m and n are of different parity; and the τ_{mn} numbers are of order unity for the basic modes ($n \sim 1$) and depend little on m.

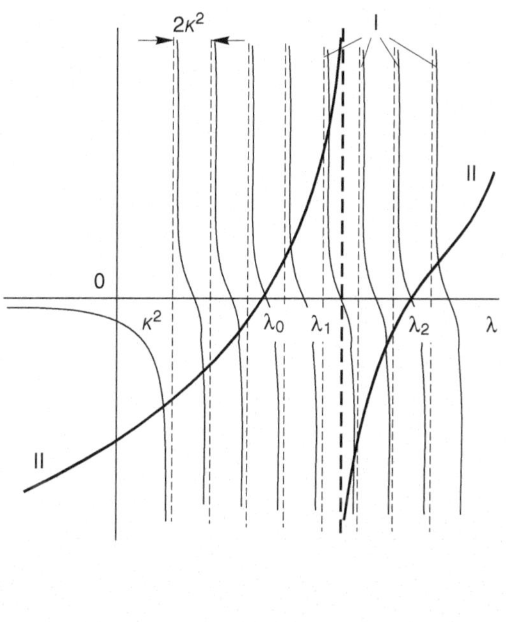

Figure 2.16 Graphic solution of dispersion Eqs. (2.122), (2.123). Curves I correspond to the right side of (2.122), and curves II to the right side of (2.123).

Let us now address the polarisation of the resulting waveguide modes. From (2.103) we have the following expressions for the oscillation magnetic field components:

$$B_x = \frac{k_z \bar{c}}{\omega}\psi' = \frac{k_z \bar{c}}{\omega}\frac{c_n y_n(\xi)}{\lambda - \lambda_n}, \quad B_y = -\frac{k_z \bar{c}}{\omega}\varphi' = -\frac{\epsilon k_z \bar{c}}{\kappa\omega}\frac{c_n \alpha_{mn} y_n(\xi/\kappa)}{(\lambda - \lambda_n)(\lambda - \kappa^2\lambda_m)}.$$

Given that $y_n(\xi) \sim 1$ for the basic modes, while $y_m(\xi/\kappa) \sim \sqrt{\kappa}$, we have, for the oscillations in question,

$$\left|\frac{B_y}{B_x}\right| \sim \frac{\epsilon}{\kappa^{1/2}}\frac{\alpha_{mn}}{\lambda - \kappa^2\lambda_m} \approx \left[\sqrt{1 + \frac{\delta^2\kappa}{4\epsilon^2\alpha_{mn}^2}} \pm \frac{\delta\kappa^{1/2}}{2\epsilon\alpha_{mn}}\right]^{-1}.$$

If $\delta^2 \gg (\epsilon\kappa)^2$, we have, for the two roots of the dispersion equation,

$$\left|\frac{B_y}{B_x}\right|^{(1)} \sim \frac{\epsilon\kappa}{|\delta|} \ll 1,$$

$$\left|\frac{B_y}{B_x}\right|^{(2)} \sim \frac{|\delta|}{\epsilon\kappa} \gg 1.$$

The first of these relations corresponds to the waveguide FMS mode with a dominant large-scale B_x magnetic field component, and the small-scale B_y component has a small amplitude. The second relation describes the waveguide Alfvén mode consisting of a dominant small-scale B_y component and large-scale B_x component of small amplitude.

In the $\delta = 0$ ($\lambda_n = \kappa^2\lambda_m$) case, the ratio of amplitudes for each mode becomes the same:

$$\left|\frac{B_y}{B_x}\right|^{(1,2)} \sim 1.$$

This can be interpreted as resonance between waveguide Alfvén and FMS modes, when the large- and small-scale components in each of them have comparable amplitude, with the modes themselves identically polarised.

2.12 Waveguides in Plasma Filaments

In Sections 2.1–2.11 dealt with waveguides for MHD waves in a medium having a 1D inhomogeneity in the Alfvén speed distribution across magnetic field lines. The above-addressed waveguides could be described using the Cartesian system of coordinates (x, y, z) in which the medium inhomogeneity is directed along one of the transverse coordinates (say, x), whereas in the two other directions the medium is homogeneous. Waveguides with such a type of inhomogeneity do exist in the Earth magnetosphere, e.g. at the plasmapause [71–74], in the inner plasmasphere [75], and in the magnetopause-adjacent part of the outer magnetosphere [76–80]. Of course, such waveguides cannot be regarded as strictly 1D-inhomogeneous. They are also inhomogeneous in the other two coordinates. However, the plasma inhomogeneity scale in them is much smaller across the magnetic shells (in the x coordinate) than the inhomogeneity scales in the other two directions (y and z), so that, in the main order of perturbation theory, the plasma can be assumed to be homogeneous in these two directions.

The Earth magnetosphere and the Sun, however, contain plasma formations with another type of inhomogeneity that are capable of being waveguides for MHD waves. They are plasma filaments extended along magnetic field lines. In the Sun they take the form of magnetic loops (see, e.g. [81, 82]), while in the Earth magnetosphere they are plasma filaments detaching themselves from the evening bulge of the plasmasphere and drifting towards the magnetopause [83] (Figure 2.17). They are termed as 'magnetospheric ducts' because whistler modes of electromagnetic oscillations, FMS and Alfvén waves travelling in the magnetosphere are channelled along these filaments [84, 85]. Thus, these regions of space play the role of waveguides for Alfvén and FMS waves. The theory of waveguide propagation of Alfvén waves was applied to good effect when investigating geomagnetic pulsations propagating in the Earth magnetosphere. It was found that the high-latitude regions of Earth, where the ionospheric feet of magnetospheric ducts are mapped along geomagnetic field lines, are related to the high-latitude peak in the occurrence of localised geomagnetic pulsations Pc1 [86]. There are also studies that observed a fine structure in geomagnetic pulsations Pc1 [87], which was interpreted as their propagation in plasma filaments [88].

Pc1 pulsations are regarded as Alfvén wave packets moving along geomagnetic field lines [11]. They are amplified as a result of a cyclotron instability due to high-energy anisotropic protons in the ring current or in the outer radiation belt each time they pass through the region of the instability located near the geomagnetic equator [89, 90].

The cyclotron instability is, however, only effective in amplifying the waves travelling in the quasi-longitudinal regime (with $k_\perp < \sqrt{u}k_\|$, see [24], p. 309]). As was shown in the Introduction, the wave packet vector increases and the packet is quick to exit this regime (this happens within one period of its bounce oscillations between the magneto-conjugated ionospheres) as it moves in a transverse-inhomogeneous medium. Thus, if the initial amplitude of the packet is not high enough, it could be concluded, based on the above, that the amplitude will not be able to increase to detectible values. The presence of a waveguide, however, makes it possible to view this process from another aspect. Alfvén waves propagate in the waveguide regime without increasing $k_\perp$. This allows the waves (upon being reflected from the ionosphere) to repeatedly travel through the instability region, in the continuous amplification regime, eventually reaching the necessary amplitude. This section will address waveguides for Alfvén waves travelling in the magnetospheric ducts.

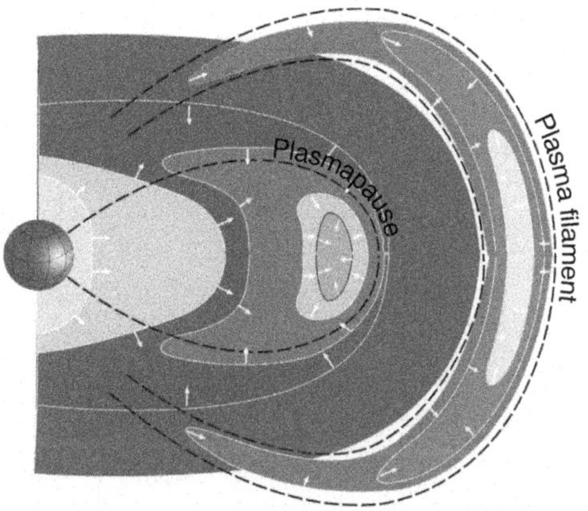

Figure 2.17 Alfvén speed distribution in the dayside magnetosphere (gray scale). v_A distribution is shown in the meridional section inside the plasmapause and in the plasma filament (duct for MHD waves) in the outer magnetosphere. Smaller v_A are shown by light gray. The white arrows indicate the direction of the FMS group velocity in the meridional section.

2.12.1 Axisymmetric Plasma Waveguide for Quasi-Longitudinal Alfvén Waves

Let us examine the possibility of waveguide propagation for quasi-longitudinal Alfvén waves in magnetospheric ducts (see [91]). We will use a model in the form of an infinite axisymmetric plasma cylinder for plasma filaments extending along magnetic field lines. To describe MHD oscillations propagating in such a waveguide, we will employ a cylindrical system of coordinates (ρ, ϕ, z), where the magnetic field is along the z axis, and the plasma density inhomogeneity is along the radius: $\rho_0 = \rho_0(\rho)$ (Figure 2.18(a)).

We will employ such a model to describe MHD oscillations using the equations for the components of a perturbed magnetic field $\mathbf{B} = (B_\rho, B_\phi, B_z)$. The plasma in the model medium is homogeneous along the ϕ and z coordinates, which is why we will seek the solutions for oscillation harmonics of the form $\exp(im\phi + ik_z z - i\omega t)$ (where $m = 0, 1, 2, 3, \ldots$ is the azimuthal wave number) and use these harmonics to decompose an arbitrary perturbation. By means of the decomposition (2.72) and the Maxwell equation (2.102) it is possible to relate the perturbed magnetic field components in the cylindric system of coordinates to the components of the scalar and vector potentials:

$$B_\rho = \frac{k_z\bar{c}}{\omega}\left(i\frac{m}{\rho}\varphi + \nabla_\rho\psi\right), \quad B_\phi = -\frac{k_z\bar{c}}{\omega}\left(\nabla_\rho\varphi - i\frac{m}{\rho}\psi\right),$$

$$B_z = i\frac{\bar{c}}{\omega}\left(\frac{1}{\rho}\nabla_\rho\rho\nabla_\rho\psi - \frac{m^2}{\rho^2}\psi\right). \tag{2.124}$$

Using (1.8) and (2.69) with the dielectric permeability tensor of the form (2.70), we obtain, for the axisymmetric harmonics ($m = 0$) of quasi-longitudinal Alfvén waves, a system of equations similar to (2.75), (2.76):

$$\left(\frac{\omega^2}{v_A^2} - k_z^2\right)B_\phi = iu\frac{\omega^2}{v_A^2}B_\rho, \tag{2.125}$$

$$\frac{\partial}{\partial\rho}\frac{1}{\rho}\frac{\partial\rho B_\rho}{\partial\rho} + \left(\frac{\omega^2}{v_A^2} - k_z^2\right)B_\rho = -iu\frac{\omega^2}{v_A^2}B_\phi. \tag{2.126}$$

Substituting B_ϕ from (2.125) into (2.126), we have

$$\frac{\partial}{\partial\rho}\frac{1}{\rho}\frac{\partial\rho B_\rho}{\partial\rho} + \left(\frac{\omega^2}{v_A^2} - k_z^2 - \frac{u^2k_z^4}{(\omega^2/v_A^2) - k_z^2}\right)B_\rho = 0. \tag{2.127}$$

To better understand the properties of the solution of (2.127), we will use the substitution $\overline{B} = \rho^{1/2}B_\rho$ in this equation to transform it into an equation having the form of the

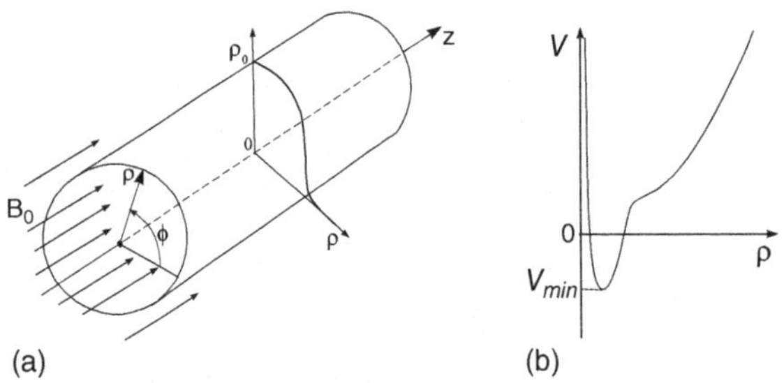

Figure 2.18 Schematic representation of a magnetospheric duct model in a cylindric system of coordinates (ρ, ϕ, z): (a) direction of outer magnetic field B_0 and plasma density distribution $\rho_0(\rho)$ in a plasma filament; (b) the form of the potential $V(\rho)$ in (2.128).

Schrödinger equation,

$$\frac{\partial^2 \overline{B}}{\partial \rho^2} - V(\rho)\overline{B} = 0, \tag{2.128}$$

with the potential

$$V(\rho) = k_z^2 - \frac{\omega^2}{v_A^2} + \frac{u^2 k_z^4}{(\omega^2/v_A^2) - k_z^2} + \frac{3}{4}\frac{1}{\rho^2}. \tag{2.129}$$

Depending on the form of the potential $V(\rho)$, Eq. (2.128) can have, given the above boundary conditions, both a continuous and discrete set of solutions. The boundary conditions in the problem are the requirement that oscillation amplitude should be finite on the plasma cylinder axis and on the asymptotic $\rho \to \infty$. A discrete set of solutions exists only if the condition

$$V_{\min} \equiv \min V(\rho) < 0 \tag{2.130}$$

is satisfied. Near the plasma cylinder axis we will use the Alfvén speed profile representation

$$v_A^2(\rho) = v_A^2(0)\left(1 + \frac{\rho^2}{R^2}\right), \tag{2.131}$$

applicable for $\rho \ll R$, where $R = \sqrt{v_A^2/|\nabla_\rho^2 v_A^2|_{\rho=0}}$ is the characteristic $v_A(\rho)$ variation scale near the plasma cylinder axis. Next, representing, in the potential (2.129), the local frequency of Alfvén oscillations travelling in the waveguide as

$$\omega^2 = k_z^2 v_A^2(0)\left(1 - \frac{\lambda}{k_z R}\right),$$

results in

$$V(\rho) = \frac{k_z}{R}\left[\overline{\rho}^2 + \lambda - \frac{\epsilon^2}{\overline{\rho}^2 + \lambda} + \frac{3}{4\overline{\rho}^2}\right], \tag{2.132}$$

where $\overline{\rho} = \rho\sqrt{k_z/R}$ is a dimensionless radial coordinate, $\epsilon = uk_z R$. The characteristic form of such a potential is shown in Figure 2.18(b). The condition that a discrete set of eigen solutions exist for (2.128) with potential (2.132) is reduced to the requirement

$$\epsilon > \frac{3}{4}. \tag{2.133}$$

Thus, there is no waveguide in the axisymmetric filament for a large-scale Alfvén wave if $\epsilon \ll 1$ (the 'shallow well' approximation). As we will see further, if $\epsilon \gg 1$, the basic waveguide modes are localised near the potential well bottom and the following approximate representation can be used for them in the potential (2.132):

$$\frac{\epsilon^2}{\overline{\rho}^2 + \lambda} \approx \frac{\epsilon^2}{\lambda}\left(1 - \frac{\overline{\rho}^2}{\lambda}\right),$$

which is applicable for $\lambda \gg k_z \Delta^2/R$, where Δ is the size of the solution localisation in the coordinate ρ. Next, Eq. (2.127) is reduced to

$$\frac{\partial}{\partial \overline{\rho}}\frac{1}{\overline{\rho}}\frac{\partial \overline{\rho} B_\rho}{\partial \overline{\rho}} + \left[\frac{\epsilon^2}{\lambda} - \lambda - \left(1 + \frac{\epsilon^2}{\lambda^2}\right)\overline{\rho}^2\right] B_\rho = 0. \tag{2.134}$$

Switching to the new independent variable $\xi = (1 + \epsilon^2/\lambda^2)^{1/2}\overline{\rho}^2$, we obtain

$$\xi\frac{\partial^2 B_\rho}{\partial\xi^2} + \frac{\partial B_\rho}{\partial\xi} + \left[D - \frac{\xi}{4} - \frac{1}{4\xi}\right] B_\rho = 0, \tag{2.135}$$

where $D = (1/4)[(\epsilon^2/\lambda) - \lambda]/\sqrt{1 + \epsilon^2/\lambda^2}$. The solution of (2.135), satisfying the boundary condition for $\xi = 0$, has the form

$$B_\rho = \xi^{1/2}\exp\left(-\xi^2/2\right)\Phi(-(D-1), 2, \xi), \tag{2.136}$$

where $\Phi(x, y, z)$ is a confluent hypergeometric function. As $\rho \to \infty$ ($\xi \to \infty$), this solution satisfies the boundary condition only if

$$D - 1 = n, \tag{2.137}$$

where $n = 0, 1, 2, \ldots$, so that $\Phi(-n, 2, \xi)$ are an nth-order polynomial and the solution (2.136) describes the eigenfunctions of the problem. Solving (2.137) relative to λ, we obtain

$$\lambda_n = \epsilon - 2\sqrt{2}(n + 1),$$

the problem eigenvalues determining the eigen frequencies of the oscillations travelling in the waveguide. The characteristic size of the eigenfunction localisation in the ξ variable is of order unity, which corresponds to $k_z\Delta^2/R \sim 1$. Thus, the above suggestions about their being localised near the potential well bottom ($\lambda \approx \epsilon \gg k_z\Delta^2/R \sim 1$) are valid for the basic harmonics of the oscillations ($n \sim 1$).

The polarisation of the waveguide modes we are considering can easily be obtained from the expression (2.125), which we will rewrite as

$$\frac{B_\phi}{B_\rho} = -i\frac{\varepsilon}{\overline{\rho}^2 + \lambda}.$$

If $\epsilon \gg 1$, we have, for the basic waveguide modes, $B_\phi/B_\rho \approx -i$, i.e. the polarisation is nearly circular and the direction of the polarisation vector rotation corresponds to Alfvén waves. The same conditions provide the possibility of effective amplification of Alfvén oscillations as they pass through the region of ion-cyclotron instability located near the geomagnetic equator.

2.12.2 Axisymmetric Plasma Waveguide for Kinetic Alfvén and FMS Waves

Let us examine the possibility of channelling kinetic Alfvén and FMS waves in a cylindrical plasma filament, the model of which was described in Section 2.12.1 (see also [92]). In the outer magnetosphere, where the plasma waveguides we discuss are located, the transverse structure of kinetic Alfvén waves is determined by dispersion effects arising when the low finite plasma pressure (here: $1 > \beta \gg m_e/m_i$) is taken into account. As in Section 2.12.1, we describe the structure of the oscillations using the components of the perturbed magnetic field in a cylindric system of coordinates (2.124), and obtain, from (1.8), (2.69) with dielectric permeability tensor (2.70), a system of coupled equations analogous to (2.115), (2.116), for axially-symmetric modes ($m = 0$):

$$\frac{\partial}{\partial\overline{\rho}}\frac{1}{\overline{\rho}}\frac{\partial\overline{\rho}B_\rho}{\partial\overline{\rho}} - (\overline{\rho}^2 - \lambda)B_\rho = -i\epsilon B_\phi, \tag{2.138}$$

$$\kappa^4 \frac{\partial}{\partial \bar{\rho}} \frac{1}{\bar{\rho}} \frac{\partial \bar{\rho} B_\phi}{\partial \bar{\rho}} - (\bar{\rho}^2 - \lambda) B_\phi = i\epsilon B_\rho, \tag{2.139}$$

where $\bar{\rho} = \rho\sqrt{k_z/R}$. For the first harmonics of the waveguide modes localised near the 'potential well' bottom, we employ a quadratic decomposition of Alfvén speed (2.131) and introduce the following notations: $\kappa^4 = \omega^2[(3/4)\rho_i^2 + \rho_s^2]/v_A^2 = u^2\beta^{*2}$ $(\beta^{*2} = v_s^2/v_A^2)$, $\epsilon = uk_zR$, $u = \omega/\omega_i$, and the parameter λ is introduced via the relation

$$\omega = k_z v_A(0)(1 + \lambda/k_zR),$$

determining the frequency of MHD eigen oscillations travelling down the waveguide. For values of $\kappa, \epsilon, u \ll 1$, characteristic of the Earth magnetosphere, we have two limiting cases for the system of Eqs. (2.138), (2.139). For $\kappa = 0$, we again arrive at the system of Eqs. (2.125), (2.126), describing a large-scale Alfvén waveguide in plasma filaments. For $\epsilon = 0$, one of the solutions of the system of Eqs. (2.138), (2.139) is a large-scale waveguide magnetosonic mode

$$B_\rho = Cy_n(\bar{\rho}), \qquad \lambda = \lambda_n = 4(n+1), \qquad B_\phi = B_z = 0, \tag{2.140}$$

where $n = 0, 1, 2, 3, \ldots$ is the waveguide FMS mode number, C is an arbitrary constant, $y_n(z) = c_n z \exp(-z^2/2) L_n^1(z^2)$ are eigen functions normalised to unity, c_n is the normalisation coefficient, $L_n^1(x)$ are generalised Laguerre polynomials. The other solution, in this case, describes a small-scale Alfvén mode with components

$$B_\rho = B_z = 0, \quad B_\phi = Cy_m\left(\frac{\bar{\rho}}{\kappa}\right), \quad \lambda = \kappa^2\lambda_m = 4(m+1)\kappa^2, \tag{2.141}$$

where $m = 0, 1, 2, 3, \ldots$ is the waveguide Alfvén mode number. Let us seek a solution to the system of Eqs. (2.138), (2.139), using the Green's function, which, for (2.138), has the form

$$G(\bar{\rho}, \bar{\rho}') = \sum_{i=0}^{\infty} \frac{y_i(\bar{\rho}) y_i(\bar{\rho}')}{\lambda - \lambda_i}.$$

The solution to (2.138) satisfying the boundary conditions (limited amplitude on the waveguide axis $\bar{\rho} = 0$ and on the asymptotic $\bar{\rho} \to \infty$) is then

$$B_\rho(\bar{\rho}) = -i\epsilon \sum_{i=0}^{\infty} \frac{c_i y_i(\bar{\rho})}{\lambda - \lambda_i}, \tag{2.142}$$

where

$$c_i = \int_0^\infty B_\phi(\bar{\rho}) y_i(\bar{\rho}) d\bar{\rho}. \tag{2.143}$$

The solution to (2.139) is obtained in a similar manner:

$$B_\phi(\bar{\rho}) = i\frac{\epsilon}{\kappa} \sum_{n=0}^{\infty} \frac{c_n y_n(\bar{\rho}/\kappa)}{\lambda - \kappa^2\lambda_n}, \tag{2.144}$$

where

$$c_n = -i\epsilon \sum_{i=0}^{\infty} \frac{c_i \alpha_{ni}}{\lambda - \lambda_i}, \quad \alpha_{ni} = \int_0^\infty y_n(\bar{\rho}/\kappa) y_i(\bar{\rho}) d\bar{\rho}. \tag{2.145}$$

Substituting (2.144) into (2.143) produces an equation for the coefficients

$$c_i = \frac{\epsilon^2}{\kappa} \sum_{n=0}^{\infty} \frac{\alpha_{ni}}{\lambda - \kappa^2 \lambda_n} \sum_{j=0}^{\infty} \frac{c_j \alpha_{nj}}{\lambda - \lambda_j}. \tag{2.146}$$

For various limiting cases, this equation can produce dispersion equations for the parameter λ, determining the waveguide mode eigen frequencies. Analysis of the dispersion equation is quite similar to the one in Section 2.11.

Let us consider, as we did in Section 2.11, the solutions for which both the relation $|\lambda - \kappa^2 \lambda_m| \ll \kappa^2$ and $|\lambda - \lambda_i| \ll 1$ hold true. In this case, the equation for coefficients (2.146) produces a quadratic dispersion equation, its solutions

$$\lambda \approx \frac{\lambda_i + \kappa^2 \lambda_m}{2} \pm \sqrt{\frac{(\lambda_i - \kappa^2 \lambda_m)^2}{4} + \frac{\epsilon^2}{\kappa} \alpha_{mi}^2} = \kappa^2 \lambda_m + \frac{\delta}{2} \pm \sqrt{\frac{\delta^2}{4} + \frac{\epsilon^2}{\kappa} \alpha_{mi}^2}$$

determining the frequencies of waveguide eigenmodes. Here, $\delta = \lambda_i - \kappa^2 \lambda_m$. If $\delta \gg \alpha_{mi}\epsilon/\sqrt{\kappa}$, we have the following relations for the amplitudes of the waveguide oscillation magnetic field components, for the two roots of the dispersion equation:

$$\left|\frac{B_\phi}{B_\rho}\right|^{(1)} \sim \frac{\epsilon \alpha_{mi}}{\kappa^{1/2}(\lambda^{(1)} - \kappa^2 \lambda_m)} \approx \frac{\epsilon \alpha_{mi}}{\kappa^{1/2}|\delta} | \ll 1,$$

$$\left|\frac{B_\phi}{B_\rho}\right|^{(2)} \sim \frac{\epsilon \alpha_{mi}}{\kappa^{1/2}(\lambda^{(2)} - \kappa^2 \lambda_m)} \approx \frac{\kappa^{1/2}|\delta|}{\epsilon \alpha_{mi}} \gg 1.$$

In this case, the waveguide modes are in a state remote **far ?** enough from internal resonance, so that the amplitude ratio of the magnetic field components differs between them. The upper relation describes large-scale waveguide FMS modes with a small contribution from the small-scale component B_ρ; whereas the lower relation refers to small-scale waveguide Alfvén modes with contribution from the large-scale component B_φ of small amplitude. If $|\lambda_i - \kappa^2 \lambda_m| \ll \alpha_{mi}\epsilon/\sqrt{\kappa}$, the waveguide Alfvén and FMS modes are in the state of internal resonance, with the amplitude ratio of the magnetic field components being the same for each solution:

$$\left|\frac{B_\phi}{B_\rho}\right|^{(1,2)} \sim 1,$$

making it impossible to determine the type (Alfvénic or FMS) of the waveguide mode. The mechanism of such resonant interaction may prove to be effective in generating small-scale Alfvén waves in magnetospheric ducts.

2.12.3 Waveguide Propagation of Geomagnetic Pulsations in the Outer Magnetosphere

Solutions to MHD equations were obtained in Sections 2.12.1 and 2.12.2, which describe Alfvén oscillations propagating along a cylindric plasma waveguide. The characteristics of such a waveguide were assumed to be homogeneous along the geomagnetic field lines. In the magnetosphere, however, the properties of plasma filaments capable of serving as waveguides for MHD waves vary along their axes. Let us examine the possibility of wave propagation in these plasma filaments when the wave parameters adiabatically adapt in accordance with the changing plasma and magnetic field characteristics.

This means that, at each given moment of time, Alfvén waves will have their spatial structure determined by the local characteristics of the waveguide duct. To estimate the waveguide properties of magnetospheric ducts, let us set the dependence of plasma filament properties on two coordinates: radial and longitudinal. The following mechanism of plasma filament formation and evolution was proposed in [83] and [93]. During magnetospheric storms, plasma filaments flake off the dusk bulge; these filaments are then carried away to the dayside magnetopause via magnetospheric convection system. Let us assume that, at the moment it leaves the plasmasphere, the longitudinal characteristics (density distribution and geomagnetic field strength) in the plasma filament are the same as the corresponding characteristics of the plasmapause. For the magnetic field, we will use the dipole approximation

$$B(\theta) = B_0\left(\frac{a_0}{a}\right)^3 P(\theta), \tag{2.147}$$

where a is the equatorial radius of the field line, $P(\theta) = \sqrt{1+3\sin^2\theta}/\cos^6\theta$, θ is latitude counted from the geomagnetic equator, and the $_0$ index corresponds to the parameters at the plasmapause, in the equatorial plane. The concentration profile will be described by the function $n(\theta) = n(0)P(\theta)$, and we will assume $k_z(\theta) = \omega/v_A(\theta)$ for the current Alfvén wave vector. Since the typical duration of plasma concentration variation, on the magnetic shells we examine in the inner magnetosphere, is long enough (about 24 hours, see [94], p. 287]) compared to geomagnetic pulsation periods, we will assume that the concentration is inversely proportional to the plasma filament volume.

The parameter characterising waveguide properties of the plasma filament is $\epsilon = uk_zR$, which can be written as

$$\epsilon(a,\theta) = \frac{\omega^2}{\omega_i}\frac{R}{v_A} = \frac{m_i\bar{c}}{eB_0^2}\omega^2 R\sqrt{4\pi n m_i}. \tag{2.148}$$

According to the condition that the number of particles is fully conserved in the plasma filament, we have

$$N \sim \pi a_0 R_0^2 n_0 = \pi a R^2 n,$$

whence

$$n^{1/2}R = n_0^{1/2}R_0\left(\frac{a}{a_0}\right)^{1/2}. \tag{2.149}$$

Substituting (2.149) into (2.148), yields

$$\epsilon(a,\theta) = \epsilon_0\left(\frac{a}{a_0}\right)^{11/2} Q(\theta), \tag{2.150}$$

where ϵ_0 is the equatorial value of parameter ϵ at the plasmapause near the dusk bulge, and function $Q(\theta) = \cos^9\theta/(1+3\sin^2\theta)^{3/4}$ describes the dependence of ϵ on latitude θ. As was shown in Section 2.12.1, the large-scale Alfvén waveguide exists in the plasma filament only if the condition (2.133) holds. If this condition is satisfied for $\theta = 0$, the conditions exist near the equatorial plane in the magnetospheric channel for channelling **the plasma filament can serve there as waveguide for the** large-scale Alfvén waves. It follows from (2.150), however, that away from the equator the value of ϵ decreases rapidly as θ increases. At certain latitude $\theta = \theta_{cr}$ the condition (2.133) no longer holds and the waveguide vanishes.

Let us estimate the value of θ_{cr} for magnetospheric ducts located in various magnetospheric regions. We will set the following medium parameter values at the plasmapause: $n_0 = 10^3\text{cm}^{-3}$, $B_0 = 200\text{nT}$, $R_0 = R_E/2$. For geomagnetic pulsations of the Pc1 frequency range with $\omega \sim 0.1$ rad/s in the region where the plasma filament breaks off the dusk bulge ($a = 5.5R_E$), we obtain $\epsilon_0 \approx 5$, which corresponds to $\theta_{cr} \approx 30°$. For ducts that are farthest from Earth ($a = 10R_E$), we have $\epsilon(a, 0) \approx 200$ and $\theta_{cr} \approx 50°$. It is thus evident that the angular sizes of the regions where the waveguide can exist exceed the size of the region where the ion-cyclotron instability mechanism is active ($\Delta\theta \approx 20°$, see [95]), which allows the Alfvén waves to cross the entire instability region, in the regime of quasiparallel propagation (the condition $\epsilon > 1$ corresponds to $k_\perp < \sqrt{u}k_\|$) and be effectively amplified.

Let us now estimate the critical frequency above which there exists a large-scale Alfvén waveguide in the plasma filament in the region where the filament breaks off the plasmasphere. The critical frequency below which the waveguide properties of the plasma filament vanish is found from condition $\epsilon = 3/4$ for $\theta = 0$. Given the above parameters, the critical value ranges from 0.1 Hz (for $a = 5, 5R_E$) to 0.03 Hz (for $a = 10R_E$), spanning the geomagnetic pulsation range of Pc1 to Pc3. Whereas the high-latitude localised geomagnetic pulsations of the Pc1 frequency range have been studied thoroughly enough [86], localised pulsations of the Pc3 range have been examined relatively poorly [92].

The condition for a small-scale Alfvén waveguide in magnetospheric ducts is the requirement $\beta > m_e/m_i$. Using a dipole model for the magnetic field, and function $n(\theta) = n(0)P(\theta)$ for the plasma concentration distribution, let us represent the parameter β as follows:

$$\beta = 8\pi n(0)(T_e + T_i)a^6/B_0^2 P^{3/2}(\theta).$$

Here, T_e, T_i are background plasma ion and electron temperatures, B_0 is the equatorial geomagnetic field strength on the Earth. Let us set the following parameter values: $m_e/m_i = 5 \cdot 10^{-4}$, $B_0 = 0.32$ G, $T_e = T_i = 5$ eV $= 8 \cdot 10^{-12}$ erg, $n(0) = 10^3$ cm^{-3}. For the critical latitude θ_{cr} above which the waveguide properties of the plasma filament vanish, we then have: $\theta_{cr} \approx 60°$ for $a = 5, 5R_E$ and $\theta_{cr} \approx 69°$ for $a = 10R_E$. The latitude where the field line crosses the ionosphere is determined as

$$\theta_i = \arccos\sqrt{r_i/a},$$

where r_i is the ionosphere radius. For $a = 5, 5R_E$ we have $\theta_i = 62°$, while for $a = 10R_E$ we have $\theta_i = 68°$. It can thus be concluded that a small-scale Alfvén waveguide exists in magnetospheric ducts, through their entire length.

2.13 FMS Wave Passing Through a Tangential Discontinuity

In Sections 2.1–2.12, we examined MHD oscillations in models of the medium with immobile plasma. In the real magnetosphere, however, the plasma is in continuous motion, which affects the properties of waves propagating in it. This influence is most essential in regions with sharp variations in plasma motion velocity, namely in shear flows. The best-known of these is plasma flow near the magnetopause, a sharp enough boundary between the solar wind and the magnetosphere. The effect of the shear flow on MHD waves in the magnetosphere can manifest itself in the following manner.

Solar wind waves (FMS waves, as a rule) can penetrate through the magnetopause into the magnetosphere. In the process, they are partially reflected from the magnetopause. The reflection coefficient depends on the spatial structure of the wave. The ray theory approximation (in optics for example) generally concerns the angle (determined by the wave field structure) of wave incidence on or reflection from the boundary dividing two media.

Another effect related to shear flows is the instability of the boundary dividing a moving and an immobile medium with respect to oscillations excited in them. Such an instability is called the Kelvin–Helmholtz instability. It can excite both surface waves, when the two media divided by the shear flow are opaque for oscillations driven at the boundary, and waves escaping from the boundary when one or both media are transparent.

It was suggested in [96] that solar wind MHD oscillations might play a significant role in the energy balance of the terrestrial magnetosphere. This was followed by a number of theoretical studies examining solar wind MHD wave penetration into the magnetosphere. We would like to note [97], where the theory was constructed of magnetosonic waves penetrating through the magnetopause, regarded as a tangential discontinuity in the plasma and magnetic field parameters. Later studies, [98] and [99], estimated the energy flux transferred by magnetosonic waves from the solar wind into the magnetosphere. These estimates suggest that no more than 1–2 % of the solar wind oscillation energy flux penetrates into the dayside magnetosphere, and the resulting flux is $\sim 10^9$ W. It was demonstrated in [100], however, that a significantly larger wave-energy flux penetrates into the geotail.

This section will address FMS waves penetrating through the magnetopause, the latter simulated by a tangential discontinuity of the parameters of the medium.

2.13.1 Model Medium and Matching MHD Equation Solutions

Let us select a flat-layer model medium (see Figure 2.19) where a boundary in the form of a tangential discontinuity ($x = 0$) separates the solar wind (region $I, x < 0$) from the magnetosphere(region $II, x > 0$). A Cartesian system of coordinates (x, y, z) will be chosen such that the x axis is normal to the separation boundary and directed into the magnetosphere, the z axis is along the background magnetic field, and the y axis completes the right-hand system. In the calculations below, the $i = I$, II index in the parameter notations indicates the number of the region in Figure 2.19.

Let us assume the plasma and magnetic field parameters to be homogeneous in each region I–II, their values changing jump-wise when crossing the boundary $x = 0$. Let us

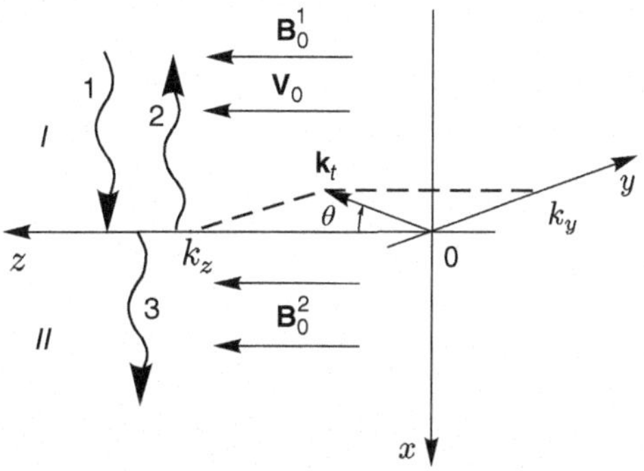

Figure 2.19 Model medium and coordinate system. Roman numbers indicate the following regions: *I* – solar wind, *II* – magnetosphere. FMS waves are labelled as: *1* – incident on the magnetosphere, *2* – reflected from the magnetopause, *3* – penetrating into the magnetosphere.

assume the background plasma velocity vector $\mathbf{v}_0^i$ and magnetic field vector $\mathbf{B}_0^i$ to be along the z axis. Inside the magnetosphere ($i = II$), the plasma is immobile: $v^{II} = 0$, and in region I, it moves at $v^I = v_0$.

The unperturbed parameters of the medium being homogeneous, we will seek a solution to the system of linearised MHD equations (1.5) for an arbitrary perturbation in the form of a decomposition into Fourier harmonics:

$$\Phi = \tilde{\Phi} \exp\left[-i\omega t + i\mathbf{k}_t\mathbf{r}_t + ik_x x\right],$$

where $\tilde{\Phi}$ is the amplitude of any of the perturbed parameters of the medium, ω is the oscillation frequency, $\mathbf{k}_t = (k_y, k_z)$ is the tangential wave vector, and k_x, is the normal component of a wave vector. A dispersion equation for magnetosonic waves in a moving medium has the form

$$(\overline{\omega}^{(i)})^4 - (\overline{\omega}^{(i)})^2 k^2 \left(v_A^{(i)^2} + v_s^{(i)^2}\right) + k^2 k_z^2 v_A^{(i)^2} v_s^{(i)^2} = 0, \tag{2.151}$$

where $\overline{\omega}^{(i)} = \omega - k_z v^{(i)}$ is the Doppler-shifted oscillation frequency, $k^2 = k_x^2 + k_y^2 + k_z^2$, $v_s^{(i)} = \sqrt{\gamma P_0^{(i)}/\rho_0^{(i)}}$ is the sound speed, and $v_A^{(i)} = B_0^{(i)}/\sqrt{4\pi\rho_0^{(i)}}$ is the Alfvén speed in region i.

Let us consider the following problem. A solar-wind magnetosonic wave incident to the magnetopause, is partly reflected and partly penetrates into the magnetosphere (see Figure 2.19). The wave energy flux penetrating into the magnetosphere from the solar wind is determined by the value of the wave vector normal component k_x. It is of special interest, therefore, to analyse the behaviour of function $k_x^2(\mathbf{k}_t, \overline{\omega}^{(i)})$.

The energy flux is only transferred through the magnetopause by waves with $k_x^2 > 0$. Let us define the set of wave parameters satisfying this condition. Obviously, the width of the transparent zones (of the parameter range determined by condition $k_x^2 > 0$) is much larger for FMS waves than for slow ones. Therefore, our further analysis will ignore the contribution from slow magnetosonic waves.

The behaviour of function $k_x^2(\mathbf{k}_t, \overline{\omega}^{(i)})$ is simplest in an immobile medium, the magnetosphere. We will describe the vector $\mathbf{k}_t$ in a polar system of coordinates, $\mathbf{k}_t = (k_t, \theta)$, where $k_t = \sqrt{k_y^2 + k_z^2}$ is the modulus of the tangential wave vector, θ is the polar angle to the axis k_z (see Figure 2.19). (2.151) then produces

$$k_x^2 = -k_t^2 + \frac{\omega^4}{\left(\omega^2 - k_t^2 c_s^2 \cos^2\theta\right) v_f^2}, \tag{2.152}$$

where the superscripts indicating the magnetospheric parameters are omitted for simplicity, and the notations are: $v_f = \sqrt{v_A^2 + v_s^2}$ is the FMS propagation speed, $c_s = v_A v_s / v_f$ is the SMS propagation speed. It can be seen from (2.152) that $|k_x^2| = \infty$ for $k_t = |\omega/c_s \cos\theta|$. Moreover, if θ is fixed, there are two points:

$$\tilde{k}_t^{(1,2)} = \frac{\omega}{\sqrt{2}c_s|\cos\theta|}\left[1 \pm \sqrt{1 - 4\frac{c_s^2}{v_f^2}\cos^2\theta}\right]^{1/2}, \tag{2.153}$$

where $k_x^2 = 0$. The '−' sign in this expression corresponds to the upper index (1) in $\tilde{k}_t^{(1,2)}$, and the '+' sign, to index (2). The dependence of $k_x^2(k_t)$ in the magnetosphere is shown in Figure 2.20(a). One transparent zone can be seen to exist for FMS waves: $0 \le k_t \le \tilde{k}_t^{(1)}$.

In a moving medium, the structure of k_x^2 is more complex. There are two singular points in Eq. (2.151) for a moving medium:

$$\bar{k}_t^{(1,2)} = \frac{\omega}{(v_0 \pm c_s)|\cos\theta|},$$

where $|k_x^2| = \infty$. The superscript (1) in $\bar{k}_t^{(1,2)}$ corresponds to the plus sign, and (2) to the minus sign, in the denominator of this expression. Moreover, there are four points in (2.151):

$$k_t^{(j)} = \frac{\omega \tilde{k}_t^{(1,2)}}{\tilde{k}_t^{(1,2)} v_0 \cos\theta \pm \omega}, \quad j = 1, 2, 3, 4, \tag{2.154}$$

where $k_x^2 = 0$. The roots in (2.154) are labelled as follows. The superscript $j = 1$ corresponds to $\tilde{k}_t^{(2)}$ in the enumerator and to the '+' sign in the denominator in (2.154), $j = 2$ to $\tilde{k}_t^{(2)}$ in the enumerator and the '−' sign in the denominator, $j = 3$ to $\tilde{k}_t^{(1)}$ and the '+' sign, $j = 4$ to $\tilde{k}_t^{(1)}$ and '−'. Here, $\tilde{k}_t^{(1,2)}$ is determined by an expression similar to (2.153), the only difference being that v_f and c_s correspond to the solar wind parameters. For $k_t = 0$ we have $k_x^2 = \omega^2/v_f^2$, and for $k_t \to \infty$

$$k_x^2 \cong k_t^2 \frac{c_s^2 - v_0^2}{(1 - v_0^2\cos^2\theta/v_f^2)v_0^2 - c_s^2}.$$

The behaviour of k_x^2 for $k_t \to \infty$ depends on the direction of $\mathbf{k}_t$, i.e. on the value of angle θ. For $|\cos\theta| > |\cos\theta^*| \equiv (v_f/v_0)\sqrt{1-(c_s/v_0)^2}$, we have $k_x^2 > 0$ (assuming $c_s < v_0$), and $k_x^2 < 0$ for $|\cos\theta| < |\cos\theta^*|$. The dependence $k_x^2(k_t)$ in the solar wind is shown in Figure 2.20(b).

There are two transparent zones for FMS waves, $0 \le k_t \le k_t^{(1)}$ and $k_t^{(2)} \le k_t$, when $0 \le \theta \le \theta^*$. For $\theta^* \le \theta \le \pi - \theta^*$, the latter transparent zone for FMS waves vanishes. The transparent zone corresponds to the interval $0 \le k_t \le k_t^{(1)}$. Same as in the magnetosphere, the width of transparent zones for FMS waves is much larger than the width of transparent zones for slow magnetosonic waves. If $\pi \ge \theta \ge \pi - \theta^*$, then $k_x^2 > 0$ for $k_t \ge 0$ and FMS waves are able to propagate for k_t in the entire range $0 \le k_t < \infty$.

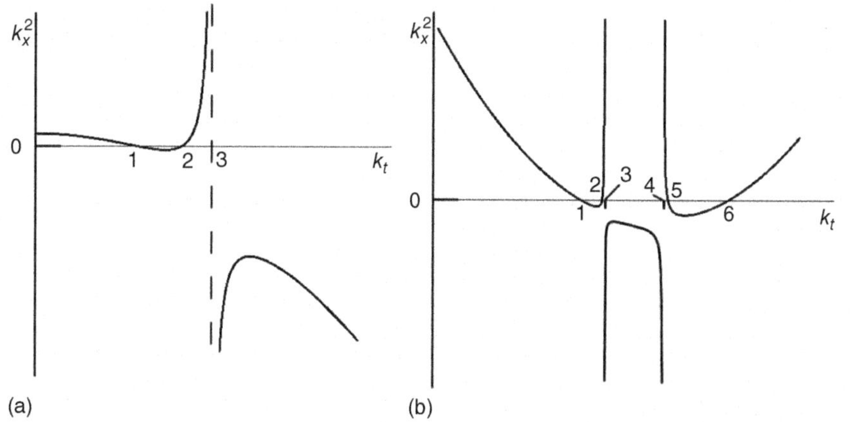

Figure 2.20 The k_t dependence of function k_x^2 for fixed ω and θ. (a) – $k_x^2(k_t)$ in the magnetosphere, the points labelled: 1 – $\tilde{k}_t^{(1)}$, 2 – $\tilde{k}_t^{(2)}$, 3 – $k_t = |\omega/c_s \cos\theta|$. (b) – $k_x^2(k_t)$ in the solar wind, the points labelled: 1 – $k_t^{(1)}$, 2 – $k_t^{(3)}$, 3 – $\bar{k}_t^{(1)}$, 4 – $\bar{k}_t^{(2)}$, 5 – $k_t^{(4)}$, 6 – $k_t^{(2)}$.

Let us now match the solutions describing FMS waves in the solar wind and in the magnetosphere. As the FMS wave falls onto the magnetopause, the latter changes its shape, which is determined by the structure of the incident and reflected waves. The matching conditions at the perturbed boundary have the form (see [97]):

$$\left\{ P + \frac{B_0^{(i)} B_z}{4\pi} \right\}_{\pm} = 0, \tag{2.155}$$

$$\left(\tilde{B}_x + B_0^{(i)} \frac{\partial \zeta}{\partial z} \right)_{\pm} = 0, \tag{2.156}$$

$$\left(\frac{\mathrm{d}\zeta}{\mathrm{d}t} - v_x \right)_{\pm} = 0, \tag{2.157}$$

where the braces indicate that the value of the enclosed expression remains conserved while the enclosed values experience a jump as the perturbed boundary is crossed, and the round brackets correspond to the enclosed expression being zero either side of the boundary (denoted here by subscripts $\pm$). Here, P is the perturbed gas-kinetic plasma pressure, B_z is the parallel (to the unperturbed magnetic field $\mathbf{B}_0$) component of the oscillation magnetic field, ζ is the normal component of the plasma shift in the wave ($v_x = d\eta/dt$ is the normal component of the perturbed speed). Equations (2.155) and (2.156) are obtained directly from linearised MHD equations (1.5) integrated across the perturbed boundary over an infinitely small interval. Equation (2.157) assumes that the wave vector tangential component is conserved when the boundary is crossed, i.e. the boundary oscillation can be represented as

$$\zeta|_{x=0} = \eta \exp\left(\mathrm{i}\mathbf{k}_t \mathbf{r}_t - \mathrm{i}\omega t\right), \tag{2.158}$$

where η is the boundary oscillation amplitude. By means of (2.158) and excluding the amplitude, η, from (2.156), (2.157), the matching conditions at the boundary produce

$$\omega(v_{x1} + v_{x2}) = \overline{\omega} v_{x3}, \tag{2.159}$$

$$B_0^{II}(B_{x1} + B_{x2}) = B_0^{I} B_{x3}, \tag{2.160}$$

$$P_1 + P_2 + B_0^{I} \frac{B_{z1} + B_{z2}}{4\pi} = P_3 + B_0^{II} \frac{B_{z3}}{4\pi}, \tag{2.161}$$

where the subscripts 1, 2, 3 correspond, respectively, to the incident, reflected and penetrating FMS wave (see Figure 2.19).

The system of perturbed ideal-MHD equations (1.5) can be used to produce

$$k_x \left(P + B_0 \frac{B_z}{4\pi} \right) = \frac{\overline{\omega}^2 - k_z^2 v_A^2}{\overline{\omega}} \rho_0 v_x,$$

which when substituted into (2.161) yields a closed system of equations producing the following relations between the components of the incident, reflected and penetrating wave (see [97]):

$$v_{x2} = R v_{x1}, \qquad v_{x3} = T v_{x1}, \tag{2.162}$$

where the notations are:

$$R = \frac{1 - Z}{1 + Z}, \tag{2.163}$$

$$T = \frac{\omega}{\overline{\omega}} \frac{2}{1 + Z} \tag{2.164}$$

are the reflection and penetration coefficients for a monochromatic FMS wave penetrating through the magnetopause, simulated by a tangential discontinuity,

$$Z = \frac{\rho_0^{II}}{\rho_0^{I}} \frac{k_x^{I}}{k_x^{II}} \frac{\omega^2 - k_z^2 v_A^{II2}}{\overline{\omega}^2 - k_z^2 v_A^{I2}}.$$

For $B_0 = 0$ and $v_0 = 0$, (2.163) and (2.164) yield the sound wave reflection coefficient and the coefficient of sound wave penetration through a flat transition layer [101].

2.13.2 Energy Flux Transferred by FMS Waves Through the Magnetopause

The energy flux density is a quadratic function of the wave amplitude. Let us use this equation for the normal component of the flux (see [97, 102]):

$$\tilde{f}_x = \rho_0 \frac{\omega(\overline{\omega}^2 - k_z^2 v_A^2)}{\overline{\omega}^2 k_x} |v_x|^2 = \frac{\omega \overline{\omega}^4}{k^4} \frac{k_x}{\overline{\omega}^2 - k_z^2 v_A^2} \frac{|\tilde{\rho}|^2}{\rho_0}, \tag{2.165}$$

where $\tilde{\rho}$ is the perturbed plasma density. The relationship between the incident, reflected and penetrating FMS wave amplitudes (2.162) can be used to demonstrate that, in the case of an incident monochromatic FMS wave on the tangential discontinuity, the following full energy flux conservation law is valid:

$$\tilde{f}_{x3} = \tilde{f}_{x1} + \tilde{f}_{x2},$$

where the subscripts 1,2,3 label the incident, reflected and penetrating wave, respectively.

Figure 2.21 shows the characteristic k_t dependence $\tilde{f}_{zi}$ $(i = 1, 2, 3)$ of the incident, reflected and penetrating FMS wave energy flux densities normalised to the incident wave energy flux density for $\theta = 0$. There are two transparent zones in the solar wind and one transparent zone in the magnetosphere, for fast magnetosound. It is evident that about 75 % of the incident wave energy flux penetrate into the magnetosphere in the first transparent zone in

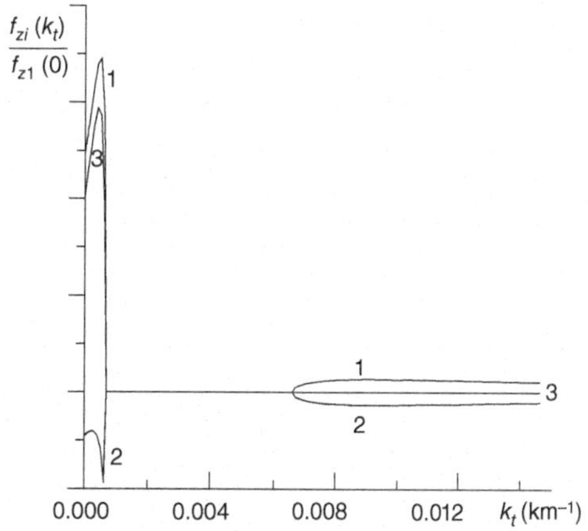

Figure 2.21 The characteristic form of the k_t dependencies of the relative densities of monochromatic FMS energy fluxes in the solar wind and in the magnetosphere, in the $0 \le \theta \le \theta^*$ sector. Energy flux densities are labelled as: 1 – FMS wave incident to the magnetopause; 2 – wave reflected from the magnetopause; 3 – wave penetrating into the magnetosphere.

the solar wind. For waves in the second zone, the magnetosphere proves opaque and they are completely reflected.

Expression (2.165) was written for the energy flux density of one wave harmonic with certain frequency ω and wave vector $\mathbf{k}_t$. To find the full energy density, an inverse Fourier transformation (2.165) is necessary over the entire frequency and wave vector spectrum of the oscillations. Let us assume that the oscillation spectrum shape does not depend on the sign of ω. An inverse Fourier transform then has the form

$$f_x = \frac{2}{(2\pi)^{3/2}} \int_0^\infty d\omega \int_0^\pi d\theta \int_0^\infty \tilde{f}_x(k_t, \theta, \omega) k_t \, dk_t. \tag{2.166}$$

(2.165) implies that the integral (2.166) is divergent if $\tilde{\rho}^2 = const$. To avoid this, one should determine the $\tilde{\rho}^2$ spectrum making sure that the integral is convergent. A model spectral function based on satellite data and satisfying the above conditions is presented in Appendix C:

$$\tilde{\rho}^2 = C\omega^{-\alpha} k_t^{-2\beta}. \tag{2.167}$$

Here, $C = \tilde{C}\overline{\rho}^2 \Phi(\mathbf{k}_t, \omega)$ is the oscillation amplitude determined from its averaged amplitude, $\overline{\rho}^2$, $\Phi(\mathbf{k}_t, \omega)$ is the filter function taking into account the spectrum cutoff thresholds and the presence of 'transparent zones' in the magnetosphere and in the solar wind, $\tilde{C}$ is the normalisation factor, and the exponents $\alpha \approx \beta \approx 5/3$ correspond to the Kolmogorov spectrum of turbulent oscillations.

Let us integrate (2.166) numerically, using, for $\tilde{\rho}^2$, the model expression (2.167). This requires the medium parameters to be set. For the solar wind (region *I*), we will use averaged parameters based on a number of observations; for the magnetosphere (region *II*), we will select parameters balanced with solar wind parameters in accordance with the full pressure conservation condition (2.59). The respective parameter sets are listed in Table 2.1.

Note that the solar wind plasma temperature in this model is somewhat above the observed value. This is due to the solar wind temperature being selected based on the balanced pressure condition (2.59) for a given temperature of the magnetospheric plasma. Our model medium lacks dynamic pressure of the solar wind flow on the magnetopause, resulting in higher solar wind plasma temperature, which affects little the final estimates, however.

Apart from the equilibrium parameters, it is necessary to set the mean amplitude of plasma density oscillations for magnetosonic oscillations in the solar wind, $\overline{\tilde{\rho}}$. Based on spacecraft observations, the amplitude of turbulent oscillations of the solar wind plasma

Table 2.1 Averaged plasma and magnetic field parameters in the solar wind (*I*) and the magnetosphere (*II*) satisfying the full pressure conservation condition (2.59).

Region	$v^{(i)}$	$B^{(i)}$	$n^{(i)}$	$T^{(i)}$
(*i*)	(km/s)	(nT)	(cm^{-3})	(K°)×10^{-5}
I	400	5	5	50.8
II	0	30	1	2

concentration is 20–50 % of its mean value [103, 104]. For the mean amplitude of FMS waves incident on the magnetosphere, let us select $\overline{n}_1 = 1\ \text{cm}^{-3}$, which is 20 % of the mean concentration of the solar wind plasma, $n^I = 5\ \text{cm}^{-3}$. The mean value of the oscillation density amplitude is determined as $\bar{\tilde{\rho}}_1 = \overline{n}_1 m_p$, where m_p is proton mass.

Various models of the solar wind/Earth magnetosphere interaction involve processes related to geotail length variation. Magnetic flux variation was determined in [105], over the substorm growth phase, in the polar cap region, via the length variation of the closed geotail, $30R_E$ to $\sim 200R_E$, due to magnetic field energy accumulation. The above model ignores the finite dimensions of the magnetosphere. To at least partially take them into account, let us determine the 'geoeffective' oscillation energy flux so that integration in (2.166) over k_t uses the value $\hat{k}_t = 2\pi/l$ as the lower threshold, where l is the characteristic maximum geotail length. Of course, this should only be done in the case when l is smaller than the characteristic scale of the solar wind oscillation parameter correlation $l \leq \hat{l}$ (see Appendix C).

It is now possible to estimate the total energy transferred into the magnetosphere by the 'geoeffective' magnetosonic wave flux over a period of order the substorm growth phase — $\Delta t \approx 3 \times 10^3$s. Let us examine the geotail as a cylinder of radius $R = 16R_E$ and length l. It is necessary to set the model variation of the geotail length, over the growth phase period, from $l = l_{\min} \approx 30R_E$ to $l = l_{\max} \approx 150R_E$. Let us accept the following model geotail length variation:

$$l = l_{\min} + \left(\frac{t}{\Delta t}\right)^{\gamma} \Delta l, \tag{2.168}$$

where $\Delta l = l_{\max} - l_{\min}$. The total energy transferred into the magnetosphere over time Δt by the 'geoeffective' FMS flux is

$$W = 2\pi R \int_0^{\Delta t} f_{x3}(l(t)) l(t) dt = 2\pi R \int_{l_{\min}}^{l_{\max}} f_{x3}(l) \left(\frac{dl}{dt}\right)^{-1} l\, dl.$$

From (2.168) we have:

$$\frac{dl}{dt} = \gamma \left(\frac{l - l_{\min}}{\Delta l}\right)^{(\gamma-1)/\gamma} \frac{\Delta l}{\Delta t}.$$

A substantial role in this estimate is played by the geotail growth rate index, γ. Various substorm events may involve different growth scenarios for the length of the closed part of the geotail – either slow ($\gamma < 1$) or fast scenarios ($\gamma > 1$). Figure 2.22 shows how the total energy transferred into the magnetosphere by the FMS flux over the characteristic duration of the substorm growth phase $\Delta t \approx 3 \cdot 10^3$s depends on the value of γ. For $\gamma = 0.5$ we have $W \approx 6 \cdot 10^{15}$J; while for $\gamma = 5$, $W \approx 10^{15}$J, which is quite comparable to the energy produced in the magnetosphere during a single substorm of medium intensity ($W \sim 10^{15}$J).

As demonstrated in [98, 99], low-effective wave energy transfer through the dayside magnetopause is determined by the subsonic speed of the flux around the magnetopause and the small size of the latter as compared to the geotail. The value of the wave energy flux through the geotail surface is much larger due to the high permeability of the night-side magnetopause resulting from the supersonic speed of the flux around the magnetopause and a significantly larger area of wave penetration.

The significant FMS flux penetrating the magnetosphere does not mean, however, that these waves can make a substantial contribution to its energy balance. For this flux to be

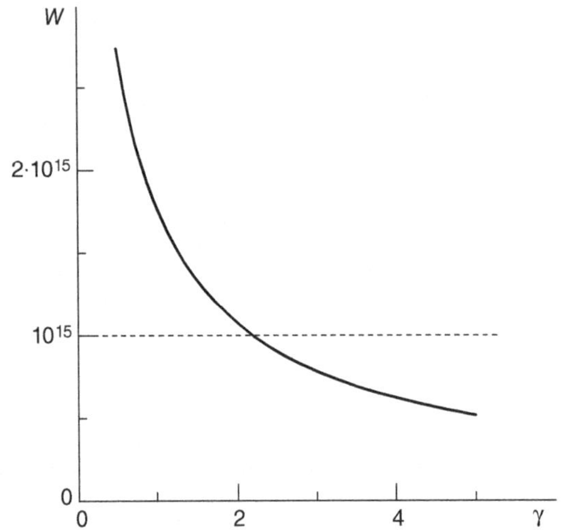

Figure 2.22 Total energy W transferred into the magnetosphere by the 'geoeffective' FMS flux over time $\Delta t \approx 3 \cdot 10^3$s vs. the geotail growth rate index γ in the model Eq. (2.168).

geoeffective, its energy must be transferred in one way or another to the magnetospheric plasma. A mechanism of such a transfer can be provided by the resonant interaction of MHD waves in an inhomogeneous magnetospheric plasma.

2.14 Unstable MHD Shear Flows in the Presence/Absence of Boundary Walls

Unstable liquid and gas shear flows (the Kelvin–Helmholtz instability) have long attracted researchers' attention. Such flows are widespread in many media, both on Earth and in space plasma (see [106, 107]). Early analytical works on this topic were, as a rule, based on models regarding the shear layer as a tangential discontinuity in the shear flow speed profile [108, 109]. Later works examined the perturbations of shear flows in models with a smooth speed profile [110, 111]. It is also worth noting [112–114], studying the instability of a shear flow in an unbounded liquid flow. A considerable number of investigations examined plasma shear flows in the presence of a background magnetic field. Those papers were chiefly stimulated by tasks related to studying the Earth magnetosphere wrapped around by the solar wind flow, comet plasma tail dynamics [115], and stability of high-speed plasma flows within the solar wind itself [107].

An important role in solving problems concerning the shear flow stability is played by the boundary conditions chosen for oscillations driven by the shear layer. In analytical problems the boundary conditions are chosen away from the shear layer, for convenient analysis of the resulting solutions, while numerical problems are guided either by the convenience of the numerical calculation or by the possibility of comparing the results to those of previous analytical studies. The boundary conditions can be selected in different ways [113, 114, 116] examined the shear flow of an unbounded liquid. A natural boundary condition for unstable modes in such a problem would be the absence of waves reaching the shear layer. In other words, only waves travelling away from the shear layer must be present far from it.

The condition was set in [117, 118] that the normal component of the oscillation velocity should become zero at a certain, identical, distance both sides of the shear layer. To put it differently, rigid walls were assumed to be present. Problem statements are also possible with a boundary condition in the form of a rigid wall on one side of the shear layer [119]. The medium characteristics and the parameters of the oscillations in question are as a rule different in different problems. It is difficult enough therefore to assess the effect the boundary conditions exert on the shear flow stability by way of comparing the findings of various papers.

In this section we will examine the stability of MHD shear flows of a conducting liquid (gas) in the presence of a background magnetic field in problems with various types of boundary conditions (see [120]). We will consider two types of oscillations. In oscillations of the first type the tangential wave vector is normal to the outer magnetic field direction. Such oscillations are similar to oscillations of the shear flow in a non-viscous compressible liquid [113, 114].

In the second type of oscillations, the angle between the tangential wave vector and the magnetic field differs from $\pi/2$. In this case, the presence of a magnetic field plays a remarkable role, which is the greater the smaller the angle. We will examine oscillations with their tangential wave vector being parallel to the magnetic field. Thus, two limiting cases will be investigated, with all possible situations realised in-between.

The numerical calculations below refer chiefly to the shear flow with a velocity profile of the form $u = \tanh x$. For a more complete understanding of the numerical results, however, we will also solve, for each type of the boundary conditions, the problem of the stability of a shear flow in the form of a tangential discontinuity in the velocity profile.

2.14.1 Model of Medium and Basic Equations

Let us consider a model of the medium presented in Figure 2.23. We will direct the coordinate y axis along the velocity of a flow stratified over the x coordinate, and introduce the z axis, completing a righthand system of coordinates. We will decompose the full perturbation field into Fourier harmonics of the form $\exp[i(k_y y + k_z z - \omega t)]$, where ω is the oscillation frequency of a single harmonic. Let us define the tangential wave vector as $\mathbf{k}_t = (k_y, k_z)$. We will choose the shear flow velocity profile along the x axis in the form

$$v_0(x) = \overline{v}_0 \tanh(x/a), \tag{2.169}$$

where $\overline{v}_0$ is half the characteristic shear flow velocity jump, a is the characteristic thickness of the shear layer. The other medium parameters: density, ρ_0, pressure, P_0, and magnetic field strength, B_0 – will be assumed to be homogeneous.

The equation for the displacement of the plasma element ζ in MHD oscillations in a moving medium has the form of (2.4):

$$\frac{\partial}{\partial x}\left(\frac{\rho_0 \Omega_A^2}{k_x^2}\frac{\partial \zeta}{\partial x}\right) + \rho_0 \Omega_A^2 \zeta = 0, \tag{2.170}$$

where $v_x = d\zeta/dt = -i\overline{\omega}\zeta$ is the x-component of the velocity vector of medium oscillations in the wave, $\overline{\omega} = \omega - k_t v_0 \cos\phi$ is the Doppler-shifted oscillation frequency,

$$\Omega_A^2 = \overline{\omega}^2 - (\mathbf{k}_t \mathbf{v}_A)^2,$$

$$k_x^2 = k_t^2 - \frac{\overline{\omega}^4}{\overline{\omega}^2 (v_s^2 + v_A^2) - v_s^2 (\mathbf{k}_t \mathbf{v}_A)^2}. \tag{2.171}$$

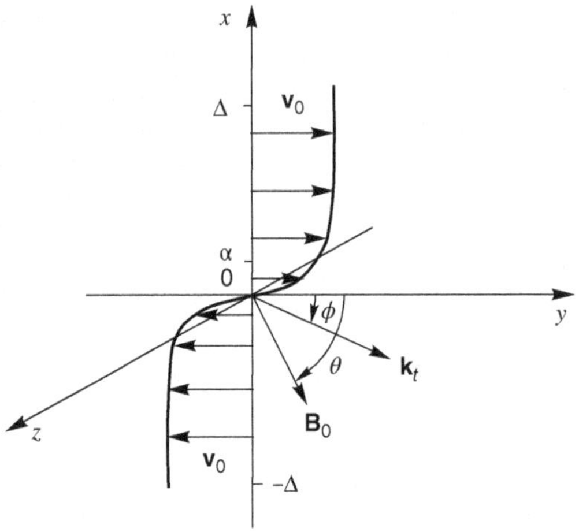

Figure 2.23 Model medium and coordinate system: a is the characteristic scale of the shear layer, $\pm\Delta$ is the location of possible boundaries in the form of rigid walls, $\mathbf{v}_0$, $\mathbf{B}_0$ are the unperturbed speed and background magnetic field vectors, $\mathbf{k}_t$ is the tangential wave vector of the oscillations.

Here, $v_s = \sqrt{\gamma P_0/\rho_0}$ is the sound speed in the medium, $\mathbf{v}_A = \mathbf{B}_0/\sqrt{4\pi\rho_0}$ is the Alfvén speed. We will make the variables dimensionless to conform to the notations in [113]. Our notations: $\alpha = k_t a$ is the dimensionless tangential wave number, $c = \omega/(k_t \overline{v}_0 \cos\phi)$ is the phase velocity of the oscillations made dimensionless by dividing by the flow speed $\mathbf{v}_0$ as projected onto the $\mathbf{k}_t$ direction, $M = \overline{v}_0 \cos\phi/v_s$ is the Mach number determined from $\mathbf{v}_0$ as projected onto the $\mathbf{k}_t$ direction (see Figure 2.23), $\overline{M} = M\sqrt{\beta^*/(1+\beta^*)}$ is the modified Mach number,

$$u(x) = v_0(x)/\overline{v}_0 = \tanh(x/a) \tag{2.172}$$

is the dimensionless background plasma velocity. The parameter $\beta^* = v_s^2/v_A^2$ characterising the relative role of the plasma compressibility and the magnetic field is proportional to the ratio between the gas-kinetic pressure P_0 of the medium and magnetic field pressure $B_0^2/8\pi$.

Let us consider two limiting cases differing in the vector $\mathbf{k}_t$ and $\mathbf{B}_0$ directions. If $\mathbf{k}_t \perp \mathbf{B}_0$ ($\theta - \phi = \pi/2$, see Figure 2.23), (2.170) is reduced to

$$\left(\frac{(c-u)^2}{\overline{M}^2(c-u)^2 - 1}\zeta'\right)' + \alpha^2(c-u)^2\zeta = 0, \tag{2.173}$$

which, up to an accuracy of the Mach number $\overline{M}$ redefinition, corresponds to the equation in [113, 114]. The prime in (2.173) denotes the derivative over the dimensionless coordinate x/a: $\zeta' = a\partial\zeta/\partial x$. As $\beta^* \to \infty$, we have $\overline{M} = M$ so that the role of magnetic field can be neglected in (2.173). Note that, as $\beta^* \to 0$, the modified Mach number $\overline{M} \to 0$, which corresponds to the incompressible medium approximation.

In the other limiting case, $\mathbf{k}_t \parallel \mathbf{B}_0$ ($\theta = \phi$), (2.170) is reduced to

$$\left[\left(1 + \beta^* + \frac{\beta^*}{M^2(c-u)^2 - 1}\right)\zeta'\right]' + \alpha^2\left[M^2\beta^*(c-u)^2 - 1\right]\zeta = 0. \tag{2.174}$$

The following calculations will require the expression for the full perturbed pressure, which, if $\mathbf{k}_t \parallel \mathbf{B}_0$, is related to the displacement ζ by

$$P + \frac{\mathbf{B}_0\mathbf{B}}{4\pi} = -\rho_0 v_A^2 \left(\frac{M^2\beta^*(c-u)^2}{M^2(c-u)^2 - 1} + 1\right)\zeta'. \tag{2.175}$$

Note that (2.173), (2.174) employ the Mach number M. When calculating oscillations in a conducting medium, in the presence of a magnetic field, the Alfvén Mach number

$M_A^2 = M^2\beta^*$ is introduced and defined as the ratio of half the shear flow velocity jump and the Alfvén speed. It can be readily seen that, as $\beta^* \to \infty$, (2.173) and (2.174) become identical and describe an ordinary hydrodynamic flow.

2.14.2 Types of Boundary Conditions

To formulate the problem of the structure of oscillations described by (2.173), (2.174), these equations need to be complemented by corresponding boundary conditions. As noted above, there are three possible types of boundary conditions that are encountered, in various combinations, in virtually all problems of shear flow stability. They have the simplest form when a rigid wall is present at a certain distance Δ from the shear layer (say from point $x = 0$). A natural condition at this wall is that the v_x-component of the perturbed oscillation velocity (and the displacement $\zeta(\Delta)$) turn to zero.

In the case of an unbounded medium, two types of boundary conditions are possible. On the asymptotics $|x| \gg a$, the medium is uniform and the solution of (2.170) can be represented as

$$\zeta = \overline{\zeta}\exp(ik_x x). \tag{2.176}$$

Let us cosider a neutrally stable oscillation mode (with zero growth rate $c_i \equiv Im\, c = 0$). If the medium is opaque ($k_x^2 < 0$) away from the layer, it would be natural to choose a solution that decreases exponentially as $|x| \to \infty$. If $k_x^2 > 0$, the medium will be transparent for the oscillations in question. The boundary condition at infinity is the absence of waves running towards the shear layer. This means that, as long as $x \to \infty$, a function of the form $\zeta = \overline{\zeta}\exp(ik_x x)$, and if $x \to -\infty$ a function of the form $\zeta = \overline{\zeta}\exp(-ik_x x)$ (assuming $k_x > 0$) will be chosen as the boundary condition.

For unstable oscillations ($c_i > 0$), the k_x-component of the full wave vector ($\mathbf{k} = (k_x, k_y, k_z)$) becomes complex on the asymptotics. Formally, a concept of waves escaping from the shear layer can be introduced for any weakly unstable oscillations, in which $\mathrm{Re}(v_{gx}) > 0$ if $x > 0$ and $\mathrm{Re}(v_{gx}) < 0$ if $x < 0$, where $v_{gx} = \partial\omega/\partial k_x$ is the group velocity, the wave energy is transferred along the x coordinate. In compliance with the energy conservation law,

$$\frac{\partial W}{\partial t} + \frac{\partial}{\partial x}(\mathbf{v}_{gx} W) = 0,$$

where W is the quadratic (over oscillation amplitude) density of wave energy. For monochromatical unstable ($c_i > 0$) oscillations, the relation $\mathrm{Im}\, k_x > 0$ is valid if $x > 0$, and $\mathrm{Im}\, k_x < 0$ if $x < 0$. This ensures that the amplitude of oscillations escaping from the shear layer decreases exponentially. This takes place for a 'well-defined' group velocity only. The concepts of 'well' and 'poorly' defined group velocity were introduced in Appendix D, examining an example when using a 'poorly defined' group velocity in the boundary condition leads to an incorrect result. In the case we are dealing with, the expression for the FMS energy density far from the shear layer (in an uniform medium) has the form (see [102])

$$W = |\zeta|^2 \rho_0 \frac{\omega k^2}{k_x^2 \overline{\omega}^3}(2\overline{\omega}^2 - k^2(v_A^2 + v_s^2))(\overline{\omega}^2 - (\mathbf{k}_t \mathbf{v}_A)^2).$$

Let us write the group velocity expressions for the two limiting cases. If $\mathbf{k}_t \perp \mathbf{B}_0$ we have

$$v_{gx} = v_s \frac{1+\beta^*}{\beta^*} \frac{k_x}{k_t} \frac{M^2}{(c-u)}, \tag{2.177}$$

where $k_x = \pm k_t \sqrt{M^2(c-u)^2\beta^*/(1+\beta^*) - 1}$, while for $\mathbf{k}_t \parallel \mathbf{B}_0$, we have

$$v_{gx} = \frac{v_s}{\beta^*} \frac{k_x}{k_t} \frac{\left[M^2(c-u)^2(1+\beta^*) - 1\right]^2}{(c-u)^3 \left[M^2(c-u)^2(1+\beta^*) - 2\right]}, \tag{2.178}$$

where

$$k_x = \pm k_t \sqrt{\frac{(M^2(c-u)^2 - 1)(M^2\beta^*(c-u)^2 - 1)}{M^2(c-u)^2(1+\beta^*) - 1}}, \tag{2.179}$$

and the signs $\pm$ are chosen according to the boundary conditions, enabling the energy to 'escape' from the shear layer. This means that, as $x \to \infty$, the sign chosen for k_x in the boundary condition (2.177) must be such as to satisfy the requirement $\mathrm{Re}\,(v_{gx}) > 0$, while satisfying the condition $\mathrm{Re}\,(v_{gx}) < 0$ in the $x \to -\infty$ case. This is the so-called causality principle: the waves carry the energy away from their source, the shear layer.

As was noted above, three types of boundary conditions are possible in shear flow stability problems. If the medium is boundless we have, for $x \to \pm\infty$,

$$\frac{\partial \zeta}{\partial x} = ik_x\zeta, \tag{2.180}$$

where the sign for k_x is chosen such that $\mathrm{Re}\,(v_{gx}) > 0$ as $x \to \infty$; or $\mathrm{Re}\,(v_{gx}) < 0$ as $x \to -\infty$.

If there is a rigid wall on one side of the shear layer (say at point $x = -\Delta$), the boundary conditions take the form

$$\begin{cases} \partial\zeta/\partial x = ik_x\zeta, & x \to \infty, \\ \zeta = 0, & x = -\Delta, \end{cases} \tag{2.181}$$

where the sign for k_x is chosen such that $\mathrm{Re}(v_{gx}) > 0$; while in the case of walls on both sides (at points $x = \pm\Delta$), the boundary conditions have the form

$$\zeta(\pm\Delta) = 0. \tag{2.182}$$

The boundary conditions for other components of the MHD oscillation field can be written based on (B.13)–(B.15).

2.14.3 Unstable Shear Flows in a Boundless Medium

It should be noted that the presence of a magnetic field manifests itself differently in the two limiting cases: $\mathbf{k}_t \perp \mathbf{B}_0$ and $\mathbf{k}_t \parallel \mathbf{B}_0$. In the former case, we used a modified, $\overline{M}$, rather than the ordinary Mach number, M, to obtain (2.173), completely coinciding in form with the equation describing the ordinary hydrodynamic flow (see, e.g. [113, 114]). In the latter case, an analogous transformation is impossible and, as we will be able to see below, the magnetic field in such flows plays a special role resulting in their stabilisation. Let us discuss the two limiting cases separately.

2.14.3.1 The $\mathbf{k}_t \parallel \mathbf{B}_0$ Case

Let the shear flow velocity profile have the form of a tangential discontinuity:

$$v_0(x) = \begin{cases} \bar{v}_0 & x > 0, \\ -\bar{v}_0 & x < 0, \end{cases} \tag{2.183}$$

which corresponds to

$$u(x) = \begin{cases} 1 & x > 0, \\ -1 & x < 0. \end{cases}$$

We will seek a solution to (2.174) in the form of (2.176) in either halfspace separated by the shear flow. The matching conditions for the solutions obtained for $x > 0$ and $x < 0$ are the requirements that the ζ displacement and the perturbed pressure should be continuous on the tangential discontinuity. Applying, at $x = 0$, the matching conditions (2.155) and (2.157) to solutions obtained in these two domains, we have the following dispersion equation:

$$\sqrt{\frac{(M^2(c-1)^2(1+\beta^*)-1)(M^2\beta^*(c-1)^2-1)}{M^2(c-1)^2-1}} \tag{2.184}$$
$$-\sqrt{\frac{(M^2(c+1)^2(1+\beta^*)-1)(M^2\beta^*(c+1)^2-1)}{M^2(c+1)^2-1}} = 0.$$

Transferring one of the terms into the left-hand side and squaring both the sides produces the following dispersion equation:

$$c^4M^4 - 2c^2M^2(1+M^2) + \left(M^4 - 2M^2 + \frac{2}{1+\beta^*}\right) = 0. \tag{2.185}$$

The solution of this equation has the form

$$c^2 = \frac{1 + M^2 \pm \sqrt{4M^2 - (1-\beta^*)/(1+\beta^*)}}{M^2}. \tag{2.186}$$

If $\beta^* > 1$ the expression under the radical is above zero. The minus sign before the radical in (2.186) corresponds to unstable oscillations ($c_i > 0$) provided the condition

$$1 - \sqrt{\frac{(\beta^*-1)}{(\beta^*+1)}} < M^2 < 1 + \sqrt{\frac{(\beta^*-1)}{(\beta^*+1)}}, \tag{2.187}$$

is met, as derived in [97, 107, 121]. As $\beta^* \to \infty$, the right-hand inequality in (2.187) gives the known instability criterion [108] for hydrodynamic shear flows: $M^2 < 2$. As we proceed to the incompressible medium limit ($M \to 0$), leaving $M_A^2 = M^2\beta^*$ finite, the left-hand inequality in (2.187) produces another known instability criterion for an incompressible conducting liquid in a magnetic field: $M_A > 1$ [109, 122].

Analysis of (2.186) shows that if $\beta^* < 1$ the expression under the radical may be below zero if $M^2 < M_0^2 = (1-\beta^*)/4(1+\beta^*)$. At first glance, it appears that one of the roots must represent an unstable oscillation mode. A more detailed analysis of calculations resulting in the solution of the dispersion Eq. (2.186), however, reveals that the solution describing the instability for $\beta^* < 1$ is spurious because it does not comply with the necessary boundary conditions for $|x| \to \infty$. If the presence of waves travelling away from the shear layer (as determined by the sign of their group velocity v_{gx}) is required on both asymptotics, then for $\beta^* < 1$ the solution exponentially decreasing on one asymptotic will be exponentially increasing on the other. This is due to a 'poorly determined' group velocity of such waves (see Appendix D). Thus, the shear flow under study is stable in a boundless medium with $\beta^* < 1$. In a medium with $\beta^* > 1$ the instability criterion for a shear flow having the form of a tangential discontinuity is given by (2.187).

Let us now find the solution to (2.174) with the boundary conditions (2.180) for a flow having the form of a smooth transition layer with the speed profile (2.169). We will integrate this equation numerically over the contour which, in the lower halfplane of the complex variable x/a, runs parallel to the real axis, below all the singular points in (2.174) that correspond to the resonant interaction between various MHD oscillation modes.

In view of the fact that the singular points in (2.174) for unstable oscillations are in the upper halfplane of the complex x, one might conclude that it is sufficient to integrate over the real axis. This approach is, however, applicable only for oscillations with a large enough value of the growth rate, when the singular points in (2.174) are far from the real axis. In weakly unstable oscillations, they are close to it, which is why integration over the real axis will produce an error comparable to the value of the growth rate c_i itself. The boundary conditions (2.180) were formulated for $x \to \pm\infty$. In numerical integration, they should be placed at a finite distance from the shear layer where the flow is virtually homogeneous and the solution of (2.174) has the form of (2.176). In numerical calculations, we set the boundary conditions for $x = \pm 5a$. A greater distance between the start and end points of the integration interval practically does not affect the final result.

A numerical solution for (2.174), in a boundless medium, for a flow in the form of a smooth transition layer with velocity profile (2.169) is presented in Figure 2.24. Figure 2.24(a) demonstrates the growth rate c_i distribution corresponding to $\beta^* = 10$; and Figure 2.24(b) does the same for $\beta^* = 1.1$. Note that the domain of existence for unstable oscillations is limited on the small M side. It is evident that, in the limit corresponding to a flow of the tangential discontinuity type ($\alpha = 0$), the domain of existence for unstable oscillations determined by condition (2.187), shrinks towards $M = 1$ as $\beta^* \to 1$. Stabilisation of oscillations of the shear flow featuring Mach numbers outside of this interval

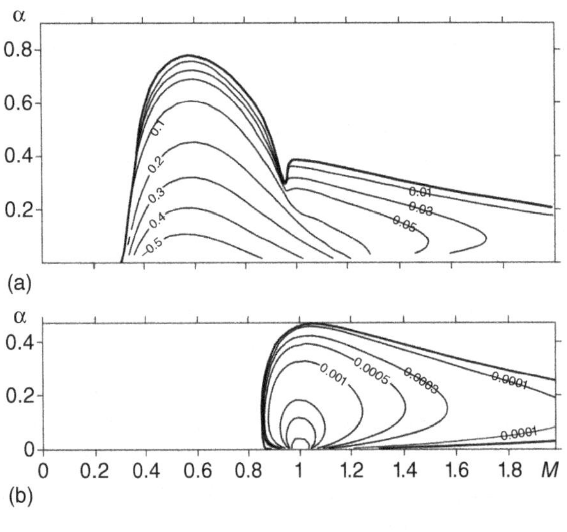

Figure 2.24 Growth rate ($c_i = \text{Im}c$) isoline distribution for MHD oscillations with $\mathbf{k}_t \parallel \mathbf{B}_0$ generated by a shear flow in the form of a smooth transition layer in a boundless medium, for two values of β^*: (a) $\beta^* = 10$, (b) $\beta^* = 1.1$.

is determined by the impact of medium compressibility and magnetic field line tension (Maxwell tensions).

2.14.3.2 The $\mathbf{k}_t \perp \mathbf{B}_0$ Case

As noted above, this problem is analogous to that of stable ordinary hydrodynamic shear flow. Its solution for a flow with a velocity profile in the form of a tangential discontinuity is analogous to solving the problem in Section 2.14.3.1 for the $\mathbf{k}_t \parallel \mathbf{B}_0$ case. The solution for unstable radiative oscillation modes as well as the domain of their existence are determined by (2.186) and (2.187) in this case, where we should set $\beta^* = \infty$.

To find a solution to the problem with a smooth velocity profile of the form (2.169) we will integrate numerically (2.173) with boundary conditions (2.180). Figure 2.25 shows the distribution in the $(\alpha, \overline{M})$ plane of growth rate c_i isolines for such oscillations. It corresponds to the distribution obtained in [112, 113]. The $\overline{M} < 1$ domain corresponds to the unstable surface mode for oscillations with exponentially decreasing amplitudes as they move away from the shear layer. The $\overline{M} > 1$ domain corresponds to the unstable radiative mode for oscillations travelling away from the shear layer [108]. It can be seen in the Figure 2.25 that, in the tangential discontinuity approximation ($\alpha \to 0$), the domain of existence for unstable oscillations related to the radiative mode is limited by the threshold value $\overline{M} < \overline{M}_c = \sqrt{2}$ obtained from (2.187) in the $\beta^* \to \infty$ limit. For hydrodynamic flows this condition for radiative oscillation modes was obtained in [108]. The fact that oscillations stabilise upon exceeding a critical value, $\overline{M}_c$, is due to the compressibility of the medium, which is determined by, among other things, the magnetic field pressure.

2.14.4 Instability of the Shear Flow Bounded by One Rigid Wall

It is evident from the general concepts that the presence of a rigid wall must cause changes in the regime of unstable oscillations. These changes should have more effect on the radiative oscillation modes reflected from such a wall. The wall location also becomes an additional factor determining the distribution and value of the oscillation growth rate. Let us consider two limiting cases as we did in the unbounded flow problem.

2.14.4.1 The $\mathbf{k}_t \parallel \mathbf{B}_0$ Case

Let us examine a shear flow having the form of a tangential discontinuity (2.183), with a rigid wall at distance $x = -\Delta$ from it. This problem is described by (2.174) with boundary conditions (2.181). In the $x > 0$ halfspace the solution will be sought in the form of an escaping wave $\zeta = \overline{\zeta} \exp(ik_x x)$, while, on the other side of the shear layer, it will have the form of the sum of the escaping wave and the one reflected from the wall:

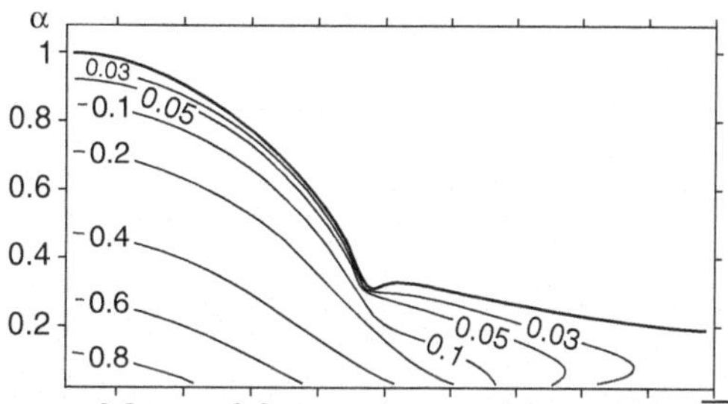

Figure 2.25 Growth rate ($c_i = \text{Im}c$) isoline distribution for MHD oscillations with $\mathbf{k}_t \perp \mathbf{B}_0$ generated by a shear flow in the form of a smooth transition layer in a boundless medium.

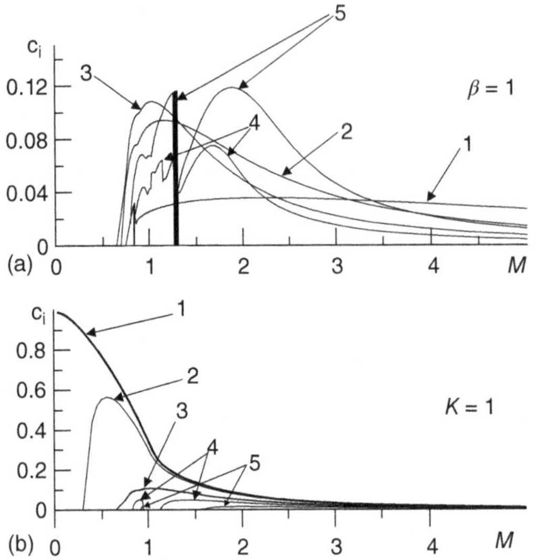

Figure 2.26 Growth rate $c_i(M)$ distribution of MHD oscillations with $\mathbf{k}_t \parallel \mathbf{B}_0$ for a shear flow in the form of a tangential discontinuity bounded by a rigid wall on one side, for different values of parameters β^* and κ: (*a*) $\beta^* = 1$, plots 1–5 refer to $\kappa = 0.1, 0.5, 1, 5, 10$; (*b*) $\kappa = 1$, plots 1–5 refer to $\beta^* = \infty, 10, 1, 0.5, 0.2$.

$$\zeta = \zeta_1 \exp(ik_x x) + \zeta_2 \exp(-ik_x x). \tag{2.188}$$

From the condition of continuous displacement ζ (2.157) and full perturbed pressure (2.155), at point $x = 0$, we obtain the dispersion equation

$$\frac{M^2(c+1)^2(1+\beta^*)-1}{M^2(c-1)^2(1+\beta^*)-1} \; \frac{M^2\beta^*(c+1)^2-1}{M^2\beta^*(c-1)^2-1} \; \frac{M^2(c-1)^2-1}{M^2(c+1)^2-1} = -\tan^2(k_x^- \Delta),$$

where k_x^- is defined by (2.179) for $u = -1$. The numerical solution of this equation is given in Figure 2.26.

Figure 2.26(a) plots the dependence $c_i(M)$ for $\beta^* = 1$ and for five different values $\kappa = k_t\Delta = 0.1, 0.5, 1, 5, 10$. All the curves have a threshold value, M_c, below which the oscillations are stable. The presence of the lower critical value, M_c, same as in the unbounded medium case, is determined by the stabilising effect of Maxwell tensions. Small-scale oscillations of $c_i(M)$ occur for such values of M when the argument of the tangent in the right-hand side of the dispersion equation is large (i.e. $|k_x^- \Delta| \gg 1$).

Varying the magnetic field strength (parameter β^*) makes it possible to trace the changes in the critical value M_c. Figure 2.26(b) shows curves $c_i(M)$ for $\kappa = 1$ and five different values of $\beta^* = \infty, 10, 1, 0.5, 0.2$. If $\beta^* = \infty$ (the hydrodynamic flow regime), the critical M_c value is absent. If $\beta^* \neq \infty$, a lower threshold appears for unstable oscillations, M_c, shifting as β^* decreases towards $M = 1$. If $\beta^* < 1$, the domain of existence for unstable oscillations splits into two domains. One, corresponding to the surface mode, is located between the two critical points $M_c < 1$. The other corresponds to the radiative mode and is left-bounded by point $M_c > 1$. It should be noted that, in contrast to an unbounded flow, the oscillations never completely stabilise for any arbitrarily small values of β^*.

The numerical solution to (2.174) with the boundary conditions (2.181) is shown in Figure 2.27 for a flow with a smooth velocity profile (2.169) (for $\Delta = -20a$). The (**a**, **b**) cases correspond to the two values of the parameter $\beta^* = 10, 1$. As $\alpha \to 0$ the domains of existence for unstable oscillations correspond to those obtained in the problem of the stability of a tangential discontinuity. Note that, for $\alpha > 0.5$, a domain of unstable

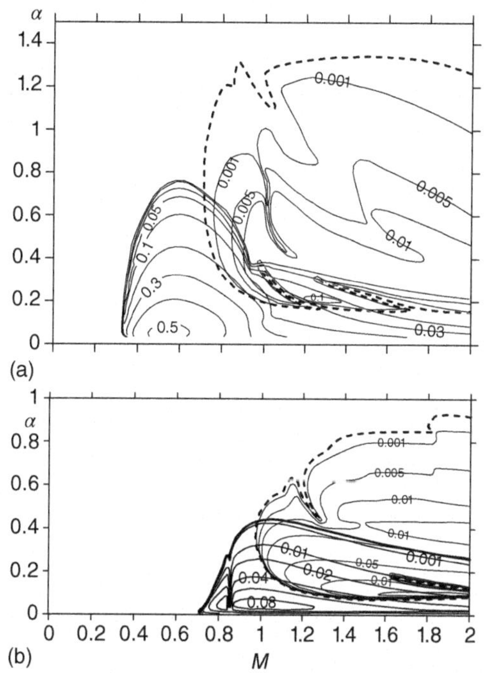

Figure 2.27 Growth rate c_i isoline distribution for MHD oscillations with $\mathbf{k}_t \parallel \mathbf{B}_0$ for a shear flow with a smooth transition layer bounded by a rigid wall ($\Delta = 20a$) on one side, for two different values of parameter β^*: (a) – $\beta^* = 10$, (b) – $\beta^* = 1$.

oscillations forms with an ultra-low growth rate, $c_i < 0.01$, near the $M = 1$. The formation of this domain is related to oscillations that were reflected from the wall and crossed the shear layer. Such oscillations are lacking in a boundless medium.

Unlike the boundary wall-free case, the domain of existence for unstable oscillations in a shear flow has no upper threshold in M in the presence of one such wall. This can be explained as follows. Along with unstable oscillation modes, a tangential discontinuity is known to generate the radiative neutral mode [123, 124]. In particular, this results in the phenomenon of over-reflection for oscillations incident to the shear layer. In other words, waves reflected from or crossing the shear layer have a higher amplitude than the incident wave. This takes place precisely in the range $M > M_c$, when the tangential discontinuity is stable in the unbounded case. The presence of a wall results in wave reflection, the wave then falls on the shear layer and is reflected from it, with a higher amplitude. This means that oscillations that were neutrally stable in the unbounded flow begin to be amplified.

2.14.4.2 The $\mathbf{k}_t \perp \mathbf{B}_0$ Case

Recall that this oscillation regime is analogous to oscillations in an ordinary hydrodynamic flow. The latter oscillations are different in that they are described by means of the modified Mach number, $\overline{M}$, which takes into account the presence of magnetic pressure. The solution of (2.173) for a flow with a velocity profile in the form of a tangential discontinuity corresponds to curve *1* in Figure 2.26(b) for $\beta^* = \infty$. For a flow with a continuous velocity profile (2.169) we will solve (2.173) numerically, with boundary conditions (2.181). Figure 2.28 shows the growth rate c_i distribution in the $(\alpha, \overline{M})$ plane for oscillations generated by the shear layer in the presence of a rigid wall at distance $\Delta = -15a$ from the layer. What makes both flows with a tangential discontinuity in their speed and flows with a velocity profile

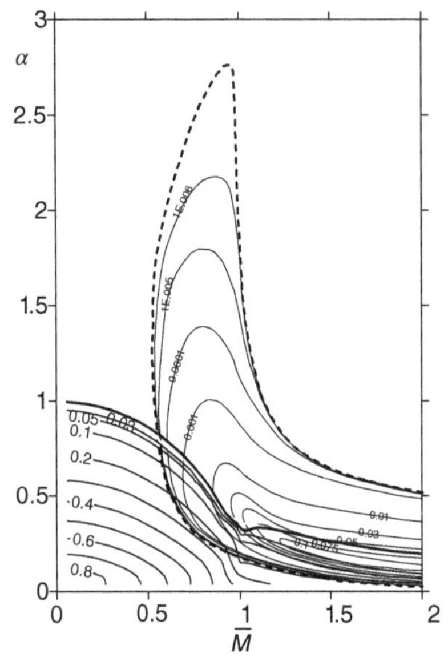

Figure 2.28 Growth rate c_i isoline distribution for MHD oscillations with $\mathbf{k}_t \perp \mathbf{B}_0$ for a shear flow with a smooth transition layer bounded by a rigid wall on one side ($\Delta = -15a$). Thick lines correspond to the surface and radiative oscillation modes, thin lines to the oscillation mode reflected from the wall.

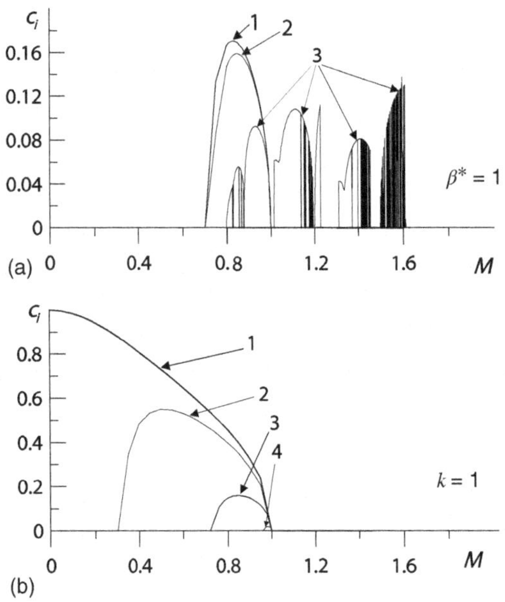

Figure 2.29 Dependence of growth rate $c_i(M)$ for MHD oscillations with $\mathbf{k}_t \parallel \mathbf{B}_0$ for a shear flow in the form of a tangential discontinuity between two rigid walls, for different values of parameters β^* and κ: (*a*) $\beta^* = 1$, plots *1–3* correspond to values $\kappa = 0.1, 1, 5$; (*b*) $\kappa = 1$, plots *1–4* correspond to values $\beta^* = \infty, 10, 1, 0.1$.

of the form (2.169) essentially different from the unbounded medium case is that the critical value $\overline{M} = M_c$, upon exceeding which the unbounded flow becomes stable, vanishes. As noted above, this is explained by the effect of a mechanism for over-reflection of oscillations from the shear layer.

Increasing magnetic field strength fails to stabilise the unstable oscillations. It will be shown in Section 2.14.5 that such a domain of unstable oscillations fails to form, in the presence of two walls either side of the shear layer.

2.14.5 Shear Flow Instability Between Two Boundary Walls

Miura and Pritchett [117, 118] studied the dependence of oscillation growth rate $c_i(\alpha)$ in shear flows with a velocity profile of the form (2.169) for certain values of the Mach number M and parameter β^*. The boundary conditions were selected in the form of two rigid walls either side of the shear layer, and integration was along the real axis x. When such an approach is used, as was noted above, domains with small growth rate are described with a large error.

2.14.5.1 The $k_t \parallel B_0$ Case

Let us examine a shear flow in the form of a tangential discontinuity (2.183) between two rigid walls ($x = \pm\Delta$). In this case, solutions to (2.174) satisfying the boundary conditions (2.182) are sought both sides of the shear layer as (2.188). Matching conditions for perturbed pressure (2.155), displacement ζ (2.157) and full perturbed pressure (2.175) at point $x = 0$ produces the dispersion equation

$$\frac{M^2(c+1)^2(1+\beta^*)-1}{M^2(c-1)^2(1+\beta^*)-1}\;\frac{M^2\beta^*(c+1)^2-1}{M^2\beta^*(c-1)^2-1}\;\frac{M^2(c-1)^2-1}{M^2(c+1)^2-1}=\frac{\tan^2(k_x^-\Delta)}{\tan^2(k_x^+\Delta)},$$

where $k_x^\pm$ are defined by (2.179) for $u = \pm 1$ respectively. A numerical solution of this equation is shown in Figure 2.29. Figure 2.29(a) plots $c_i(M)$ for $\beta^* = 1$ for three different values $\kappa = k_t\Delta = 0.1,\ 1,\ 5$. The domains of existence for both unstable oscillations are limited on two sides. Moving the walls far enough away from the shear layer results in very conspicuous periodicity in the oscillation growth rate distribution thanks to standing waves forming between the reflecting surfaces. As in the one-wall case, small-scale growth rate oscillations are observed due to tangent arguments increasing: $|k_x^\pm\Delta| \gg 1$.

This is how periodicity in the oscillation growth rate distribution can be explained. Standing waves form between the walls, their eigenfrequency depending on the value of the tangential wave vector and the distance between the walls. Instability of the radiative mode is due to the resonance between the neutral mode generated by the shear layer and the wave reflected from the wall. Varying the flow parameters (Mach numbers M) makes various standing wave harmonics resonate with the neutral oscillation mode. A peak is then observed in the oscillation growth rate distribution. If the walls are near enough to each other, the frequency of even the fundamental eigen harmonic of the standing waves is higher than the frequency of the neutral mode radiated by the shear layer. In this case, unstable radiative oscillation modes fail to be excited and the oscillation growth rate is due to the unstable surface mode only.

Let us examine the behaviour of the growth rate for various values of parameter β^*. This will give us the opportunity to trace the effect the magnetic field strength exerts on the value of the critical Mach number M_c. Figure 2.29(b) plots the dependences of growth rate $c_i(M)$ for $\kappa = 1$ for various values $\beta^* = \infty,\ 10,\ 1,\ 0.1$. If $\beta^* = \infty$, which corresponds to the hydrodynamic flow case, the lower critical Mach number is lacking. As β^* diminishes, it shifts to $M = 1$, and the domain of existence for the surface mode of unstable oscillations narrows down. The walls in this case are near enough to each other, which fact manifests itself in complete stabilisation of the radiative oscillation mode (in the $M > 1$ domain). However small are the β^* values, the flow never completely stabilises.

The numerical solution of (2.174) for a flow with a smooth velocity profile (2.169) between two rigid walls ($\Delta = \pm 10a$) is shown in Figure 2.30. The distribution of growth

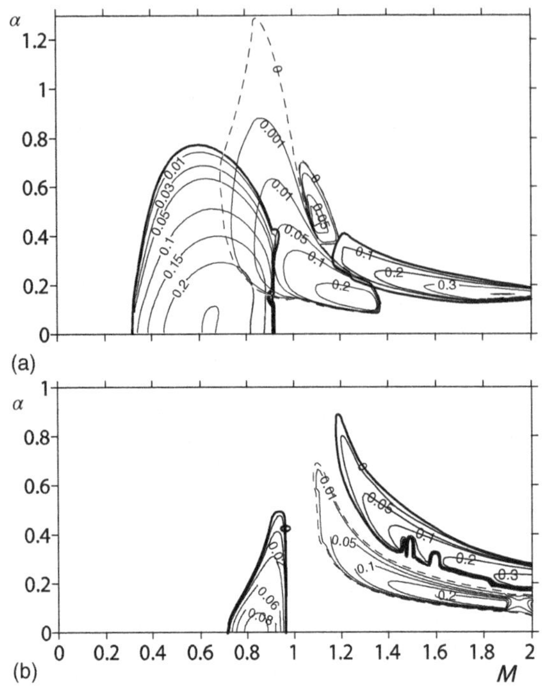

Figure 2.30 Growth rate c_i isoline distribution for MHD oscillations with $\mathbf{k}_t \parallel \mathbf{B}_0$ for a shear flow in the form of a smooth transition layer between two rigid walls ($\Delta = \pm 10a$) for two different values of parameter β^*: (*a*) $\beta^* = 10$, (*b*) $\beta^* = 1$.

rate c_i isolines in Figure 2.30(a), (b) corresponds to $\beta^* = 10$, 1. As β^* decreases, the domain of existence for the unstable surface mode collapses, and the absolute growth rate diminishes. The domain of existence for unstable oscillations related to the radiative mode ($M > 1$) has a characteristic quasiperiodic, patchy structure. This is due to resonance between various harmonics of standing waves between the walls and the neutral mode of oscillations radiated by the shear flow. As β^* decreases, the number of patches, their size and growth rate decrease. If $\beta^* < 1$, the shear flow is close to being stable even though never reaching complete stabilisation for any values of β^*.

2.14.5.2 The $\mathbf{k}_t \perp \mathbf{B}_0$ Case

The solution to (2.173) for a flow with a velocity profile in the form of a tangential discontinuity is given by curve 1 in Figure 2.29(a), corresponding to $\beta^* = \infty$. It describes the unstable surface mode right-limited by a critical value of the Mach number, $M_c = 1$.

Let us solve (2.173) numerically, with boundary conditions (2.182), for a flow with a smooth velocity profile (2.169). It describes, accurate up to Mach number $\overline{M}$ redefinition, a hydrodynamic flow. The distribution of oscillation growth rate isolines for a shear layer located between two walls ($\Delta = \pm 20a$), is shown in Figure 2.31. Moving the walls far enough away, the growth rate distribution exhibits a patchy, quasiperiodic structure in the domain where unstable oscillations of the radiative mode ($\overline{M} > 1$) are observed in the boundless flow. This is explained by standing unstable oscillations appearing between the reflecting walls. As the walls move closer to the shear layer, first the patches of unstable radiative oscillations disappear, and then the domain of surface mode instability decreases. As the walls move closer than $\Delta \approx \pm 1.2a$, the shear flow becomes completely stable, even in the incompressible medium limit ($\overline{M} \to 0$).

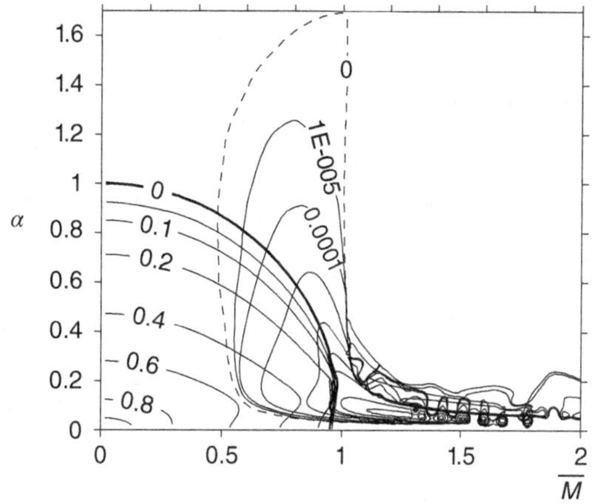

Figure 2.31 Growth rate c_i isoline distribution for MHD oscillations with $\mathbf{k}_t \perp \mathbf{B}_0$ for a shear flow in the form of a smooth transition layer bounded by rigid walls on both sides ($\Delta = \pm 20a$). The thick lines correspond to the surface and radiative oscillation modes, and the thin lines to oscillation modes reflected from the walls.

It can thus be concluded that the presence of walls has a stabilising effect on the shear flow. The instability of the radiative mode stabilising as the walls move closer is due to the changing eigenfrequencies of the harmonics of the standing waves between the walls. Stabilisation of the surface mode can be explained as follows. In the presence of walls, the characteristic scale of unstable surface mode across the shear layer is determined by the distance between the walls. If they are far enough from the shear layer, both the characteristic scale and the domain of unstable surface mode remain virtually the same as they are in the boundless flow. As the walls move closer, this scale diminishes. The range of the transverse wave number α and the Mach number $\overline{M}$ making the surface mode oscillations unstable also decreases. Once the distance between the walls reaches the characteristic scale of the shear flow velocity profile, the shear flow is completely stabilised.

2.15 Geotail Instability Due to Shear Flow at the Magnetopause

Kelvin–Helmholtz (K-H) instability of a shear flow was examined in Section 2.14 in a homogeneous conducting liquid, for various boundary conditions away from the shear layer. The situation in the real magnetospheric plasma is much more complex. As a rule, these plasma are strongly inhomogeneous. Instability develops in the shear layer near the resonance surface where the tangential component of the MHD wave phase velocity coincides with the velocity of the plasma flow enwrapping the magnetosphere. A number of papers regarded observable geomagnetic pulsations as unstable modes of magnetopause oscillations [125–128] and provided a theoretical interpretation of such oscillations [129–134]. The recent appearance of multi-spacecraft observations has made it possible to examine in much detail MHD oscillations generated and propagating in the Earth's magnetosphere [135–137].

Theoretical studies of MHD oscillations driven by the K-H instability in the Earth's magnetosphere have a long history (see [21, 138, 139]). First investigations in this field relied on the simplest models of the medium including two plasma-filled homogeneous half-spaces [140, 141]. The plasma in one of these half-spaces is separated from the other by a sharp boundary and moves relative to it. Subsequent investigations used models with a smooth transition layer [117, 129, 131, 142]. Analytical solutions cannot be obtained for such models but numerical solutions of MHD equations allow us to determine not only the spectrum of unstable oscillations generated by the shear flow at the magnetopause, but also the solar wind velocity range required for their instability [143, 144].

More complex magnetosphere models were also examined having the form of a flat layer with two sharp boundaries enwrapped by the solar wind flow [100, 145, 146]. Similar models were also used for studying magnetosonic oscillations generated by high-speed solar wind jet flows [147]. Models with a sharp boundary allow algebraic dispersion equations to be obtained for unstable oscillation modes [130, 148]. In some special cases it is possible to find analytical solutions of these equations for determining the frequencies and growth rates of unstable oscillations [120].

The shear layer is unstable for MHD oscillations with a transverse wavelength that is greater than its thickness and becomes stable for oscillations whose transverse wavelength is comparable to or smaller than the layer thickness [118, 149]. The growth rate of unstable oscillations in magnetospheric models with a smooth boundary is smaller than in models with a sharp boundary. Moreover, resonance surfaces appear in the boundary layer for the Alfvén and slow magnetosonic (SMS) waves [35, 37, 61]. Here, the energy of unstable FMS waves generated by the magnetopause shear flow is absorbed [150]. The stability of the magnetopause is determined by a trade-off between two effects: instability developing in the shear layer and MHD oscillation energy absorbed on resonance surfaces. There are papers studying both these effects [54, 151, 152].

Models in which regions adjacent to the shear layer are also inhomogeneous have yet another feature. FMS waves have turning points inside the magnetosphere. Eigen-oscillations of the magnetospheric waveguide are localised between the magnetopause and the turning points [153, 154]. It is easiest to simulate this situation by placing a reflecting wall in a homogeneous plasma layer describing the magnetosphere [146]. In more realistic models where the layer adjacent to the magnetopause is inhomogeneous, the turning points of different eigenmodes are at different distances from the magnetopause [155, 156]. In magnetospheric models with finite cross-section, the existence range of unstable oscillations extends to much shorter characteristic wavelengths (see [157, 158] for the cylindrical geomagnetic tail model).

In this section we solve the problem of geomagnetic tail (geotail) instability using the inhomogeneous plasma cylinder model. As we will see further, various types of the geotail FMS oscillations can become unstable thanks to the K-H instability. Such oscillations can serve, for example, as a source of pumping for the magnetospheric resonator for ultralow-frequency FMS oscillations in the near-Earth part of the plasma sheet (see [159, 160] and Section 3.6).

2.15.1 Model Medium and Basic Equations

Let us consider a model geotail in the form of an inhomogeneous plasma cylinder, as shown in Figure 2.32. The plasma distribution over radius in this model refers to geotail lobes. The plasma sheet is not included explicitly. Its presence is simulated by the Alfvén and SMS speed distribution over radius. Away from the cylinder axis they vary from the values characteristic of the plasma sheet to values typical of the geotail lobes. The Alfvén

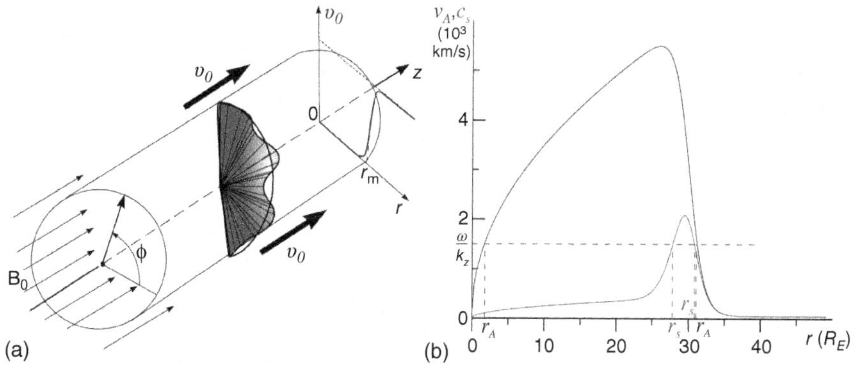

Figure 2.32 (a) cylindrical model of the geotail wrapped around by the solar wind plasma flow and a schematic structure of the unstable "global mode" of the geotail. (b) Alfvén speed $v_A(\rho)$ and SMS speed $c_s(\rho)$ distributions in the geotail and the solar wind. On the resonance surfaces $\rho = \rho_s$ (points 2 and 3) and $\rho = \rho_A$ (points 1 and 4), the parallel phase velocity $\overline{\omega}/k_z$ of the monochromatic wave coincides with, respectively, the local SMS speed c_s and the Alfvén speed v_A.

speed distribution over radius was calculated in [160] based on spacecraft data on the magnetospheric plasma concentration and magnetic field strength distribution [161, 162]. Since the basic results of that paper refer to the region of open field lines, the presence of the plasma sheet should not be an essential element in our calculations.

We will employ a cylindric system of coordinates (ρ, ϕ, z), where the origin of coordinates $\rho = 0$ coincides with the plasma cylinder axis. The background magnetic field is along the z axis. We will assume the plasma in the magnetosheath to be moving along the z axis at speed v_0, and the plasma in the geotail to be immobile (Figure 2.32(a)).

The transition from magnetospheric plasma parameters to magnetosheath parameters occurs in a narrow transition layer of thickness $\Delta_\rho \ll \rho_m$, where ρ_m is the characteristic radius of the geotail. The plasma density distribution over radius is assumed to be such that the maximum density is reached at the plasma cylinder axis and drops to a minimum towards the cylinder boundary. The magnetic field strengths in the geotail is higher than in the solar wind. The Alfvén speed distribution over radius $v_A(\rho) = B_0/\sqrt{4\pi\rho_0}$ in the model geotail is shown in Figure 2.32(b). Also shown is the SMS speed distribution $c_s(\rho) = v_A v_s/\sqrt{v_A^2 + v_s^2}$, where the sound speed in plasma $v_s = \sqrt{\gamma P_0/\rho_0}$ is determined from the background plasma pressure P_0, satisfying the plasma equilibrium condition (2.59). We will assume that the magnetic field strengths are virtually homogeneous inside and outside the plasma cylinder and vary within the thin transition layer of thickness $\Delta_\rho \ll \rho_m$ only. The numerical calculations below will use the following parameter values: $\rho_m = 30R_E$, $\Delta_\rho = 2R_E$, where $R_E = 6370$ km is the Earth radius. Such a distribution is typical of the plasma parameters in the geotail lobes.

It follows from the equilibrium condition (2.59) that the plasma pressure varies within the transition layer only. The values for the main plasma and magnetic field parameters at the geotail boundary as used in the numerical calculations are listed in Table 2.2. These parameters enable the plasma configuration equilibrium condition to be satisfied.

Let us define the spatial structure of monochromatic MHD wave in the geotail model. We will denote $v_\rho = d\zeta/dt = \partial\zeta/\partial t + (\mathbf{v}_0\nabla)\zeta$, the radial component of the plasma speed vector

Table 2.2 Main parameters of the model medium at the geotail boundary.

Parameter\region	Tail lobes	Magnetosheath
B_0(nT)	20	5
v_A(km/s)	6000	50
v_s(km/s)	420	177
$\beta^* = v_s^2/v_A^2$	0.005	12.6

in the wave, where ζ is the radial shift of the plasma element. Let us examine a monochromatic wave of the form $\exp(ik_z z + im\phi - i\omega t)$, where k_z is the wave vector component along the z axis, $m = 0, 1, 2, 3, \ldots$ is the azimuthal wave number, ω is wave frequency. Writing out the system of linearised MHD equations (1.5) componentwise in the cylindric system of coordinates as in Appendix B produces the following equation for the radial component of the plasma element shift:

$$\frac{\partial}{\partial \rho}\frac{\rho_0 \Omega_A^2}{k_\rho^2}\frac{1}{\rho}\frac{\partial \rho\zeta}{\partial \rho} + \rho_0 \Omega_A^2 \zeta = 0, \tag{2.189}$$

analogous to (2.4), where the notations are $\Omega_A^2 = \overline{\omega}^2 - k_z^2 v_A^2$, $\overline{\omega} = \omega - k_z v_0$ is the Doppler-shifted oscillation frequency,

$$k_\rho^2 = \frac{\overline{\omega}^4}{\overline{\omega}^2(v_A^2 + v_s^2) - k_z^2 v_A^2 v_s^2} - k_z^2 - \frac{m^2}{\rho^2} = k_z^2\left(\frac{\overline{\omega}_A^4/(1+\beta^*)}{(\overline{\omega}_A^2 - \overline{\omega}_s^2)} - 1 - \frac{m^2}{k_z^2\rho^2}\right)$$
$$= \frac{k_z^2}{1+\beta^*}\frac{(\overline{\omega}_A^2 - \overline{\omega}_{A1}^2)(\overline{\omega}_A^2 - \overline{\omega}_{A2}^2)}{(\overline{\omega}_A^2 - \overline{\omega}_s^2)} \tag{2.190}$$

is the squared radial component of the wave vector in the WKB approximation when the solution of (2.189) can be written as $\zeta \sim \exp(i \int k_\rho d\rho)$, $\beta^* = v_s^2/v_A^2$, $\overline{\omega}_A = \overline{\omega}/k_z v_A(\rho)$, $\overline{\omega}_s = \sqrt{\beta^*/(1+\beta^*)}$, and $\overline{\omega}_{A1}^2, \overline{\omega}_{A2}^2$ are the roots of the biquadratic (for $\overline{\omega}_A$) equation $k_\rho^2 = 0$. Note that β^*, at an accuracy up to a factor close to unity, coincides with the plasma parameter $\beta = 8\pi P_0/B_0^2$ – the gas-kinetic plasma pressure to magnetic pressure ratio.

The other MHD oscillation field components are expressed in terms of ζ as follows:

$$v_\rho = -i\overline{\omega}\zeta, \quad v_\phi = -\frac{1}{K_s^2}\left(v_A^2 + \frac{K_A^2 v_s^2}{\chi_s^2}\right)\frac{m}{\overline{\omega}\rho^2}\frac{\partial \rho\zeta}{\partial \rho}, \tag{2.191}$$

$$v_z = -\frac{k_z K_A^2 v_s^2}{\overline{\omega}\chi_s^2 \rho}\frac{\partial \rho\zeta}{\partial \rho} - \zeta\frac{dv_0}{d\rho},$$

$$B_\rho = ik_z B_0 \zeta, \quad B_\phi = -\frac{k_z B_0}{\overline{\omega}} v_\phi, \tag{2.192}$$

$$B_z = -\frac{K_A^2 B_0}{\chi_s^2}\left(1 - \frac{k_z^2 v_s^2}{\overline{\omega}^2}\right)\frac{1}{\rho}\frac{\partial \rho\zeta}{\partial \rho} - \zeta\frac{dB_0}{d\rho},$$

$$P = -\overline{\gamma} P_0 \frac{K_A^2}{\chi_s^2}\frac{1}{\rho}\frac{\partial \rho\zeta}{\partial \rho} + \zeta\frac{d}{d\rho}\left(\frac{B_0^2}{8\pi}\right), \tag{2.193}$$

where the notations are:

$$K_A^2 = 1 - \frac{k_z^2 v_A^2}{\overline{\omega}^2}, \quad K_S^2 = K_A^2 - \frac{m^2 v_A^2}{\rho^2 \overline{\omega}^2}, \quad \chi_S^2 = K_S^2 - K_A^2 \frac{k_z^2 + m^2/\rho^2}{\overline{\omega}^2} v_S^2.$$

The FMS turning points are determined by the zeros in function $k_\rho^2(\rho)$, and the resonance surfaces for Alfvén and SMS waves, by the singular points in (2.189), where the coefficient for the higher derivative becomes zero. At the Alfvén resonance point $\rho = \rho_A$, we have $\Omega_A^2(\rho_A) = 0$. At the magnetosonic resonance point $\rho = \rho_s$, the denominator in the definition of k_ρ^2 (2.190) becomes zero, producing a local dispersion equation for SMS waves as $|k_\rho^2| \to \infty$: $\overline{\omega}^2 = k_z^2 c_s^2(\rho_s)$. On the resonance surfaces, the longitudinal phase velocity $\overline{\omega}/k_z$ of the MHD wave coincides with the local velocity of Alfvén or SMS waves (see Figure 2.32(b)).

We will use the following boundary conditions in our numerical solution of (2.189). As $\rho \to 0$, we will select a finite-amplitude solution of (2.189) approximately representable in this limit as:

$$\rho^2 \zeta'' + \sigma \rho \zeta' + (k_{\rho 0}^2 \rho^2 - 2 + \sigma)\zeta = 0, \tag{2.194}$$

where $\zeta' = \nabla_\rho \zeta$, $k_{\rho 0}^2 \equiv k_\rho^2(\rho \to 0)$ (for $m \neq 0$, we have $k_{\rho 0}^2 \approx -m^2/\rho^2$). Here, $\sigma = 1$ for $m = 0$, and $\sigma = 3$ for $m \neq 0$. The finite solution of (2.189), for $\rho \to 0$, has the form:

$$\zeta = C \begin{cases} \rho, & \text{for} \quad m = 0, \\ \rho^{m-1}, & \text{for} \quad m \neq 0, \end{cases} \tag{2.195}$$

where C is an arbitrary constant. The second boundary condition will be determined in the magnetosheath for $\rho = 2\rho_m$. It prescribes that the oscillation amplitude here is selected to be equal to the amplitude of the harmonic given by the model FMS spectrum in the magnetosheath (see Section 2.13.2 and Appendix C), which is used to determine the constant C and the oscillation amplitude in the entire space.

The solution of (2.189) near the resonance surface for SMS waves $\rho = \rho_s$ is rather similar to (2.63), the solution of (2.62). As $\rho \to \rho_s$, it takes the form

$$\zeta = -C \ln\left(\frac{\rho - \rho_s}{a_s} + i\varepsilon_s\right),$$

where C is the integration constant, a_s is the characteristic variation scale of plasma parameters near the resonance surface, ε_s is a regularising factor due to oscillation dissipation near the resonance surface. When $\varepsilon_s = 0$, the solution possesses a logarithmic singularity on the resonance surface for SMS waves. To regularise the singularities in (2.189), we will redefine the expressions for oscillation frequency by adding imaginary additions to include Alfvén and SMS oscillation dissipation on resonance surfaces, in the definition of Ω_A^2 and in the denominator in (2.190). We will set $\overline{\omega} = \omega - k_z v_0 + i\gamma_A$ in the definition of Ω_A^2, and $\overline{\omega} = \omega - k_z v_0 + i\gamma_s$ in the denominator in (2.190), where $\gamma_{A,s}$ are Alfvén and SMS decrements near their respective resonance surfaces.

The decrements determine the amplitude and characteristic scale of Alfvén and SMS wave localisation at the resonance surfaces. For Alfvén waves, the decrement is small ($\gamma_A \sim 10^{-3}\omega$), enabling a large amplitude of resonant oscillations and their narrow localisation region over radius. The SMS decrement strongly depends on the plasma ion to electron temperature ratio (see Figure 1.3 in Chapter 1). The plasma electrons are hotter than the ions in the solar wind ($T_e \approx 3T_i$), therefore, we set $\gamma_s \approx 10^{-2}\omega$ in the magnetosheath. In contrast, the plasma ions in the geotail lobes are hotter than the

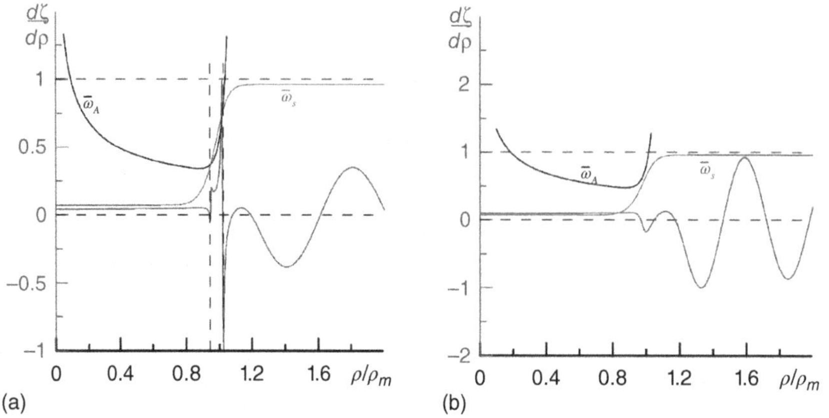

Figure 2.33 Spatial structure of monochromatic MHD waves with $m = 1$ for two different values of the parallel phase speed $\overline{\omega}/k_z$: (a) oscillations with resonant surfaces for SMS waves in the geotail, $\overline{\omega}_A(\rho_s) = \overline{\omega}_s(\rho_s)$, (b) oscillations with no resonance surfaces in the geotail.

electrons ($T_i \approx 8T_e$), implying $\gamma_s \approx 0,8\omega$. The transition from the SMS decrement selected for the magnetosheath to the decrement typical of the magnetosphere takes place in the same transition layer as for the other plasma parameters.

Figure 2.33 shows the radial structure of two monochromatic waves, with resonance surfaces for SMS oscillations in the magnetosphere for one (Figure 2.33(a)), but no surfaces for the other (Figure 2.33(b)). These figures illustrate the unity-normalised structure of the derivative $d\zeta/d\rho$, determining the oscillation amplitude on resonance surfaces. The resonance surfaces for SMS waves are determined by the intersection points of functions $\mathrm{Re}\,(\overline{\omega}_A(\rho))$ and $\mathrm{Re}\,(\overline{\omega}_s(\rho))$, where the real part of the denominator in (2.190) becomes zero.

2.15.2 Calculating the Magnetopause MHD Instability Growth Rate in the Tangential Discontinuity Model

For the geotail, we will use a model in the form of an inhomogeneous plasma cylinder as described in Section 2.15.1 and shown in Figure 2.32. Plasma distribution over radius in this model corresponds to the geotail lobes. It does not explicitly take into account the plasma sheet of the real magnetosphere. Its presence is simulated by the Alfvén and sound speed distribution over radius. The local instability of the geotail boundary is determined by the parameters of the immediately adjoining magnetospheric and solar wind regions. The instability of global modes is, on the contrary, determined by the integral characteristics of the geotail plasma and does not depend on details of the plasma distribution over radius. Therefore the plasma sheet cannot have much impact on these processes.

To describe the structure of oscillations along the radial coordinate we will use (2.189) for the radial component of the plasma displacement. The other wave field components are described by (2.191)–(2.193). The radial component of the MHD wave vector is described by (2.190) in the WKB approximation, when the solution of (2.189) can be represented as $\zeta \sim \exp(i \int k_\rho d\rho)$.

In the WKB approximation, the solution of the problem of MHD oscillation stability in the cylindric model with a boundary in the form of a tangential discontinuity examined

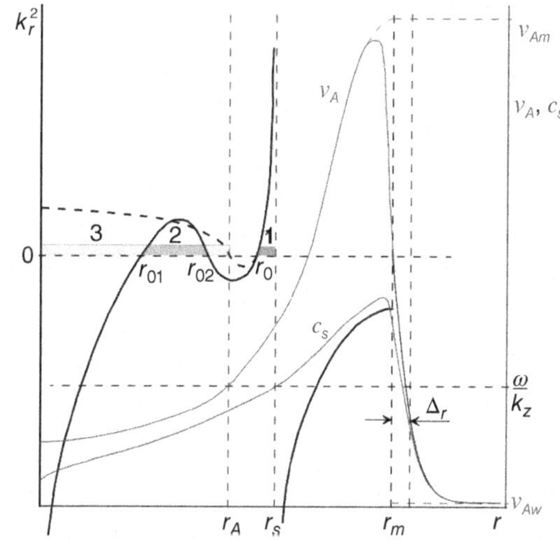

Figure 2.34 Alfvén $v_A(\rho)$ and SMS wave speed distribution $c_s(\rho)$ (light and dark gray lines; the vertical axis is to the right) inside and outside the geotail in the plasma cylinder model. Squared wave-vector WKB component distribution over radius, $k_\rho^2(\rho)$ (black lines; the vertical axis is to the left, the thick dashed lines correspond to $m = 0$). Coordinates ρ_A and ρ_s correspond to resonance surfaces for Alfvén and SMS oscillations, $\rho_{00}, \rho_{01}, \rho_{02}$ are the turning points for magnetosonic waves. Numbers and shades of grey denote the transparent regions: *1* – for SMS waves, *2* – for FMS waves with $m \neq 0$, and *3* – for FMS waves with $m = 0$.

here is determined by the magnitude of the wave vector radial component $k_\rho(\rho)$ on both sides of the boundary. Let us analyse the behaviour of $k_\rho^2(\rho)$ in the above geotail model. Qualitatively, the distribution of k_ρ^2 within the plasma cylinder is shown in Figure 2.34 for such values of m, k_z and ω that enable the presence of all possible resonance surfaces and turning points for MHD waves within the geotail.

The ordinary turning points are determined by the zeros of function $k_\rho^2(\rho)$. In the distribution in Figure 2.34, their number can vary from one (ρ_{00}) to three ($\rho_{00}, \rho_{01}, \rho_{02}$). The number of turning points is determined by the values of parameters m, k_z and ω. Thus, for the axisymmetric mode, $m = 0$, the turning point ρ_{02} is absent, and ρ_{01} coincides with ρ_A, the latter point determining, for oscillations with $m \neq 0$, the location of the resonance surface for Alfvén waves. The resonance surfaces are determined by the singular points in (2.189) where the coefficient before the higher derivative turns to zero. The Alfvén resonance point ρ_A, determined by $\Omega_A^2(\rho_A) = 0$, is located in the opaque region (ρ_{02}, ρ_{00}). If $m = 0$, then ρ_A is a turning point (the coefficient before the higher derivative does not turn to zero at it). The magnetosonic resonance point ρ_s is determined by the denominator turning to zero in (2.190), producing a local dispersion equation for SMS waves as $|k_\rho^2| \to \infty$: $\omega^2 = k_z^2 c_s^2(\rho_s)$.

A transparent region for FMS waves (where $k_\rho^2(\rho) > 0$) can exist within the geotail - located within the interval $\rho_{01} \leq \rho \leq \rho_{02}$ if $m \neq 0$, and within the interval $0 \leq \rho \leq \rho_A$ if $m = 0$. The interval $\rho_{00} \leq \rho \leq \rho_s$ is a transparent region for SMS waves (where ρ_s determine the resonance surface for SMS). As can be seen from (2.190), the behaviour of $k_\rho^2(\rho)$ in the $0 < \rho < \rho_m$ interval depends on the value of $\overline{\omega}_A(\rho)$ for the endpoints of the interval. You can get the idea of how the distribution $k_\rho^2(\rho)$ varies as the phase speed of the wave in question ω/k_z increases by shifting the function $k_\rho^2(\rho)$ in Figure 2.34 to the right. As $\omega/k_z \to 0$ there is an opaque region covering the entire magnetosphere corresponding to the $\rho_s < \rho < \rho_m$ part of the plot; as $\omega/k_z \to \infty$ there is a transparent region corresponding to the $\rho_{01} < \rho < \rho_{02}$ part of the plot as $\rho_{01} \to 0$ and $\rho_{02} > \rho_m$. Depending on the magnitudes of $\overline{\omega}_{Am}^2$ and $\overline{\omega}_{Aw}^2$ determined by the wave phase speed, the boundary can be adjoined by both the transparent and opaque regions of the oscillations in question. Hereafter the subscripts $_{m,w}$ indicate that the relevant values are determined by the parameters in the magnetospheric and solar wind regions adjoining the separating boundary.

The matching condition for solutions at the boundary can be easily obtained by integrating (2.189) in the narrow interval $(\rho_m - \varepsilon, \rho_m + \varepsilon)$ for $\varepsilon \to 0$:

$$\left.\frac{\rho_0 \Omega_A^2}{k_\rho^2} \frac{\partial \ln \zeta}{\partial \rho}\right|_{\rho_m - \varepsilon} = \left.\frac{\rho_0 \Omega_A^2}{k_\rho^2} \frac{\partial \ln \zeta}{\partial \rho}\right|_{\rho_m + \varepsilon}. \tag{2.196}$$

Using (2.191) and (2.193), it can be demonstrated that the matching condition (2.196) is analogous to the requirement that the plasma should be equally shifted on both sides of the boundary ($\zeta_{\rho_m - \varepsilon} = \zeta_{\rho_m + \varepsilon}$, or the impermeability condition) and the full perturbed pressure across the boundary ($(P + B_z B_0/4\pi)_{\rho_m - \varepsilon} = (P + B_z B_0/4\pi)_{\rho_m + \varepsilon}$) should be conserved.

Let us now define the boundary conditions in the problem under study. As $\rho \to 0$, the boundary condition is that the amplitude of oscillations described by (2.189) should be finite. As for the boundary conditions for $\rho \to \infty$, its definition is related to the causality principle. In this problem, we will be interested in such solutions of (2.189) that describe unstable modes of the oscillations. Such solutions, according to the causality principle, describe waves escaping from the shear layer that has generated them. In other words, the energy flux of these waves must be directed away from the shear layer.

As can be seen from Section 2.15, the group velocity must be positive for unstable modes that, as $\rho \to \infty$, carry energy away from the shear layer: $\mathrm{Re}(v_{g\rho}) > 0$. Differentiating (2.190) over ω yields

$$v_{g\rho} = v_{A\infty} \frac{1 + \beta_\infty^*}{k_z} \mathrm{Re} k_{\rho\infty} \frac{\left[\overline{\omega}_{A\infty}^2 - \overline{\omega}_{s\infty}^2\right]^2}{\overline{\omega}_{A\infty}^3 \left[\overline{\omega}_{A\infty}^2 - 2\overline{\omega}_{s\infty}^2\right]}. \tag{2.197}$$

The boundary condition for the wave escaping from the shear layer, as $\rho \to \infty$, has the form

$$\frac{\partial \zeta}{\partial \rho} = i k_{\rho\infty} \zeta, \tag{2.198}$$

where the sign of $k_{\rho\infty} \equiv k_\rho(\rho \to \infty) = \pm\sqrt{k_{\rho\infty}^2}$ is determined by the requirement $\mathrm{Re}\,(v_{g\rho}) > 0$.

To form a qualitative idea of how the growth rate of the unstable oscillations at the geotail boundary behaves depending on the solar wind flow velocity, let us solve the above problem using the WKB approximation in the ρ coordinate. We will consider such parameter sets for unstable oscillations that enable the resonance surfaces and turning points for MHD waves to be far from the boundary $\rho = \rho_m$, while making the WKB approximation applicable for describing oscillations near that boundary. If the oscillations are weakly unstable ($|\mathrm{Im}\,(\omega)| \ll |\mathrm{Re}\,(\omega)|$), we will regard the boundary-adjacent region to be an opaque region if $\mathrm{Re}\, k_\rho^2(\rho_m) < 0$, but we will regard it as a transparent region if $\mathrm{Re}\, k_\rho^2(\rho_m) > 0$.

If a transparent region adjoins the geotail boundary on the solar wind side ($\rho > \rho_m$), then the wave escaping from the magnetosphere will be the solution of (2.189) for the unstable mode in the solar wind. Its WKB solution has the form

$$\zeta = C_w \sqrt{\frac{k_\rho}{\rho_0 \Omega_A^2 \rho}} \exp\left(i \int_{\rho_m}^{\rho} k_\rho d\rho'\right), \tag{2.199}$$

where C_w is an arbitrary constant. If the solar wind is opaque, however, the solution will have the form of a surface wave decreasing in amplitude with distance from the boundary:

$$\zeta = C_w \sqrt{\frac{k_\rho}{\rho_0 \Omega_A^2 \rho}} \exp\left(-\int_{\rho_m}^{r} \sqrt{-k_\rho^2 d\rho'}\right). \tag{2.200}$$

Analogously, if an opaque region adjoins the boundary on the magnetosphere side ($\rho < \rho_m$), the WKB solution of (2.189) has the form of a surface wave whose amplitude decreases into the magnetosphere:

$$\zeta = C_m \sqrt{\frac{k_\rho}{\rho_0 \Omega_A^2 \rho}} \exp\left(\int_{\rho_m}^{\rho} \sqrt{-k_\rho^2 d\rho'}\right). \tag{2.201}$$

If a transparent region adjoins the boundary on the magnetosphere side, the WKB solution here has the form

$$\zeta = C_m \sqrt{\frac{k_\rho}{\rho_0 \Omega_A^2 \rho}} \cos\left(\int_{\rho_m}^{\rho} k_\rho d\rho' + \Psi\right), \tag{2.202}$$

where $\Psi = \int_{\overline{\rho}}^{\rho_m} k_\rho d\rho + \pi/4$ is the integral phase from a turning point $\rho = \overline{\rho}$ to the geotail boundary $\rho = \rho_m$. If a transparent region for SMS waves adjoins the boundary, then $\overline{\rho} = \rho_{00}$; if a transparent region for FMS waves does, then, we have $\overline{\rho} = 0$ for $m = 0$, and $\overline{\rho} = \rho_{01}$ for $m \neq 0$.

Let us match the solution in the magnetosphere to the solution describing the oscillation structure in the solar wind. We will regard the geotail boundary as a tangential discontinuity at $\rho = \rho_m$. Note that, in this approximation, which can be defined as local, the dispersion properties of the oscillations are determined by the parameters of the medium directly adjoining the boundary from both inside and outside. In this case the result does not depend on variations in the medium properties far from the tangential discontinuity. Matching according to (2.196), we obtain a dispersion equation which we will write as

$$b \frac{c^2 - 1}{\frac{(c-M_A)^2}{\epsilon^2} - 1} = \begin{cases} -\sqrt{k_{\rho m}^2 / k_{\rho w}^2}, & \text{for } \operatorname{Re} k_{\rho m}^2, \operatorname{Re} k_{\rho w}^2 < 0, \\ i\sqrt{-k_{\rho m}^2 / k_{\rho w}^2}, & \text{for } \operatorname{Re} k_{\rho m}^2 < 0, \operatorname{Re} k_{\rho w}^2 > 0, \\ -\cot\Psi \sqrt{-k_{\rho m}^2 / k_{\rho w}^2}, & \text{for } \operatorname{Re} k_{\rho m}^2 > 0, \operatorname{Re} k_{\rho w}^2 < 0, \\ i\cot\Psi \sqrt{k_{\rho m}^2 / k_{\rho w}^2}, & \text{for } \operatorname{Re} k_{\rho m}^2, \operatorname{Re} k_{\rho w}^2 > 0, \end{cases} \tag{2.203}$$

where the notations are: $b = B_{0m}^2 / B_{0w}^2$, $c = \omega / k_z v_{Am}$ is dimensionless phase velocity, $M_A = v_{0w}/v_{Am}$ is the Alfvén Mach number determined from velocity v_{Am},

$$k_{\rho m}^2 = k_z^2 \left(\frac{c^4}{c^2(1+\beta_m^*) - \beta_m^*} - 1 - \kappa_m^2\right),$$

$$k_{\rho w}^2 = k_z^2 \left(\epsilon^{-2} \frac{(c - M_A)^4}{(c - M_A)^2 (1 + \beta_w^*) - \epsilon^2 \beta_w^*} - 1 - \kappa_m^2\right),$$

$\beta_{m,w}^* = v_{Sm,w}^2 / v_{Am,w}^2$, $\kappa_m = m / k_z \rho_m$, $\epsilon = v_{Aw}/v_{Am}$ (assuming $v_{Aw} \ll v_{Am}$). Let us seek a solution for the dispersion equation (2.203) using the perturbation method with respect to small parameter ($\epsilon \ll 1$), while assuming

$$c = c_0 + \epsilon c_1 + \cdots \tag{2.204}$$

In the zero order of perturbation theory we have $c_0 = M_A$. In the first order of perturbation theory, we square the left- and right-hand sides of (2.203) to obtain the equation for c_1:

$$\overline{b}^2 (M_A^2 - 1)^2 \left(\frac{c_1^4}{c_1^2(1 + \beta_w^*) - \beta_w^*} - 1 - \kappa_m^2 \right) = \pm (c_1^2 - 1)^2 k_{\rho m0}^2, \tag{2.205}$$

where $k_{\rho m0}^2 \equiv k_{\rho m}^2(c = M_A)$. The plus sign in the right-hand side and $\overline{b} = b$ correspond to $\mathrm{Re}\,(k_{\rho m}^2) < 0$, the minus sign and $\overline{b} = b\tan(\Psi + \pi/4)$ correspond to $\mathrm{Re}\,(k_{\rho m}^2) > 0$. Equation (2.205) is of the sixth order relative to c_1, and its solution can be sought numerically. In the $|c_1| \gg 1$ (but $\epsilon |c_1| \ll c_0$) case, however, it can be roughly reduced to a biquadratic equation,

$$c_1^4 \pm c_1^2 \frac{\overline{b}^2 (M_A^2 - 1)^2}{k_{\rho m0}^2 (1 + \beta_w^*)} \pm \frac{\overline{b}^2}{k_{\rho m0}^2} (1 + \kappa_m^2)(M_A^2 - 1)^2 = 0. \tag{2.206}$$

The solution of (2.206) for $\mathrm{Re}\,(k_{\rho m}^2) < 0$ has the form

$$c_1^2 = \frac{b^2(M_A^2 - 1)^2}{2k_{\rho m0}^2(1 + \beta_w^*)} \pm \sqrt{\frac{b^4(M_A^2 - 1)^4}{4k_{\rho m0}^4(1 + \beta_w^*)^2} - \frac{b^2(M_A^2 - 1)^2(1 + \kappa_m^2)}{k_{\rho m0}^2}}. \tag{2.207}$$

Obviously, the condition $|c_1| \gg 1$ holds for $b \gg 1$ and $|M_A^2 - 1| \gtrsim 1$. The value of

$$k_{\rho m0}^2 = k_z^2 \left(\frac{M_A^4}{M_A^2(1 + \beta_m^*) - \beta_m^*} - 1 - \kappa_m^2 \right)$$

is real and, hence, there are no unstable oscillations for $k_{\rho m0}^2 > 0$ $(c_1^2 > 0)$, while for $k_{\rho m0}^2 < 0$, the solution for an unstable mode is obtained by choosing the minus sign before the radical in (2.207). It is easily verifiable that $k_{\rho m0}^2 < 0$ for $M_A < M_0$ and $M_1 < M_A < M_2$, where $M_0^2 = \beta_m^*/(1 + \beta_m^*)$, and $M_{1,2}^2$ are the roots of the biquadratic (relative to M_A) equation $k_{\rho m0}^2 = 0$:

$$M_{1,2}^2 = \frac{(1 + \kappa_m^2)(1 + \beta_m^*)}{2} \pm \sqrt{\frac{(1 + \kappa_m^2)^2(1 + \beta_m^*)^2}{4} - \beta_m^*(1 + \kappa_m^2)}.$$

If $\beta_m^* \ll 1$, we obtain approximate expressions $M_1^2 \approx M_0^2 + M_0^4/M_2^2 < 1$ and $M_2^2 \approx (1 + \kappa_m^2)(1 + \beta_m^*) > 1$.

It follows from the second equation in (2.203), for $\mathrm{Re}\,(k_{\rho w}^2) > 0$, which corresponds to a transparent region in the solar wind, that the value of c_1^2 cannot be real, which contradicts the solution (2.207) for $k_{\rho m0}^2 < 0$. In this case, there are no unstable oscillations. In the case of $\mathrm{Re}\,(k_{\rho w}^2) < 0$, corresponding to the first equation in (2.203), the signs are the same for the left- and right-hand sides of the equations only if $M_A > 1$. Hence, for $\mathrm{Re}(k_{\rho m}^2) < 0$ unstable oscillations are driven on the geotail boundary within the range of shear flow parameters

$$1 < M_A < M_2. \tag{2.208}$$

This conclusion is consistent with the one in [141].

The solution of (2.206) for $\mathrm{Re}(k_\rho^2) > 0$ away from the function $\overline{b} = b\tan(\Psi + \pi/4)$ poles $(\overline{b}^2 = \infty)$ and zeros $(\overline{b}^2 = 0)$ has the form

$$c_1^2 = -\frac{\overline{b}^2(M_A^2 - 1)^2}{2k_{\rho m0}^2(1 + \beta_w^*)} \pm \sqrt{\frac{\overline{b}^4(M_A^2 - 1)^4}{4k_{\rho m0}^4(1 + \beta_w^*)^2} + \frac{\overline{b}^2(M_A^2 - 1)^2(1 + \kappa_m^2)}{k_{\rho m0}^2}}. \tag{2.209}$$

As in the $k^2_{\rho m0} < 0$ case, the solutions corresponding to the transparent region on the solar wind side ($\mathrm{Re}(k^2_{\rho w}) > 0$) describe stable oscillations only. Unstable solutions are obtained for $k^2_{\rho m0} > 0$ corresponding to the parameter ranges $M_0 < M_A < M_1$ and $M_A > M_2$. The unstable-oscillation growth rate structure is described by a discrete set of the phase velocity values $c = c_{0n}$ defined by the eigenmode frequencies of FMS waves travelling along the FMS waveguide in the geotail lobes. Their dispersion equation has the form

$$\Psi(c_{0n}) = \pi\left(n - \frac{1}{4}\right),$$

where $n = 1, 2, 3, \ldots$ is the number of the waveguide eigen harmonic.

If $M_A > M_2$, only positive (right of the dispersion equation roots) branches of the functions correspond to unstable solutions: $\bar{b}(M_A) = b\tan(\Psi(M_A) + \pi/4) > 0$; while for $M_0 < M_A < M_1 < 1$, only the negative (lefthand) branches: $\bar{b}(M_A) = b\tan(\Psi(M_A) + \pi/4) < 0$ do. The former range ($M_A > M_2$) corresponds to a transparent region for FMS waves adjoining the geotail boundary, while the latter ($M_0 < M_A < M_1$) to a transparent region for SMS waves. It can be shown that the resulting solutions describe stable oscillations ($\mathrm{Im}(c) < 0$) only as we approach the poles and zeros of function $\bar{b}(M_A)$.

Figure 2.35(a) shows an example of a numerical solution of the dispersion equation (2.206) for the azimuthal harmonic $m = 1$ with the longitudinal component of the wave vector $k_z\rho_m = 2$ and the following parameters: $\epsilon = v_{Aw}/v_{Am} = 0.08$, $\beta^*_m = 0.005$, $b = B^2_{0m}/B^2_{0w} = 16$. Not a single eigen mode fits into the SMS waveguide. Therefore the geotail is stable if $M_A < 1$. If $M_A > 1$, the oscillations under scrutiny are unstable in the (2.208) range, as well as in the intervals corresponding to the above-discussed roots of (2.209). Each of these roots corresponds to one of the eigen harmonics of the magnetospheric FMS waveguide adjoining the magnetopause. As M_A increases, higher and higher harmonics become unstable. As can be seen from Figure 2.35(a), there is no noticeable

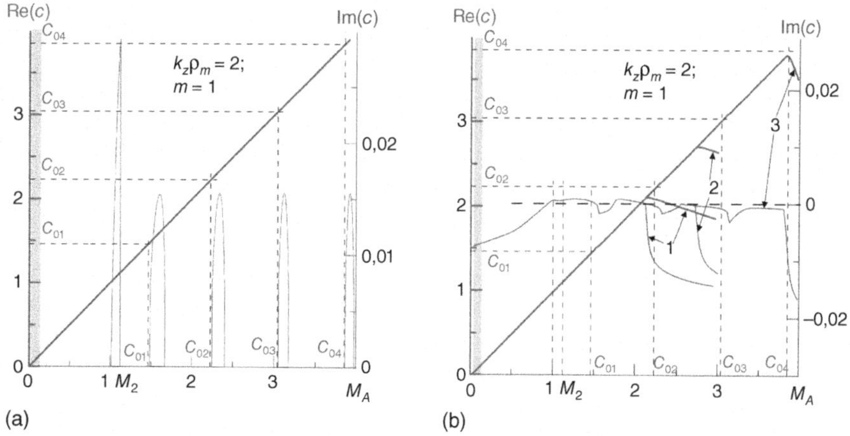

Figure 2.35 The Mach number M_A dependence of the frequency (Re (c), dark gray line) and growth rate (Im (c) light gray lines) for unstable oscillations driven at the geotail boundary. (a) WKB solution for the model with a boundary in the form of a tangential discontinuity, where $c_{01,02,03,04}$ are the roots of the dispersion equation $\tan(\Psi(c_{0n}) + \pi/4) = 0$ determining, in the WKB approximation, the FMS-waveguide eigenfrequencies (c_{0n}) in the geotail lobes. (b) solution for the model with a boundary in the form of a smooth transition layer of characteristic thickness $\Delta \equiv \Delta_\rho/\rho_m = 0,066$ for the same parameters as in panel (a).

M_A dependence of the maximum values of the oscillation growth rate, but the width of the unstable oscillation ranges decreases as M_A (and the eigenharmonic number n) increases.

2.15.3 Geotail Instability in a Smooth-Boundary Model

Let us examine an MHD instability problem for the geotail wrapped around by the solar wind in the model with a boundary in the form of a smooth transition layer. In this case the solution of (2.189) can only be found numerically. For convenient search of the numerical solutions and their comparison to the above WKB approximation results, let us rewrite (2.189) in a dimensionless form:

$$\frac{\partial}{\partial \xi}\frac{\tilde{b}(\xi)[\overline{\omega}_A^2(\xi)-1]}{\xi\kappa^2(\xi)}\frac{\partial \xi\zeta}{\partial \xi}+(k_z\rho_m)^2\tilde{b}(\xi)[\overline{\omega}_A^2(\xi)-1]\zeta=0, \tag{2.210}$$

where the notations are: $\xi=\rho/\rho_m$, $\overline{\omega}_A(\xi)=[c-M_A\tilde{v}_0(\xi)]/\tilde{v}_A(\xi)$, $\tilde{v}_A(\xi)=v_A(\xi)/v_{Am}$, $\tilde{v}_0(\xi)=v_0(\xi)/v_{0m}$, $\tilde{b}(\xi)=B_0^2(\xi)/B_{0m}^2$,

$$\kappa^2(\xi)=\frac{\overline{\omega}_A^4}{\overline{\omega}_A^2(\xi)(1+\beta^*(\xi))-\beta^*(\xi)}-1-\frac{\kappa_m^2}{\xi^2},$$

$\beta^*(\xi)=v_A^2(\xi)/v_s^2(\xi)$. We will simulate the shear flow velocity $\tilde{v}_0(\xi)$, Alfvén speed $\tilde{v}_A(\xi)$ and squared magnetic field strength $\tilde{b}(\xi)$ profiles using the following functions:

$$\tilde{v}_0(\xi)=\frac{1}{2}\left[1+\tanh\frac{\xi-1}{\Delta}\right],$$

$$\tilde{v}_A(\xi)=\frac{1}{2}\left[\epsilon+\epsilon_0+(1-\epsilon_0)\sqrt{\xi}+(\epsilon-\epsilon_0-(1-\epsilon_0)\sqrt{\xi})\tanh\frac{\xi-1}{\Delta}\right],$$

$$\tilde{b}(\xi)=\frac{1}{2}\left[1+b^{-1}-(1-b^{-1})\tanh\frac{\xi-1}{\Delta}\right],$$

where $\Delta=\Delta_\rho/\rho_m$ (Δ_ρ is the characteristic thickness of the magnetopause transition layer), $\epsilon=v_{Aw}/v_{Am}$, $\epsilon_0=v_A(0)/v_{Am}$, $b=B_{0m}^2/B_{0w}^2$, while the function $\beta^*(\xi)$ will be defined from the plasma configuration equilibrium condition (2.59), which can be written as

$$\beta^*(\xi)=\frac{\beta_m^*}{\tilde{b}(\xi)}+\frac{\overline{\gamma}}{2}\left(\frac{1}{\tilde{b}(\xi)}-1\right).$$

The following values of dimensionless parameters were used in the numerical calculations: $\Delta=0.066$, $b=16$, $\epsilon_0=0.016$, $\epsilon=0.008$, $\beta_m^*=0.005$. A boundary-value problem was solved in order to find the frequency of oscillations (the parameter c in dimensionless variables), satisfying the boundary conditions for $\xi\to\infty$ (2.198) and $\xi\to 0$. The latter requirement means that, for $\xi\to 0$, the solution of (2.210) must coincide with the finite-in-amplitude solution (2.195) of the approximate equation (2.194).

The results for the numerical calculations of the unstable oscillation growth rate for the azimuthal harmonic $m=1$ with the longitudinal wavenumber $k_z\rho_m=2$ are presented in Figure 2.35(b). Comparison with Figure 2.35(a) which presents the local approximation solution of the same problem for oscillations in the geotail model with a boundary in the form of a tangential discontinuity reveals considerable differences in the oscillation growth rate distribution.

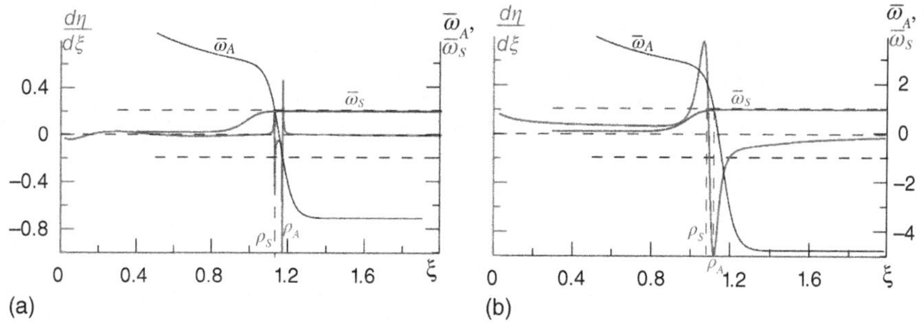

Figure 2.36 Radial structure of unstable oscillations in the geotail for the azimuthal harmonic $m = 1$, normalised to the maximum value of $|d\zeta/d\xi|_{max}$: (a) oscillations close to the second harmonic, $n = 2$, of the eigenmodes propagating in the FMS waveguide in the geotail lobes ($k_z\rho_m = 2$), (b) 'global mode' oscillations for small values of $k_z\rho_m \to 0$.

First, it should be noted that the solution for the geotail with a smooth boundary is, for $M_A > M_2$, a bunch of curves on the $c(M_A)$ plot which, upon passing through the eigenvalues $\text{Re}(c) = c_{0n}$, diverge from the zero approximation value $c \approx M_A$ (the solution obtained in Section 2.15.2 in the zero approximation). The solutions were obtained using the numerical integration method for (2.210) and searching for eigenvalues of c corresponding to the set boundary conditions.

Figure 2.35(b) plots the solutions in the interval $0 < M_A < 4$ for the harmonics $n = 1, 2, 3$ of the magnetospheric FMS waveguide. Comparing it to Figure 2.35(a) reveals a manifold decrease of the oscillation growth rate. Only the first few harmonics ($n = 1$, 2, in our case) remain unstable. This is explained by blurred boundary layer and competition between the effects of oscillation dissipation on resonance surfaces and shear flow instability. The eigenvalue points c_{0n} (the same c_{0n} and M_2 points are shown in Figure 2.35(b) as those obtained in the WKB approximation in Figure 2.35(a) are displaced, and the first region of the unstable oscillations expands.

Figure 2.36(a) plots the spatial structure of unstable oscillations close to the second harmonic, $n = 2$. As follows from our analysis of the WKB solution, the solar wind is an opaque region for the unstable mode of the oscillations. The resonance surfaces for Alfvén and SMS waves, determined by the conditions $\overline{\omega}_A(\rho_A) = \pm 1$ and $\overline{\omega}_A(\rho_s) = \pm\overline{\omega}_s(\rho_s)$, respectively, are located in the region of the magnetopause transition layer.

Note that the range $M_A > 1$ is never reached for the actually observed speed of the solar wind along the Earth magnetosphere. The above analysis implies that the geotail boundary always remains locally stable. As will be shown below, however, another type of unstable geotail oscillations exist that prove to be unstable for any, arbitrarily small values of the solar wind flow speed.

2.15.4 K-H Instability of Global Modes in the Geotail

Figure 2.37 plots the growth rate distribution for another type of unstable MHD oscillations in the geotail. Their existence in no way follows from the above calculations of the local instability of the magnetopause based on the WKB approximation. They have a spatial structure shown in Figure 2.36(b), its characteristic feature being the practically constant value of the first derivative $d\zeta/d\rho$ across the geotail. They can be defined as

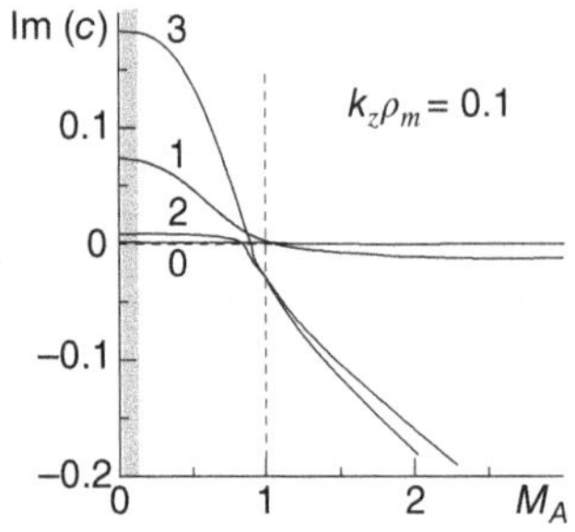

Figure 2.37 The Mach number, M_A, dependence of the growth rate of the 'global' modes in the geotail for the first azimuthal harmonics $m = 0, 1, 2, 3$ and $k_z\rho_m = 0.1$.

'global' modes of the geotail oscillations. The following features of these oscillations are of note:

1. For harmonics with $m \neq 0$ as $k_z\rho_m \to 0$, the $\mathrm{Im}c(M_A)$ distribution plots do not depend on $k_z\rho_m$.
2. As M_A increases, the oscillation growth rate decreases, turning to zero for harmonics with $m \neq 0$ at a certain $M_A = M_{Ac}$ (the M_{Ac} value differs for different azimuthal harmonics m). If $M_A > M_{Ac}$ these harmonics become stable ($\mathrm{Im}\,(c) < 0$).
3. The $\mathrm{Im}\,c(M_A)$ plots for the harmonic $m = 0$ differ very much for oscillations with various $k_z\rho_m$ and have no limiting value of M_{Ac} that would limit the domain of existence for the unstable oscillations.
4. The absolute values of the growth rate for azimuthal harmonics with $m \neq 0$ are much larger than for oscillations with $m = 0$.

To qualitatively analyse these features of the global modes for $k_z\rho_m \to 0$, let us inspect the following simplified model. The derivative $d\zeta/d\rho$ being nearly constant inside the plasma cylinder, we can neglect the second derivative of ζ in (2.189):

$$\left(\nabla_\rho \ln\left(\frac{\rho_0\Omega_A^2}{k_\rho^2}\right) + \rho^{-1}\right)\frac{\partial\zeta}{\partial\rho} + k_\rho^2\zeta \approx 0.$$

The solution of this equation is

$$\zeta = C\exp\left(-\int_{\rho_m}^{\rho} \frac{k_\rho^2\rho'}{\rho'\nabla_{\rho'}\ln\left(\rho_0\Omega_A^2/k_\rho^2\right) + 1}d\rho'\right),$$

where C is an arbitrary constant. If $k_z\rho_m \ll 1$ the solar wind is an opaque region for global modes with $m \neq 0$. Considering the magnetopause as a tangential discontinuity, we obtain the following dispersion equation from the matching condition (2.196):

$$-\frac{\rho_{0w}\Omega_{Aw}^2}{\sqrt{-k_{\rho w}^2}} = \frac{\rho_{0m}\Omega_{Am}^2}{k_{\rho m}^2}\left(\rho_m^{-1} - \frac{k_{\rho m}^2}{\nabla_\rho \ln\left(\rho_{0m}\Omega_{Am}^2/k_{\rho m}^2\right) + \rho_m^{-1}}\right).$$

Or, in the dimensionless form,

$$\frac{(c - M_A)^2/\epsilon^2 - 1}{\sqrt{-k_{\rho w}^2/k_z^2}} = \frac{b(c^2 - 1)}{k_z\rho_m}\left[(k_z\rho_m)^2 - \frac{k_z^2}{k_{\rho m}^2}\right], \qquad (2.211)$$

where

$$k_{\rho w}^2 = k_z^2 \left(\frac{(c - M_A)^4/\epsilon^4}{(c - M_A)^2(1 + \beta_w)/\epsilon^2 - \beta_w} - 1 - \kappa_m^2 \right),$$

$$k_{\rho m}^2 = k_z^2 \left(\frac{c^4}{c^2(1 + \beta_m) - \beta_m} - 1 - \kappa_m^2 \right),$$

$b = B_{0m}^2/B_{0w}^2 > 1$, $\epsilon = v_{Aw}/v_{Am} \ll 1$, $\kappa_m = m/k_z\rho_m$. To estimate the solution, $|\nabla_\rho \ln (\rho_{0m}\Omega_{Am}^2/k_{rm}^2)| \lesssim \rho_m^{-1}$ is set in (2.211).

We will seek the solution of (2.211) according to perturbation theory, as a decomposition (2.204) in small parameter ϵ. In the zero approximation, we have $c_0 = M_A$. In the first order of perturbation theory we obtain

$$c_1^2 \approx b(M_A^2 - 1)(m + m^{-1}) + 1. \tag{2.212}$$

This solution describes the unstable mode ($c_1^2 < 0$) if

$$M_A < M_{Ac} = \sqrt{1 - \frac{m}{b(m^2 + 1)}}.$$

Thus, it can be seen that, in the $k_z\rho_m \to 0$ limit, the instability growth rate (c_1) does not depend on $k_z\rho_m$. The instability region is bounded by the condition $0 < M_A < M_{Ac}$, which is consistent with the features of unstable global modes with $m \neq 0$ listed above in the items 1–2.

The solar wind is a transparent region for the global mode $m = 0$, in the same limiting case, $k_z\rho_m \ll 1$. The following dispersion equation is obtained from the matching condition (2.196):

$$i\frac{(c - M_A)^2/\epsilon^2 - 1}{\sqrt{k_{\rho w}^2/k_z^2}} = \frac{b(c^2 - 1)}{k_z\rho_m} \left[(k_z\rho_m)^2 - \frac{k_z^2}{k_{\rho m}^2} \right]. \tag{2.213}$$

As before, we have $c_0 = M_A$ in the zeroth order of perturbation theory. In the first order, we obtain the following approximate expression for c_1:

$$c_1 \approx i\frac{b}{k_z\rho_m} \frac{(M_A + 1)(M_A^2 - M_0^2)}{(M_A - \beta_m^*)\sqrt{1 + \beta_w^*}},$$

where $M_0^2 = \beta_m^*/(1 + \beta_m^*)$.

It is easily verifiable that the mode in question is unstable ($\mathrm{Im}(c_1) > 0$) throughout the M_A variation range, except for a narrow interval, $M_0 < M_A < \beta_m^*$. This is consistent with the features of unstable global modes with $m = 0$ listed above in the items 3 and 4. Note that the thus obtained solutions should be treated as merely illustrative of the qualitative behaviour of the oscillation global modes. The precise values of their growth rate found numerically may differ considerably from these simplified estimates.

Figure 2.38 shows a sample dependence of the growth rate of azimuthal harmonics with $m = 0$ and $m = 1$ on the unstable global mode frequency for various speeds of the solar wind flow around the magnetosphere. The calculations relied on numerically integrating (2.210) as the parameter $k_z\rho_m$ varied $0 < k_z\rho_m < 8$. What stands out is both the difference in the qualitative behaviour of the oscillation harmonics, and their absolute values – the harmonic $m = 1$ growth rates are many times larger than those for the $m = 0$ harmonic. When the maximum speed of the enwrapping flow is $v_0 = 800$ km/s, the

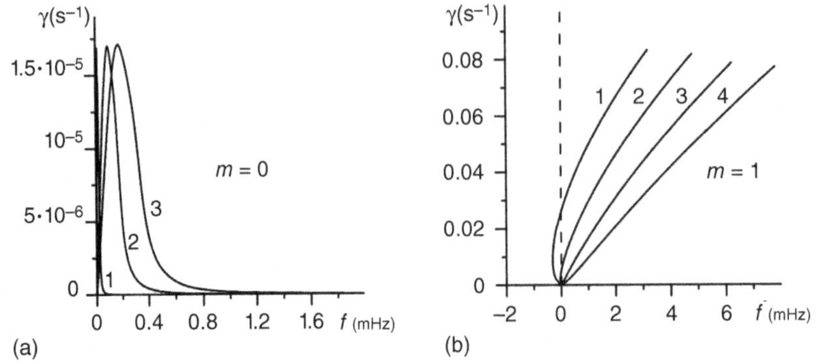

Figure 2.38 The dependence of the growth rate $\gamma \equiv \text{Im}\omega$ of unstable azimuthal harmonics $m = 0$ and $m = 1$ on the global mode frequency $f = \text{Re}(\omega)/2\pi$ for various speeds of the solar wind flow around the magnetosphere: *1* - $v_0 = 200$ km/s, *2* - $v_0 = 400$ km/s, *3* - $v_0 = 600$ km/s, *4* - $v_0 = 800$ km/s.

harmonic $m = 0$ becomes fully stable. Negative values of the frequency $f = \text{Re}(\omega)/2\pi < 0$ correspond to unstable waves the z-component of the phase velocity of which $\text{Re}(\omega)/k_z$ is directed against the solar wind plasma flow around the magnetosphere, while positive $f = \text{Re}(\omega)/2\pi > 0$ refer to waves travelling down the stream.

It is easy to infer that, in the $k_z\rho_m \to 0$ limit, the above-discussed 'global modes' of the geotail unstable oscillations correspond to oscillations with $\mathbf{k}_t \perp \mathbf{B}_0$ discussed in Section 2.15. As we observed, such oscillations are driven in the hydrodynamic regime, where there is no lower threshold for their instability in the speed of the solar wind flowing around the magnetosphere. In the $k_z = 0$ limit, resonance surfaces for Alfvén and SMS waves disappear for such oscillations and they are not absorbed by the background plasma. These oscillations propagate chiefly along the azimuthal direction as surface waves, while the entire geotail cross-section can be regarded as a 'thin' layer for them (see Figure 2.32(a)). Unstable 'global' modes can serve as a permanent source exciting oscillations in the largest magnetospheric FMS resonator located in the near-Earth part of the geotail plasma sheet (see Section 3.6.3).

2.16 Kelvin–Helmholtz Instability in the Geotail Low-Latitude Boundary Layer

In this section, we consider the Kelvin–Helmholtz instability in the low-latitude boundary layer (LLBL) of the geotail. This study relies on a cylindrical model of the geotail, with a smooth boundary enwrapped by a helical solar wind flow. In Section 2.15, we chose a geotail model with a magnetic field directed along the plasma cylinder axis. This model is suitable for most of the geotail lobes. However it is incorrect for the LLBL, where the magnetic field direction is almost perpendicular to the geotail axis. In this section, we use another cylindrical geotail model where the magnetic field is directed azimuthally.

The system of ideal MHD equations describing the plasma dynamics in an inhomogeneous medium has the form

$$\frac{d}{dt}\bar{\rho}\bar{\mathbf{v}} = -\nabla\bar{P} - \frac{1}{4\pi}\bar{\mathbf{B}} \times [\nabla \times \bar{\mathbf{B}}], \tag{2.214}$$

$$\frac{\partial}{\partial t}\overline{\mathbf{B}} = \nabla \times [\overline{\mathbf{v}} \times \overline{\mathbf{B}}], \quad (2.215)$$

$$\frac{d}{dt}\overline{\rho} + \nabla(\overline{\rho}\,\overline{\mathbf{v}}) = 0, \quad (2.216)$$

$$\frac{d}{dt}\frac{\overline{P}}{\overline{\rho}^{\overline{\gamma}}} = 0, \quad (2.217)$$

$$\overline{\mathbf{j}} = \frac{\overline{c}}{4\pi}[\nabla \times \overline{\mathbf{B}}], \quad (2.218)$$

where $d/dt = \partial/\partial t + (\overline{\mathbf{v}} \cdot \nabla)$ is the Lagrangian derivative, $\overline{\mathbf{j}}$, $\overline{\mathbf{B}}$ and $\overline{\mathbf{v}}$ are full current, magnetic field and plasma velocity vectors (including the background plasma parameters and disturbances). $\overline{P}$ and $\overline{\rho}$ are the plasma pressure and density, $\overline{\gamma}$ is the adiabatic index, $\overline{c}$ is the velocity of light in vacuum.

2.16.1 Cylindrical Model of the Geotail Near LLBL

The LLBL area has the form of a 'belt' encircling the magnetopause within $\pm(3 \div 5)\ R_E$ from the equatorial surface (see Figure 2.39(a)). One of its features is the fact that the inclination angle of the geomagnetic field lines to the direction of the solar wind velocity is maximum here. The geotail LLBL is where the cross-tail current spreads out and transforms into magnetopause currents.

We use the cylindrical geotail model in Figure 2.39(b) to describe the distribution of the medium parameters in the LLBL. We will use the cylindrical coordinate system (ρ, ϕ, z), in which the z axis coincides with the plasma cylinder axis, the magnetic field $\mathbf{B}_0 = (0, B_{0\phi} \equiv B_0, 0)$ is along the azimuthal ϕ-coordinate and its strength, like the other model parameters, varies over the radial ρ-coordinate. Hereafter, the subscripts $_0$ denote the values of background medium parameters.

We will find the distribution of the background plasma and magnetic field equilibrium parameters from a steady-state system of Eqs. (2.214) – (2.218) where all time derivatives are zero. Differences in the plasma concentration between the geotail current sheet ($n_0 \sim 1 \div 3\ \text{cm}^{-3}$) and the solar wind ($n_0 \sim 3 \div 10\ \ \text{cm}^{-3}$) regions adjacent to the LLBL

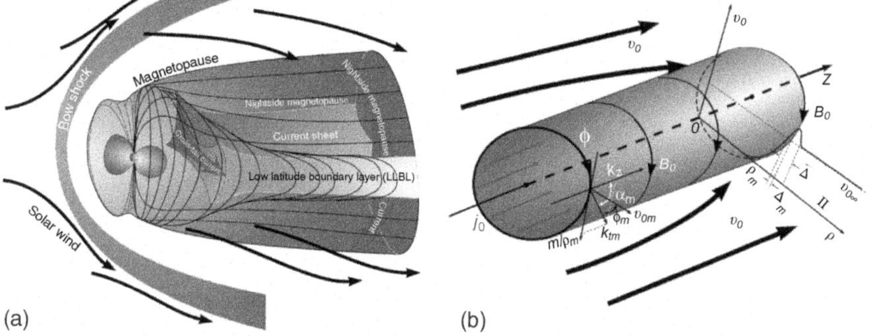

Figure 2.39 The low-latitude boundary layer (LLBL): (a) schematic of the magnetic field lines and the electric current configuration in the geotail LLBL, (b) cylindrical model of the geotail enwrapped by a helical plasma flow. Here: I is magnetosphere, II is solar wind.

are small enough [163, 164]. Therefore, to simplify further calculations, we will assume $n_0 = const$. The geotail current sheet is simulated by current along the plasma cylinder axis. In the medium model in question we have $\mathbf{j}_0 = (0, 0, j_{0z})$ in the background plasma. To describe the radial distribution of the longitudinal current density we use this expression

$$j_{0z}(\rho) = \frac{\bar{j}_0}{2}\left[1 + \tanh\frac{\rho_m^2 - \rho^2}{\Delta_m^2}\right], \tag{2.219}$$

where $\bar{j}_0 = const$, ρ_m is the characteristic radius of the magnetosphere, Δ_m is the typical LLBL thickness. Here and below, the subscripts $_m$ denote the values of background plasma parameters at the conditional mean magnetosphere boundary.

We use the z-component of the vector equation (2.218) to obtain

$$\begin{aligned} B_0(\rho) &= \frac{2\pi\bar{j}_0}{\bar{c}\rho}\int_0^\rho \rho'\left(1 + \tanh\frac{\rho_m^2 - \rho'^2}{\Delta_m^2}\right)d\rho' \\ &= \frac{\pi\bar{j}_0}{\bar{c}\rho}\left[\rho^2 - \Delta_m^2 \ln\frac{\cosh(\rho^2 - \rho_m^2)/\Delta_m^2}{\cosh(\rho_m^2/\Delta_m^2)}\right] \end{aligned} \tag{2.220}$$

for the background magnetic field strength: $\mathbf{B}_0 = (0, B_0, 0)$.

When $\Delta_m \to 0$ (2.220) transforms into

$$B_0 = \frac{2\pi}{\bar{c}}\bar{j}_0\begin{cases} \rho, & \text{for } 0 \le \rho < \rho_m, \\ \rho_m^2/\rho, & \text{for } \rho \ge \rho_m, \end{cases} \tag{2.221}$$

the expression obtained in [157] for a geotail model with a boundary in the form of a tangential discontinuity. Specifying density ρ_0 allows the radial distribution of the Alfvén speed to be determined:

$$v_A(\rho) = B_0/\sqrt{4\pi\rho_0} = v_{Am}\sqrt{a_A(\rho)}, \tag{2.222}$$

where

$$a_A(\rho) = \frac{1}{(2\rho\,\rho_m)^2}\left[\rho^2 - \Delta_m^2 \ln\frac{\cosh(\rho^2 - \rho_m^2)/\Delta_m^2}{\cosh(\rho_m^2/\Delta_m^2)}\right]^2,$$

and $v_{Am} = (\bar{j}_0\rho_m/\bar{c})\sqrt{\pi/\rho_0}$ is the Alfvén speed at the magnetopause.

To find the equilibrium plasma pressure it is necessary to set the plasma motion velocity. Let us assume that this velocity has only two components: $\mathbf{v}_0 = (0, v_{0\phi}, v_{0z})$. If we introduce the angular velocity of the plasma rotation $\Omega(\rho)$, then $v_{0\phi} = \rho\,\Omega$. Let us define the radial distribution of Ω and the velocity component v_{0z} by the following model expressions:

$$\Omega(\rho) = \frac{\overline{\Omega}}{2\sqrt{1 + (\rho/\overline{\Delta})^4}}\left(1 - \tanh\frac{\rho_m^2 - \rho^2}{\Delta_m^2}\right),\quad v_{0z}(\rho) = \frac{v_{0\infty}}{2}\left(1 - \tanh\frac{\rho_m^2 - \rho^2}{\Delta_m^2}\right), \tag{2.223}$$

which, for $\Delta_m \to 0$, transform into

$$\Omega(\rho) = \overline{\Omega}\begin{cases} 0, & \rho < \rho_m, \\ 1/\sqrt{1 + (\rho/\overline{\Delta})^4}, & \rho \ge \rho_m, \end{cases} \quad \text{and} \quad v_{0z}(\rho) = v_{0\infty}\begin{cases} 0, & \rho < \rho_m, \\ 1, & \rho \ge \rho_m, \end{cases} \tag{2.224}$$

respectively. Here $\overline{\Delta}$ is the characteristic magnetosheath thickness, $\overline{\Omega}$ is the characteristic angular velocity at the magnetopause, and $v_{0\infty}$ is the value of the velocity z– component at infinity. We define $\overline{\Omega}$ as follows. Let us set the flow helicity index at the magnetopause as $S_m = v_{0\phi,m}/v_{0\infty}$ (see Figure 2.39(b)). Given $v_{0\phi,m} = \rho_m \Omega(\rho_m)$, we then have $\overline{\Omega} = S_m v_{0\infty}\sqrt{1+(\rho_m/\overline{\Delta})^4}/\rho_m$.

Let us now determine the equilibrium pressure distribution of the background plasma $P_0(\rho)$. It is easy to verify that both the ϕ – and z – components of the vector equation (2.214) are identically equal to zero. Using the ρ – component of that equation we get the equilibrium condition for the background plasma:

$$\frac{d}{d\rho}\left(P_0 + \frac{B_0^2}{8\pi}\right) = \rho_0 \Omega^2 \rho - \frac{B_0^2}{4\pi\rho}. \tag{2.225}$$

Integrating (2.225) from ρ to ∞ produces

$$P_0(\rho) = P_0(\infty) - \frac{B_0^2(\rho)}{8\pi} + \frac{1}{4\pi}\int_\rho^\infty \frac{B_0^2(\rho')}{\rho'}\,d\rho' - \rho_0 \int_\rho^\infty \rho'\Omega^2(\rho')\,d\rho'. \tag{2.226}$$

Here $B_0(\infty) = 0$ was taken into account. The integrands in (2.226) are quite complex and the corresponding integrals can be calculated numerically. However, one can simulate the results of such an integration. Using (2.221) and (2.224) for $B_0(\rho)$ and $\Omega(\rho)$ and integrating in (2.226), we get

$$P_0(\rho) = \begin{cases} P_{0m} + \dfrac{B_{0m}^2}{4\pi}\left(1 - \dfrac{\rho^2}{\rho_m^2}\right), & \rho < \rho_m, \\[2ex] P_0(\infty) - \dfrac{\rho_0}{2}(\overline{\Omega\Delta})^2\left[\dfrac{\pi}{2} - \arctan\left(\dfrac{\rho}{\overline{\Delta}}\right)^2\right], & \rho \geq \rho_m, \end{cases} \tag{2.227}$$

where the plasma pressure at infinity is related to the pressure on the magnetopause P_{0m} by

$$P_0(\infty) = P_{0m} + \frac{\rho_0}{2}(\overline{\Omega\Delta})^2\left[\frac{\pi}{2} - \arctan\left(\frac{\rho_m}{\overline{\Delta}}\right)^2\right].$$

(2.226) can be used to set, by analogy with (2.219) and (2.223), the following model for $P_0(\rho)$

$$P_0(\rho) = P_{0m} + \frac{\rho_0}{4}(\overline{\Omega\Delta})^2\left[1 + \tanh\frac{\rho^2 - \rho_m^2}{\Delta_m^2}\right]\arctan\frac{\overline{\Delta}^{-2}(\rho^2 - \rho_m^2)}{\overline{\Delta}^{-4} + \rho^2\rho_m^2} + \frac{B_{0m}^2}{8\pi}\left(1 - \frac{\rho^2}{\rho_m^2}\right)\left[1 - \tanh\frac{\rho^2 - \rho_m^2}{\Delta_m^2}\right].$$

This model expression is in very good agreement with the exact numerical integration results.

The radial distribution of the sound velocity is defined as

$$v_s(\rho) = \sqrt{\gamma P_0/\rho_0} = v_{sm}\sqrt{a_s(\rho)}, \tag{2.228}$$

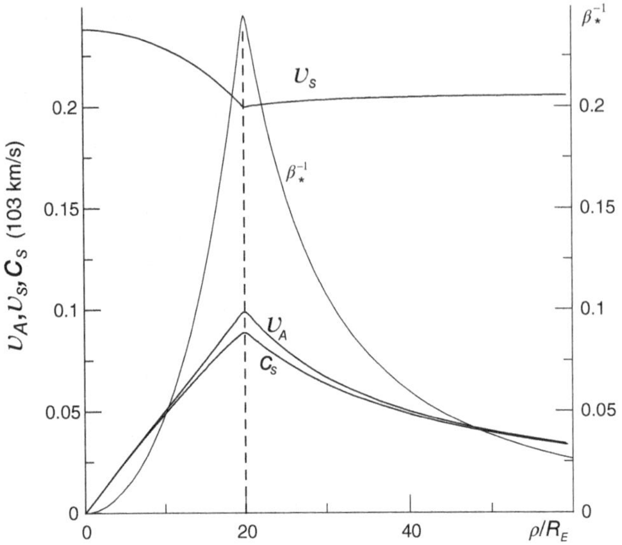

Figure 2.40 Radial distribution of Alfvén speed v_A, sound speed v_s, SMS speed c_s (left vertical axis) and the β_*^{-1} parameter (right vertical axis) in the equilibrium geotail model for $v_{0z} = 400$ km/s and $S_m = 0.2$. Here (see Fig. 2.39(b)): **I** – magnetosphere, **II** – solar wind.

where

$$a_s(\rho) = 1 + \frac{(\overline{\Omega\Delta})^2}{4v_{sm}^2}\left[1 + \tanh\frac{\rho^2 - \rho_m^2}{\Delta_m^2}\right]\arctan\frac{\overline{\Delta}^{-2}(\rho^2 - \rho_m^2)}{\overline{\Delta}^{-4} + \rho^2\rho_m^2}$$
$$+\frac{\overline{\gamma}}{2\beta_m^*}\left(1 - \frac{\rho^2}{\rho_m^2}\right)\left[1 - \tanh\frac{\rho^2 - \rho_m^2}{\Delta_m^2}\right],$$

$v_{sm} = \sqrt{\overline{\gamma}P_{0m}/\rho_0}$ is the sound velocity at the magnetopause, and $\beta_{*m} = v_{sm}^2/v_{Am}^2$ is a dimensionless parameter characterising the background plasma pressure to magnetic pressure ratio. This value coincides, to within a factor of $\overline{\gamma}/2 \approx 1$, with the plasma parameter $\beta_m = 8\pi P_{0m}/B_{0m}^2$.

To describe the medium parameters in the geotail LLBL the following constants are used in the numerical calculations: $\rho_m = 20R_E$ (where $R_E = 6370$ km is the Earth's radius); $\Delta_m = 1.5R_E$, $\overline{\Delta} = 5R_E$; $v_{Am} = 100$ km/s; $v_{sm} = 200$ km/s. The flow velocity v_{0z} varied from 200 km/s to 1000 km/s, and the flow helicity index S_m from 0 to 1. Figure 2.40 shows the model distributions (for $v_{0z} = 400$ km/s and $S_m = 0.2$): the Alfvén speed (v_A), the sound speed (v_s), the SMS wave velocity $\left(c_s = v_A v_s/\sqrt{v_A^2 + v_s^2}\right)$ and the parameter β_*^{-1}. Note that the maxima in the radial distribution of the Alfvén speed, the SMS wave velocity and the β_*^{-1} parameter are reached at the magnetopause, while the maximum in the sound velocity distribution, on the plasma cylinder axis. Similar behaviour of the medium parameters is observed in the real current sheet region of the geotail [164, 165].

2.16.2 Basic Equation and Boundary Conditions for MHD Oscillations in the Cylindrical Coordinate System

To describe the MHD oscillation structure in the geotail model in question we linearise the system of Eqs. (2.214) – (2.217) with respect to small disturbances of the medium parameters. To do this, we use the following representations of the parameters: $\overline{\mathbf{B}} = \mathbf{B}_0 + \mathbf{B}$, $\overline{\mathbf{v}} = \mathbf{v}_0 + \mathbf{v}$, $\overline{P} = P_0 + P$ and $\overline{\rho} = \rho_0 + \tilde{\rho}$, where the second terms in each of these expressions describe small disturbances related to the MHD oscillations. The system of linearised equations has the form

$$\rho_0 \left[\frac{\partial}{\partial t}\mathbf{v} + (\mathbf{v}_0\nabla)\mathbf{v} + (\mathbf{v}\nabla)\mathbf{v}_0\right] + \tilde{\rho}(\mathbf{v}_0\nabla)\mathbf{v}_0$$
$$-\nabla P - \frac{1}{4\pi}\left(\mathbf{B}_0 \times [\nabla \times \mathbf{B}] + \mathbf{B} \times [\nabla \times \mathbf{B}_0]\right), \tag{2.229}$$

$$\frac{\partial}{\partial t}\mathbf{B} = \nabla \times [\mathbf{v}_0 \times \mathbf{B}] + \nabla \times [\mathbf{v} \times \mathbf{B}_0], \tag{2.230}$$

$$\frac{\partial}{\partial t}\tilde{\rho} + \nabla(\tilde{\rho}\mathbf{v}_0) + \rho_0 \,\mathrm{div}\,\mathbf{v} = 0, \tag{2.231}$$

$$\frac{\partial}{\partial t}P + (\mathbf{v}_0\nabla)P + (\mathbf{v}\nabla)P_0 + \gamma P_0 \,\mathrm{div}\,\mathbf{v} = 0. \tag{2.232}$$

Here $\rho_0 = const$ and $\mathrm{div}\,\mathbf{v}_0 = 0$ are taken into account. Assuming the medium to be homogeneous over the ϕ and z coordinates we will seek the solutions of (2.229) – (2.232) as an expansion in harmonics of the form $\exp(-i\omega t + im\phi + ik_z z)$, where ω is the oscillation frequency, $m = \pm 0, 1, 2\ldots$ and k_z are the wave numbers in the ϕ and z coordinates, respectively. The following expressions for the components of the MHD oscillation wave field are obtained in the Appendix R:

$$v_\phi = -i\frac{m}{\rho}\frac{\Omega_A^2 v_s^2}{\overline{\omega}^2\Omega_d^2}\frac{1}{\rho}\frac{\partial}{\partial\rho}\rho v_\rho \tag{2.233}$$
$$-i\left[\frac{m\Omega^2}{\overline{\omega}^2} + \frac{\rho\Omega' + 2\Omega}{\overline{\omega}} + \frac{k_z m v_s^2}{\rho\overline{\omega}^2}\frac{\Omega_c^2}{\Omega_d^2}\right]\frac{v_\rho}{\varkappa_s^2},$$

$$v_z = -i\frac{k_z}{\kappa_\rho^2}\frac{1}{\rho}\frac{\partial}{\partial\rho}\rho v_\rho - i\frac{\Omega_c^2}{\Omega_d^2}v_\rho, \tag{2.234}$$

$$B_\rho = -\frac{mB_0}{\rho\overline{\omega}}v_\rho, \tag{2.235}$$

$$B_\phi = -i\frac{B_0}{\overline{\omega}}\left[\varkappa_s^2\frac{\Omega_A^2}{\Omega_d^2}\frac{1}{\rho}\frac{\partial}{\partial\rho}\rho v_\rho + \left(k_z\frac{\Omega_c^2}{\Omega_d^2} + \frac{m\Omega'}{\overline{\omega}} + \left(\ln\frac{B_0}{\rho}\right)'\right)v_\rho\right], \tag{2.236}$$

$$B_z = i\frac{mB_0}{\rho\overline{\omega}}\left[\frac{k_z}{\kappa_\rho^2}\frac{1}{\rho}\frac{\partial}{\partial\rho}\rho v_\rho + \left(\frac{\Omega_c^2}{\Omega_d^2} - \frac{v_{0z}'}{\overline{\omega}}\right)v_\rho\right], \tag{2.237}$$

$$P = -i\frac{\rho_0\overline{\omega}}{k_z\Omega_d^2}\left[\frac{k_z\Omega_A^2 v_s^2}{\overline{\omega}^2}\frac{1}{\rho}\frac{\partial}{\partial\rho}\rho v_\rho + \Omega_b^2 v_\rho\right], \tag{2.238}$$

where $\overline{\omega} = \omega - m\Omega - k_z v_{0z}$ is the Doppler-shifted oscillation frequency in a moving medium, the prime denotes the derivative with respect to ρ (e.g. $\Omega' = \nabla_\rho\Omega$), and the following notations are used:

$$\Omega_A^2 = \overline{\omega}^2 - \frac{m^2}{\rho^2} v_A^2, \tag{2.239}$$

$$\Omega_b^2 = \Omega_c^2 \left(1 - \frac{k_t^2 v_A^2}{\overline{\omega}^2}\right) - \frac{\Omega_d^2}{\overline{\omega}^2}\left[\frac{\Omega_A^2 v'_{0z}}{\overline{\omega}} + k_z v_A^2 \left(\left(\ln\frac{B_0}{\rho}\right)' + \frac{m\Omega'}{\overline{\omega}}\right)\right], \tag{2.240}$$

$$\Omega_c^2 = \varkappa_s^2 \Omega_A^2 \frac{v'_{0z}}{\overline{\omega}} + k_z \left[\frac{v_s^2}{\gamma}(\ln P_0)'\right. \tag{2.241}$$

$$\left. + \frac{m}{\overline{\omega}}\left(\Omega'(v_s^2 + \varkappa_s^2 v_A^2) + 2v_s^2 \frac{\Omega}{\rho}\right) + v_A^2 \left(\left(\ln\frac{B_0}{\rho}\right)' + 2\frac{m^2 v_s^2}{\rho^3 \overline{\omega}^2}\right)\right],$$

$$\Omega_d^2 = \kappa_\rho^2 (v_s^2 + \varkappa_s^2 v_A^2), \tag{2.242}$$

$$\kappa_\rho^2 = \frac{\overline{\omega}^2}{(v_s^2 + \varkappa_s^2 v_A^2)} - k_t^2, \quad k_t^2 = \frac{m^2}{\rho^2} + k_z^2, \tag{2.243}$$

$$\varkappa_s^2 = 1 - \frac{m^2 v_s^2}{\rho^2 \overline{\omega}^2}. \tag{2.244}$$

Here, the coefficients (2.239)–(2.242) have the dimension of frequency squared, k_t is the tangential wave number, κ_ρ is the ρ component of the wave vector in the WKB approximation as $\rho \to \infty$ (when the components of the wave field can be represented as $\sim \exp(i \int \kappa_\rho d\rho)$), and $\varkappa_s$ is a dimensionless parameter.

If the radial component of the oscillation velocity is expressed in terms of $\zeta = i v_\rho / \overline{\omega}$ (the ζ parameter coincides with the plasma radial displacement in the wave when $\Omega = 0$: $v_\rho = d\zeta/dt$), then the equation of linear oscillations resulting from the equations system (2.229) – (2.232), with the expressions (2.235) substituted into it, has the form

$$\frac{\partial}{\partial\rho}\frac{\Omega_A^2}{\kappa_\rho^2 \rho}\frac{\partial \rho\zeta}{\partial\rho} + \left(\Omega_A^2 + \tilde{\Omega}^2\right)\zeta = 0, \tag{2.245}$$

where

$$\tilde{\Omega}^2 = \frac{2v_A^2}{\rho}(\ln \rho B_0)' - \frac{2}{\rho}\left(\frac{\Omega_A^2 v_A^2 \varkappa_s^2}{\Omega_d^2}\right)' + \frac{4k_z^2 v_A^2 v_s^2}{\rho^2 \Omega_d^2}\left(1 + \frac{2m\Omega}{\overline{\omega}}\right)$$
$$+\rho\left[\Omega^2\left(\frac{\Omega_A^2}{\Omega_d^2} - 1\right) + \frac{2m\,\Omega_A^2\,\Omega}{\rho^2\,\Omega_d^2\,\overline{\omega}} v_s^2\right]' \tag{2.246}$$
$$-4\frac{\Omega^2}{\Omega_d^2}\left[(\overline{\omega}^2 - k_t^2 v_A^2)\left(1 + \frac{m\Omega}{\overline{\omega}}\right) - k_z^2(v_A^2 + v_s^2) + \frac{\Omega^2\Omega_A^2}{4\overline{\omega}^2} k_t^2 \rho^2\right].$$

If the azimuthal velocity of the flow around the plasma cylinder is zero ($\Omega = 0$), then we have $\tilde{\Omega}^2 = 0$ as $\rho \to \infty$, and (2.245) transforms into a well-known equation (2.4) for plane-stratified flows (see also [120, 129]).

To find the solution of (2.245) the boundary conditions for ζ should be specified. When $\rho \to 0$, the natural requirement is that the oscillation amplitude should be finite. The solutions satisfying this requirement differ between symmetric ($m = 0$) and asymmetric ($m \neq 0$) modes. Equation (2.245) for symmetric modes, when $\rho \to 0$, can be written as

$$\frac{\partial}{\partial \rho}\frac{1}{\rho}\frac{\partial \rho \zeta}{\partial \rho} + \kappa_{\rho 0}^2 \zeta = 0,$$

where $\kappa_{\rho 0} = \kappa_\rho(\rho \to 0)$ is constant. Its solution, finite as $\rho \to 0$, is

$$\zeta = C_0 J_1(\kappa_{\rho 0}\rho) \approx C_0\, \kappa_{\rho 0}\rho, \tag{2.247}$$

where C_0 is an arbitrary constant, $J_1(x)$ is the Bessel function. For modes with $m \neq 0$, as $\rho \to 0$, (2.245) has the form

$$\frac{\partial}{\partial \rho}\rho\frac{\partial \rho \zeta}{\partial \rho} - m^2 \zeta = 0,$$

and its solution, finite as $\rho \to 0$, is

$$\zeta = C_0 \rho^{m-1}. \tag{2.248}$$

As $\rho \to \infty$, the boundary condition for unstable oscillation modes is the presence of a wave carrying the energy away from the shear layer in the absence of an incident wave. This condition expresses the causality principle: the wave source on the $\rho \to \infty$ asymptotic is the shear flow layer at the magnetopause. We especially underscore that this condition is true even for unstable surface waves because, thanks to the imaginary part of their frequency ($\text{Im}\,\omega > 0$), they also have a nonzero radial component of the group velocity, $v_{g\rho} = \text{Re}\,\partial\omega/\partial k_\rho$. That is, unstable surface waves, like radiative modes, transfer energy in the radial direction, in contrast to the neutrally stable surface waves (in which $\text{Im}\,\omega = 0$).

Assuming that the WKB approximation is applicable on the $\rho \to \infty$ asymptotic for describing the wave field, we choose the following asymptotical solution of (2.245),

$$\zeta = C_1\sqrt{\frac{\kappa_\rho}{\rho\Omega_A^2}}\exp\left(i\int_{\rho_m}^{\rho}\kappa_\rho\, d\rho'\right),$$

which ensures that the above requirement is fulfilled. Thus, the boundary condition as $\rho \to \infty$ can be represented as:

$$\frac{\partial \zeta}{\partial \rho} = i\kappa_\rho \zeta, \tag{2.249}$$

which also ensures that the amplitude of unstable oscillations decreases when they run away from the shear layer (see [120]).

2.16.3 Numerical Solution of the Basic Equation and Discussion

It is more convenient to use dimensionless parameters for numerically solving equation (2.245):

$$c = \frac{\omega}{k_{tm} v_{0m} \cos\phi_m} \tag{2.250}$$

– the ratio of the tangential component of the phase velocity of the wave at the magnetopause to the velocity of the flow around the magnetosphere as projected onto the tangential wave vector direction ($\mathbf{k}_t\mathbf{v}_{0m} = k_{tm}v_{0m}\cos\phi_m$, see Figure 2.39(b)) and

$$M = \frac{v_{0m}\cos\phi_m}{v_{sm}} \tag{2.251}$$

is the Mach number determined from the ratio of the same the flow velocity projection $v_{0m} \cos \phi_m$ to the sound speed at the magnetopause. Using the earlier-introduced flow helicity index at the magnetopause, we write

$$\cos \phi_m = \frac{k_z + mS_m/\rho_m}{\sqrt{(1+S_m^2)(k_z^2 + m^2/\rho_m^2)}}.$$

In this notation, we have, for example, the following expression for the coefficient Ω_A^2 in (2.245):

$$\Omega_A^2 = k_{tm}^2 v_{sm}^2 \left[M^2 (c - u(\rho))^2 - \frac{a_A(\rho)}{\beta_m} \frac{\rho_m^2}{\rho^2} \sin^2 \alpha_m \right],$$

where

$$u(\rho) = \frac{\mathbf{k}_t \cdot \mathbf{v}_0}{\mathbf{k}_{tm} \cdot \mathbf{v}_{0m}} \approx \frac{1}{2} \left[1 + \frac{\rho_m}{v_{sm}} \frac{\Omega(\rho) - \Omega(\rho_m)}{M} \sin \alpha_m \right] \left(1 - \tanh \frac{\rho_m^2 - \rho^2}{\Delta_m^2} \right)$$

is the profile of the velocity $\mathbf{v}_0$ projected on the $\mathbf{k}_t$ vector direction, $\sin \alpha_m = m/(k_{tm}\rho_m)$ (see Figure 2.39(b)), and the function $a_A(\rho)$ is defined by (2.222). The expressions for κ_ρ^2 and $\tilde{\Omega}^2$ are more cumbersome so we will not write them out explicitly here.

The numerical solutions of (2.245) with the boundary conditions (2.247), (2.248) are shown in Figures 2.41–2.44. Figure 2.41 shows the structures of various modes of unstable oscillations generated by the shear layer. Figure 2.41(a) shows the radial structures of the surface mode (curve 1) and of the radiative mode (curve 2). The same figure shows the real part distribution of the squared radial wave number in the WKB approximation, $\mathrm{Re}\, k_\rho^2(\rho) = \mathrm{Re}\, \kappa_\rho^2(\rho)(1 + \tilde{\Omega}^2/\Omega_A^2)$ (curve 3). Note the sharp change in k_ρ^2 in the shear layer characteristic of resonant oscillations.

The surface mode is an FMS wave for which the magnetospheric and the solar wind regions adjacent to the shear layer are opaque ($\mathrm{Re}\, k_\rho^2 < 0$). The radiative mode is an FMS

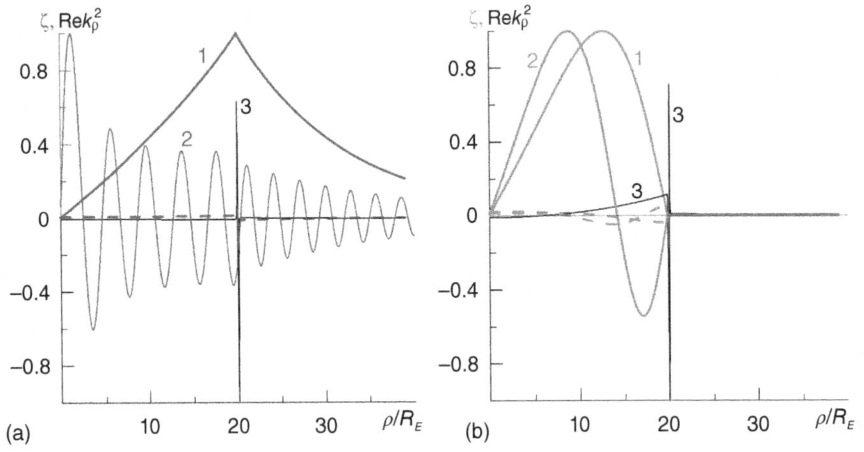

Figure 2.41 Radial structure of unstable MHD oscillations generated by the magnetopause shear flow: (a) – surface mode structure (1), radiative mode structure (2) ; (b) – structure of the first (1) and second (2) eigen-mode harmonics of the FMS waveguide in the geomagnetic tail. On both panels: solid lines are $\mathrm{Re}\,\zeta/(\mathrm{Re}\,\zeta)_{\max}$, dotted lines are $\mathrm{Im}\,\zeta/(\mathrm{Re}\,\zeta)_{\max}$ (in panel (a) for the surface mode only); black curves (3) are the squared wave number, $\mathrm{Re}\,k_\rho^2/(\mathrm{Re}\,k_\rho^2)_{\max}$.

wave for which the solar wind region adjacent to the shear layer is transparent ($\mathrm{Re}\,k_\rho^2 > 0$). Figure 2.41(a) gives an example where the magnetosphere is also transparent for the radiative mode. Figure 2.41(b) shows the structures of the two first harmonics of the FMS waveguide existing in the geotail cylindrical model under study. The magnetospheric region adjacent to the shear layer is transparent (here, $\mathrm{Re}\,k_\rho^2 > 0$), while the solar wind is opaque ($\mathrm{Re}\,k_\rho^2 < 0$) for these waves.

Note that the amplitude of unstable FMS waves reaches its maximum: at the magnetopause, for a surface wave; near the axis of the plasma cylinder simulating the geomagnetic tail, for a radiative mode; and inside the plasma cylinder, for the waveguide eigenmodes. Such a spatial distribution of the amplitude of unstable oscillations can be explained simply enough.

The amplitude of a surface FMS wave rapidly decreases as it moves from the shear layer into the adjacent opaque regions. For the radiative mode in Figure 2.41(a), its only turning point (where the radial component of its phase velocity vanishes) is near the cylinder axis. The energy of unstable oscillations generated by the shear layer 'accumulates' near it. Here, the oscillations 'slow down', 'stop' and begin to slowly travel in the opposite direction. We could expect such a temporal behaviour of oscillations when solving the problem of an evolving initial perturbation amplified by the Kelvin–Helmholtz instability. In our, stationary, problem this manifests itself in increasing oscillation amplitude near the turning point.

The maximum amplitudes of the waveguide eigenmodes are located at the antinodes of their radial distribution. This can also be explained by the 'accumulation' of the energy of the unstable waveguide FMS waves generated by the shear layer. The solar wind is an opaque region for them and the shear layer is close to their turning points. In this case the eigenmodes are captured between the turning points located near the axis and near the outer boundary of the plasma cylinder, and their energy 'accumulates' in the waveguide.

Figure 2.42 gives examples of the growth rate distribution for the surface wave and for three first FMS-waveguide harmonics over k_z (rather, over dimensionless $k_z\Delta_m$). The azimuthal wave number for all these waves is $m = 1$, and the Mach number is $M = 1.4$. Note that the surface wave remains unstable even if its characteristic radial scales, determined by the k_z-component of the wave vector, are much less than Δ_m. This is the fundamental distinction between the magnetospheric models of finite cross-section and the models with Cartesian geometry, where the only characteristic scale is the shear layer thickness.

Recall that in plane-parallel shear flow models those oscillations are unstable whose characteristic transverse wavelengths are greater than the shear layer thickness [120, 129, 131]. Once the wavelength becomes comparable with, or smaller than, the shear layer thickness, the oscillations stabilise. The finite shear layer thickness of the cylindrical model at hand manifests itself in the fact that the growthrate maximum in the $\mathrm{Im}c$ distribution is on the $k_z\Delta_m \sim 1$ scale, whereas the width of the region of existence for unstable oscillations is much wider (over the $k_z\Delta_m$ parameter).

The eigen-mode growth rates of the FMS waveguide are much smaller than the surface wave growth rate. Their maximum is attained at $k_z\Delta_m = 0$, and their distribution is particular in that they asymptotically tend to zero as $k_z\Delta_m \to \infty$. Such behaviour for the eigen-mode growth rates of the magnetospheric FMS waveguide has been studied in detail in [152, 156] where a model with Cartesian geometry was used.

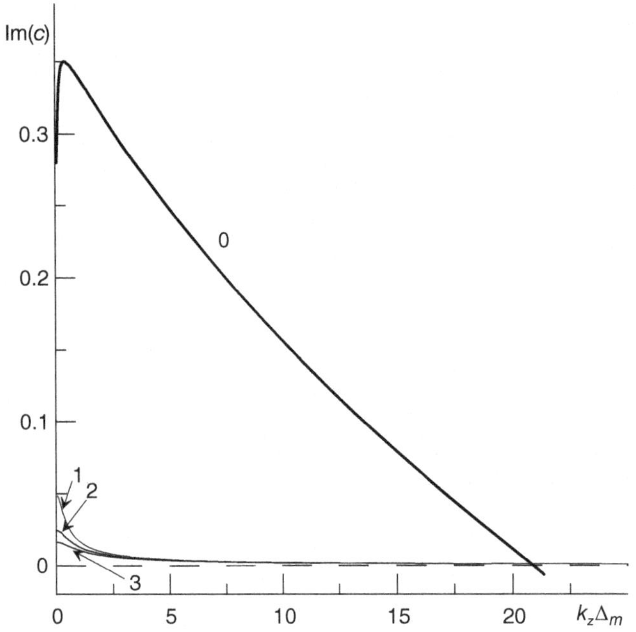

Figure 2.42 $k_z\Delta_m$ dependence of the oscillation growth rate Im (c) for the given azimuthal wave number $m = 1$ and the Mach number $M = 1.4$: (0) – for the surface mode, (1,2,3) – for three harmonics of the FMS waveguide in the geotail.

Figure 2.43 presents growth rate isoline maps (in the $(k_z\Delta_m, M)$ plane) for an MHD oscillation driven by solar wind shear flows at the magnetopause. Two limiting cases are considered: flows parallel to the z axis (for which $S_m = 0$), and strongly twisted flows (with $S_m = 1$). Calculations were made for two azimuthal harmonics, $m = 0, 1$. Here the restricted area is shown near the local maximum in the oscillation growth rate distribution ($0 \le k_z\Delta_m \le 5$, $0 \le M \le 5$). The left hand part of each plot shows the growth rate distribution for the surface and radiative modes. Each of these distributions has a local maximum at $M \approx 0.5$ and $0.5 \le k_z\Delta_m \le 1.5$. This maximum corresponds to the maximum for the surface wave growth rate. Indeed, the observed surface waves amplified by the Kelvin–Helmholtz instability more often than not exhibit much larger tangential wavelengths ($\sim 2 - 15\,R_E$, see [22, 166, 167]) than the typical magnetopause thickness in observations ($\sim 0.1 - 1R_E$, see [168, 169]).

When shifted right along the M axis, the structure of unstable oscillations is gradually transformed from surface-type waves into radiative modes. The instability of the harmonic $m = 1$ in the longitudinal shear flow ($S_m = 0$) has a peculiar feature: when $k_z\Delta_m \le 0.7$ the instability areas for the surface and radiative modes are separated by an area of stable oscillations. Quasi-equidistant 'islands' with a small growth rate exist in the right part of the plots corresponding to the magnetospheric waveguide eigen-modes. These 'islands' are also present in the area of unstable surface and radiative modes, but are not visible against the background distributions with a large growth rate. Note also that in strongly twisted shear flows ($S_m = 1$) the boundary between unstable and stable oscillations shifts to the area of large Mach numbers. The range of the maximum growth rate remains localised at the same value, $M \approx 0.5$, determined by the shear layer thickness.

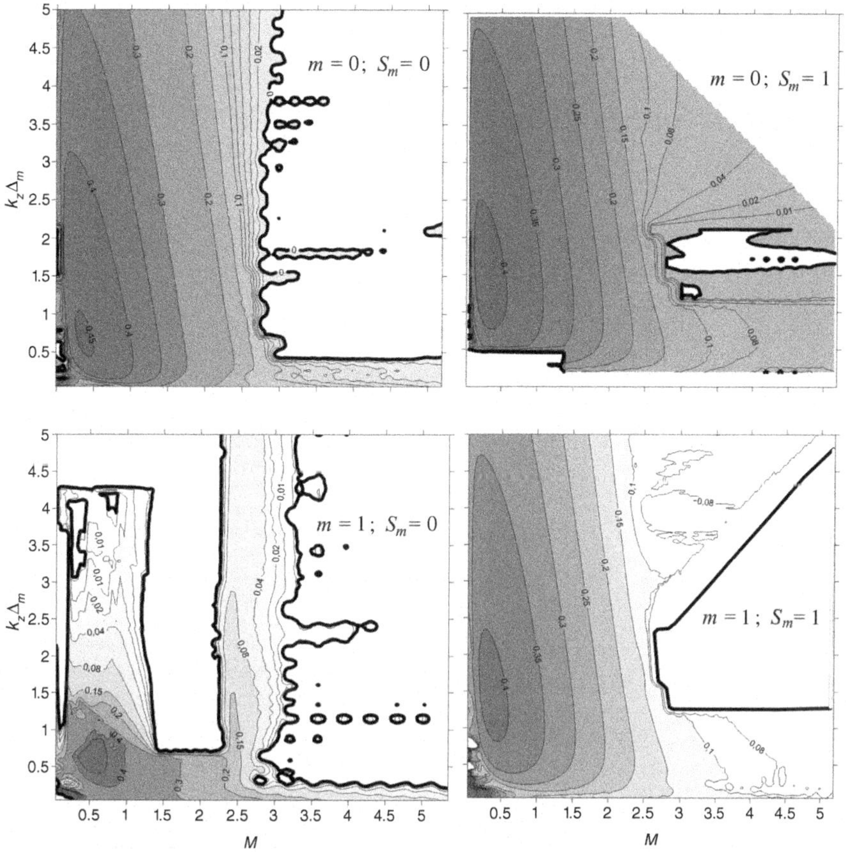

Figure 2.43 Growth rate (Im c) isoline maps for the basic $m = 0$ and first ($m = 1$) azimuthal harmonics of MHD oscillations in the $(M, k_z\Delta_m)$ plane in the longitudinal solar wind flows (left-hand panels, helicity index $S_m = 0$) and in the strongly twisted flows (right-hand panels, $S_m = 1$). The bold line depicts the boundary $\mathrm{Im}\,(c) = 0$ separating the areas of unstable ($\mathrm{Im}\,(c) > 0$) and stable ($\mathrm{Im}\,(c) < 0$) oscillations.

Let us consider the frequency ($f = \mathrm{Re}\,\omega/2\pi$) dependence of the unstable oscillation growth rate ($\gamma = \mathrm{Im}\,\omega$) in order to determine the spectrum of oscillations driven at the magnetopause by different solar wind flows. Such distributions for the first four azimuthal harmonics, $m = 0, 1, 2, 3$, of oscillations driven in the slow- ($v_m = 200$ km/s), medium- ($v_m = 400$ km/s) and high-speed ($v_m = 800$ km/s) flows are shown in the Figure 2.44 plots. Longitudinal ($S_m = 0$) and strongly twisted ($S_m = 1$) shear flows are considered. The unstable oscillation frequency variation is related to the wave vector k_z-component varying from $k_z = 0.01/\Delta_m$ to the limiting values, when the oscillation growth rate passes through zero and becomes negative.

Note that in the low- and medium-speed flows these distributions are practically independent of the azimuthal wave number m. Differences in the distributions of the longitudinal and strongly twisted flows are also minimal. The unstable oscillation frequencies are in the range $0 < f < 0.04$ Hz, corresponding to the Pc3 to Pc6 geomagnetic pulsation range. The growth rate maximum, $\gamma \approx 0.04\ \mathrm{s}^{-1}$, corresponds to the frequency $f \approx 0.02$ Hz (pulsations in the Pc4 range).

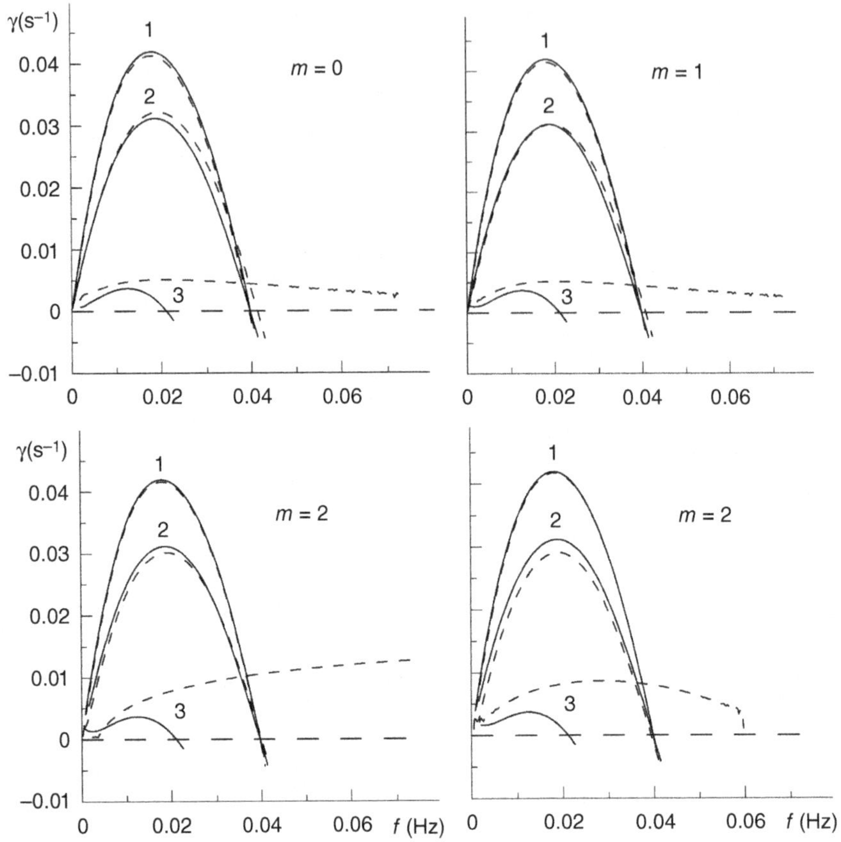

Figure 2.44 Dependence of the growth rate of unstable MHD oscillations $\text{Im}(\omega)$ on their frequency $f = \text{Re}\omega/2\pi$ (for azimuthal harmonics $m = 0, 1, 2, 3$) for different velocities of the solar wind flowing around the magnetosphere: $v_{0m} = 200$, 400, 800 km/s (curves 1,2,3). Solid lines depict the oscillations generated by a longitudinal flow (the helicity index $S_m = 0$), dashed lines stand for the oscillations generated by a strongly twisted flow ($S_m = 1$).

Distributions in the high-speed longitudinal flows are also practically identical and cover the Pc4–Pc6 geomagnetic pulsation range, their maximum localised at frequency $f \approx 0.01$ Hz (pulsations in the Pc5 range). The unstable oscillations growth rate in the high-speed flows is much smaller than in the slow- and medium-speed flows. This is due to the fact that only the radiative mode is unstable in high-speed flows, whereas surface waves with large growth rates are driven at the magnetopause in slower flows. This can explain the fact that global Pc5 pulsations associated with the MHD waveguide in the geomagnetic tail are observed only when the magnetosphere passes through high-speed solar wind flows [170].

The distributions in strongly twisted high-speed flows differ most. The unstable oscillation growth rate in such flows is much higher than in the longitudinal flows, and their frequency ranges are much wider. For the harmonics $m = 0, 1, 3$ the limiting values of the frequency reach $f \approx 0.1$ Hz, and for the $m = 2$ harmonic, $f \approx 1$ Hz. This means that the magnetopause is unstable in high-speed strongly twisted flows for oscillations of the entire geomagnetic pulsation range, Pc1 to Pc6.

Let us list the main results of the above.

1. It is shown that three types of unstable MHD waves exist in the geotail: (i) surface waves at the magnetopause, (ii) waves radiated into the solar wind, and (iii) eigen-modes of the waveguide within the geotail.
2. Unstable surface waves generated in the low- to medium-speed solar wind have the largest growth rate. They are driven in the Pc3–Pc6 geomagnetic pulsation range, and have a local maximum at Pc4 frequencies. In the cylindrical model in question the minimum transverse wavelengths of the unstable oscillations are much smaller than the shear layer thickness. This differentiates it from models with Cartesian geometry, where the minimum transverse wavelengths of unstable oscillations are larger than the boundary-layer thickness.
3. In high-speed solar wind flows the magnetopause is stable to surface waves, but unstable to radiative modes. The growth rate of such oscillations is an order of magnitude smaller than that for surface waves generated in the low- to medium-speed solar wind. If the helicity of high-speed solar wind flows enwrapping the magnetosphere is small the oscillations are unstable in the Pc4–Pc6 geomagnetic pulsation range, their maximum growth rate being in the Pc5 range.
4. If the helicity of high-speed solar wind flow enwrapping the magnetosphere is large enough, the oscillations are unstable throughout the Pc1–Pc6 geomagnetic pulsation range.

2.17 Cherenkov Radiation of the Fast Magnetoacoustic Waves

Impulses of the solar wind dynamic pressure are considered as one of the main sources of the ULF waves in the magnetosphere [171–175]. This impulse produces the fast magnetoacoustic mode propagating across the magnetic shells deep into the magnetosphere. Reaching the resonance shell, it generates the shear Alfvén mode. In theoretical studies, the case is usually considered when the impulse drops on the magnetosphere normal to the magnetopause [176–179]. However, it is more natural to suppose that the impulse drops onto the magnetopause on some finite angle θ [180]. In this case, the disturbance will run along the magnetopause with velocity $u = u_{sw}/\sin\theta$, where u_{sw} is the solar wind velocity. This velocity can be larger than the Alfvén speed inside the magnetosphere.

The source moving in medium with velocity larger than the wave's phase speed in this medium emits the wave which concentrated on the Mach cone expanding with the phase speed on the normal to its surface. This mechanism is called the Cherenkov emission. The Cherenkov emission of the fast magnetoacoustic mode in the homogeneous plasma was for the first time considered in [181].

In this section we consider the spatial structure of the fast magnetoacoustic mode generated by the super-Alfvénic source moving on the magnetopause [182] when the plasma inhomogeneity is taken into account. The one-dimensionally inhomogeneous model of the magnetosphere (box model) is adapted where the field lines are straight and directed along the z-axis. The mass density ρ_0 to vary only in the x direction. The x, y and z coordinates play the role of the radial, azimuthal and field-aligned coordinate in the magnetosphere, respectively (Figure 2.45). The magnetopause is assigned with the coordinate x_M. The Alfvén speed

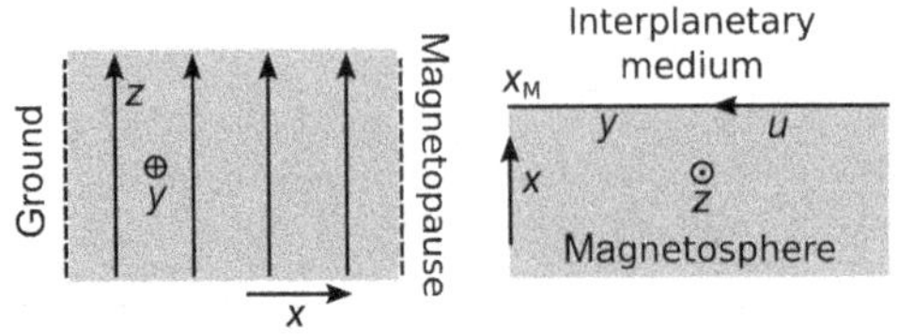

Figure 2.45 The box model of the magnetosphere.

v_A is supposed to decrease with the radial coordinate, as typical in the major part of the magnetosphere.

Along the magnetopause (in the y-direction) the current flows which bounds the magnetosphere and separates it from the interplanetary medium. The source is running in the same y-direction with the velocity u. The Mach number on the magnetopause is $M = u/v_{AM} > 1$ where v_{AM} is the Alfvén speed value on the magnetopause.

The moving source is identified with the indentation on the place where the solar wind imhomogeneity touches the magnetopause. Here the localised imbalance of pressure equilibrium appears accompanied by a change of the magnetopause current system [183]. It is this current which generates the wave propagating into the magnetosphere

2.17.1 The Single Fourier-Harmonic

First, let us consider the wave with given frequency ω, azimuthal k_y and parallel k_z wave vector. In the chosen coordinate system, the system of MHD equations yields the following single equation for the wave's $\tilde{B}_z(\omega)$ component:

$$\tilde{B}_z'' + \frac{(v_A^2)'}{v_A^2}\frac{\omega^2}{\omega^2 - k_z^2 v_A^2}\tilde{B}_z' + \left(\frac{\omega^2}{v_A^2} - k_y^2 - k_z^2\right)\tilde{B}_z = 0. \tag{2.252}$$

Here $v_A(x)$ is the Alfvén speed, the prime denotes the derivative with respect to the radial coordinate ($[...]' = d[...]/dx$).

Later on, we will assume that the parallel (z) component of the wave vector is small. Then, the fast mode's frequency is $\omega \sim k_\perp v_A$, where $k_\perp$ is the perpendicular component of the wave's vector. Owing to this, the separation between the fast mode localisation region and the Alfvén resonance (characterised by the condition $\omega^2 = k_z^2 v_A^2(x)$) is large, and the Alfvén resonance exerts only little influence on the fast mode. In this case, at large distance from the Alfvén resonance, Eq. (2.252) is written in the form

$$\tilde{B}_z'' + \frac{(v_A^2)'}{v_A^2}\tilde{B}_z' + \left(\frac{\omega^2}{v_A^2} - k_y^2 - k_z^2\right)\tilde{B}_z = 0. \tag{2.253}$$

To find the boundary condition on the magnetopause (x_M), we will assume that the wave vanishes in the interplanetary medium. Then, in the presence of the point-like moving source, the boundary condition can be written in the form

$$B_z(y,t) = \frac{4\pi}{c} I_0 \delta(y - ut) e^{ik_z z}, \tag{2.254}$$

or, for the Fourier component,

$$\tilde{B}_z = \frac{2I_0}{c}\delta(\omega - k_y u) e^{ik_z z}. \tag{2.255}$$

The second boundary condition is decay at $x \to -\infty$.

Equation (2.253) can be solved in the WKB approximation. The wave vector's radial component is

$$k_x(x) = \sqrt{\frac{\omega^2}{v_A^2(x)} - k_y^2 - k_z^2}. \tag{2.256}$$

In the transparent region, $k_x^2 > 0$. The boundary of the transparent region where $k_x = 0$ is denoted x_0.

The solution of Eq. (2.253) with boundary condition (2.255) is

$$\tilde{B}_z(x, k_y, \omega, z) = A\delta(\omega - k_y u)\frac{\sin(\psi_0 - \psi + \frac{\pi}{4})}{\sin(\psi_0 + \frac{\pi}{4})} \tag{2.257}$$

where

$$A = \frac{2I_0}{c}\sqrt{\frac{k_{xM}}{k_x}\frac{v_{AM}}{v_A}}e^{ik_z z}, \tag{2.258}$$

$$\psi = \int_x^{x_M} k_x dx', \tag{2.259}$$

ψ_0 is the value on the reflection shell x_0. Taking into account that k_z is considered small, we can rewrite the latter equation in a simpler form: $\psi = |k_y|\xi$ where

$$\xi = \int_x^{x_M} dx' \sqrt{\frac{u^2}{v_A^2} - 1}. \tag{2.260}$$

2.17.2 Summation of the Fourier-Harmonics

Let us now perform the inverse Fourier-transform of (2.257):

$$B_z(x, y, t) = \int_{-\infty}^{\infty} dk_y e^{ik_y y} \int_{-\infty}^{\infty} d\omega e^{-i\omega t}\tilde{B}_z(x, k_y, \omega). \tag{2.261}$$

The result is

$$B_z = \pi A(I_1 + I_2 + I_3 + I_4) \tag{2.262}$$

where

$$\begin{aligned}
I_1 &= \Theta(\tau + \xi_0 - \xi)2\sum_{N=0}^{\infty}\delta\left(\frac{\tau - \xi}{\xi_0} - 2N\right)\cos\left(\frac{\pi}{4}\frac{\tau - \xi}{\xi_0}\right),\\
I_2 &= -\Theta(\tau - \xi_0 + \xi)\left[\cot\left(\frac{\pi}{2}\frac{\tau + \xi}{\xi_0}\right)\sin\left(\frac{\pi}{4}\frac{\tau + \xi}{\xi_0}\right) - \cos\left(\frac{\pi}{4}\frac{\tau + \xi}{\xi_0}\right)\right],\\
I_3 &= \Theta(\tau + \xi_0 - \xi)\left[\cot\left(\frac{\pi}{2}\frac{\tau - \xi}{\xi_0}\right)\sin\left(\frac{\pi}{4}\frac{\tau - \xi}{\xi_0}\right) - \cos\left(\frac{\pi}{4}\frac{\tau - \xi}{\xi_0}\right)\right],\\
I_4 &= -\Theta(\tau - \xi_0 + \xi)2\sum_{N=0}^{\infty}\delta\left(\frac{\tau + \xi}{\xi_0} - 2N\right)\cos\left(\frac{\pi}{4}\frac{\tau + \xi}{\xi_0}\right).
\end{aligned} \tag{2.263}$$

This solution is depicted on Figure 2.46. The spatial variation of the Alfvén speed is chosen in the form

$$\frac{1}{v_A^2(x)} = \frac{1}{v_{AM}^2}\left(1 + \frac{x - x_M}{l}\right) \tag{2.264}$$

where l is characteristic inhomogeneity scale.

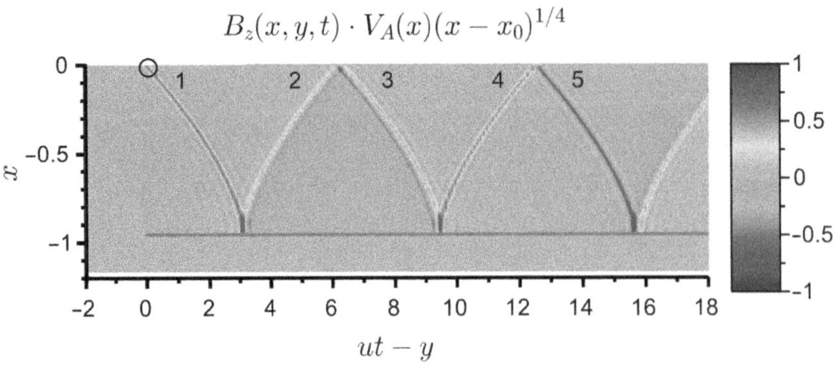

Figure 2.46 The spatial structure of the fast mode wave field generated by the Cherenkov mechanism in the source reference frame (normalised amplitude). The parameters chosen in Eq. (2.264) are $l = 1$, $x_M = 0$. The Mach number on the magnetopause is chosen $M = 5$. The sum of the first 20 harmonics is depicted. The source is denoted by the circle on the top left, the grey line denotes the general reflection surface. The numbers denote different wave branches.

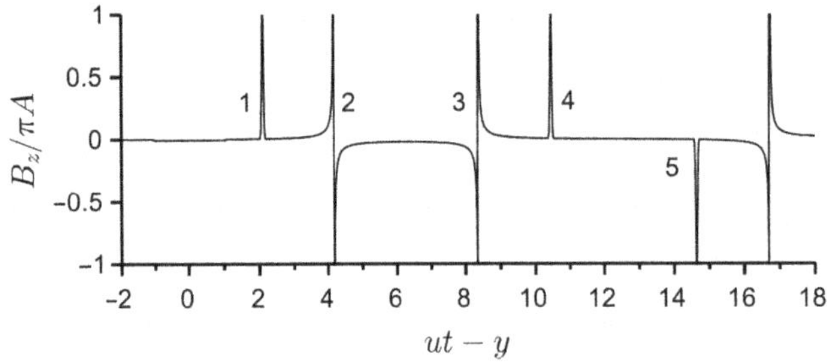

Figure 2.47 The section of the wave field along the azimuthal coordinate (exact solution at $x = -0.5$). The numbers denote wave branches, as in Figure 2.46.

Due to the δ-functions and cotangents in the solution (2.263), all wave energy is concentrated along the thin arched lines. The section of the wave field along the azimuthal coordinate is shown in Figure 2.47. As one can see, all field is sharply concentrated in a series of peaks. Thus, a satellite moving in the magnetosphere will see the fast mode as spikes of the magnetic field strongly resembling Pi2 pulsations.

An interesting feature of the solution is the boundedness of region where the fast mode can propagate. This boundary can be coined as the general reflection surface. In Figure 2.46, it is depicted by the grey line at $x \approx -0.95$. On this boundary, the branches of the solution are directed on the right angle to the y-axis. It is easy to see that this location corresponds to the Alfvén speed value equal to the source's velocity u (the local Mach number equals 1). The width of the mode's propagation region Δx can be evaluated from Eq. (2.264):

$$\Delta x = l\left(1 - \frac{1}{M^2}\right). \tag{2.265}$$

Remind that $M = u/v_{AM}$ is the Mach number on the boundary. Thus, the width grows with the increase of the M value. For the source with M just slightly larger than 1, the wave field is narrowly concentrated near the magnetopause (our solution cannot be applied for the sub-Alfvénic source when $M < 1$). On the other hand, the width grows with the increase of the inhomogeneity scale l. The weaker the inhomoheneity, the deeper the wave can penetrate into the magnetosphere. At the homogeneous plasma limit, $l \to \infty$, the wave propagates in plasma without any hindrance, presenting ordinary Cherenkov emission.

Let us discuss the main features of this solution. The moving source generates fast magnetoacoustic wave with different harmonics of the cavity mode. Each harmonic is a cavity mode, that is, a wave standing between the magnetopause and a reflection surface corresponding to the harmonic's eigenfrequency. Thus, in contrast to the homogeneous plasma case, in the non-uniform plasma the wave's spatial structure is formed by standing rather than running waves. The summation of these harmonics results in formation of the fast magnetoacoustic Mach cone. Its every point moves with the local Alfvén speed. Since the angle between the source's velocity and the wave's speed is determined by the relation $\alpha(x) = \arccos \frac{v_A(x)}{u}$, and the increase of the Alfvén speed into of the magnetosphere, the cone rotates with the distance from the magnetopause. On the magnetic shell where the local Mach number becomes equal 1, the angle becomes equal 90°. Thus, the wave cannot penetrate into the magnetosphere deeper than this shell. The magnetic shell with $v_A(x) = u$ was coined as the general reflection surface for the fast mode generated by the Cherenkov mechanism.

It is the plasma inhomogeneity being the reason for the emergence of the boundedness of the wave's propagation region. Increase of the inhomogeneity scale leads to deeper penetration of the wave into the magnetosphere. On the other hand, the propagation region's width grows with the increase of the source's speed u.

2.18 MHD Oscillation Field Penetrating from the Magnetosphere to Ground

In all of the above-addressed problems the gradient of the background plasma inhomogeneity was assumed to be across the magnetic field lines. Such an approximation is fully justifiable in the main order of perturbation theory, for most problems concerning the structure of magnetospheric MHD oscillations. The fact is magnetic field does not stop charged plasma particles from freely flowing along the magnetic field lines, but does restrict considerably their transverse movement. The main component of the plasma density gradient in the magnetosphere is therefore across the magnetic shells.

A completely different situation is found near the ionosphere, where the plasma density gradient is chiefly determined by the gravitation field. Here, plasma stratification is nearly vertical, and the geomagnetic field lines are generally inclined to the Earth surface. Near the poles, the plasma density gradient and the geomagnetic field vectors are nearly parallel, while near the equator they are normal to each other. Thus, the main component in the polar and middle latitudes is the geomagnetic field-aligned component of the ionospheric plasma density gradient.

Most geomagnetic pulsations observable at Earth surface are related to Alfvén waves penetrating from the magnetosphere. The electromagnetic field of these waves suffers considerable transformation while moving from the magnetosphere to the ground, through the ionosphere and atmosphere [184–186]. The effect the ionosphere exerts on the Alfvén wave field that reaches the ground manifests itself differently in different frequency ranges. The most essential element of the process are waves exciting currents in the ionospheric conducting layer. These currents generate FMS waves in the ionosphere that are responsible for the major contributions to the electromagnetic oscillations induced at Earth surface.

There is a waveguide for the shortest-period oscillations (the Pc1 frequency range: $f \sim 10^{-1}$–1 Hz) in the ionospheric F2-layer where FMS waves can propagate along the ionosphere. A considerable number of theoretical and experimental papers are devoted to investigating the relationship between the field of these waves and electromagnetic oscillations on the ground. Theoretical investigations of waveguide FMS waves and oscillations they induce on Earth were carried out in [187–191].

For oscillations in a lower-frequency range ($f \sim 10^{-2}$– 10^{-1} Hz) conditions are created in the ionospheric E-layer for a certain ionospheric whistler to be generated. This phenomenon was first investigated theoretically in [192] and, in more detail, in [193]. In the same frequency range, conditions are created in the upper ionosphere (F2-layer and above) for partial confinement of Alfvén waves along the geomagnetic field lines – a so-called ionospheric Alfvén resonator (IAR) appears [194, 195].

Somewhat simpler are the conditions for the lowest-frequency ($f \sim 10^{-2}$–10^{-3} Hz) magnetospheric Alfvén oscillations penetrating to the ground [196]. The wavelength of such oscillations is much larger than any of the characteristic scales of ionospheric plasma inhomogeneity along the geomagnetic field lines. Thanks to this fact, the ionosphere can in many cases be regarded as a thin layer, mathematically. This allows a far enough advancement of the analytical study of the field of such waves penetrating from the magnetosphere to the ground. One of the first studies examining this process theoretically was [197]. That paper addressed the problem of waves with $k_\perp \ll k_\parallel$ that come from the magnetosphere and are reflected from the ionosphere, where $k_\parallel$ and $k_\perp$ are the longitudinal and transverse components of the Alfvén wave vector in the magnetosphere (which was regarded as homogeneous in that paper). The theory was further developed for the $k_\perp \gg k_\parallel$ case in [184]. A case was examined when the geomagnetic field is normal to Earth surface, in contrast to [197], where it was assumed to be inclined. Similar calculations for waves with arbitrary $k_\perp$ and $k_\parallel$ were made in [198] for a model where geomagnetic field is normal to Earth surface.

The theory was later developed in papers examining Alfvén waves passing through a horizontally inhomogeneous ionosphere. The horizontal inhomogeneity of the ionosphere was simulated in [199–202] by a plane containing a limited area where the conductivity differs from that of the remaining ionosphere. The conductivity both inside and outside this area was assumed to be homogeneous. Another type of inhomogeneous ionosphere was discussed in [203, 204]. It featured continuously varying conductivity in the horizontal direction. Those papers demonstrated that the angle to which the polarisation ellipsis of the geomagnetic pulsation field rotates as these pulsations penetrate from the magnetosphere to the ground can differ considerably from $\pi/2$ of the horizontally homogeneous ionospheric model.

A next step in the studies of Alfvén oscillation field penetrating through the ionosphere to the ground was made in [205, 206]. A model which included arbitrary inclination of geomagnetic field lines as well as Alfvén speed and conductivity tensor component distribution over height was used to inspect how Alfvén waves with arbitrary transverse structure penetrate to the ground.

Practically all the above-mentioned papers regard the ionosphere as a region where incident Alfvén waves from the magnetosphere that pass through it suffer a considerable transformation in their field. Part of the wave energy is absorbed due to wave dissipation in the ionospheric conducting layer [186, 207, 208]. Various physical processes are considered as the sources of Alfvén waves in those papers, both inside and outside the magnetosphere (see reviews [15, 209]). The ionosphere is assumed to be a passive element of the medium. As an exception, [196, 199, 210, 211] (see also the monographs [212, 213]) consider the ionosphere as the Alfvén wave source. The model employed is that of an optically thin ionosphere with a vertical geomagnetic field.

We will next examine the process by which the MHD wave field penetrates from the magnetosphere to the ground using two models. Let us first consider the problem in a model with a vertical geomagnetic field. It is the simplest case providing the opportunity to find the boundary conditions at the upper ionospheric boundary both for Alfvén and magnetosonic waves, in the presence of external currents in the ionosphere. This will be of use in the following problems about the structure of transverse small-scale MHD waves in the magnetosphere the source of which are external currents in the ionosphere. The same problem will then be solved for Alfvén waves in the model with an inclined geomagnetic field.

2.18.1 Boundary Conditions for MHD Waves at the Upper Ionospheric Boundary in a 'Thin Layer' Model with a Vertical Magnetic Field

An important role in the process of MHD-wave field penetrating from the magnetosphere to Earth is played by the Alfvén speed and medium conductivity distribution over height. The typical height distributions of these parameters are shown in Figure 2.48(a). Two layers can be distinguished in the ionosphere: conducting layer (III) and upper ionosphere

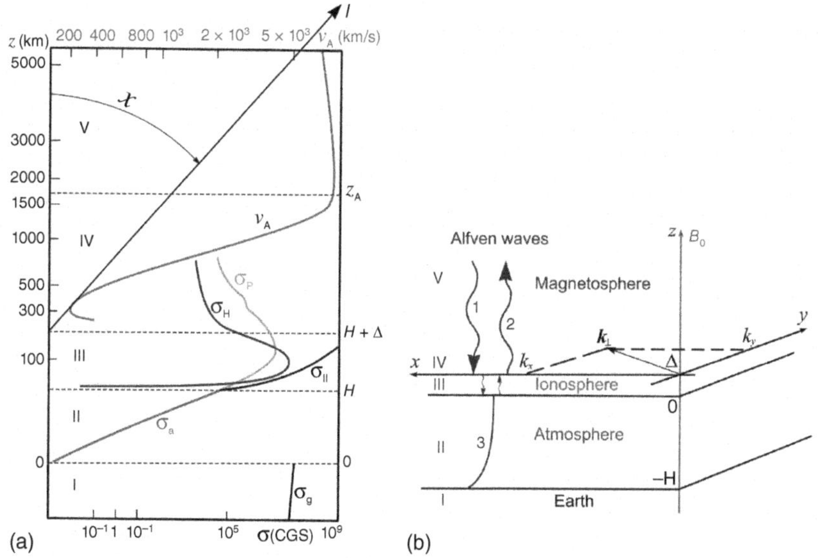

Figure 2.48 (a) typical height profiles of the conductivity tensor components $\hat{\sigma}$ and the Alfvén speed v_A. Roman numerals indicate the following layers: I – ground with isotropic conductivity σ_g, II – atmosphere with conductivity σ_a, III – lower ionosphere with transverse Pedersen σ_P and Hall conductivities σ_H and parallel conductivity $\sigma_\parallel$, IV – upper ionosphere, where $\sigma_P, \sigma_H \to 0$, V – magnetosphere. (b) model of near-Earth medium with a vertical magnetic field; Alfvén wave penetrating to Earth is shown schematically: 1 – incident wave from the magnetosphere, 2 – wave reflected from the ionosphere, 3 – field of the wave penetrating to ground.

(IV), up to $z_A \approx (1.5 \div 2) \cdot 10^3$ km. At this height, the Alfvén speed profile changes sharply from fast-increasing in the upper ionosphere to slow-decreasing in the magnetosphere. This height is the upper boundary of the ionospheric Alfvén resonator (IAR).

Let us consider the problem of a monochromatic MHD wave of the form $\exp(i\mathbf{kr} - i\omega t)$, arriving from the magnetosphere and reflected by the ionosphere, where $\mathbf{k}$ is the wave vector, ω is wave frequency. The wave type will be determined below. Let us examine the simplest flat-layer model with vertical geomagnetic field. We will use a Cartesian system of coordinates (x, y, z), where the z axis is along the magnetic field lines (see Figure 2.48(b)). Let us assume the magnetosphere and the atmosphere to be homogeneous, and the wavelength along the magnetic field to be large enough for the ionosphere to be capable of being regarded as a thin layer ($k_z\Delta \ll 1$, where Δ is the thickness of the ionospheric conducting layer).

Since the main role in defining the structure of low-frequency MHD oscillations is played by the conductivity of the medium, it is most convenient to use electric field components to describe them. When external currents are present in the medium, monochromatic MHD oscillations are described by the Maxwell equations of the form

$$\text{curl } \mathbf{E} = ik_0\mathbf{B}, \quad \text{curl } \mathbf{B} = -ik_0\mathbf{E} + \frac{4\pi}{\bar{c}}\left[\mathbf{j} + \mathbf{j}^{\text{ext}}\right], \tag{2.266}$$

where $k_0 = \omega/\bar{c}$, $\bar{c}$ is speed of light in vacuum, $\mathbf{E}$ and $\mathbf{B}$ are the electric and magnetic fields of the oscillations, $\mathbf{j}^{\text{ext}}$ is the external current unrelated to the field of the oscillations. For low-frequency MHD waves we are interested in, the conduction current in layers with anisotropic conductivity (in the ionosphere and magnetosphere, in our model) can be represented as

$$\mathbf{j} = \sigma_{\|}\mathbf{E}_{\|} + \sigma_P\mathbf{E}_{\perp} + \sigma_H\left[\frac{\mathbf{B}_0}{B_0} \times \mathbf{E}_{\perp}\right], \tag{2.267}$$

where $\mathbf{B}_0$ is the geomagnetic field vector, $\mathbf{E}_{\|}$ is the parallel (along $\mathbf{B}_0$) and $\mathbf{E}_{\perp}$ transverse components of the disturbed electric field, $\sigma_{\|}$, σ_P and σ_H are the parallel, Pedersen and Hall conductivities of the medium.

Equations (2.266) have the simplest form for field components $E_{\perp}$ and $E_b = [\mathbf{B}_0 \times \mathbf{E}_{\perp}]/B_0$, where the E_b component is a projection of field $\mathbf{E}$ on the direction perpendicular to $\mathbf{B}_0$ and $\mathbf{k}_{\perp}$ completing the right-hand system of coordinates. For homogeneous layers (conducting ground and the atmosphere, where $\sigma_{\|} = \sigma_{\perp} = \sigma_H = \sigma$) lacking any external currents, we obtain, from (2.266),

$$\frac{\partial^2 E_{\perp,b}}{\partial z^2} - [k_{\perp}^2 - k_0(k_0 + i\varkappa)]E_{\perp,b} = 0, \tag{2.268}$$

where $\varkappa = 4\pi\sigma/\bar{c}$. Let us examine a model where the ground conductivity is infinite. The solutions of these equations in the atmosphere ($-H < z < 0$) that would satisfy the boundary conditions $E_{\perp,b}(z = -H) = 0$ then have the following form, for oscillations with $k_0 \ll k_{\perp}, \varkappa$

$$E_{\perp,b} = C_{\perp,b} \sinh\left[(z + H)\sqrt{k_{\perp}^2 - ik_0\varkappa}\right], \tag{2.269}$$

where $C_{\perp,b}$ are arbitrary constants.

In anisotropic layers (ionosphere and magnetosphere) the parallel conductivity of plasma is much larger than its transverse conductivity $\sigma_{\|} \gg \sigma_P, \sigma_H$. In the $\sigma_{\|} \to \infty$ limit, we have

$E_{\parallel} = 0$, from (2.266). For the other two components of the electric field we obtain a system of coupled equations:

$$\frac{\partial^2 E_\perp}{\partial z^2} + ik_0\varkappa_P E_\perp = ik_0\varkappa_H E_b - i\frac{4\pi k_0}{\bar{c}} j_\perp^{\text{ext}}, \tag{2.270}$$

$$\frac{\partial^2 E_b}{\partial z^2} - (k_\perp^2 - ik_0\varkappa_P)E_b = -ik_0\varkappa_H E_\perp - i\frac{4\pi k_0}{\bar{c}} j_b^{\text{ext}}. \tag{2.271}$$

The small terms proportional to $k_0 \ll k_\perp, \varkappa_{P,H}$ (where $\varkappa_{P,H} = 4\pi\sigma_{P,H}/\bar{c}$) are ignored here. Let us integrate these equations over the thickness of the ionospheric conducting layer ($0 < z < \Delta$), assuming it to be thin for the oscillations. This means that the oscillation field is practically invariable inside this layer, $\mathbf{E}(\Delta) \approx \mathbf{E}(0)$, and its components can be taken out of the integral. The result is

$$\left.\frac{\partial E_\perp}{\partial z}\right|_\Delta = \left.\frac{\partial E_\perp}{\partial z}\right|_0 - i\frac{4\pi\omega}{\bar{c}^2}\Sigma_P E_\perp + i\frac{4\pi\omega}{\bar{c}^2}\Sigma_H E_b - i\frac{4\pi\omega}{\bar{c}^2} J_\perp^{\text{ext}},$$
$$\left.\frac{\partial E_b}{\partial z}\right|_\Delta = \left.\frac{\partial E_b}{\partial z}\right|_0 + k_\perp^2 \Delta E_b - i\frac{4\pi\omega}{\bar{c}^2}\Sigma_P E_b - i\frac{4\pi\omega}{\bar{c}^2}\Sigma_H E_\perp - i\frac{4\pi\omega}{\bar{c}^2} J_b^{\text{ext}},$$

where the notations are:

$$\Sigma_{P,H} = \int_0^\Delta \sigma_{P,H}(z)dz$$

are the integral Pedersen an Hall conductivity of the ionosphere,

$$J_{\perp,b}^{\text{ext}} = \int_0^\Delta j_{\perp,b}^{\text{ext}}(z)dz,$$

are densities of external currents in the conducting layer of the ionosphere integrated over height. Substituting the solutions for the lower edge of the ionosphere as (2.269) into these equations and proceeding to the (x, y, z) system of coordinates where

$$E_x = (k_x E_\perp + k_y E_b)/k_\perp, \qquad E_y = (k_x E_b - k_y E_\perp)/k_\perp$$

and the analogous relations for the vector components $(J_x^{\text{ext}}, J_y^{\text{ext}})$ are valid, produces the following equations for the upper boundary of the current sheet:

$$\left.\frac{\partial E_x}{\partial z}\right|_\Delta = (k_\perp + k_y^2\Delta)E_x - k_x k_y \Delta E_y - i\frac{4\pi\omega}{\bar{c}^2}(\Sigma_P E_x - \Sigma_H E_y + J_x^{\text{ext}}), \tag{2.272}$$

$$\left.\frac{\partial E_y}{\partial z}\right|_\Delta = (k_\perp + k_x^2\Delta)E_x - k_x k_y \Delta E_x - i\frac{4\pi\omega}{\bar{c}^2}(\Sigma_P E_y + \Sigma_H E_x + J_y^{\text{ext}}). \tag{2.273}$$

Here we used the condition $k_0^2, k_0\varkappa \ll k_\perp^2$, natural for low-frequency MHD waves. We will now distinguish the boundary conditions for different branches of the MHD oscillations using, instead of the transverse vector of the oscillation electric field, its decomposition (2.72) into the potential and vortical components:

$$E_x = -ik_x\varphi + ik_y\psi, \qquad E_y = -ik_y\varphi - ik_x\psi.$$

Recall that the scalar potential φ of the oscillations describes the Alfvén wave field, and the parallel component ψ of the vector potential describes the magnetosonic wave field. Substituting these expressions in (2.272), (2.273), we obtain the boundary conditions for

Alfvén and magnetosonic waves at the upper boundary of the ionospheric conducting layer, which we will write as

$$\varphi(\Delta) = i\frac{v_P}{\omega}\frac{\partial\varphi}{\partial z}\bigg|_\Delta + \frac{\Sigma_H}{\Sigma_P}\psi(\Delta) - i\frac{k_xJ_x^{\text{ext}} + k_yJ_y^{\text{ext}}}{k_\perp^2\Sigma_P}, \tag{2.274}$$

$$\psi(\Delta) = i\frac{v_P}{\omega}\frac{\partial\psi}{\partial z}\bigg|_\Delta - \frac{\Sigma_H}{\Sigma_P}\varphi(\Delta) + i\frac{k_yJ_x^{\text{ext}} - k_xJ_y^{\text{ext}}}{k_\perp^2\Sigma_P}, \tag{2.275}$$

where $v_p = \overline{c}^2/4\pi\Sigma_P$ is the characteristic speed of low-frequency whistler in the ionosphere (see [214]). Note that we omitted from these equations some terms that provide small corrections to the phase of the reflected wave for oscillations with a large wavelength along the magnetic field (and it is these oscillations that we are going to address hereafter). The remaining terms in (2.274), (2.275) describe the following physical effects. The first terms in the right-hand sides describe decay in the corresponding branches of MHD waves due to their dissipation related to the finite Pedersen conductivity of the ionospheric current layer.

The second terms are related to the field of the wave which is incident on the ionosphere generating another branch of MHD oscillations by exciting Hall currents. In other words, the Alfvén wave incident to the ionosphere does not only produce a reflected Alfvén wave, but also a magnetosonic wave generated by Hall currents in the ionosphere. Analogously, the magnetosonic wave falling onto the ionosphere is reflected from the ionosphere as the sum of the magnetosonic and Alfvén waves.

The third terms in the right-hand sides of (2.274), (2.275) describe external currents in the ionosphere capable of being the source of corresponding MHD waves in the magnetosphere. Using the condition that the current is closed (div$\mathbf{j}$ = 0) and the atmosphere is current-free, the last term in (2.274) can be written as

$$j_z^{\text{ext}}(\Delta) = -i(k_xJ_x^{\text{ext}} + k_yJ_y^{\text{ext}}),$$

i.e. oscillations of parallel external currents at the upper boundary of the ionospheric current layer can serve as the source of Alfvén waves in the magnetosphere. Analogously, the last term in (2.275) can be represented as

$$\int_0^\Delta [\text{curl}\mathbf{j}^{\text{ext}}]_z dz = i(k_xJ_y^{\text{ext}} - k_yJ_x^{\text{ext}}),$$

i.e. vortical external currents in the ionospheric conducting layer can be the source of magnetosonic waves in the magnetosphere.

2.18.2 Alfvén Waves Penetrating to Ground from the Magnetosphere in a Model Geospace with an Inclined Magnetic Field

The problem of MHD wave field penetrating from the magnetosphere to ground in the model with an inclined magnetic field is much more complex than the above addressed problem using a vertical field model. Let us examine it in accordance with [205, 215]. We will choose the following systems of coordinates for solving the above obtained equations (2.268) as well as equations analogous to (2.270), (2.271) for a model with an inclined geomagnetic field. We will assume, with necessary accuracy, the Earth surface to be flat when studying electromagnetic oscillation field. Let us introduce the Cartesian system of coordinates (x, y, z), where the x axis is south-to-north along a magnetic meridian, the

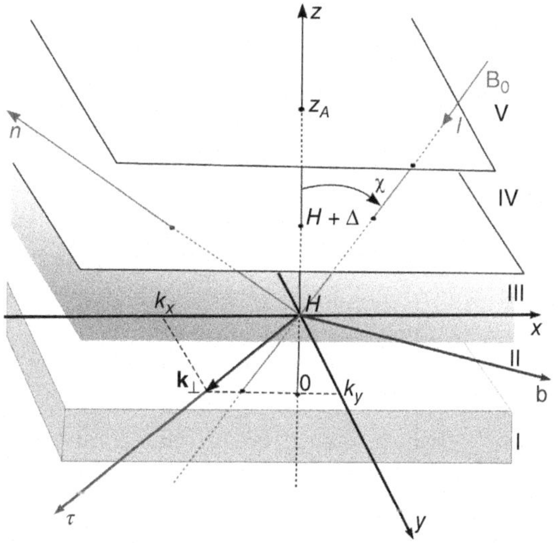

Figure 2.49 Mutual positions of three systems of coordinates used in the problem of MHD wave field penetrating from the magnetosphere to Earth, for a geospace model with inclined geomagnetic field: (x, y, z), (τ, b, z) and (n, y, l). Roman numerals denote the following layers: I – ground with conductivity σ_g, II – atmosphere with conductivity σ_a, III – lower ionosphere (E-layer) and IV – upper ionosphere with anisotropic conductivities, V – magnetosphere.

y axis is west-to-east along a parallel, and the z axis is directed vertically upwards along the normal to the Earth surface (Figure 2.49).

The components of a disturbed electromagnetic field are functions of the coordinates and time (e.g. $B_x = B_x(x, y, z, t)$). Assuming the medium to be stationary, these components can be represented as Fourier decompositions over harmonics of certain frequency ω:

$$B_x(x, y, z, t) = \int_{-\infty}^{\infty} \tilde{B}_x(x, y, z, \omega) \exp(-i\omega t)\, d\omega.$$

Given the fact that the medium is horizontally homogeneous, Fourier decompositions are possible over spatial harmonics with certain values of the horizontal components of the wave vector $\mathbf{k}_\perp = (k_x, k_y)$:

$$\tilde{B}_x(x, y, z, \omega) = \int_{-\infty}^{\infty} dk_x \int_{-\infty}^{\infty} dk_y\, \overline{B}_x(k_x, k_y, z, \omega) \exp(ik_x x + ik_y y). \tag{2.276}$$

We will use these Fourier harmonics in further calculations, without writing out their dependence on k_x, k_y and ω, for brevity.

To solve (2.268) in isotropic media – in the ground and in the atmosphere – we will use the system of coordinates (τ, b, z) obtained by rotation about the z axis relative to the (x, y, z) system. We will direct the τ axis along the horizontal wave vector $\mathbf{k}_\perp$, and the b axis, perpendicular to the τ axis, in the same horizontal plane. The following relations exist between horizontal vector components of the disturbed wave magnetic field:

$$\overline{B}_\perp = (k_x/k_\perp)\overline{B}_x + (k_y/k_\perp)\overline{B}_y,\ \overline{B}_b = -(k_y/k_\perp)\overline{B}_x + (k_x/k_\perp)\overline{B}_y.$$

Analogous relations exist for the components of a disturbed electric field.

For anisotropic media – the ionosphere and magnetosphere – we will employ the system of coordinates (n, y, l), obtained by rotation to angle χ about the y axis relative to the (x, y, z) system so that the l axis is along $\mathbf{B}_0$ (see Figure 2.49). The n axis lies in the meridional plane and is perpendicular to the l and y axes. In this system of coordinates,

$$B_n = B_x \cos\chi + B_z \sin\chi,\ B_l = -B_x \sin\chi + B_z \cos\chi \tag{2.277}$$

and analogous relations are valid for the electric components of the wave field.

2.18.2.1 Low-Frequency Electromagnetic Oscillation Field in the Ground and Atmosphere

Let us consider a model homogeneous ground with finite conductivity σ_g. In the ground, the solutions of (2.268) for the disturbed electromagnetic field components satisfying the boundary condition (finite oscillation amplitudes) are

$$\bar{E}_b(z) = \bar{E}_b(0)\exp(k_g z),\ \overline{B}_b(z) = \overline{B}_b(0)\exp(k_g z),$$

where the $z = 0$ point corresponds to the ground/atmosphere boundary, and $k_g = \sqrt{k_\perp^2 - ik_0\kappa_g}$, $\kappa_g = 4\pi\sigma_g/\bar{c}$. The other components of the electromagnetic field in isotropic layers are expressed as follows:

$$\begin{aligned} \bar{E}_\perp &= -\frac{1}{\kappa - ik_0}\frac{\partial \overline{B}_b}{\partial z}, & \bar{E}_z &= i\frac{k_0}{\kappa - ik_0}\overline{B}_b, \\ \overline{B}_\perp &= \frac{i}{k_0}\frac{\partial \bar{E}_b}{\partial z}, & \overline{B}_z &= \frac{k_\perp}{k_0}\bar{E}_b. \end{aligned} \tag{2.278}$$

For solutions in the ground, one has to set $\kappa = \kappa_g$. Thanks to the high conductivity of the ground we will use, for simplicity, the limit $\kappa_g \to \infty$ in the calculations below, which yields $E_\perp(0) = E_b(0) = 0$.

The atmospheric conductivity σ_a is much smaller than the ground conductivity σ_g, therefore we will assume that the condition $k_\perp^2 \gg k_0\kappa_a$ holds, where $\kappa_a = 4\pi\sigma_a/\bar{c}$ across the atmosphere ($0 < z < H$). The result will be the solutions for the components $B_\perp$ and E_b in the form

$$\overline{B}_\perp(z) = B_\perp(0)\cosh(k_\perp z),\ \bar{E}_b(z) = -i\frac{k_0}{k_\perp}B_\perp(0)\sinh(k_\perp z). \tag{2.279}$$

Thus, for $z = H$, we have

$$\overline{B}_\perp(H) = i\frac{k_\perp}{k_0}E_b(H)\coth(k_\perp H). \tag{2.280}$$

With satisfactory accuracy we can assume that, for $z = H$,

$$\overline{B}_b(H) = 0.$$

This equation is used as the second boundary condition for solving the problem in the ionosphere.

2.18.2.2 Low-Frequency Electromagnetic Oscillation Field in the Ionosphere

It is easiest to write a system of equation analogous to (2.270), (2.271), in the ionospheric model with an inclined magnetic field, using the system of coordinates (n, y, l). If a four-component column vector is introduced,

$$\alpha = \begin{pmatrix} \bar{E}_n \\ \bar{E}_y \\ \overline{B}_y \\ \overline{B}_n \end{pmatrix},$$

this system of equations can be represented as

$$-i\frac{\partial \alpha}{\partial z} = \hat{Q}\alpha + \hat{q}\alpha + ig. \tag{2.281}$$

The matrix $\hat{Q}$ here consists of the components of the horizontal wave vector $k_\perp = (k_x, k_y)$, and the matrix $\hat{q}$ consists of the components $\kappa_{P,H} = 4\pi\sigma_{P,H}/\overline{c}$:

$$\hat{Q} = \begin{pmatrix} k_x \tan\chi & 0 & k_0/\cos\chi & 0 \\ 0 & k_x \tan\chi & 0 & -k_0/\cos\chi \\ -\dfrac{k_y^2}{k_0\cos\chi} & \dfrac{k_x k_y}{k_0 \cos^2\chi} & k_x\tan\chi & -k_y\dfrac{\tan\chi}{\cos\chi} \\ -\dfrac{k_x k_y}{k_0} & \dfrac{k_x^2}{k_0\cos\chi} & -k_y\sin\chi & -k_x\tan\chi \end{pmatrix}$$

$$\hat{q} = \begin{pmatrix} 0 & 0 & 0\ 0 \\ 0 & 0 & 0\ 0 \\ \dfrac{\kappa_P}{\cos\chi} & -\dfrac{\kappa_H}{\cos\chi} & 0\ 0 \\ -\kappa_H\cos\chi & -\kappa_P\cos\chi & 0\ 0 \end{pmatrix}$$

Moreover, a column vector of external currents is present in (2.281)

$$g = \frac{4\pi}{\overline{c}\cos\chi}\begin{pmatrix} 0 \\ 0 \\ j_n^{ext} \\ -j_y^{ext} \end{pmatrix}$$

As will be seen from the following calculations, the components of this vector are much smaller than the components of the vector $\hat{Q}\alpha$. Thus, we can apply the perturbation method when searching for the solution of the system (2.281). In the zero approximation, we have this system of equations:

$$-i\frac{\partial \alpha^{(0)}}{\partial z} = \hat{Q}\alpha^{(0)}, \tag{2.282}$$

the solutions of which have the form

$$\alpha^{(0)}(z) = \overline{\psi}\exp(ik_z z). \tag{2.283}$$

Substituting (2.283) in (2.282) produces a system of algebraic equations for k_z, the solutions of which have the form

$$k_z^{(1)} = k_z^{(2)} = k_x \tan\chi \equiv k_{zA}, \quad k_z^{(3)} = ik_\perp \equiv k_{zf},$$

$$k_z^{(4)} = -ik_\perp \equiv k_{zf}^*.$$

The roots $k_z^{(1)}$ and $k_z^{(2)}$ correspond to the Alfvén wave, and $k_z^{(3)}$ and $k_z^{(4)}$ to the magnetosonic wave of frequency $\omega = 0$. In the following calculations we will also need the transverse wave-vector components for the Alfvén

$$k_{nA} \equiv k_x/\cos\chi, \quad k_{\perp A} = \sqrt{k_{nA}^2 + k_y^2}$$

and magnetosonic waves

$$k_{nF} = k_x \cos\chi + ik_\perp \sin\chi, \quad k_{\perp F} = k_\perp \cos\chi + ik_x \sin\chi .$$

Each root $k_z^{(i)}$ $(i = 1, 2, 3, 4)$ has a corresponding column vector of coefficients $\overline{\psi}^i$ defined by (2.282) relating the amplitudes of various components of the electric field for each wave. The full solution of the system (2.282) is an arbitrary combination of linearly independent vectors $\overline{\psi}^i \exp(ik_z^{(i)} z)$. In the four-vector space, the set of $\overline{\psi}^i$ forms a full system, which is why the solution of (2.281) can be sought as

$$\alpha(z) = \overline{\psi}^i F_i(z) \equiv \hat{\psi} F(z), \qquad (2.284)$$

where $F(z)$ is the column vector of the desired coefficients, and the matrix $\hat{\psi}$ is composed of the column vectors $\overline{\psi}^i$:

$$\hat{\psi} = \begin{pmatrix} \dfrac{k_{nA}}{k_{\perp A}} & -\dfrac{(k_{nA}^2 - k_y^2)\tan\chi}{k_{\perp A}^2} & -\dfrac{k_y}{k_{\perp F}} & -\dfrac{k_y}{k_{\perp F}^*} \\ \dfrac{k_y}{k_{\perp A}} & -\dfrac{2k_{nA}k_y \tan\chi}{k_{\perp A}^2} & \dfrac{k_{nF}}{k_{\perp F}} & \dfrac{k_{nF}^*}{k_{\perp F}^*} \\ 0 & \dfrac{k_{nA}}{k_0} & -i\dfrac{k_y}{k_0} & i\dfrac{k_y}{k_0} \\ 0 & -\dfrac{k_y}{k_0} & -i\dfrac{k_{nF}}{k_0} & i\dfrac{k_{nF}^*}{k_0} \end{pmatrix}$$

Substituting the solution (2.284) into the system of Eq. (2.281) and left-multiplying it by $\hat{\psi}^{-1}$, an inverse matrix to $\hat{\psi}$, will produce a system of equations for the coefficients $F(z)$:

$$-\frac{\partial F}{\partial z} = \hat{\Lambda} F + \hat{P} + \overline{r}. \qquad (2.285)$$

The expressions for the matrix $\hat{\Lambda}$ and the first column of the matrix $\hat{P}$ we are interested in have the form

$$\hat{\Lambda} = \begin{pmatrix} k_{zA} & k_{\perp A}/\cos\chi & 0 & 0 \\ 0 & k_{zA} & 0 & 0 \\ 0 & 0 & k_{zF} & 0 \\ 0 & 0 & 0 & k_{zF}^* \end{pmatrix}$$

$$P^1 = \frac{k_0}{k_{\perp A}\cos\chi} \begin{Bmatrix} i\tan\chi \left(\dfrac{k_{nA}}{k_{\perp A}}\kappa_P - \dfrac{k_y}{k_{\perp A}}\kappa_H \right) \\ i\kappa_P \\ \dfrac{1}{2}\left(\kappa_H + i\dfrac{k_y \sin\chi}{k_\perp}\kappa_P \right) \\ \dfrac{1}{2}\left(-\kappa_H + i\dfrac{k_y \sin\chi}{k_\perp}\kappa_P \right) \end{Bmatrix},$$

and, for the column vector $\bar{r}$,

$$\bar{r} = \frac{4\pi\omega}{\bar{c}^2}\begin{pmatrix} -i(j_n \tan\chi)/(k_{\perp A}\cos\chi) \\ i(k_{nA}j_n + k_y j_y)/(k_{\perp A}^2\cos\chi) \\ -(k_y j_n - k_{nF} j_y)/(2k_{\perp F}k_\perp) \\ (k_y j_n - k_{nF}^* j_y)/(2k_{\perp F}^* k_\perp) \end{pmatrix}$$

In the zero approximation, the solutions (2.284) with coefficients $F_i(z)$ satisfy Eq. (2.282), therefore we will seek these coefficients as

$$F_i(z) = f_i(z)\exp(ik_z^{(i)}(z - H)),$$

where the phase is counted from the $z = H$ boundary, for convenience. Equations (2.285) must be supplemented by the boundary conditions at the lower ionospheric boundary ($z = H$). The boundary conditions for functions $f_i(z)$ that we obtain from (2.278) and (2.280) have the form

$$f_2(H) = 0, \tag{2.286}$$

$$(1 + \cot(k_\perp H))f_3(H) + (1 - \cot(k_\perp H))f_4(H) = 0. \tag{2.287}$$

A natural requirement in the upper ionosphere and the magnetosphere would be the absence of solutions with amplitudes increasing as $z \to \infty$. This results in the following boundary condition at the upper ionospheric boundary

$$f_4(z_A) = 0. \tag{2.288}$$

It will be seen from further calculations that this inequality holds across the ionosphere:

$$|f_1| \gg |f_2|, |f_3|, |f_4|.$$

This allows us, in the first order of perturbation theory, to leave only terms proportional to f_1 in the right-hand side of Eq. (2.285). As a result, the first couple of equations in (2.285) is separated from the other two and has the form

$$f_1' = i\frac{k_{\perp A}}{\cos\chi}f_2 - i\frac{k_0\tan\chi}{k_{\perp A}\cos\chi}\left(\frac{k_{nA}}{k_{\perp A}}\kappa_P - \frac{k_y}{k_{\perp A}}\kappa_H\right)f_1 + \frac{4\pi k_0\tan\chi}{\bar{c}k_{\perp A}\cos\chi}j_n\exp(-ik_{zA}(z - H)), \tag{2.289}$$

$$f_2' = -\frac{k_0}{k_{\perp A}\cos\chi}\kappa_P f_1 - \frac{4\pi k_0}{\bar{c}k_{\perp A}^2\cos\chi}(k_{nA}j_n + k_y j_y)\exp(-ik_{zA}(z - H)), \tag{2.290}$$

where the prime denotes derivative d/dz. As will be seen later, the function $f_1(z)$ varies little over the interval $H \le z \le z_a$, and one can set

$$f \equiv f_1(H) \approx f_1(z). \tag{2.291}$$

Let us introduce the notations

$$X_{P,H}(z) = \frac{4\pi}{c}\int_H^z \sigma_{P,H}(z')dz', \quad K_{P,H} \equiv X_{P,H}(H+\Delta),$$
$$\bar{J}_{n,y}(z) = \frac{4\pi}{c}\int_H^z \bar{j}_{n,y}(z')\exp(-ik_{zA}(z'-H))dz', \quad \bar{I}_{n,y} \equiv \bar{J}_{n,y}(H+\Delta).$$

Integrating (2.290) with the boundary condition (2.286) will produce

$$f_2(z) = -f\frac{k_0}{k_{\perp A}\cos\chi}X_P(z) - \frac{k_0}{k_{\perp A}^2\cos\chi}(k_{nA}\bar{J}_n(z) + k_y\bar{J}_y(z)).$$

In the upper ionosphere ($z \to \infty$), we have

$$f_2 = -f\frac{k_0}{k_{\perp A}\cos\chi}K_P - \frac{k_0}{k_{\perp A}^2\cos\chi}(k_{nA}\bar{I}_n + k_y\bar{I}_y). \tag{2.292}$$

For the condition $|f_2| \ll |f_1|$ to be satisfied, it is necessary that each term in the right-hand side of (2.292) should be much smaller than $|f|$. As for the terms proportional to the currents in the ionosphere, the condition of their being small will be formulated in Section 2.18.2.3. For the first term to be small, the following inequality must be satisfied:

$$\frac{k_0K_P}{k_{\perp A}} \ll 1. \tag{2.293}$$

It allows us to omit the term proportional to f_1 in the right-hand side of (2.289). Integrating (2.289) will then produce

$$f_1(z) = f\left[1 - i\frac{k_0}{\cos^2\chi}\int_H^z X_P(z')dz'\right] - \frac{k_0\sin\chi}{k_{\perp A}\cos^2\chi}\bar{J}_n(z). \tag{2.294}$$

As will be seen from findings in Section 2.18.2.3, the last term in this equation is small. For the condition (2.291) to be satisfied, therefore, we need

$$k_0K_P\Delta \ll 1. \tag{2.295}$$

The conditions (2.293) and (2.295) are equivalent to the requirement that the frequency of the oscillations in question, ω, should be small compared to the frequency of a low-frequency whistler in the ionosphere, $\omega_{PH} = \bar{c}^2/4\pi\Sigma_P\Delta$. To be able to completely ignore (as we did above) the transverse conductivity in the upper ionosphere, the frequency ω must be much smaller than the eigenfrequency of the ionospheric Alfvén resonator (IAR):

$$\omega\int_{H+\Delta}^{z_A}\frac{dz}{v_A(z)} \ll 1.$$

Knowing the expression for $f_1(z)$, it is easy to integrate the second couple of equations (2.285) with the boundary conditions (2.287), (2.288), as well:

$$f_3(z) = -i\frac{fk_0}{2k_{\perp A}\cos\chi}\left[\left(\bar{K}_H - i\frac{k_y\sin\chi}{k_\perp}\bar{K}_P\right)e^{-2k_\perp H}\right.$$
$$\left. - \int_H^z\left(\varkappa_H + i\frac{k_y\sin\chi}{k_\perp}\varkappa_P\right)\exp[(k_\perp + ik_x\tan\chi)(z'-H)]dz'\right], \tag{2.296}$$

$$f_4(z) = i\frac{fk_0}{2k_{\perp A}\cos\chi}\int_z^{\infty}\left(\varkappa_H - i\frac{k_y\sin\chi}{k_\perp}\varkappa_P\right)$$
$$\times\exp[(-k_\perp + ik_x\tan\chi)(z' - H)]\,dz', \tag{2.297}$$

where

$$\overline{K}_{P,H} = \int_H^{\infty}\varkappa_{P,H}(z)\exp[(-k_\perp + ik_x\tan\chi)(z-H)]\,dz.$$

The full solution of the system (2.281) in the upper ionosphere has the form

$$\alpha = \overline{\psi}^1F_1 + \overline{\psi}^2F_2 = \begin{pmatrix} F_1\ k_{nA}/k_{\perp A} \\ F_1\ k_y/k_{\perp A} \\ F_2\ k_{nA}/k_0 \\ -F_2\ k_y/k_0 \end{pmatrix}. \tag{2.298}$$

This solution only includes components related to the Alfvén wave in the magnetosphere. The electromagnetic field of a magnetosonic wave generated in the E-layer by Hall currents fails to penetrate into the upper ionosphere because the ionosphere is an opaque region for it. With satisfactory accuracy, the solutions $F_1(z)$ and $F_2(z)$ in the upper ionosphere can be represented as

$$F_1(z) = \left[f - i\frac{k_0(z-H)}{\cos^2\chi}(fK_P + \overline{I}_\perp)\right]\exp(ik_x(z-H)\tan\chi), \tag{2.299}$$

$$F_2(z) = -\frac{k_0}{k_{\perp A}\cos\chi}\left[(fK_P + \overline{I}_\perp)\right]\exp(ik_x(z-H)\tan\chi), \tag{2.300}$$

where

$$\overline{I}_\perp = \frac{k_{nA}\overline{I}_n + k_y\overline{I}_y}{k_{\perp A}}.$$

Using the condition of closed current $\operatorname{div}\mathbf{j} = 0$, we find:

$$\overline{I}_\perp = \frac{4\pi}{\overline{c}k_{\perp A}}\overline{j}_\parallel,$$

where $\overline{j}_\parallel$ is the density oscillation harmonic of external parallel current at the upper boundary of the ionospheric conducting layer ($z = H + \Delta$).

It follows from the solutions (2.298)–(2.300) that the transverse components of disturbed magnetic field can be written as

$$\begin{aligned}\overline{B}_y(z) &= B_A\frac{k_{nA}}{k_{\perp A}}\exp(ik_x(z-H)\tan\chi),\\ \overline{B}_n(z) &= -B_A\frac{k_y}{k_{\perp A}}\exp(ik_x(z-H)\tan\chi),\end{aligned} \tag{2.301}$$

where

$$B_A = -\frac{fK_P + \overline{I}_\perp}{\cos\chi}$$

is the Alfvén oscillation amplitude in the upper ionosphere. The second equation can be rewritten as

$$f = -\frac{B_A \cos\chi + \bar{I}_\perp}{K_P}. \tag{2.302}$$

If $k_{nA} \gg k_y$ in the magnetosphere (at the upper boundary of the ionosphere, to be more precise), then

$$|\bar{B}_y| \sim B_A, \qquad |\bar{B}_n| \ll B_A. \tag{2.303}$$

These are Alfvén waves with toroidal polarisation. They include, for example, Alfvén waves generated in the magnetosphere, at resonant surfaces, by FMS waves penetrating from the solar wind, or driven at the magnetopause by the Kelvin-Helmholtz instability. Conversely, if $k_{nA} \ll k_y$, then

$$|\bar{B}_y| \ll B_A, \qquad |\bar{B}_n| \sim B_A. \tag{2.304}$$

These are poloidal Alfvén waves, their possible source being external currents in the ionosphere, for example. These waves will be dealt with in more detail in the Chapter 3.

2.18.2.3 Boundary Conditions for Alfvén Waves at the Upper Boundary of the Ionosphere

Expression (2.301) can be used to obtain the boundary conditions for Alfvén waves at the upper boundary of the ionosphere ($z = z_a$). Note that the derivative of the wave field components along the field line as written in the system of coordinates (x, y, z) has the form

$$\frac{\partial}{\partial l} = \cos\chi \frac{\partial}{\partial z} + ik_x \sin\chi.$$

Using equations (2.289), (2.290) for the derivative $\nabla_z f_{1,2}$, while taking into account expressions (2.292), (2.294) and (2.302), we obtain, at satisfactory accuracy, the boundary condition at the upper boundary of the ionosphere ($z = z_A$) for the Alfvén wave magnetic field components:

$$\left[\frac{\partial \bar{B}_{n,y}}{\partial l} + i\frac{\omega}{v_A^2} v_P \bar{B}_{n,y} \pm \frac{k_{y,nA}}{k_{\perp A}^2} \frac{\omega \bar{j}_\parallel}{v_A^2 \Sigma_P} \exp[ik_x(z-H)\tan\chi]\right]_{z=z_A} = 0, \tag{2.305}$$

where $v_P = \bar{c}^{-2}\cos\chi/4\pi\Sigma_P$ is the propagation speed of the low-frequency ionospheric whistler. Here, the upper '+' corresponds to the n index, and the lower '−' to y. Note that some terms are omitted from (2.305) which fail to describe new physical effects, but only contribute a small correction to the phase of the reflected Alfvén wave in the magnetosphere. The remaining terms describe the Alfvén wave source (external currents) and the wave dissipation in the ionosphere. Let us use the Maxwell equation (2.266) in order to write analogous boundary conditions for the transverse components of the wave electric field:

$$\left[\bar{E}_{n,y} + i\frac{v_P}{\omega}\frac{\partial \bar{E}_{n,y}}{\partial l} - i\frac{k_{nA,y}}{\bar{c}k_{\perp A}^2}\frac{\bar{j}_\parallel}{\Sigma_P} \exp[ik_x(z-H)\tan\chi]\right]_{z=z_A} = 0. \tag{2.306}$$

By means of a Fourier transform inverse to (2.276), it is possible to render (2.305) and (2.306) as the boundary conditions for monochromatic Alfvén waves with arbitrary transverse structure of their field in the magnetosphere:

$$\left.\frac{\partial \tilde{B}_{n,y}}{\partial l}\right|_{l=l_+} = i\frac{\omega v_{P+}}{v_{A+}^2}\left[\tilde{B}_{n,y} \pm \frac{4\pi}{c}\nabla_{y,n}\tilde{J}_{\parallel}\right]_{l=l_+}, \tag{2.307}$$

$$\left.\tilde{E}_{n,y}\right|_{l=l_+} = \left[-i\frac{v_P}{\omega}\frac{\partial \tilde{E}_{n,y}}{\partial l} + \frac{\nabla_{n,y}\tilde{J}_{\parallel}}{V_P}\right]_{l=l_+}, \tag{2.308}$$

where the subscript '+' corresponds to values taken at the point $l = l_+$ where the field line intersects the upper boundary of the ionosphere ($z = z_a$), and the notations are

$$\nabla_{n,y} \equiv \frac{\partial}{\partial n}, \frac{\partial}{\partial y}; \quad v_P = \frac{\bar{c}^2 \cos\chi}{4\pi\Sigma_P}; \quad V_P = \frac{\Sigma_P}{\cos\chi}.$$

The function $\tilde{J}_{\parallel}$ is related to the density of parallel external currents at the upper boundary of the current layer $\tilde{j}_{\parallel}$ by

$$\Delta_{\perp}\tilde{J}_{\parallel} = \tilde{j}_{\parallel},$$

where $\Delta_{\perp} = \nabla_n^2 + \nabla_y^2$ is the transverse Laplacian. Using (2.308) to represent the electric field of Alfvén waves in terms of the scalar potential (2.72) yields the following boundary condition:

$$\varphi|_{l=l_{\pm}} = \left[\pm i\frac{v_P}{\omega}\frac{\partial\varphi}{\partial l} - \frac{\tilde{J}_{\parallel}}{V_P}\right]_{l=l_{\pm}}, \tag{2.309}$$

where the ± signs correspond to the upper ionospheric boundary in Earth's Northern (l_+) or Southern (l_-) hemispheres. The boundary conditions (2.309) for $\chi = 0$ completely correspond to the boundary conditions (2.274) obtained above for the 'thin layer' model. It should be noted that the field of magnetosonic wave generated by Hall currents in the ionosphere can be neglected at the upper ionospheric boundary because their amplitude exponentially decreases upwards, for the low-frequency oscillations under study.

As we will see in the Chapter 3, the amplitude of poloidal Alfvén wave excited in the magnetosphere by external currents in the ionospheric E-layer can be written as

$$B_a = \frac{2\bar{c}}{\omega t_A v_{A0}}\frac{k_y a_N}{\lambda_{PN}}\frac{\tilde{J}_{\parallel +}}{V_P}, \tag{2.310}$$

where v_{A0} is the Alfvén speed at the magnetic shells we examine at the equatorial plane, t_A is the Alfvén transit time along the field line between the magneto-conjugated ionospheres. Moreover, (2.310) includes two characteristic scales: characteristic transverse scale of magnetospheric plasma inhomogeneity a_N and characteristic transverse length of poloidal Alfvén wave λ_{PN} in the neighbourhood of the resonance magnetic shell it is generated on. Exact expressions for these parameters will be obtained in the Chapter 3. For order-of-magnitude estimates, we will note that

$$k_y a_N \sim m,$$

where $m \gg 1$ is the azimuthal wave number of poloidal Alfvén wave,

$$\omega t_A \sim N,$$

N is the harmonic number of standing Alfvén waves in the magnetosphere (for fundamental harmonics $N \sim 1$) and

$$\lambda_{PN} \sim a_N^{1/3}/k_y^{2/3}.$$

Substituting (2.310) in (2.302), and later in (2.292) and (2.294), we find that the terms in these equations that are proportional to the currents are small if

$$\frac{m^{2/3}}{N}\frac{\overline{c}^2}{v_{A0}\Sigma_P} \gg 1. \tag{2.311}$$

For the characteristic values of the parameters of the medium under investigation — $v_{A0} \sim 10^3$km/s, $\Sigma_P \sim 10^8$ km/s, the condition (2.311) is reduced to the requirement that $m^{2/3} \gg N$, which is necessarily satisfied for poloidal Alfvén waves.

2.18.2.4 Electromagnetic Oscillations Induced on the Earth Surface by Magnetospheric Alfvén Waves

Let us define the field of electromagnetic oscillations induced at the Earth surface by incident Alfvén waves from the magnetosphere which are then reflected by the ionosphere. The condition (2.311) allows the direct effect of external currents in the ionosphere on electromagnetic oscillations at the Earth surface to be neglected because monochromatic Alfvén waves are 'tied' to a certain resonance magnetic shell. With time the waves 'accumulate' amplitude on this shell so that the amplitude of electromagnetic oscillations induced by them on Earth's surface becomes significantly larger than that directly linked to external currents.

Based on (2.286), (2.294), (2.296), (2.297) and (2.302), we have, at the lower ionospheric boundary,

$$F_1(H) = -\frac{\cos\chi}{K_P}B_A,$$

$$F_2(H) = 0,$$

$$F_3(H) = i\frac{k_0}{2k_{\perp A}}e^{-2k_\perp H}\left(\frac{\overline{K}_H}{K_P} + i\frac{k_y\sin\chi}{k_\perp}\frac{\overline{K}_P}{K_P}\right)B_A,$$

$$F_4(H) = -i\frac{k_0}{2k_{\perp A}}\left(\frac{\overline{K}_H}{K_P} - i\frac{k_y\sin\chi}{k_\perp}\frac{\overline{K}_P}{K_P}\right)B_A.$$

Substituting these expressions in (2.284) allows the components of the oscillation field $\bar{E}_n, \bar{E}_y, \overline{B}_y, \overline{B}_n$ at the lower ionospheric boundary, and, through simple geometric transformations, the horizontal components of the field to be obtained:

$$\overline{B}_\perp(H) = \frac{k_\perp}{k_{\perp A}}\cosh(k_\perp H)e^{-k_\perp H}\left(\frac{\overline{K}_H}{K_P} - i\frac{k_y\sin\chi}{k_\perp}\frac{\overline{K}_P}{K_P}\right)B_A,$$

$$\bar{E}_b(H) = -i\frac{k_0}{k_{\perp A}}\sinh(k_\perp H)e^{-k_\perp H}\left(\frac{\overline{K}_H}{K_P} - i\frac{k_y\sin\chi}{k_\perp}\frac{\overline{K}_P}{K_P}\right)B_A,$$

$$\bar{E}_\perp(H) = -i\frac{k_\perp\cos\chi}{k_{\perp A}K_P}B_A,$$

$$\overline{B}_b(H) = 0.$$

Based on (2.279), we have, at the Earth surface,

$$\overline{B}_\perp(0) = \frac{k_\perp}{k_{\perp A}} e^{-k_\perp H} \left(\frac{\overline{K}_H}{K_P} - i\frac{k_y \sin\chi}{k_\perp} \frac{\overline{K}_P}{K_P} \right) B_A,$$

$$\bar{E}_\perp(0) = \bar{E}_b(0) = \overline{B}_b(0) = 0.$$

or, for the field components in the coordinate system (x, y, z), we obtain the following equations:

$$\begin{aligned} \overline{B}_x(0) &= \overline{B}_y(z_a)\overline{R}(k_x, k_y)\cos\chi, \\ \overline{B}_y(0) &= -\overline{B}_n(z_a)\overline{R}(k_x, k_y), \end{aligned} \tag{2.312}$$

relating Alfvén wave amplitudes at the upper boundary of the ionosphere $\overline{B}_{n,y}(z_a)$ to the amplitudes of the magnetic field at Earth's surface, where

$$\overline{R}(k_x, k_y) = \frac{1}{\Sigma_P} \int_0^\infty \left(\sigma_H(z) - i\frac{k_y}{k_\perp}\sigma_P(z)\sin\chi \right) \times \exp(-k_\perp + ik_x(z - z_a)\tan\chi)dz. \tag{2.313}$$

Here, the upper limit of the integration extends to infinity because, outside the conducting layer, the transverse conductivities of the ionospheric plasma decrease exponentially, therefore the integrals themselves change little. Function $\overline{R}(k_x, k_y)$ takes its simplest form for large-scale Alfvén waves with $k_\perp \Delta \ll 1$. In this case the ionosphere can be regarded as a thin layer for Alfvén oscillations. For them, we have

$$\overline{R}(k_x, k_y) \approx \frac{\Sigma_H - i(k_y/k_\perp)\Sigma_P \sin\chi}{\Sigma_P}.$$

A consequence of equations (2.312) is that the dominant component of the magnetic field of the oscillations changes as they penetrate to the ground. If we have a toroidal Alfvén wave (with dominant azimuthal component $\overline{B}_y(z_a) \gg \overline{B}_n(z_a)$) in the upper ionosphere, after they penetrate to the ground the poloidal component $\overline{B}_x(0) \gg \overline{B}_y(0)$ becomes dominant. If we have a poloidal Alfvén wave ($\overline{B}_y(z_a) \ll \overline{B}_n(z_a)$) in the upper ionosphere, the azimuthal component of the field $\overline{B}_y(0) \gg \overline{B}_x(0)$ becomes dominant on the ground. This effect is known as a $\pi/2$ turn of the polarisation vector [184, 197]. Note that this is only valid for a horizontally homogeneous model ionosphere. For a horizontally inhomogeneous ionosphere, the rotation of the polarisation ellipse may differ significantly from $\pi/2$ [203, 204].

Additionally, the oscillation hodograph undergoes direction reversal. For a poloidal Alfvén wave in the magnetosphere, the oscillation hodograph in the plane (B_n, B_y) rotates clockwise. In contrast, the hodograph for a toroidal Alfvén wave rotates counter-clockwise. The situation is reversed at Earth's surface. For oscillations induced at Earth's surface by a poloidal Alfvén wave, the hodograph in the (B_x, B_y) plane rotates counter-clockwise, while rotating clockwise for oscillations induced by a toroidal Alfvén wave.

Another consequence of the general formulae (2.312), (2.313) is that, in the general case, it is not only the field of oscillations caused by the Hall conductivity of the ionosphere that penetrates to Earth surface, as is the case in the vertical magnetic field problem, but so does the field caused by the Pedersen conductivity. This effect is only present in problems relying on a model medium with an inclined geomagnetic field when the horizontal wave vector of the oscillations in question is outside the magnetic meridian plan ($k_y \neq 0$). Most papers examining the MHD wave field penetration to Earth solve problems where this effect cannot be detected. They either consider models with vertical geomagnetic field ($\chi = 0$), or study the meridional propagation of waves ($k_y = 0$) in models with an inclined magnetic field. In all these cases, only oscillations caused by Hall currents in the ionosphere penetrate to Earth surface.

3

MHD Oscillations in 2D-Inhomogeneous Models

This chapter discusses magnetospheric MHD oscillations in models where the plasma is inhomogeneous, not only across magnetic shells but along the magnetic field lines as well. These models are chiefly axisymmetric, the parameters remaining inhomogeneous in the azimuthal coordinate. The full wave field can be decomposed in such models into azimuthal harmonics of the form $\exp(im\phi)$, where ϕ is the azimuthal angle, and $m = 0, 1, 2, 3, \ldots$ are the azimuthal wave numbers. All wave processes characteristic of 1D-inhomogeneous models also take place in a 2D-inhomogeneous medium. They have certain peculiarities, though.

Theoretical investigation of MHD waves in the magnetosphere was first started in [19], where equations were derived describing the longitudinal structure and spectrum of azimuthally symmetric waves ($m = 0$) which were called 'toroidal'. Plasma and magnetic field oscillations in these waves are in the azimuthal direction (Figure 3.1, left). The other limiting case was later examined in [215] and [216] – that of a $m \to \infty$ wave. Alfvén waves of this type were called 'poloidal'. The plasma and magnetic field in such waves oscillate in the radial direction, in the magnetic meridian plane (Figure 3.1, right). The eigenfrequencies and longitudinal structure differ comparatively little between poloidal and toroidal standing Alfvén waves [217, 218].

The results of those papers were later generalised for the case of models involving not only geomagnetic field line curvature, but the transverse inhomogeneity of plasma, as well. A magnetospheric model was used in [219, 220] having the form of a half-cylinder, while a more complex magnetospheric model with a dipole magnetic field was employed in [217, 221, 222]. It was shown that a characteristic resonant singularity appears in the Alfvén wave field distribution across magnetic shells in those models.

The above approaches in theoretical research have been found to describe, from different standpoints, the same physical phenomenon – the Alfvén resonance, a process where toroidal standing Alfvén waves with $m \sim 1$ are excited in an axisymmetric model magnetosphere [223–225]. The structure of these waves across magnetic shells has a characteristic singularity on the resonance surface which is analogous to the one observed in unbounded plasma. The amplitude of the Alfvén waves excited on the resonance surface is determined by competing effects: their energy dissipation in the ionosphere and transverse dispersion, causing these waves to escape from the resonance surface across the magnetic shells [48, 49]. Both these effects are small in the real magnetosphere, resulting in a characteristic resonant peak in the distribution of the Alfvén wave field components across the magnetic shells. At the same time, these oscillations have the form of standing waves

Magnetospheric MHD Oscillations: A Linear Theory, First Edition. Anatoly Leonovich, Vitalii Mazur, and Dmitri Klimushkin.

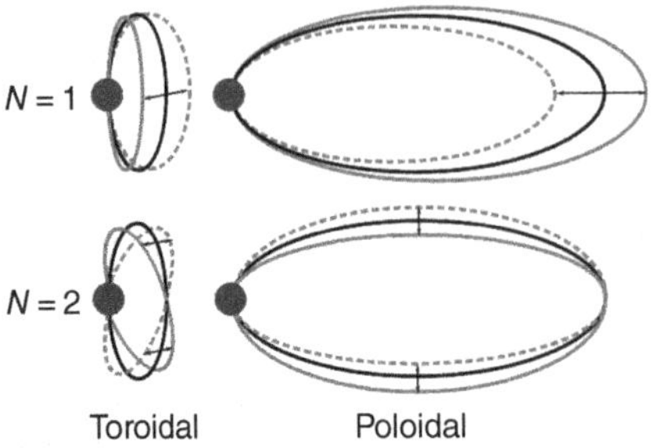

Figure 3.1 Toroidal and poloidal oscillations of field lines. The fundamental ($N = 1$) and second ($N = 2$) harmonics are shown.

along the geomagnetic field lines between the magnetoconjugated ionospheres. Alfvén resonance theory for an axisymmetric dipole-like model magnetosphere is discussed in Sections 3.1–3.3 of this monograph.

The main difference between model magnetospheres with curvilinear magnetic field and those with straight magnetic field lines lies in the structure of the magnetosonic waves involved in the Alfvén resonance. The problem of magnetosonic oscillations for a realistic 2D (in the geomagnetic meridian plane) plasma distribution proves to be strongly non-1D, rendering its exact analytical solution impossible due to inherent mathematical difficulties. Two different approaches are employed in order to overcome them.

The first consists of applying numeric simulation methods, allowing one to describe the structure of the full MHD oscillation field inside the magnetospheric cavity [177, 226, 227]. The framework of this approach made it possible to establish the fact that magnetosonic oscillation energy is concentrated in a narrow neighbourhood near the equatorial plane, which was confirmed from spacecraft observations [228–230]. This effect is absent from the magnetospheric model with straight geomagnetic field lines. This approach, while producing the general picture of MHD waves excited in the magnetosphere, unfortunately fails to explore the regularities and features of the interaction between the Alfvén and magnetosonic oscillations. In this respect, numerical simulation should be treated as a numerical experiment the results of which are themselves to be analytically interpreted.

The other approach consists in employing a combined (analytical and numerical) method for solving a system of specially derived equations describing MHD oscillations in the magnetosphere. Inhomogeneous plasma lacks independent modes of MHD oscillations, their wave field possessing a complex structure. In the Alfvén resonance region, it resembles the Alfvén wave field, while resembling magnetosonic wave field away from it. Therefore subdividing MHD oscillation field into different modes analogous to those in a homogeneous plasma, becomes a matter of convention in an inhomogeneous plasma.

A number of papers employ a method of describing MHD oscillation field in an inhomogeneous plasma by way of subdividing it into the potential and the vortical components [64, 65]. They demonstrate that the potential component of the field in a homogeneous plasma corresponds to Alfvén waves, while the vortical, to magnetosonic waves. This method was used in [231, 232] to explore magnetosonic eigen-oscillation decay process in a waveguide caused by the resonant interaction between these oscillations and Alfvén waves, while in [233, 234], it enabled the spatial structure to be described of magnetosonic eigen-oscillations in the magnetospheric cavity using a dipole magnetosphere model.

To date, magnetosonic oscillations in the magnetosphere have been less well studied theoretically than Alfvén waves. This is evidenced even by the number of theoretical papers addressing magnetosonic oscillations in the magnetosphere being much smaller than those dealing with Alfvén waves. This ratio is determined by sizeable mathematic difficulties inherent in analytical investigations of fast magnetosonic mode (FMS) oscillation field. Certain achievements can be found in this area, however, resulting from simple 1D-inhomogeneous model magnetospheres. Note primarily [235, 236], examining the possibility fast magnetosonic oscillations confined inside a magnetospheric resonator formed by the plasma-density gradient near the plasmapause. This resonator, first discovered theoretically, was later found in spacecraft observations [237]. A simple 1D-inhomogeneous model magnetosphere (box model) was used in [238, 239] to analyse the structure of magnetosonic eigenmodes of magnetospheric oscillations. An analogous approach to studying magnetosonic eigen oscillations in the magnetosphere was also employed in [76, 79, 240]. Papers [182, 241] are devoted to the magnetosonic oscillations in 1D-inhomogeneous model magnetosphere generated by non-stationary sources. A theoretical investigation into the FMS oscillation structure and spectrum in a dipole-like magnetosphere can be found in Sections 3.4–3.6 of this monograph.

A special variety of magnetospheric MHD oscillations are transversely small-scale Alfvén waves [46, 139]. 'Transversely small-scale' means small characteristic scales of wave field variation across magnetic field lines as compared to typical scale of magnetospheric parameter variation and to the parallel wavelength. This means, in particular, that the azimuthal wave number $m \gg 1$ for such oscillations. Transverse small scale is a natural property of Alfvén oscillations in a transverse-inhomogeneous plasma [242, 243] due to small transverse dispersion of Alfvén waves. Small dispersion means that neighbouring field lines oscillate almost independently of each other and, transverse inhomogeneity present, with different frequency. As a result, the oscillation phases are mixed, i.e. their transverse structure becomes small-scale [36, 217, 244]. It is only transverse dispersion that can stop or prevent it from becoming small scale, but its being small results in a small transverse wavelength.

This reasoning, however, only proves that Alfvén waves are small-scale in the normal direction to the magnetic shells. In the azimuthal direction, the wave can be small- or large-scale. Which of the two possibilities is realised is determined by the wave excitation mechanism. Alfvén resonance is an example of how azimuthally large-scale wave with $m \sim 1$ is excited.

Magnetosonic waves with $m \gg 1$ cannot penetrate deep into the magnetosphere from the outside because it is an opaque region for such waves. Thanks to this, effective generation of Alfvén oscillations with $m \gg 1$ requires that their source should be at the same magnetic shells where they are excited. Such waves can be generated by energetic particles drifting in the magnetosphere the in the azimuthal direction [245–250]. Another example of the local sources is external currents in the conducting ionospheric layer caused by neutrals moving due to various processes of both natural and artificial origin (see Section 2.18 in this monograph).

Whereas the parallel structure of Alfvén waves with $m \gg 1$ has been studied long enough, comparatively little is known of their structure across magnetic shells. Such research was done in [251–253]. They demonstrated that transverse-small scale Alfvén waves have a more complex structure than resonant Alfvén waves with $m \sim 1$. A monochromatic source excites a poloidal standing Alfvén wave on a resonance magnetic shell where the source frequency coincides with the local frequency of poloidal Alfvén oscillations. This wave then escapes across magnetic shells towards another, 'toroidal', resonance surface, where it is fully absorbed due to dissipation. During this transit, the polarisation of the oscillations changes from 'poloidal' to 'toroidal'. If the movement of the Alfvén wave near the 'poloidal' resonance surface is restricted by turning points in the transverse coordinate (due to non-monotonic distribution of the Alfvén speed across magnetic shells), this wave can be captured in the resonator and remains poloidal. The field structure of monochromatic Alfvén waves with $m \gg 1$ is addressed in this chapter of the monograph, Sections 3.7–3.10.

Research into monochromatic oscillations allows their non-trivial spatial structure to be revealed but is not enough for describing real MHD oscillations in the magnetosphere having, as a rule, a more or less wide frequency spectrum [254–257]. A characteristic example of such oscillations are daytime geomagnetic pulsations Pc3 excited by magnetosonic waves of extra-magnetospheric origin [254, 258–260]. As a rule, these oscillations are stochastic in nature.

Another type includes oscillations excited in the magnetosphere by broadband correlated sources. The most illustrative example is a source of the sudden pulse type [176, 219, 261, 262], its time dependence described by δ-function. Oscillations of this class include all waves caused by transition processes in the magnetosphere. Characteristic examples include sudden storm commencement (SSC) or Pi2 pulsations [171, 263]. Such oscillations are less frequent than stochastic ones but do play an important role in magnetospheric physics.

Various phenomena of both natural and artificial origin can serve as the source of transverse-small scale Alfvén oscillations in the magnetosphere. External currents in the conducting layer of the ionosphere can excite transverse-small scale Alfvén waves in the magnetosphere. The external current themselves can be caused by various natural processes. Considerable electromagnetic perturbations are known to be observed in zones where large earthquakes are imminent [264, 265]. Penetrating into the ionosphere, these oscillations generate external currents in the conducting layer that serve as the source of Alfvén waves in the magnetosphere. Some perturbations are related to the regions of high thunderstorm activity. A certain part of discharges in the thunderstorm clouds is not directed Earthward but into the ionosphere [266]. This also must lead to intense external currents generated in the ionosphere.

Noteworthy artificial sources of external currents are rocket flying through the conducting layer of the ionosphere during spacecraft launches [267]. Analogous perturbations are also registered during active experiments involving a charged-particle cloud ejected into the ionosphere [268] or when they are artificially generated by means of large electromagnetic contours [188]. Papers discussing observations of disturbances related to surface explosions deserve a special note. Concerning these last sources, there is a detailed observation material enabling a detailed comparison of theory-based predictions to observation data. Galperin et al. [75, 269] present data on a strong pulse of Alfvén oscillations (an amplitude of $\sim$100 nT, in the magnetic component) observed during the active MASSA-1 experiment onboard Aureole-3 satellite. It was demonstrated in [270] that, in the direction along the geomagnetic field lines, the shape of the nonlinear Alfvén wave probably observed in the experiment resembles the pulse observed in the experiment. In [271], the shape of the registered pulse is explained by the transverse structure of the blast-generated Alfvén wave packet (see Section 3.15 of this monograph). Transverse-small scale Alfvén oscillations excited in the dipole-like magnetosphere by broadband and localised sources are studied in Sections 3.11–3.15 of this chapter of the monograph.

A separate area contains investigations of MHD oscillations in the geotail. The common trait of planetary magnetospheres possessing their own magnetic field is a magnetotail forming under the effect of the solar wind flow [272, 273]. A typical magnetotail structure forms a current sheet dividing it into two lobes (for example, the geotail). This current, closing through the magnetopause, forms a magnetic field which becomes nearly dipole near Earth, while the curvature radius of the field lines extending into the tail is very small in their apex. This creates specific conditions for propagation and interaction of Alfvén and slow magnetosonic waves.

The propagation direction of FMS waves is determined by their spatial structure and can be any direction relative to the background magnetic field. The phase velocities of Alfvén and slow magnetosound (SMS) waves are directed nearly along the background magnetic field, enabling their confinement on magnetic shells bounded by a high-conducting ionosphere in the Earth's magneto-conjugated hemispheres. These modes can interact on the field lines extending into the magnetotail, forming a complex wave-field pattern [274–278]. Under certain conditions, these oscillations can switch into the ballooning instability regime [279–282]. This instability, along with the tearing instability (see [283, 284]), is often regarded as the reconnection mechanism of the geomagnetic field in the current sheet, during substorms. These two types of instabilities are sometimes invoked to explain flapping oscillations of the geotail current sheet [285, 286].

Studies of the ballooning instability often employ the simplest approximation to analyse the local dispersion equation [281, 287, 288]. This equation is obtained for a medium that is homogeneous along the magnetic field lines, its parameters corresponding to the current sheet parameters. It was shown in [289], however, that MHD equation solutions obtained for the same geotail model produce different results for the local and WKB approximations, the latter taking into account parameter variations along magnetic field lines (see Section 3.16 of this Chapter). Moreover, Alfvén and SMS waves on field lines crossing the current sheet suffer mutual linear transformation resulting in a peculiar coupled mode of these oscillations. It also can switch to the ballooning instability regime if the geotail current is intense enough. The spatial structure and oscillation spectrum of such coupled modes in the geotail is examined in Section 3.17 of this Chapter.

3.1 Resonance Between FMS and Kinetic Alfvén Waves in a Dipole-Like Magnetosphere

In the ideal MHD approximation, solution equations describing resonance between Alfvén and FMS waves have a singularity at resonance magnetic shells. Here, the frequency of a monochromatic FMS wave playing the role of the source of resonant oscillations coincides with the frequency of a harmonic of standing toroidal Alfvén waves. This singularity can be eliminated by taking into account effects that are outside the ideal MHD framework, such as finite Larmor radius of plasma ions or electron inertia (see Section 2.7). Below we will examine the process of resonance for kinetic Alfvén waves in an axisymmetric, dipole-like model magnetosphere. Mathematically, it is a 2D-inhomogeneous model where the magnetic field and plasma are inhomogeneous, both across magnetic shells and along magnetic field lines.

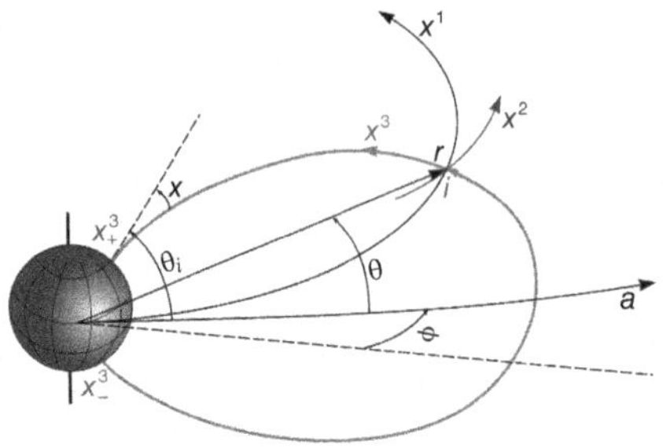

Figure 3.2 Coordinate systems related to geomagnetic field lines in an axisymmetric model magnetosphere: curvilinear orthogonal system of coordinates (x^1, x^2, x^3), curvilinear non-orthogonal system of coordinates (a, ϕ, θ).

We assume the plasma and magnetic field to be homogeneous in the azimuthal direction. Let us introduce an orthogonal system of curvilinear coordinates (x^1, x^2, x^3) where the coordinate x^3 is along the magnetic field line, the coordinate x^1 is across the magnetic shells and the azimuthal coordinate x^2 completes the right-hand system of coordinates (Figure 3.2). An arbitrary perturbation in such a model can be sought as a decomposition into Fourier harmonics $\exp(ik_2x^2 - i\omega t)$, where ω is the frequency of the oscillations, k_2 is the azimuthal wave number (if the azimuthal angle ϕ is used as the x^2, then $k_2 = m = 0, 1, 2, 3, \ldots$).

It is easiest to obtain the system of MHD equations relating the field of FMS oscillations and kinetic Alfvén waves from the approximate vector equation (2.69), where the dielectric permeability tensor has the form:

$$\hat{\varepsilon} = \frac{\bar{c}^2}{v_A^2}\begin{pmatrix} 1 - \dfrac{3k_1^2\rho_i^2}{4g_1} & -\dfrac{3k_1k_2\rho_i^2}{4\sqrt{g_1g_2}} & 0 \\ -\dfrac{3k_1k_2\rho_i^2}{4\sqrt{g_1g_2}} & 1 - \dfrac{3k_2^2\rho_i^2}{4g_2} & 0 \\ 0 & 0 & v_A^2\dfrac{\tilde{G}(s_e/\rho_s)}{\omega^2\rho_s^2} \end{pmatrix}, \tag{3.1}$$

where g_1, g_2, g_3 are the diagonal components of the metric tensor, and the k_1 component of the wave vector should be understood as a derivative, $k_1 = -i\nabla_1 \equiv -i\partial/\partial x^1$, acting on the wave-field components. Recall that here, $s_e = \bar{c}/\omega_{pe}$ is the electron skin depth in plasma, $\rho_s = \rho_i\sqrt{T_e/T_i}$, where $\rho_i = v_i/\omega_i$ is the ion Larmor radius, and the function $\tilde{G}(s_e/\rho_s)$ has the following property (see Section 2.7):

$$\frac{\rho_s^2}{\tilde{G}(s_e/\rho_s)} \approx \begin{cases} \rho_s^2, & s_e \ll \rho_s, \\ -s_e^2, & s_e \gg \rho_s. \end{cases}$$

The length element squared in the orthogonal curvilinear system of coordinates has the form

$$ds^2 = g_1(dx^1)^2 + g_2(dx^2)^2 + g_3(dx^3)^2,$$

while the covariant components of the curl**E** are

$$(\text{rot}\mathbf{E})_1 = \frac{g_1}{\sqrt{g}}\left(\frac{\partial E_3}{\partial x^2} - \frac{\partial E_2}{\partial x^3}\right), \quad (\text{rot}\mathbf{E})_2 = \frac{g_2}{\sqrt{g}}\left(\frac{\partial E_1}{\partial x^3} - \frac{\partial E_3}{\partial x^1}\right),$$

$$(\text{rot}\mathbf{E})_3 = \frac{g_3}{\sqrt{g}}\left(\frac{\partial E_2}{\partial x^1} - \frac{\partial E_1}{\partial x^2}\right),$$

where $g = g_1 g_2 g_3$. For the transverse electric field of oscillations, we will use the decomposition (2.72), where the field components are expressed in terms of the scalar and vector potentials as

$$E_1 = -\nabla_1\varphi + ik_2\frac{g_1}{\sqrt{g}}\psi, \qquad E_2 = -ik_2\varphi - \frac{g_2}{\sqrt{g}}\nabla_1\psi. \tag{3.2}$$

Writing out (2.69) componentwise produces

$$E_3 \approx -i\frac{\rho_s^2}{\tilde{G}(s_e/\rho_s)}\frac{\omega\sqrt{g_3}}{v_A}\Delta_\perp\varphi, \tag{3.3}$$

where $\Delta_\perp = g_1^{-1}\nabla_1^2 - g_2^{-1}k_2^2$ is the transverse Laplacian. Substituting this expression into the two other equations yields

$$\frac{\omega^2}{v_A^2}\Lambda^2 p\Delta_\perp\varphi' + \hat{L}_T\varphi' = i\frac{k_2}{\sqrt{g_3}}\left(\tilde{\Delta} + \frac{\omega^2}{v_A^2}\right)\psi, \tag{3.4}$$

$$\nabla_1\tilde{\Delta}_\perp\psi + \sqrt{g_3}\hat{L}_P\frac{p}{\sqrt{g_3}}\nabla_1\psi = -i\frac{k_2\sqrt{g}}{g_2}\left(\frac{\omega^2}{v_A^2}\Lambda^2\Delta_\perp\varphi + p\hat{L}_P\varphi\right), \tag{3.5}$$

where $\varphi' \equiv \nabla_1\varphi, p = \sqrt{g_2/g_1}, \Lambda^2 = (3/4)\rho_i^2 + s_e^2/\tilde{G}(s_e/\rho_s)$ (in the $s_e \sim \rho_s$ region, the Λ^2 function is complex, see Figure 1.4). The following operators are also used:

$$\hat{L}_T = \frac{1}{\sqrt{g_3}}\nabla_3\frac{p}{\sqrt{g_3}}\nabla_3 + p\frac{\omega^2}{v_A^2}, \tag{3.6}$$

$$\hat{L}_P = \frac{1}{\sqrt{g_3}}\nabla_3\frac{p^{-1}}{\sqrt{g_3}}\nabla_3 + p^{-1}\frac{\omega^2}{v_A^2}, \tag{3.7}$$

$$\tilde{\Delta} = \tilde{\Delta}_\perp + \nabla_3\frac{p}{\sqrt{g_3}}\nabla_3\frac{p^{-1}}{\sqrt{g_3}}, \tag{3.8}$$

$$\tilde{\Delta}_\perp = \frac{g_3}{\sqrt{g}}\nabla_1\frac{g_2}{\sqrt{g}}\nabla_1 - \frac{k_2^2}{g_2}, \tag{3.9}$$

where $\nabla_i \equiv \partial/\partial x^i$ ($i = 1, 2, 3, \ldots$). The operators (3.6) and (3.7) describe the parallel structure of toroidal and poloidal Alfvén waves, while (3.8) and (3.9) are analogous to the Laplace operator in the curvilinear orthogonal system of coordinates. It is easy to verify that switching to the Cartesian system of coordinates in the 1-D-inhomogeneous plasma model produces the system of Eqs. (2.77) and (2.78), describing Alfvén resonance in non-ideal plasma ($g_1 = g_2 = g_3 = 1$ and $\nabla_1 = \nabla_x, \nabla_3 = ik_z, k_2 = k_y$).

For axisymmetric oscillations ($m = 0$) the right-hand sides of (3.4) and (3.5) turn to zero. Equation (3.4) then describes axisymmetric Alfvén wave uncoupled to FMS wave field. This wave cannot be resonance-driven. Equation (3.5) in this case describes axisymmetric FMS wave lacking resonant magnetic shells. Given the plasma gyrotropy effect (non-zero small parameter $u = \omega/\omega_i \ll 1$), these oscillations are coupled, even if the coupling is weak. We will, however, ignore this effect for the time being.

For $m \neq 0$ oscillations, (3.4) describes the field of kinetic Alfvén waves excited on resonance magnetic shells by large-scale monochromatic FMS waves. The solution of the second equation, describing the field of large-scale FMS wave, will be found later in Sections 3.5 and 3.6. Note here that the magnetosphere is an opaque region for FMS waves with $m \gg 1$, their amplitude decreasing exponentially into the magnetosphere, at a scale of $\sim L/m$, where L is the characteristic transverse size of the magnetosphere. Alfvén resonance inside the magnetosphere is thus effective enough only for the first azimuthal harmonics of MHD oscillations ($m \sim 1$).

As we will see later, the expression in the right-hand side of (3.4) varies little, on the scale of the resonant Alfvén wave localisation. We will therefore deem it constant, in the main order of perturbation theory, when solving the problem of resonant Alfvén wave field structure. We will use (2.309) differentiated over x^1 as the boundary condition for Alfvén waves on the ionosphere, where the presence of external currents in the conducting layer is ignored:

$$\varphi'\big|_{x^3=x^3_\pm} = \left[\pm i\frac{v_P}{\omega\sqrt{g_3}}\frac{\partial\varphi'}{\partial x^3}\right]_{x^3=x^3_\pm}. \qquad (3.10)$$

Here points $x^3_\pm$ are the coordinates of the field-line intersection points with the upper boundary of the ionosphere in the Northern (+) and Southern (−) hemispheres, respectively. Recall that we assume the upper boundary to be at a height of $z = z_A \approx 1500$ km, where the Alfvén speed gradient suffers a dramatic change. Moreover, we will assume the parameter $\delta_\pm = (v_p/v_A)_\pm \ll 1$ to be small, using it to develop the perturbation theory below.

The following preliminary remark should be made in order to find the solution of the partial differential equation (3.4). As will be seen later, it describes Alfvén standing waves between the magneto-conjugated ionospheres, along geomagnetic field lines (along the x^3 coordinate). The wavelength of their fundamental harmonics is comparable to the field-line length ($\sim L$) on the resonance magnetic shell. If the solution of (3.4) is sought in the x^3 coordinate, in the Wentzel-Kramers-Brillouin (WKB) approximation ($\varphi \sim \exp(\int k_3 dx^3)$), then $k_3 \sim L^{-1}$ for the oscillations in question. Across magnetic shells (along the x^1 coordinate) the characteristic localisation scale of resonant Alfvén waves is small being determined by the small parameters of the problem. Thus the ratio $|k_1| \gg |k_3|$ holds true for the WKB components of the wave vector.

If the WKB solution of (3.4) is sought in the general form as

$$\varphi' = \exp\left(i\Theta(x^1, x^3)\right),$$

where $\Theta(x^1, x^3)$ is a large quasi-classical phase, then

$$k_1 = \nabla_1\Theta(x^1, x^3), \qquad k_3 = \nabla_3\Theta(x^1, x^3).$$

The condition $|k_1| \gg |k_3|$ means that the phase Θ depends on x^1 much more strongly than on x^3. In the main order of the parameter $|k_3/k_1| \ll 1$ decomposition it should be assumed that Θ does not depend on x^3 and it is possible to extract the main component Θ_0 of the phase depending on x^1 only and to represent the full phase as

$$\Theta(x^1, x^3) = \Theta_0(x^1) + \Theta_1(x^1, x^3),$$

where $\Theta_0 \gg \Theta_1$. In the main order of perturbation theory, the wave vector component $k_1 = \nabla\Theta_0(x^1)$ then does not depend on x^3 and is a function of the x^1 coordinate alone. Hence, the full solution of (3.4) can be represented as

$$\varphi'(x^1, x^3) = V'(x^1)H(x^1, x^3), \qquad (3.11)$$

where function $V'(x^1) = \exp\left(i\Theta_0(x^1)\right)$ describes, in the WKB approximation, the small-scale structure of the oscillations, across the magnetic shells and $H(x^1, x^3) = \exp\left(i\Theta_1(x^1, x^3)\right)$ describes their structure along magnetic field lines. The x^1 dependence of $H(x^1, x^3)$ is weak enough, however, meaning that, in the main order, function $V'(x^1)$ alone is responsible for the x^1 derivative. The solution of (3.4) can also be sought in the form of (3.11) in the general case when the WKB approximation does not apply. This approach is a version of the multiple-scale method in problems for a two-variable function.

3.1.1 Longitudinal Structure of Toroidal Alfvén Waves

We will seek the general solution of (3.4) with boundary conditions (3.10) in this form:

$$\varphi'(x^1, x^3) = V'(x^1)(T(x^1, x^3) + h(x^1, x^3)), \tag{3.12}$$

where function $T(x^1, x^3)$ is the solution of the zero-approximation problem

$$\hat{L}_T T = \frac{1}{\sqrt{g_3}} \nabla_3 \frac{p}{\sqrt{g_3}} \nabla_3 T + p\frac{\omega^2}{v_A^2} T = 0 \tag{3.13}$$

with homogeneous boundary conditions on the ionosphere $T(x^1, x^3_\pm) = 0$. Equation (3.13) is derived from (3.4) if the small terms are neglected: the first term, proportional to the small parameter Λ^2, and the right-hand side, describing the source of resonant oscillations – magnetosonic wave field.

This equation can be regarded as an equation of the zero order of perturbation theory determining the structure of Alfvén waves along magnetic field lines. The x^1-coordinate dependence of function T is determined by the equation coefficients – Alfvén speed $v_A(x^1, x^3)$ and metric tensor components $g_i(x^1, x^3)$ that depend on x^1 as parameter. Hence, the characteristic variation scale of the solution $T(x^1, x^3)$ in the x^1 coordinate is determined by the background plasma inhomogeneity scale, which is much larger than the localisation scale of resonant Alfvén oscillations in this coordinate.

The second term in the right-hand side of (3.12) is a small correction ($|h(x^1, x^3)| \ll |T(x^1, x^3)|$) which should be included in the next, first, order of perturbation theory. The non-zero right-hand side of (3.4) and the small correction in the boundary conditions (3.10) describing oscillation decay in the ionosphere will also be taken into account in this first order. The solutions of (3.13) with homogeneous boundary conditions on the ionosphere are eigenfunctions and the corresponding eigenvalues of the frequency

$$T = T_N(x^1, x^3), \quad \omega = \Omega_{TN}(x^1), \quad N = 1, 2, 3 \ldots .$$

They describe the parallel (along magnetic field lines) structure of toroidal standing Alfvén waves (N is the parallel wavenumber) on the resonance magnetic shell. In the main order of perturbation theory, the full two-dimensional structure of such waves in a meridional section can be represented as

$$\varphi' = \varphi'_N(x^1, x^3) = C\delta(x^1 - x^1_{TN})T_N(x^1, x^3),$$

where C is an arbitrary constant, $\delta(x)$ is the delta-function, x^1_{TN} is the coordinate of the resonance magnetic shell. This means that the oscillations are concentrated on the resonance magnetic shell $x^1 = x^1_{TN}$, while having a structure described by eigen-functions $T_N(x^1_{TN}, x^3)$ along the magnetic field lines.

It follows from the general theory of ordinary differential equations (see [290]) that the set of the solutions of (3.13) forms a complete system, and the corresponding eigenfunctions can be chosen to be orthonormal:

$$\int_{x_-^3}^{x_+^3} \frac{\sqrt{g}}{v_A^2 g_1} T_N T_{N'} dx^3 = \delta_{NN'}.$$

An arbitrary solution of (3.4) can be sought in the form of an expansion into the functions:

$$\varphi'(x^1, x^3) = \sum_{N=1}^{\infty} V_N'(x^1) T_N(x^1, x^3). \tag{3.14}$$

For the basic harmonics of standing waves ($N \sim 1$), the solution of (3.13) can only be found numerically. To form an idea of the qualitative behaviour of eigenfunctions $T_N(x^1, x^3)$, however, let us obtain this equation in the WKB approximation, applicable to harmonics with $N \gg 1$. We will seek the solution of (3.13) in the form of

$$T = \exp(i\overline{\theta}).$$

Here $\overline{\theta}$ is the large quasiclassical phase, which we will represent as

$$\overline{\theta} = \overline{\theta}_0 + \overline{\theta}_1 + \cdots,$$

where $|\overline{\theta}_0| \gg |\overline{\theta}_1| \gg \cdots$. Substituting this solution into (3.13) yields the following equation for the phase:

$$-\overline{\theta}'^{\,2} + i\overline{\theta}'' + i(\ln \frac{g_2}{g})'\overline{\theta}' + g_3 \frac{\omega^2}{v_A^2} = 0,$$

where the prime denotes the derivative: $\theta' = \nabla_3 \theta = \partial\theta/\partial x^3$. In the zero order of the WKB approximation we have

$$\overline{\theta}_0' = \pm\sqrt{g_3} \frac{\omega}{v_A},$$

while in the first order:

$$\overline{\theta}_1' = \frac{i}{2}\left(\ln \frac{g_2}{\sqrt{g}} \theta_0'\right)' = \frac{i}{2}\left(\ln \frac{p}{v_A}\right)'.$$

The orthonormal eigenfunctions, therefore, have the following form, in the WKB approximation:

$$T_N(x^1, x^3) = \sqrt{\frac{2v_A}{pt_A}} \sin\left(\Omega_{TN} \int_{x_-^3}^{x^3} \frac{\sqrt{g_3} dx^3}{v_A}\right), \tag{3.15}$$

where $\Omega_{TN} = \pi N/t_A$ are the eigen values of the problem (3.13) in the WKB approximation,

$$t_A = \int_{x_-^3}^{x_+^3} \frac{\sqrt{g_3} dx^3}{v_A} \tag{3.16}$$

is the Alfvén transit time along a magnetic field line between the magneto-conjugated ionospheres, on the resonance magnetic shell $x^1 = x_{TN}^1$. The solutions (3.15) describe standing waves between points x_+^3, x_-^3, with an $N-1$ node on the field line. Note that, in the general case, the functions (3.15) can be asymmetric about the equatorial plane.

3.1.2 Structure of Resonant Kinetic Alfvén Waves Across Magnetic Shells

Substituting a solution in the form of (3.12) into (3.4) yields, in the first order of perturbation theory,

$$\frac{\omega^2}{v_A^2}\frac{p\Lambda^2}{g_1}T_N\nabla_1^2V_N' + V_N'\left(\hat{L}_T(\Omega_{TN})h_N + p\frac{\omega^2-\Omega_{TN}^2}{v_A^2}T_N\right) = i\frac{k_2}{\sqrt{g_3}}\left(\tilde{\Delta} + \frac{\omega^2}{v_A^2}\right)\psi, \tag{3.17}$$

an equation determining the structure of the oscillations across magnetic shells (function $V_N'(x^1)$). Note that, in (3.17), we ignore the small term $\sim k_2^2\Lambda^2$ responsible for a small correction to the eigenfrequency of the oscillations in question.

In the boundary conditions on the ionosphere, we will include the non-zero right-hand side in (3.10) describing oscillation dissipation in the ionosphere:

$$h_N|_{x^3-x_\pm^3} = \left[\pm i\frac{v_P}{\omega\sqrt{g_3}}\frac{\partial T_N}{\partial x^3}\right]_{x^3=x_\pm^3}. \tag{3.18}$$

Integrating (3.17) along a field line between the magneto-conjugated ionospheres of the Northern and Southern hemispheres, with boundary conditions (3.18), produces

$$\sigma_N(x^1)\frac{\partial^2 V_N'}{\partial x^{1\,2}} + \frac{(\omega+i\gamma_{TN})^2-\Omega_{TN}^2}{\omega^2}V_N' = i\frac{k_2}{\omega^2}\mu_N(x^1), \tag{3.19}$$

where

$$\sigma_N(x^1) = \int_{x_-^3}^{x_+^3}\frac{p\Lambda^2}{g_1v_A^2}T_N^2(x^1,x^3)\sqrt{g_3}dx^3. \tag{3.20}$$

If the WKB approximation is applicable to the oscillations, in the x^3 coordinate ($N \gg 1$),

$$\sigma_N(x^1) = \frac{1}{t_A}\int_{x_-^3}^{x_+^3}\frac{\Lambda^2}{g_1}\frac{\sqrt{g_3}dx^3}{v_A}$$

is the average value of Λ^2/g_1 on the field line: $\sigma_N(x^1) = \langle\Lambda^2/g_1\rangle$, where the $\langle f\rangle$ brackets denote the magnetic field line-averaged value of function $f(x^3)$. Given its general properties (see Chapter 1), function Λ^2 can be represented as $\sigma_N = \rho_N^2\exp(-i\alpha_N)$, where ρ_N has the dimension of length, and the phase $0 \le \alpha_N \le \pi$.

Moreover, the boundary conditions on the ionosphere (3.18) produced an imaginary addition to the frequency in (3.19):

$$\gamma_{TN} = \frac{1}{\Omega_{TN}^2}\left[v_{p+}P_+\left(\frac{\partial T_N}{\partial l}\right)_+^2 + v_{p-}P_-\left(\frac{\partial T_N}{\partial l}\right)_-^2\right],$$

which is a decrement of standing toroidal Alfvén waves in the conducting layer of the ionosphere, where $\pm$ indicate the ionosphere of the Northern or Southern hemisphere, respectively. The derivative is taken along the physical length of the field line ($dl = \sqrt{g_3}dx^3$).

Function

$$\mu_N(x^1) = \int_{x_-^3}^{x_+^3}T_N\left(\tilde{\Delta} + \frac{\omega^2}{v_A^2}\right)\psi dx^3$$

in the right-hand side of (3.19) describes the source of resonant Alfvén waves which is related to the FMS field (function ψ).

If the functions $\Omega_{TN}(x^1)$ vary monotonically in the neighbourhood of the resonance surface, the eigenfrequencies of the resonant Alfvén oscillations in their localisation region can be written in the linear approximation as

$$\Omega_{TN}^2(x^1) \approx \omega^2 \left(1 - \frac{x^1 - x_{TN}^1}{a_N}\right), \tag{3.21}$$

where $a_N = (\nabla_1 \ln \Omega_{TN}^2)_{x_{TN}^1}^{-1}$ is the characteristic variation scale of function $\Omega_{TN}^2(x^1)$ near the resonance shell. In the neighbourhood of the resonance surface (3.18) can therefore be approximately written as

$$\rho_N^2 e^{-i\alpha_N} \frac{\partial^2 V_N'}{\partial x^{1\,2}} + \frac{x^1 - x_{TN}^1 + i\varepsilon_{TN}}{a_N} V_N' = i\frac{k_2}{\omega^2}\mu_N(x^1), \tag{3.22}$$

where $\varepsilon_{TN} = 2a_N\gamma_{TN}/\omega$ is the characteristic scale of resonant oscillation decay in the x^1 coordinate, due to their dissipation in the ionosphere. This equation has the same form as (2.86), which describes the structure of kinetic Alfvén waves in 1D-inhomogeneous plasma. Setting the right-hand side of (3.22) as constant, let us write down its solution satisfying the boundary conditions (finite amplitude on the asymptotics), as

$$V_N'(x^1, \omega) = i\frac{k_2}{\omega^2}\mu_N \left(\frac{a_N}{\rho_N}\right)^{2/3} e^{i\alpha_N/3} G\left(e^{i\alpha_N/3}\frac{x^1 - x_{TN}^1 + i\varepsilon_{TN}}{a_N^{1/3}\rho_N^{2/3}}\right), \tag{3.23}$$

where function $G(z)$ is the solution of inhomogeneous Airy equation (2.88), its integral representation being (2.89) and asymptotics (2.90). In the two limiting cases $\alpha_N = \pi$ ($\beta \ll m_e/m_i$ is the 'cold plasma' approximation) and $\alpha_N = 0$ ($\beta \gg m_e/m_i$ is the 'warm plasma' approximation) the qualitative behaviour of function (3.23) for small decay ($\gamma_{TN} \ll (a_N/\rho_N)^{2/3}\omega$) is shown in Figure 2.13. The exact distribution of the real and imaginary parts of function $G(z)$ for $\alpha_N = 0$ and $\gamma_{TN} = 0$ is shown in Figure 2.12.

In the real magnetosphere, however, standing Alfvén waves decay rather strongly, thanks to their dissipation in the ionospheric conducting layer. The function $G(z)$ asymptotic is therefore more likely to be determined by the dissipative scale ε_{TN} and has the form $G(z) \approx z^{-1}$, which for $|z| \to \infty$ produces

$$V_N'(x^1, \omega) \approx i\frac{k_2}{\omega^2}\frac{\mu_N a_N}{x^1 - x_{TN}^1 + i\varepsilon_{TN}}. \tag{3.24}$$

This approximation is applicable if the characteristic wavelength l_N of kinetic Alfvén waves is smaller than their dissipative scale ε_{TN} (which is analogous to the condition $\gamma_{TN} \gtrapprox (a_N/\rho_N)^{2/3}\omega$). Expression (3.24) in this case can be used even on the resonance surface, whereas the parallel component of the electric field becomes negligibly small (see (3.3).

The solution (3.23) can be generalised using the expansion (3.21). Expressing the argument of the function G through frequency ω and replacing the wave vector component k_2 with the derivative of the FMS oscillation field with respect to x^2 yields

$$V_N'(x^1, \omega) = \frac{\nabla_2 \tilde{\mu}_N(x^1, x^2, \omega)}{\omega^2}\left(\frac{a_N}{\rho_N}\right)^{2/3} e^{i\alpha_N/3} \times G\left[2\left(\frac{a_N}{\rho_N}\right)^{2/3} e^{i\alpha_N/3}\frac{\omega - \Omega_{TN} + i\gamma_{TN}}{\Omega_{TN}}\right]. \tag{3.25}$$

This form of the expression can be used for describing the field of monochromatic Alfvén waves both near the resonance surface, where the linear decomposition (3.21) is applicable, and away from it, where that decomposition ceases to apply. What is more important, however, is the fact that this expression can be used for studying broadband resonant Alfvén oscillations by doing an inverse Fourier transformation over the source frequency spectrum at observation point.

The full spatial structure of resonant kinetic Alfvén waves excited by a monochromatic FMS wave in a dipole-like magnetosphere is therefore a set of standing Alfvén waves (each harmonic on its own resonance magnetic shell). The parallel structure of such waves is described by (3.13), the WKB solution of which has the form of (3.15). The structure of each such wave across magnetic shells is described by (3.22), its solution (3.23) describing waves escaping from the resonance surface. The direction in which these waves propagate depends on plasma temperature. In a $\langle\beta\rangle \gg m_e/m_i$ plasma, kinetic Alfvén waves escape along the x^1 coordinate one side of the resonance magnetic shell, while an opaque region is located on the other side. In a $\langle\beta\rangle \ll m_e/m_i$ plasma, the propagation pattern is the complete opposite – the transparent and opaque regions change places with respect to the above case, along the x^1 coordinate. The wave escapes in the side which in a $\langle\beta\rangle \gg m_e/m_i$ plasma was an opaque region.

In the region where $\langle\beta\rangle \sim m_e/m_i$, however, the behaviour of the amplitude of oscillations escaping from the resonance surface differs drastically from that in those two limiting cases. Using the asymptotic expressions (2.90) away from the resonance surface yields, for oscillation amplitude distribution across magnetic shells away from the resonance surface ($x^1 - x^1_{TN} \gg l_N$):

$$|V'_N(x^1)| \approx \sqrt{\pi}\left|\frac{k_2\mu_N}{\omega^2}\right|\left(\frac{\alpha_N}{\rho_N}\right)^{2/3}\left(\frac{l_N}{x^1 - x^1_{TN}}\right)^{1/4}$$
$$\times \exp\left[-\frac{2}{3}\left(\frac{x^1 - x^1_{TN}}{l_N}\right)^{3/2}\sin\frac{\alpha_N}{2} - \varepsilon_{TN}\frac{\sqrt{x^1 - x^1_{TN}}}{l_N^{3/2}}\cos\frac{\alpha_N}{2}\right], \tag{3.26}$$

where $l_N = \alpha_N^{1/3}\rho_N^{2/3}$ is the characteristic wavelength in the x^1 coordinate. Thus, the decay of oscillations escaping from the resonance surface is only due to their (weak) absorption in the ionospheric conducting layer for $\alpha_N = 0$, but the oscillations decay strongly at the scale of order wavelength l_N if $\alpha_N \sim 1$. As we saw in Chapter 1, this is due to collisionless decay of Alfvén waves thanks to background plasma electrons (Landau damping) resulting in their strong absorption.

A region in Earth's magnetosphere where Alfvén waves are strongly absorbed is located in the transition layer of the plasmapause, where the plasma parameters change from values characteristic of a cold plasmasphere ($\beta \ll m_e/m_i$) to values typical of the outer magnetosphere ($\beta \gg m_e/m_i$). The same region exhibits a sharp gradient in background plasma density (and Alfvén speed) across magnetic shells, resulting in high density of the resonance magnetic shells. The projection of this region onto the ionosphere is where such optical phenomena as 'stable auroral red (SAR) arcs' are registered most frequently [291, 292]. They are triggered by fluxes of epithermal electrons precipitating into the ionosphere from the magnetosphere. One explanation of why these fluxes appear is the above-mentioned interaction between Alfvén waves and background plasma electrons. This mechanism was suggested in [293] to explain the magnetospheric ring current decay during the substorm recovery phase.

3.1.3 Feedback from Resonant Alfvén Oscillations to FMS Wave Field

Above, we found the spatial distribution of the field of resonant standing Alfvén waves excited in the magnetosphere by a large-scale monochromatic FMS wave. Let us now estimate the degree to which the resonant Alfvén waves feed back to the generating FMS wave. We assume that the large-scale FMS wave has characteristic spatial scales comparable to the scales $\sim L$ of magnetospheric plasma inhomogeneity. We will integrate (3.5) with respect to the variable x^1 across the resonance magnetic shell x_{TN}, on scale ε which is much larger than the Alfvén wave localisation scale but is much smaller than the FMS wave scale: $l_N \ll \varepsilon \ll L$.

Taking into account the term with maximum amplitude ($\sim \nabla_1\varphi$) only in the right-hand side of (3.5) results in the following approximate expression:

$$\{\tilde{\Delta}_\perp \psi\}_{x^1_{TN}\pm\varepsilon} \approx i\frac{k_2\sqrt{g}}{g_2}\frac{\omega^2}{v_A^2}\frac{\Lambda^2}{g_1}\{\varphi'\}_{x^1_{TN}\pm\varepsilon}, \tag{3.27}$$

where the braces denote the difference between the values of the enclosed expression at points $x^1_{TN} \pm \varepsilon$, and the coefficients in the right-hand side refer to the resonance surface. Using (3.14) for φ', where the characteristic parallel function is determined as (3.15), and the transverse function as (3.23), and given $|G(z)| \sim 1$ in the transparent region, we obtain the following estimate:

$$\{\varphi'\} \sim \frac{1}{\sqrt{g_2 g_3}}\frac{k_2 a_N}{\omega^2}\left(\frac{a_N}{\rho_N}\right)^{2/3}\frac{v_A}{t_A}\tilde{\Delta}_\perp\psi,$$

where $a_N \sim L$ is the characteristic scale of v_A variation scale in the x^1 coordinate and all the right-hand side parameters refer to the resonance magnetic shell. Substituting this into (3.27) yields the following estimate:

$$\frac{\{\tilde{\Delta}_\perp\psi\}_{x^1_{TN}\pm\varepsilon}}{(\tilde{\Delta}_\perp\psi)_{x^1_{TN}}} \sim \frac{(k_2\rho_N)^2}{g_1 g_2}\left(\frac{a_N}{\rho_N}\right)^{2/3} = \frac{(k_2 l_N)^2}{g_1 g_2} \ll 1,$$

showing that the second derivative with respect to x^1 of function ψ changes relatively little when passing through the resonance surface. Iterated integration of (3.5) easily demonstrates that, on the scales in question, the variations of the function ψ itself and its first derivative $\nabla_1\psi$ are also small. Consequently, the right-hand side of (3.4) can be assumed to be practically constant on the resonant Alfvén wave localisation scale, and the right-hand side of (3.5) can be assumed to be zero when determining the FMS wave field in the main order of perturbation theory.

3.2 Alfvén Resonance in a Dipole-Like Magnetosphere

In Section 3.1, we determined the spatial structure of kinetic Alfvén waves excited by FMS waves on resonance magnetic shells in a dipole-like axisymmetric model magnetosphere. The same model will be used in this section to examine Alfvén resonance in the case when

Alfvén waves are strongly absorbed in the ionosphere. In this case, the transverse structure of Alfvén waves is determined by their dissipation, and the kinetic corrections in the previous section are small and can be ignored. We will determine the field structure of both resonant Alfvén waves and the monochromatic FMS waves they are excited by. We will also estimate the feedback from Alfvén waves to the FMS oscillation field.

3.2.1 Model of the Medium and Basic Equations

Same as in Section 3.1, we will examine a model magnetosphere with a dipole-like geomagnetic field (see Figure 3.2). Our numerical calculations will assume a dipole field, while our analytical calculations will assume merely a dipole-like field, with similar magnetic field-line geometry. This allows the results of the analytical calculations to be applied to field configurations that are analogous to those corresponding to Earth's dayside magnetosphere (see also simulation in [294]). We will not include solar wind plasma movement. This will definitely place certain restrictions on the applicability of the model. Calculation results in such a model are applicable to the dayside magnetosphere. A solar wind stagnation point is located in the subsolar region of the magnetopause, therefore, plasma movement-related effects make a comparatively small contribution to the field structure of the MHD oscillations in question.

Let us define the following model for numerical calculations. We will use a curvilinear orthogonal system of coordinates (x^1, x^2, x^3) related to the geomagnetic field lines, where the coordinate x^1 identifies the magnetic shell, x^2 the field line on the magnetic shell and x^3 the point on the field line. We will assume a dipole geomagnetic field in the numerical calculations. We will use the angle θ between the point on the field line and the equatorial plane (see Figure 3.2) as the longitudinal coordinate x^3. The length element along the field line is given by

$$dl = \sqrt{g_3}dx^3 = a\overline{\eta}(\theta)\cos\theta\, d\theta,$$

where a is the equatorial radius of the field line, which we will use instead of the coordinate x^1, $\overline{\eta}(\theta) = \sqrt{1 + 3\sin^2\theta}$. We will use the azimuthal angle ϕ ($k_2 = m = 0, \pm1, \pm2, \ldots$ is the azimuthal wave number) as the coordinate x^2. Note that despite the system of curvilinear coordinates (a, ϕ, θ) not being orthogonal, each of these coordinates is uniquely related to the corresponding curvilinear coordinates (x^1, x^2, x^3). The transverse components of the metric tensor in this system of coordinates have the form

$$g_1 = \cos^6\theta/\overline{\eta}^2(\theta), \qquad g_2 = a^2\cos^6\theta.$$

The latitude where the field line intersects the ionosphere is determined as

$$\theta^* = \arccos\sqrt{r_i/a},$$

where r_i is the radius of the ionosphere.

Another element of the model is the Alfvén speed v_A. We will use the following model for it:

$$v_A = \frac{v_{Am} + v_{Asw}}{2} - \frac{v_{Am} - v_{Asw}}{2}\tanh\frac{a\cos^2\theta - a_m}{\Delta_m}, \tag{3.28}$$

where v_{Asw} is Alfvén speed in the solar wind, Δ_m is the effective thickness of the magnetopause. Our numerical calculations will assume the magnetopause to be a sphere of radius a_m, even though the above model allows the magnetopause to be set in any axisymmetric form. Function $v_{Am}(a,\theta)$ describes a 2D distribution of Alfvén speed inside the magnetosphere. We will use the following representation for this function:

$$v_{Am}(a,\theta) = \frac{1}{2}\left[\left(v_{A1}\left(\frac{a_1}{a}\right)^{\mu_1} + v_{A2}\left(\frac{a_2}{a}\right)^{\mu_2}\right) - \left(v_{A1}\left(\frac{a_1}{a}\right)^{\mu_1} - v_{A2}\left(\frac{a_2}{a}\right)^{\mu_2}\right)\tanh\frac{a-a_p}{\Delta_p}\right]\left(\frac{\overline{\eta}(\theta)}{\cos^6\theta}\right)^{\nu}, \tag{3.29}$$

where v_{A1} and v_{A2} are the characteristic Alfvén speed values in the inner and outer magnetospheric regions, respectively; a_p, Δ_p are the equatorial radius and characteristic thickness of the plasmapause. The last term in this relation describes the Alfvén speed variation along geomagnetic field lines. Formula (3.29) simulates the Alfvén speed distribution in the moderately disturbed dayside magnetosphere, for the following values of its parameters: $v_{A1} = 250$ km/s, $v_{A2} = 500$ km/s, $v_{Asw} = 50$ km/s, $a_m = 10R_E$ ($R_E = 6370$ km is Earth radius), $\Delta_m = 0.5R_E$, $a_p = 4R_E$, $\Delta_p = 0.5R_E$, $a_1 = 2.5R_E$, $a_2 = 5R_E$, $\mu_1 = 1.5$, $\mu_2 = 1$, $\nu = 0.25$. Figure 3.3 illustrates the equatorial distribution of $v_A(a,0)$ across magnetic shells in this magnetosphere model. The same figure shows the distribution of the basic period of magnetospheric Alfvén eigen-oscillations in the WKB approximation (3.16) depending on magnetic shell parameter $L = a/R_E$ (McIlwain parameter).

We will seek the MHD oscillation field in the axisymmetric model of the magnetosphere as a decomposition into Fourier harmonics of the form $\exp(ik_2x^2 - i\omega t)$. We will make use of the cold plasma approximation ($T_i = T_e = 0$). In this case SMS is absent and only Alfvén waves and fast magnetosonic oscillations are present in the plasma. Decomposing the oscillation electric field into the potential and vortical components (2.72), we will obtain, from the system of linearised ideal MHD equations (1.5), the following system of equations (see Appendix E):

$$\nabla_1\sqrt{g_3}\,\hat{L}_T\nabla_1\varphi - k_2^2\sqrt{g_3}\,\hat{L}_P\varphi = ik_2\left(\nabla_1\sqrt{g_3}\,\hat{L}_T\frac{p^{-1}}{\sqrt{g_3}}\psi - \sqrt{g_3}\,\hat{L}_P\frac{p}{\sqrt{g_3}}\nabla_1\psi\right), \tag{3.30}$$

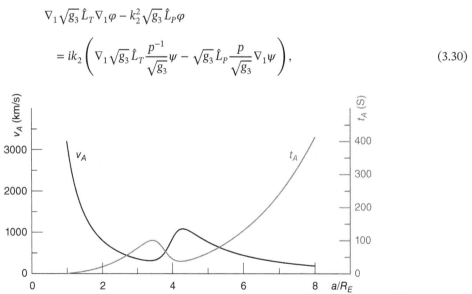

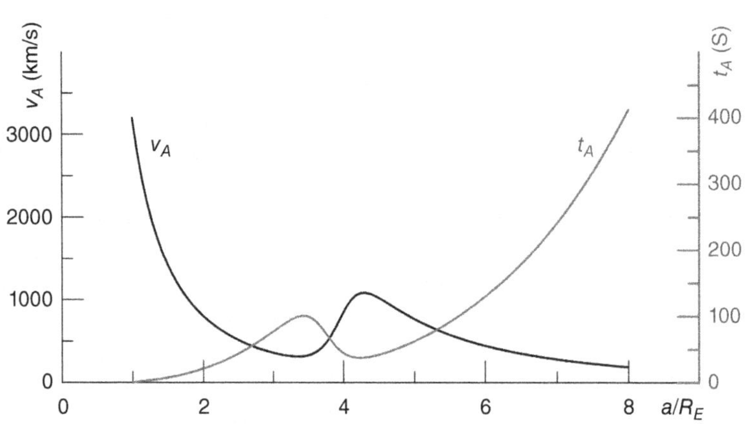

Figure 3.3 Equatorial dependence of Alfvén speed $v_A(L,0)$ in the model under study and the corresponding dependence of the basic period $t_A(L)$ of magnetospheric Alfvén eigen oscillations on the magnetic shell parameter L.

$$\nabla_1 \frac{p}{\sqrt{g_3}} \nabla_1 \psi - k_2^2 \frac{p^{-1}}{\sqrt{g_3}} \psi + \sqrt{\frac{g}{g_3}} \hat{L}_T \frac{p^{-1}}{\sqrt{g_3}} \psi = -\frac{i}{k_2} \sqrt{\frac{g}{g_3}} \hat{L}_T \nabla_1 \varphi, \tag{3.31}$$

where the notations are: $\nabla_i = \partial/\partial x^i$ $(i = 1, 2, 3)$, g_i are the metric tensor components, $g = g_1 g_2 g_3$, $p = \sqrt{g_2/g_1}$, operators $\hat{L}_T$ and $\hat{L}_P$ are determined by the expressions (3.6) and (3.7), v_A is the Alfvén speed. Equations (E.20) and (E.21) (Appendix E) link the covariant components of the electric (**E**) and magnetic (**B**) field of the MHD oscillations with potentials φ and ψ. The physical components of the vectors are related to the covariant components by $\tilde{E}_i = E_i/\sqrt{g_i}$, $\tilde{B}_i = B_i/\sqrt{g_i}$ and the length element along the field line is determined as $dl = \sqrt{g_3}\, dx^3$.

Equation (3.30) describes the field of Alfvén waves, and (3.31) that of magnetosonic waves. When applied to homogeneous plasma, operators in the left-hand sides of these equations produce dispersion equations for, respectively, Alfvén, $\omega^2 = k_\parallel^2 v_A^2$, and fast magnetosonic waves, $\omega^2 = k^2 v_A^2$. Interaction between these waves in an inhomogeneous plasma is described by the right-hand sides of (3.30) and (3.31), vanishing when applied to homogeneous plasma.

Equation (3.30) can also be obtained directly from the system (3.4) and (3.5) in the $\Lambda^2 \to 0$ limit. Let us differentiate (3.4) with respect to x^1, multiply (3.5) by $-ik_2/\sqrt{g_3}$ and sum the resulting equations. The result is

$$\begin{aligned} &\frac{\omega^2}{v_A^2} \Lambda^2 \sqrt{g} \Delta_\perp^2 \varphi + \nabla_1 \sqrt{g_3} \hat{L}_T \nabla_1 \varphi - k_2^2 \sqrt{g_3} \hat{L}_P \varphi \\ &= ik_2 \left(\nabla_1 \sqrt{g_3} \hat{L}_T \frac{p^{-1}}{\sqrt{g_3}} \psi - \sqrt{g_3} \hat{L}_P \frac{p}{\sqrt{g_3}} \nabla_1 \psi \right). \end{aligned} \tag{3.32}$$

If $\Lambda^2 = 0$, this will produce (3.30). Note again that this approximation can be used for strong enough dissipation of Alfvén waves when their dissipative scale is much larger than the dispersion scale in the x^1 coordinate ($\varepsilon_{TN} \gg l_N$). Below, we will examine the problem of monochromatic FMS wave falling from the solar wind onto the magnetosphere and reflected from the magnetopause, in the presence of resonance surfaces for Alfvén waves inside the magnetosphere.

3.2.2 Resonant Alfvén Wave Field Structure

In the neighbourhood of the magnetic shell where its field is concentrated, the localisation scale of the resonant Alfvén wave in the x^1 coordinate is much smaller than the wavelength along the field line. As was demonstrated in Section 3.1.1, the solution of (3.30) for potential φ can be sought in this case as

$$\varphi = [V(x^1) T(x^1, x^3) + h(x^1, x^3)] \exp(ik_2 x^2 - i\omega t). \tag{3.33}$$

Here, function $V(x^1)$ describes the small-scale structure of Alfvén wave across magnetic shells; $T(x^1, x^3)$ the wave structure along the field line; and function $h(x^1, x^3)$ is a small correction in the higher orders of perturbation theory. Detailed solution of an equation of the form (3.30) using the perturbation method is given in Section 3.1.1.

In the main order of perturbation theory, an equation of the form (3.13) is obtained for function $T(x^1, x^3)$ with homogeneous boundary conditions on the ionosphere: $T(x^1, x_\pm^3) = 0$. Its solutions satisfy the boundary conditions on the ionosphere are a set of eigenfunctions $T_N(x^1, x^3)$ and the corresponding set of eigenfrequencies $\omega = \Omega_{TN}(x^1)$,

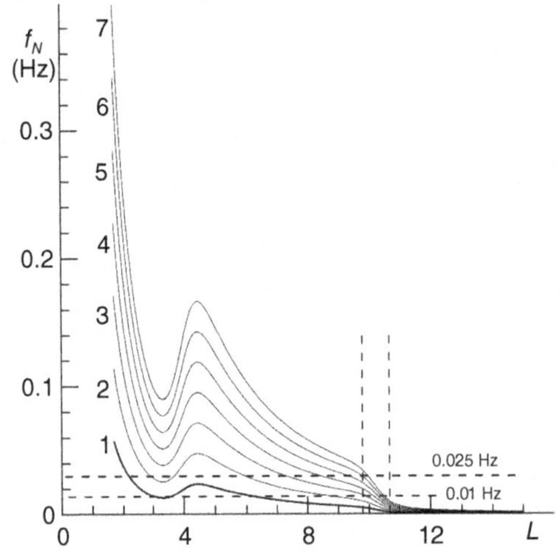

Figure 3.4 Eigenfrequency distribution for the first seven harmonics of standing toroidal Alfvén waves across magnetic shells in a dipole model magnetosphere. Horizontal dashed lines indicate the frequencies of magnetosonic waves incident on the magnetosphere from the solar wind. Vertical dashed lines are conditional boundaries of the transition layer of the magnetopause.

where $N = 1, 2, \ldots$ is the longitudinal eigen number identifying the harmonics of standing Alfvén waves. These solutions describe standing Alfvén waves of the toroidal type on the magnetic shell corresponding to the coordinate x^1. In the WKB approximation, the solutions of this equation in the x^3 coordinate have the form (3.15), and the corresponding eigenvalues of the standing Alfvén wave frequencies have the form (3.16).

Rigorously, the WKB approximation in the x^3 coordinate is inapplicable to the fundamental harmonics of standing Alfvén waves. It can only be used in this case for a qualitative representation of the oscillation structure. To examine these oscillations more accurately, their eigenfunctions and eigenfrequencies were numerically calculated. Graphs in Figure 3.4 illustrate the dependence of eigenfrequencies $\Omega_{TN}(x^1)$ for the first seven harmonics of standing toroidal Alfvén waves on the magnetic shell parameter L in the above-described model dipole magnetosphere. These distributions feature a characteristic 'knee' in the neighbourhood of the plasmapause ($L = 4$) and a sharp decrease in v_A (and $\Omega_{TN}(x^1)$) in the transition region of the magnetopause ($L \approx 10$).

In the next order of perturbation theory, (3.30) yields an equation for function V_N analogous to (3.17). If left-multiplied by T_N and integrated along the field line this equation acquires this form

$$\nabla_1[(\omega + i\gamma_{TN})^2 - \Omega_{TN}^2]\nabla_1 V_N - k_y^2[(\omega + i\gamma_{TN})^2 - \Omega_{TN}^2]V_N = i\frac{k_y}{\beta_N}\mu_N, \tag{3.34}$$

where the following notations are:

$$k_y = k_2\sqrt{\alpha_{PN}/\alpha_{TN}}, \quad \beta_N = \sqrt{\alpha_{PN}\alpha_{TN}},$$

$$\alpha_{PN} = \oint \frac{T_N^2 d\ell}{pA^2}, \quad \alpha_{TN} = \oint \frac{pT_N^2 d\ell}{A^2},$$

$$\mu_N = \oint T_N \left[\nabla_1\sqrt{g_3}\hat{L}_T\frac{g_1}{\sqrt{g}}\psi - \sqrt{g_3}\hat{L}_P\frac{g_2}{\sqrt{g}}\nabla_1\psi\right] dx^3.$$

When deriving (3.34), we ignored any distinction between operators $\hat{L}_P$ and $\hat{L}_T$ in the left-hand side of the equation. This is permissible for harmonics of Alfvén oscillations with $m \sim 1$ because their characteristic scale in the x^1 coordinate is much larger than the distance between the poloidal and toroidal resonance surfaces making the wave field structure on them practically identical (see Sections 3.7 and 3.13). Standing Alfvén wave decrement γ_{TN} originates from the boundary conditions on the ionosphere and is related to wave energy dissipation in the ionospheric conducting layer (see Section 3.1.2).

If the Alfvén speed varies monotonously in the magnetospheric region in question, functions $\Omega_{TN}(x^1)$ can be linearised and represented as (3.21), in the narrow neighbourhoods of resonance shells $x^1 = x^1_{TN}$ (where $\omega = \Omega_{TN}(x^1_{TN})$). For the dimensionless coordinate $\xi = k_y(x^1 - x^1_N)$, we will rewrite (3.34) as follows:

$$\nabla_\xi(\xi + i\varepsilon_N)\nabla_\xi V_N - (\xi + i\varepsilon_N)V_N = i\frac{a_N\mu_N}{\beta_N\omega^2},$$

where $\varepsilon_N = 2\gamma_{TN}k_y a_N/\omega$. As was shown in Section 3.1, the right-hand side of this equation changes across magnetic shells on much larger scales than the resonant Alfvén wave localisation scale. The Fourier method can be used to find the solution in this case (see Section 3.13). The solution of (3.34) satisfying the boundary conditions (finite oscillation amplitude on the asymptotics in the x^1 coordinate) can be represented as

$$V_N = -\frac{a_N\mu_N}{\beta_N\omega^2}F_N(\xi + i\varepsilon_N), \tag{3.35}$$

where the function $F_N(\xi + i\varepsilon_N)$ has the following integral representation:

$$F_N(\xi + i\varepsilon_N) = \int_0^\infty \frac{\exp[ik(\xi + i\varepsilon_N)]}{\sqrt{1+k^2}}\,dk. \tag{3.36}$$

This function describes the structure of resonant Alfvén wave across magnetic shells. If $\varepsilon_N \ll 1$, then $k \gg 1$ is responsible for the bulk of the integral (3.36) at $|\xi| \to 0$. In this case, we have $F_N \sim \ln(\xi + i\varepsilon_N)$. This means that a known logarithmic singularity exists on the toroidal resonance surface for Alfvén waves ($\xi = 0$) for $\varepsilon_N \to 0$. Such a singularity can be observed, for example, in field components E_2 and B_1 of the resonant Alfvén oscillations, while components E_1 and B_2, proportionate to $\nabla_\xi F_N(\xi)$, have a singularity of the form ξ^{-1}. On the asymptotics ($|\xi| \gg \varepsilon_N$), $k \ll 1$ is responsible for the bulk of the integral (3.36), resulting in the following asymptotic representation: $F_N(\xi) \approx i(\xi + i\varepsilon_N)^{-1}$.

Thus, we have now determined the structure of resonant Alfvén waves near a resonance magnetic shell. To determine the full field of MHD oscillations in the entire model magnetosphere, however, we also have to determine the field structure of the fast magnetosonic wave driving Alfvén waves on resonance magnetic shells.

3.2.3 Field Structure of Monochromatic FMS Oscillations in a Dipole-Like Magnetosphere

Magnetosonic wave field is determined by (3.31). Unlike (3.30), it lacks parameters that could be used to strictly justify employing the multiple-scale method we used earlier to examine the field structure of resonant Alfvén waves. Nevertheless, we will use this method for a qualitative (and quantitative, to a certain degree of accuracy) examination of the magnetosonic wave structure.

Comparing the spatial structure of FMS oscillations obtained using this method to that obtained by numerically solving the system of MHD equations shows their good qualitative consistence (see [226]). Moreover, a special investigation has shown that the WKB solution of equations having no singularities in the coefficients obtained at the WKB method applicability limit (when the characteristic wavelength is of order the inhomogeneity scale), demonstrates a rather good (~10–15% error) coincidence with the precise numeric solution [295].

Based on this, we will seek the solution of (3.31) as

$$\psi = w(x^1)[H(x^1, x^3) + h(x^1, x^3)] \exp(ik_2 x^2 - i\omega t),$$

where function $w(x^1)$ describes magnetosonic wave structure across magnetic shells, $H(x^1, x^3)$ describes the longitudinal structure, and $h(x^1, x^3)$ is a small correction to the solution in the higher orders of perturbation theory. We will seek the solution for function $w(x^1)$ in the WKB approximation as

$$w(x^1) = \exp[i\Theta(x^1)],$$

where Θ is the large quasiclassical phase, which can be represented as a series, $\Theta = \Theta^{(0)} + \Theta^{(1)} + \Theta^{(2)} + \cdots$, $(|\Theta^{(0)}| \gg |\Theta^{(1)}| \gg |\Theta^{(2)}| + \cdots)$. In the main order of perturbation theory, we obtain from (3.31) an equation for function H, the solution of which is a set of eigenfunctions, $H_n(x^1, x^3)$ ($n = 1, 2, \ldots$ is the parallel wavenumber for FMS waves) and the corresponding set of eigenvalues of the wave vector k_{1n} ($k_1 \equiv \nabla_1 \Theta^{(0)}(x^1)$). The method of solving this equation (in the WKB approximation or numerically) is detailed in Section 3.5.

In the next, first, order of perturbation theory, we obtain this equation

$$\begin{aligned} \hat{L}_f h_n - \frac{2k_{1n}k_{1n}^{(1)}}{g_1} H_n + i\frac{k_{1n}}{g_1}\nabla_1 H_n + i\frac{g_3}{\sqrt{g}}\nabla_1 \frac{g_2}{\sqrt{g}} k_{1n} H_n \\ + \frac{\sqrt{g}}{k_2 g_1} \sum_{N=1}^{\overline{N}} \frac{k_{1n}\delta_{nN}^{(1)}}{\beta_N} \frac{T_N}{v_A^2} \xi F_N'(\xi) = 0, \end{aligned} \tag{3.37}$$

where $\hat{L}_f$ (left-hand side operator in (3.31), where $\nabla_1 = ik_{1n}$, left-multiplied by $(g_1 g_2)^{-1/2}$) is a zero-approximation operator for FMS waves, $\overline{N}$ is the number of resonance surfaces for Alfvén waves for a given FMS frequency, and the following notations are introduced: $k_{1n}^{(1)} \equiv \nabla_1 \Theta_n^{(1)}(x^1)$,

$$\begin{aligned} \delta_{nN}^{(1)} &= \oint T_N \left(\hat{L}_T \frac{g_1}{\sqrt{g}} - \hat{L}_P \frac{g_2}{\sqrt{g}} \right) H_n \sqrt{g_3}\, dx^3 \\ &= 2 \oint H_n (\nabla_l \ln p)(\nabla_l T_N)\, dx^3, \end{aligned} \tag{3.38}$$

where $\nabla_l \equiv \partial/\sqrt{g_3}\partial x^3$ is a derivative with respect to the parallel coordinate, and the closed-contour integral presumes a two-way integration along the field line between the magnetoconjugated ionospheres. The last term in (3.37) describes the feedback from resonant Alfvén waves to the exciting magnetosonic wave. Notably, the number of resonance surfaces can be large enough in the model magnetosphere we examine which includes an adjacent solar-wind region. In view of the fact that the model is only applicable near the magnetosphere, however, let us restrict ourselves to taking into account only resonances in the region covering the magnetosphere proper and the transition region of the magnetopause ($L \leq 10.5$ in our model). Moreover, it will be seen from the following calculations that the effect of resonances in the solar wind region is negligibly small.

The total number of resonances, $\overline{N}$, excited in the magnetosphere by a monochromatic magnetosonic wave is therefore limited.

Dashed horizontal lines in Figure 3.4 indicate the frequency of FMS wave incident on the magnetosphere. The intersection points of these lines and the $\Omega_{TN}(x^1)$ curves determine the resonance surface coordinates $x^1 = x^1_{TN}$, their number $\overline{N}$ depending on magnetosonic wave frequency. Let us left-multiply (3.37) by $H_n g_1/\sqrt{g}$ and integrate along the field line. The result will be an equation which when solved with respect to function $k^{(1)}_{1n}$ will yield

$$k^{(1)}_{1n} = \frac{i}{2}\nabla_1 \ln k_{1n} + i\tilde{k}_{1n} + \overline{k}_{1n}, \tag{3.39}$$

where the notations are

$$\tilde{k}_{1n} = \frac{1}{2}\oint \nabla_1 \left(\frac{g_2 H_n^2}{\sqrt{g}}\right)\frac{dx^3}{g_2}, \tag{3.40}$$

$$\overline{k}_{1n} = \frac{1}{2k_2}\sum_{N=1}^{\overline{N}} \frac{a_{nN}\delta^{(1)}_{nN}}{\beta_N}\xi F'_N(\xi), \tag{3.41}$$

$$a_{nN} = \oint \frac{T_N H_n}{v_A^2} dx^3, \tag{3.42}$$

and function H_n is assumed to be normalised as follows:

$$\oint \frac{H_n^2}{\sqrt{g}} dx^3 = 1.$$

Function $k^{(1)}_{1n}$ describes the magnetosonic wave amplitude variation. The first term in (3.39) is a pre-exponent in the WKB approximation, the second – $\tilde{k}^{(1)}_{1n}$ is a geometrical factor related to wave propagation in a magnetic field with lines converging at the origin of coordinates, and the last term – $\overline{k}^{(1)}_{1n}$ is a correction due to feedback from resonant Alfvén waves to the magnetosonic wave. The structure of the nth longitudinal harmonic of magnetosonic oscillations across magnetic shells can be represented in the WKB approximation as follows (see Section 3.5):

$$w_n = C_n \frac{\exp\left[\int_{x_n^1}^{x^1} (\tilde{k}_{1n} - i\overline{k}_{1n})dx^{1\prime}\right]}{\sqrt{|k_{1n}|}} \begin{cases} \exp\left[\int\limits_{x_n^1}^{x^1} |k_{1n}|dx^{1\prime}\right], & x^1 < x_n^1, \\ \sin\left[\int\limits_{x_n^1}^{x^1} k_{1n}dx^{1\prime} + \frac{\pi}{4}\right], & x^1 > x_n^1, \end{cases}$$

where x_n^1 is the FMS turning point determined by condition $k^2_{1n}(x^1_n) = 0$, C_n is oscillation amplitude independent of x^1. Note that this solution is inapplicable in the turning point neighbourhood, where an approximate analytical solution should be employed matching the WKB solutions on the asymptotics.

It was assumed in Section 3.1 that the presence of resonant Alfvén waves little affects the structure and spectrum of the exciting magnetosonic oscillations. To verify this assumption, let us calculate numerically the transverse structure of magnetosonic wave given the feedback from the resonant Alfvén waves. Note that if the magnetosphere is symmetrical about the equatorial surface, only those functions a_{nN} and $\delta^{(1)}_{nN}$ that have the same parity of

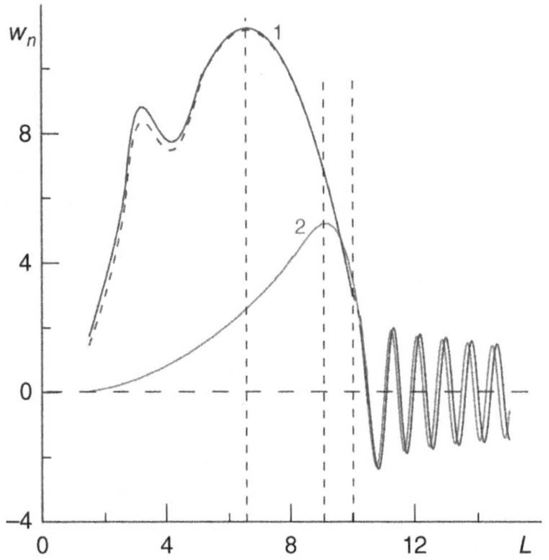

Figure 3.5 The structure of monochromatic ($f = \omega/2\pi = 0.01$ Hz) FMS waves across magnetic shells for two first longitudinal harmonics ($n = 1, 2$), with (dashed lines) or without (solid lines) feedback from resonant Alfvén waves. The vertical dashed lines denote FMS turning points ($L = 6.1$ and $L = 9.2$) and the magnetopause ($L = 10$).

the longitudinal wave numbers n and N will be non-zero. Figure 3.5 displays the plots of two functions w_1 and w_2 with (dashed lines) or without (solid lines) the feedback from the Alfvén waves to the magnetosonic oscillation field.

The calculations of functions in Figure 3.5 presumed a small enough value of Alfvén wave dissipation in the ionosphere: $\varepsilon_N = 5 \times 10^{-3}$. For the characteristic value $k_y a_N \sim 0.2$ in the selected model of the magnetosphere, this corresponds to a small decrement in Alfvén waves: $\gamma_{TN} \sim 10^{-2}\omega$. For function w_2 both curves virtually coincide, while for w_1 the difference is noticeable but also small. For large decrements ($\varepsilon_N \geq 10^{-1}$) this difference becomes negligibly small even for w_1. Thus the feedback from resonant Alfvén oscillations to the structure of the exciting magnetosonic wave is indeed small. This feedback can be neglected in the main order of perturbation theory when studying magnetosonic wave field.

This fact is notable in itself being related to the quite definite subdivision of the MHD oscillation field into the potential and vortical components. The coefficients in the left-hand side of (3.31) have no singularities even on resonance magnetic shells, making it possible to use the WKB approximation in the x^1 coordinate for describing FMS wave field everywhere except the small neighbourhoods of the turning points. The absolute maxima in the function w_1 and w_2 distributions in Figure 3.5 correspond to these points.

3.2.4 MHD Oscillation Magnetic Field Amplitude Distribution in the Meridional Plane

Let us first determine the MHD oscillation magnetic field distribution across magnetic shells. We will numerically calculate the amplitude in the meridional plane along some lines crossing the magnetic shells. For Alfvén waves, it is convenient to use the ionospheric upper boundary, where the magnetic field transverse components $B_{rA} = B_{1A}/\sqrt{g_1}$ and $B_{\phi A} = B_{2A}/\sqrt{g_2}$ have an antinode, as such a line. It is more difficult to select such a line for FMS waves. For the longitudinal component $B_{lf} = B_{3f}/\sqrt{g_3}$ we will draw the line in such a manner that it should cross the maximum of functions $\nabla_l H_n$, while for the components B_{rf} and $B_{\phi f}$ it should cross the maximum of functions H_n determining the FMS wave structure

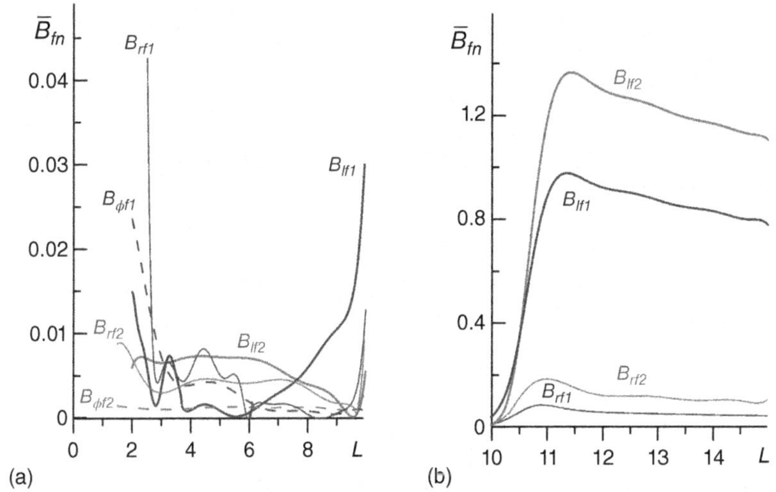

Figure 3.6 Amplitude distributions of magnetic field components, B_{rf}, $B_{\phi f}$ and B_{lf}, across magnetic shells for monochromatic ($f = 0.01$ Hz) magnetosonic oscillations: (a) inside the magnetosphere, (b) in the solar wind region. Numbers (1) and (2) in the panels correspond to the two first longitudinal harmonics of FMS oscillations ($n = 1, 2$).

along magnetic field lines. For functions $H_1(x^1, l)$ and $\nabla_l H_2(x^1, l)$ this line lies in the plane of the magnetic equator, while for functions $H_2(x^1, l)$ and $\nabla_l H_1(x^1, l)$ it runs along the transparent region boundary.

We will restrict ourselves to calculating oscillations of low enough frequency $f = \omega/2\pi = 0.01$ Hz, for which only one resonance surface exists inside the magnetosphere (see Figure 3.4). As will be seen below, this frequency is much lower than the frequencies of magnetosonic eigen oscillations in the magnetosphere. Figure 3.6 contains amplitude distributions for the three magnetic-field components of oscillations with azimuthal wave number $m = 1$, for two first longitudinal harmonics of magnetosonic waves incident on the magnetosphere ($n = 1, 2$).

The amplitudes of the incident and reflected FMS wave in this figure are equal. Their difference due to partial energy dissipation in the Alfvén resonance region can only be found in the second order of the WKB method. The constants C_n are selected such that the amplitude of the longitudinal component B_{lf} of the oscillation field on magnetic shell $L = 11$ equal unity ($|B_{lf}|$=1 nT).

Figure 3.6a is the distribution of the components $B_{rf}, B_{\phi f}$ and B_{lf} inside the magnetosphere ($L < 10$), while Figure 3.6b refers to the components B_{rf} and B_{lf} in the solar wind region ($L > 10$). Component $B_{\phi f}$ is so small in the solar wind region as to make it invisible in the plot. Such a distribution of the amplitudes is caused by their different dependence on the wave vector component k_{1n}. It follows from (E.2) (Appendix E) that $B_{lf} \sim k^2_{\perp fn}$, where $k^2_{\perp fn} = (k^2_{1n}/g_1 + k^2_2/g_2)$ is the FMS wave vector transverse component squared, $B_{rf} \sim k_{1n}$ and $B_{\phi f} \sim k_2$, assuming $k_{1n} \gg k_2$ in the solar wind region. This also causes sharply decreased amplitudes of B_{lf} and B_{rf} of the oscillation field components when passing from the solar wind region into the magnetosphere, because k_{1n} decreases by several orders of magnitude in the process. Inside the magnetosphere the amplitudes of the three magnetosonic wave field components are comparable in value (Figure 3.6a).

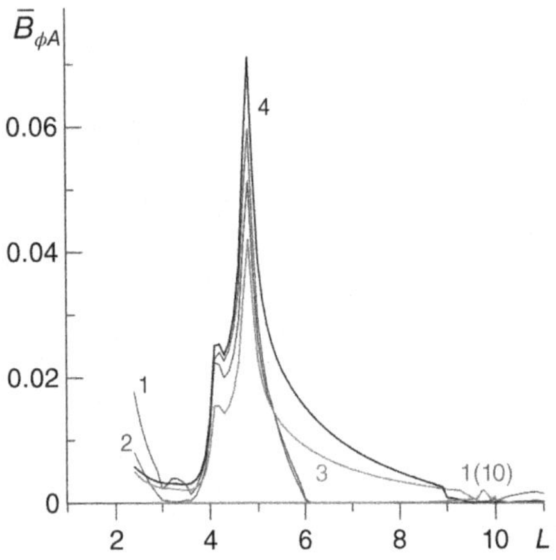

Figure 3.7 Mean amplitude distribution for the field component $B_{\phi A}$ of the two first harmonics ($N = 1, 2$) of resonant Alfvén waves across magnetic shells. These harmonics are excited by a monochromatic ($f = 0.01$ Hz, $m = 1$) magnetosonic waves with $n = 1$ (lines 1,2) and $n = 2$ (lines 3,4). The amplitude distribution of the first harmonic ($N = 1$) is shown (line 1(10)) for comparison, as excited by magnetosonic wave with $m = 10$, $n = 1$ which practically does not penetrate into the magnetosphere.

As can be seen in Figure 3.4, even slight changes in the frequency of the FMS wave incident on the magnetosphere (see the vertical shift of horizontal dashed lines) causes significant changes in the location of the resonance surfaces. A similar effect would take place for small changes in the Alfvén speed distribution. It especially concerns the first longitudinal harmonics $N \sim 1$ of the resonant Alfvén waves. To trace the amplitude variations in the excited Alfvén wave depending on the location of the resonance magnetic shell, Figure 3.7 shows the distributions of the mean amplitudes of the component $B_{\phi A}$ for the two first harmonics of standing Alfvén waves ($N = 1, 2$), exited by monochromatic ($f = 0.01$ Hz) FMS waves with $n = 1, 2$. Note that the mean amplitudes $\overline{B}_{rA}$ and $\overline{B}_{\phi A}$ are determined as follows:

$$B_{rA} = \overline{B}_{rA} F(\xi), \qquad B_{\phi A} = \overline{B}_{\phi A} F'(\xi),$$

where

$$\overline{B}_{rA} = -\sqrt{2}\frac{\Omega_{TN}}{\omega^3 \beta_N} \frac{\bar{c} m a_N}{\sqrt{t_A v_{Ai} p_i g_{2i}}} \delta_{nN}^{(1)}(\nabla_1 w_n),$$

$$\overline{B}_{\phi A} = i\beta_N p_i \overline{B}_{rA},$$

t_A is the Alfvén speed transit time between the magnetoconjugated ionospheres, $v_{Ai} = 2 \times 10^3$ km/s, and the i index means that the values are taken on the upper boundary of the ionosphere. To obtain the value of the amplitude on the resonance surface ($\xi = 0$) requires that the mean amplitude $\overline{B}_{\phi A}$ should be multiplied by $F'(0)$, i.e. as follows from (3.36), the amplitude should be multiplied by ε_N^{-1} if $\varepsilon_N \ll 1$.

Two peaks in the $\overline{B}_{\phi A}$ distribution, in the neighbourhood of the plasmapause, are related to the parameter a_N variations near the extrema of the function $\Omega_{TN}(x^1)$. Using the decomposition (3.21) makes a_N become infinite at the extremum points. Decomposition of the next, second, order accuracy should be used in this case, which was done in calculating the plots in Figure 3.7. It should be noted, however, that the above-presented solution for resonant Alfvén waves is inapplicable in the immediate neighbourhood of these points (see the horizontal straight line $f = 0.025$ Hz crossing the curve $\Omega_{T2}(L)$ in Figure 3.4). Another analytical research method analogous to the one in [224] should be used for describing the resonant interaction between MHD waves near the extrema of $\Omega_{TN}(x^1)$. It is near these points that the largest amplitude of resonant Alfvén oscillations is to be expected.

Another, paradoxical, feature of the plots in Figure 3.7 is the sharp increase in the amplitude of the resonant Alfvén oscillations from the transparent region into the opaque region, through the magnetosonic wave turning point, determined by the condition $k_{1n}(x_n^1) = 0$. For the $n = 1$ harmonic this point is located at $L = 6.1$, and for the $n = 2$ harmonic, at $L = 9.2$. This is why the outer magnetosphere ($6 < L < 10$) is dominated by the amplitudes of two even Alfvén resonant wave harmonics ($N = 2, 4$) excited by the FMS $n = 2$ harmonic. Upon passing into the opaque region, the magnetosonic wave structure along geomagnetic field lines changes dramatically. In the opaque region this structure resembles the longitudinal structure of FMS-excited standing Alfvén waves. The value of the correlator $\delta_{nN}^{(1)}$ determining the amplitude of resonant Alfvén oscillations (see (3.38) increases sharply in the process resulting in the observed effect.

Figure 3.8 is the transverse structure of the $B_{\phi A}$ component of the field of the resonant Alfvén oscillations excited by magnetosonic oscillations with frequency $f = 0.01$ Hz. For comparison, this figure shows the structure of the parallel B_{lf}-component of the magnetosonic wave field with $n = 2$. Recall that the amplitude distribution of resonant Alfvén oscillations is along the upper ionospheric boundary, and that of the magnetosonic wave, along a specially selected line in the meridional plane where its amplitude is highest. It can be seen that the amplitudes of the resonant peaks inside the transition layer are much smaller than the amplitude of the only resonant harmonic inside the magnetosphere.

The amplitude of the resonant harmonic inside the magnetosphere is comparable to the amplitude of magnetosonic oscillations in the solar wind and is much higher than the amplitude of magnetosonic oscillations in the magnetosphere. It follows from Figure 3.7 that, inside the magnetosphere, the resonant harmonic ($N = 1$) is located in the region ($L \approx 7.1$) where its amplitude is highest. If the resonance surface happens to be in the opaque region ($L < 6$) for the exciting magnetosonic wave, the magnitude of the resonant Alfvén oscillations increases considerably. This means that the amplitude of resonant Alfvén oscillations inside the magnetosphere can be comparable to or even exceed one of the solar wind magnetosonic waves incident on the magnetosphere. This can be observed even at a large enough decrement of Alfvén waves due to their energy dissipation in the ionosphere (e.g. $\gamma_{TN} \sim 0.1\omega$).

Figure 3.9 shows, in the meridional plane, the amplitude distribution of the full magnetic field of resonant Alfvén oscillations excited by monochromatic FMS wave from the solar

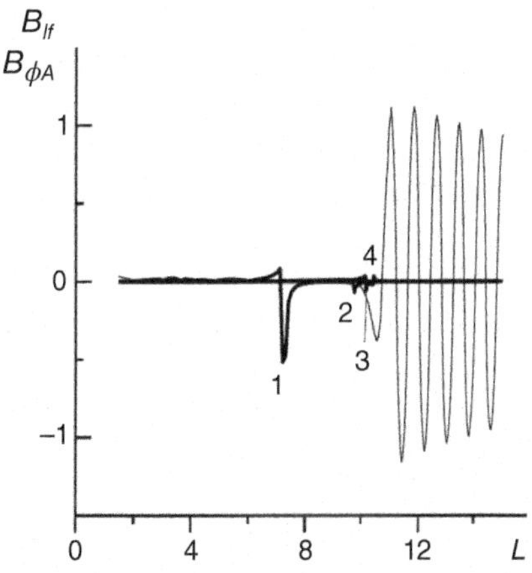

Figure 3.8 Distribution of the $B_{\phi A}$ component (thick lines) of the field of resonant Alfvén oscillations excited by monochromatic ($f = 0.01$ Hz) FMS waves incident on the magnetosphere. The numbers indicate the first four harmonics of resonant Alfvén waves ($N = 1, 2, 3, 4$). The distribution of the parallel B_{lf} component (thick line) of the field of the second parallel harmonic ($n = 2$) of the FMS wave is given for comparison.

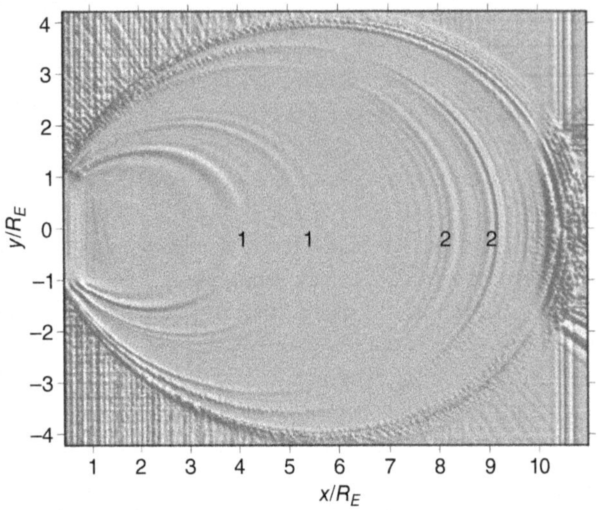

Figure 3.9 Amplitude distribution in the meridional plane for the full field of resonant Alfvén oscillations excited by monochromatic (frequency 0.01 Hz) magnetosonic wave in a dipole magnetosphere. Numbers 1, 2 denote the first ($N = 1$) and the second ($N = 2$) harmonics of standing Alfvén waves.

wind incident on the magnetosphere. The FMS wave frequency is 0.01 Hz, and its spatial structure corresponds to a harmonic with parallel wave number $n = 1$ and azimuthal wave number $m = 1$. It can be seen that three basic harmonics of standing Alfvén waves ($N = 1$ in the inner magnetosphere), three second harmonics ($N = 2$ – in the outer magnetosphere) and higher harmonics – in the transition layer region – are excited inside the magnetosphere at various resonance magnetic shells. The fact that several identical harmonics of standing Alfvén waves are excited is due to the non-monotonic distribution of the Alfvén speed in the model. The amplitude of resonant oscillations reaches its maximum in the antinodes of their parallel structure – for example, on the ionospheric upper boundary and, for the even harmonics $N = 2, 4, \ldots$, near the equatorial plane.

3.3 Resonant Alfvén Waves Excited in a Dipole-Like Magnetosphere by Broadband Sources

Sections 3.1 and 3.2 addressed the field structure of resonant standing Alfvén waves excited in a dipole-like magnetosphere by monochromatic FMS oscillations. In the real magnetosphere, FMS waves are as a rule excited by broadband sources [294]. We will therefore examine here the behaviour of the field of resonant Alfvén waves excited by various types of broadband sources in the magnetosphere, following [150]. We will trace the field dynamics of resonant Alfvén waves using as an example the behaviour of the main component of their magnetic field, B_2, which is implied by (E.21) (Appendix E), to be related to the scalar potential as follows:

$$B_2 = i\frac{\bar{c}}{\omega}\frac{g_2}{\sqrt{g}}\nabla_3\varphi',$$

where $\varphi' = \nabla_1\varphi$. Representing the full field potential φ' of the resonant oscillations as the sum with respect to all standing Alfvén wave harmonics (3.14), we can write

$$B_2(x^1,x^2,x^3,t) = \sum_{N=1}^{\infty} F_N(x^1,x^2,t)\nabla_3 T_N(x^1,x^3), \tag{3.43}$$

where the notations are

$$F_N(x^1,x^2,t) = \frac{1}{2\pi}\int_{-\infty}^{\infty} \tilde{\mu}_N(x^1,x^2,\omega)\tilde{Q}_N(x^1,\omega)e^{-i\omega t}d\omega, \tag{3.44}$$

$$\tilde{\mu}_N = i\frac{g_2}{\sqrt{g}}\frac{c\nabla_2\mu_N(x^1,x^2,\omega)}{\Omega_{TN}^3},$$

$$\tilde{Q}_N(x^1,\omega) = \left(\frac{a_N}{\rho_N}\right)^{2/3} e^{i\alpha_N/3} G\left[2\left(\frac{a_N}{\rho_N}\right)^{2/3} e^{i\alpha_N/3}\frac{\omega - \Omega_{TN} + i\gamma_{TN}}{\Omega_{TN}}\right]. \tag{3.45}$$

Here, expression (3.25) is used for the function $V_N'(x^1,\omega)$ describing the transverse structure of the Nth harmonic of standing Alfvén waves, function $G(z)$ has an integral representation, (2.89), and its asymptotics on the real axis have the form (2.90). The function $\tilde{\mu}_N$ describes the Alfvén wave source determined by the magnetospheric FMS oscillation field.

Using their inverse Fourier transforms to express the functions in the integrand in (3.44) allows us to write

$$F_N(x^1,x^2,t) = \int_{-\infty}^{\infty} \overline{\mu}_N(x^1,x^2,t')Q_N(x^1,t-t')dt',$$

where

$$Q_N(x^1,\tau) = \frac{1}{2\pi}\int_{-\infty}^{\infty} \tilde{Q}_N(x^1,\omega)e^{-i\omega\tau}d\omega = Q_N^{(+)}(x^1,\tau) + Q_N^{(-)}(x^1,\tau),$$

$$Q_N^{(+)}(x^1,\tau) = \frac{1}{2\pi}\int_0^{\infty} \tilde{Q}_N(x^1,\omega)e^{-i\omega\tau}d\omega,$$

$$Q_N^{(-)}(x^1,\tau) = \frac{1}{2\pi}\int_{-\infty}^{0} \tilde{Q}_N(x^1,\omega)e^{-i\omega\tau}d\omega. \tag{3.46}$$

Evidently, $Q_N^{(-)}(x^1,\tau) = [Q_N^{(+)}(x^1,\tau)]^*$, where * denotes a complex conjugate. The function $\tilde{Q}_N(x^1,\omega)$ being analytical in the upper halfplane of complex ω, we have $Q_N(x^1,\tau) = 0$ for $\tau < 0$, allowing us to write

$$\begin{aligned} F_N(x^1,x^2,t) &= \textstyle\int_{-\infty}^{t}\overline{\mu}_N(x^1,x^2,t')Q_N(x^1,t-t')dt' \\ &= \textstyle\int_0^{\infty}\overline{\mu}_N(x^1,x^2,t-\tau)Q_N(x^1,\tau)d\tau. \end{aligned} \tag{3.47}$$

The representation (3.47) expresses the causality principle: the resonant oscillation field at time t is determined by the source acting over the previous time interval.

Let us examine several examples of the behaviour of resonant standing Alfvén waves excited by the FMS oscillation field, using for it certain model representations.

3.3.1 Monochromatic Source of FMS Waves

Let the source function have the form

$$\tilde{\mu}_N(x^1,x^2,\omega) = M_N(x^1,x^2)\delta(\omega - \omega_0),$$

where $M_N(x^1,x^2)$ describes the spatial structure of the field of a monochromatic source with frequency ω_0 determined by the FMS oscillation field. Substituting this expression into (3.44) produces

$$\begin{aligned} F_N(x^1,x^2,t) &= \frac{1}{2\pi} M_N(x^1,x^2)\tilde{Q}_N(x^1,\omega_0)e^{-i\omega_0 t} \\ &\approx \frac{1}{2\pi} M_N(x^1,x^2)\tilde{Q}_N(x^1_{TN}(\omega_0),\omega_0)e^{-i\omega_0 t}. \end{aligned}$$

Here, the function $M_N(x^1,x^2)$ variation scale over variable x^1 is assumed to be much wider than for function $\tilde{Q}_N(x^1,\omega_0)$. Thus, we obtained the same result as in Sections 3.1 and 3.2, examining Alfvén waves excited by monochromatic FMS oscillations.

3.3.2 Pulse Source of FMS Waves

Let us examine a source function of the form

$$\overline{\mu}_N(x^1,x^2,t) = M_N(x^1,x^2)\delta(t)$$

– a sudden pulse of FMS oscillations at time $t = 0$. Substituting this expression into (3.47) yields

$$F_N(x^1,x^2,t) = M_N(x^1,x^2)Q_N(x^1,t).$$

Hence, the function $Q_N(x^1,t)$ describes the response of the Nth harmonic of standing Alfvén waves to a sudden pulse of FMS oscillations. Let us inspect this function in more detail. We will substitute for $\tilde{Q}_N(x^1,\omega)$ in (3.45) (for $\alpha_N = 0$ as an example) into the function $Q_N^{(+)}(x^1,\tau)$ definition and proceed to the integration variable in the integrand $\xi = 2(a_N/\rho_N)^{2/3}(\omega - \Omega_{TN})/\Omega_{TN}$:

$$Q_N^{(+)}(x^1,\tau) = \frac{\Omega_{TN}}{4\pi} e^{-i\Omega_{TN}\tau} \int_{-\infty}^{\infty} G(\xi + i2^{1/3}\gamma_{TN}\tau_N)\exp\left(-i\xi\frac{\tau}{2^{1/3}\tau_N}\right)d\xi,$$

where $\tau_N = (2a_N/\rho_N)^{2/3}/\Omega_{TN}$ is the characteristic time interval during which effects due to the ‘kinetic’ dispersion of Alfvén waves are manifest. Here, the lower integration limit extends to $-\infty$ because we assume the condition $\Omega_{TN}\tau_N \gg 1$ to hold, meaning strong localisation of the integrand near $\omega = \Omega_{TN}$. Using the integral representation (2.89) for function $G(z)$ and successively integrating with respect to ξ (which produces $\sim 2\pi\delta(\nu - \tau/2^{1/3}\tau_N)$ under the integral over ν) and ν produces

$$Q_N^{(+)}(x^1,\tau) = -i\Theta(\tau)\frac{\Omega_{TN}}{2}\exp\left(-i\Omega_{TN}\tau - \gamma_{TN}\tau - i\frac{\tau^3}{6\tau_N^3}\right),$$

where $\Theta(\tau)$ is the Heaviside step function ($\Theta(x) = 0$ for $x < 0$, $\Theta(x) = 1$ for $x \geq 0$). Thus the full response of the Nth harmonic of standing Alfvén waves can be represented as

$$Q_N(x^1,t) = -\Theta(t)\Omega_{TN}(x^1)e^{-\gamma_{TN}t}\sin\left(\Omega_{TN}t + \frac{t^3}{6\tau_N^3}\right),$$

which are oscillations with variable frequency $\Omega_{TN} + t^2/6\tau_N^3$ emerging immediately after the sudden pulse (at time $t = 0$) and decaying at decrement γ_{TN}. Oscillation frequency variation at observation point is due to Alfvén wave dispersion. It is explained by the fact that the oscillations are excited simultaneously over a wide range of resonance magnetic shells (with its own resonant frequency at each shell). Thanks to the transverse dispersion, the

Alfvén waves slowly (much slower than the Alfvén speed) move across the magnetic shells. As a result, waves with frequencies corresponding to the resonance shell from which they came reach the observation point at time t. Of course, these dispersion effects can only be noticeable if their dispersion time does not exceed the Alfvén wave decrement time: $\gamma_{TN}\tau_N \lesssim 1$.

3.3.3 FMS Wave Source in the Form of a Wave Packet (Substorm Pi2 Model)

Let us examine a source in the form of a wave packet (Figure 3.10a):

$$\overline{\mu}_N(x^1, x^2, t) = M_N(x^1, x^2)e^{-\Gamma|t|}\sin\omega_0 t.$$

For $\omega_0 \gtrapprox \Gamma$ its shape resembles a wave packet of substorm geomagnetic pulsations Pi2. Let us see what response such a FMS wave packet causes in standing Alfvén waves. Substituting this expression into (3.47) yields

$$F_N(x^1, x^2, t) = M_N(x^1, x^2)\int_0^\infty \left[\cos\left(\omega_0 t - (\omega - \Omega_{TN})\tau + \frac{\tau^3}{6\tau_N^3}\right)\right.$$
$$\left. - \cos\left(\omega_0 t - (\omega + \Omega_{TN})\tau - \frac{\tau^3}{6\tau_N^3}\right)\right]\exp\left[-\gamma_{TN}\tau - \Gamma|t-\tau|\right]d\tau.$$

If $\Gamma \gg \gamma_{TN}$, the bulk of the integral is near $\tau = t$ and we can easily set $\tau^3 \approx t^3$ in the cosine arguments, after which the integral is easily calculated. We obtain

$$F_N(x^1, x^2, t) \approx \frac{M_N(x^1, x^2)\Omega_{TN}^2}{(\Gamma^2 + (\omega_0 - \Omega_{TN})^2)(\Gamma^2 + (\omega_0 + \Omega_{TN})^2)}$$
$$\times \left[\left(2\Gamma\omega_0\cos\omega_0 t + (\omega_0^2 - \Omega_{TN}^2 - \Gamma^2)\sin\omega_0 t\right)e^{-\Gamma|t|}\right.$$
$$\left. + 4\Gamma\omega_0\Theta(t)e^{-\gamma_{TN}t}\cos\left(\Omega_{TN}t + \frac{t^3}{6\tau_N^3}\right)\right].$$

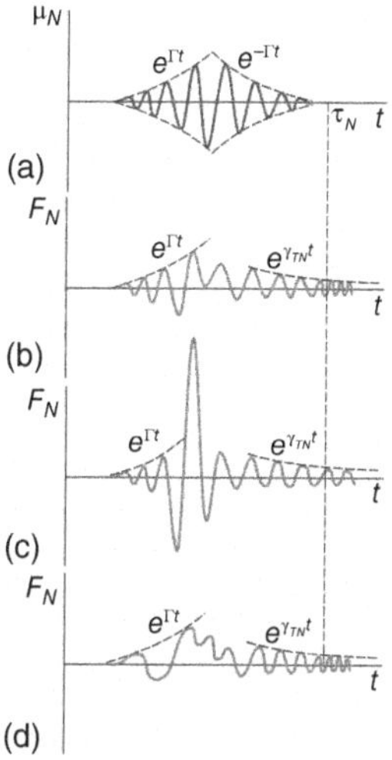

Figure 3.10 (a) FMS wave packet; (b–d) behaviour of the Nth harmonic of standing Alfvén waves excited by FMS wave packet: (b) exciting FMS wave frequency is higher than the frequency of Alfvén eigen oscillations, $\omega_0 \gg \Omega_{TN}$; (c) $|\omega_0 - \Omega_{TN}| \ll \Gamma$ – standing Alfvén waves and FMS oscillations are in resonance; (d) $\omega_0 \ll \Omega_{TN}$.

Here, the first term in the square brackets describes forced Alfvén oscillations at observation point caused by FMS wave packet passage. These oscillations' life-time is limited by the wave packet passage interval. The second term describes, same as in the pulse source case, eigen oscillations of standing Alfvén waves of variable frequency decaying at γ_{TN}. Figure 3.10 provides a qualitative picture of the behaviour of Alfvén waves excited by such a source. Cases are shown when the wave-packet carrier frequency is much larger than the frequency of magnetospheric Alfvén eigen oscillations ($\omega_0 \gg \Omega_{TN}$ – Figure 3.10b), when it is much smaller than that frequency ($\omega_0 \ll \Omega_{TN}$ – Figure 3.10) and when these frequencies are close ($|\omega_0 - \Omega_{TN}| \ll \Gamma$ – Figure 3.10c). The last case corresponds to resonance of a wave packet with carrier frequency coinciding with the frequency of a standing Alfvén wave harmonic, on the magnetic shell in question.

3.3.4 Stochastic Source of FMS Waves (Dayside Pc3 Model)

In many cases, the source of magnetospheric FMS oscillations has a stochastic character. For example, FMS oscillations penetrating into Earth's dayside magnetosphere from the solar wind are such a source. In this case the value of $\overline{\mu}_N(x^1, x^2, t)$ cannot be assumed as given, but it is possible to set the statistical characteristics of the ensemble of random functions describing the behaviour of the field of stochastic FMS oscillations. For the simplest case, let us think of waves with different frequencies as uncorrelated ('white noise'). It is then possible to write

$$\left\langle \tilde{\mu}_N(x^1, x^2, \omega)\mu_{N'}(x^1, x^2, \omega') \right\rangle = m_N(x^1, x^2, \omega)m_{N'}(x^1, x^2, \omega)\delta(\omega - \omega') \tag{3.48}$$

pair correlator of the functions describing the sources of FMS oscillations with frequencies ω and ω' that excite standing Alfvén wave harmonics with parallel wave numbers N and N'. The angle brackets here denote phase averaging of random oscillations.

For the pair correlator of the basic components of the standing Alfvén wave field we then have

$$\begin{aligned}
&\left\langle \tilde{B}_2^*(x^1, x^2, x^3, \omega)\tilde{B}_2(x^1, x^2, x^3, \omega') \right\rangle \\
&\quad = \sum_{N,N'} m_N^*(x^1, x^2, \omega)m_{N'}(x^1, x^2, \omega)\tilde{Q}_N^*(x^1, \omega)\tilde{Q}_{N'}(x^1, \omega) \\
&\quad \times \nabla_3 T_N(x^1, x^3)\nabla_3 T_{N'}(x^1, x^3)\delta(\omega - \omega').
\end{aligned} \tag{3.49}$$

Given the functions $Q_N(x^1, \omega)$ being narrow-localised with respect to frequency near $\omega = \Omega_{TN}(x^1)$, correlations with $N' \neq N$ can be neglected, in the main order of perturbation theory, in the sum (3.49). The mean square of the Alfvén oscillation amplitude at observation point will then be

$$\left\langle \left| \tilde{B}_2(x^1, x^2, x^3, \omega) \right|^2 \right\rangle = P(x^1, x^2, x^3, \omega)\delta(\omega - \omega'),$$

where

$$P(x^1, x^2, x^3, \omega) = \sum_N \left| m_N(x^1, x^2, \Omega_{TN}) \right|^2 |\tilde{Q}_N(x^1, \omega)|^2 \left| \nabla_3 T_N(x^1, x^3) \right|^2.$$

An inverse Fourier transform will produce

$$\begin{aligned}
&\left\langle \left| \tilde{B}_2(x^1, x^2, x^3, t) \right|^2 \right\rangle \\
&= \int_{-\infty}^{\infty} d\omega \int_{-\infty}^{\infty} d\omega' \exp[i(\omega - \omega')] \left\langle \left| \tilde{B}_2^*(x^1, x^2, x^3, \omega) \right|^2 \right\rangle \\
&= \int_{-\infty}^{\infty} P(x^1, x^2, x^3, \omega) d\omega.
\end{aligned}$$

Using the representation (3.45) (for $\alpha_N = 0$) for $\tilde{Q}_N(x^1, \omega)$ yields

$$\begin{aligned}
\int_{-\infty}^{\infty} |\tilde{Q}_N(x^1, \omega)|^2 d\omega = \frac{\Omega_{TN}}{2} \left(\frac{a_N}{\rho_N} \right)^{2/3} \int_{-\infty}^{\infty} d\xi \int_{-\infty}^{\infty} dv \int_{-\infty}^{\infty} dv' \\
\times \exp\left[i\left(\frac{v^3}{3} - \frac{v'^3}{3} \right) - i(v - v')\xi - (v + v')\frac{\varepsilon_{TN}}{l_n} \right],
\end{aligned}$$

where $\xi = 2(a_N/\rho_N)^{2/3}(\omega - \Omega_{TN})/\Omega_{TN}$, $\varepsilon_{TN} = 2\gamma_{TN}a_N/\Omega_{TN}$, $l_N = a_N^{1/3}\rho_N^{2/3}$. Successively integrating over ξ, v' and v yields

$$\int_{-\infty}^{\infty} |\tilde{Q}_N(x^1,\omega)|^2 d\omega = \pi\left(\frac{a_N}{\rho_N}\right)^{2/3}\frac{l_N\Omega_{TN}}{\varepsilon_{TN}} = \frac{\pi}{2}\frac{\Omega_{TN}^2}{\gamma_{TN}},$$

whence

$$\left\langle\left|\tilde{B}_2(x^1,x^2,x^3,t)\right|^2\right\rangle = \frac{\pi}{2}\sum_N \frac{\Omega_{TN}^2}{\gamma_{TN}}\left|m_N(x^1,x^2,\Omega_{TN})\right|^2\left|\nabla_3 T_N(x^1,x^3)\right|^2. \tag{3.50}$$

The expression (3.50) completely determines the amplitude distribution of standing Alfvén waves excited in an axisymmetric magnetosphere by stochastic FMS oscillations featuring a 'white noise' spectrum (3.48).

Let us discuss the amplitude distribution of geomagnetic Pc3 pulsations continually observed in the dayside magnetosphere, while assuming them to be a set of harmonics of standing Alfvén waves excited by stochastic FMS oscillations. As a source of such FMS oscillations, instability of solar wind protons reflected from the bow shock wave front is considered in [296]. FMS oscillations penetrate into the magnetosphere and excite standing Alfvén waves on resonance magnetic shells.

An important role in the Alfvén wave amplitude distribution in (3.50) is played by the source spectral function $m(x^1,x^2,\omega)$, describing the FMS oscillation field. Let us assume that it has a sufficiently conspicuous local maximum at certain frequency $\omega = \overline{\omega}$ determined by the FMS oscillation generation mechanism. This is also supported by spacecraft observations of geomagnetic Pc3 pulsations [255, 256].

Figure 3.11 is a schematic representation of standing Alfvén waves excited by such a source. As the number of the parallel harmonic of standing Alfvén waves increases, their amplitude decreases. Function $m_1(x^1,x^2,\omega)$ has two, and function $m_2(x^1,x^2,\omega)$ three local maxima in the amplitude distribution across magnetic shells, which is determined by non-monotonically changing Alfvén speed near the plasmapause. The right-hand maxima of these functions are higher than the left-hand ones because we assume the magnetosphere to be an opaque region for the FMS oscillations in question, their amplitude decreasing from the magnetopause into the magnetosphere. Therefore the inner magnetosphere is dominated by oscillations of the first harmonic $N = 1$, while its outer part by those of higher

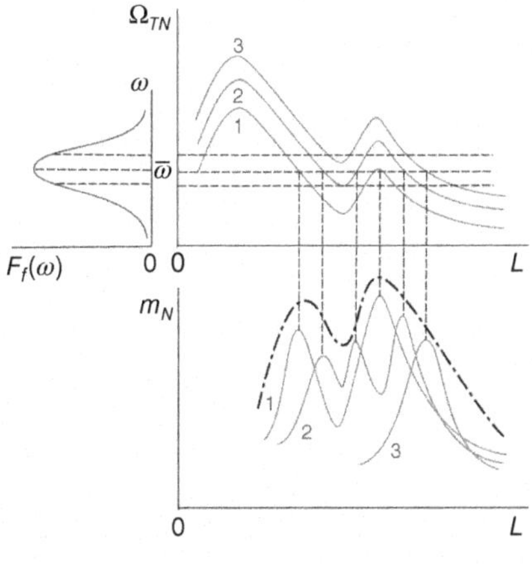

Figure 3.11 Standing Alfvén waves excited in the magnetosphere by a stochastic source of FMS oscillations. Top left – the spectrum of stochastic FMS oscillations, $F_f(\omega)$. Top right – eigenfrequency Ω_{TN} distribution for three first harmonics ($N = 1, 2, 3$) of resonant Alfvén waves across magnetic shells ($L = a/R_E$ – McIlwain parameter). Bottom – distribution, across magnetic shells, of the oscillation amplitude m_N for the first three harmonics of the excited standing Alfvén waves and their envelope (dash-dotted line).

harmonics $N = 2, 3, \ldots$ of standing Alfvén waves, their frequencies Ω_{TN} approximately equal to the central frequency of the source, $\overline{\omega}$. This pattern of the amplitude distribution of daytime geomagnetic pulsations is also supported by ground-based observations [12].

3.4 Magnetosonic Resonance in a Dipole-Like Magnetosphere

In Sections 3.1–3.3 of this chapter addressed Alfvénic resonance in a dipole-like model magnetosphere. We employed a 'cold plasma' model. Effects determining the transverse dispersion of kinetic Alfvén waves can be included as small corrections to this model. Finite-pressure plasma features another branch of MHD oscillations – SMS. SMS waves propagate almost along magnetic field lines. This fact makes SMS waves similar to Alfvén waves.

Same as resonant Alfvén waves, SMS oscillations with small azimuthal wave numbers ($m \sim 1$) can be generated as a result of resonance with fast magnetosound, as demonstrated in [297]. To differentiate between the resonant interaction processes between FMS oscillations and Alfvén or SMS waves, we will hereafter call them 'Alfvén resonance' and 'magnetosonic resonance', respectively. The process of magnetosonic resonance consists in monochromatic FMS wave exciting SMS oscillations on those magnetic shells where the FMS frequency coincides with the local frequency of SMS eigen oscillations. The structure and spectrum of resonant SMS oscillations were not calculated in [297]. Such calculations are carried out in Section 2.6 of this monograph, using a 1D-inhomogeneous model magnetosphere (see also [51, 61, 298]).

Below we will examine magnetosonic resonance in a 2D-inhomogeneous model magnetosphere with a dipole-like magnetic field.

3.4.1 Self-Consistent Model of a Dipole Magnetosphere with Rotating Plasma

To describe the plasma dynamics in magnetic field let us employ the system of ideal MHD equations (1.1)–(1.4). We will assume the magnetic field in the model magnetosphere to be a dipole, its field-line equation for the meridional plane being

$$r = a \cos^2 \theta, \tag{3.51}$$

where r is the radius of a point on a field-line in a spherical system of coordinates, a is the equatorial radius of the field line, θ is the magnetic latitude as measured from the equatorial plane (see Figure 3.12). The length element along the field line has the form

$$dl = a \cos\theta \sqrt{1 - 3 \sin^2 \theta} d\theta,$$

and the dipole magnetic field strength is determined by

$$B_0(a, \theta) = \overline{B}\left(\frac{\bar{a}}{a}\right)^3 \frac{\sqrt{1 + 3 \sin^2 \theta}}{\cos^6 \theta}. \tag{3.52}$$

It is possible to select $\bar{a} = R_E$ for Earth's magnetosphere, $\overline{B} \approx 0.32$ G is then the equatorial strength of the magnetic field on Earth's surface.

MHD equations have the simplest form in a system of coordinates related to magnetic field lines. Our calculations rely on an orthogonal system of curvilinear coordinates (x^1, x^2, x^3) (see Figure 3.12). Any other functions uniquely linked to each of the coordinates

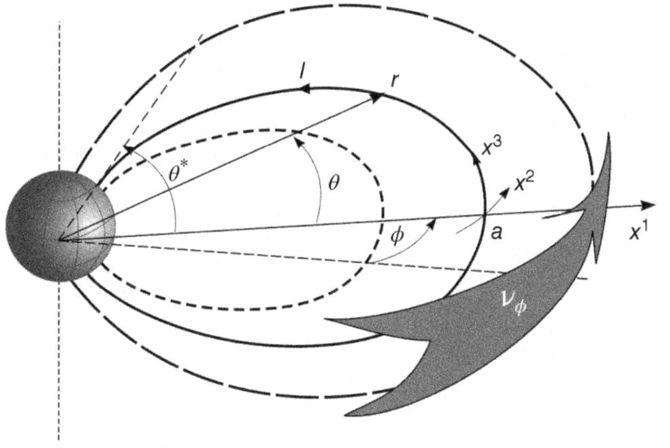

Figure 3.12 Model axisymmetric magnetosphere and systems of coordinates: (x^1, x^2, x^3) is an orthogonal system of curvilinear coordinates related to magnetic field lines, (a, ϕ, θ) is a non-orthogonal system of curvilinear coordinates. Dashed lines indicate the plasmapause ($a = a_p$) and magnetopause ($a = a_m$).

(x^1, x^2, x^3) can be used as coordinates. It was found to be convenient to use in numeric calculations the field-line equatorial radius a instead of coordinate x^1, the azimuthal angle ϕ instead of x^2, and latitude θ (or field-line length l), as counted from the equatorial plane instead of the parallel coordinate x^3. Note that the coordinates (a, ϕ, θ) are of course not orthogonal.

Metric tensor components determining the length elements along the coordinate lines in variables a, θ have the form (see Appendix P)

$$g_1 = \frac{\cos^6\theta}{1 + 3\sin^2\theta}, \qquad g_2 = a^2\cos^6\theta. \tag{3.53}$$

The third metric-tensor component can be determined from the ratio of the segments of two field lines of equatorial radius a and a_0 between infinitely close coordinate surfaces x^3 =const and $x^3 + dx^3$ =const. If the field line of equatorial radius a crosses the coordinate surface x^3 = const at latitude θ, and the equatorial radius a_0 field line, at latitude θ_0, the expression for g_3 is

$$\frac{g_3(a,\theta)}{g_3(a_0,\theta_0)} = \left(\frac{a}{a_0}\right)^6 \left(\frac{\cos\theta}{\cos\theta_0}\right)^{12} \frac{1 + 3\sin^2\theta_0}{1 + 3\sin^2\theta} = \frac{B_0^2(a_0,\theta_0)}{B_0^2(a,\theta)}. \tag{3.54}$$

It is easily verifiable that the equation $\text{div}\mathbf{B}_0 = 0$, reducible to

$$\frac{\partial}{\partial x^3}\frac{\sqrt{g}}{g_3}B_3 = \frac{\partial}{\partial x^3}\sqrt{g_1 g_2}B_0 = 0,$$

in the selected system of coordinates, holds for a dipole magnetic field. The condition $\text{curl}\mathbf{B}_0 = 0$ also holds meaning a current-free plasma. The magnetic field is called 'force-free' in this case.

We will next construct a stationary ($\partial/\partial t = 0$) solution to the system of equations (1.1)–(1.4), which will determine the spatial distribution of equilibrium plasma parameters. For the sake of simplicity, we will assume the plasma to move in the azimuthal direction only $\mathbf{v}_0 = (0, v_\phi = v_2/\sqrt{g_2}, 0)$ (see Figure 3.12). We will also assume the model to be azimuthally symmetric. Both (1.3) and (1.4) are valid, because the first term is zero in either. The first and third components in the vector equation (1.2) are also identically equal to zero, while the third yields

$$\frac{B_3}{g_3}\nabla_3\frac{v_2}{g_2} = 0. \tag{3.55}$$

The ϕ component of the metric tensor g_2 in any axisymmetric system of coordinates using the azimuthal angle as the x^2 coordinate can be represented as $g_2 = \rho^2$, where $\rho = a\cos^3\theta$ is the radius in the cylindric system of coordinates (ρ, ϕ, z) as counted from the symmetry axis. Given that $\Omega = v_\phi/\rho = v_2/g_2$ is the angular velocity of the plasma rotation, and $\mathbf{B}_0 = (0, 0, B_3/\sqrt{g_3})$, Eq. (3.55) can be rewritten as

$$\mathbf{B}_0\nabla\Omega = 0. \tag{3.56}$$

It follows from this equation that the angular velocity of the plasma motion in the model magnetosphere under study is constant on each magnetic shell and only changes across magnetic shells – in the x^1 coordinate: $\Omega \equiv \Omega(x^1)$. Therefore the azimuthal rotation of plasma on each magnetic shell is at a constant angular velocity without changing the magnetic field-line geometry. This means, in particular, that the conditional boundaries separating regions with sharply differing velocities of plasma motion (such as the plasmapause and magnetopause) must coincide with one of the magnetic shells, as shown in Figure 3.12.

Our numeric calculations rely on the following model $\Omega(x^1)$:

$$\Omega(a) = \frac{d(a)}{2}\left[(\Omega_m + \Omega_{sw}) - (\Omega_m - \Omega_{sw})\tanh\frac{a - a_m}{\Delta_{mp}}\right], \tag{3.57}$$

where Ω_m, Ω_{sw} are the characteristic values of the plasma angular velocity in the neighbourhood of the magnetopause, in the magnetosphere and solar wind, respectively, a_m is the equatorial radius of the magnetopause, Δ_{mp} is the characteristic thickness of the magnetopause transition layer. The coefficient $d(a)$ is necessary for modelling the plasma motion character inside the magnetosphere and in the solar wind. Let us select the following model for this:

$$d(a) = \begin{cases} \exp\left(-\alpha(a)(a - a_m)^2/\Delta_m^2\right), & a \le a_m, \\ a_m/a, & a > a_m, \end{cases}$$

where Δ_m is the characteristic scale of plasma velocity variation in the magnetosphere, $\alpha(a)$ is a coefficient of order unity. Let us employ the following numerical values of the parameters: $a_m = 10R_E$, $\Delta_{mp} = 0.5R_E$, where R_E is Earth's radius, $\Omega_m = v_{\phi m}/a_m$, $\Omega_{sw} = v_{\phi sw}/a_m$, where $v_{\phi m} = 50$ km/s, $v_{\phi sw} = 400$ km/s are the characteristic azimuthal velocities of plasma motion in the neighbourhood of the magnetopause, in the magnetosphere and solar wind, $\Delta_m = 10R_E$.

For the selected models of magnetic field and plasma motion velocity, the second component in the vector equation (1.1) identically turns to zero, while the other two equations can be written as

$$\frac{\rho_0\Omega^2}{2}\nabla_1 g_2 = \nabla_1 P_0, \tag{3.58}$$

$$\frac{\rho_0\Omega^2}{2}\nabla_3 g_2 = \nabla_3 P_0. \tag{3.59}$$

Equilibrium in the rotating plasma is sustained by its gas-kinetic pressure gradient.

To solve the system of Eqs. (3.58) and (3.59) in the general form, let us differentiate (3.58) with respect to x^3, (3.59) with respect to x^1 and subtract one from the other. The result will be

$$(\nabla_1 g_2)\nabla_3(\rho_0\Omega^2) - (\nabla_3 g_2)\nabla_1(\rho_0\Omega^2) = 0. \tag{3.60}$$

We will seek a solution to this equation using the characteristics method. The characteristics of the operator in the left-hand side of (3.60) are determined by

$$\frac{dx^1}{dx^3} = -\frac{\nabla_3 g_2}{\nabla_1 g_2},$$

reducible to the total differential $dg_2 = 0$, implying that the metric tensor component g_2 does not change along the characteristic. In the axisymmetric system of coordinates $g_2 = \rho^2$, hence, the characteristics in the plane $x^2 = \text{const}$ are straight lines running parallel to the symmetry axis. Two of them are shown in Figure 3.12 as vertical dashed lines.

If a coordinate s is introduced which changes along the characteristic, by parameterising the coordinates x^1 and x^3:

$$\frac{dx^1}{ds} = 1, \qquad \frac{dx^3}{ds} = -\frac{\nabla_3 g_2}{\nabla_1 g_2},$$

Eq. (3.60) is reduced to

$$\frac{\partial \rho_0\Omega^2}{\partial s} = 0. \tag{3.61}$$

This implies that the value of $\rho_0\Omega^2$ remains the same along the characteristic. In particular, an expression can be written determining the plasma density at an arbitrary point in the meridional plane via the value at the point of its vertical projection onto the equatorial plane:

$$\rho_0(a,\theta) = \rho_0(\rho,0)\frac{\Omega^2(\rho)}{\Omega^2(a)}. \tag{3.62}$$

One of the most important parameters determining the MHD-oscillation structure in plasma is the Alfvén speed $v_A = B_0/\sqrt{4\pi\rho_0}$. Using expression (3.52) for B_0, and (3.62) for ρ_0 results in

$$v_A(a,\theta) = v_A(\rho,0)\frac{\Omega(a)}{\Omega(\rho)}\cos^3\theta\sqrt{1+3\sin^2\theta}. \tag{3.63}$$

The equilibrium plasma pressure for given magnetic field $B_0(a,\theta)$ and density distribution $\rho_0(a,\theta)$ can also be found from (3.58) and (3.59). Integrating (3.59) along the field line yields

$$\begin{aligned} P_0(a,\theta) &= P_0(a,0) + \frac{\Omega^2(a)}{2}\int_0^\theta \rho_0(a,\theta')\frac{\partial g_2(a,\theta')}{\partial\theta'}d\theta' \\ &= P_0(a,0) - 3a^2\Omega^2(a)\int_0^\theta \rho_0(a,\theta')\sin\theta'\cos^5\theta'\,d\theta', \end{aligned} \tag{3.64}$$

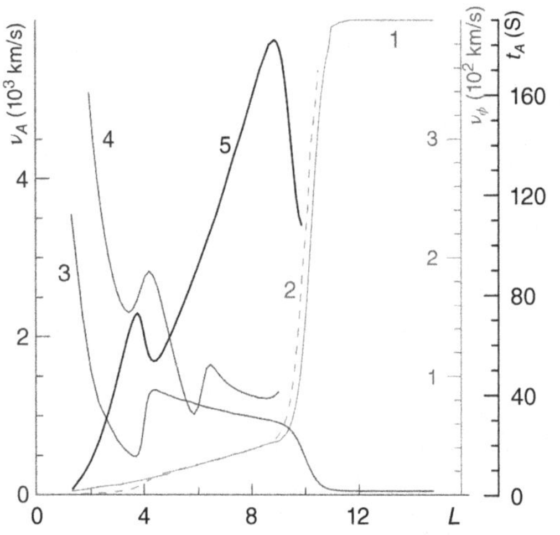

Figure 3.13 Distribution across magnetic shells of plasma equatorial velocity v_ϕ(curves 1 and 2), Alfvén speed $v_A(a, \theta)$ (curve 3 in the equatorial plane $\theta = 0$, curve 4 along radius **r** at angle $\theta = 30°$ to the equator) and the basic harmonic period t_A of magnetospheric standing Alfvén waves (curve 5).

where $P_0(a, 0)$ is pressure at the point where the field line crosses the equatorial plane, to determine which we will integrate (3.58) for $\theta = 0$:

$$P_0(a, 0) = P_0(\bar{a}, 0) + \int_{\bar{a}}^{a} \rho_0(a', 0)\Omega^2(a')a' da', \tag{3.65}$$

where $\bar{a}$ is the equatorial radius of the boundary magnetic shell, which we will choose to be equal to the ionospheric radius $\bar{a} = r_i$.

Figure 3.13 is the equatorial distribution of plasma velocity for two different $\Omega(a)$ distribution models given by function $\alpha(a)$. Curve 1 corresponds to $\alpha = 1$ inside the entire magnetosphere, while curve 2 was obtained for $\alpha(a) = (3 - \tanh(a - a_p)/\Delta_p)/4$, where $a_p = 4R_E$ is the plasmapause equatorial radius, and $\Delta_p = 0.5R_E$ is the characteristic thickness of the plasmapause transition layer. This model $\alpha(a)$ allows simulation of the sharp change in v_ϕ at the plasmapause. Such a distribution of v_ϕ results in a sharp jump in plasma density when passing through the plasmapause.

Sharp changes in Alfvén speed distribution when passing through the plasmapause and magnetopause are due to plasma density $\rho_0(a, \theta)$ variation. Curve 3 in Figure 3.13 shows the equatorial distribution of Alfvén speed $v_A(a, 0)$, in accordance with model [299], as given by analytical formulas (3.28) and (3.29) for $\theta = 0$. Also shown is the distribution of v_A along radius **r** at inclination $\theta = 30°$ to the equatorial plane (curve 4). An interesting feature in this distribution is the 'double plasmapause' – a phenomenon which is not infrequently observed by spacecrafts. Within the model in question, one of the local maxima is formed due to the sharply changing density in the plasmapause transition layer. The second maximum is due to a similarly sharp density change in the cylindric layer formed by the plasmapause projected along the characteristics (see (3.62)). Curve 5 in the same figure illustrates the value of the basic period, t_A, of the magnetospheric standing Alfvén waves found from (3.16) in the WKB approximation.

Figure 3.14 shows an isoline distribution of parameter $\beta = 8\pi P_0(a, \theta)/B_0^2(a, \theta)$, determining the plasma gas-kinetic to magnetic pressure ratio. The pressure value $P_0(\bar{a}, 0) = 0.2$ nPa at height $h = 1500$ km above Earth surface were chosen as the boundary condition at the ionosphere. For plasma concentration $n_0 = 10^4$ cm^{-3} this corresponds to temperature $T_0 = 1500$ K. What is often taken as the conditional boundary of the magnetosphere is

Figure 3.14 Parameter $\beta = 8\pi P_0(a,\theta)/B_0^2(a,\theta)$ isolines in the meridional plane (x,z). Coordinates $x = a\cos^3\theta$ and $z = a\cos^2\theta\sin\theta$ are in Earth radius units.

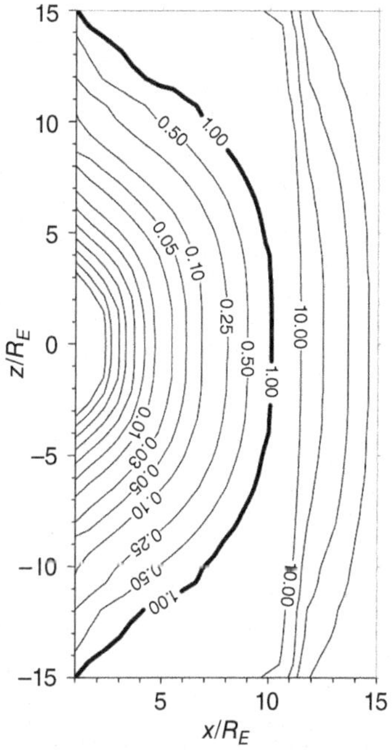

the surface where the sum of the dynamic and gas-kinetic plasma pressure of the solar wind becomes equal to the geomagnetic field pressure ($\beta = 1$ in our model). It can be seen in the model magnetosphere in question, that this surface is very different from the magnetopause, determined earlier as the boundary separating the regions of fast-moving solar-wind plasma from those of slow convection of the magnetospheric plasma. The surfaces only coincide in the near-equatorial region.

In this manner, we have constructed a self-consistent model of a dipole magnetosphere with rotating plasma balanced by its gas-kinetic pressure gradient.

3.4.2 Basic Equations for Magnetosonic Waves

To describe the magnetosonic oscillation field in a finite-pressure plasma, let us employ Eq. (E.2) (Appendix E), neglecting the fact that its right-hand side, describing the effect the Alfvén wave field has on magnetosonic oscillations, is non-zero. As could be seen in Sections 3.1–3.2, the feedback from Alfvén waves to FMS oscillation is small even on resonance surfaces. Interaction between Alfvén and SMS waves can be significant in oscillations with large azimuthal wave numbers $m \gg 1$. For the oscillations with $m \sim 1$ we examine here, this interaction is non-essential.

In a homogeneous plasma, the operator in the left-hand side of (E.7) gives a dispersion equation for magnetosonic waves:

$$\omega^4 - \omega^2 k^2 (v_A^2 + v_s^2) + k^2 k_\parallel^2 v_A^2 v_s^2 = 0, \tag{3.66}$$

where $k^2 = k_\parallel^2 + k_\perp^2$, $k_\parallel^2 = k_3^2/g_3$, $k_\perp^2 = k_1^2/g_1 + k_2^2/g_2$. The solution of (3.66) can be written as

$$\omega^2 = \frac{k^2}{2}(v_A^2 + v_s^2) \pm \sqrt{\frac{k^4}{4}(v_A^2 + v_s^2)^2 - k^2 k_\parallel^2 v_A^2 v_s^2}.$$

The plus sign refers to the dispersion equation for FMS waves, and the minus sign, to the one for SMS waves. If one of the inequalities $v_s \ll v_A$, $v_A \ll v_s$, $|k_{\|}| \ll |k_{\perp}|$ holds true, the following approximate dispersion equations can be obtained:

$$\omega^2 \approx k^2 c_f^2$$

– for FMS waves, where $c_f^2 = v_A^2 + v_s^2$ and,

$$\omega^2 \approx k_{\|}^2 c_s^2,$$

– for SMS waves, where $c_s^2 = v_A^2 v_s^2/(v_A^2 + v_s^2)$. In the model magnetosphere we are dealing with at least one of the above inequalities holds true ($v_s \ll v_A$).

When describing both fast and slow magnetosonic waves, potential ψ can be represented, in the linear approximation, as the sum $\psi = \psi_f + \psi_s$, where the component ψ_f is related to the FMS wave, and ψ_s to the SMS wave. Demonstrably, the following conditions are realised in an inhomogeneous plasma for the full perturbed pressure of SMS waves:

$$\left(\tilde{P} + P_m\right)/\tilde{P} \sim \left(\tilde{P} + P_m\right)/P_m = \frac{k_{\|}^2}{k^2}\frac{v_A^2}{v_A^2 + v_s^2},$$

where $P_m = B_0 B_{\|}/4\pi$ is perturbed magnetic pressure, and $\tilde{P}$ is perturbed gas-kinetic pressure described by (E.5) (Appendix E) if $\varphi = 0$. This implies that, for oscillations with $k_{\perp} \gg k_{\|}$, the full pressure in a SMS wave is practically unperturbed:

$$\tilde{P} + \frac{B_0 B_{\|}}{4\pi} \approx 0. \tag{3.67}$$

As will be seen from further calculations, the resonant SMS-wave field is narrowly localised across magnetic shells near the resonance surface. Away from this surface, the potential ψ is determined by the FMS-oscillation field ($\psi \approx \psi_f$). If we ignore, in the operator $\hat{L}_0$ in (E.7) (Appendix E), the small (of order $v_s/v_A \ll 1$) component related to the derivatives with respect to the longitudinal coordinate x^3, the result will be an equation describing the FMS wave field away from the resonance surface

$$v_A^2 \tilde{\Delta}\psi_f + v_s^2 \overline{\Delta}\psi_f + \omega^2 \psi_f = 0, \tag{3.68}$$

where

$$\tilde{\Delta} = \frac{g_3}{\sqrt{g}}\nabla_1 \frac{g_2}{\sqrt{g}}\nabla_1 - \frac{k_2^2}{g_2} + \nabla_3 \frac{p}{\sqrt{g_3}}\nabla_3 \frac{p^{-1}}{\sqrt{g_3}},$$

$$\overline{\Delta} = \frac{B_0}{P_0^\sigma}\frac{1}{\sqrt{g_1 g_2}}\left(\nabla_1 \frac{pP_0^\sigma}{B_0}\nabla_1 - \frac{k_2^2}{p}\frac{P_0^\sigma}{B_0} + \nabla_3 \frac{\sqrt{g}}{g_3}\frac{P_0^\sigma}{\rho_0}\nabla_3 \frac{\rho_0}{B_0\sqrt{g_3}}\right),$$

$\sigma = 1/\overline{\gamma}$, $\overline{\gamma} = 5/3$ is the adiabatic index.

Near the resonance surface, the full field of magnetosonic oscillations is the sum of the FMS and SMS wave fields: $\psi = \psi_f + \psi_s$. Substituting this expression for potential ψ into (E.7) and taking (3.68) into account yield this equation

$$\frac{B_0\sqrt{g_3}}{4\pi\rho_0}\hat{L}_0 \frac{B_0}{\sqrt{g_3}}\tilde{\Delta}\psi_s + v_s^2\overline{\Delta}\psi_s + \omega^2\psi_s$$
$$= -\frac{v_A^2 v_s^2}{\omega^2}\frac{\rho_0/\sqrt{g_1 g_2}}{B_0 P_0^\sigma}\nabla_3 \frac{\sqrt{g}}{g_3}\frac{P_0^\sigma}{\rho_0}\nabla_3 \frac{B_0}{\sqrt{g_3}}\tilde{\Delta}\psi_f, \tag{3.69}$$

describing resonant SMS oscillations, where

$$\hat{L}_0 = \frac{v_s^2}{\omega^2}\frac{\rho_0}{P_0^\sigma \sqrt{g}}\nabla_3 \frac{\sqrt{g}}{g_3}\frac{P_0^\sigma}{\rho_0}\nabla_3 + 1.$$

The right-hand side in (3.69) describes the source of resonant SMS waves – the field of monochromatic FMS wave. We will assume it to be known from the solution of (3.68) (see Section 3.5).

The boundary conditions for SMS waves on the ionosphere are given by (2.275), where we will neglect external currents and Alfvén wave generation due to finite Hall conductivity in the ionosphere. As was shown in Section 3.5, the near-Earth magnetosphere is an opaque region for the FMS waves discussed here, which is why their contribution into the full field of magnetosonic waves on the ionosphere will also be neglected. Let us assume that the vector potential ψ component in the boundary condition (2.275) is only related to the SMS wave field:

$$\psi_s\big|_{l=l_\pm} = \mp i\frac{v_{p\pm}}{\omega}\frac{\partial \psi_s}{\partial l}\bigg|_{l=l_\pm}, \tag{3.70}$$

where the derivative $\partial/\partial l \equiv \partial/\sqrt{g_3}\partial x^3$ is along the field line, and the longitudinal coordinate is represented by field line length l counted from the equatorial plane ($l_\pm$ are the coordinates of the field line intersection with the ionosphere of the Northern or Southern Hemisphere). The right-hand side in (3.70) describes SMS wave decay due to the Joule dissipation of their energy in the conducting layer of the ionosphere.

3.4.3 Structure of Standing SMS Waves Along Magnetic Field Lines

The calculations that follow will concern the first few harmonics of SMS waves standing along magnetic field lines. Longitudinally, the characteristic wavelength of such oscillations is of order field-line length. We will find that the characteristic scale of resonant SMS oscillations across magnetic shells is much smaller than their longitudinal wavelength: $|\nabla_1\psi_s/\psi_s| \gg |\nabla_3\psi_s/\psi_s|$. This allows one to use the multiple-scale method for finding a solution to (3.69), by representing the potential ψ_s in the form

$$\psi_s = U(x^1)[S(x^1,x^3) + h(x^1,x^3)]\exp(ik_2x^2 - i\omega t), \tag{3.71}$$

where function $U(x^1)$ describes the small-scale structure of the oscillations in the x^1 coordinate, and $S(x^1,x^3)$ the oscillation structure along magnetic field lines, in the main order of perturbation theory. The characteristic scale of the $S(x^1,x^3)$ variation in x^1 is much larger than the $U(x^1)$ variation scale. The small correction $h(x^1,x^3)$ describes the oscillations in higher orders of perturbation theory.

An equation for the longitudinal structure of SMS waves is obtained if (3.69) retains the terms of only the main order ($\sim \nabla_1^2\psi_s$) of perturbation theory:

$$\frac{\rho_0 p^{-1}}{B_0 P_0^\sigma}\nabla_3 \frac{\sqrt{g}P_0^\sigma}{g_3\rho_0}\nabla_3 \frac{B_0}{g_1\sqrt{g_3}}S + \frac{\omega^2}{c_s^2}S = 0. \tag{3.72}$$

We will assume that, in the main order, the functions $S(x^1,x^3)$ satisfy the homogeneous boundary conditions on the ionosphere: $S(x^1,x^3_\pm) = 0$. The solution of (3.72) are the eigenfunctions $S_N(x^1,x^3)$, ($N = 1, 2, 3, \ldots$ is the longitudinal wave number) and the corresponding eigenvalues of the oscillation frequency $\omega^2 = \Omega_{sN}^2$. We will assume the eigenfunctions to be normalised by the following condition

$$\int_{l_-}^{l_+} \frac{pP_0^\sigma}{g_1 g_3} \frac{v_A^2}{c_s^2} S_N^2 dl = 1. \tag{3.73}$$

For a qualitative idea of the structure of the eigenfunctions $S_N(x^1, x^3)$, let us solve (3.72) in the WKB approximation, applicable to harmonics with $N \gg 1$. We will use the standard WKB method analogous to the one in Section 3.1.1 and write the solution of (3.72) normalised by the condition (3.73), in the two first orders of the WKB approximation, as

$$S_N(x^1, x^3) = \sqrt{\frac{2}{t_s} \frac{g_3 g_1^{3/2}}{P_0^\sigma g_2^{1/2}} \frac{c_s}{v_A^2}} \sin\left(\Omega_{sN} \int_{l_-}^{l} \frac{dl}{c_s}\right), \tag{3.74}$$

where $\Omega_{sN} = 2\pi N / t_s$,

$$t_s = \int_{l_-}^{l_+} \frac{dl}{c_s} \tag{3.75}$$

is the transit time along the magnetic field line between the magnetoconjugated ionospheres, at slow magnetosonic wave speed c_s. The solution to (3.72) for the basic harmonics ($N \sim 1$) can only be found numerically, same as in Section 3.4.6.

3.4.4 Structure of Resonant SMS Oscillations Across Magnetic Shells

In the first order of perturbation theory, we will assume the functions $h_N(x^1, x^3)$ in (3.71) to satisfy the following boundary conditions on the ionosphere:

$$h_N\big|_{l=l_\pm} = \mp i U_N(x^1) \frac{v_{p\mp}}{\omega} \frac{\partial S_N}{\partial l}\bigg|_{l=l_\pm}. \tag{3.76}$$

Let us left-multiply (3.69) by $S_N\, pP_0^\sigma / \sqrt{g_3} v_s^2$ and integrate along the field line between the magnetoconjugated ionospheres. Given (3.72) and the boundary condition (3.76), we obtain the following equation for function $U_N(x^1)$:

$$\frac{(\omega^2 + i\gamma_N)^2 - \Omega_{sN}^2}{\omega^2} \nabla_1^2 U_N^2 - \left[\beta_{1N} + (k_2^2 \beta_{2N} + \beta_{3N}) \frac{(\omega^2 + i\gamma_N)^2 - \Omega_{sN}^2}{\omega^2}\right] U_N = D_N, \tag{3.77}$$

where the notations are:

$$\beta_{1N} = \int_{l_-}^{l_+} \frac{pP_0^\sigma}{g_3} S_N \left(\nabla_3 \frac{g_2}{\sqrt{g}} \nabla_3 \frac{g_1}{\sqrt{g}} S_N + \frac{\Omega_{sN}^2}{c_s^2} S_N\right) dl,$$

$$\beta_{2N} = \int_{l_-}^{l_+} \frac{pP_0^\sigma}{g_2 g_3} \frac{v_A^2}{c_s^2} S_N^2 dl,$$

$$\beta_{3N} = -\int\limits_{l_-}^{l_+} \frac{pP_0^\sigma}{g_3} \frac{v_A^2}{v_s^2} S_N \nabla_3 \frac{g_2}{\sqrt{g}} \nabla_3 \frac{g_1}{\sqrt{g}} S_N dl,$$

$$D_N = \frac{\Omega_{sN}^2}{\omega^2} \int\limits_{l_-}^{l_+} \frac{pP_0^\sigma}{g_3} \frac{v_A^2}{c_s^2} S_N \tilde{\Delta}\psi_f \, dl,$$

and the expression for the decrement:

$$\gamma_N = \frac{1}{2\Omega_{sN}^2} \left[\frac{pP_0^\sigma v_A^2}{g_1 g_3} v_{p+} \left. \frac{\partial S_N}{\partial l} \right|_{l=l_+} + \frac{pP_0^\sigma v_A^2}{g_1 g_3} v_{p-} \left. \frac{\partial S_N}{\partial l} \right|_{l=l_-} \right], \tag{3.78}$$

owing to the Joule dissipation of SMS-wave energy in the ionospheres of the Northern ($l = l_+$) and Southern ($l = l_+$) hemispheres.

Expressions for the coefficients in (3.77) are significantly simplified for oscillations with $N \gg 1$, when the WKB approximation is applicable, in the longitudinal coordinate:

$$\beta_{1N} \approx 0, \quad \beta_{2N} \approx \frac{1}{t_s} \int\limits_{l_-}^{l_+} \frac{g_1}{g_2} \frac{dl}{c_s}, \quad \beta_{3N} \approx \frac{\Omega_{sN}^2}{t_s} \int\limits_{l_-}^{l_+} \frac{g_1 dl}{c_s v_s^2},$$

$$\gamma_N = \frac{1}{t_s} \left[\frac{v_{p+}}{c_s^+} + \frac{v_{p-}}{c_s^-} \right].$$

Interestingly, the decrement in this approximation is independent of the harmonic number of standing SMS waves, meaning that it does not depend on the oscillation frequency.

Let us solve (3.77) near the resonance magnetic shell $x^1 = x_{sN}^1$, determined by the condition $\omega = \Omega_{sN}(x_{sN}^1)$. We will examine the magnetospheric region where function $\Omega_{sN}(x^1)$ changes monotonically. Near the resonance surface, $\Omega_{sN}(x^1)$ can be decomposed accurate to linear terms:

$$\Omega_{sN} \approx \omega \left(1 - \frac{x^1 - x_{sN}^1}{a_N} \right). \tag{3.79}$$

This expansion applies when $|x^1 - x_{sN}^1| \ll L$, where $a_N = (\partial \ln \Omega_{sN} / \partial x^1)^{-1}$ is the characteristic Ω_{sN} variation scale at point $x^1 = x_{sN}^1$. Substituting (3.79) into (3.77) and switching to the dimensionless variable $\xi = (x^1 - x_{sN}^1)/\lambda_{sN}$, where $\lambda_{sN} = 1/\sqrt{\beta_{3N}}$, yields

$$(\xi + i\varepsilon) \frac{\partial^2 U_N}{\partial \xi^2} - [c_N + (1 + d_N)(\xi + i\varepsilon)] U_N = G_N, \tag{3.80}$$

an equation describing the transverse structure of resonant SMS wave near the resonance surface. The coefficients in this equation, $\varepsilon = \gamma_N a_N / \omega \lambda_{sN}$, $c_N = \beta_{1N} a_N \lambda_{sN}$, $d_N = \beta_{2N} k_2^2 \lambda_{sN}^2$, $G_N = D_N \lambda_{sN} a_N$, can be regarded as constant because they change little on the localisation scale of the desired solution $U_N(\xi)$.

The solution to (3.80) can be found by substituting the desired function $U_N(\xi)$ as a Fourier integral:

$$U_N(\xi) = \frac{1}{\sqrt{2\pi}} \int\limits_{-\infty}^{\infty} \overline{U}_N(k) e^{ik\xi} dk.$$

Substituting this expression into (3.80) produces an easily solvable first-order differential equation for function $\overline{U}_N(k)$ (see Section 3.13). A reverse Fourier transformation results in solving the initial equation as

$$U_N(\xi) = iG_N(0) \int_0^\infty \frac{\exp[ik(\xi + i\varepsilon) + i\zeta(k)]}{k^2 + 1 + d_N} dk, \tag{3.81}$$

where

$$\zeta(k) = \frac{c_N}{\sqrt{1 + d_N}} \arctan \frac{k}{\sqrt{1 + d_N}}.$$

Let us examine the behaviour of $U_N(\xi)$ near the resonance surface ($\xi \to 0$) and on the asymptotics ($|\xi| \to \infty$). As $\xi \to 0$ the bulk of the integral (3.81) is taken for $k \gg 1$ making it possible to set $\zeta(k) \approx \zeta(\infty)$ in the exponent of the integrand, and neglect all the terms except k^2, in the denominator. An expression for the second derivative is easily found:

$$U_N''(\xi)|\xi| \to 0^{\approx} \frac{G_N(0)e^{i\zeta(\infty)}}{(\xi + i\varepsilon)},$$

which we integrate to obtain

$$U_N(\xi)|\xi| \to 0^{\approx} G_N(0)e^{i\zeta(\infty)}(\xi + i\varepsilon)\ln(\xi + i\varepsilon). \tag{3.82}$$

Such a behaviour of $U_N(\xi)$ as $\xi \to 0$ can be found directly from (3.80) using Frobenius' method.

On the asymptotics, $|\xi| \to \infty$, the bulk of the integral in (3.81) is taken for $k \ll 1$ making it possible to set $k = 0$ in both $\zeta(k)$ and the denominator of the integrand. The integral is then easily calculated:

$$U_N(\xi)|\xi| \to \infty^{\approx} - \frac{G_N(0)}{1 + d_N} \frac{1}{\xi + i\varepsilon}. \tag{3.83}$$

As the resonant SMS oscillations move farther from the resonance surface, their amplitude decreases linearly on the asymptotics. Such a behaviour satisfies the boundary conditions in the x^1 coordinate – limited amplitude of the oscillations in their entire domain of existence.

3.4.5 The Field Component Structure of Resonant SMS Oscillations Near the Resonance Surface

Let us examine the behaviour of different field components of resonant SMS oscillations near the resonance surface. For the magnetic field components we will use the expressions (E.2) (Appendix E), where we will set $\varphi = 0, \psi = \psi_{sN}$. Using its approximate representation (3.82) for function $U_N(\xi)$ produces

$$B_{1N} \approx -i\frac{\overline{B}_N}{\lambda_{sN}} \ln(\xi + i\varepsilon) \left(\frac{g_1}{\sqrt{g}} \nabla_3 \frac{g_2}{\sqrt{g}} S_N \right),$$

$$B_{2N} \approx k_2 \overline{B}_N (\xi + i\varepsilon)\ln(\xi + i\varepsilon) \left(\frac{g_2}{\sqrt{g}} \nabla_3 \frac{g_1}{\sqrt{g}} S_N \right),$$

$$B_{3N} \approx i\frac{\overline{B}_N}{\lambda_{sN}^2} \frac{1}{(\xi + i\varepsilon)} \frac{S_N}{g_1},$$

where $\overline{B}_N = (\overline{c}/\omega)G_N(0)e^{i\zeta(\infty)}$. As $\varepsilon \to 0$ the longitudinal field component B_{3N} has a higher-order singularity ($\sim\xi^{-1}$). Near the ionosphere, however, where function $S_N(x^1, x^3)$ has a node, the radial field component B_{1N} dominates, which has a logarithmic singularity. The azimuthal component B_{2N} has no singularities in the amplitude distribution.

For the electric field components we will use the expressions (E.1) (Appendix E). Substituting into them U_N as (3.82) yields

$$E_{1N} \approx ik_2\overline{E}_N(\xi + i\varepsilon)\ln(\xi + i\varepsilon)\left(\frac{g_1}{\sqrt{g}}S_N\right),$$

$$E_{2N} \approx -\frac{\overline{E}_N}{\lambda_{sN}}\ln(\xi + i\varepsilon)\left(\frac{g_2}{\sqrt{g}}S_N\right),$$

where $\overline{E}_N = G_N(0)e^{i\zeta(\infty)}$. The azimuthal component E_{2N} has a logarithmic singularity as $\varepsilon \to 0$.

For the velocity field components we will use (E.3) (Appendix E), as well as Eq. (E.4), relating them to perturbed pressure. Near the resonance surface, we have

$$v_{1N} \approx -\frac{\overline{v}_N}{\lambda_{sN}}\ln(\xi + i\varepsilon)\frac{S_N}{\sqrt{g_3}},$$

$$v_{2N} \approx -ik_2\overline{v}_N(\xi + i\varepsilon)\ln(\xi + i\varepsilon)\frac{S_N}{\sqrt{g_3}},$$

$$v_{3N} \approx -\frac{\overline{v}_N/\rho_0}{\lambda_{sN}^2\omega^2}\frac{1}{(\xi + i\varepsilon)}\nabla_3\frac{v_A^2\rho_0}{g_1}\frac{S_N}{\sqrt{g_3}},$$

where $\overline{v}_N = (\overline{c}/B_0)G_N(0)e^{i\zeta(\infty)}$. Plasma velocity oscillations are somewhat different in character from the electromagnetic field component oscillations. Both the transverse components v_{1N} and v_{2N} have a node near the ionosphere. The longitudinal component v_{3N}, with singularity $\sim \xi^{-1}$, thus dominates in the entire domain of existence of resonant SMS oscillations.

The expression for perturbed pressure near the resonance surface is

$$\tilde{P}_N \approx -i\frac{\overline{B}_N B_0}{4\pi\lambda_{sN}^2}\frac{1}{(\xi + i\varepsilon)}\frac{S_N}{g_1\sqrt{g_3}}.$$

It is easily verifiable that the condition (3.67) that full perturbed pressure should be zero holds for resonant SMS wave.

3.4.6 Numerical Solutions of Equations for Resonant SMS Waves

Exact solutions to (3.72) and (3.80), the equations describing the resonant SMS oscillation structure, can be found only numerically for the fundamental harmonics of standing SMS waves. Our numerical solutions will use a system of coordinates (a, ϕ, θ) related to the dipole magnetic field lines (see Figure 3.12).

The plasma and magnetic field parameters in the calculations below are within the framework of a self-consistent model dipole magnetosphere developed in Section 3.4.1. Their distribution in the magnetic meridian plane correspond to a moderately perturbed dayside magnetosphere. The calculations concern the magnetic shell $L = 6.6$, which corresponds to

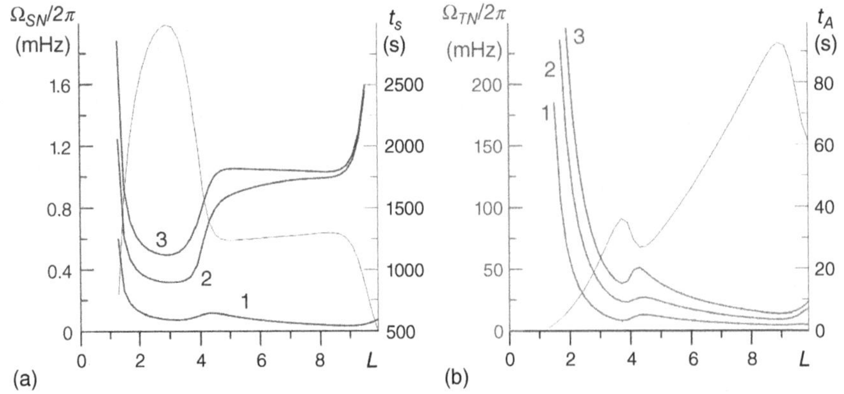

Figure 3.15 (a) Distribution across magnetic shells of the eigenfrequencies of the first three harmonics of standing SMS waves ($\Omega_{sN}/2\pi$) and the SMS transit time t_s along the field line. (b) Frequencies of the first three harmonics of standing toroidal Alfvén waves ($\Omega_{TN}/2\pi$) and Alfvén speed transit time between magneto-conjugated ionospheres, t_A.

the location of geostationary satellites. Note that these calculations neglect the effects due to the background plasma motion. Special calculations dedicated to resonant Alfvén waves have demonstrated that these effects are small [300].

Figure 3.15a displays the eigenfrequency distributions of the first three harmonics of standing SMS waves. They resulted from a numerical solution of (3.72) with homogeneous boundary conditions on the ionosphere. For comparison, Figure 3.15b shows analogous distributions for the first three harmonics of standing toroidal Alfvén waves in the same model magnetosphere. Each figure indicates the transit time along the field-line between the magneto-conjugated ionospheres, at SMS wave speed (c_s) and Alfvén speed (v_A), which determine the value of the oscillation eigen-frequencies in the WKB approximation. Unlike Alfvén waves, the frequencies of the basic harmonics of SMS waves differ very much from their WKB values. Thus, the precise value of the frequency of the first harmonic of SMS waves differs nearly fivefold from its WKB value.

Figure 3.16b depicts the structures of the first three harmonics of Alfvén and SMS waves standing along the field line. We find about the same picture for Alfvén waves as what we saw in the WKB approximation. The structure of the main harmonics of standing SMS waves differ radically from their WKB version (3.74). Note the nonmonotonous character of function $S_N(x^1, x^3)$ as $\theta \approx \pm 28°$. This is due to a particular feature of the model magnetosphere employed, namely a double plasmapause. In the neighbourhood of $\theta \approx \pm 28°$ the field line crosses the plasmapause of the second kind, which does not coincide with the magnetic shell surface. The main feature of the basic harmonics of SMS waves in the model magnetosphere in question is their fast amplitude decay as they approach the ionosphere.

For comparison, Figure 3.16b shows the structure of three first harmonics of SMS waves on short magnetic field lines at magnetic shell $L = 1.3$. A different model magnetosphere was used for these calculations (see [94]). They were aimed at determining the structure of SMS waves generated in the plasmasphere by the solar terminator moving along the ionosphere [28, 301]. The structure of the standing SMS waves in question is asymmetric about the equatorial plane – the field lines of the magnetic shell in question cross the ionosphere in the sunlit region in the Northern hemisphere and in the shaded region in the Southern.

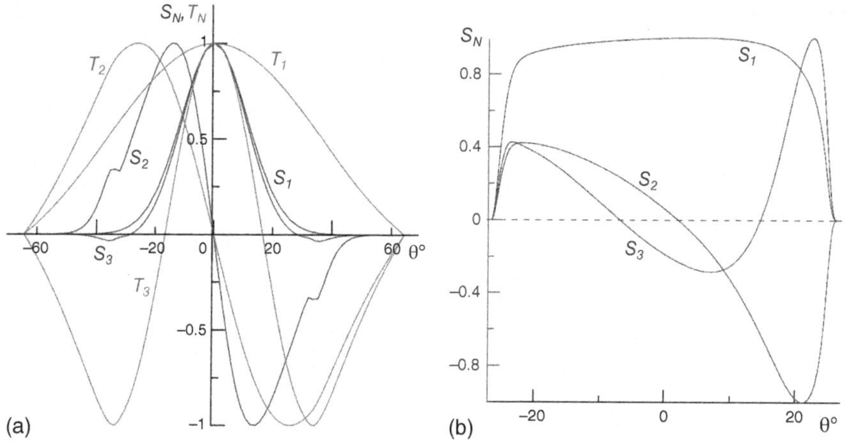

Figure 3.16 (a) Structure along magnetic field lines of the first three harmonics of standing Alfvén waves ($T_N(\theta)$, $N = 1, 2, 3$) and SMS waves ($S_N(\theta)$, $N = 1, 2, 3$) at magnetic shell $L = 6.6$. (b) Structure of the first three harmonics of standing SMS waves ($S_N(\theta)$, $N = 1, 2, 3$) at magnetic shell $L = 1.3$.

Notably, the inner plasmasphere is the only region in the terrestrial magnetosphere where the background plasma electron temperature is considerably higher than the ion temperature: $T_e \approx 2T_i$. This makes the SMS waves decay weakly allowing them to be observed here with noticeable amplitude.

Figure 3.17 illustrates the structure of the magnetic field physical components ($B_x \equiv B_1/\sqrt{g_1} = |B_x|e^{i\alpha_x}$, $B_y \equiv B_2/\sqrt{g_2} = |B_y|e^{i\alpha_y}$, $B_z \equiv B_3/\sqrt{g_3} = |B_z|e^{i\alpha_z}$) for the fundamental harmonic of resonant SMS waves ($N = 1$) in the neighbourhood of the resonance magnetic shell $L = 6.6$. Figure 3.17a refers to the oscillation structure in the near-equatorial region (away from the function S_N and $\partial S_N/\partial l$ nodes). The amplitude of monochromatic FMS wave playing the role of a source for the resonant SMS oscillations is chosen such that $|B_z| = 1$ at the amplitude distribution maximum. The initial phase of the oscillations

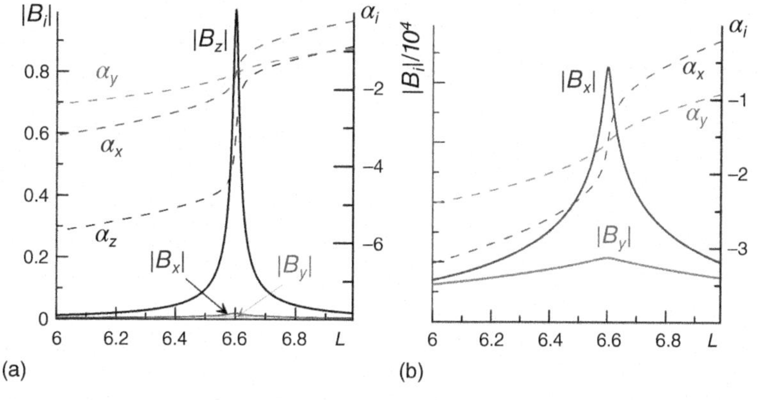

Figure 3.17 Transverse structure of the magnetic field components for the first harmonic ($N = 1$) of resonant SMS oscillations near the resonant shell $L = 6.6$: (a) standing SMS wave amplitude $|B_i|$ and phase α_i distribution ($i = x, y, z$) near the equatorial surface; (b) oscillation amplitude and phase distributions near the ionosphere (longitudinal component B_z turns zero).

is chosen as being zero in the asymptotically remote region right of the resonance shell. Our numerical calculations relied on the decay value for which $\varepsilon = 10^{-2}$. The resonant structure of the oscillations in this case is quite distinct. As γ_N (and ε) increase the maximum amplitude decreases and the resonance peak width increases. Passing through the resonance peak the phase of the parallel B_z-component changes by about π, that of the B_x-component by $\sim \pi/2$, while remaining practically unchanged in the B_y-component.

The corresponding distributions of the B_x-, B_y-components of the oscillation field near the ionosphere are shown in Figure 3.17b. Recall that the B_z-component turns zero at the ionosphere. The amplitudes of the transverse field components near the ionosphere are, however, much smaller than their equatorial value. This is due to the rapid decrease of function $\partial S_N/\partial l$, describing the longitudinal structure of the oscillation field, as we move closer to the ionosphere (see Figure 3.16a). Thus the Alfvén waves are the main component of the MHD oscillations registered at Earth surface. For FMS waves, Earth is in the opaque region, while the amplitude of SMS waves decays strongly towards the ionosphere.

3.5 FMS Oscillations in a Dipole-Like Magnetosphere

Axial symmetry in the dipole-like model presumes the geomagnetic field and plasma to be inhomogeneous both along the geomagnetic field lines and across the magnetic shells. The field lines are curvilinear and cross the ionosphere in the Northern and Southern hemispheres (a typical example is a dipole magnetic field). The magnetosphere has a sharp enough boundary separating it from the solar wind – the magnetopause. The magnetopause is a thin transition layer where the plasma parameters vary from characteristic values for the magnetosphere to values typical of the solar wind. Magnetosonic oscillations are somewhat effective in penetrating through this boundary from the solar wind into the magnetosphere and back. At the same time, the magnetosphere can also serve as a resonator for FMS waves – FMS eigen oscillations can form inside it. Below, we will examine the spatial structure and frequency spectrum of such oscillations.

3.5.1 Longitudinal Structure of FMS Oscillations

As was shown in Sections 3.1–3.4, studies of fast magnetosonic oscillations can neglect, in the main order of perturbation theory, their interaction with resonant Alfvén and SMS waves, even on resonance surfaces. To examine the spatial structure and spectrum of FMS oscillations in the axisymmetric magnetosphere, we will therefore employ a homogeneous equation (3.68) in curvilinear orthogonal coordinates (x^1, x^2, x^3) tied to the geomagnetic field lines. Most of the magnetosphere satisfies the condition $v_s^2 \ll v_A^2$. For the sake of simplicity, we will therefore restrict ourselves to a model magnetosphere with a 'cold' plasma (see Section 3.2.1). In such a medium, FMS wave propagation is mostly determined by the spatial distribution of the Alfvén speed. When necessary, the solutions obtained below are easily generalised for the 'warm' plasma case ($v_s^2 \lesssim v_A^2$).

Moreover, we will assume that the wavelength of the FMS oscillations in question is much smaller than the characteristic scale of a magnetospheric plasma inhomogeneity, across magnetic shells. To describe such oscillations away from turning points it is possible to use the WKB approximation in the x^1 coordinate. Such an approach is qualitatively applicable to oscillations with wavelength of order the magnetospheric inhomogeneity scale. There are no oscillations with an inverse ratio of wavelength and inhomogeneity scale, in the magnetosphere. Let us also assume that, along the field line, the wavelength of the oscillations is of order the characteristic scale of the magnetospheric plasma inhomogeneity. Such an approach enables the multiple-scale method to be used for solving (3.68). Let us represent the desired solution as

$$\psi_f = u(x^1)\exp[i(\tilde{\Phi}(x^1)]\left[H(x^1,x^3) + h(x^1,x^3\right]\exp[i(k_2x^2 - \omega t)], \tag{3.84}$$

where $\tilde{\Phi}(x^1)$ is the large quasiclassical phase, $u(x^1)$ is the function describing the slow variation of the oscillation amplitude across magnetic shells, $H(x^1,x^3)$ is the function describing the oscillation structure along the geomagnetic field lines, in the main order of perturbation theory, $h(x^1,x^3)$ is a small correction to $H(x^1,x^3)$ in higher orders.

Substituting this expression into (3.68), in the main order of perturbation theory, results in an expression for function $H(x^1,x^3)$:

$$\nabla_3\frac{p}{\sqrt{g_3}}\nabla_3\frac{g_1}{\sqrt{g}}H + \left(\frac{\omega^2}{v_A^2} - k_\perp^2\right)H = 0, \tag{3.85}$$

where $p = \sqrt{g_2/g_1}$, $k_\perp^2 = (k_1^2/g_1 + k_2^2/g_2)$ is the transverse wave vector squared ($k_1 = \nabla_1\tilde{\Phi}$). Equation (3.85) should be complemented by boundary conditions at the ionosphere. As will be shown further, the ionosphere lies deep inside an opaque region for the magnetosonic waves in question and does not play a significant role in determining their structure. Thanks to this, we will assume the ionosphere to be ideally conducting, for the sake of simplicity. This means that $E_1 = E_2 = 0$ at the ionosphere, whence

$$H|_{x^3=x^3_\pm} = 0, \tag{3.86}$$

where $x^3_\pm$ are the points where the field line crosses the ionosphere in the Northern or Southern hemisphere. The problem (3.85) and (3.86) can be regarded as a problem of the eigenvalues of the wave number squared, k_1^2, for given FMS frequency ω. The solution of this problem is a set of eigenfunctions H_n (where $n = 0, 1, 2, 3, \ldots$ is the ordinal number of the longitudinal harmonic of FMS waves) and a corresponding set of eigenvalues, k_{1n}^2. The set H_n is complete making it possible to decompose any perturbation of the monochromatic FMS oscillation field in the magnetosphere into these eigenfunctions.

To represent the eigenmode structure qualitatively, let us use the WKB approximation in the parallel coordinate x^3, i.e. write the solution of (3.85) in the following form

$$H = \exp[i\Theta(x^1,x^3)],$$

where $\Theta(x^1,x^3) = \Theta_0(x^1,x^3) + \Theta_1(x^1,x^3) + \cdots$ is the quasiclassical phase in the parallel coordinate ($|\Theta_0| \gg |\Theta_1| \gg \cdots$). Substituting this solution into (3.85) yields the solution for the phase:

$$-(\Theta')^2 + i\Theta'' + i\Theta'\left(2\nabla_3\ln\frac{g_1}{\sqrt{g}} + \nabla_3\ln\frac{g_2}{\sqrt{g}}\right) + g_3\nabla_3\frac{g_2}{\sqrt{g}}\nabla_3\frac{g_1}{\sqrt{g}} + k_3^2 = 0,$$

where $k_3^2 = g_3\left(\omega^2/v_A^2 - k_\perp^2\right)$. In the main order of perturbation theory we have $\Theta_0' = \pm k_3$, whence

$$\Theta_0 = \pm\int_{x^3_-}^{x^3} k_3 dx^3.$$

In the first order, we obtain

$$\Theta_1 = \frac{i}{2}\ln\left(\frac{|k_3|g_1}{g_3\sqrt{g}}\right).$$

The general WKB solution of (3.85) can thus be represented as

$$H = f[A\exp(i\Theta_0) + B\exp(-i\Theta_0)], \tag{3.87}$$

where $f = \sqrt{g_3\sqrt{g}/|k_3|g_1}$.

To further investigate the resulting solution, we will use the model introduced in Section 3.2.1. As follows from the general form of the solution (3.87), its behaviour is determined by function $k_3 = \pm\sqrt{g_3(F(a,\theta) - k_{1n}^2)/g_1}$, where

$$F(a,\theta) = g_1\left(\frac{\omega^2}{v_A^2} - \frac{m^2}{g_2}\right).$$

In the transparent regions, where $F > k_{1n}^2$, the solution is a periodic function, while in the opaque regions, where $F < k_{1n}^2$, it consists of the sum of the decreasing and increasing exponentials. The form of the solution is thus determined by the behaviour of function $F(a,\theta)$.

Figure 3.18 shows the dependence of function $F(a,\theta)$ on the longitudinal coordinate (angle θ) for four different values of parameter m at magnetic shell $L = 6.6$. As follows from the boundary conditions on the ionosphere (3.86), a non-trivial solution of (3.85) exists only if the field line has at least one interval where $F > k_{1n}^2$. Four variants (I, II, III, IV) of the difference $F - k_{1n}^2$ are shown in Figure 3.18 responsible for four different types of solutions for eigenmodes. The simplest solution is type IV, for which $F > k_{1n}^2$ along the entire field line. In this case, the WKB solution has the form

$$H_n = Cf\sin\left(\int_{x_-^3}^{x^3} k_3(x^1, x^{3'})dx^{3'}\right), \tag{3.88}$$

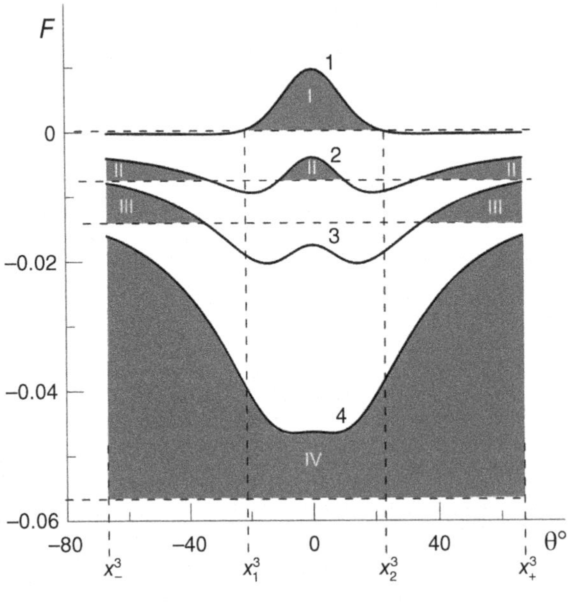

Figure 3.18 Dependence of function $F(L,\theta)$ on geomagnetic latitude θ and azimuthal wave number m at magnetic shell $L = 6.6$. Curve 1 refers to $m = 1$, curve 2 to $m = 5$, curve 3 to $m = 7$, curve 4 to $m = 10$. Horizontal dashed lines are possible eigenvalues of parameter k_{1n}^2. Areas where $F(a,\theta) - k_{1n}^2 > 0$ labelled as I, II, III, IV (shown in grey) correspond to the four types of longitudinal structure of FMS eigen oscillations.

where C is an arbitrary constant, and eigenvalues k_{1n}^2 are determined by the quantisation condition

$$\int_{x_-^3}^{x_+^3} k_3 dx^3 = \pi(n+1),$$

where $n = 0, 1, 2, \ldots$ is parallel wave number. Solutions of type I have the form (see Appendix F)

$$H_n = Cf \begin{cases} \sinh\left(\int_{x_-^3}^{x^3} |k_3'| dx^{3'}\right), & x_-^3 \le x^3 < x_1^3, \\ e^{\overline{\psi}} \sin\left(\int_{x_1^3}^{x^3} k_3' dx^{3'} + \frac{\pi}{4}\right), & x_1^3 < x^3 < x_2^3, \\ \mp \sinh\left(\int_{x_2^3}^{x^3} |k_3'| dx^{3'} - \overline{\psi}\right), & x_2^3 < x^3 \le x_+^3, \end{cases} \tag{3.89}$$

where the two different $\mp$ signs correspond to even or odd n in the quantisation condition

$$\int_{x_1^3}^{x_2^3} k_3 dx^3 = \pi\left(n + \frac{1}{2}\right).$$

The coordinates x_1^3, x_2^3 are turning points, where $k_{1n}^2 = F(x^1, x_{1,2}^3)$. Moreover, the following notation is introduced for the integral phase:

$$\overline{\psi} = \int_{x_-^3}^{x_1^3} |k_3| dx^3 = \int_{x_2^3}^{x_+^3} |k_3| dx^3.$$

Expressions for type II and III WKB solutions being more cumbersome, we will not write them out here, even though their qualitative behaviour is quite obvious from the two above examples. It should only be noted that if there is more than one transparent region on the field line, where $F > k_{1n}^2$, the spectrum of eigenvalues k_{1n}^2 (and corresponding eigenfunctions H_n) splits up.

To determine the structure of the main harmonics of FMS oscillations more accurately, let us integrate (3.85) numerically. Figure 3.19 displays the structure of the four first harmonics

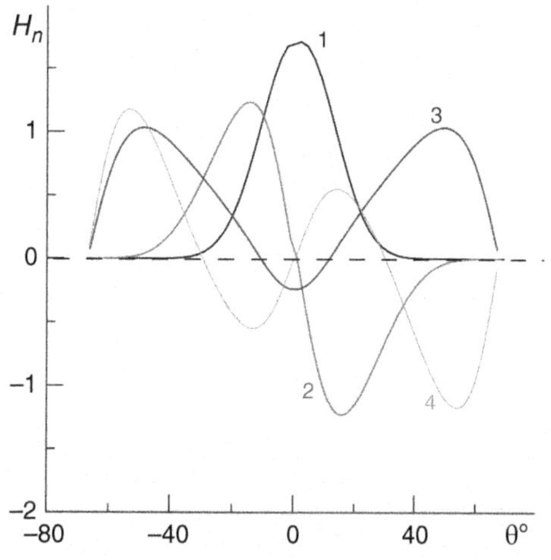

Figure 3.19 Longitudinal structure of first four harmonics ($n = 0, 1, 2, 3$) of the FMS eigenmodes in the dipole-like magnetosphere, at magnetic shell $L = 6.6$.

$H_n(L, \theta)$ at magnetic shell $L = 6.6$. They are normalised as follows:

$$\oint H_n^2 dx^3 \equiv 2\int_{x_-^3}^{x_+^3} H_n^2 dx^3 = 1.$$

These calculations employ the following values of the wave parameters: $m = 1, f = \omega/2\pi = 0.025$ Hz is the frequency equalling the frequency of the second harmonic of the Alfvén eigen oscillations (Ω_{T2}) in the model magnetosphere we are studying, at magnetic shell $L = 6.6$. The bold line in this figure and the ones that follow marks the functions corresponding to the eigenmode ($n = 0$, $m = 1, f = f_2 = \Omega_{T2} = 0.025$ Hz), which we will select as reference in comparing to other modes of the FMS oscillations. Harmonics H_1, H_2 are the solutions of (3.85) of the (3.89) form (type I in Figure 3.18), i.e. they are periodic functions near the equatorial plane and exponentially decrease in amplitude when approaching the ionosphere. Harmonics H_3, H_4 are solutions of the (3.88) form (type IV in Figure 3.18), which have the form of periodic functions along the entire field line.

Dependencies of eigenvalues k_{1n}^2 on the magnetic shell parameter L are shown in Figure 3.20 for five first harmonics of the FMS eigenmodes ($n = 0, 1, 2, 3, 4$). Dependence $k_{1n}^2(L)$ is shown in Figure 3.20b for the entire range of the magnetic shells under examination ($1.5 \leq L \leq 15$), while dependence $k_{1n}^2(L)$ in Figure 3.20a refers to the inner magnetosphere ($1.5 \leq L \leq 10$). Evidently, $k_{10}^2 > 0$ for $L > 2.2$, and $k_{10}^2 < 0$ for $L < 2.2$, for the reference harmonic. For the second harmonic, inequality $k_{11}^2 > 0$ holds in the outer magnetosphere for $L > 6.1$, as well as in the $3.1 < L < 3.6$ region inside the plasmasphere. For harmonics $n = 2, 3, 4$, inequality $k_{1n}^2 > 0$ holds in the outer magnetosphere only.

Let us define the transparent region for the magnetosonic modes under study using the requirement that these three inequalities hold simultaneously: $k_{1n}^2 > 0$, $k_2^2 > 0$, $k_{3n}^2 > 0$. This means that the magnetosonic wave can freely propagate in any direction, in the

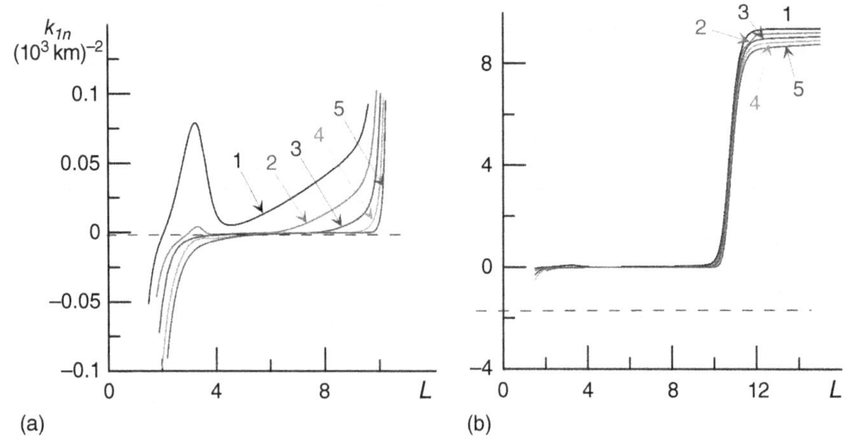

Figure 3.20 Squared quasi-classical wave-vector component k_{1n}^2 for the first five harmonics ($n = 0, 1, 2.3, 4$) of the FMS eigenmodes vs. the magnetic shell parameter L: (a) $k_{1n}^2(L)$ distribution inside the magnetosphere and (b) $k_{1n}^2(L)$ distribution in the magnetic shell range $1.5 < L < 15$, including the solar wind region.

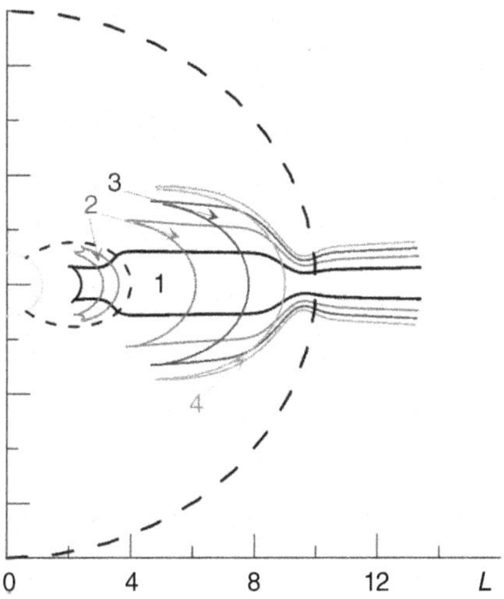

Figure 3.21 The boundaries of transparent regions (in the meridional plane) for the first four parallel harmonics of FMS oscillations ($n = 0, 1, 2, 3$), in a dipole-like model magnetosphere, for frequency $\omega = 2\pi f_2$ and azimuthal wavenumber $m = 1$.

region in question. The boundaries of the transparent regions for the above-discussed four harmonics of FMS modes are shown in Figure 3.21. The boundary of the reference mode is closest to the ionosphere and singly connected. This mode is strongly localised near the equatorial plane. For the second mode ($n = 1$, $m = 1$, $f = f_2$) the inner boundary of the transparent region is farther from the ionosphere, while the region itself consists of two unconnected regions: external, for $L > 6.1$, and internal, for $3.1 < L < 3.6$. The latter transparent region is a resonator for FMS oscillations as first described in [235, 236].

For harmonics $n = 2, 3, 4$, the inner boundaries of the transparent regions approach the magnetopause. Near the equatorial plane, all the oscillation modes have a narrow exit channel from the magnetosphere into the solar wind. This enables the FMS oscillations to penetrate from the solar wind into the magnetosphere. Instability of solar wind protons reflected from the bow shock front can serve as the source of such oscillations [1, 302]. Note that the shape of the transparent regions in the solar wind can differ substantially from what has been obtained in this monograph because of plasma motion and the magnetic field configuration differing from a dipole here. Hopefully, this difference is not too big a short distance from the magnetopause, however, because the stagnation point is located here, near the equator, where the solar wind speed $v_{sw} = 0$. The size of the transparent regions in the magnetosphere decreases with increasing n.

The boundaries of the transparent regions are shown in Figure 3.22 for mode ($n = 0, m = 1$), for different frequencies $f = f_1 = f_2/2$; $f = f_2$; $f = f_3 = 3f_2/2$. The distances between the equator and the lateral boundaries are close for the eigen modes with these three frequencies, while the inner boundary shifts towards the magnetopause as frequency f increases. Note that, for frequency $f = f_3$, the transparent region consists of two unconnected parts – the inner part, forming a resonator below the plasmapause; and the outer part, open into the solar wind. The boundaries of transparent regions behave in an analogous manner for fixed frequency, but varying azimuthal wave number $m = 1, 2, 3, \ldots$. The lateral boundaries for the first azimuthal harmonics practically coincide, while the inner boundary moves closer to the magnetopause as m increases.

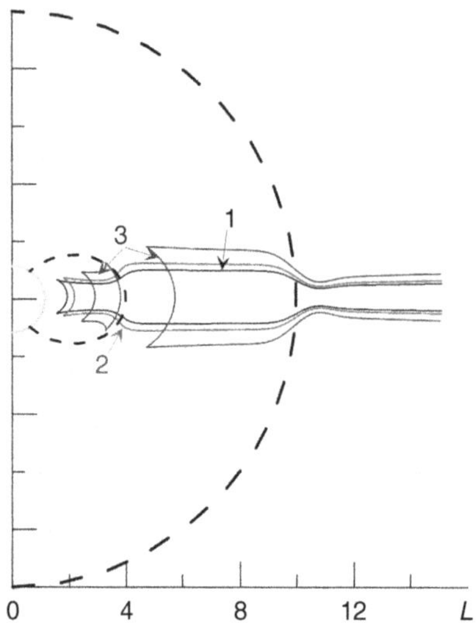

Figure 3.22 The boundaries of the transparent regions for FMS oscillation harmonic ($n = 0, m = 1$), for different frequencies: (1) $f = f_1$, (2) $f = f_2$, (3) $f = f_3$.

3.5.2 FMS Oscillation Structure Across Magnetic Shells

In the first order of perturbation theory the following equation

$$\sqrt{g_3}u_n\hat{L}_T\frac{g_1}{\sqrt{g}}h_n - k_{\perp n}^2u_nh_n + i\left[\frac{k_{1n}}{g_1}\nabla_1u_nH_n + \frac{g_3}{\sqrt{g}}\nabla_1\frac{g_2}{\sqrt{g}}k_{1n}u_nH_n\right] = 0$$

is derived from (3.68) for the function $u(x^1)$ describing the slow variation of the oscillation amplitude across magnetic shells (see (3.84)). Left-multiplying this equation by $g_1u_nH_n/\sqrt{g}$ and integrating along the field line between the magneto-conjugated ionosphere 'there and back' produces

$$\oint \nabla_1\left(\frac{g_2}{\sqrt{g}}k_{1n}u_n^2H_n^2\right)\frac{dx^3}{g_2} = 0. \tag{3.90}$$

This takes into account the fact that the integral of the summands proportional to h_n turns zero because the operator $\hat{L}_T$ is Hermitian. Equation (3.90) can be re-written as

$$a_{1n}\nabla_1(k_{1n}u_n^2) = a_{2n}k_{1n}u_n^2, \tag{3.91}$$

where

$$a_{1n} = \oint \frac{H_n^2}{\sqrt{g}}dx^3, \qquad a_{2n} = \oint \nabla_1\left(\frac{g_2H_n^2}{\sqrt{g}}\right)\frac{dx^3}{g_2}.$$

Normalising the function H_n such that $a_{1n} = 1$ and denoting $\tilde{k}_{1n} \equiv a_{2n}(x^1)/2$, let us write the solution of (3.91) as follows:

$$u_n(x^1) = \frac{C}{\sqrt{|k_{1n}|}}\exp\left(-\int \tilde{k}_{1n}dx^1\right),$$

where C is an arbitrary constant. The general solution of (3.68) can thus be represented as

$$\psi_f = \frac{C}{\sqrt{|k_{1n}|}}\exp\left(i\int(k_{1n} + i\tilde{k}_{1n})dx^1 + ik_2x^2 - i\omega t\right)H_n(x^1, x^3). \tag{3.92}$$

The specific form of this solution depends on the boundary conditions in the x^1 coordinate. In Section 3.4.4 we discussed the problem of Alfvén resonances excited in the magnetosphere by a FMS wave from the solar wind falling onto the magnetosphere. Below we will examine the problem of magnetospheric FMS eigen oscillations in resonators formed by inhomogeneously distributed Alfvén speed.

3.6 FMS Resonators in Earth's Magnetosphere

The hypothesis that the terrestrial magnetosphere can play the role of a resonator for FMS waves was first advanced a long enough time ago [238, 239, 303]. The eigenmodes of such a resonator that were called global modes in [238] should be the largest-scale, and hence, the lowest-frequency, geomagnetic oscillations, occupying the entire magnetosphere. As we saw in Sections 3.2.3–3.2.4, the near-Earth part of the magnetosphere (being an opaque region) is unavailable for such oscillations. However, the oscillations of such a resonator can be transported to Earth by resonant Alfvén waves excited on field lines passing through the resonator region. Such oscillations are particular in that their frequency is independent of the point where they are registered on Earth's surface.

Such ultra low frequency (ULF) oscillations with discrete spectrum were indeed discovered in observations using both HF radars [304, 305], and ground-based magnetometer networks [306, 307]. The spectra of these oscillations contain very pronounced maxima at frequencies 1.3, 1.9, 2.6, 3.4, … mHz. These frequencies hardly change, either event to event, or over the course of one event. This caused them to be called 'magic frequencies'. Such oscillations are as a rule recorded on Earth surface in the midnight-dawn sector, at latitudes 60°–80°.

To regard these observations as supporting the theory of a global magnetospheric FMS resonator, has proven somewhat premature [308, 309]. A critical analysis in Section 3.6.1 demonstrates that the available versions of the theory not only fail to describe the basic characteristics of the observable oscillations but are even unable to provide an acceptable explanation of the mere possibility that a global magnetospheric resonator could exist.

3.6.1 Qualitative Proof that FMS Resonators Exist in the Magnetosphere

If a FMS wave is localised in a rectangular homogeneous resonator with reflecting walls, with sides length l_x, l_y, l_z (see Figure 3.23), its wave-vector components are quantised:

$$k_x = \frac{\pi n_x}{l_x}; \quad k_y = \frac{\pi n_y}{l_y}; \quad k_z = \frac{\pi n_z}{l_z}, \tag{3.93}$$

where n_x, n_y, n_z are wave numbers capable of taking values from the sequence of natural numbers. On order of magnitude, the basic frequency of the resonator

$$f = \frac{\omega}{2\pi} \sim \frac{v_A}{l}, \tag{3.94}$$

where l is the smallest of the resonator box dimensions. The approximate formula (3.94) remains valid for resonators of arbitrary shape with inhomogeneous plasma and magnetic field. Meanwhile, v_A should be regarded as the characteristic value of Alfvén speed, and l as

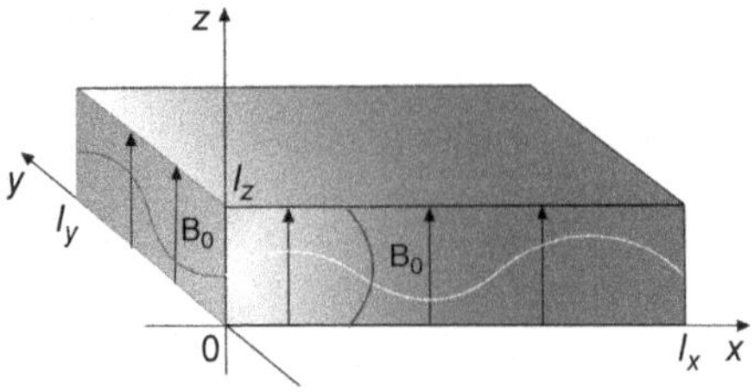

Figure 3.23 Model FMS resonator in the form of a rectangular box with ideally reflecting walls ('box model').

the smallest of its three mutually perpendicular dimensions. More strictly, l is the smallest distance between parallel planes that could contain the resonator.

The role of reflecting walls in natural – including magnetospheric – resonators is played by an inhomogeneous medium. Two significantly differing cases can be distinguished. In the first case, reflection is from wave turning points (or rather, surfaces). This concept is strictly defined if the WKB approximation is applicable. Let the above resonator be inhomogeneous in one coordinate, e.g. x, the inhomogeneity being such that the WKB approximation is applicable. A quasi-classical wave vector can then be found for the FMS wave:

$$k_x^2(x) = \frac{\omega^2}{v_A^2(x)} - k_y^2 - k_z^2. \tag{3.95}$$

This assumes that ω, k_y and k_z are constants and the Alfvén speed depends on x. The wave propagates in a transparent region, where $k_x^2 > 0$, but is unable to propagate in an opaque region, where $k_x^2 < 0$. The Alfvén speed can be easily seen to have smaller values in the transparent region than in the opaque region. Planes where $k_x^2 = 0$ play the role of reflecting walls. Depending on the width and height of the potential barrier formed by the opaque region, reflection can be partial or complete. A resonator emerges if the transparent region in the x coordinate is bounded by turning points. In the three-dimensional case, a resonator can exist if the transparent region located in a region of relatively small values of the Alfvén speed is bounded by a turning surface on all sides.

The other version of reflecting walls of a resonator are sharp boundaries separating abrupt changes in the parameters of the medium (Alfvén speed, in our case). The WKB approximation is inapplicable to such a boundary. The wave suffers partial reflection (the larger the Alfvén speed jump the more pronounced the reflection) even if the boundary separates two transparent regions. The two above cases of a resonator with different types of reflecting walls are not mutually exclusive. The situation is quite possible when part of the resonator boundaries are turning surfaces, while the other part are sharp, separating boundaries.

Let us now proceed from general remarks to discussing various versions of the theory of a magnetospheric resonator. In all these versions, the outer boundary of the resonator is a sharp magnetospheric boundary – the magnetopause [177, 233, 238], in some versions – the bow shock front [305]. Below the magnetopause there is a transparent region for FMS waves that forms a resonator for them. The solar wind, where the Alfvén speed is much smaller, is also a transparent region for such waves. The magnetopause therefore cannot be an ordinary reflecting surface, but is capable of playing the role of a partially reflecting boundary due to a sharp Alfvén speed jump on it.

Currently, theoretical works are available calculating the structure and spectrum of the 'global modes' in various model magnetospheres. This was done in [238, 239] for a box model of the magnetosphere, in [219] for a half-cylinder model, and [226, 233, 310] used an axisymmetric model with a dipole geomagnetic field. All these models describe the FMS resonator eigenmodes confined between the magnetopause and the turning surface separating the transparent region in the outer magnetosphere from the opaque region in the inner magnetosphere. The opaque region is present thanks to the Alfvén speed increasing significantly when approaching Earth.

As will be seen in the calculations below, the eigen frequencies of such modes ($f > 5$ mHz) were found to be much higher than those observed ($\sim$1 mHz). This result could have been expected even based on a simple estimate. The characteristic values are $v_A \sim 10^3$ km/s, $l \sim 10^5$ km, for the magnetosphere, yielding $f \sim 10$ mHz. The result of an exact calculation can of course differ 2–3-fold from an estimate, but hardly by as much as an order of magnitude. The value $f > 5$ mHz, obtained in the above-mentioned papers, confirms this.

However, difficulties concerning the eigen frequency values can be regarded as second-degree compared to the following circumstance of principal importance. The existing magnetotail casts doubt on the entire concept of a global magnetospheric resonator. The above-discussed FMS wave appears to be able to freely escape into the opened magnetotail. This is the reason why [76, 240, 305] proposed to treat the observed ULF oscillations with a discrete spectrum as the geotail waveguide eigenmodes. This interpretation, however, can hardly be accepted as successful too. It is very difficult to explain the discrete oscillation spectrum within its framework. The above authors employ a model waveguide in the form of a rectangular channel, denoting the transverse coordinates y and z, and the coordinate along the channel, running along the waveguide and into the geotail, x. That k_y and k_z are quantised follows from geometrical reasoning, but the k_x component can take a continuous series of values, as can the frequency ω. The above authors' suggestion that the observed oscillations correspond to values $k_x \ll k_y, k_z$ (in this case, the particular value of k_x is of no importance when calculating the eigen frequency) is not supported by any arguments. It is completely impossible to understand why the waveguide excitation mechanism, while generating a wave with $k_y, k_z \sim 1/l$ (l is the transverse magnetospheric scale), cannot excite a wave with $k_x \sim 1/l$, but only with $k_x \ll 1/l$. Given that there is an inhomogeneity of approximately the same scale l, along the x coordinate, it is completely unjustified to suggest the existence of a wave with $k_x \ll 1/l$. The its wavelength cannot be much larger than the inhomogeneity scale, because its source size is much larger than the dimensions of the plasma configuration, and this cannot be an eigen-oscillation.

In the remote geotail, where the scale of inhomogeneity along the tail is much larger than the transverse scale, waves with $k_x \ll k_y, k_z$ are possible. Even in this case, however, the question remains as to the oscillation excitation mechanism. Moreover, it becomes even more difficult to match the theoretical values of oscillation eigen frequencies to what is actually observed. At characteristic transverse scale ($l \sim (1-3) \times 10^5$ km) of the waveguide, the Alfvén speed $v_A \sim (3-10) \times 10^3$ km/s in the tail lobes, producing frequency $f \approx 30$ mHz for the main harmonic, too remote from the frequencies of observable oscillations. Finally, a FMS waveguide can possibly exist in the neutral plasma sheet [311]. The same difficulties persist for this waveguide. In particular, we have $f \sim 10$ mHz for the frequency of the main harmonic, for $l \sim 10^4$ km, $v_A \sim 10^2$ km/s.

It should be admitted therefore that the concept of magnetospheric 'global modes' is currently in an unsatisfactory form. Unknown are the localisation region of the eigenmodes, the causes of their being trapped inside the magnetosphere and their spatial structure. The difference between the theoretical and observed values of the oscillation eigen frequencies is too big.

As was noted above, FMS resonators and waveguides are located in the regions where there is a minimum in the Alfvén speed v_A (or rather FMS propagation speed $c_f = \sqrt{v_A^2 + v_s^2}$) distribution. In most of the magnetosphere $\beta^* = v_s^2/v_A^2 \ll 1$ and $c_f \approx v_A$. In some regions parameter β^* can take values $\beta^* \sim 1$, but even in this case $c_f \sim v_A$. There is only one region in the magnetosphere where $\beta^* \gg 1$, – the thin near-equatorial part of the plasma sheet, playing no significant role for large-scale FMS oscillations of the magnetosphere. Thus, a decisive role in investigations of FMS resonators and waveguides is played by the global Alfvén speed distribution in the magnetosphere. No less important is this distribution for standing Alfvén waves. Their longitudinal structure is determined by the Alfvén speed distribution along the field line, and their transverse structure by certain integral (along the field line) characteristics, also related to this distribution.

Figure 3.24 provides a schematic representation of the Alfvén speed distribution in the magnetosphere. This distribution reflects the main structural elements of the magnetosphere: the magnetopause, cusps, the plasmasphere, the geotail lobes and the plasma sheet.

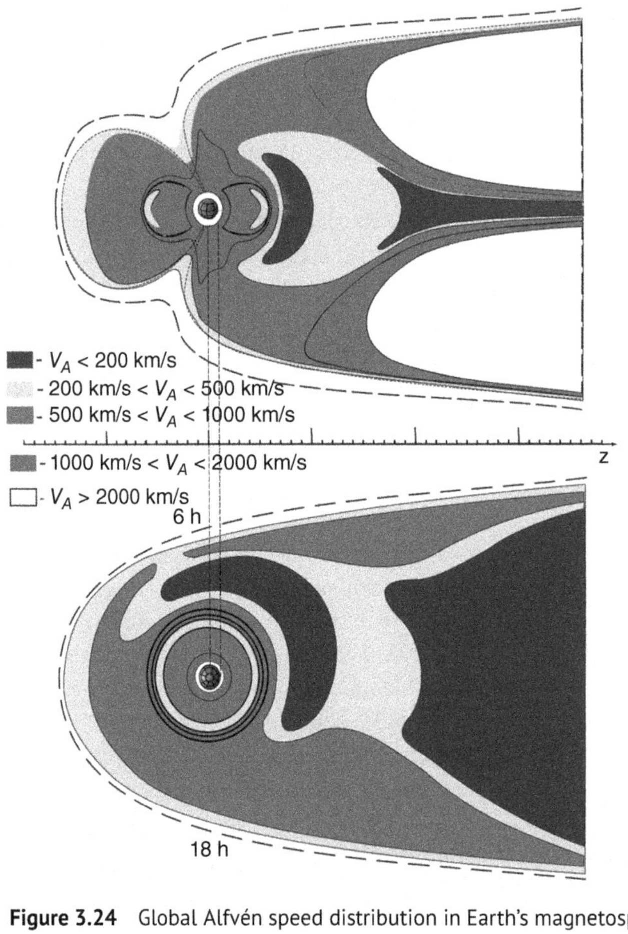

Figure 3.24 Global Alfvén speed distribution in Earth's magnetosphere.

The distribution in Figure 3.24 was plotted using the characteristic values of the magnetic field and plasma parameters in [161, 162].

To describe the plasma density and magnetic field distribution in the $-10R_E > z > -80R_E$ range, the following empirical dependencies were used: $B_0 \approx 3.5 \cdot (20/z)^{1.95}$ nT, $n_0 \approx (10/z)^{1.3}$ cm^{-3} in the plasma sheet and $B_0 \approx 75/\sqrt{z}$ nT, $n_0 \approx (10/z)^{2.6}$ cm^{-3} in the geotail lobes, where z is in Earth radii. In this case we have v_A (km/s) $\approx 2.18B_0(\text{nT})/\sqrt{n_0(\text{cm})^{-3}}$ for the Alfvén speed. The maximum value of the Alfvén speed, $\sim$ 5000 km/s, is reached in the geotail lobes. Outside the magnetosphere, the Alfvén speed tends to $\sim$ 50–100 km/s, typical of the solar wind.

The Alfvén speed in the plasma sheet is over an order of magnitude smaller than in the geotail lobes. Based on the data in [162], we can estimate the value of v_A in the near-Earth part of the plasma sheet (NEPS) for different distances from Earth: $v_A \approx 240$ km/s at $z = -10$ R_E, $v_A \approx 70$ km/s at $z = -15R_E$ and $v_A \approx 200$ km/s at $z = -25\,R_E$. It is thus possible to accept $v_A \approx 200$ km/s as the typical value in the NEPS.

As can be seen in Figure 3.24, there are minima in the v_A distribution in the outer dayside magnetosphere and inside the plasmasphere. It is here that the above- mentioned magnetospheric resonators for FMS waves are localised. The largest-scale and deepest minimum in the Alfvén speed, however, is located in the NEPS. The comparatively small value of the geomagnetic field and the high plasma density in this region guarantee the smallest Alfvén speed of the entire magnetosphere, $v_{A0} \sim 100$ km/s. The characteristic size of this region $L_0 \sim 10^5$ km. Using dimension-based concepts, the eigenmode frequency of the possible resonator is estimated $f_0 = \omega_0/2\pi \sim v_{A0}/L_0 \sim 1$ mHz.

The fact that v_A increases manifold (tens of times) when moving away from the NEPS towards Earth and into the geotail lobes ensures that FMS waves are reliably trapped in these directions. Towards the magnetopause, v_A increases by a much smaller factor (2–5 times), which can prove insufficient for a turning surface to exist. In this case, FMS wave is reflected thanks to the Alfvén speed jump at the magnetopause bounding this resonator. The question remains whether FMS waves can escape from the NEPS into its flat and thin ($a_0 \sim 10^4$ km thick) remote part, where v_A is of the same order as in the NEPS (Figure 3.25). Simple reasoning shows that such an escape is impossible. Let us introduce a Cartesian system of coordinates (x, y, z) such that the z coordinate be directed from Earth into the geotail, and the x coordinate across the plasma sheet. For FMS wave, $k_x \sim \pi/a_0$, $k_y \sim \pi/L$ in the remote part of the plasma sheet, where a_0 is the characteristic thickness of the plasma

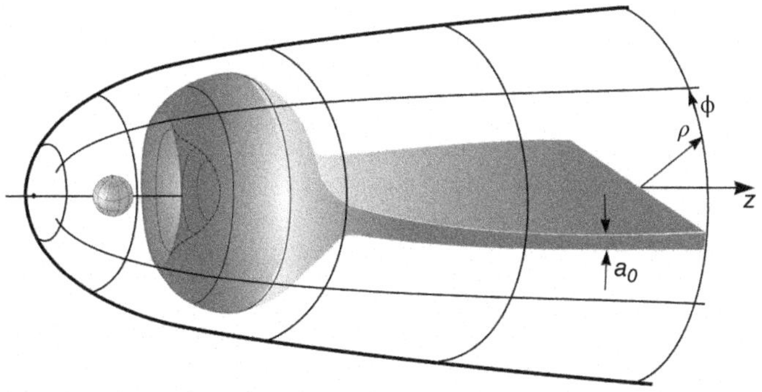

Figure 3.25 Axisymmetric model magnetosphere with a plasma sheet.

sheet, L is the characteristic size of the magnetosphere along the y axis. Given that $v_A \sim v_{A0}$ and $a_0 \ll L_0, L$ in this region,

$$k_z^2 = \omega_0^2 / v_A^2 - k_x^2 - k_y^2 \sim -\pi^2 / a_0^2,$$

i.e. the remote part of the plasma sheet is an opaque region for $\sim \omega_0$ modes.

The NEPS can thus serve as an FMS resonator, being encircled on all sides by a boundary reflecting FMS waves. The characteristic frequency of the eigenmodes of such a resonator, $f \sim 1$ mHz, which corresponds to the observed oscillations in the magnetosphere. The FMS-resonator modes should manifest themselves on Earth via standing Alfvén waves excited by the Alfvén resonance mechanism (field line resonance). The field lines passing through the NEPS cross the Earth surface at latitudes 60°–80°, which is consistent with the latitudinal localisation of the observable oscillations. Certain asymmetry in their longitudinal localisation can also be explained. The convective flow of the magnetospheric plasma (from the nightside through the dawn sector to the dayside) is known to shift the NEPS from the symmetric midnight position into the dawn sector [162]. As a result, the resonator for low-frequency FMS oscillations is shifted into the same sector, in full accordance with the observations.

Below, we will investigate the eigenmode structure and spectra for each of the above-noted magnetospheric FMS resonators.

3.6.2 FMS Resonators in the Dayside Magnetosphere

Let us examine the problem of magnetosonic wave incident to the magnetosphere from the solar wind ($x^1 > x^1_{mp}$, where x^1_{mp} is the magnetopause radial coordinate) in a model with a dipole-like magnetic field. Recall that our model does not include solar wind plasma motion, which can affect the result to a certain degree. Since the transparent regions for the eigenmodes in question are localised near the equatorial plane, however, containing a stagnation point for the solar wind near the dayside magnetosphere, these effects will hopefully not be too large. Outside the magnetosphere, the solution (3.92) can be rewritten as

$$\psi_{fn} = \overline{\psi}_{fn} H_n(x^1, x^3) \exp(im\phi - i\omega t),$$

where m is the azimuthal wave number, ϕ is the azimuthal angle,

$$\overline{\psi}_{Fn} = k_{1n}^{-1/2} \exp\left(-\int \tilde{k}_{1n} dx^1\right) \times \left[C_{in} \exp\left(-i\int k_{1n} dx^1\right) + C_{rn} \exp\left(i\int k_{1n} dx^1\right)\right], \tag{3.96}$$

where C_{in}, C_{rn} are the amplitudes of, respectively, the incident and reflected waves. It was demonstrated in Section 3.5.1 that there is a magnetic shell for each monochromatic FMS wave ($x^1 = x^1_n$) inside the magnetosphere which separates the transparent ($x^1 > x^1_n$) and opaque ($x^1 < x^1_n$) regions for oscillations with fixed wave numbers: longitudinal – n and azimuthal – m.

Let us represent the solution in the opaque region as

$$\overline{\psi}_{Fn} = \frac{C}{\sqrt{|k_{1n}|}} \exp\left(\int_{x^1_n}^{x^1} (|k_{1n}| - \tilde{k}_{1n}) dx^{1'}\right), \quad x^1 < x^1_n.$$

Using the Swan method to match the WKB solutions obtained on both sides of the turning point (see [69]), we will bypass the turning point $x^1 = x_n^1$ in the complex x^1 plane and will obtain, in the transparent region, a solution of the form

$$\overline{\Psi}_{Fn} = \frac{C}{\sqrt{|k_{1n}|}} \exp\left(\int_{x_n^1}^{x^1} \tilde{k}_{1n} dx^{1'}\right) \sin\left(\int_{x_n^1}^{x^1} k_{1n} dx^{1'} - \frac{\pi}{4}\right), \quad x^1 > x_n^1. \tag{3.97}$$

Matching the logarithmic derivatives of solutions (3.96) and (3.97) at the magnetopause $(x^1 = x_{mp}^1)$ yields

$$k_{1n}^- \cot\left(\int_{x_n^1}^{x_{mp}^1} k_{1n} dx^1 - \frac{\pi}{4}\right) = k_{1n}^+ \frac{C_{in} - C_{rn}}{C_{in} + C_{rn}}, \tag{3.98}$$

where $k_{1n}^{\pm} \equiv k_{1n}(x_{\pm}^1)$ are the values of the transverse wave number near the magnetopause, respectively, inside and outside the magnetosphere $(x_{\pm}^1 = \lim_{\varepsilon \to 0}(x_{mp}^1 \pm \varepsilon))$. In this problem, the magnetopause is represented as a tangential discontinuity where the Alfvén speed has a jump from its value in the magnetosphere to values characteristic of the solar wind.

If the reflection coefficient is defined as $R_n = C_{rn}/C_{in}$, we can obtain for it, from (3.98)

$$R_n = -\frac{1 - i(k_{1n}^-/k_{1n}^+) \cot\left(\int_{x_n^1}^{x_{mp}^1} k_{1n} dx^1 - \frac{\pi}{4}\right)}{1 + i(k_{1n}^-/k_{1n}^+) \cot\left(\int_{x_n^1}^{x_{mp}^1} k_{1n} dx^1 - \frac{\pi}{4}\right)}. \tag{3.99}$$

If the incident wave frequency coincides with a resonator eigenmode, the introduced reflection coefficient becomes infinite because the denominator of this expression is zero. This should be regarded as the presence of solutions to (3.68) satisfying the boundary conditions in question (limited amplitude in the entire domain of existence for the oscillations) in the absence of an FMS wave incident on the magnetosphere $(C_{in} = 0, C_{rn} \neq 0)$. Using the quantum mechanics analogy, it is possible to say that the desired solutions describe energy levels (in our case, oscillation frequencies) in a 'potential well' formed by an inhomogeneity in the Alfvén speed distribution. Equating the denominator in (3.99) to zero produces the following quantisation condition:

$$\int_{x_n^1}^{x_{mp}^1} k_{1nj} dx^1 = \pi\left(j + \frac{1}{4}\right) - \frac{i}{2} \ln \frac{k_{1nj}^+ + k_{1nj}^-}{k_{1nj}^+ - k_{1nj}^-}, \tag{3.100}$$

determining the transverse wave number in the WKB approximation, where $j = 0, 1, 2, \ldots$. Since $k_{1nj} \equiv k_{1nj}(\omega)$, it is possible to find from (3.100) the eigen oscillation frequencies of the magnetospheric resonator ω_{mnj}, defined by three wave numbers: azimuthal – m, longitudinal – n and transverse – j. The basic harmonics of just these FMS eigen oscillations $(n \sim m \sim j \sim 1)$ should be regarded as the 'global modes' of the dipole-like model magnetosphere under study.

Since $C_{rn} \neq 0$, the magnetosphere is not an ideal resonator: part of the energy of these eigen oscillations escapes into the solar wind. The form of (3.100) implies that the eigen frequencies are complex-valued: $\omega_{mnj} = \mathrm{Re}(\omega_{mnj}) + i\mathrm{Im}(\omega_{mnj})$, where $\mathrm{Re}(\omega_{mnj})$ is the real part of the frequency, and $\mathrm{Im}(\omega_{mnj})$ is the decrement due to part of the energy of the oscillation eigenmodes escaping into the solar wind. If the jump the parameters suffer at the magnetopause is very large $(k_{1nj}^+ \gg k_{1nj}^-)$, the logarithm in (3.100) becomes zero and the magnetosphere becomes an ideal resonator $(\mathrm{Im}(\omega_{mnj}) = 0)$.

Table 3.1 Eigen frequencies $\mathrm{Re}\omega_{mnj}$ (rad/s) × 10^2 and decrements $\mathrm{Im}\,\omega_{mnj}(\mathrm{s}^{-1}) \times 10^2$ for several first harmonics of a FMS resonator in the outer part of the dayside magnetosphere.

		$j = 0$	
$m \backslash n$	0	1	2
0	(4.6; −0.15)	(9.4; −0.15)	(13.8; −0.14)
1	(5.0; −0.16)	(9.7; −0.16)	(14.0; −0.15)
2	(5.7; −0.17)	(10.2; −0.17)	(14.4; −0.15)
		$j = 1$	
$m \backslash n$	0	1	2
0	(9.1; −0.1)	(17.4; −0.14)	(23; −0.15)
1	(10.8; −0.15)	(17.6; −0.15)	(23.1; −0.15)
2	(11.8; −0.16)	(18; −0.16)	(23.4; −0.14)
		$j = 2$	
$m \backslash n$	0	1	2
0	(13.3; −0.13)	(23.3; −0.14)	(30.2; −0.16)
1	(13.5; −0.15)	(24.0; −0.15)	(30.3; −0.16)
2	(14.1; −0.18)	(24.6; −0.16)	(30.7; −0.15)

Let us solve (3.100) numerically relative to the eigen frequencies ω_{mnj}, using the model dayside magnetosphere described in Section 3.2.1. Table 3.1 lists the eigen frequencies of an FMS resonator in the outer part of the dayside magnetosphere, for the first three values of the wavenumbers: $m, n, j = 0, 1, 2$. The first number in the parentheses shows the real part of the frequency, and the second (negative) number, its imaginary part – the decrement due to part of the eigenmode energy escaping into the solar wind.

Note that the Q factor of the oscillations is rather high ($\mathrm{Re}(\omega_{mnj}) \gg |\mathrm{Im}(\omega_{mnj})|$). However, the frequency of the lowest-frequency eigen oscillations is significantly higher than that of the ULF-oscillations with discrete spectrum observable in the magnetosphere ($\omega \sim 10^{-3}$ (rad/s), see [76]). The observed oscillations therefore cannot be the eigen oscillations of a FMS resonator in the outer part of the dayside magnetosphere, but other mechanisms should be found to explain them.

It was shown in [235, 236] that a transparent region for FMS oscillations also exists in the plasmasphere bounded on two sides by opaque regions, across the magnetic shells. The same is evident from the calculations of the transparent region configuration in Section 3.5.1. The eigen mode structure of this resonator was studied in [227, 231, 310] for various models of magnetospheric plasma distribution. Section 3.5.1 of this monograph uses a model dipole-like magnetosphere to inspect the structure of FMS oscillations along magnetic field lines. Let us now examine the eigen frequency spectrum of this resonator.

We will use x^1_{n1} to denote the turning point that determine the magnetic shell separating the resonator transparent region below the plasmapause from the inner (close to the ionosphere) opaque region, while using x^1_{n2} to denote the coordinate of the shell separating the transparent region (resonator) from the opaque region in the outer magnetosphere. The solution inside the transparent region satisfying the boundary condition for $x^1 < x^1_{n1}$, is given by (3.97). An analogous formula for a solution satisfying the boundary condition for $x^1 > x^1_{n2}$ can be written by adding $\pi/2$ to the phase and replacing the lower limit of integration by x^1 and the upper limit by x^1_{n2} in (3.97). For the oscillation eigen mode both these solutions must coincide, which is only possible when the quantisation condition holds:

$$\int_{x^1_{n1}}^{x^1_{n2}} k_{1nj} dx^1 = \pi \left(j + \frac{1}{2} \right), \tag{3.101}$$

where $j = 0, 1, 2, \ldots$. This equation defines the resonator eigen frequencies ω_{mnj} in the zero approximation, when the opaque regions bounding the resonator across the magnetic shells can be assumed to be of infinite extent.

In the real magnetospheric conditions, however, this statement is not quite accurate. The outer opaque region separating the resonator from the outer magnetosphere is a potential barrier of finite height and width. This results in part of oscillations escaping into the outer magnetosphere, i.e. the eigen frequencies of such a resonator are also complex-valued. The imaginary part of the frequency is oscillation decrement due to part of the oscillation energy leaking through the potential barrier. The value of this decrement depends on the values of the quantum numbers m, n and j, determining the eigen frequency ω_{mnj} spectrum. When the azimuthal wave number is large, $m \gg 1$, this decrement is small ($|\text{Im}\omega_{mnj}| \ll \text{Re}\omega_{mnj}$). In this case, the potential well where the resonator eigenmode is localised is deep and energy leak from it is small. If $m \sim 1$ the decrement value becomes comparable to the mode eigen frequency ($|\text{Im}\omega_{mnj}| \sim \text{Re}\omega_{mnj}$), which means smaller potential-well depth and larger leaks of the oscillation energy into the outer magnetosphere. We will here restrict ourselves to examining only the frequency spectrum of the eigen oscillations of such a resonator, while ignoring their decay thanks to leaking through the potential barrier.

Let us solve numerically (3.101) with respect to the resonator eigen frequencies. The frequency spectrum of several first harmonics of the eigen frequencies is presented in Table 3.2. Note that not all table cells are filled in. This is due to the fact that if wavenumbers n and j are fixed, the potential well becomes deep enough for an eigenmode to exist starting from a certain value $m = \overline{m}_{nj}$ only. Thus we have, e.g. $\overline{m}_{00} = 2, \overline{m}_{10} = 1, \overline{m}_{20} = 0 \ldots$. When $m < \overline{m}_{nj}$ the potential well is too shallow and an eigenmode with given n and j cannot fit into it.

The calculated frequencies of the resonator eigenmodes can be seen to belong in the Pc1–2 geomagnetic pulsation range. A number of studies suggest that this resonator plays a special role in generating storm-time geomagnetic pulsations Pi2 (see [312–314]).

3.6.3 FMS Resonator in the Near-Earth Plasma Sheet

Let us now examine the eigen oscillations of an FMS resonator localised in the NEPS. The basic structural elements of the magnetosphere necessary for this research are shown in Figure 3.25: the magnetopause and the plasma sheet subdividing the geotail into two lobes. In its near-to-Earth part, the plasma sheet widens substantially and occupies most of the magnetospheric cavity. This is an important feature in terms of the following theoretical investigation.

Table 3.2 Eigen frequencies Re ω_{mnj} (rad/s), of several first harmonics of the FMS resonator below the plasmapause.

	$j = 0$		
$m \backslash n$	0	1	2
0	—	—	0.32
1	—	0.23	0.327
2	0.136	0.238	0.334

	$j = 1$		
$m \backslash n$	0	1	2
3	—	—	0.473
5	—	0.39	0.5
7	0.317	0.42	0.527

	$j = 2$		
$m \backslash n$	0	1	2
9	—	—	0.662
11	—	0.586	0.694
13	0.523	0.623	0.728

3.6.3.1 Model of the Medium

An important role in this investigation is played by the FMS wave speed distribution $c_f = \sqrt{v_A^2 + v_s^2}$, where v_A is Alfvén speed, v_s is sound speed in plasma. Of most interest here is the relationship between the v_A and v_s values. The Alfvén speed distribution in the magnetosphere is known well enough (see Figure 3.24). It is much less known about the sound speed distribution. Since $v_s^2/v_A^2 \approx \beta$, where $\beta = 8\pi P_0/B_0^2$ is the plasma gas-kinetic to magnetic pressure ratio, it is possible to examine the β distribution in the magnetosphere.

In most of the magnetosphere, $\beta \ll 1$. There are two regions in the magnetosphere with a high enough plasma density – the plasmasphere and the plasma sheet. The plasma in the plasmasphere is cold and the geomagnetic field strength high, so that $\beta \ll 1$. The situation is completely different in the plasma sheet. The plasma here is hot and the geomagnetic field strength is much smaller than in the plasmasphere, making large values of β possible, including $\beta \gg 1$ [165]. Let us examine this issue in more detail.

To understand the situation more clearly, two substantially different parts should be distinguished in the plasma sheet – the near (NEPS) and far part (distant plasma sheet [DPS]). Both are filled with plasma of practically the same density and temperature so that $P_{0,\mathrm{NEPS}} \sim P_{0,\mathrm{DPS}}$. The plasma is transferred by the magnetospheric convective system from the far to the near-Earth plasma sheet, where, on the inner edge of the plasma sheet, it partially precipitates into the ionosphere resulting in diffuse aurorae, auroral arcs and other geomagnetic activity-related processes. The magnetic field is, however, substantially different between these two parts of the plasma sheet.

The far plasma sheet can be regarded as a nearly flat layer where the magnetic field varies from $-B_{0T}$ in one geotail lobe to B_{0T} in the other. The geomagnetic field strength decreases to nearly zero in the centre of the plasma sheet. The equilibrium condition for the far plasma sheet is $P_{0,\mathrm{DPS}} = B_{0T}^2/8\pi$. This means that plasma pressure in the plasma sheet is balanced by magnetic pressure in the geotail lobes. When calculating β near the centre of the plasma sheet, it should be taken into account that magnetic field here has a comparatively small vertical component B_{0n}. The presence of this component results in magnetic field lines being closed in the plasma sheet, but very elongated into the geotail. On the order of magnitude, $B_{0n} \lesssim 0.1B_{0T}$. In the centre of the far plasma sheet, therefore, $\beta_{DPS} = 8\pi P_{0,\mathrm{DPS}}/B_{0n}^2$, which is fully consistent with data in [165].

The geomagnetic field strength in the near-Earth plasma sheet is high enough, at least $B_{0,\mathrm{NEPS}} \gtrsim B_{0T}$. Assuming $P_{0,\mathrm{NEPS}} \sim P_{0,\mathrm{DPS}}$, we have $\beta_{\mathrm{NEPS}} \lesssim 1$. The value $\beta_{\mathrm{NEPS}} \gg 1$ is impossible. This means that plasma pressure in the near plasma sheet is not balanced. The dimensions of this part, as well as of the entire magnetosphere, depend substantially on the geomagnetic disturbance level. Under quiet conditions, $K_p = 1$–2, the inner boundary of the near plasma sheet is located at a distance of approximately 10 R_E, while its outer boundary, at a distance of 20–30 R_E from Earth. Under perturbed conditions, $K_P \gtrsim 5$, these distances decrease to 5 R_E and 10 R_E, respectively. ULF-oscillations with a discrete spectrum are observable under quiet geomagnetic conditions only, $K_P < 3$. Thus, we will assume the typical value of the dimensions of the near plasma sheet under these conditions: $\sim 15R_E \approx 10^5$ km. The value $\beta_{\mathrm{NEPS}} \lesssim 1$ means that $c_f \sim v_A$ here. Therefore, we can apply the 'cold' plasma approximation here for a qualitative investigation and order-of-magnitude estimates.

Plasma density in the plasma sheet is much higher and the Alfvén speed much lower than in the geotail lobes. Thus, the plasma sheet boundaries in Figure 3.25 should be regarded as boundaries separating magnetospheric regions with high vs. low value of the Alfvén speed. Such a boundary fails to reach Earth because both the magnetic field strength and Alfvén speed increase when approaching Earth.

3.6.3.2 Coordinate System and Basic Equations

Let us describe the spatial structure of monochromatic FMS waves in an inhomogeneous plasma, in the 'cold' plasma approximation, by

$$\Delta\Phi + \frac{\omega^2}{v_A^2}\Phi = 0, \tag{3.102}$$

where Φ is any component of the oscillation field. This should be regarded as a model equation. Different systems of curvilinear coordinates will feature different coefficients for derivatives. However, when the characteristic spatial scales of the oscillations are much smaller than the parameter variation scales (the WKB approximation being applicable), (3.102) describes, with good accuracy, FMS oscillations in any orthogonal system of coordinates. At the applicability limit, it also describes oscillations with wavelengths of order the medium inhomogeneity scales.

Solution of (3.102) is substantially simplified if the problem under study has a symmetry in a certain direction. We are interested in oscillations localised in the near plasma sheet. As can be seen in Figure 3.25, there is a certain axial symmetry (about the z axis) in this part of the magnetosphere. Such a symmetry is absent from the middle and DPS. For the

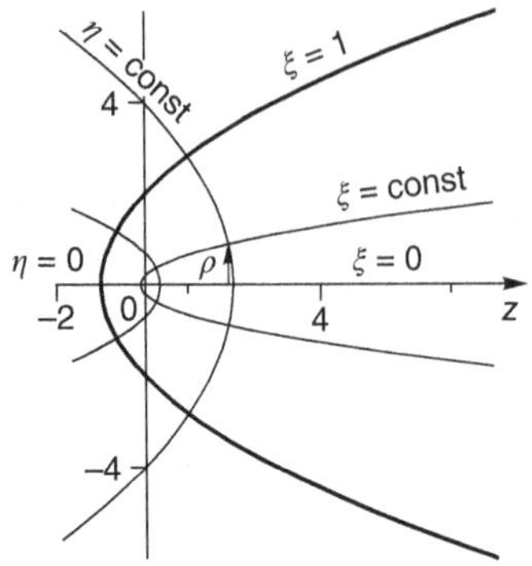

Figure 3.26 Orthogonal system of dimensionless parabolic coordinates (ξ, η), in the meridional section, including the z axis. The focus, $z = 0$, is at Earth's centre. The surface $\xi = 1$ coincides with the magnetopause. The semiaxes $z = (0, \infty)$ and $z = (0, -\infty)$ correspond to the coordinate surfaces $\xi = 0$ and $\eta = 0$, respectively.

problem we are dealing with, however, the far plasma sheet plays no significant role which is why its absence can be utterly neglected. Both the geotail lobes and the far plasma sheet are opaque regions for the large-scale FMS waves we are examining.

This allows an axisymmetric model magnetosphere to be used for transiting from a 3D-inhomogeneous to a 2D-inhomogeneous problem, making the research much simpler. Since the magnetospheric cavity resembles a paraboloid, it is natural to choose parabolic coordinates for describing it (see [315]). Let us introduce two sets of mutually orthogonal parabolic surfaces as the coordinates, denoting them as ξ and η, focused at Earth centre. These surfaces are obtained by rotating the generating parabolic curve about the z axis, passing through the Earth and Sun centres (Figure 3.26).

The dimensionless parabolic coordinates ξ and η are related to the ordinary cylindric coordinates (z being the common axis) by the following relations:

$$\xi = (\sqrt{\rho^2 + z^2} - z)/2\sigma,$$
$$\eta = (\sqrt{\rho^2 + z^2} + z)/2\sigma,$$

where σ is the distance between the focus and the paraboloid vertex, $\xi = 1$, corresponding to the magnetopause. In other words, σ is the distance between the Earth centre and the subsolar point on the magnetopause. Assuming an axial symmetry in the problem, we will seek the solution in the form of an expansion into azimuthal harmonics of the form $\Phi(\xi, \eta, \varphi) = \overline{\Phi}(\xi, \eta) \exp(im\phi)$, where $m = 0, 1, 2, \ldots$ is the azimuthal wavenumber. Equation (3.102), rewritten in the parabolic coordinates ξ and η, has the form

$$\frac{\partial}{\partial \xi} \xi \frac{\partial \overline{\Phi}}{\partial \xi} + \frac{\partial}{\partial \eta} \eta \frac{\partial \overline{\Phi}}{\partial \eta} + \left[(\xi + \eta) \frac{\omega^2 \sigma^2}{v_A^2(\xi, \eta)} - \frac{m^2}{4} \left(\frac{1}{\xi} + \frac{1}{\eta} \right) \right] \overline{\Phi} = 0. \tag{3.103}$$

It is easily verifiable that choosing an Alfvén speed distribution of the form

$$v_A^2(\xi, \eta) = \frac{v_{A0}^2}{\sigma} \frac{\xi + \eta}{a(\xi) + b(\eta)},$$

where v_{A0} is a constant having the dimension of velocity, and $a(\xi)$ and $b(\eta)$ are arbitrary functions of variables ξ and η, will turn (3.103) into an equation with separable variables.

Functions $a(\xi)$ and $b(\eta)$ can be chosen such as to simulate the basic properties of the v_A distribution in the magnetospheric region in question. Let us consider oscillations localised in the magnetosphere ignoring their small leakage into the solar wind via the magnetopause. Thus, choosing functions $a(\xi)$ and $b(\eta)$ we will not require that $v_A^2(\xi, \eta)$

should behave correctly in the solar wind region. The main features of the v_A distribution in the near plasma sheet and in the geotail lobes can be simulated by the following functions:

$$a(\xi) = [\alpha_0 + (\alpha_1 - \alpha_0)\xi]\xi, \tag{3.104}$$

$$b(\eta) = \begin{cases} [\alpha_0 + (\alpha_2 - \alpha_0)\eta]\eta, & \eta \le \eta_1, \\ \beta_1[1 - (\eta - \eta_{\min})^2/\Delta_1^2], & \eta_1 < \eta < \eta_2, \\ \beta_2[1 + (\eta - \eta_{\max})^2/\Delta_2^2], & \eta \ge \eta_2, \end{cases} \tag{3.105}$$

where $\eta_{\min}$ is the coordinate corresponding to the Alfvén speed minimum localisation in the near plasma sheet. The parameters of the model (3.105) are chosen such that function $b(\eta)$ is continuous and smooth. Points η_1 and η_2 can be regarded as the near and far (from Earth) boundaries of the near plasma sheet, in the η coordinate. The turning points of the FMS oscillations are located between these boundaries. The Alfvén speed maximum in the middle part of the geotail is a potential barrier confining the low-frequency oscillations inside the near plasma sheet. For higher-frequency oscillations, this maximum is not a potential barrier that would prevent their leaking into the geotail. This may be the reason why higher-frequency FMS oscillations with a discrete spectrum are not observed in the magnetosphere.

If the desired function is represented as $\overline{\Phi}(\xi, \eta) = \zeta(\xi)\mu(\eta)$, the solution of the original partial differential equation can be reduced to solving a system of two ordinary differential equations:

$$\xi\frac{\partial^2\zeta}{\partial\xi^2} + \frac{\partial\zeta}{\partial\xi} + \left[\Omega^2 a(\xi) - \frac{m^2}{4\xi} + Q\right]\zeta = 0, \tag{3.106}$$

$$\eta\frac{\partial^2\mu}{\partial\eta^2} + \frac{\partial\mu}{\partial\eta} + \left[\Omega^2 b(\eta) - \frac{m^2}{4\eta} - Q\right]\mu = 0. \tag{3.107}$$

Here, Q is the separation constant, and $\Omega = \omega\sigma/v_{A0}$ is a dimensionless frequency. They are the eigenvalues of the problem to be solved.

To numerically calculate the eigen frequency spectrum for the FMS oscillations localised in the near plasma sheet, let us set the parameters defining the model of the Alfvén speed (3.104) and (3.105). We will choose the following values of the constants: $v_{A0} = 750$ km/s, $\sigma = 10R_E$ ($R_E = 6370$ km is the Earth radius), $\alpha_0 = 0.04$, $\alpha_1 = 10$, $\alpha_2 = 5$, $\eta_{\min} = 1.5$, $\eta_{\max} = 6$, $\Delta_1 = 1$, $\Delta_2 = 2.5$, $\eta_1 = 0.6$, and η_2 will be determined from the condition of equality of the second and third expressions in (3.105) at point $\eta = \eta_2$. The corresponding 2D (in the meridional plane) Alfvén speed distribution is shown in Figure 3.27. This model describes well enough both the v_A distribution in the near plasma sheet, and its maximum in the middle part of the geotail, as well as qualitatively correctly describing its variation towards Earth.

Moreover, solving the eigenvalue problem requires that boundary conditions be formulated for functions $\zeta(\xi)$ and $\mu(\eta)$. We do not take into account the small leakage of the oscillations into the solar wind, therefore, the boundary condition for $\zeta(\xi)$ can be formulated at the magnetopause for $\xi = 1$. Let us assume this boundary to be ideally reflecting, which yields the first boundary condition $\zeta(1) = 0$. We will set the second boundary condition at the symmetry axis $\xi = 0$. As $\xi \to 0$, the first term in the square

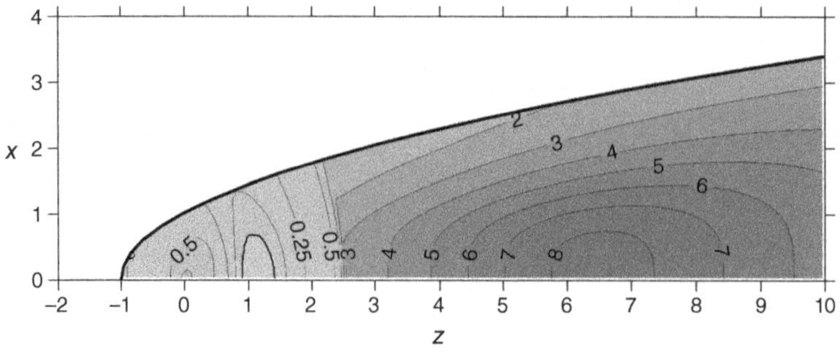

Figure 3.27 Alfvén speed $v_A(10^3$ km/s) distribution isolines in the meridional plane, in the parabolic model magnetosphere.

brackets in (3.106) can be neglected, and the remainder is reduced to the Bessel equation, its solution expressed in terms of the Bessel functions:

$$\zeta(\xi) = CJ_{2m}(2\sqrt{Q\xi}) + DY_{2m}(2\sqrt{Q\xi}),$$

where C and D are arbitrary constants. We will require that the solution be finite throughout its definition domain. Since function $Y_\nu(x)$ is singular as $x \to 0$, let us set $D = 0$. As $\eta \to 0$, the boundary condition for function $\mu(\eta)$ is thus

$$\mu(\eta) = CI_{2m}(2\sqrt{Q\eta}).$$

In the WKB approximation, the oscillations in question are localised between two turning points: $\overline{\eta}_1$ and $\overline{\eta}_2$ $(0 < \overline{\eta}_1 < \overline{\eta}_2 < \eta_2)$. If $\eta > \overline{\eta}_2$, the function $\mu(\eta)$ decreases exponentially into the opaque region. Moving on along the geotail, the Alfvén speed reaches a maximum and starts to decrease. The opaque region is again replaced with a transparent region. This also enables the oscillations to escape into the solar wind through a potential barrier.

This barrier is so wide and high, however, that the oscillation leakage through it is small (smaller than that through the magnetopause) and can be neglected. Thus, the second boundary condition in this approximation can be formulated as an impermeable wall located deep inside the opaque region. If this wall is located much farther than the characteristic scale of the function $\mu(\eta)$ decrease into the opaque region, the desired solution is practically independent of where in particular it is localised. Our calculations presumed it to be at Alfvén speed distribution maximum point $\mu(\eta_{\max}) = 0$.

These boundary conditions are only satisfied by such solutions of (3.106) and (3.107) that correspond to the eigenvalues of parameters $\Omega = \Omega_{mnl}$ and $Q = Q_{mnl}$, where $m, n, l = 0, 1, 2, \ldots$ are the wavenumbers determining the number of nodes in the eigenfunctions $\Phi_{mnl}(\phi, \xi, \eta)$ in the ϕ, ξ and η coordinates, respectively. The set of eigenvalues Ω_{mnl} determines the set of FMS resonator eigen frequencies $f_{mnl} = \omega_{mnl}/2\pi$. These frequencies are listed in Table 3.3 for the first three harmonics, for each wavenumber.

Evidently, the calculated eigen frequencies coincide well enough with the spectrum of ULF-oscillations (0.8, 1.3, 1.9, 2.6, … mHz) observable in the magnetosphere. This calculated frequency spectrum exhibits one interesting feature. The eigen frequencies are not distributed uniformly along the spectrum, but are grouped into certain clusters. Thus, frequencies $f_{000} = 0.73$ mHz and $f_{100} = 1.04$ mHz represent clusters consisting of one frequency only. The sets of frequencies ($f_{001} = 1.41$; $f_{010} = 1.36$; $f_{200} = 1.32$ mHz)

Table 3.3 Eigen frequencies $f_{mnl}(\text{mHz}) = \omega_{mnl}/2\pi$ of several first harmonics of a FMS resonator in the near-Earth plasma sheet.

	$m = 0$		
$n\backslash l$	0	1	2
0	0.73	1.41	2.13
1	1.36	1.96	2.65
2	1.97	2.55	3.2

	$m = 1$		
$n\backslash l$	0	1	2
0	1.04	1.66	2.42
1	1.66	2.24	2,91
2	2.29	2.84	3.47

	$m = 2$		
$n\backslash l$	0	1	2
0	1.32	1.91	2.7
1	1.96	2.52	3.17
2	2.59	3.13	3.75

and ($f_{101} = 1.66; f_{110} = 1.66; f_{300} = 1.59$ mHz) are clusters consisting of three frequencies of mean values $\bar{f} \approx 1.35$ mHz and $\bar{f} \approx 1.6$ mHz, respectively. Other oscillation harmonics with mean frequencies $\bar{f} \approx 1.95$ mHz, $\bar{f} \approx 2.2$ mHz, $\bar{f} \approx 2.6$ mHz, $\bar{f} \approx 3.1$ mHz can be grouped together into clusters consisting of 5–7 harmonics (including harmonics with $m, n, l > 2$). Since we neglect corrections of order $v_s^2/v_A^2 \sim \beta$, these values may be corrected by a few to 10–20%.

The number of harmonics in each cluster can, under otherwise equal conditions, be regarded as indicative of a relative probability to observe oscillations with mean frequency corresponding to a given cluster. It can hardly be expected, however, that the conditions for the excitation of different oscillation harmonics will be identical. The corresponding frequency must be present in the oscillation source spectrum. The Kelvin–Helmholtz instability at the magnetopause can exemplify such a source. As we have seen in Section 2.15.3, the resonator frequency spectrum falls within the range of FMS oscillations excited by this instability at the geotail boundary (see also [316, 317]). These oscillations can also be excited in the magnetosphere under the impact of solar wind pressure pulses [318, 319].

The oscillations in question are transported to Earth by Alfvén waves excited in the Alfvén resonance (field line resonance [FLR]) process [320, 321]. The characteristic eigen frequencies of toroidal Alfvén oscillations at the magnetic shells in question ($10 < L < 20$) are just within the range of the main modes of the FMS resonator [322]. Obviously, the spectrum of the calculated frequencies of the resonator in question is very close to the spectrum of the observed 'magic frequencies'. It is equally easy to explain the localisation of the oscillations registered on Earth at latitudes 60°–80°. It is this ionospheric region that the resonator in the near plasma sheet is mapped onto along magnetic field lines. Thanks to magnetospheric convection, this resonator is shifted into the magnetospheric midnight-dawn sector.

In its localisation region, the lateral walls of the resonator are not located far from the plasmapause. Thanks to this, it cannot be regarded as an ideal resonator. It is partially permeable to oscillations coming from the solar wind, and part of the energy of its eigen oscillations also escapes into the solar wind. The Q-factor of this resonator is, however, high enough. Simple estimates show that the ratio between the decrement γ of the eigenmodes, related to their leakage into the solar wind through the magnetopause, and the oscillation frequency is, on the order of magnitude, $\gamma/\omega \sim v_{Asw}/v_{Am}$, where $v_{Am} \sim$ (300–500) km/s, $v_{Asw} \sim$ (50–100) km/s are the characteristic Alfvén speed values in the solar wind and the magnetosphere near the magnetopause (see [323]). Thus, we have $\gamma \sim (0.1\text{–}0.3)\omega$. This is enough to make the continually operating source excite corresponding oscillations in the resonator.

The following reasonable suggestion is possible regarding the stability of the observable frequencies. The oscillations in question are as a rule observed under quiet enough conditions of geomagnetic disturbance ($K_p < 3$), therefore, the parameters of the near part of the plasma sheet are always approximately identical. Hence, the FMS resonator formed under such conditions features practically identical characteristics.

3.7 Monochromatic Transverse-Small-Scale Alfvén Waves with $m \gg 1$ in a Dipole-Like Magnetosphere

In Sections 3.1–3.6 of this chapter dealt with MHD oscillations that are large-scale in the azimuthal direction ($m \sim 1$). Let us now examine the spatial structure of Alfvén waves that are small-scale both across magnetic shells and in the azimuthal direction ($m \gg 1$). We will carry out both a complete analytical investigation of this structure and a numerical calculation of its main characteristics using a dipole-like model magnetosphere with 'cold' plasma. Fast magnetosonic waves with $m \gg 1$ cannot penetrate deep into the magnetosphere because the magnetosphere is an opaque region for such waves (see Section 3.5). Thanks to this, generation of Alfvén waves with $m \gg 1$ requires their source to be located at the magnetic shells where they are excited. External currents in the ionospheric conducting layer may serve as such a source. They are generated by neutrals moving in the ionospheric conducting layer due to various processes of both natural and artificial origin (see Section 2.18).

To describe the structure of such Alfvén oscillations, let us use (3.30), where we will set the right-hand side, describing the FMS wave field, as zero:

$$\nabla_1 \hat{L}_T \nabla_1 \varphi + \nabla_2^2 \hat{L}_P \varphi = 0. \tag{3.108}$$

For operators $\hat{L}_T$, $\hat{L}_P$ we will use the following representation:

$$\hat{L}_T = \frac{\partial}{\partial l} p \frac{\partial}{\partial l} + p \frac{\omega^2}{v_A^2}, \tag{3.109}$$

$$\hat{L}_P = \frac{\partial}{\partial l} p^{-1} \frac{\partial}{\partial l} + p^{-1} \frac{\omega^2}{v_A^2}, \tag{3.110}$$

$p = \sqrt{g_2/g_1}$, and the derivatives are taken along the physical length l of the magnetic field lines ($dl = \sqrt{g_3}dx^3$).

3.7.1 Formulating the Problem of the Alfvén Oscillation Structure in the WKB Approximation

The transverse-small-scale character of the oscillations under study makes it natural to use the WKB approximation in the x^1 coordinate, directed across the magnetic shells (see Figure 3.2). Let us choose the dependence of an individual azimuthal harmonic on the transverse coordinates in the form

$$\varphi = \exp(iQ + ik_2x^2),$$

where $Q = Q(x^1, l)$ is the large quasi-classical phase, $k_2 = m/\rho$ is the azimuthal wavenumber (ρ is the radius counted from the symmetry axis, in a cylindrical system of coordinates, see Figure 3.12). Note that examining individual azimuthal harmonics with fixed m loses any sense if $m \gg 1$, therefore we will hereafter use the general designation of the azimuthal coordinate x^2 and azimuthal wavenumber k_2; the change of the latter value can be regarded as continuous for $m \to \infty$. The condition of a small-scale potential in the x^1 coordinate

$$\left|\frac{1}{\sqrt{g_1}}\frac{\partial\varphi}{\partial x^1}\right| \gg \left|\frac{\partial\varphi}{\partial l}\right|$$

yields an analogous inequality for phase Q:

$$\left|\frac{1}{\sqrt{g_1}}\frac{\partial Q}{\partial x^1}\right| \gg \left|\frac{\partial Q}{\partial l}\right|,$$

i.e. the dependence of function Q on the coordinate x^1 is much stronger than its dependence on the coordinate l. This means that a dominant term can be extracted depending on x^1 only. In other words, the main order term Q_0 can be regarded as independent of the l coordinate when expanding the phase into an asymptotic series of the WKB approximation,

$$Q = Q_0 + Q_1 + Q_2 + \cdots .$$

Denote

$$\vartheta = Q_0; \quad \exp[i(Q_1 + Q_2 + \cdots)] = H + h + \cdots ,$$

or, in other words, write the solution of (3.108) in the WKB approximation as

$$\varphi = \exp(i\vartheta(x^1, \omega) + ik_2x^2)[H(x^1, l, \omega) + h(x^1, l, \omega) + \cdots]. \tag{3.111}$$

Note also that function

$$\tilde{\vartheta}(x^1, x^2, \omega) = \vartheta(x^1, \omega) + k_2x^2$$

is a full quasi-classical phase in transverse coordinates. The quasiclassical covariant components of the wave vector are determined by the ratios

$$k_1 = \partial\tilde{\vartheta}/\partial x^1, \qquad k_2 = \partial\tilde{\vartheta}/\partial x^2.$$

The latter equality implies that

$$k_1 = k_1(x^1, \omega) = \partial\tilde{\vartheta}(x^1, \omega)/\partial x^1,$$

and, as was to be expected, the identity $k_2 = \partial\tilde{\vartheta}(x^1, \omega)/\partial x^2 = k_2$ takes place. Despite k_1 and k_2 not depending on l, the wave-vector squared

$$k_\perp^2 = \frac{k_1^2}{g_1} + \frac{k_2^2}{g_2}$$

does depend on l.

Substituting expression (3.111) into the homogeneous equation (3.108) produces, in the main order of the WKB approximation,

$$\hat{L}H = 0, \tag{3.112}$$

where

$$\hat{L} \equiv \hat{L}(x^1, l, \omega) = k_1^2\hat{L}_T + k_2^2\hat{L}_P = \frac{\partial}{\partial l}q\frac{\partial}{\partial l} + q\frac{\omega^2}{A^2},$$
$$q = \sqrt{g_\perp}k_\perp^2 = pk_1^2 + p^{-1}k_2^2, \quad g_\perp = g_1g_2.$$

The boundary condition for Alfvén waves at the ionosphere has the form of (2.309). In the main order of perturbation theory, we will consider an ideally conducting ionosphere, i.e. assume

$$H|_{l_\pm} = 0, \tag{3.113}$$

where $l_\pm$ are the points where the field line crosses the ionosphere in the Northern and Southern Hemispheres, respectively. For given x^1 and ω, the relations (3.112) and (3.113) can be regarded as an eigenvalue problem for $\kappa = k_1/k_2$. To be convinced that this problem can be treated in this manner, let us represent these relations as

$$(\kappa^2\hat{L}_T + \hat{L}_P)H = 0, \quad H|_{l_\pm} = 0. \tag{3.114}$$

Let

$$\kappa = \kappa_N(x^1, \omega), \quad H = H_N(x^1, l, \omega) \tag{3.115}$$

be the solution of the eigenvalue problem. Here, $N = 1, 2, \ldots$ is the number of the eigenvalue equalling the number of the function H_N half-waves on the field line. For given k_2, Eqs. (3.112) and (3.113) can be regarded as the eigenvalue problem for k_1. Clearly,

$$k_1 = k_{1N}(x^1, \omega) = k_2\kappa_N(x^1, \omega). \tag{3.116}$$

Hence

$$\vartheta = \vartheta_N(x^1, \omega) = \int k_{1N}(x^1, \omega)dx^1 = k_2\int \kappa_N(x^1, \omega)dx^1. \tag{3.117}$$

Thus, the solution of the eigenvalue problem determines the main order of the WKB approximation, in the x^1 coordinate.

To conclude this section, let us list the formulas for the components of perturbed electric and magnetic fields, in the WKB approximation. Substituting (3.111) into the expressions for the MHD oscillation electromagnetic field components (E.1) and (E.2) (Appendix E), yields, in the main order,

$$\begin{aligned} &E_1 = -ik_1He^{i\tilde{\vartheta}}, \qquad E_2 = -ik_2He^{i\tilde{\vartheta}}, \quad E_3 = 0, \\ &B_1 = \frac{\bar{c}}{\omega}\frac{k_2}{p}\frac{\partial H}{\partial l}e^{i\tilde{\vartheta}}, \quad B_2 = -\frac{\bar{c}}{\omega}k_1p\frac{\partial H}{\partial l}e^{i\tilde{\vartheta}}, \\ &B_3 = -2i\frac{\bar{c}}{\omega}\sqrt{g_3}\frac{k_1k_2}{q}\frac{\partial \ln p}{\partial l}\frac{\partial H}{\partial l}e^{i\tilde{\vartheta}}. \end{aligned} \tag{3.118}$$

3.7.2 Qualitative Investigation of the Eigenvalue Problem

Problems (3.112) and (3.113) define H_N as a function of l accurate to an arbitrary factor. To define this factor, let us examine the normalised solutions of this problem, $R_N(x^1, l, \omega)$, determined by the following relations:

$$\hat{L}R_N = 0, \quad R_N|_{l_\pm} = 0,$$
$$\oint \frac{q_N}{v_A^2} R_N^2 \, dl = 1. \tag{3.119}$$

Here

$$q_N = pk_{1N}^2 + p^{-1}k_2^2 = k_2^2(p\kappa_{1N}^2 + p^{-1}),$$

and the curvilinear integral means 'two-way' integration along the field line between the two magneto-conjugated ionospheres.

The relation $\kappa(x^1, \omega) = k_1/k_2$, for given x^1, determines the functional dependence between frequency ω and parameter $\kappa = k_1/k_2$. Let us introduce an inverse function,

$$\omega = \omega(x^1, k_1/k_2). \tag{3.120}$$

Parameter ω_N can be regarded as the solution of the eigenvalue problem (3.112) and (3.113) for parameter ω, for given values of parameters k_1 and k_2, and the equality (3.120) as a local dispersion equation. With such a statement of the problem, the eigenfunctions are $R_N[x^1, l, \omega_N(x^1, k_{1N}/k_2)]$.

A special role in the following is played by the two limiting cases: $\kappa = 0$ and $\kappa = \infty$, corresponding to poloidal and toroidal modes of Alfvén oscillations. Let us inspect them more closely. When $\kappa = 0$ ($k_1 = 0$), the eigenvalue problem (3.112) and (3.113) has the form

$$\hat{L}_P(\omega)H = 0, \quad H|_{l_\pm} = 0.$$

Let us denote its solutions as

$$\omega = \Omega_{PN}(x^1), \quad H = P_N(x^1, l)$$

calling them poloidal eigenfrequencies and poloidal eigenfunctions, respectively. We will assume the latter to be the normalised condition

$$\oint \frac{1}{pv_A^2} P_N^2 dl = 1.$$

It is easy to see that

$$\Omega_{PN}(x^1) = \omega_N(x^1, 0), \quad P_N(x^1, l) = k_2 R_N[x^1, l, \Omega_{PN}(x^1)].$$

When $\kappa \to \infty$ ($k_1 \to \infty$) we have, from (3.112) and (3.113)

$$\hat{L}_T(\omega)H = 0, \quad H|_{l_\pm} = 0.$$

The solutions of this problem are toroidal eigenfrequencies and eigenfunctions, which we will denote as

$$\omega = \Omega_{TN}(x^1), \quad H = T_N(x^1, l).$$

We will assume that the eigenfunction normalisation condition

$$\oint \frac{p}{v_A^2} T_N^2 dl = 1$$

holds, resulting in

$$\Omega_{TN}(x^1) = \omega_N(x^1, \infty), \quad T_N(x^1, l) = k_1 R_N[x^1, l, \Omega_{TN}(x^1)].$$

The latter equality should be understood as follows: as $k_1 \to \infty$, function $R_N = T_N/k_1$ tends to zero.

Of key value for the theory propounded here is the difference between poloidal and toroidal eigenfrequencies. Their difference $\Delta\Omega_N = \Omega_{TN} - \Omega_{PN}$ is called polarisation spectral splitting [321]. A useful analytical expression can be derived for it. We rely on the identities

$$\frac{\partial}{\partial l} p \frac{\partial T_N}{\partial l} + p \frac{\Omega_{TN}^2}{A^2} T_N = 0, \quad \frac{\partial}{\partial l} \frac{1}{p} \frac{\partial P_N}{\partial l} + \frac{1}{p} \frac{\Omega_{PN}^2}{A^2} P_N = 0.$$

Let us left-multiply the first one by P_N/p, and the second by pT_N, subtract the latter from the former and integrate along the field line. The transformation and integration by parts produces the equality

$$\Omega_{TN}^2 - \Omega_{PN}^2 = \oint \frac{\partial^2 \ln p}{\partial l^2} P_N T_N dl \Big/ \oint \frac{1}{v_A^2} P_N T_N dl. \tag{3.121}$$

Hence, the polarisation spectral splitting is determined by geomagnetic field-line curvature. Indeed, $p = \sqrt{g_2/g_1}$ is independent of l in a magnetic field with straight lines, making the right-hand side in (3.121) equal zero. It will be demonstrated in Section 3.7.3 that the difference $\Delta\Omega_N$ is small compared to the frequencies.

Of special importance for a mode with fixed frequency ω are surfaces defined (for fixed N) by equations

$$\Omega_{TN}(x^1) = \omega, \qquad \Omega_{PN}(x^1) = \omega. \tag{3.122}$$

Let us call them 'poloidal' and 'toroidal' resonance surfaces, their coordinates x_{PN}^1 and x_{TN}^1 found by solving Eq. (3.122) with respect to x^1. The distance between them can be characterised by the difference $\Delta_N = x_{TN}^1 - x_{PN}^1$. If $\Delta_N > 0$ and the functions Ω_{PN} and Ω_{TN} decrease monotonously, the poloidal surface is left of the toroidal one, $x_{TN}^1 > x_{PN}^1$. Given $\Delta\Omega_N \ll \Omega_{TN}, \Omega_{PN}$, it is easy to obtain the explicit expression for Δ_N. We will decompose functions Ω_{TN}, Ω_{PN} near the resonance surfaces as

$$\Omega_{TN} = \omega \left(1 - \frac{x^1 - x_{TN}^1}{2a_N}\right), \quad \Omega_{PN} = \omega \left(1 - \frac{x^1 - x_{PN}^1}{2a_N}\right). \tag{3.123}$$

Here, $a_N = (\nabla_1 \ln \Omega_{(T,P)N})^{-1}$ is the characteristic scale of the function $\Omega_{TN}(x^1), \Omega_{PN}(x^1)$ variation near the resonance surfaces. This scale can be regarded as being the same for both functions when $\Delta\Omega_N \ll \Omega_{TN}, \Omega_{PN}$. This inequality implies that the applicability ranges of the expansions of (3.123) overlap. Subtracting one from the other produces

$$\Delta_N = 2\alpha_N a_N, \tag{3.124}$$

where $\alpha_N = \Delta\Omega_N/\omega$, assuming $\alpha_N \ll 1$. The definition of resonance surfaces implies that function $k_{1N}(x^1, \omega)$ tends to zero on the poloidal surface, while tending to infinity on the toroidal surface. Let us examine its behaviour in the neighbourhood of these surfaces. We will use perturbation theory for this purpose. When $|x^1 - x^1_{PN}| \ll \Delta_N$ the values of $\omega^2 - \Omega^2_{PN}$ and k^2_{1N} can be assumed to be small, near the poloidal surface. Let also $H = P_N + h_N$, where h_N is a small correction. Linearising the problem (3.112) and (3.113) with respect to small parameters yields

$$k_2^2\hat{L}_P(\Omega_{PN})h_N + k^2_{1N}\hat{L}_T(\Omega_{PN})P_N + k_2^2\frac{\omega^2 - \Omega^2_{PN}}{pv_A^2}P_N = 0,$$
$$h_N|_{l_\pm} = 0.$$

We will left-multiply this equation by P_N and integrate along the field line. Given that the operator $\hat{L}_P$ is Hermitian (combined with the boundary condition) we obtain

$$k^2_{1N} = k_2^2\frac{\omega^2 - \Omega^2_{PN}}{w_{PN}},$$
$$w_{PN} = -\oint P_N\hat{L}_T(\Omega_{PN})P_N dl = \oint \frac{\partial^2 p}{\partial l^2}P_N^2 dl. \tag{3.125}$$

Analogously, we have, near the toroidal surface, when $|x^1 - x^1_{TN}| \ll \Delta_N$:

$$k^2_{1N} = k_2^2\frac{w_{TN}}{\omega^2 - \Omega^2_{TN}},$$
$$w_{TN} = -\oint T_N\hat{L}_P(\Omega_{TN})T_N dl = -\oint \frac{\partial^2 p^{-1}}{\partial l^2}T_N^2 dl. \tag{3.126}$$

The analytical estimates and results of the numerical calculations in Section 3.7.3 demonstrate that constants w_{PN} and w_{TN} are positive.

If the expansions (3.123) are applicable, we have, from (3.125) and (3.126), respectively,

$$k^2_{1N} = k_2^2\frac{\omega^2}{w_{PN}}\frac{x^1 - x^1_{PN}}{a_N}, \qquad k^2_{1N} = -k_2^2\frac{w_{TN}}{\omega^2}\frac{a_N}{x^1 - x^1_{TN}}. \tag{3.127}$$

It is evident from this that the poloidal resonance surface is an ordinary turning point, in the x^1 coordinate, where k_1^2 tends to zero, while the toroidal resonance surface is a singular turning point, where k_1^2 has a pole. The transparent region is located near the poloidal surface if $x^1 > x^1_{PN}$, and near the toroidal surface, if $x^1 < x^1_{TN}$, i.e. the transparent region is located in the range $x^1_{PN} < x^1 < x^1_{TN}$.

The opaque regions (where k_1^2 are negative) are located outside this range. The asymptotical values of k^2_{1N} in these regions can be found analytically. We have

$$k^2_{1N} \to \begin{cases} -k_2^2/p^2_{\min}, & x^1 - x^1_{TN} \gg \Delta_N, \\ -k_2^2/p^2_{\max}, & x^1_{PN} - x^1 \gg \Delta_N, \end{cases} \tag{3.128}$$

where $p^2_{\min}$ and $p^2_{\max}$ are the smallest and the largest values of function $p^2 = p^2(l)$ on the field line. In simple models of the geomagnetic field (e.g. in a dipole field) the values $p^2_{\min}$ and $p^2_{\max}$ are reached at, respectively, the equator and the ionosphere. The complete distribution of $k_1^2(x^1)$ is shown in Figure 3.28.

To conclude this section, let us address the issue of the transverse group velocity of the oscillations under study. We will define its contravariant components in the usual manner:

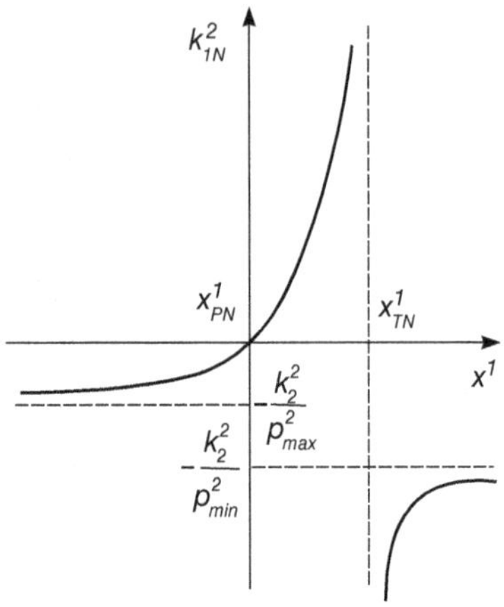

Figure 3.28 Distribution of the square of the wave-vector WKB component k^2_{1N} for Alfvén waves with $m \gg 1$, in the transverse coordinate x^1.

$v^i_N = \partial\omega/\partial k_i$. We have

$$v^1_N = \frac{\partial\omega_N(x^1, k_1/k_2)}{\partial k_1} = \frac{1}{k_2}\frac{\partial\omega_N}{\partial\kappa} = \frac{1}{k_2}\left(\frac{\partial\kappa_N}{\partial\omega}\right)^{-1} = \left(\frac{\partial k_{1N}}{\partial\omega}\right)^{-1}. \tag{3.129}$$

To obtain the expression for $\partial k_{1N}/\partial\omega$, we will differentiate (3.119) with respect to ω. Given

$$\frac{\partial\hat{L}}{\partial\omega} = 2k_{1N}\frac{\partial k_{1N}}{\partial\omega}\hat{L}_T(\omega) + 2\omega\frac{q_N}{v_A^2},$$

we obtain the equation

$$\hat{L}\frac{\partial R_N}{\partial\omega} + 2k_{1N}\frac{\partial k_{1N}}{\partial\omega}\hat{L}_T(\omega)R_N + 2\omega\frac{q_N}{v_A^2}R_N = 0.$$

Left-multiply this equation by R_N and integrate along the field line. Thanks to the operator $\hat{L}$ being Hermitian and the normalisation condition (3.119), we obtain

$$v^1_N = \left(\frac{\partial k_{1N}}{\partial\omega}\right)^{-1} = -\frac{k_{1N}}{\omega}\oint R_N\hat{L}_T(\omega)R_N dl$$
$$= \frac{k_{1N}}{\omega}\oint p\left[\left(\frac{\partial R_N}{\partial l}\right)^2 - \frac{\omega^2}{v_A^2}R_N^2\right] dl. \tag{3.130}$$

The following equation is obtained in a similar manner:

$$v^2_N = -\frac{k_2}{\omega}\oint R_N\hat{L}_P(\omega)R_N dl = \frac{k_2}{\omega}\oint\frac{1}{p}\left[\left(\frac{\partial R_N}{\partial l}\right)^2 - \frac{\omega^2}{v_A^2}R_N^2\right] dl. \tag{3.131}$$

We will use the relation

$$\frac{\omega^2}{v_A^2}R_N^2 = -\frac{R_N}{q_N}\frac{\partial}{\partial l}q_N\frac{\partial R_N}{\partial l},$$

to reduce the expressions for group velocities to

$$v^i_N = \frac{\tilde{k}^i k_1 k_2}{\omega}\oint\left(\frac{p'/p}{q_N}\right)' R_N^2 dl. \tag{3.132}$$

The notations are $\tilde{k}^1 = k_2$, $\tilde{k}^2 = -k_1$.

In the neighbourhood of the resonance surfaces, these formulas are simplified. Near the poloidal surface,

$$v_N^1 = \frac{k_{1N} w_{PN}}{k_2^2 \omega}, \quad v_N^2 = -\frac{k_{1N}^2 w_{PN}}{k_2^3 \omega}, \tag{3.133}$$

while near the toroidal surface,

$$v_N^1 = \frac{k_2^2 w_{TN}}{k_{1N}^3 \omega}, \quad v_N^2 = \frac{k_2 w_{TN}}{k_{1N}^2 \omega}. \tag{3.134}$$

It is evident from these relations and (3.127) that the group velocity tends to zero at the poloidal surface according to the law

$$v_N^1 \sim (x^1 - x_{PN}^1)^{1/2}, \quad v_N^2 \sim x^1 - x_{PN}^1, \tag{3.135}$$

and, on the toroidal surface, according to the law

$$v_N^1 \sim (x_{TN}^1 - x^1)^{3/2}, \quad v_N^2 \sim x_{TN}^1 - x^1. \tag{3.136}$$

It follows from (3.132) that

$$k_{1N} v_N^1 + k_2 v_N^2 = 0, \tag{3.137}$$

i.e. the transverse group velocity v_N^i is perpendicular to the phase gradient $k_i = \nabla_i \vartheta$ and directed along the characteristic (line of constant phase) (see Figure 3.29). Using the concept of group velocity, it is possible to introduce a new variable, τ, the time it takes the wave to travel along the characteristic line. Let us set

$$d\tau = \frac{dx^1}{v_N^1} = \frac{dx^2}{v_N^2}, \tag{3.138}$$

implying that the differentials of the coordinates dx^1 and dx^2 are taken along the characteristic line, i.e. are related by $k_{1N} dx^1 + k_2 dx^2 = 0$. Let us assume that the wave propagation duration τ is counted from the moment the wave is generated at the poloidal resonance surface. Hence,

$$\tau = \int_{x_{PN}^1}^{x^1} \frac{dx^{1\prime}}{v_N^1}. \tag{3.139}$$

Using the relations (3.135) and (3.136), it is easy to verify that the integral (3.139) converges at the lower limit and diverges when $x^1 \to x_{TN}^1$. This means that the time required for the wave to approach the toroidal surface is infinite.

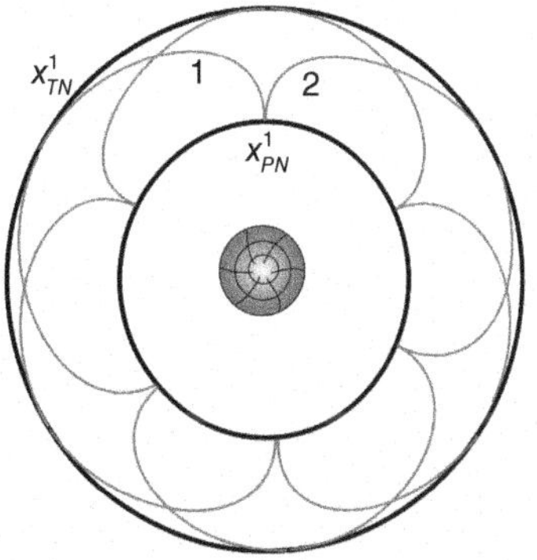

Figure 3.29 Lines of constant phase $\vartheta(x^1, x^3) = \text{const}$ (characteristic lines) in the transverse section of an axisymmetric magnetosphere. Curves 1 correspond to $k_2 > 0$, and curves 2 to $k_2 < 0$. The circles are the transverse sections of the resonance surfaces: the inner circle is the poloidal ($x^1 = x_{PN}^1$), the outer the toroidal surface ($x^1 = x_{TN}^1$).

It follows from (3.138) that

$$v_N^1 = \frac{dx^1}{d\tau}, \qquad v_N^2 = \frac{dx^2}{d\tau},$$

where the differentials of the coordinates are also assumed to be taken along the characteristic line. We will introduce the full derivative of the function of the coordinates x^1, x^2 determined by

$$\frac{\partial}{\partial \tau} = v_N^1 \frac{\partial}{\partial x^1} + v_N^2 \frac{\partial}{\partial x^2}. \tag{3.140}$$

The relation (3.137) can then be represented as

$$\frac{\partial \tilde{\vartheta}}{\partial \tau} = 0,$$

i.e. the wave phase is constant along the characteristic line.

3.7.3 Structure of High-*m* Alfvén Waves Along Magnetic Field Lines

Solving the longitudinal equation (3.119) for any realistic models of geomagnetic field and plasma is undoubtedly a numerical problem. An analytical WKB method can however be used along the parallel coordinate l, for harmonics with large numbers N. Let us consider the two solution methods and compare their results. We will start with the WKB method.

Assume that $R_N(l) = \exp[iS(l)]$, where $S(l)$ is the large quasi-classical phase along the longitudinal coordinate. Equation (3.119) produces, in the WKB approximation, an equation for the phase

$$-S'^2 + iS'' + i(\ln q)'S' + \omega^2/v_A^2 = 0.$$

Hereafter, the prime denotes an l derivative. It will be shown below that, to warrant the desired accuracy of the result, it is necessary to keep the first three terms in the asymptotical decomposition of the phase S

$$S = S_0 + S_1 + S_2 + \cdots$$

In the main (zero) order,

$$-S_0'^2 + \omega^2/v_A^2 = 0; \quad S_0(l) = \pm\omega \int \frac{dl}{v_A(l)}.$$

In the next (first) order,

$$-2S_0'S_1' + iS_0'' + i(\ln q)'S_0' = 0,$$

whence

$$S_1 = -i \ln C + \frac{i}{2} \ln \frac{q}{v_A}, \quad e^{iS_1} = C\left(\frac{v_A}{q}\right)^{1/2},$$

where C is an arbitrary constant. In the second order,

$$-2S_0'S_2' - S_1'^2 + iS_1'' + i(\ln q)'S_1' = 0.$$

Simple transformations produce

$$S_2 = \pm\frac{1}{8\omega} \int v_A[(\ln v_A)'^2 - (\ln q)'^2 + 2(\ln v_A)'' - 2(\ln q)'']dl'.$$

We will denote $\overline{S} = S_0 + S_2$. The general solution of (3.119) can then be written as

$$R_N = \left(\frac{v_A}{q}\right)^{1/2} (c_+ e^{i\overline{S}} + c_- e^{-i\overline{S}}) = \left(\frac{v_A}{q}\right)^{1/2} (c_1 \sin \overline{S} + c_2 \cos \overline{S}).$$

The boundary condition (3.119) implies $c_2 = 0$, whence the quantisation condition

$$\omega \oint \frac{dl}{v_A} + \frac{1}{8\omega} \oint v_A[(\ln v_A)'^2 - (\ln q)'^2 + 2(\ln v_A)'' - 2(\ln q)'']dl = 2\pi N.$$

follows. Assuming large N and using the iteration method to solve this equation yields

$$\omega = \omega_N \equiv \frac{2\pi N}{t_A} - \frac{1}{16\pi N} \oint v_A[(\ln v_A)'^2 - (\ln q)'^2 + 2(\ln v_A)'' - 2(\ln q)'']dl, \tag{3.141}$$

where

$$t_A = t_A(x^1) = \oint \frac{dl}{v_A(x^1, l)} \tag{3.142}$$

is the forward-and-backward transit time at local Alfvén speed along a field line between the magnetoconjugated ionospheres. Since

$$(\ln q)' = \frac{p^2\kappa - 1}{p^2\kappa + 1}(\ln p)', \tag{3.143}$$

equality (3.141) defines $\omega = \omega_N$ as a function of parameter κ. It can also be treated as an equation defining function $\kappa = \kappa_N(\omega)$. It is evident from (3.141) that mode dispersion is only manifest in the second order of the WKB approximation. With respect to the eigenmodes under study, it is sufficient to restrict ourselves to these two orders. We will use the normalisation condition (3.119) to determine the constant C and obtain

$$R_N = \left(\frac{2v_A}{qt_A}\right) \sin\left(\frac{2\pi N}{t_A} \int_{l_-}^{l} \frac{dl'}{v_A}\right). \tag{3.144}$$

Equalities (3.141) and (3.143) imply relations

$$\Omega_{PN} \equiv \frac{2\pi N}{t_A} + \frac{1}{16\pi N} \oint v_A[(\ln p)'^2 - (\ln v_A)'^2 - 2(\ln p)'' - 2(\ln v_A)'']dl, \tag{3.145}$$

$$\Omega_{TN} \equiv \frac{2\pi N}{t_A} + \frac{1}{16\pi N} \oint v_A[(\ln p)'^2 - (\ln v_A)'^2 + 2(\ln p)'' - 2(\ln v_A)'']dl \tag{3.146}$$

as special cases, and we have, from (3.144):

$$\begin{aligned} P_N &= \left(\frac{2pv_A}{t_A}\right)^{1/2} \sin\left(\frac{2\pi N}{t_A} \int_{l_-}^{l} \frac{dl'}{v_A}\right), \\ T_N &= \left(\frac{2v_A}{pt_A}\right)^{1/2} \sin\left(\frac{2\pi N}{t_A} \int_{l_-}^{l} \frac{dl'}{v_A}\right). \end{aligned} \tag{3.147}$$

For the polarisation splitting of the spectrum in (3.145) and (3.146), we have

$$\Delta\Omega_N = \frac{1}{4\pi N} \oint v_A(\ln p)'' dl. \tag{3.148}$$

The same formula results from (3.121) upon substituting (3.147). Analogously, from (3.125) and (3.126) we obtain

$$w_{PN} = \frac{1}{t_A} \oint v_A p p'' dl, \quad w_{TN} = \frac{1}{t_A} \oint v_A \frac{1}{p} \left(\frac{1}{p} \right)'' dl. \tag{3.149}$$

Note that expressions (3.125) and (3.126) for k_{1N}^2 can be obtained from (3.141) by expanding it, in the former case, with respect to small parameters $\omega - \Omega_{PN}$ and κ^2, and, in the latter case, with respect to parameters $\omega - \Omega_{TN}$ and κ^{-2}. The constants $w_{(P,T)N}$ are directly obtained in the form (3.149). Finally, let us use the approximation in question to write the expression for the group velocity:

$$v_N^i = \frac{\tilde{k}^i k_1 k_2}{2\pi N} \oint \left(\frac{p'/p}{q_N} \right)' \frac{v_A}{q_N} dl. \tag{3.150}$$

Let us now address the numerical solution of the above-obtained equations. We will restrict ourselves to a relatively simple model magnetosphere. We will assume a dipole geomagnetic field. In this case, we have for parameter p:

$$p = \sqrt{g_2/g_1} = a(1 + 3\sin^2\theta)^{1/2}, \tag{3.151}$$

where a is the equatorial radius of the field line, θ is latitude counted from the equator (see Figure 3.2). The length element along the field line is given by

$$dl = a\cos\theta(1 + 3\sin^2\theta)^{1/2} d\theta.$$

The latitudes of the points where the field line crosses the ionospheres of the conjugated hemispheres are $\theta_+ = \theta^*$ and $\theta_- = -\theta^*$, where

$$\theta^* = \arccos\sqrt{r_i/a},$$

r_i is the radius of the upper boundary of the ionosphere (as counted from the Earth centre). Section 2.18 explains why, for the basic harmonics of standing Alfvén waves, this boundary should be chosen at a height of 1000–2000 km from Earth surface, where the Alfvén speed ceases to increase rapidly with height. Using the variables a and θ as coordinates in the meridional plane, the geomagnetic field strength can be represented as

$$B(a,\theta) = B_0(a_0/a)^3 v(\theta); \qquad v(\theta) = (1 + 3\sin^2\theta)^{1/2}/\cos^6\theta.$$

Here, B_0 is the geomagnetic field strength in the equatorial plane, at certain magnetic shell $a = a_0$. To describe the Alfvén speed in the meridional plane, we will use the following model function:

$$v_A(a,\theta) = v_{A0}(a_0/a)^\mu [v(\theta)]^\nu. \tag{3.152}$$

Selecting appropriate values of the constants v_{A0}, μ and ν, this expression can be used to simulate a wide class of Alfvén speed distributions in the magnetosphere. Using the (3.152) model is restricted by magnetospheric regions with monotonous v_A variation across magnetic shells. In particular, this model fails to account for the presence of the plasmapause. Let us carry out numerical calculations for the following values of constants $a_0 = 4R_E = 2.5 \times 10^4$ km, $r_i = 7.9 \times 10^3$ km, $v_{A0} = 10^3$ km/s, $\mu = 3/2$, and $\nu = 1/4$. The calculation results are displayed in Figures 3.30–3.33.

Figure 3.30 illustrates the longitudinal structure of the first three harmonics of poloidal and toroidal standing Alfvén waves of unit amplitude, at magnetic shell $L = 6.6$. The role

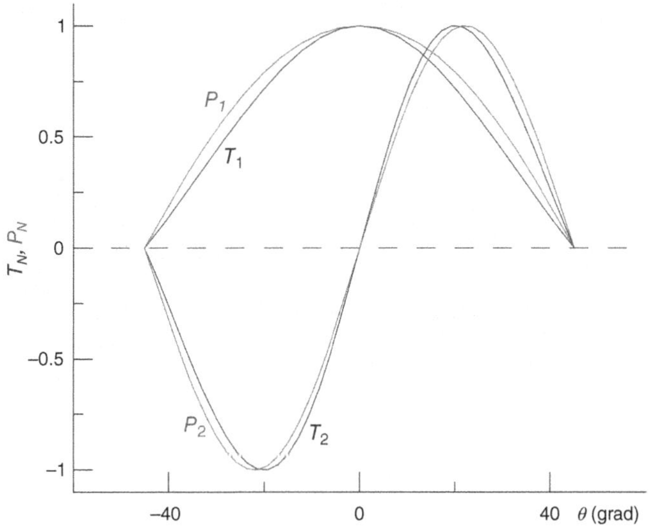

Figure 3.30 Structure of the first two harmonics ($N = 1, 2$) of standing toroidal (dark lines) and poloidal (light lines) Alfvén waves, along field-lines, at magnetic shell $L = 6.6$.

of the longitudinal coordinate is played in this figure by magnetic latitude θ, as counted from the equatorial plane. Evidently, the difference between the longitudinal structure of the oscillations in question is only minimal.

Figure 3.31 shows the eigenfrequency $\Omega_{(P,T)N}$ distribution for the first two harmonics of standing poloidal and toroidal Alfvén waves, over various magnetic shells. What catches our attention is the fact that polarisation splitting of the spectrum is large enough for the basic harmonic only (relation $\Delta\Omega_1/\Omega_{T1} = (\Omega_{T1} - \Omega_{P1})/\Omega_{T1} \approx 0.2$ at magnetic shell $L = 2$). It is negligibly small for the higher harmonics ($\Delta\Omega_2/\Omega_{T2} \approx 0.02$; $\Delta\Omega_3/\Omega_{T3} \approx 0.005\ldots$). It could be expected from the WKB formula (3.148) that the splitting value could only decrease severalfold.

Strictly speaking, the WKB approximation is however inapplicable to the basic harmonics, but generally does also produce the correct result for them on order of magnitude. We thus arrive at an important conclusion. A large enough value of the difference $\Delta\Omega_N$ necessary for the theory in question to be successfully applied is always reached for the

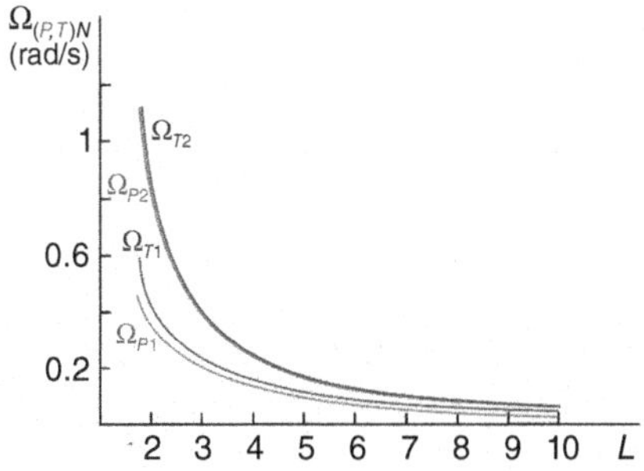

Figure 3.31 Poloidal Ω_{PN} and toroidal Ω_{TN} eigenfrequencies ($N = 1, 2$) vs. magnetic shell (the McIlwain) parameter $L = a/R_E$.

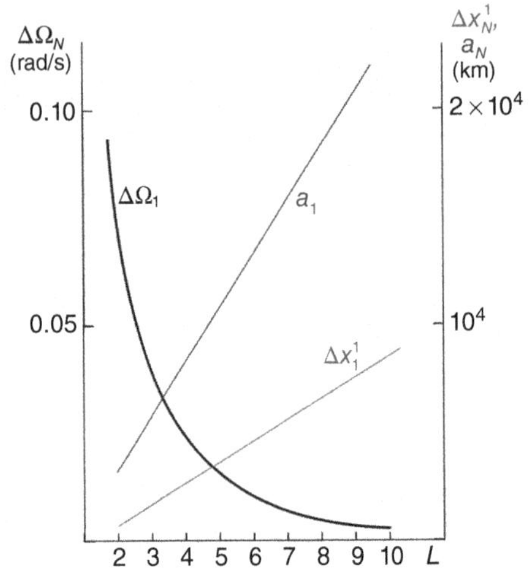

Figure 3.32 The polarisation splitting $\Delta\Omega_1$ of the spectrum, equatorial splitting Δx^1_1 of resonance surfaces and the characteristic scale a_1 of the transverse inhomogeneity of the Alfvén speed for the basic harmonic of Alfvén oscillations ($N = 1$) vs. magnetic shell parameter $L = a/R_E$.

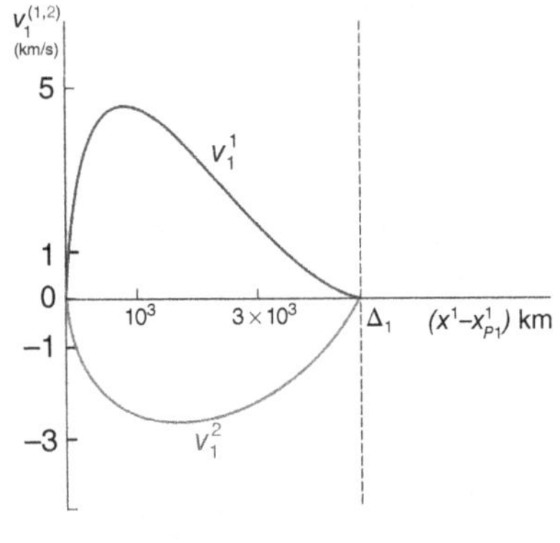

Figure 3.33 The group velocity components v^1_1 and v^2_1 for the main harmonic of standing Alfvén waves vs. the radial coordinate in the equatorial plane ($\theta = 0$), the toroidal resonance surface $x^1_{T1} = x^1_{P1} + \Delta_1$ for the waves is at magnetic shell $L = 6.6$.

basic harmonic ($N = 1$), but may not be reached for the $N > 2$ harmonics. Note that this feature only concerns Alfvén oscillations in a 'cold' plasma. In model magnetospheres with a 'warm' or moving plasma [321, 324], as well as featuring a magnetic field shear [325], the polarisation splitting of the spectrum is significant for other basic harmonics ($N \sim 1$) of standing Alfvén waves as well. This is due to a significant change in the eigenfrequency of poloidal Alfvén oscillations at the poloidal resonance shell, while the toroidal frequency and the location of the toroidal resonance surface remain practically unchanged.

Let us define a_N as

$$\frac{1}{2a_N} = \frac{d \ln \sqrt{\Omega_{PN}\Omega_{TN}}}{dx^1} = \frac{1}{2}\left(\frac{d \ln \Omega_{PN}}{dx^1} + \frac{d \ln \Omega_{TN}}{dx^1}\right). \tag{3.153}$$

By order of magnitude, $a_N \sim a$. It depends little on N. In contrast, $\Delta_N = 2\alpha_N a_N$ does strongly depend on N. The characteristic equatorial dependencies of these parameters and polarisation splitting $\Delta\Omega_1$ of the spectrum for the basic harmonic ($N = 1$) on the magnetic shell parameter are shown in Figure 3.32. The width Δ_1 of the transparent region is large enough (10^3 to 10^4 km in the magnetosphere) for the basic mode, $N = 1$. Mapped onto the ionosphere along magnetic field lines, this results in several hundred kilometres. The value of Δ_N is less than a hundred kilometres for $N > 1$ harmonics.

Analysing the definitions (3.125) and (3.126) for w_{PN} and w_{TN} can produce the following estimates:

$$w_{PN} \sim \alpha_N v_{A0}^2, \qquad w_{TN} \sim \alpha_N v_{A0}^2/a^4, \tag{3.154}$$

where v_{A0} is the characteristic value of the Alfvén speed in the magnetosphere, at the magnetic shell under study. An analogous estimate of the characteristic equatorial values of the physical components of the group velocity,

$$\hat{v}_N^1 \sim \hat{v}_N^2 \sim \alpha_N v_{A0}/m \tag{3.155}$$

follows from (3.130) and (3.131), which agrees well enough with the plots in Figure 3.33. Numerical simulation of the propagation of a narrow packet of poloidal (initially) Alfvén waves also confirms such a dependence of their transverse group velocity on the radial coordinate [326].

3.7.4 Dissipation of Standing Alfvén Waves in the Ionosphere

To account for dissipation of Alfvén wave energy in the ionosphere, it is necessary to keep the first term in the right-hand side of the boundary condition (2.309). Instead of the problem (3.112) and (3.113), we then arrive at the following eigenvalue problem:

$$\hat{L}(x^1, k_1, k_2, \omega)H = 0, \quad H|_{l_\pm} = \mp i(v_{p\pm}/\omega)(\partial H/\partial l)|_{l_\pm}. \tag{3.156}$$

Its solutions will differ from the above eigenvalues k_{1N} in certain corrections δk_{1N}:

$$k_1 = k_{1N} + \delta k_{1N}.$$

The value δk_{1N} is known to be related to the local decrement $\gamma_N = \gamma_N(x^1, \omega)$ of the mode as follows:

$$\delta k_{1N} = i\gamma_N/v_N^1. \tag{3.157}$$

Indeed, the correction to the quasi-classical phase can in this case be written as

$$\delta\tilde{\vartheta} = \int_{x_{PN}^1}^{x^1} \delta k_{1N} dx^{1'} \equiv i\Gamma(x^1), \tag{3.158}$$

where the following notation is introduced:

$$\Gamma(x^1) = \int_{x_{PN}^1}^{x^1} \frac{\gamma_N}{v_N^1} dx^{1'} = \int_0^\tau \gamma_N d\tau'. \tag{3.159}$$

The last equality makes use of the definition (3.138). Thus, in accordance with (3.111), we have the factor

$$\exp(i\delta\tilde{\vartheta}) = \exp(-\Gamma),$$

describing the decay of the wave as it propagates along the characteristic line. Note that, approaching this issue strictly formally, the equality (3.157) must be regarded as the definition of the decrement γ_N.

To actually define the correction δk_{1N}, let us use perturbation theory. Let $H = f_N(R_N + h_N)$, where f_N is a constant independent of l. We will linearise the problem (3.156) with respect to small parameters δk_{1N} and h_N:

$$\begin{aligned} \hat{L}(x^1, k_1, k_2, \omega)h_N + \delta k_{1N}\frac{\partial \hat{L}(x^1, k_1, k_2, \omega)}{\partial k_1}R_N = 0, \\ h_N|_{l_\pm} = \mp i\frac{v_{p\pm}}{\omega}\left.\frac{\partial R_N}{\partial l}\right|_{l_\pm}. \end{aligned} \tag{3.160}$$

We will multiply the first of these relations by R_N and integrate along the field-line. The result will be

$$\delta k_{1N} = -\oint R_N \hat{L}\, h_N dl \Big/ \oint R_N \frac{\partial \hat{L}}{\partial k_1} R_N dl. \tag{3.161}$$

Since $\partial \hat{L}/\partial k_1 = 2k_1\hat{L}_T$, we then have, in accordance with (3.130),

$$\oint R_N \frac{\partial \hat{L}}{\partial k_1} R_N dl = 2k_1 \oint R_N \hat{L} R_N dl = -2\omega v_N^1.$$

Comparing (3.157)–(3.161) produces this equality:

$$\oint R_N \hat{L}\, h_N dl = 2i\omega\gamma_N. \tag{3.162}$$

Transforming the left-hand side by partial integration and making use of the boundary condition yields the following expression for the decrement:

$$\gamma_N = \frac{1}{\omega^2}\left[q_N^+ v_{p+}\left(\frac{\partial R_N}{\partial l}\right)_+^2 + q_N v_{p-}\left(\frac{\partial R_N}{\partial l}\right)_-^2\right]. \tag{3.163}$$

The notations are $q_N^\pm = q_N(l_\pm) = p_\pm k_{1N}^2 + p_\pm^{-1}k_2^2$, where $p_\pm = p(l_\pm)$. Combined with equality (3.157), this expression defines the correction δk_{1N}. Note also that the local decrement γ_N can be regarded as a correction to the eigenfrequency (3.120) related to decay in the ionosphere: $\omega = \omega_N - i\gamma_N$. In this case, the decrement should be regarded as a function of variables x^1 and κ: $\gamma_N = \gamma_N[x^1, \omega_N(x^1, \kappa)]$. Near the poloidal surface, we have, from (3.163),

$$\gamma_N \equiv \gamma_{PN}(\omega) = \frac{1}{\omega^2}\left[\frac{v_{p+}}{p_+}\left(\frac{\partial P_N}{\partial l}\right)_+^2 + \frac{v_{p-}}{p_-}\left(\frac{\partial P_N}{\partial l}\right)_-^2\right], \tag{3.164}$$

and near the toroidal surface,

$$\gamma_N \equiv \gamma_{TN}(\omega) = \frac{1}{\omega^2}\left[p_+ v_{p+}\left(\frac{\partial T_N}{\partial l}\right)_+^2 + p_- v_{p-}\left(\frac{\partial T_N}{\partial l}\right)_-^2\right]. \tag{3.165}$$

For large N, when the longitudinal WKB approximation formulas are applicable, we substitute (3.144) into (3.163) to obtain

$$\gamma_N = \frac{2}{t_A}\left[\frac{v_{p+}}{v_{A+}} + \frac{v_{p-}}{v_{A-}}\right], \tag{3.166}$$

where $v_{A\pm} = v_A(l_\pm)$ is the Alfvén speed value at the upper ionospheric boundary. In this approximation, γ_N is effectively independent of N. Expression (3.166) can be used to estimate γ_N. The typical values are: $v_{A\pm} = (3\times10^3 - 3\times10^4)$ km/s, $v_{p\pm} \sim 10^2$ km/s , for the dayside ionosphere; and $v_{p\pm} \sim 10^3$ km/s, for the nightside ionosphere.

Hence, $v_{p\pm}/v_{A\pm} \sim 0.03$–$0.003$ for the dayside ionosphere and $v_{p\pm}/v_{A\pm} \sim 0.3$–$0.03$ for the nightside ionosphere. Parameters $v_{p\pm}/v_{A\pm}$ being small would guarantee a weak decay of the mode, i.e. small decrement γ_N compared to the difference of neighbouring eigenfrequencies (say ω_N and ω_{N+1}), which are equal to a/t_A on the order of magnitude. Expression (3.166) adequately corresponds to the spatial distribution of the decrement of geomagnetic pulsations observed in mid-latitudes [327].

3.7.5 Amplitude Distribution of High-m Alfvén Oscillations Across Magnetic Shells

In the main order of the WKB approximation in the x^1 coordinate, (3.112) and (3.113) determine the function $H = H_N$ accurate to an arbitrary factor that can depend on x^1 and ω. In other words, the solution in this order can be written as

$$H_N(x^1, l, \omega) = f_N(x^1, \omega) R_N(x^1, l, \omega). \tag{3.167}$$

Function f_N should be regarded as the amplitude of a standing wave at a given magnetic shell. The equation defining f_N is the solvability condition for the correction of the next order of the WKB approximation in the x^1 coordinate. Let us set

$$H = H_N + \tilde{h}_N = f_N(R_N + h_N). \tag{3.168}$$

Here, the next order correction $\tilde{h}_N$ is in the form $f_N h_N$, without loss of generality. Let us assume that the function h_N includes, not only corrections related to the next order of the WKB approximation, but also corrections due to decay in the ionosphere. This means that function h_N satisfies the boundary condition (3.160).

Substituting (3.168) into homogeneous equation (3.108) produces, in the order following the main order of the WKB approximation,

$$-f_N \hat{L} h_N + i[\nabla_1(k_1 f_N \hat{L}_T R_N) + k_1 \hat{L}_T \nabla_1(f_N R_N)] = 0.$$

Left-multiply this equation by $f_N R_N$ and integrate along the field line. Given the fact that operator $\hat{L}_T$ is Hermitian to functions that become zero when $l = l_\pm$, the following relation is obtained:

$$\nabla_1 k_1 f_N^2 \oint R_N \hat{L}_T R_N dl + i f_N^2 \oint R_N \hat{L}_T h_N dl = 0.$$

Finally, we use equalities (3.130), (3.131) and (3.162) to obtain

$$\nabla_1 v_N^1 f_N^2 = -2\gamma_N f_N^2. \tag{3.169}$$

The relation have a simple physical sense. Let us denote Alfvén oscillation energy density as w. It consists of the perturbed magnetic field energy and the kinetic energy of plasma particles:

$$w = \frac{B^2}{8\pi} + \frac{\rho_0 v_E^2}{2},$$

where $\mathbf{v}_E = \bar{c}[\mathbf{E}\mathbf{B}_0]/B_0^2$ is the velocity of the electric drift of plasma in the wave field. Substituting relations (3.118) and (3.167) produces

$$\begin{aligned} w &= \frac{1}{8\pi}\left(\frac{B_1 B_1^*}{g_1} + \frac{B_2 B_2^*}{g_2}\right) + \frac{\rho_0 \bar{c}^2}{2B_0^2}\left(\frac{E_1 E_1^*}{g_1} + \frac{E_2 E_2^*}{g_2}\right) \\ &= \frac{\bar{c}^2}{8\pi\omega^2}\frac{f_N^2}{\sqrt{g_\perp}}\left[q_N\left(\frac{\partial R_N}{\partial l}\right)^2 + q_N \frac{\omega^2}{v_A^2} R_N^2\right]. \end{aligned}$$

Let us now calculate the energy $\overline{w}$ of the oscillations contained in a thin flux tube of unit size, in the x^1 and x^2 coordinates (i.e. $dx^1 = 1$ and $dx^2 = 1$). The transverse section of such a tube, $\sigma = \sqrt{g_1 g_2} = \sqrt{g_\perp}$. We have

$$\overline{w} = \frac{1}{2}\oint w\sigma dl = \frac{\overline{c}^2}{16\pi} f_N^2 \oint \left[\frac{q_N}{\omega^2}\left(\frac{\partial R_N}{\partial l}\right)^2 + \frac{q_N}{v_A^2} R_N^2 \right] dl = \frac{\overline{c}^2}{8\pi} f_N^2.$$

On the other hand, we will integrate the transverse contravariant components of the Poynting vector

$$S^1 = \frac{\overline{c}}{4\pi}\frac{1}{\sqrt{g}} E_2^* B_3, \quad S^2 = -\frac{\overline{c}}{4\pi}\frac{1}{\sqrt{g}} E_1^* B_3$$

over the volume of the same flux tube to find the contravariant components of the transverse flux vector of the tube energy:

$$\overline{S}^i = \frac{1}{2}\oint S^i \sigma dl = \frac{\overline{c}^2}{8\pi} v_N^i f_N^2 = v_N^i \overline{w}. \tag{3.170}$$

Here, we use relations (3.118), (3.130), (3.131) and (3.167). It is evident from the formulas that equality (3.169) is the balance equation for the flux tube oscillation energy:

$$\nabla_i \overline{S}^i = -2\gamma_N \overline{w}.$$

This of course takes into account the fact that $\nabla_2 \overline{S}^2 = 0$ thanks to the axial symmetry. The solution of (3.169) is

$$f_N = C v_N^{-1/2} \exp(-\Gamma), \tag{3.171}$$

where C is a constant independent of x^1. It should be noted that the exponential part of this expression was actually defined in Section 3.7.4 (see (3.158)). Combining expressions (3.111), (3.117), (3.167) and (3.171) yields the solution in the two main orders of the WKB approximation:

$$\varphi(x^1, l, \omega) = \frac{C}{\sqrt{v_N^1}} \exp\left(i \int_{x_{PN}^1}^{x^1} k_{1N} dx^{1'} - \int_{x_{PN}^1}^{x^1} \frac{\gamma_N}{v_N^1} dx^{1'} \right) R_N(x^1, l, \omega). \tag{3.172}$$

It is evident from relations (3.135) that the integral in the exponential function of formula (3.171) converges at the lower limit. Consequently, the amplitude of f_N, due to the presence of a pre-exponential factor, has a singularity, $f_N \sim (x^1 - x_{PN}^1)^{-1/4}$, on the poloidal surface. Away from the toroidal surface, f_N decreases due to both increasing group velocity v_N^1, and decay in the ionosphere. On the toroidal surface, the pre-exponential factor in (3.171) becomes infinite, in accordance to the law $(x_{TN}^1 - x^1)^{-3/4}$. Since the integral in the exponential function also tends to infinity (due to the infinite time necessary for the wave to approach the toroidal surface), however, the full amplitude f_N tends to zero.

The behaviour of the amplitude will be different in the presence of instability in the Alfvén waves discussed here. If the influence of the instability mechanism is stronger than the decay in the ionosphere, the wave is more likely to grow than decay. Such a situation can be described if γ_N is assumed to be negative. The exponential factor in (3.172) then increases, from the poloidal to the toroidal surface, becoming infinite on the latter [328]. The instability of the oscillations in question can be related to e.g. the presence in the plasma of high-energy charged particles (see [328–335]). A detailed review of other types of instability in the waves in question can be found in [26]. A review of the experimental data on the generation of different classes of the ULF waves by kinetic instabilities is presented in [336].

To conclude this section, let us discuss the issue of the WKB-approximation applicability condition in the x^1 coordinate. The original form of this condition is known to be

$$\left|\frac{d}{dx^1}\frac{1}{k_1}\right| \ll 1. \tag{3.173}$$

To analyse this inequality, let us make use of the model expression

$$\hat{k}_1^2 = (\hat{k}_2^0)^2 \frac{x^1 - x^1_{PN}}{x^1_{TN} - x^1}. \tag{3.174}$$

Here, $\hat{k}^0_{1,2} = k_{1,2}/\sqrt{g_{1,2}(0)}$ are the physical components of the wave vector in the equatorial plane, $l = 0$. Apart from producing a correct qualitative behaviour of the $\hat{k}_1^2(x^1)$ function, the expression (3.174), is also correct in order of magnitude. It follows from expression (3.174) that the inequality (3.173) can only be satisfied if

$$\hat{k}_1 \hat{\Delta}_N \gg 1, \tag{3.175}$$

where $\hat{\Delta}_N = \sqrt{g_1}(x^1_{TN} - x^1_{PN})$ is the characteristic physical distance between the resonance surfaces (e.g. at the equatorial surface). Since $\hat{k}_1 \sim \hat{k}_2$ between the resonance surfaces, the condition (3.175) implies that many transverse wavelengths fit in between these surfaces. In order of magnitude, $\hat{k}_2^0 = m/a$, therefore the inequality (3.175) can be rewritten as $m \gg a/\hat{\Delta}_N \sim 1/\alpha_N$. For the basic harmonic ($N = 1$), this leads to a condition which effectively is no different from the originally accepted $m \gg 1$. Even for the next harmonic ($N = 2$), however, the condition for the azimuthal wave number becomes more strict: $m \gg 10$–10^2. We emphasise that this only refers to the 'cold' plasma model. In a 'warm' and moving plasma, as well as in a magnetic field with a shear, the condition $m \gg 1$ is sufficient for this theory to be applied to all basic harmonics ($N \sim 1$) of standing Alfvén waves.

Even if the inequality (3.175) is satisfied, the condition (3.173) is broken near the resonance surfaces. Assuming (3.175) to hold, it is easy to see that the WKB approximation is applicable if

$$|x^1 - x^1_{PN}| \gg \lambda_{PN}, \quad |x^1 - x^1_{TN}| \gg \lambda_{TN}, \tag{3.176}$$

where

$$\lambda_{PN} \sim \hat{\Delta}_N/(\hat{k}_2\hat{\Delta}_N)^{2/3}, \quad \lambda_{TN} \sim \hat{\Delta}_N/(\hat{k}_2\hat{\Delta}_N)^2. \tag{3.177}$$

It should be emphasised that $\lambda_{PN}, \lambda_{TN} \ll \hat{\Delta}_N$. Therefore, in the neighbourhood of the resonance surfaces, where $|x^1 - x^1_{PN}| \lesssim \lambda_{PN}$ and $|x^1 - x^1_{TN}| \lesssim \lambda_{TN}$, the formulas of the transverse WKB approximation are inapplicable. In particular, the conclusion that amplitude f_N becomes infinite (or tends to zero) on resonance surfaces proves to be wrong. To investigate the solutions in the neighbourhood of these surfaces, it is necessary to discard the WKB approximation and return to the (3.108).

3.7.6 Solution Near the Poloidal Resonance Surface

For perturbations depending on the x^2 coordinate as $e^{ik_2x^2}$, the homogeneous equation (3.108) takes the form

$$[\nabla_1 \hat{L}_T(\omega)\nabla_1 - k_2^2 \hat{L}_P(\omega)]\varphi = 0. \tag{3.178}$$

Let us find its solution near the poloidal surface using perturbation theory based on the fact that the desired solution is close to the poloidal mode. This implies

$$|\nabla_1/\sqrt{g_1}| \ll k_2/\sqrt{g_2}, \tag{3.179}$$

i.e. the first term in (3.178) is small. In the main order of perturbation theory, omitting this term and using the zeroth order approximation for the boundary condition yields

$$\hat{L}_P(\omega)\varphi = 0, \quad \varphi|_{l_\pm} = 0.$$

The solution of this eigenvalue problem is known:

$$\varphi = u_N P_N, \quad \omega^2 = \Omega_{PN}^2, \tag{3.180}$$

where u_N is a factor independent of l. Comparing (3.180) to (3.111) and (3.167), it can be seen that, in the region where the WKB approximation is applicable over the transverse coordinate,

$$u_N = (1/k_2) f_N \exp(i\vartheta). \tag{3.181}$$

In the main order of perturbation theory presented here, the factor u_N is not defined because the solution (3.180) is degenerate. The equation for u_N, as is usual in this case, is the solvability condition for the next-approximation correction.

Let us set

$$\varphi = u_N(x^1,\omega)P_N(x^1,l) + h_N$$

and linearise (3.178) assuming the first term of this equation, as well as function h_N and difference $\omega^2 - \Omega_{PN}^2$ to be small. Multiplying the resulting equality by P_N and integrating along the field-line yields

$$\begin{aligned}\nabla_1^2 u_N \oint P_N \hat{L}_T(\Omega_{PN})P_N dl - k_2^2(\omega^2 - \Omega_{PN}^2)U_N \\ -k_2^2 \oint P_N \hat{L}_P(\Omega_{PN})h_N dl = 0.\end{aligned} \tag{3.182}$$

To calculate the last summand in these relations, we will linearise the boundary condition (2.309):

$$h_N|_{l_\pm} = \mp i\frac{v_{p\pm}}{\omega}\frac{\partial P_N}{\partial l}\bigg|_{l_\pm} - \frac{J_\parallel^\pm}{V_{p\pm}}. \tag{3.183}$$

Here, the term with external currents in the ionosphere is finally taken into account. As will be shown below, it should actually be taken into account in the immediate proximity to the poloidal surface only. Integrating by parts and making use of (3.164) produces

$$\oint P_N \hat{L}_P(\Omega_{PN})h_N dl = 2i\gamma_N u_N + I_\parallel,$$

where

$$I_\parallel = 2\left[-\left(\frac{\partial P_N}{\partial l}\right)_{l+}\frac{J_\parallel^+}{P_+ V_{p+}} + \left(\frac{\partial P_N}{\partial l}\right)_{l-}\frac{J_\parallel^-}{P_- V_{p-}}\right]. \tag{3.184}$$

Taking notice also of the equality (3.125), we obtain from (3.182) the final equation for function u_N:

$$w_{PN}\nabla_1^2 u_N + k_2^2[(\omega + i\gamma_N)^2 - \Omega_{PN}^2]u_N = k_2^2 I_\parallel. \tag{3.185}$$

It should be emphasised that this is an inhomogeneous equation, therefore it defines the solution including the amplitude.

The coefficients in (3.185) are functions of the x^1 coordinate. The strongest dependence on x^1 is bracketed, before u_N. On the poloidal surface, the bracketed expression tends to zero. The other values, $w_{PN}, I_\parallel$ and γ_N, can as a rule be assumed constant in the region of interest. If the poloidal surface is not too close to the function $\Omega_{PN}(x^1)$ extremum, the expansion (3.123) is applicable near this surface. Hence,

$$(\omega + i\gamma_N)^2 - \Omega_{PN}^2 \approx \omega^2 \left(\frac{x^1 - x_{PN}^1}{a_N} + 2i\frac{\gamma_N}{\omega} \right).$$

Let us introduce a dimensionless variable

$$\xi_{PN} = \frac{x^1 - x_{PN}^1}{\lambda_{PN}},$$

where the constant λ_{PN} is defined by

$$\lambda_{PN} = \left(\frac{w_{PN} a_N}{k_2^2 \omega^2} \right)^{1/3}. \tag{3.186}$$

Equation (3.185) can be reduced to

$$\frac{d^2 u_N}{d\xi_{PN}^2} + (\xi_{PN} + i\varepsilon_{PN})u_N = \frac{a_N}{\lambda_{PN}} \frac{I_\parallel}{\omega^2}, \tag{3.187}$$

where

$$\varepsilon_{PN} = 2\frac{a_N}{\lambda_{PN}} \frac{\gamma_N}{\omega}.$$

It is easy to see that λ_{PN} has the same dimension as the x^1 coordinate and is the characteristic scale of the solution in this coordinate (for $\varepsilon_{PN} \lesssim 1$). Parameter ε_{PN} characterises the role of decay at the ionosphere. If $\varepsilon_{PN} \ll 1$, this role is insignificant, but if $\varepsilon_{PN} \gg 1$, this decay, conversely, defines the form of the solution, while the first term in (3.185) can in this case be omitted. Given (3.154), the following estimate is possible for the model magnetosphere described in Section 3.7.3:

$$\lambda_{PN} \sim \alpha_N^{1/3} a / m^{2/3}, \tag{3.188}$$

which fully complies with the definition (3.177). This estimate implies that the mode poloidality condition (3.179) is equivalent to the inequality $\hat{k}_2 \hat{\lambda}_{PN} \gg 1$, satisfied if $m \gg 1/\alpha_N$. We will assume the latter inequality to hold.

A full definition of the solution to (3.187) requires the boundary conditions to be set in the x^1 coordinate. In the opaque region, $\xi \to -\infty$, the condition that the solution be finite seems natural. In the transparent region, $\xi \to \infty$, we will require that the solution have the form of a travelling wave carrying the energy away from the poloidal surface. These conditions are based on the remarkable property of the solution near the toroidal surface. In Section 3.7.7 will demonstrate that wave incident to the toroidal surface is completely absorbed close to it. This means that there is no wave reflected from the toroidal surface and travelling to the poloidal surface.

The desired solution of (3.187) can be written in terms of the standard function $G(z)$, which satisfies the inhomogeneous Airy equation (2.88) and the above-mentioned boundary conditions. This function has an integral representation (2.89) and asymptotics (2.90). Important properties of function $G(z)$ become evident if it is compared, in the asymptotic region $z \to \infty$, to the solution of the homogeneous Airy equation

$$\overline{G}'' + z\overline{G} = 0.$$

Using the WKB approximation in the last equation, for $z \gg 1$, yields the general solution:

$$\overline{G}(z) = \frac{A}{z^{1/4}} \exp\left(i\frac{2}{3}z^{3/2}\right) + \frac{B}{z^{1/4}} \exp\left(-i\frac{2}{3}z^{3/2}\right).$$

Comparing this to (2.90) one can see that function $G(z)$ coincides, in the asymptotic region, with one of the solutions of the homogeneous equation. In other words, the presence of the right-hand side in (2.88) is only essential if $z \sim 1$, but can be omitted entirely in the $z \gg 1$ region, without affecting the form of the solution. It should be kept in mind, of course, that the amplitude and phase in the solution for the asymptotical region are defined by the right-hand side in the $z \sim 1$ region. Studies using Green's function demonstrate that this property of function $G(z)$ is due to the characteristic spatial scale of the solution decreasing when moving into the $z \gg 1$ region. The decreased scale of the solution is the main reason why the inhomogeneous term of the equation was neglected. Applying the function $G(z)$ properties to solving (3.187) below will find that the inhomogeneous term in the boundary condition (3.183) should only be taken into account in the $|x^1 - x^1_{PN}| \sim \lambda_{PN}$ region. This means neglecting it when using the WKB approximation in the x^1 coordinate.

Making use of function $G(z)$, the solution of (3.187) can be written as

$$u_N = \frac{a_N}{\lambda_{PN}} \frac{I_\parallel}{\omega^2} G(\xi_{PN} + i\varepsilon_{PN}) = \frac{a_N}{\lambda_{PN}} \frac{I_\parallel}{\omega^2} G\left(\frac{x^1 - x^1_{PN}}{\lambda_{PN}} + i\varepsilon_{PN}\right). \tag{3.189}$$

In accordance with (3.180), for the potential of a perturbed electric field,

$$\varphi_N = \frac{a_N}{\lambda_{PN}} \frac{I_\parallel}{\omega^2} G\left(\frac{x^1 - x^1_{PN}}{\lambda_{PN}} + i\varepsilon_{PN}\right) P_N(x^1, l). \tag{3.190}$$

Recall that these formulas are only applicable near the poloidal surface, for $|x^1 - x^1_{PN}| \ll \Delta_N$.

3.7.7 Solution Near the Toroidal Resonance Surface

Same as in Section 3.7.6, we will use perturbation theory to find the solution assuming it to be close to a toroidal mode. This means that the following inequality holds true:

$$|\nabla_1/\sqrt{g_1}| \gg k_2/\sqrt{g_2}, \tag{3.191}$$

which is opposite to (3.179), and the second term in brackets in (3.178) should be regarded as small. In the main order of perturbation theory, we have

$$\varphi = V_N T_N, \qquad \omega^2 = \Omega^2_{TN}, \tag{3.192}$$

where V_N is a factor independent of l. In the region where the WKB approximation in the x^1 coordinate is applicable,

$$V_N = (1/k_1)f_N \exp(i\vartheta). \tag{3.193}$$

Let us assume, in the next order, that

$$\varphi = V_N(x^1, \omega)T_N(x^1, l) + h_N.$$

Linearising the boundary condition produces

$$h_N|_{l_\mp} = \mp i \frac{v_{p\pm}}{\omega} \frac{\partial T_N}{\partial l}\bigg|_{l_\pm} V_N. \tag{3.194}$$

The term with external currents is omitted here because of the extremely small scale of the solution (see below). Let us linearise (3.178), multiply the resulting relation by T_N and integrate along the field line. The result will be

$$\oint T_N \hat{L}_T(\Omega_{TN})\nabla_1^2 h_N dl + \nabla_1(\omega^2 - \Omega_{TN}^2)\nabla_1 V_N$$
$$-k_2^2 V_N \oint T_N \hat{L}_P(\Omega_{TN})T_N dl = 0.$$

Integrating by parts, using relations (3.194) and formula (3.165) yields

$$\oint T_N \hat{L}_T(\Omega_{TN})\nabla_1^2 h_N dl = 2i\gamma_N \nabla_1^2 V_N.$$

Given the definition in (3.126), we obtain the desired equation for V_N:

$$\nabla_1[(\omega + i\gamma_N)^2 - \Omega_{TN}^2]\nabla_1 V_N - k_2^2 w_{TN} V_N = 0. \tag{3.195}$$

Let us find the solution of this equation for the case when the decomposition (3.123) can be applied to function $\Omega_{TN}(x^1)$. We will introduce a dimensionless variable,

$$\xi_{TN} = \frac{x^1 - x_{TN}^1}{\lambda_{TN}},$$

defining the constant λ_{TN} by this relation:

$$\lambda_{TN} = \frac{\omega^2}{k_2^2 w_{TN} a_N}. \tag{3.196}$$

Equation (3.195) can then be written as

$$\frac{d}{d\xi_{TN}}(\xi_{TN} + i\varepsilon_{TN})\frac{d}{d\xi_{TN}}V_N - V_N = 0, \tag{3.197}$$

where

$$\varepsilon_{TN} = 2\frac{a_N}{\lambda_{TN}}\frac{\gamma_N}{\omega}. \tag{3.198}$$

As in Section 3.7.6, λ_{TN} is the characteristic scale of the solution in the neighbourhood of the toroidal surface, while parameter ε_{TN} characterises the role of dissipation in the ionosphere, in this neighbourhood. For the model magnetosphere used here, we obtain, in the order of magnitude, given the estimate (3.154),

$$\lambda_{TN} \sim a/(\alpha_N m^2). \tag{3.199}$$

This relation also coincides with the definition (3.177). Using (3.199), the toroidality condition (3.191) again results in the inequality $m \gg 1/\alpha_N$.

Let us introduce the finite amplitude (as $z \to \infty$) function $g(z)$, satisfying the equation

$$(zg')' - g = 0$$

and bounded at $z \to \infty$. This function is expressed via the modified zero-order Hankel function:

$$g(z) = K_0(2z^{1/2}). \tag{3.200}$$

For $z \gg 1$, it has the following asymptotic representation:

$$g(z) \approx (\sqrt{\pi}/2)z^{-1/4}\exp(-2z^{1/2}), \tag{3.201}$$

yielding, for small z,

$$g(z) \approx -(1/2)\ln z. \tag{3.202}$$

The point $z = 0$ is a singular turning point. The behaviour of function $g(z)$ for negative z is determined by bypassing the singular point. In our case, this path, due to decay, is determined in accordance with the rule $z = \eta + i\varepsilon_{TN}$. This leads to the following asymptotic representation for $z \to -\infty$:

$$g(z) \approx (\sqrt{\pi}/2)(-z)^{-1/4}\exp[-2i(-z)^{1/2} - i\pi/4]. \tag{3.203}$$

The solution of (3.197) has the form

$$V_N = Dg(\xi_{TN} + i\varepsilon_{TN}) = Dg\left(\frac{x^1 - x^1_{TN}}{\lambda_{TN}} + i\varepsilon_{TN}\right), \tag{3.204}$$

where D is an arbitrary constant. Hence, in accordance with (3.192),

$$\varphi_N = Dg\left(\frac{x^1 - x^1_{TN}}{\lambda_{TN}} + i\varepsilon_{TN}\right)T_N(x^1, l). \tag{3.205}$$

This solution for the transparent region is a wave arriving at the toroidal surface and absorbed in its neighbourhood, at scale $|x^1 - x^1_{TN}| \sim \lambda_{TN}$. The reflected wave is completely absent. In the opaque region, the solution decreases exponentially when moving away from the toroidal surface. It should be emphasised that formulas (3.204) and (3.205) are applicable for $|x^1 - x^1_{TN}| \ll \Delta_N$.

3.7.8 Global Structure of High-*m* Alfvén Wave (Matching the Solutions for Different Regions)

To obtain a full description of the spatial structure of the mode, it is necessary to match the solutions obtained for different regions in x^1. It is convenient to use for this purpose the dimensionless function $r_N(x^1, l)$, determined by the equality

$$r_N = (q_N t_A/v_A)^{1/2} R_N.$$

This function satisfies the relation

$$\langle r_N^2 \rangle = 1,$$

where

$$\langle F \rangle = \frac{1}{t_A} \oint F(l) \frac{dl}{v_A} \tag{3.206}$$

is the field-line-averaged value of a function $F = F(l)$. Near the poloidal and toroidal surfaces, respectively,

$$r_N = (t_A/pv_A)^{1/2} P_N, \qquad r_N = (t_A p/v_A)^{1/2} T_N.$$

For large N, when the WKB approximation is applicable in the longitudinal coordinate l,

$$r_N = \sqrt{2} \sin\left(\frac{2\pi N}{t_A} \int_{l_-}^{l} \frac{dl'}{v_A}\right).$$

Making use of this definition, the solution (3.190) for the perturbed potential near the poloidal surface can be written as

$$\varphi_N = \tilde{\varphi} \left(\frac{pv_A}{p_0 v_{A0}}\right)^{1/2} G\left(\frac{x^1 - x^1_{PN}}{\lambda_{PN}} + i\varepsilon_{PN}\right) r_N(x^1, l), \tag{3.207}$$

where

$$\tilde{\varphi} = \left(\frac{p_0 v_{A0}}{t_A}\right)^{1/2} \frac{a_N}{\lambda_{PN}} \frac{I_\parallel}{\omega^2},$$

is the characteristic value of the perturbed potential. The '0' index denotes the equatorial values of the corresponding parameters. The solution (3.207) and the WKB approximation applicability ranges overlap in the transverse coordinate. In the overlap region $\lambda_{PN} \ll x^1 - x^1_{PN} \ll \Delta_N$, we have from (3.207), in accordance with (2.90),

$$\varphi_N = -\sqrt{\pi}\tilde{\varphi} \left(\frac{pv_A}{p_0 v_{A0}}\right)^{1/2} \left(\frac{\lambda_{PN}}{x^1 - x^1_{PN}}\right)^{1/4}$$
$$\times \exp\left[\frac{2}{3} i \left(\frac{x^1 - x^1_{PN}}{\lambda_{PN}}\right)^{3/2} - \varepsilon_{PN} \left(\frac{x^1 - x^1_{PN}}{\lambda_{PN}}\right)^{1/2} + i\frac{\pi}{4}\right] r_N. \tag{3.208}$$

On the other hand, we obtain from (3.127) and (3.133), given (3.186),

$$k_{1N} = \frac{(x^1 - x^1_{PN})^{1/2}}{\lambda_{PN}^{3/2}}, \qquad v^1_N = v^1_{PN} \left(\frac{x^1 - x^1_{PN}}{\lambda_{PN}}\right)^{1/2},$$

where $v^1_{PN} = \omega \lambda^2_{PN}/a_N$ is the characteristic value of the transverse group velocity in the neighbourhood of the poloidal surface. Hence,

$$\vartheta = \int_{x^1_{PN}}^{x^1} k_{1N} dx^{1\prime} = \frac{2}{3} \left(\frac{x^1 - x^1_{PN}}{\lambda_{PN}}\right)^{3/2},$$

$$\Gamma = \int_{x^1_{PN}}^{x^1} \frac{\gamma_N}{v^1_N} dx^{1\prime} = \varepsilon_{PN} \left(\frac{x^1 - x^1_{PN}}{\lambda_{PN}}\right)^{1/2}.$$

Comparing (3.172) to (3.208) in the overlap region, it can be seen that they completely coincide, functionally, in both the x^1 and l coordinates. This makes it possible to determine the constant C. Substituting these expressions into the general formula (3.172) produces the solution in the WKB approximation applicability region matched to the solution in the neighbourhood of the poloidal surface:

$$\begin{aligned}\varphi_N &= -\sqrt{\pi}\tilde{\varphi}\left(\frac{v^1_{PN}}{v^1_N}\frac{p^{-1}k_2^2}{pk_{1N}^2+p^{-1}k_2^2}\frac{pv_A}{p_0v_{A0}}\right)^{1/2} \\ &\times\exp\left(i\int_{x^1_{PN}}^{x^1}k_{1N}dx^{1'}-\int_{x^1_{PN}}^{x^1}\frac{\gamma_N}{v^1_N}dx^{1'}+i\frac{\pi}{4}\right)r_N.\end{aligned} \tag{3.209}$$

Analogously, the solution (3.209) is matched to the solution in the neighbourhood of the toroidal surface. In the region where they are both applicable, $\lambda_{TN} \ll x^1_{TN} - x^1 \ll \Delta_N$, we have, from (3.127) and (3.134), given the definition (3.196),

$$k_{1N} = \frac{1}{\lambda_{TN}^{1/2}(x^1_{TN}-x^1)^{1/2}}, \qquad v^1_N = v^1_{TN}\left(\frac{x^1_{TN}-x^1}{\lambda_{TN}}\right)^{3/2},$$

where $v^1_{TN} = \omega\lambda_{TN}^2/a_N$ is the characteristic value of v^1_N in the neighbourhood of the toroidal surface. Let us denote

$$\overline{\vartheta} = \int_{x^1_{PN}}^{x^1_{TN}} k_{1N}dx^1 \tag{3.210}$$

the full incursion of the quasiclassical phase between the resonance surfaces. It should be emphasised that the integral (3.210) converges at the upper limit. In order of magnitude, $\overline{\vartheta} \sim \hat{k}_2\Delta_N \sim m\alpha_N \gg 1$. Thanks to this definition, we have, for $0 \le x^1_{TN} - x^1 \ll \Delta_N$,

$$\vartheta(x^1) = \overline{\vartheta} - \int_{x^1}^{x^1_{TN}} k_{1N}dx^{1'} = \overline{\vartheta} - 2\left(\frac{x^1_{TN}-x^1}{\lambda_{TN}}\right)^{1/2}.$$

Unfortunately, the value of $\Gamma(x^1)$ cannot be calculated in the same manner, because the corresponding integral diverges for $x^1 \to x^1_{TN}$. Let us introduce an auxiliary coordinate, $\overline{x}^1$, located within the same limits: $0 < x^1_{TN} - \overline{x}^1 \ll \Delta_N$. It is then possible to write

$$\begin{aligned}\Gamma(x^1) &= \int_{x^1_{PN}}^{\overline{x}^1}\frac{\gamma_N}{v^1_N}dx^{1'} + \int_{\overline{x}^1}^{x^1}\frac{\gamma_N}{v^1_N}dx^{1'} \\ &= \Gamma(\overline{x}^1) - \varepsilon_{TN}\left(\frac{\lambda_{TN}}{x^1_{TN}-\overline{x}^1}\right)^{1/2} + \varepsilon_{TN}\left(\frac{\lambda_{TN}}{x^1_{TN}-x^1}\right)^{1/2}.\end{aligned}$$

It is easy to verify that the value of

$$\overline{\Gamma} = \Gamma(\overline{x}^1) - \varepsilon_{TN}\left(\frac{\lambda_{TN}}{x^1_{TN}-\overline{x}^1}\right)^{1/2} \tag{3.211}$$

is effectively independent of the coordinate $\overline{x}^1$ if the latter is within the above limits. Thus, near the toroidal surface,

$$\Gamma(x^1) = \overline{\Gamma} + \varepsilon_{TN}\left(\frac{\lambda_{TN}}{x^1_{TN}-x^1}\right)^{1/2}.$$

The relations allow the WKB approximation formula (3.209), near the toroidal surface, to be written as

$$\varphi_N = -\sqrt{\pi}\hat{k}_2^0\hat{\lambda}_{PN}^0\tilde{\varphi}\left(\frac{\lambda_{TN}}{x_{TN}^1 - x^1}\right)^{1/2}\left(\frac{p_0 v_A}{p v_{A0}}\right)^{1/2}$$
$$\times\exp\left[i\left(\overline{\vartheta} + \frac{\pi}{4}\right) - \overline{\Gamma} - 2i\left(\frac{x_{TN}^1 - x^1}{\lambda_{TN}}\right)^{1/2} - \varepsilon_{TN}\left(\frac{\lambda_{TN}}{x_{TN}^1 - x^1}\right)^{1/2}\right] r_N. \tag{3.212}$$

Using the asymptotic representation (3.201) it can be ascertained that the solution (3.205) for $x_{TN}^1 - x^1 \gg \lambda_{TN}$ functionally coincides with solution (3.212). Comparing them, it is possible to determine the constant D. After that, (3.205) takes the form

$$\varphi_N = -2i\hat{k}_2^0\hat{\lambda}_{PN}^0 e^{i\overline{\vartheta}-\overline{\Gamma}}\tilde{\varphi}\left(\frac{p_0 v_A}{p v_{A0}}\right)^{1/2} g\left(\frac{x^1 - x_{TN}^1}{\lambda_{TN}} + i\varepsilon_{TN}\right) r_N. \tag{3.213}$$

Formulas (3.207), (3.209) and (3.213) combined describe the perturbed potential of the mode throughout the range of its existence, except the intervals deep within the opaque regions, i.e. $x^1 - x_{TN}^1 \gg \lambda_{TN}$ and $x_{PN}^1 - x^1 \gg \lambda_{PN}$. The solution in them is given by the WKB approximation (asymptotical values k_{1N} are determined by relations (3.128)). We do not write out the corresponding formulas, because the mode amplitude in these regions is negligibly small, i.e. oscillation is effectively absent in them.

The full structure of monochromatic standing Alfvén wave with $m \gg 1$ is qualitatively shown in Figure 3.34. The monochromatic source (external currents in the ionosphere) excites a poloidal standing Alfvén wave (with dominant electromagnetic-field components E_2 and B_1) at poloidal resonance surface, $x^1 = x_{PN}^1$. The wave travels across magnetic shells to the toroidal resonance surface $x^1 = x_{TN}^1$, where it is completely absorbed thanks to its energy dissipating in the ionospheric conducting layer. The wave remains standing along

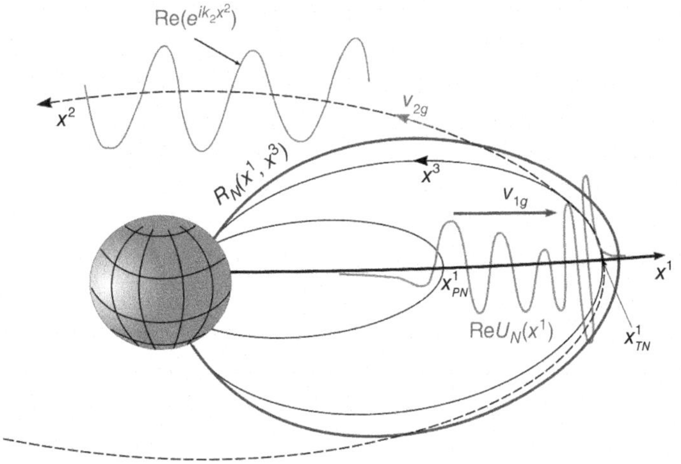

Figure 3.34 Schematic representation of the structure of standing Alfvén wave with $m \gg 1$ in a dipole-like magnetosphere. Function $R_N(x^1, x^3)$ describes the wave structure along magnetic field lines, and $U_N(x^1)$ across magnetic shells.

magnetic field lines as it propagates across magnetic shells, but its polarisation gradually changes, becoming toroidal as it approaches the toroidal surface (with dominant components E_1 and B_2).

3.8 Electromagnetic Oscillations Induced at Earth Surface by Magnetospheric Standing High-*m* Alfvén Waves

If the characteristic transverse scales of the Alfvén waves are much larger than the atmosphere thick they keep their spatial structure intact when the oscillations penetrate from the magnetosphere to ground [185, 206]. The polarisation ellipse turns in the process due to Hall currents excited in the ionosphere, which generate a FMS wave, its wave field penetrate to ground (see Section 2.18). If the characteristic transverse wavelength is smaller than the thickness of the atmosphere, the wave's structure expands to a scale comparable to the thickness of the atmosphere when it penetrate to ground.

Let us now see how the field of transverse-small-scale Alfvén wave with $m \gg 1$ penetrates from the magnetosphere to ground. The condition (2.311), certainly valid for the waves in question, allows us to neglect the direct effect of èxternal currents in the ionosphere on the electromagnetic oscillations at Earth surface. This is explained by the fact that Alfvén waves are 'tied' to a certain resonance magnetic shell. Currents in the ionosphere drive Alfvén waves on it, which induce electromagnetic oscillations at Earth surface, with much larger amplitudes than the oscillations directly related to the currents. As a result, it becomes possible to use for the electromagnetic oscillations induced at Earth surface ($z = 0$) the asymptotic expressions (2.312), relating them to the Alfvén oscillation field at the upper ionospheric boundary ($z = z_a$).

To describe the Alfvén field structure at the upper ionospheric boundary, we will use magnetic field components in the system of coordinates (n, y, l) (see Figure 2.49), giving the relationship between the components $B_n = B_1/\sqrt{g_1}$ and $B_l = B_3/\sqrt{g_3}$ on the one hand, and the components B_x and B_z on the other by (2.277). We will use Eq. (E.2) (Appendix E) and the expressions (3.207), (3.209) and (3.213) for the potential φ in various regions between the resonance surfaces for the magnetic field components of Alfvén oscillations to write the expressions for the magnetic field components of the oscillations at the upper ionospheric boundary $z = z_a$ (or $l = l_*$) as:

$$\begin{aligned} \tilde{B}_{nN}(x, k_y, l_*, \omega) &= B_A G(\xi_{PN}(x, \omega)), \\ \tilde{B}_{yN}(x, k_y, l_*, \omega) &= i\frac{B_A}{k_y \lambda_{PN} \cos\chi} G'(\xi_{PN}(x, \omega)) \end{aligned} \tag{3.214}$$

for $|x - x_{PN}| \ll \Delta x_N$,

$$\begin{aligned} \tilde{B}_{nN}(x, k_y, l_*, \omega) &= -\sqrt{\pi} B_A \left(\frac{v_{PN}^x}{v_N^x} \frac{k_y^2 \cos^2\chi}{k_{xN}^2 + k_y^2 \cos^2\chi} \right)^{1/2} \\ &\quad \times \exp(i\vartheta_N(x) - \Gamma_N(x) + i\pi/4), \\ \tilde{B}_{yN}(x, k_y, l_*, \omega) &= -\frac{k_{xN}}{k_y \cos\chi} \tilde{B}_{nN}(x, k_y, l_*, \omega) \end{aligned} \tag{3.215}$$

for $x - x_{PN} \gg \lambda_{PN}$ and $x_{TN} - x \gg \lambda_{TN}$, and

$$\tilde{B}_{nN}(x, k_y, l_*, \omega) = -2ik_y\lambda_{PN}\cos\chi B_A\, g(\xi_{TN}(x,\omega)) \times \exp(i\overline{\vartheta}_N - \overline{\Gamma}_N),$$
$$\tilde{B}_{yN}(x, k_y, l_*, \omega) = 2\frac{\lambda_{PN}}{\lambda_{TN}} B_A\, g'(\xi_{TN}(x,\omega))\exp(i\overline{\vartheta}_N - \overline{\Gamma}_N) \tag{3.216}$$

for $|x - x_{TN}| \ll \Delta x_N$, where $B_A = (k_y\overline{c}/\omega)(pv_A/p_0v_{A0})^{1/2}\tilde{\varphi}$ is the characteristic amplitude of the magnetic field of Alfvén oscillations at the upper ionospheric boundary, $k_y = k_2/\sqrt{g_2}$. Note that a tilde in these expressions denotes a separate Fourier harmonic in decompositions of the oscillations, not only with respect to frequencies ω, but to azimuthal wavenumbers k_y as well. The function arguments in (3.214) and (3.216) have the form

$$\xi_{PN}(x,\omega) = \frac{x - x_{PN}(\omega)}{\lambda_{PN}} + i\varepsilon_{PN}, \quad \xi_{TN}(x,\omega) = \frac{x - x_{TN}(\omega)}{\lambda_{TN}} + i\varepsilon_{TN},$$

where $\varepsilon_{(P,T)N} = \gamma_{(P,T)N}a_N/\Omega_{(P,T)N}\lambda_{(P,T)N}$, $\gamma_{(P,T)N}$ are the wave decrement values at the ionosphere near the respective resonance surfaces (the decrement is assumed to be small: $\gamma_{(P,T)N} \ll \Omega_{(P,T)N}$). In the expressions (3.214)–(3.216), λ_{PN} and λ_{TN} are the characteristic Alfvén wavelength across magnetic shells, in the neighbourhood of, respectively, the poloidal and the toroidal resonance surface. Functions $G(z)$ and $G'(z)$, $g(z)$ and $g'(z)$ describe the wave field structure in x near these surfaces and have the following integral representations:

$$G(z) = -\int_0^\infty \exp(isz - is^3/3)ds, \quad G'(z) = \frac{\partial G}{\partial z},$$
$$g(z) = \frac{1}{2}\int_0^\infty s^{-1}\exp(isz - i/s)ds, \quad g'(z) = \frac{\partial g}{\partial z}. \tag{3.217}$$

Expressions (3.215) describe the wave field in the region between the resonance surfaces, where the WKB approximation is applicable in x. In these formulas, $k_{x,N}(x,\omega)$ is the x component of the horizontal wave vector, in the WKB approximation. To define the function $k_{xN}(x,\omega)$ precisely is difficult enough (see Section 3.7.2), but its behaviour is aptly simulated by the following expression:

$$k_{x,N}(x,\omega) = k_y\left(\frac{x - x_{PN}(\omega)}{x_{TN}(\omega) - x}\right)^{1/2}. \tag{3.218}$$

Function $v_N^x(x,\omega)$ is the x component of the oscillation group velocity defined as

$$v_N^x(x,\omega) = \frac{\partial k_{xN}(x,\omega)}{\partial\omega},$$

and v_{PN}^x is the characteristic value of $v_N^x(x,\omega)$ near the poloidal resonance surface, where it is permissible to write

$$v_N^x(x,\omega) \approx v_{PN}^x\sqrt{\frac{x - x_{PN}(\omega)}{\lambda_{PN}}}.$$

Function

$$\vartheta_N(x) = \int_{x_{PN}}^x k_{xN}(x',\omega)dx'$$

describes the phase incursion, and

$$\Gamma_N(x) = \int_{x_{PN}}^{x} \frac{\gamma_N(x',\omega)}{v_N^x(x',\omega)} dx'$$

describes the integral decrement, which are both acquired by the wave as it moves across magnetic shells from the poloidal resonance surface to point x within the interval (x_{PN}, x_{TN}). Correspondingly, $\overline{\vartheta} = \vartheta(x_{TN})$ is the full phase incursion within the interval Δx_N, and $\overline{\Gamma} = \Gamma(\tilde{x})$ is the integral decrement as the wave moves from $x = x_{PN}$ to an arbitrary point $x = \tilde{x}$, near the toroidal surface: $|\tilde{x} - x_{TN}| \ll \Delta x_N$.

To find the oscillation field structure on the ground we will Fourier-transform the expressions (3.214)–(3.216) with respect to wave numbers k_x. Next we will use the linking formulas (2.312) to subject the oscillation field at Earth surface to an inverse Fourier transformation. We will use the integral representations (3.217) to find inverse Fourier transforms of functions g and G in the form

$$\begin{aligned}
\overline{G}(l_*) &= -i2\pi\lambda_{PN}\theta(k_x) \\
&\quad \times \exp\left(-i\frac{(k_x\lambda_{PN})^3}{3} - ik_x x_{PN} - \varepsilon_{PN}k_x\lambda_{PN}\right), \\
\overline{G}'(l_*) &= -ik_x\lambda_{PN}\overline{G}(l_*), \\
\overline{g}(l_*) &= \pi k_x^{-1}\theta(k_x)\exp\left(-\frac{i}{k_x\lambda_{TN}} - ik_x x_{TN} - \varepsilon_{TN}k_x\lambda_{TN}\right), \\
\overline{g}'(l_*) &= ik_x\lambda_{TN}\overline{g}(l_*).
\end{aligned} \tag{3.219}$$

To conduct an analogous decomposition of functions (3.215), we will make use of the fact that the exponent contains a large quasiclassical phase, $|\vartheta(x)| \gg 1$. This enables us to use the saddle-point method when calculating the integrals of the form

$$\overline{B} = \int_{-\infty}^{\infty} A(x)\exp\left(i\vartheta(x) - ik_x x + i\frac{\pi}{4}\right) dx,$$

where $A(x)$ is a slowly varying pre-exponential factor. Equating the first derivative of the exponent to zero, we will obtain an equation determining the saddle-point $\overline{x}$:

$$\left.\frac{\partial\vartheta}{\partial x}\right|_{x=\overline{x}} = k_{xN}(\overline{x},\omega) = k_x. \tag{3.220}$$

The second derivative of the phase at the saddle point has the form

$$\left.\frac{\partial^2\vartheta}{\partial x^2}\right|_{x=\overline{x}} = \left.\frac{\partial k_{xN}(x,\omega)}{\partial x}\right|_{x=\overline{x}} = -\frac{\omega}{2a_N}\frac{1}{v_N^x(\overline{x},\omega)}.$$

This equation makes use of the fact that the wave in question is localised in the interval $\Delta x_N \ll a_N$. The dependence $k_{xN}(x,\omega)$ can be written in the general form as

$$k_{xN}(x,\omega) \equiv k_{xN}(x - x_{PN}(\omega)).$$

Hence

$$\frac{\partial k_{xN}}{\partial\omega} = -\frac{\partial x_{PN}}{\partial\omega}\frac{\partial k_{xN}}{\partial x}.$$

Thanks to the wave being localised within a narrow interval of magnetic shells, a linear approximation can be used for $x_{PN}(\omega)$

$$x_{PN}(\omega) \approx x - 2\frac{\omega - \Omega_{PN}(x)}{\omega} a_N,$$

which produces the above-obtained result. We will use the standard formulas of the saddle-point method (see [337]) to obtain

$$\overline{B}_{nN}(l_*) = -2i\pi B_A \theta(k_x) \frac{k_y \lambda_{PN} \cos\chi}{\sqrt{k_{xN}^2 + k_y^2 \cos^2\chi}} \times \exp\left(i\vartheta(x,\omega) - ik_x x - \Gamma(x,\omega)\right),$$
$$\overline{B}_{yN}(l_*) = -\frac{k_x}{k_y \cos\chi} \overline{B}_{nN}(l_*). \tag{3.221}$$

An inverse Fourier transformation at Earth surface reduces the functions $\overline{G}$ and $\overline{g}$ back into G and g (same as it does $\overline{G}'$ and $\overline{g}'$ into G' and g') resulting in:

$$\tilde{B}_{xN}(0) = i\frac{B_A}{k_y \lambda_{PN} \Sigma_P} \int_0^\infty \left[\sigma_H(z) - i\frac{k_y}{|k_y|}\sigma_P(z)\sin\chi\right] \times G'(\tilde{\xi}_{PN}(x_z,\omega))\, e^{-k_y z} dz,$$
$$\tilde{B}_{yN}(0) = -\frac{B_A}{\Sigma_P} \int_0^\infty \left[\sigma_H(z) - i\frac{k_y}{|k_y|}\sigma_P(z)\sin\chi\right] \times G(\tilde{\xi}_{PN}(x_z,\omega))\, e^{-k_y z} dz \tag{3.222}$$

for $|x - x_{PN}| \ll \Delta x_N$,

$$\tilde{B}_{xN}(0) = 2\frac{\lambda_{PN}}{\lambda_{TN}} \frac{B_A \cos\chi}{\Sigma_P} \exp(i\overline{\vartheta}_N - \overline{\Gamma}_N) \times \int_0^\infty \sigma_H(z)\, g'(\tilde{\xi}_{TN}(x_z,\omega)) dz,$$
$$\tilde{B}_{yN}(0) = 2ik_y \lambda_{PN} \frac{B_A \cos\chi}{\Sigma_P} \exp(i\overline{\vartheta}_N - \overline{\Gamma}_N) \times \int_0^\infty \sigma_H(z) g(\tilde{\xi}_{TN}(x_z,\omega)) dz \tag{3.223}$$

for $|x - x_{TN}| \ll \Delta x_N$. Here, σ_H and σ_P are the Hall and Pedersen conductivities of plasma, Σ_P is the integral Pedersen conductivity, $x_z = x + (z - z_a)\tan\chi$ is the x coordinate mapped along the field line, from height z onto the upper ionospheric boundary (see Figure 3.35). The arguments of functions $g(z)$ and $G(z)$ have the form

$$\tilde{\xi}_{PN}(x_z,\omega) = \frac{x_z - x_{PN}}{\lambda_{PN}} + i\varepsilon_{PN},$$
$$\tilde{\xi}_{TN}(x_z,\omega) = \frac{x_z - x_{TN}}{\lambda_{TN}} + i\left(\varepsilon_{TN} + \frac{z}{\lambda_{TN}}\right).$$

Expressions (3.222) were obtained making use of the fact that the main contribution to the Fourier integral of functions G and G' is from harmonics with $k_x \ll k_y$ and it is possible to set $k_t \approx k_y$ in (2.312). Analogously, calculations of (3.223) relied on the fact that the main contribution to the Fourier integral of functions g and g' is from harmonics with $k_x \gg k_y$ and it is possible to assume $k_t \approx k_x$ in (2.312).

For doing an inverse Fourier transform of functions (3.221) at Earth surface, we will again make use of a large quasiclassical phase $|\vartheta(\overline{x}(k_x), \omega)| \gg 1$ present in the exponent. To calculate integrals of the form

$$\tilde{B} = \int_0^\infty \tilde{A}(k_x)\exp[i\vartheta(\overline{x}(k_x)) - ik_x(x_z - \overline{x}(k_x)) - k_t z]dk_x$$

we will employ the saddle-point method. Here $k_t = (k_x^2 + k_y^2)^{1/2}$, and the saddle-point $\overline{x}(k_x)$ is defined by (3.220). Equating the first derivative of the exponent to zero produces

$$x_z - \overline{x}(\overline{k}_x) + i\frac{\overline{k}_x}{\overline{k}_t}z = 0,$$

an equation determining the saddle-point $\overline{k}_x$ (where $\overline{k}_x = k_{xN}(\overline{x}(\overline{k}_x, \omega)$ according to (3.220)). This equation can be regarded as defining $\overline{\overline{x}} = \overline{x}(\overline{k}_x)$ and be rewritten as

$$x_z - \overline{\overline{x}} + i\frac{k_{xN}(\overline{\overline{x}}, \omega)}{\overline{k}_t}z = 0,$$

where $\overline{k}_t = (k_{xN}^2(\overline{\overline{x}}, \omega) + k_y^2)^{1/2}$. The second derivative of the exponent, at the saddle point is equal to

$$\frac{2a_N}{\omega}\overline{v}_N^x \equiv \frac{2a_N}{\omega}v_N^x(\overline{\overline{x}}, \omega) + i\frac{k_y^2}{\overline{k}_t^3}z.$$

Using the standard formulas of the saddle-point method (see [337]) we obtain, at Earth surface, for $x_z - x_{PN} \gg \lambda_{PN}$ and $x_{TN} - x_z \gg \lambda_{TN}$:

$$\begin{aligned}\tilde{B}_{nN}(0) = {} & \pi\frac{B_A}{\Sigma_P}\int_0^\infty \left[\sigma_H(z) - i\frac{k_y}{\overline{k}_t}\sigma_P(z)\sin\chi\right] \\ & \times\left(\frac{v_{PN}^x}{|\overline{v}_N^x|}\frac{\overline{k}_{xN}^2\cos^2\chi}{\overline{k}_{xN}^2 + k_y^2\cos^2\chi}\right)^{1/2} \\ & \times\exp\left(i\vartheta(\overline{x}) - \Gamma(\overline{x}) - \frac{k_y^2 z}{\overline{k}_t} + i\left(\frac{\pi}{4} - \frac{\arg\,\overline{v}_N^x}{2}\right)\right)dz, \qquad (3.224) \\ \tilde{B}_y(0) = {} & \sqrt{\pi}\frac{B_A}{\Sigma_P}\int_0^\infty \left[\sigma_H(z) - i\frac{k_y}{\overline{k}_t}\sigma_P(z)\sin\chi\right] \\ & \times\left(\frac{v_{PN}^x}{|\overline{v}_N^x|}\frac{\overline{k}_y^2\cos^2\chi}{\overline{k}_{xN}^2 + k_y^2\cos^2\chi}\right)^{1/2} \\ & \times\exp\left(i\vartheta(\overline{x}) - \Gamma(\overline{x}) - \frac{k_y^2 z}{\overline{k}_t} + i\left(\frac{\pi}{4} - \frac{\arg\,\overline{v}_N^x}{2}\right)\right)dz.\end{aligned}$$

Formulas (3.222)–(3.224) completely solve the problem of the field of monochromatic transverse-small-scale Alfvén oscillations penetrating from the magnetosphere to the Earth surface. These formulas are written in the simplest form if functions $\sigma_{P,H}(z)$ are localised (at e.g. height H) at a much smaller scale than the characteristic vertical scale of the wave in the ionosphere. In this case, the ionosphere can be regarded mathematically as a thin layer, which allows all functions (excluding $\sigma_{P,H}(z)$) to be taken out of the integral sign at point $z = H$. As a result, we obtain:

$$\tilde{B}_{xN}(0) = i\frac{\overline{B}}{k_y\lambda_{PN}}\, G'(\tilde{\xi}_{PN})\exp(-k_yH),$$
$$\tilde{B}_{yN}(0) = -\overline{B}\, G(\tilde{\xi}_{PN})\exp(-k_yH); \qquad (3.225)$$

for $|x_H - x_{PN}| \ll \Delta x_N$

$$\tilde{B}_{xN}(0) = \sqrt{\pi}\overline{B}\left(\frac{v_{PN}}{|\overline{v}_N^x|}\frac{\overline{k}_{xN}^{2}\cos^2\chi}{\overline{k}_{xN}^{2} + k_y^2\cos^2\chi}\right)^{1/2}$$
$$\times\exp\left[i\vartheta\,(\overline{\overline{x}}) - \Gamma(\overline{\overline{x}}) - \frac{k_y^2H}{k_t} + i\left(\frac{\pi}{4} - \frac{\arg\overline{v}_N^x}{2}\right)\right], \qquad (3.226)$$
$$\tilde{B}_{yN}(0) = \frac{k_y}{\overline{k}_{xN}}\tilde{B}_{xN}(0);$$

for $x_H - x_{PN} \gg \lambda_{PN}$ and $x_{TN} - x_H \gg \lambda_{TN}$

$$\tilde{B}_{xN}(0) = 2\frac{\lambda_{PN}}{\lambda_{TN}}\overline{B}\, g'(\tilde{\xi}_{TN})\exp(i\overline{\vartheta}_N - \overline{\Gamma}_N),$$
$$\tilde{B}_{yN}(0) = 2ik_y\lambda_{PN}\cos\chi\overline{B}\, g(\tilde{\xi}_{TN})\exp(i\overline{\vartheta}_N - \overline{\Gamma}_N). \qquad (3.227)$$

for $|x_H - x_{TN}| \ll \Delta x_N$. Here

$$\overline{B} = \frac{B_A}{\Sigma_P}\left(\Sigma_H - i\frac{k_y}{|k_y|}\Sigma_P\sin\chi\right),$$

$$x_H = x + (H - z_A)\tan\chi, \quad \overline{\overline{x}} = x_H + i\frac{\overline{k}_{xn}H}{k_t},$$

$$\tilde{\xi}_{PN} = \frac{\overline{\overline{x}} - x_{PN}}{\lambda_{PN}} + i\varepsilon_{PN}, \quad \tilde{\xi}_{TN} = \frac{\overline{\overline{x}} - x_{TN}}{\lambda_{TN}} + i\varepsilon_{TN},$$
$$\overline{v}_N^x = v_N^x(\overline{\overline{x}}, \omega) + i\frac{\omega k_y^2H}{2\overline{k}_t^{3}a_N}.$$

A qualitative scheme of the penetration of the electromagnetic field of Alfvén oscillations is shown in Figure 3.35. From the upper ionosphere the Alfvén wave field penetrates into the lower ionosphere, where it induces currents generating a fast magnetosonic wave. The field of the magnetosonic oscillations is re-emitted into the atmosphere and reaches the Earth surface.

Figure 3.35a is a scheme of how Alfvén oscillations with a much larger wavelength across magnetic shells than the neutral atmosphere thick ($\lambda_{(P,T)N} > H$) penetrate to ground. In this case, the oscillation structure in x remains the same when they penetrate to ground as it was at the upper ionospheric boundary ($z = z_A$). The oscillation amplitude of course diminishes in the process as $\exp(-k_tH)$. Figure 3.35b is a schematic representation of the penetration of wave with $\lambda_{TN} < H$. As the oscillations penetrate to ground, their structure can be seen

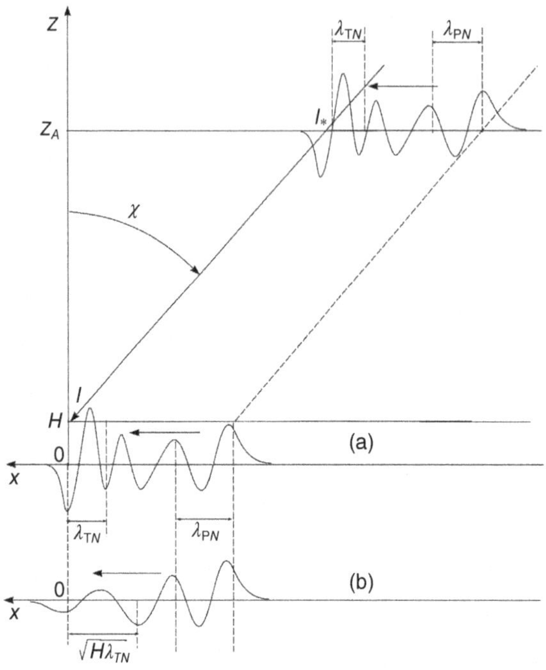

Figure 3.35 Penetration of the high-m Alfvén wave field from the magnetosphere to ground: (a) in the $\lambda_{(P,T)N} > H$ case, (b) in the $\lambda_{(P,T)N} \lesssim H$ case. Shown are the hodographs of oscillations in the plane (B_x, B_y), at various points inside the transparent region (x_{PN}, x_{TN}) at the upper ionospheric boundary and at Earth surface.

to expand and their characteristic wavelength in x to reach $\sqrt{\lambda_{TN}H}$ near the projection of the toroidal resonance surface.

3.9 Linear Transformation of Standing High-*m* Alfvén Waves Near the Toroidal Resonance Surface

Section 3.7 discussed the structure of monochromatic standing Alfvén wave with $m \gg 1$ generated at the poloidal resonance surface and travelling across magnetic shells towards the toroidal resonance surface, where it is fully absorbed. Ionospheric external currents were discussed as a source of such a wave, and its decay in the conducting layer of the ionosphere as its dissipation mechanism. If no dissipation mechanisms existed, the Alfvén oscillation field would have a singularity at the toroidal resonance surface. As was shown in Section 3.1, however, if the Alfvén oscillation decay is small, the wave field structure near the toroidal resonance surface is determined, not by dissipation, but by the kinetic Alfvén wave generated here. These wave travel away from the toroidal resonance surface across magnetic shells and carry the energy away from the resonance magnetic shell.

Alfvén waves discussed in Section 3.7 also propagate across magnetic shells. Their transverse dispersion is determined by the magnetic field line curvature, and their wavelength across magnetic shells (in the x^1 coordinate) is much larger than what can be found in kinetic Alfvén waves [338]. Therefore we will call them 'large-scale' Alfvén waves in this section, in contrast to small-scale kinetic Alfvén waves. When their dissipation near the toroidal resonance surface is small such a wave can transform into kinetic Alfvén waves.

Let us consider the mechanism of linear transformation of large-scale Alfvén waves (their dispersion being determined by the magnetic field line curvature) into kinetic Alfvén waves.

For this, we will use (3.32), where we will neglect the right-hand side, describing the field of FMS wave with $m \gg 1$, which is negligibly small in the inner magnetosphere:

$$\frac{\omega^2}{v_A^2}\Lambda^2\sqrt{g}\Delta_\perp^2\varphi + \nabla_1\sqrt{g_3}\hat{L}_T\nabla_1\varphi - k_2^2\sqrt{g_3}\hat{L}_P\varphi = 0, \tag{3.228}$$

where the operators $\hat{L}_T$, $\hat{L}_P$, $\Delta_\perp$ are defined as (3.6), (3.7) and (3.9) respectively. We will use

$$\varphi|_{l_\pm} = \mp i\frac{v_{P\pm}}{\omega}\frac{\partial\varphi}{\partial l}\bigg|_{l_\pm}, \tag{3.229}$$

as the boundary condition at the ionosphere. This equation takes into account oscillation decay due to the finite conductivity of the ionosphere, but neglects the presence of external currents (see (2.309)). As we did in Section 3.7, we will seek the solution to (3.228) using the multiple-scale method, representing it as

$$\varphi = V_N(x^1,\omega)T_N(x^1,l) + h_N,$$

where function $V_N(x^1,\omega)$ describes the small-scale structure of Alfvén oscillations across magnetic shells, $T_N(x^1,l)$ describes their structure along magnetic field lines, in the main order, and $h_N(x^1,l,\omega)$ is a small correction in higher orders of perturbation theory. Functions $T_N(x^1,l)$ are the eigenfunctions of the zeroth-order problem

$$\hat{L}_T(\Omega_{TN})T_N(x^1,l) = 0, \qquad T_N(x^1,l_\pm) = 0,$$

which describes the structure of standing Alfvén waves at the toroidal resonance shell. Functions $\Omega_{TN}(x^1)$ are the eigenvalues of the problem and the frequencies of the harmonics of toroidal standing Alfvén waves at the magnetic shell in question ($N = 1, 2, 3, \ldots$ is the parallel wave number).

Left-multiplying (3.228) by T_N and integrating along the field line produces, in the first order of perturbation theory, an equation for the function $V_N(x^1)$:

$$\omega^2\sigma_N(x^1)\frac{\partial^4 V_N}{\partial x^{1\,4}} + \nabla_1[(\omega+i\gamma_N)^2 - \Omega_{TN}^2]\nabla_1 V_N - k_2^2 w_{TN}V_N = 0, \tag{3.230}$$

where the parameter σ_N, describing the transverse dispersion of small-scale kinetic Alfvén waves, is determined by (3.20); the parameter w_{TN}, describing the dispersion of a large-scale Alfvén wave by (3.126); and the Alfvén wave decrement at the ionosphere γ_N by (3.165). Recall (see Chapter 1 and Section 3.1) that the parameter σ_N is positive ($\sigma_N > 0$) in the external magnetosphere with a 'warm' plasma, negative ($\sigma_N < 0$) inside the plasmasphere filled by a 'cold' plasma, and complex in the transition region of the plasmapause (with $\mathrm{Im}\sigma_N < 0$).

If the WKB approximation is applicable, the solution to (3.230) can be represented as

$$V_N \sim \exp\left(i\int k_1 dx^1\right),$$

while for the quasiclassical wave vector (neglecting the oscillation decay at the ionosphere) we obtain

$$\omega^2\sigma_N k_1^4 - (\omega^2 - \Omega_{TN}^2)k_1^2 - k_2^2 w_{TN} = 0. \tag{3.231}$$

This equation can also be regarded as determining the local frequency ω of the oscillations from their wave vector k_1. Solving this equation with respect to ω, given the condition

$|k_1^2\sigma_N| \ll 1$, produces a local dispersion equation,

$$\omega^2 = \Omega_{TN}^2 + k_1^2\sigma_N\Omega_{TN}^2 - \frac{k_2^2 w_{TN}}{k_1^2}. \tag{3.232}$$

The last term in the right-hand side is the dispersion correction related to the large-scale Alfvén wave. For large-scale Alfvén waves with

$$k_1^2 \ll \frac{k_2 w_{TN}^{1/2}}{\sigma_N^{1/2}\omega}, \tag{3.233}$$

we can set

$$\omega^2 \approx \Omega_{TN}^2 - \frac{k_2^2 w_{TN}}{k_1^2}. \tag{3.234}$$

This equation can be regarded as the dispersion equation for large-scale Alfvén waves their transverse dispersion determined by the magnetic field line curvature. It should be noted that (3.233) cannot address the limiting case $k_1^2 \to 0$, where the WKB approximation becomes inapplicable. For small-scale Alfvén waves, the opposite limiting case is valid:

$$k_1^2 \gg \frac{k_2 w_{TN}^{1/2}}{\sigma_N^{1/2}\omega}.$$

In this case,

$$\omega^2 \approx \Omega_{TN}^2(1 + k_1^2\sigma_N) \tag{3.235}$$

is the dispersion equation analogous to that (1.24) for kinetic Alfvén waves in an inhomogeneous plasma. It is possible to determine from (3.235) the x^1 component of the group velocity for small-scale standing Alfvén waves:

$$v_{gN}^1 = \frac{\partial\omega}{\partial k_1} = \Omega_{TN}k_1\sigma_N. \tag{3.236}$$

Note that, in a 'warm' plasma ($\sigma_N > 0$) the signs of the group (3.236) and the phase velocity ω/k_1 coincide, while being opposite in a 'cold' plasma ($\sigma_N < 0$).

Let us return to (3.231). Its solution with respect to k_1^2 is

$$k_1^2 = \frac{\omega^2 - \Omega_{TN}^2 \pm \sqrt{(\omega^2 - \Omega_{TN}^2)^2 + 4k_2^2\sigma_N\omega^2 w_{TN}}}{2\omega^2\sigma_N}. \tag{3.237}$$

This equation defines function $k_1^2(x^1)$ based on given dependence $\Omega_{TN}^2(x^1)$. Let us restrict ourselves to the case when, near the toroidal resonance surface ($x^1 = x_{TN}^1$), the linear decomposition can be used:

$$\Omega_{TN}^2 = \omega^2\left(1 - \frac{x^1 - x_{TN}^1}{a_N}\right), \tag{3.238}$$

where a_N is the characteristic scale of the function $\Omega_{TN}^2(x^1)$ inhomogeneity and assuming $|x^1 - x_{TN}^1| \ll a_N$. Substituting (3.238) into (3.237), produces

$$k_1^2 = \frac{1}{2\sigma_N}\left[\frac{x^1 - x_{TN}^1}{a_N} \pm \sqrt{\left(\frac{x^1 - x_{TN}^1}{a_N}\right)^2 + 4k_2^2\sigma_N\frac{w_{TN}}{\omega^2}}\right]. \tag{3.239}$$

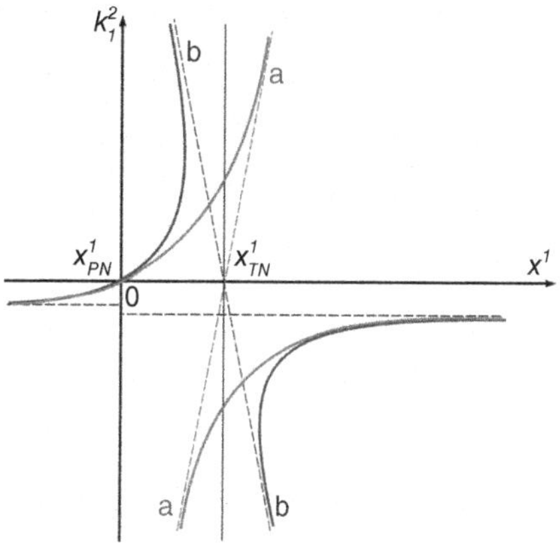

Figure 3.36 Distribution in x^1 of the quasiclassical wave vector squared k_1^2. Slanting dashed lines show the asymptotics $k_1^2 = (x^1 - x^1_{TN})/\sigma_N a_N$. The (a) case corresponds to $\sigma_N > 0$, and the (b) case to $\sigma_N < 0$.

This dependence is shown in Figure 3.36. Away from the resonance surface, when $|x^1 - x^1_{TN}| \gg k_2(\sigma_N w_{TN})^{1/2} a_N$, the two roots in (3.239) can be represented as

$$\begin{aligned} k_1^2 &= -\frac{k_2^2 w_{TN}}{\omega^2}\frac{a_N}{x^1 - x^1_{TN}}, \\ k_1^2 &= \frac{x^1 - x^1_{TN}}{a_N \sigma_N}. \end{aligned} \tag{3.240}$$

The first describes a large-scale Alfvén wave, its dispersion defined by (3.234). Its transparent region occupies $x^1 < x^1_{TN}$. The second root is a small-scale (kinetic) Alfvén wave with dispersion law (3.235). Its transparent region occupies $x^1 < x^1_{TN}$ if $\sigma_N < 0$, and $x^1 > x^1_{TN}$, if $\sigma_N > 0$.

Near the toroidal resonance surface Alfvén waves of one type are transformed into the other type. Let us examine this process qualitatively. We will return to (3.230). Substituting (3.238) into it produces

$$\sigma_N(x^1)\frac{\partial^4 V_N}{\partial x^{14}} + \frac{\partial}{\partial x^1}\left(\frac{x^1 - x^1_{TN}}{a_N} + 2i\frac{\gamma_N}{\omega}\right)\frac{\partial V_N}{\partial x^1} - \frac{k_2^2 w_{TN}}{\omega^2} V_N = 0. \tag{3.241}$$

Let us introduce a dimensionless coordinate, $\eta = (x^1 - x^1_{TN})/\lambda_{TN}$, where λ_{TN} is the characteristic length of large-scale Alfvén wave in x^1, near the toroidal resonance surface (3.196). Then introducing a complex variable in (3.241), $z = \eta + i\varepsilon_{TN}$, where ε_{TN} is defined as (3.198), produces the following equation:

$$\alpha^2 V_N^{IV} + zV_N'' + V_N' - V_N = 0, \tag{3.242}$$

where $\alpha = \sigma_N a_N/\lambda^3_{TN}$, and the superscript ^{IV} and primes denote corresponding derivatives with respect to z. The dimensionless parameter α^2 is small: $|\alpha^2| \ll 1$. Given the fact that it is complex, let us set $\alpha^2 = |\alpha^2|e^{-i\phi}$, where $0 \le \phi \le \pi$.

Equation (3.242) can be solved by means of the Laplace transform method (see [339]). The full set of linearly independent solutions is determined by the integrals along trajectories C_k:

$$F_k(z) = \frac{1}{\sqrt{\pi}}\int_{C_k}\frac{dt}{t}\exp\left(\frac{\alpha^2}{3}t^3 + \frac{1}{t} + zt\right). \tag{3.243}$$

Each integration trajectory C_k in the plane of the complex variable t is determined by the requirement that function

$$Z(t) = \frac{1}{t}\exp\left(\frac{\alpha^2}{3}t^3 + \frac{1}{t}\right)$$

should have identical values (or return to the original value with closed trajectory) at the trajectory endpoints. It is easily verifiable that $|Z(t)| \to 0$ for $|t| \to \infty$ in sectors

$$\frac{2\pi n + \phi}{3} + \frac{\pi}{6} < \arg t < \frac{2\pi n + \phi}{3} + \frac{\pi}{2},$$

where n is an arbitrary integer. These sectors are white in Figure 3.37. Moreover, $|Z(t)| \to 0$ as $t \to 0$, if $\mathrm{Re}\, t < 0$ (e.g. along the negative semiaxis of the real t). Consequently, the solutions of (3.242) are the contour integrals (3.243) along one of the contours $C_1, C_2, \ldots, C_7$ in Figure 3.37. Since only four linearly independent solutions exist, three relations link the above solutions F_k:

$$F_4 = F_2 - F_1, \qquad F_5 = F_3 - F_2, \qquad F_6 = F_1 - F_3 + F_7.$$

The general solution of (3.242) is a superposition of any four linearly independent functions F_k.

The form of this superposition depends on the boundary conditions chosen for $z \to \pm\infty$. They can be formulated as follows, for the solution we are interested in. First, this solution must be of finite amplitude. This means that growing asymptotics must be absent in the opaque regions, both for large-scale and small-scale Alfvén wave. Second, small-scale wave must carry the energy away, in the transparent region, far from the transformation region (i.e. its group velocity must be directed away from the toroidal resonance surface). Physically, this means that kinetic Alfvén wave is generated as a result of linear transformation of large-scale Alfvén wave. On the other hand, absent are kinetic Alfvén waves

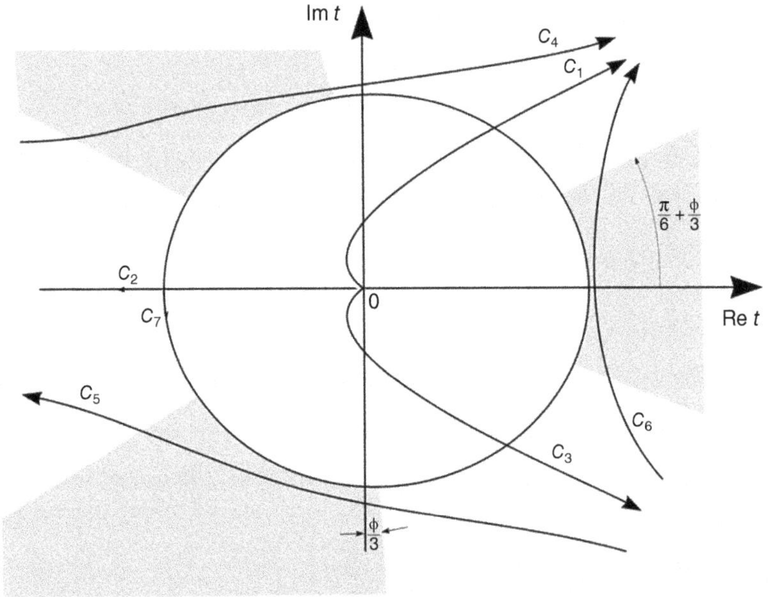

Figure 3.37 Possible integration contours in integrals (3.242). Sectors with exponentially growing asymptotics are in grey.

importing the energy from the infinity, i.e. waves generated by other sources. These requirements determine the desired solution accurate to an arbitrary factor given by the amplitude of the incident large-scale Alfvén wave.

Verifiably, the solution $F_1(z)$ satisfies the above conditions. Let us examine the asymptotic representation of function $F_1(z)$. We will use the standard calculation techniques based on the saddle-point method (see [337]) to obtain the following result:

$$F_1(z) = (-z)^{-1/4} \exp\left[-2i(-z)^{1/2} - i\frac{\pi}{4}\right] + \left(-\frac{z}{\mu}\right)^{-3/4} \times \exp\left[-\frac{2}{3}\left(\cos\frac{\phi}{2} + i\sin\frac{\phi}{2}\right)\left(-\frac{z}{\mu}\right)^{3/2} + i\frac{\pi}{2} - i\frac{\phi}{4}\right], \quad z \to -\infty, \tag{3.244}$$

$$F_1(z) = z^{-1/4} \exp\left(-2z^{1/2}\right) + \left(\frac{z}{\mu}\right)^{-3/4} \times \exp\left[-\frac{2}{3}\left(\sin\frac{\phi}{2} - i\cos\frac{\phi}{2}\right)\left(\frac{z}{\mu}\right)^{3/2} - i\frac{\pi}{4} + i\frac{\phi}{4}\right], \quad z \to \infty, \tag{3.245}$$

where $\mu = |\alpha|^{2/3}$. It can be seen from this that, in z, large-scale Alfvén wave has a characteristic scale equal to unity, and small-scale wave has a scale equal to μ. In x^1, this corresponds to scales λ_{TN} and $\mu\lambda_{TN} = |\sigma_N a_N|^{1/3}$. This of course is only valid for a small enough decay of Alfvén waves. This means $\varepsilon_{TN} \ll 1$ for the large-scale mode, and

$$\delta_N \equiv \frac{\varepsilon_{TN}}{\mu} \cong 2\frac{a_N}{\mu\lambda_{TN}}\frac{\gamma_N}{\omega} \ll 1. \tag{3.246}$$

for the small-scale mode. If the inverted inequalities $\varepsilon_{TN} \gg 1$ and $\delta_N \gg 1$ are valid, expressions (3.244) and (3.245) do not apply, and the characteristic scales of the oscillations are determined by mode decay (see below).

It follows from (3.244) and (3.245) that, for $z < 0$, large-scale mode in the transformation region is a wave travelling towards the toroidal resonance surface, with no reflected wave. Thus, the result in Section 3.7 is fully reproduced, the only difference being that large-scale Alfvén wave is not absorbed near the toroidal resonance surface, but is fully transformed into kinetic Alfvén wave. Let us examine the $\sigma_N > 0$ ($\phi = 0$) case and assume Alfvén oscillation dissipation to be weak in the ionosphere. Leaving only the basic terms and using dimensionless variable z to coordinate x^1 we will obtain, from (3.244) and (3.245), on the asymptotics:

$$F_1(x^1) = \left(\frac{\lambda_{TN}}{x^1_{TN} - x^1}\right)^{-1/4} \times \exp\left[-2i\left(\frac{x^1_{TN} - x^1}{\lambda_{TN}}\right)^{1/2} - \varepsilon_{TN}\left(\frac{\lambda_{TN}}{x^1_{TN} - x^1}\right)^{1/2} - i\frac{\pi}{4}\right], \tag{3.247}$$

for $x^1_{TN} - x^1 \gg \lambda_{TN}$,

$$F_1(z) = \left(\frac{\mu\lambda_{TN}}{x^1 - x^1_{TN}}\right)^{-1/4} \times \exp\left[-\frac{2}{3}i\left(\frac{x^1 - x^1_{TN}}{\mu\lambda_{TN}}\right)^{3/2} - \delta_N\left(\frac{x^1 - x^1_{TN}}{\mu\lambda_{TN}}\right)^{1/2} - i\frac{\pi}{4}\right], \tag{3.248}$$

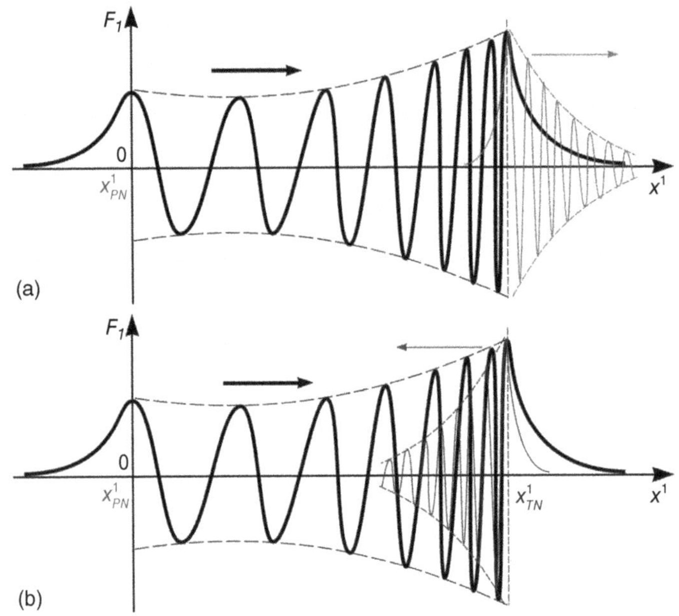

Figure 3.38 Spatial structure of Alfvén waves with $m \gg 1$ across magnetic shells. The thick (black) line is 'large-scale' Alfvén wave, the thin line is kinetic Alfvén wave: (a) $\sigma_N > 0$ and (b) $\sigma_N < 0$.

for $x^1 - x^1_{TN} \gg \mu\lambda_{TN}$. The typical scale of the decay of small-scale wave $\mu\lambda_{TN}/\delta_N^2 \sim (\sigma_N/a_N)(\omega/\gamma)^2$ is much larger than the wavelength $\mu\lambda_{TN}$ but is much smaller (for the parameter values typical of the magnetosphere) than the distance between the toroidal and the poloidal resonance surfaces. Thus, Alfvén wave is absorbed in a narrow neighbourhood of the toroidal resonance surface, upon the linear transformation into the small-scale mode. If parameter σ_N is complex (i.e. argument ϕ is far from both $\phi = 0$ and $\phi = \pi$), the scale of the decay of the small-scale mode coincides with the wavelength $\mu\lambda_{TN}$.

Analogous calculations can also be made for the $\sigma_N < 0$ case. In this case, the transparent regions for the large-scale and small-scale mode coincide, and their phase velocities have the same direction – towards the toroidal resonance surface. However, the group velocity of the small-scale mode has an opposite direction – it carries the energy away from the toroidal surface. Qualitatively, the Alfvén oscillation structure in these two cases is shown in Figure 3.38.

3.10 Magnetospheric Resonator for Standing High-*m* Alfvén Waves

In Sections 3.7 and 3.9 we examined the structure of standing Alfvén waves with $m \gg 1$ in magnetospheric regions with monotonic distribution of Alfvén speed across magnetic shells. More precisely, we examined regions with monotonic distribution of functions $\Omega_{PN}(x^1)$, $\Omega_{TN}(x^1)$, describing the frequencies of standing poloidal and toroidal Alfvén waves at different magnetic shells. The situation near the extrema of these functions is, however, completely different.

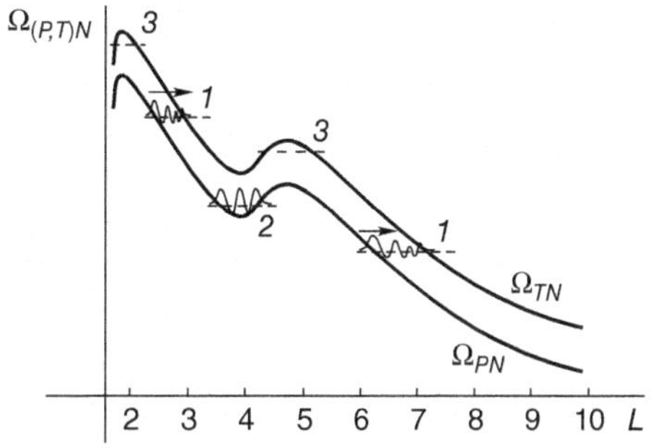

Figure 3.39 Schematic plots of functions $\Omega_{(P,T)N}(x^1)$ in the dayside magnetosphere of Earth. The McIlwain parameter $L = a/R_E$ is used as the x^1 coordinate. *1* – the transparent region is located between the poloidal and toroidal resonant surfaces; *2* – Alfvén resonator; *3* – the presence of two toroidal turning points makes Alfvén oscillations impossible.

Figure 3.39 illustrates schematically the distribution of functions $\Omega_{PN}(x^1)$, $\Omega_{TN}(x^1)$ in the dayside magnetosphere of Earth. Horizontal dashed lines in the figure correspond to the Alfvén oscillation frequency. The points where these lines intersect the function $\Omega_{PN}(x^1)$, $\Omega_{TN}(x^1)$ plots determine the turning points for Alfvén waves in the x^1 coordinate – the ordinary x^1_{PN} (poloidal) and singular x^1_{TN} (toroidal), respectively. The transparent region for Alfvén oscillations is determined by inequalities $\Omega_{PN}(x^1) < \omega < \Omega_{TN}(x^1)$. Let us analyse qualitatively the nature of these oscillations.

The most typical case is when the transparent region is limited on one side by the ordinary turning point x^1_{PN}, while on the other – by a singular turning point x^1_{TN}. If $\Omega_{PN}(x^1)$, $\Omega_{TN}(x^1)$ decrease as the magnetic shell radius increases, as is the case for most of the magnetosphere, $x^1_{TN} > x^1_{PN}$. This case (*1* in Figure 3.39) is considered in Sections 3.7 and 3.9.

However, another possibility exists. Case *2* in Figure 3.39 corresponds to solving MHD equations describing Alfvén oscillations bounded on two sides by ordinary turning points. This case lacks the singular turning point where the Alfvén wave energy is completely absorbed. Eigen oscillations, i.e. oscillations with no outer source, can exist in such a region. The frequencies of such oscillations are determined by corresponding quantisation conditions. If the WKB approximation applies, it will be the well-known Bohr–Sommerfeld quantisation condition

$$\oint k_{1N}(x^1, \omega)dx^1 = 2\pi(n + \frac{1}{2}), \tag{3.249}$$

where $n = 1, 2, 3 \dots$ is the transverse wave number. The domain of existence for such oscillations limited in the longitudinal coordinate by the points where the field line intersects the ionosphere can be called an Alfvén resonator. The quantisation condition determines the frequency spectrum ω_{Nn} of such a resonator, which depends on two wave numbers: longitudinal N and transverse n.

Let us now derive the exact formulas describing the eigen oscillations of this Alfvén resonator. Near the minimum of function $\Omega_{PN}(x^1)$ we will use the following decomposition:

$$\Omega_{PN} = \overline{\Omega}_{PN}\left(1 + \frac{(x^1 - \overline{x}^1_{PN})^2}{2a^2_{PN}}\right), \tag{3.250}$$

where $\overline{\Omega}_{PN}$ is the value of function $\Omega_{PN}(x^1)$ at the minimum point $x^1 = \overline{x}_{PN}$, and $a_{PN} = (\nabla_1^2\Omega_{PN}/\overline{\Omega}_{PN})^{-1/2}_{x^1=\overline{x}_{PN}}$ is the characteristic scale of its variation here. Let us consider oscillations with frequencies satisfying the condition $\omega - \overline{\Omega}_{PN} \ll \Delta\Omega_N$. Such oscillations are close to the poloidal mode, throughout their localisation domain, and Eq. (3.185) can be used to describe their spatial structure across magnetic shells.

Let us introduce dimensionless variable

$$\xi = (x^1 - \overline{x}^1_{PN})/\overline{\lambda}_{PN},$$

where $\overline{\lambda}_{PN} = (w_{PN} a^2_{PN}/k_2^2 \omega^2)^{1/4}$ is the characteristic scale in x^1 of the oscillations in question. On order of magnitude, $\overline{\lambda}_{PN} \sim \alpha_N^{1/2} a/m^{1/2}$. In the region of the minimum, at the outer edge of the plasmapause, we obtain an estimate, $\overline{\lambda}_{PN} \sim$ 500–1000 km, for Alfvén oscillations with $N = 1$ and azimuthal wave numbers $m = 20$–50. This variable can be used to represent the equation for the transverse structure of the oscillations in question, near the local minimum in the $\Omega_{PN}(x^1)$ distribution, as follows:

$$\frac{d^2 u_N}{d\xi^2} + (\sigma - \xi^2) u_N = \frac{a^2_{PN}}{\overline{\lambda}^2_{PN}} \frac{I_\|}{\omega^2}, \tag{3.251}$$

where

$$\sigma = \frac{a^2_{PN}}{\overline{\lambda}^2_{PN}} \frac{(\omega + i\gamma_N)^2 - \overline{\Omega}^2_{PN}}{\omega^2}.$$

If no external currents are present in the ionosphere, the right-hand side of (3.251) becomes zero and, if the solution at the asymptotics $\xi \to \pm\infty$ is required to be zero, we obtain the well-known eigenvalue problem for a quantum oscillator (see Section 2.8). Its solutions are

$$u_n = C_n y_n(\xi), \qquad \sigma = \sigma_n = 2n + 1, \tag{3.252}$$

where

$$y_n(\xi) = \pi^{-1/4} 2^{-n/2} (n!)^{-1/2} e^{-\xi^2/2} H_n(\xi),$$

$H_n(\xi)$ are the Hermitian polynomials. Functions $y_n(\xi)$ are orthonormalised:

$$\int_{-\infty}^{\infty} y_n(\xi) y_{n'}(\xi) d\xi = \delta_{nn'}.$$

The solution (3.252) determines the eigen functions and eigenfrequencies of the Alfvén resonator:

$$\varphi = C_n y_n \left(\frac{x^1 - \overline{x}^1_{PN}}{\overline{\lambda}_{PN}} \right) P_N(x^1, l), \qquad \omega = \omega_{Nn} - i\gamma_N, \tag{3.253}$$

$$\omega_{Nn} = \overline{\Omega}_{PN} \left[1 + \frac{\overline{\lambda}^2_{PN}}{a^2_{PN}} \left(n + \frac{1}{2} \right) \right].$$

Note that the eigenfrequencies in (3.253) can be obtained directly from the quantisation condition (3.249) even if n are small ($n \sim 1$), a well-known fact for the quantum oscillator.

It is now easy to obtain the solution of the inhomogeneous equation (3.251) using the Green function of its left-hand side

$$G(\xi, \xi', \sigma) = \sum_{n=0}^{\infty} \frac{y_n(\xi) y_n(\xi')}{\sigma - \sigma_n},$$

satisfying the equation

$$\frac{d^2 G}{d\xi^2} + (\sigma - \xi^2) G = \delta(\xi - \xi'). \tag{3.254}$$

We have

$$\varphi = u_N P_N = P_N(x^1, l) \sum_{n=0}^{\infty} \frac{c_n}{(\omega + i\gamma_N)^2 - \omega_{Nn}^2} y_n \left(\frac{x^1 - \overline{x}_{PN}^1}{\overline{\lambda}_{PN}} \right), \qquad (3.255)$$

where

$$c_n = \int_{-\infty}^{\infty} y_n(\xi) I_{\parallel}(\overline{x}_{PN}^1 + \overline{\lambda}_{PN}\xi) d\xi.$$

This solution is a superposition of the eigenmodes excited by a monochromatic source of frequency ω (for example, by external currents in the ionosphere), in the region of the Alfvén resonator. If frequency ω is close to one of the resonator eigenfrequencies ω_{Nn}, making the difference $\omega - \omega_{Nn}$ much smaller than the eigenfrequency splitting $(\overline{\lambda}_{PN}^2 / a_{PN}^2)\overline{\Omega}_{PN}$, and the decrement γ_N small enough, one term corresponding to the resonant frequency will be dominant in the sum (3.255).

As for case *3* in Figure 3.39, simple reasoning shows that there is no solution corresponding to the given boundary conditions. Indeed, the oscillation transparent region is in this case limited by two singular turning points. As we saw in Section 3.7, the finite-amplitude solution limited in the opaque region is in this case a wave travelling in the transparent region to the singular turning point. However, the wave cannot travels in two opposite sides the same time.

Finally, it should be noted that presence of the warm (finite-β) plasma favors the transverse Alfvén resonator in different locations [340–342]. Possible observational manifestations of the Alfven resonator are discussed in [343–345].

3.11 High-*m* Alfvén Waves Generated in the Magnetosphere by Stochastic Sources

In Sections 3.7–3.10 dealt with the field structure of a separate harmonic of transverse-small-scale Alfvén oscillations generated in a dipole-like magnetosphere by a monochromatic source. External currents in the ionosphere were discussed as the source. Let us now examine generation of Alfvén waves with $m \gg 1$ by broadband sources.

Broadband oscillations can be conditionally subdivided into two large classes. One includes stochastic oscillations excited by noise sources, i.e. sources with amplitudes that are a random function of time. External currents in the ionosphere suggested as a possible source of standing Alfvén waves with $m \gg 1$ are also likely to be noise in nature. Stochastic oscillations are characterised by their correlation properties (see [346]). The simplest example are oscillations of the 'white noise' type almost completely lacking correlation at various frequencies. Apparently, most geomagnetic pulsations can be classed as stochastic oscillations. The other class are non-stationary oscillations excited by broadband correlated sources. Let us first examine oscillations generated in the magnetosphere by stochastic sources.

3.11.1 Expressions for Physical Components of Alfvén Oscillation Magnetic Field

Let us examine the behaviour of Alfvén oscillations by addressing the physical components of their magnetic field: $\hat{B}_i(x^1, l, \omega) = \tilde{B}_i(x^1, l, \omega)/\sqrt{g_i}$, where the tilde means a separate harmonic with given frequency ω, and indices $i = 1, 2, 3$ correspond to components in the system of curvilinear coordinates (x^1, x^2, x^3) (see Figure 3.2). To obtain expressions for these components, we will represent the Fourier harmonic of the oscillation field scalar potential as the product of three functions:

$$\varphi(x^1, l, \omega) = \tilde{\varphi}_N(x^1, \omega)\tilde{Q}_N(x^1, \omega)\tilde{Z}_N(x^1, l, \omega). \tag{3.256}$$

Unlike notations in Sections 3.7–3.10, the oscillation amplitude $\tilde{\varphi}_N(x^1, \omega)$ is explicitly extracted here determined by the source (external currents in the ionosphere) and comparatively weakly depending on x^1. Function $\tilde{Q}_N(x^1, \omega)$ describes the small-scale structure of the wave across magnetic shells. Function $\tilde{Z}_N(x^1, l, \omega)$ describes the longitudinal structure of standing Alfvén wave. It is the solution of the same longitudinal eigenvalue problem (3.119) as the earlier-used functions $R_N(x^1, l, \omega)$ and $r_N(x^1, l, \omega)$, but differs from them in normalisation:

$$\left\langle \frac{q_N}{v_A} \tilde{Z}_N^2 \right\rangle = \frac{q_N^{(0)}}{v_{A0}},$$

where the angle brackets denote the field-line averaged value (3.206). Here $q_N = pk_{1N}^2 + p^{-1}k_2^2$, and the zero index denotes the equatorial values of the respective parameters (for $\theta = 0$). If the poloidal and toroidal longitudinal eigenfunctions are normalised as

$$\left\langle \frac{1}{pv_A} P_N^2 \right\rangle = \frac{1}{p_0 v_{A0}}, \qquad \left\langle \frac{p}{v_A} T_N^2 \right\rangle = \frac{p_0}{v_{A0}},$$

their relation to $\tilde{Z}_N(x^1, l, \omega)$ is defined by the following:

$$P_N(x^1, l) = \tilde{Z}_N(x^1, l, \Omega_{PN}(x^1)), \qquad T_N(x^1, l) = \tilde{Z}_N(x^1, l, \Omega_{TN}(x^1)).$$

For function $\tilde{Q}_N(x^1, \omega)$, describing the oscillation structure across magnetic shells, we obtain the following expressions:

(1) near the poloidal resonance surface, for $|x^1 - x_{PN}^1| \ll \Delta x_N^1$ (and $|\omega - \Omega_{PN}| \ll \Delta\Omega_N$)

$$\tilde{Q}_N(x^1, \omega) = G\left(\frac{x^1 - x_{PN}^1}{\lambda_{PN}} + i\varepsilon_{PN}\right) = G\left(\frac{\omega - \Omega_{PN}}{\omega_{PN}} + i\varepsilon_{PN}\right), \tag{3.257}$$

where $\omega_{PN} = \Omega_{PN}\lambda_{PN}/2a_N$;

(2) in the $x_{PN}^1 < x^1 < x_{TN}^1$ region (for $x^1 - x_{PN}^1 \gg \lambda_{PN}, (\gamma_N/\omega)a_N$ and $x_{TN}^1 - x^1 \gg \lambda_{TN}, (\gamma_N/\omega)a_N$), where (and if) the WKB approximation in x^1 applies,

$$\tilde{Q}_N(x^1, \omega) = \left(\frac{v_{PN}^1}{v_N^1} \frac{p_0^{-1}k_2^2}{p_0 k_{1N}^2 + p_0^{-1}k_2^2}\right)^{1/2} \times \exp\left[i\vartheta_N(x^1, \omega) - \Gamma_N(x^1, \omega) + i\frac{\pi}{4}\right], \tag{3.258}$$

(3) near the toroidal resonance surface, for $|x^1 - x_{TN}^1| \ll \Delta x_N^1$ (and $|\omega - \Omega_{TN}| \ll \Delta\Omega_N$)

$$\begin{aligned}\tilde{Q}_N(x^1, \omega) &= \frac{k_2\lambda_{TN}}{p_0} \exp\left[i\bar{\vartheta}_N(\omega) - \bar{\Gamma}_N(\omega) + i\frac{\pi}{2}\right] g\left(\frac{x^1 - x_{TN}^1}{\lambda_{TN}} + i\varepsilon_{TN}\right) \\ &= \frac{k_2\lambda_{TN}}{p_0} \exp\left[i\bar{\vartheta}_N(\omega) - \bar{\Gamma}_N(\omega) + i\frac{\pi}{2}\right] g\left(\frac{\omega - \Omega_{TN}}{\omega_{TN}} + i\varepsilon_{TN}\right),\end{aligned} \tag{3.259}$$

where $\omega_{TN} = \Omega_{TN}\lambda_{TN}/2a_N$. Here, function $G(z)$, describing the oscillation structure near the poloidal surface, has the following integral representation:

$$G(z) = \frac{i}{\sqrt{\pi}}\int_0^\infty \exp\left(isz - i\frac{s^3}{3}\right) ds, \tag{3.260}$$

and its asymptotics (accurate to the constant factor of order unity) are defined by (2.90). Function $g(z)$, describing the oscillation structure near the toroidal surface, is expressed via the modified zeroth-order Bessel function (3.200) and has the integral representation

$$g(z) = \frac{1}{\sqrt{\pi}}\int_0^\infty \exp\left(isz + \frac{i}{s}\right)\frac{ds}{s}. \tag{3.261}$$

Its asymptotics are defined by (3.201) and (3.203). The characteristic wavelength in the x^1 coordinate, near the poloidal (λ_{PN}) and toroidal (λ_{TN}) resonance surfaces is defined as (3.186) and (3.196), respectively. The standing Alfvén wave group velocity v_N^1 (in the x^1 coordinate) is defined as (3.130), and its characteristic value near the poloidal resonance surface is $v_{PN}^1 = \lambda_{PN}^2\Omega_{PN}/a_N$. The integral phase ϑ_N of oscillations is determined by (3.117), and the integral decrement Γ_N by (3.159). The full oscillation phase incursion $\overline{\vartheta}_N$ between resonance surfaces is defined as (3.210), and the full oscillation decrement $\overline{\Gamma}_N$ as (3.211).

Next, we will use expressions (E.2) for the oscillation magnetic field components (Appendix E) to write, for the physical components of the field of standing Alfvén oscillations,

$$\hat{B}_i(x^1, l, \omega) = \tilde{B}_N(x^1, \omega)\tilde{Q}_N^{(i)}(x^1, \omega)\tilde{Y}_N^{(i)}(x^1, l, \omega). \tag{3.262}$$

The following notations are introduced here:

$$\tilde{B}_N(x^1, \omega) = \frac{\bar{c}k_2}{v_{A0}\sqrt{g_2^{(0)}}}\tilde{\varphi}_N(x^1, \omega) \tag{3.263}$$

is the characteristic amplitude of the oscillation magnetic field,

$$\tilde{Y}_N^{(1)} = \sqrt{\frac{g_2^{(0)}}{g_2}}\tilde{Y}_N, \qquad \tilde{Y}_N^{(2)} = \sqrt{\frac{g_1^{(0)}}{g_1}}\tilde{Y}_N,$$

$$\tilde{Y}_N^{(3)} = \frac{\sqrt{g_2^{(0)}}}{k_2}\frac{\partial \ln p}{\partial l}\frac{q_N^{(0)}}{q_N}\tilde{Y}_N \tag{3.264}$$

are the functions describing the structure of its components along the background magnetic field lines, where

$$\tilde{Y}_N(x^1, l, \omega) = \frac{v_{A0}}{\omega}\frac{\partial \tilde{Z}_N(x^1, l, \omega)}{\partial l}. \tag{3.265}$$

For $\omega = \Omega_{PN}(x^1)$ we will denote the corresponding functions

$$Y_{PN}^{(1)} = \sqrt{\frac{g_2^{(0)}}{g_2}}Y_{PN}, \quad Y_{PN}^{(2)} = \sqrt{\frac{g_1^{(0)}}{g_1}}Y_{PN}, \quad Y_{PN}^{(3)} = \frac{\sqrt{g_2^{(0)}}}{k_2}\frac{\partial \ln p}{\partial l}\frac{p}{p_0}Y_{PN},$$

$$Y_{PN}(x^1,l)=\tilde{Z}_N(x^1,l,\Omega_{PN}(x^1))=\frac{v_{A0}}{\omega}\frac{\partial P_N(x^1,l)}{\partial l}.$$

Analogously, for $\omega=\Omega_{TN}(x^1)$

$$Y_{TN}^{(1)}=\sqrt{\frac{g_2^{(0)}}{g_2}}Y_{TN},\quad Y_{TN}^{(2)}=\sqrt{\frac{g_1^{(0)}}{g_1}}Y_{TN},\quad Y_{TN}^{(3)}=\frac{\sqrt{g_2^{(0)}}}{k_2}\frac{\partial\ln p}{\partial l}\frac{p_0}{p}Y_{TN},$$

$$Y_{TN}(x^1,l)=\tilde{Z}_N(x^1,l,\Omega_{TN}(x^1))=\frac{v_{A0}}{\omega}\frac{\partial T_N(x^1,l)}{\partial l}.$$

Moreover, we will denote

$$\tilde{Q}_N^{(1)}=\tilde{Q}_N,\qquad \tilde{Q}_N^{(2)}=i\frac{p_0}{k_2}\nabla_1\tilde{Q}_N \tag{3.266}$$

and function $\tilde{Q}_N^{(3)}$, satisfying the equation

$$\Delta_\perp^{(0)}\tilde{Q}_N^{(3)}=\frac{2k_2}{\sqrt{g_1^{(0)}g_2^{(0)}}}\nabla_1\tilde{Q}_N.$$

In the region where the WKB approximation applies,

$$\tilde{Q}_N^{(2)}=-p_0\frac{k_{1N}}{k_2}\tilde{Q}_N,\qquad \tilde{Q}_N^{(3)}=-\frac{2ik_{1N}k_2}{q_N^{(0)}}\tilde{Q}_N. \tag{3.267}$$

Near the poloidal surface,

$$\tilde{Q}_N^{(3)}=-\frac{2p_0}{k_2}\nabla_1\tilde{Q}_N,$$

and near the toroidal surface,

$$\tilde{Q}_N^{(3)}(x^1,\omega)=\frac{2k_2}{p_0}\int_{-\infty}^{x^1}Q_N(x^{1\prime},\omega)dx^{1\prime}.$$

Note that all functions $\tilde{Q}_N^{(i)}$ are dimensionless and of order unity, functions $\tilde{Y}_N^{(i)}$ are also dimensionless, and $\tilde{Y}_N^{(1)}$ and $\tilde{Y}_N^{(2)}$ are of order unity, and $\tilde{Y}_N^{(3)}\sim\tilde{Y}_N^{(1,2)}/m$.

3.11.2 Statistical Properties of the Oscillation Source

Let us address the Alfvén oscillations, the source of which is an external current in the ionosphere. The full density of the current $\mathbf{j}$ in the ionosphere can be represented as

$$\mathbf{j}=\hat{\sigma}\mathbf{E}+\mathbf{j}^{(\text{ext})},$$

where $\hat{\sigma}$ is the conductivity tensor, and $\mathbf{j}^{(\text{ext})}$ is the density of an external current of non-electromagnetic origin. It can be generated by neutrals moving in acoustic-gravity wave (AGW) or internal gravity wave (IGW) in the ionospheric E-layer. We will use $\tilde{j}_\parallel^{(\pm)}$ to denote the Fourier component of the longitudinal constituent of $\mathbf{j}^{(\text{ext})}$ in the conjugated ionospheres (plus for the Northern, and minus for the Southern ionosphere). Let us introduce functions $\tilde{J}^{(\pm)}(x^1,\omega)$ by defining them

$$\Delta_\perp^{(\pm)}\tilde{J}^{(\pm)}=\tilde{j}_\parallel^{(\pm)}. \tag{3.268}$$

Here,

$$\Delta_\perp^{(\pm)} = \frac{1}{g_1^{(\pm)}}\nabla_1^2 - \frac{k_2^2}{g_2^{(\pm)}}$$

is the transverse Laplacian for Alfvén waves at the ionospheric level. The standing Alfvén wave amplitude from formulas (3.262) and (3.263) is expressed in terms of $\tilde{J}^{(\pm)}$ as follows:

$$\tilde{B}_N = \frac{2\pi k_2}{\sqrt{g_1^{(0)}}\bar{c}\omega v_{A0} t_A}\frac{a_N}{\lambda_{PN}} \times \left[-\frac{\Sigma_P^{(+)}}{p_+}\tilde{J}^{(+)}\tilde{Y}_N(l_+) + \frac{\Sigma_P^{(-)}}{p_-}\tilde{J}^{(-)}\tilde{Y}_N(l_-)\right]. \tag{3.269}$$

Here $\Sigma_P^{(\pm)}$ are the integral Pedersen conductivities of the conjugated ionospheres.

For a stochastic source of the oscillations, functions $\tilde{J}^{(\pm)}(x^1, \omega)$ are random functions of frequency ω. Their statistical properties are characterised by setting correlation functions. Let us assume that external currents of the conjugated hemispheres are not correlated:

$$\left\langle \tilde{j}_\parallel^{(+)*}(x^1, \omega)\tilde{j}_\parallel^{(-)}(x^{1'}, \omega')\right\rangle = 0. \tag{3.270}$$

Hereafter the angle brackets denote statistical ensemble averaging. For autocorrelators we will take the simplest, 'white noise', model. In other words, we will set

$$\left\langle \tilde{j}_\parallel^{(\pm)*}(x^1, \omega)\tilde{j}_\parallel^{(\pm)}(x^{1'}, \omega')\right\rangle = f^{(\pm)}(x^1, x^{1'}, \omega)\delta(\omega - \omega'), \tag{3.271}$$

where $f^{(\pm)}(x^1, x^{1'}, \omega)$ are correlation functions.

Equation (3.268) implies that functions $\tilde{J}^{(\pm)}(x^1, \omega)$ are also random functions of ω and possess the statistical properties of 'white noise'. Indeed, let $\Upsilon^{(\pm)}(x^1, y^1, \omega)$ be a Green's function of (3.268), then

$$\tilde{J}^{(\pm)}(x^1, \omega) = \int_{-\infty}^{\infty} \Upsilon^{(\pm)}(x^1, y^1, \omega)\tilde{j}_\parallel^{(\pm)}\left(y^1, \omega\right) dy^1. \tag{3.272}$$

Integrating with respect to y^1 formally extends to the interval $(-\infty, \infty)$, but it is kept in mind that, in reality, it is realised within the localisation region of the external current $\tilde{j}_\parallel$. It follows from (3.270)–(3.272) that

$$\left\langle \tilde{J}^{(+)*}(x^1, \omega)\tilde{J}^{(-)}(x^{1'}, \omega')\right\rangle = 0, \tag{3.273}$$

$$\left\langle \tilde{J}^{(\pm)*}(x^1, \omega)\tilde{J}^{(\pm)}(x^{1'}, \omega')\right\rangle = F^{(\pm)}(x^1, x^{1'}, \omega)\delta(\omega - \omega'), \tag{3.274}$$

where

$$F^{(\pm)}(x^1, x^{1'}, \omega) = \int_{-\infty}^{\infty} dy^1 \int_{-\infty}^{\infty} dy^{1'}\,\tilde{\Upsilon}^{(\pm)}(x^1, y^1, \omega)\Upsilon^{(\pm)}(x^{1'}, y^{1'}, \omega)f^{(\pm)}(y^1, y^{1'}, \omega).$$

Using (3.269), (3.273), and (3.274), it is easy to verify whether the amplitude function $\tilde{B}_N(x^1, \omega)$ also possess 'white noise' properties. Below we will need a correlator,

$$\left\langle \tilde{B}_N^*(x^1, \omega)\tilde{B}_N(x^1, \omega')\right\rangle = \frac{1}{|\omega|}B_N^2(x^1, \omega)\delta(\omega - \omega'). \tag{3.275}$$

Here,

$$B_N^2(x^1,\omega) = \frac{1}{g_1^{(0)}\,|\omega|}\left(\frac{8\pi k_2}{\bar{c}v_{A0}t_A}\right)^2\left(\frac{a_N}{\lambda_{PN}}\right)^2 \times\left[\left(\frac{\Sigma_P^{(+)}\tilde{Y}_N(l_+)}{p_+}\right)^2 F^{(+)}(x^1,x^1,\omega) + \left(\frac{\Sigma_P^{(-)}\tilde{Y}_N(l_-)}{p_-}\right)^2 F^{(-)}(x^1,x^1,\omega)\right].$$

Note that function $B_N(x^1,\omega)$ has the dimension of magnetic field induction.

3.11.3 Spectral and Polarisation Properties of Alfvén Noise

The stochastic nature of the source leads to the stochastic character of the standing Alfvén waves it generates. To describe their statistical properties, we will examine the correlators of the components of the perturbed magnetic field. From formulas (3.262) and (3.275) we have

$$\left\langle \hat{B}_i^*(x^1,l,\omega)\hat{B}_j(x^1,l,\omega')\right\rangle = \mathscr{P}_{ij}(x^1,l,\omega)\delta(\omega-\omega'), \tag{3.276}$$

where

$$\mathscr{P}_{ij}(x^1,l,\omega) = \frac{1}{|\omega|}B_N^2(x^1,\omega) \times \tilde{Q}_N^{(i)*}(x^1,\omega)\tilde{Q}_N^{(j)}(x^1,\omega)\tilde{Y}_N^{(i)}(x^1,l,\omega)\tilde{Y}_N^{(j)}(x^1,l,\omega). \tag{3.277}$$

Thus, standing Alfvén waves are also 'white noise', and functions $\mathscr{P}_{ij}(x^1,l,\omega)$ have the sense of spectral densities of oscillations at a given point in space (x^1,l). In what follows, we will restrict ourselves to inspecting functions $\mathscr{P}_{11}$, $\mathscr{P}_{12}$, $\mathscr{P}_{22}$ and $\mathscr{P}_{33}$. The first three describe the transverse components and the last the longitudinal component of perturbed magnetic field of Alfvén oscillations.

Arbitrary functions of time can be represented as a Fourier integral:

$$\Phi(x^1,x^2,x^3,t) = \int_{-\infty}^{\infty}\tilde{\Phi}(x^1,x^3,\omega)e^{ik_2x^2-i\omega t}d\omega. \tag{3.278}$$

The amplitude of Alfvén waves at point (x^1,l) is determined by the correlators

$$P_{ij}(x^1,l) = \left\langle \hat{B}_i(x^1,l,t)\hat{B}_j(x^1,l,t)\right\rangle.$$

It is easy to obtain from (3.276) and (3.278)

$$P_{ij}(x^1,l) = 2\int_0^{\infty}\mathscr{P}_{ij}(x^1,l,\omega)d\omega. \tag{3.279}$$

Thanks to the chosen model random source being stationary, these correlators do not depend on time t.

The properties of spectral densities as functions of frequency ω are mainly determined by functions $\tilde{Q}_N^{(i)*}(x^1,\omega)$ and $\tilde{Q}_N^{(j)}(x^1,\omega)$, contained in (3.277). Their presence results in functions $\mathscr{P}_{ij}$ exhibiting sharp peaks near frequencies Ω_{PN} and Ω_{TN}. They are mainly concentrated in the interval $(\Omega_{PN},\Omega_{TN})$ and decrease fast away from it. A much smoother dependence of functions $\tilde{Y}_N^{(i)}$ on ω does not reflect qualitatively on the behaviour of $\mathscr{P}_{ij}$. To an even larger degree, this concerns function B_N^2. We will assume the source spectrum band to be much larger than $\Delta\Omega_N$, therefore function $B_N^2(x^1, \omega)$ can be assumed to be constant in the interval $(\Omega_{PN}, \Omega_{TN})$.

Of special interest is the issue of the frequencies the Alfvén noise spectrum is concentrated at (near Ω_{PN}, or near Ω_{TN}, or throughout the interval in between). In other words, which frequency region contributes most to the integral (3.279)? As will be seen below, the answer to this question depends on the magnitude of the standing Alfvén wave decay in the ionosphere. There are two limiting cases, of large or small decay. In the former case, monochromatic waves constituting Alfvén noise and travelling from the poloidal to the toroidal surface decay near the poloidal surface at a much smaller scale than the distance between the resonance surfaces. According to (3.139) and (3.155), the characteristic transit time between these surfaces, $\tau \sim m/\omega$, hence the wave attenuation coefficient, $\gamma_N \tau_N \sim m(\gamma_N/\omega)$. Strong decay means that $m(\gamma_N/\omega) \gg 1$. In this case, Alfvén oscillation energy is concentrated near Ω_{PN} and the oscillations are of poloidal character, i.e. the radial component of perturbed magnetic field is much larger than its azimuthal component. When decay is small, $m(\gamma_N/\omega) \ll 1$, monochromatic waves reach the neighbourhood of the toroidal surface without suffering noticeable decay and accumulate there thanks to a rapidly decreasing group velocity v_N^1 when approaching to the toroidal surface (see (3.136)). As a result, the noise spectrum is concentrated near frequency Ω_{TN} and their polarisation is of toroidal character.

Let us evaluate the correlators (3.279) in these limiting cases. When decay is large, that is the oscillations at a given magnetic shell are concentrated near the poloidal frequency, it is enough to use (3.257), for $\tilde{Q}_N(x^1, \omega)$. We have

$$\mathscr{P}_{11}(x^1, l, \omega) = \frac{1}{\Omega_{PN}} B_N^2(x^1, \Omega_{PN}) \left| G\left(\frac{\omega - \Omega_{PN}}{\omega_{PN}} + i\varepsilon_{PN}\right)\right|^2 Y_{PN}^{(1)^2}(x^1, l).$$

Here, for slowly varying functions ω, their values at point $\omega = \Omega_{PN}$ are used. Applying definition (3.279) and results in the Appendix G, we have

$$\begin{aligned} P_{11}(x^1, l) &= \frac{2\omega_{PN}}{\Omega_{PN}} B_N^2(x^1, \Omega_{PN}) Y_{PN}^{(1)^2}(x^1, l) \int_{-\infty}^{\infty} \left| G\left(\eta + i\varepsilon_{PN}\right)\right|^2 d\eta \\ &= \frac{\mu_N}{v_N} B_N^2(x^1, \Omega_{PN}) Y_{PN}^{(1)^2}(x^1, l). \end{aligned} \tag{3.280}$$

Here,

$$\mu_N = \frac{k_2 \lambda_{PN}^2}{p_0 a_N}, \qquad v_N = \frac{2k_2 a_N}{p_0} \frac{\gamma_{PN}}{\Omega_{PN}}.$$

On order of magnitude,

$$\mu_N \sim \frac{\alpha_N^{2/3}}{m^{1/3}}, \qquad v_N = m\frac{\gamma_{PN}}{\omega}.$$

In the case at hand, $v_N \gg 1$.

Remarkably, an utterly identical result can be obtained by using the WKB approximation (3.258) for $\tilde{Q}_N(x^1, \omega)$ despite the fact that it does not apply in the smaller neighbourhood of point $\omega = \Omega_{PN}$. In this approximation,

$$\mathscr{P}_{11}(x^1, l, \omega) = \frac{1}{\omega} B_N^2(x^1, \omega) \frac{v_{PN}^1}{v_N^1(x^1, \omega)} \frac{k_2^2/p_0^2}{k_{1N}^2(x^1, \omega) + k_2^2/p_0^2} e^{-2\Gamma_N(x^1,\omega)} \tilde{Y}_N^{(1)^2}(x^1, l, \omega).$$

Hence, using in (3.279) the integration variable

$$k_1 = k_{1N}(x^1, \omega) \tag{3.281}$$

and using relation (3.130), we have

$$P_{11}(x^1, l) = 2\mu_N B_N^2(x^1, \Omega_{PN}) \int_0^\infty dk_1 \frac{k_2/p_0}{k_1^2 + k_2^2/p_0^2} \times e^{-2\Gamma_N[x^1, \omega_N(x^1, k_1)]} \tilde{Y}_N^{(1)^2}(x^1, l, \omega_N(x^1, k_1)), \tag{3.282}$$

where function $\omega_N(x^1, k_1)$ is the solution of (3.281) relative to ω for given k_1. Using definitions (3.159) and (3.129) (see also [347]) it is possible to demonstrate that

$$\Gamma_N\left[x^1, \omega_N(x^1, k_1)\right] \approx \begin{cases} \frac{\gamma_{PN}}{|\nabla_1 \Omega_{PN}|} k_1, & k_1 \ll k_2/p_{0,} \\ \overline{\Gamma}_N + \frac{\gamma_{PN}}{|\nabla_1 \Omega_{PN}|} k_1, & k_1 \gg k_2/p_{0.} \end{cases} \tag{3.283}$$

Using this expression it is easy to verify that, for $v_N \gg 1$, thanks to rapidly decaying exponent, the integral (3.282) gains most from

$$k_1 \lesssim \frac{|\nabla_1 \Omega_{PN}|}{\gamma_{PN}} \sim \frac{1}{v_N} \frac{k_2}{p_0} \ll \frac{k_2}{p_0}.$$

In terms of variable ω, according to relations (3.125) and (3.127), this means that the spectrum width

$$\Delta\omega \sim \omega - \Omega_{PN} \lesssim \frac{\Delta\Omega_N}{v_N^2}.$$

The integral (3.282) is in this case easily calculated and we again arrive at (3.280).

When calculating the other correlators, the WKB approximation and expression (3.258) also produce identical results. We have

$$P_{12} = -\frac{\mu_N}{2v_N^2} B_N^2(x^1, \Omega_{PN}) Y_{PN}^{(1)}(x^1, l) Y_{PN}^{(2)}(x^1, l),$$

$$P_{22} = \frac{\mu_N}{2v_N^3} B_N^2(x^1, \Omega_{PN}) Y_{PN}^{(2)^2}(x^1, l),$$

$$P_{33} = \frac{\mu_N}{2v_N^3} B_N^2(x^1, \Omega_{PN}) Y_{PN}^{(3)^2}(x^1, l).$$

Using the findings in Appendix H, we calculate the semiaxes of the ellipse of the transverse polarisation of Alfvén oscillations and its inclination to the coordinate x^1 line:

$$\left\langle \hat{B}_{\perp\max}^2 \right\rangle = \frac{\mu_N}{v_N} B_N^2(x^1, \Omega_{PN}) Y_{PN}^{(1)^2}(x^1, l),$$

$$\left\langle \hat{B}_{\perp\min}^2 \right\rangle = \frac{\mu_N}{4v_N^3} B_N^2(x^1, \Omega_{PN}) Y_{PN}^{(2)^2}(x^1, l),$$

$$\alpha_0 = -\frac{1}{2v_N} \frac{Y_{PN}^{(2)}(x^1, l)}{Y_{PN}^{(1)}(x^1, l)}.$$

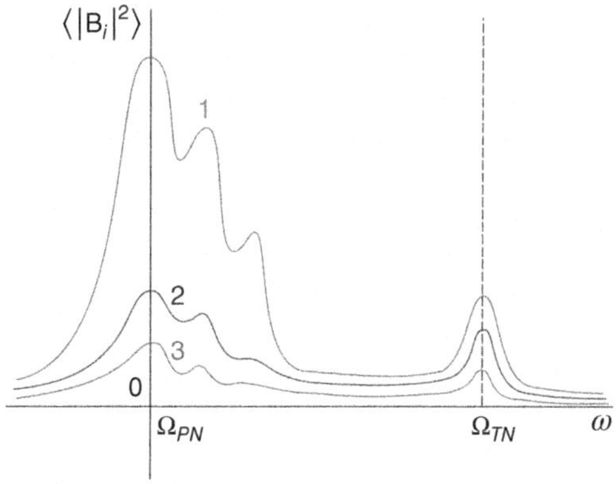

Figure 3.40 Schematic plots of spectral density of the components$\langle |B_i|^2 \rangle$ ($i = 1, 2, 3$) of perturbed magnetic field of standing Alfvén waves for strong decay of oscillations ($\nu_N \gg 1$).

Thus, for $\nu_N \gg 1$, Alfvén noise has a poloidal character. Its spectrum is concentrated near frequency $\omega = \Omega_{PN}$, the characteristic spectrum width $\Delta\omega \sim \Delta\Omega_N/\nu_N^2 \ll \Delta\Omega_N$. The polarisation ellipse is strongly elongated,

$$\left\langle \hat{B}^2_{\perp\min} \right\rangle \Big/ \left\langle \hat{B}^2_{\perp\max} \right\rangle \sim \frac{1}{\nu_N^2} \ll 1,$$

and oriented practically along the x^1 coordinate. Note also that

$$\left\langle \hat{B}^2_{\parallel} \right\rangle \Big/ \left\langle \hat{B}^2_{\perp\max} \right\rangle \sim \frac{1}{m^2\nu_N^2} \ll 1.$$

The schematic plots of the spectral densities for various components of perturbed magnetic field, for $\nu_N \gg 1$, are shown in Figure 3.40.

Let us address the $\nu_N \ll 1$ case. It can be seen from the relations in (3.283) that the exponent in the integrand of (3.282) in this case, decreases very slowly, the characteristic scale of its decrease

$$k_1 \sim \frac{|\nabla_1 \Omega_{PN}|}{\gamma_{PN}} \sim \frac{1}{\nu_N}\frac{k_2}{p_0} \gg \frac{k_2}{p_0}.$$

This exponential can then be omitted, because the integral remains converging:

$$P_{11}(x^1, l) = 2\mu_N B_N^2(x^1, \Omega_{PN}) \int_0^\infty dk_1 \frac{k_2/p_0}{k_1^2 + k_2^2/p_0^2} \tilde{Y}_N^{(1)^2}(x^1, l, \omega_N(x^1, k_1)). \quad (3.284)$$

This integral gains most from $k_1 \sim k_2/p_0$, that is, in terms of variable ω, for

$$\omega - \Omega_{PN} \sim \Delta\Omega_N.$$

It is impossible to find the integral in (3.284) in the explicit form because of the presence of $\tilde{Y}_N^{(1)^2}$, but it is not difficult to obtain an order of magnitude estimate,

$$P_{11} \sim \mu_N B_N^2. \quad (3.285)$$

If the WKB approximation is also used for correlator $\mathscr{P}_{22}$, we obtain, instead of (3.282),

$$P_{22} = \frac{2p_0\mu_N}{k_2} B_N^2(x^1, \Omega_{TN})$$
$$\times \int_0^\infty dk_1 \frac{k_1^2}{k_1^2 + k_2^2/p_0^2} e^{-2\Gamma_N[x^1, \omega_N(x^1, k_1)]} \tilde{Y}_N^{(2)^2}(x^1, l, \omega_N(x^1, k_1)). \quad (3.286)$$

Here, the exponential cannot be omitted in the integrand, or the integral will become diverging. Using (3.283), it is easy to find out that, in this case, the characteristic domain of integration with respect to k_1

$$k_1 \sim \frac{|\nabla_1 \Omega_{PN}|}{\gamma_{TN}} \sim \frac{1}{\nu_N}\frac{k_2}{p_0} \gg \frac{k_2}{p_0}.$$

In terms of the variable ω this means that the spectrum is concentrated near $\omega = \Omega_{TN}$ and it follows from (3.126) and (3.127) that its characteristic width

$$\Delta\omega \sim \Omega_{TN} - \omega - \nu_N^2 \Delta\Omega_N \ll \Delta\Omega_N.$$

Given the above, the integral (3.286) is easily calculated:

$$P_{22} = \frac{\mu_N}{\nu_N}\frac{\gamma_{PN}}{\gamma_{TN}} B_N^2(x^1, \Omega_{TN}) Y_{TN}^{(2)^2}(x^1, l). \tag{3.287}$$

It is taken into account here that $\overline{\Gamma}_N \sim \nu_N \ll 1$.

The spectrum being concentrated within a small neighbourhood of toroidal frequency Ω_{TN} in the last case, the question arises whether it is correct to use the WKB approximation, since this approximation is inapplicable for $|\omega - \Omega_{TN}| \lesssim 1$. For this reason, we will calculate P_{22} using (3.259) for $\tilde{Q}_N$. From (3.266) we have

$$P_{22} = \frac{2}{\Omega_{TN}} B_N^2(x^1, \Omega_{TN}) Y_{TN}^{(2)^2}(x^1, l) \int_{-\infty}^{\infty} \left| g'\left(\frac{\omega - \Omega_{TN}}{\omega_{TN}} + i\varepsilon_{TN}\right)\right|^2 d\omega.$$

Using the value of the integral in Appendix G we again arrive at (3.287).

It is analogously established that the spectral density $\mathscr{P}_{12}(x^1, l, \omega)$ is also concentrated near frequency $\omega = \Omega_{TN}(x^1)$ over the interval $\Delta\omega = \nu_N^2 \Delta\Omega_N$. The corresponding correlator

$$P_{12} = -2\mu_N \left(\ln \frac{1}{\nu_N}\right) B_N^2(x^1, \Omega_{TN}) Y_{TN}^{(1)}(x^1, l) Y_{TN}^{(2)}(x^1, l), \tag{3.288}$$

and, notably, WKB calculations and those based on (3.259) produce the same result.

Relations (3.280), (3.287) and (3.288) can be used to find the semi-axes of the ellipse of the transverse polarisation of the oscillations and the inclination of the ellipse to the x^1 axis:

$$\left\langle \hat{B}_{\perp\max}^2 \right\rangle = P_{22} = \frac{\mu_N}{\nu_N}\frac{\gamma_{PN}}{\gamma_{TN}} B_N^2(x^1, \Omega_{TN}) Y_{TN}^{(2)^2}(x^1, l),$$

$$\left\langle \hat{B}_{\perp\min}^2 \right\rangle = P_{11} \sim \mu_N B_N^2(x^1, \Omega_{TN}),$$

$$\alpha_0 = \frac{\pi}{2} + 2\nu_N \frac{Y_{TN}^{(1)}(x^1, l)}{Y_{TN}^{(2)}(x^1, l)} \ln \frac{1}{\nu_N}.$$

As for the correlator P_{33}, we have, in the WKB approximation, and given (3.267),

$$P_{33} = 8\mu_N B_N^2(x^1, \Omega_{TN}) \int_0^{\infty} dk_1 \frac{k_1^2 (k_2/p_0)^3}{(k_1^2 + k_2^2/p_0^2)^3} e^{-2\Gamma_N[x^1, \omega_N(x^1, k_1)]} \tilde{Y}_N^{(3)^2}(x^1, l, \omega_N(x^1, k_1)).$$

Here, the exponential factor can be omitted same as in P_{11} calculations. The spectral density is distributed throughout the interval $(\Omega_{PN}, \Omega_{TN})$. The integral is not taken explicitly, but an order-of-magnitude estimate is easily obtained:

$$P_{33} \equiv \left\langle \hat{B}_{\parallel}^2 \right\rangle \sim \mu_N B_N^2 \tilde{Y}_N^{(3)^2} \sim \frac{\mu_N}{m^2} B_N^2.$$

Figure 3.41 Schematic plots of spectral density of the components $\langle |B_i|^2 \rangle$ ($i = 1, 2, 3$) of perturbed magnetic field of standing Alfvén waves for weak decay of oscillations ($\nu_N \ll 1$).

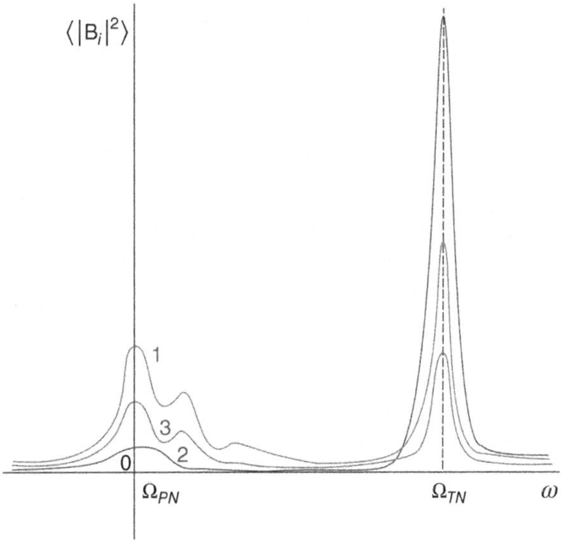

Thus, the oscillations are toroidal in character for $\nu_N \ll 1$. The semi-major axis of the polarisation ellipse is practically along the azimuth. Oscillations along this axis have a narrow spectrum concentrated near frequency $\omega = \Omega_{TN}(x^1)$, with characteristic width $\Delta\omega \sim \nu_N^2 \Delta\Omega_N$. Oscillations along the semi-minor axis of the transverse polarisation ellipse, as well as along the geomagnetic field have a wider spectrum confined within the interval $(\Omega_{PN}, \Omega_{TN})$, their amplitudes being relatively small:

$$\frac{\left\langle \hat{B}^2_{\perp\min} \right\rangle}{\left\langle \hat{B}^2_{\perp\max} \right\rangle} \sim \nu_N, \qquad \frac{\left\langle \hat{B}^2_{\parallel} \right\rangle}{\left\langle \hat{B}^2_{\perp\max} \right\rangle} \sim \frac{\nu_N}{m^2}.$$

Schematic plots of spectral densities for different components of perturbed magnetic field when $\nu_N \ll 1$ are shown in Figure 3.41.

3.12 Broadband Standing High-*m* Alfvén Waves Generated by Correlated Sources

Let us now examine standing Alfvén waves generated in the magnetosphere by non-stationary correlated sources. This means that source function $\tilde{\varphi}_N(x^1, \omega)$ in (3.256) will be assumed to be a given function of coordinate and frequency. Consequently, its Fourier transform

$$\varphi_N\left(x^1, t\right) = \int_{-\infty}^{\infty} d\omega \tilde{\varphi}_N\left(x^1, \omega\right) e^{-i\omega t} \tag{3.289}$$

is a given function of coordinate and time. As noted above, such a treatment is justified for oscillations related to magnetospheric restructuring processes, the reaction of the magnetosphere to sharp impacts of both external and internal origin. Characteristic examples are geomagnetic pulsations Pi2 and sudden commencement (SC)/SSC oscillations.

The specific task of this section is reduced to an inverse Fourier transformation

$$\varphi\left(x^1, l, t\right) = \int_{-\infty}^{\infty} d\omega \tilde{\varphi}\left(x^1, l, \omega\right) e^{-i\omega t}, \tag{3.290}$$

by means of the expression for function $\tilde{\varphi}_N(x^1, l, \omega)$ obtained earlier in Sections 3.7.5–3.7.7 and written in the general form in Section 3.11.1. Substituting (3.256) into (3.290) produces

$$\varphi\left(x^1, l, t\right) = \int_{-\infty}^{\infty} d\omega \tilde{\varphi}_N\left(x^1, \omega\right) \tilde{Q}_N\left(x^1, \omega\right) \tilde{Z}_N\left(x^1, l, \omega\right) e^{-i\omega t}. \tag{3.291}$$

By means of (3.289), this expression can be represented as

$$\varphi\left(x^1, l, t\right) = \frac{1}{2\pi} \int_{-\infty}^{\infty} dt' \varphi_N\left(x^1, t - t'\right) \int_{-\infty}^{\infty} d\omega \tilde{Q}_N\left(x^1, \omega\right) \tilde{Z}_N\left(x^1, l, \omega\right) e^{-i\omega t'}.$$

We will use the saddle-point method to calculate the integral with respect to frequency in this expression because function $\tilde{Q}_N(x^1, \omega)$ varies rapidly with respect to variable ω. The saddle point is a function of parameters x^1 and t':

$$\omega = \Omega_N\left(x^1, t'\right). \tag{3.292}$$

It will be demonstrated below that function $\Omega_N(x^1, t)$ has a simple physical sense. If the saddle-point method is used the slowly varying function $\tilde{Z}_N(x^1, l, \omega)$ can be taken out of the integrand, taking its value at the saddle point. Introducing notations

$$Z_N\left(x^1, l, t\right) = \tilde{Z}_N\left(x^1, l, \Omega_N\left(x^1, t\right)\right), \tag{3.293}$$

$$Q_N\left(x^1, t\right) = \int_{-\infty}^{\infty} d\omega \tilde{Q}_N\left(x^1, \omega\right) e^{-i\omega t}, \tag{3.294}$$

we have

$$\varphi\left(x^1, l, t\right) = \frac{1}{2\pi} \int_{-\infty}^{\infty} dt' \varphi_N\left(x^1, t - t'\right) Q_N\left(x^1, t'\right) Z_N\left(x^1, l, t'\right). \tag{3.295}$$

Formula (3.295) expresses the Alfvén wave potential in terms of source function $\varphi_N(x^1, t)$. It is easy to see that functions $Q_N(x^1, t)$ and $Z_N(x^1, l, t)$ describe the Alfvén wave excited by a source of the instantaneous pulse type. Assuming

$$\varphi_N\left(x^1, t\right) = \varphi_N\left(x^1\right) \delta(t),$$

we have

$$\varphi\left(x^1, l, t\right) = \frac{1}{2\pi} \varphi_N\left(x^1\right) Q_N\left(x^1, t\right) Z_N\left(x^1, l, t\right). \tag{3.296}$$

Function $Z_N(x^1, l, t)$ describes the longitudinal structure of standing Alfvén wave, and the main dependence on the x^1 coordinate and time t results from the rapidly varying function $Q_N(x^1, t)$.

The response to a source of the instantaneous pulse type is in a certain sense an opposite case to the monochromatic wave. Studying the wave structure in these two extreme limits also allows a general idea to be formed about all the intermediate cases. Therefore we will restrict ourselves to examining the response to the source of the instantaneous pulse type. Studying the general case (3.295) requires that a particular model source $\varphi_N(x^1, t)$ be set. It should be noted that the above-mentioned examples (SC, SSC and substorm Pi2) can very well be treated as the response of the magnetosphere to a source of the instantaneous pulse type.

3.12.1 Response of Magnetospheric Alfvén Oscillations to Instantaneous Pulse

Let us first assume that the saddle point is inside the interval $(\Omega_{PN}, \Omega_{TN})$, far enough from its ends (we will formulate this condition more strictly below). Using expression (3.258), we then have

$$Q_N\left(x^1,t\right) = \int\limits_{-\infty}^{\infty} d\omega \left[\frac{v^1_{PN}}{v^1_N\left(x^1,\omega\right)} \frac{p_0^{-1}k_2^2}{p_0 k_{1N}^2\left(x^1,\omega\right) + p_0^{-1}k_2^2}\right]^{1/2} \times \exp\left[i\Psi_N\left(x^1,\omega,t\right) - \Gamma_N(x^1,\omega) + i\frac{\pi}{4}\right]. \tag{3.297}$$

Here,

$$\Psi_N\left(x^1,\omega,t\right) = \vartheta_N(x^1,\omega) - \omega t.$$

The saddle point is defined by

$$\frac{\partial \Psi_N\left(x^1,\omega,t\right)}{\partial\omega} = 0,$$

which, given relations (3.117) and (3.139), can be represented as

$$\tau\left(x^1,\omega\right) = t. \tag{3.298}$$

This equation has a simple physical sense. It determines the frequency of the monochromatic wave reaching shell x^1 a certain time t after it was generated at its poloidal surface. It is this wave that determines the oscillation at shell x^1 at time t.

It is not difficult to obtain the limiting expressions for small and large t, for function $\omega = \Omega_N(x^1, t)$, determined by (3.298). In the two limiting cases we have

$$\Omega_N\left(x^1,t\right) = \Omega_{PN}\left(x^1\right) + \omega^3_{PN}(x^1)t^2, t \ll m/\Omega, \tag{3.299}$$

$$\Omega_N\left(x^1,t\right) = \Omega_{TN}\left(x^1\right) - \frac{1}{\omega_{TN}(x^1)t^2}, t \gg m/\Omega. \tag{3.300}$$

Here, Ω is a value of order Ω_{PN} or Ω_{TN}, and functions $\omega_{PN}(x^1)$ and $\omega_{TN}(x^1)$ are defined in (3.257) and (3.259). Thus, as t changes in the interval $(0, \infty)$ function $\Omega_N(x^1, t)$ changes in the interval $(\Omega_{PN}(x^1), \Omega_{TN}(x^1))$.

Let us determine the time-dependent quasiclassical wave vector

$$\overline{k}_{1N}\left(x^1,t\right) = k_{1N}\left[x^1, \Omega_N\left(x^1,t\right)\right] \tag{3.301}$$

and phase

$$\overline{\Psi}_N\left(x^1,t\right) = \Psi_N\left[x^1, \Omega_N\left(x^1,t\right), t\right] \equiv \vartheta_N\left[x^1, \Omega_N\left(x^1,t\right)\right] - \Omega_N\left(x^1,t\right) t. \tag{3.302}$$

It is easy to verify that

$$\frac{\partial \overline{\Psi}_N\left(x^1,t\right)}{\partial x^1}=\overline{k}_{1N}\left(x^1,t\right);\qquad \frac{\partial \overline{\Psi}_N\left(x^1,t\right)}{\partial t}=-\Omega_N\left(x^1,t\right). \tag{3.303}$$

Hence,

$$\frac{\partial \overline{k}_{1N}(x^1,t)}{\partial t}=-\frac{\partial \Omega_N\left(x^1,t\right)}{\partial x^1}.$$

Using these relations and limiting expressions (3.299) and (3.300) for $\Omega_N(x^1_, t)$ corresponding expressions for $\overline{k}_{1N}(x^1,t)$ and $\overline{\Psi}_N(x^1,t)$ can be obtained:

$$\overline{k}_{1N}(x^1,t)=\begin{cases} -\nabla_1\Omega_{PN}(x^1)t, & t\ll m/\Omega,\\ -\nabla_1\Omega_{TN}(x^1)t, & t\gg m/\Omega,\end{cases} \tag{3.304}$$

$$\overline{\Psi}_N(x^1,t)=\begin{cases} -\Omega_{PN}(x^1)t-\frac{\omega^3_{PN}(x^1)t^3}{3}, & t\ll m/\Omega,\\ \tilde{\Psi}_N(x^1)-\Omega_{TN}(x^1)t-\frac{1}{\omega_{TN}(x^1)t}, & t\gg m/\Omega.\end{cases} \tag{3.305}$$

Here,

$$\tilde{\Psi}_N(x^1)=\int_0^\infty\left[\Omega_{TN}\left(x^1\right)-\Omega_N\left(x^1,t\right)\right]dt. \tag{3.306}$$

It can be shown that

$$\tilde{\Psi}_N(x^1)=\overline{\vartheta}_N\left[\Omega_{TN}\left(x^1\right)\right]. \tag{3.307}$$

We will also need the time-dependent wave-decay coefficient

$$\overline{\Gamma}_N(x^1,t)=\Gamma_N[(x^1,\Omega_N\left(x^1,t\right)], \tag{3.308}$$

having the following limiting expressions

$$\overline{\Gamma}_N(x^1,t)=\begin{cases}\gamma_{PN}(x^1)t, & t\ll m/\Omega,\\ \tilde{\Gamma}_N(x^1)+\gamma_{TN}(x^1)t, & t\gg m/\Omega.\end{cases} \tag{3.309}$$

Here, $\tilde{\Gamma}_N(x^1)\equiv\tilde{\Gamma}_N(x^1,\Omega_{TN}(x^1))$, $\gamma_{TN}(x^1)\equiv\gamma_{TN}(\Omega_{TN}(x^1))$, $\gamma_{PN}(x^1)\equiv\gamma_{PN}(\Omega_{TN}(x^1))$, where function $\tilde{\Gamma}_N(x^1,\omega)$ is defined as

$$\tilde{\Gamma}_N(x^1,\omega)=\int_{x^1_{PN}(\omega)}^{x^1_{TN}(\omega)}\frac{\gamma_N(x^1,\omega)-\tilde{\gamma}_{TN}(\omega)}{v^1_N(x^1,\omega)}dx^1+\frac{\gamma_{TN}}{\omega_{TN}}\left(\frac{\omega_{TN}}{\Omega_{TN-\omega}}\right)^{1/2},$$

functions $\gamma_N(x^1,\omega)$, $\gamma_{PN}(\omega)$ and $\gamma_{TN}(\omega)$, respectively, as (3.163)–(3.165), and $\tilde{\gamma}_{TN}(\omega)=\gamma(x^1_{TN}(\omega),\omega)$.

Returning to the saddle-point method for calculating the integral (3.297), we have

$$\frac{\partial^2\Psi_N}{\partial\omega^2}=\frac{\partial\tau_N}{\partial\omega}=-\frac{1}{\nabla_1\Omega_{PN}}\frac{\partial k_{1N}}{\partial\omega}=-\frac{1}{\nabla_1\Omega_{PN}}\frac{1}{v^1_N}.$$

Therefore, near the saddle point,

$$\Psi_N(x^1,\omega,t)\approx\overline{\Psi}_N(x^1,t)-\frac{\left[\omega-\Omega_N\left(x^1,t\right)\right]^2}{2v^1_N(x^1,\Omega_N)\nabla_1\Omega_{PN}}.$$

When saddle-point integrating with respect to ω, the peak width is, on order of magnitude,

$$\Delta\omega \sim \left|v_N^1 \nabla_1 \Omega_{PN}\right|^{1/2} \sim \frac{\Delta\Omega_N}{(\alpha_N m)^{1/2}}.$$

Recall that $\alpha_N = \Delta\Omega_N/\Omega_{TN}$. It is assumed that $\alpha_N m \gg 1$, therefore $\Delta\omega \ll \Delta\Omega_N$ and applying the saddle-point method is justified if $\Omega_N(x^1,\omega)$ is not too close to the end of the interval $(\Omega_{PN}, \Omega_{TN})$. After the above, calculating the integral (3.297) is not difficult. We have

$$Q_N(x^1,t) = 2\sqrt{\pi}\omega_{PN}\left[\frac{k_2^2/p_0^2}{\overline{k}_{1N}^{\,2}(x^1,t) + k_2^2/p_0^2}\right]^{1/2} \times \exp\left[i\overline{\Psi}_N(x^1,t) - \overline{\Gamma}_N(x^1,t) + i\pi/2\right]. \tag{3.310}$$

If the saddle point $\omega = \Omega_N(x^1,t)$ is close to the poloidal or toroidal frequency, the above calculation is invalid. Let us specify the corresponding conditions. For the above-discussed integration technique to be justified near the poloidal frequency the following inequalities must hold:

$$\omega - \Omega_{PN} \gg \omega_{PN}, \qquad \Delta\omega \sim \left(\frac{\Omega_{PN} v_N^1}{a_N}\right)^{1/2} \ll \omega - \Omega_{PN}. \tag{3.311}$$

The former warrants the WKB approximation to be applied to $\tilde{Q}_N(x^1,\omega)$ near the saddle point, and the latter the saddle-point technique itself. Near the poloidal frequency, we have

$$\Delta\omega \sim \omega_{PN}\left(\frac{\omega - \Omega_{PN}}{\omega_{PN}}\right)^{1/4},$$

whence the latter inequality (3.311) is reduced to the former. Analogously, the following inequalities need to be satisfied near the toroidal frequency

$$\Omega_{TN} - \omega \gg \omega_{TN}, \qquad \Delta\omega \ll \Omega_{TN} - \omega. \tag{3.312}$$

Near the toroidal frequency we have

$$\Delta\omega \sim \omega_{TN}\left(\frac{\Omega_{TN} - \omega}{\omega_{TN}}\right)^{3/4},$$

and the latter inequality (3.312) is also reduced to the former.

If inequality (3.311) does not hold, i.e. $\Omega_N - \Omega_{PN} \lesssim \omega_{PN}$, expression (3.257) can be substituted into (3.294). Using (3.260) and changing integration order in the resulting multiple integral yields

$$Q_N(x^1,t) = 2i\sqrt{\pi}\omega_{PN}\theta(t)\exp\left(-i\Omega_{PN}t - \frac{i}{3}\omega_{PN}^3 t^3 - \gamma_{PN}t\right), \tag{3.313}$$

where $\theta(t)$ is the Heaviside function

$$\theta(t) = \begin{cases} 0, & t < 0, \\ 1, & t > 0. \end{cases}$$

It follows from the inequality $\Omega_N - \Omega_{PN} \lesssim \omega_{PN}$ that

$$t \lesssim \frac{1}{\omega_{PN}} \ll \frac{m}{\Omega}. \tag{3.314}$$

This implies, however, that

$$\overline{k}_{1N}\left(x^1,t\right) \sim \nabla_1\Omega_{PN}t \lesssim \frac{1}{\lambda_{PN}} \ll \frac{k_2}{p_0},$$

and the approximate expressions in the upper lines of (3.305) and (3.309) are valid for functions $\overline{\Psi}_N(x^1,t)$ and $\overline{\Gamma}_N(x^1,t)$. Given the above we can see that formulas (3.310) and (3.313) coincide.

Analogously, for $\Omega_{TN} - \Omega_N \lesssim \omega_{TN}$, expression (3.259) must be substituted into (3.294),

$$Q_N(x^1,t) = i\frac{k_2\lambda_{PN}}{p_0} \times \int_{-\infty}^{\infty} d\omega g\left(\frac{\omega - \Omega_{TN} + i\gamma_{TN}}{\omega_{TN}}\right)\exp\left[-i\omega t + i\overline{\vartheta}_N(\omega) - \tilde{\Gamma}_N(\omega)\right]. \quad (3.315)$$

$\exp[-\tilde{\Gamma}_N(\omega)]$ can be taken out of the integrand at point $\omega = \Omega_{TN}$ (resulting in $\exp[-\tilde{\Gamma}_N(x^1)]$), but dependence on ω in term $\exp[i\overline{\vartheta}_N(\omega)]$ must be taken into account, because $\overline{\vartheta}_N(\omega)$ is a large phase. For this purpose, it is sufficient to linearly expand this phase near point $\omega = \Omega_{TN}$

$$\overline{\vartheta}_N(\omega) = \overline{\vartheta}_N(\Omega_{TN}(x^1)) + \left.\frac{\partial\overline{\vartheta}_N}{\partial\omega}\right|_{\omega=\Omega_{TN}} \times (\omega - \Omega_{TN})$$
$$= \tilde{\Psi}_N(x^1) + \tau_N(\Omega_{TN})(\omega - \Omega_{TN}(x^1)).$$

After this, using the integral representation (3.261) and changing the integration order in the resulting multiple integral will produce

$$Q_N(x^1,t) = 2i\sqrt{\pi}\frac{k_2\lambda_{PN}}{p_0}\frac{1}{t} \times \exp\left(-i\Omega_{TN}t + i\tilde{\Psi}_N(x^1) - \frac{i}{\omega_{TN}t} - \Gamma_N(x^1) - \gamma_{TN}t\right). \quad (3.316)$$

The inequality $\Omega_{TN} - \Omega_N \lesssim \omega_{TN}$ means that

$$t \gtrsim \frac{1}{\omega_{TN}} \gg \frac{m}{\Omega},$$

and hence expressions in the lower lines of (3.304), (3.305) and (3.309) are applicable, and the condition $\overline{k}_{1N}(x^1,t) \gg k_2/p_0$ holds. It can be easily seen, however, that in this case expression (3.310) transforms into (3.316).

Thus, formula (3.310), originally obtained in the interval $(1/\omega_{PN}) \ll t \ll (1/\omega_{TN})$, proves to be applicable to any values of variable t, and formulas (3.313) and (3.316) are its limiting expressions for, respectively, $t \ll m/\Omega$ and $t \gg m/\Omega$.

Having obtained an expression for the transverse potential $\varphi(x^1,l,t)$ it is not difficult to evaluate the components of perturbed electric and magnetic fields of the Alfvén oscillations. Let us restrict ourselves to calculating the perturbed magnetic field. Analogously to definition (3.293), we will introduce time-dependent longitudinal functions for magnetic field

$$Y_N^{(i)}\left(x^1,l,t\right) = \tilde{Y}_N^{(i)}[x^1,l,\Omega_N(x^1,t)]$$

and time-dependent amplitude of magnetic field

$$B_N(x^1,t) = \tilde{B}_N(x^1,\Omega_N(x^1,t)),$$

where functions $\tilde{B}_N(x^1,\omega)$ and $\tilde{Y}_N^{(i)}(x^1,\omega)$ are defined by (3.263)–(3.265). Using these definitions we will obtain the following expression for physical components of perturbed magnetic field:

$$\hat{B}_i\left(x^1,l,t\right) = B_N(x^1,t)Q_N^{(i)}(x^1,t)Y_N^{(i)}\left(x^1,l,t\right).$$

Here (compare with (3.266)):

$$Q_N^{(1)}(x^1,t) = Q_N(x^1,t);$$

$$Q_N^{(2)}(x^1,t) = -\frac{\bar{k}_{1N}\left(x^1,t\right)}{k_2/p_0}Q_N(x^1,t);$$

$$Q_N^{(3)}(x^1,t) = -2i\frac{\bar{k}_{1N}\left(x^1,t\right)\cdot(k_2/p_0)}{\bar{k}_{1N}^2(x^1,t)+(k_2/p_0)^2}Q_N(x^1,t).$$

The obtained results can be given a simple and illustrative physical interpretation. An instantaneous pulse-type source has a very wide (formally, infinite) spectrum and excites Alfvén waves at all magnetic shells simultaneously. Monochromatic wave, for which this shell is poloidal, is excited at given magnetic shell x^1. Correspondingly, wave frequency ω is equal to the poloidal frequency $\Omega_{PN}(x^1)$ of this shell, and the wave polarisation is poloidal in character (magnetic field oscillates in the radial direction, in the x^1 coordinate). After this, each of the monochromatic waves travels in the radial direction, i.e. towards its toroidal surface, and as it travels, its polarisation gradually transforms from poloidal to toroidal.

It is not difficult to understand how this pattern will look at fixed magnetic shell x^1. At the first moment, a poloidal type oscillation will be excited at this shell, its frequency equal to the poloidal frequency $\Omega_{PN}(x^1)$ of the given magnetic shell. This oscillation will presently escape the given magnetic shell, starting to move towards its toroidal surface. Oscillations from other, more and more remote, shells, where they were generated as poloidal oscillations, will reach magnetic shell x^1. The more remote is such a shell from x^1, the stronger is the change of the oscillation polarisation from poloidal to toroidal over the oscillation transit time. Finally, shell x^1 will be reached by the oscillation of frequency $\Omega_{TN}(x^1)$, for which this shell is toroidal.

Thus, an oscillation slowly varying from the poloidal to the toroidal type will be observed at this shell: its frequency changes from $\Omega_{PN}(x^1)$ to $\Omega_{TN}(x^1)$, and its polarisation, from radial to azimuthal. The characteristic time $\sim m/\omega$ of such a variation, i.e. the oscillation is poloidal in character when $t \ll m/\omega$, and toroidal when $t \gg m/\omega$.

The above-described pattern features a small enough decay, when the wave decay coefficient $\Gamma \sim \gamma m/\omega$ is small over the characteristic time m/ω:

$$m\frac{\gamma}{\omega} \ll 1.$$

In the opposite case,

$$m\frac{\gamma}{\omega} \gg 1$$

the wave has no time to transform into a toroidal wave and decays completely at a stage when it is still poloidal.

It was noted above that an instantaneous-pulse type source and a monochromatic source are, in a certain sense, opposite limiting cases: the former corresponds to the δ-function of time t, and the latter to the δ-function of frequency ω. It is completely clear that the source can also be regarded as instantaneous when the pulse duration Δt is finite but small, and as monochromatic when the spectrum width $\Delta\omega$ is small. Let us formulate the corresponding requirements explicitly.

When integrating with respect to frequency in (3.297) using the saddle-point method, we could see that, near the saddle point, function $\tilde{Q}_N(x^1, \omega)$ has the form of a sharp peak in variable ω, with characteristic width

$$\Delta\Omega \sim \frac{\Delta\Omega_N}{(\alpha_N m)^{1/2}} \sim \left(\frac{\alpha_N}{m}\right)^{1/2} \Omega_N. \tag{3.317}$$

Formula (3.310) was obtained when examining an instantaneous pulse, when the source function $\varphi_N(x^1, \omega) = \text{const}$. Calculating the integral (3.297) will also remain unchanged in the case when $\tilde{\varphi}_N(x^1, \omega)$ depends on ω, but the characteristic scale $\Delta\omega$ of its change in this variable is much larger than (3.317):

$$\Delta\omega \gg \Delta\Omega. \tag{3.318}$$

This condition is at least satisfied if the source is a pulse of duration

$$\Delta t \ll \frac{1}{\Delta\Omega} \sim \left(\frac{m}{\alpha_N}\right)^{1/2} \frac{1}{\Omega_N}. \tag{3.319}$$

For characteristic values $m \approx 40, \alpha \approx 0.2$, the value $(m/\alpha_N)^{1/2} \approx 15$, the condition (3.319) then means that the pulse duration must be less than 15 oscillation periods. During the SC phenomenon or substorm explosion, this condition is more likely to hold than not. When the inequality opposite to the inequality (3.319) holds,

$$\Delta\omega \ll \Delta\Omega \sim \left(\frac{\alpha_N}{m}\right)^{1/2} \Omega_N,$$

the source can be regarded as monochromatic.

3.12.2 Conclusions vs. Observations

Based on statistical observations onboard the geostationary satellite AMPTE/CCEB it was demonstrated in [237, 348] that about 50% of MHD oscillations observed in the magnetosphere are unstructured stochastic oscillations. Of the remaining 50%, about 30% are harmonic toroidal oscillations, 10% are oscillations with a large portion of the compressible component ($B_3 \sim B_1, B_2$), and less than 5% are transverse oscillations of the poloidal type. It was shown in [255, 349, 350] that the transverse poloidal MHD oscillations ($B_1 > B_2 \gg B_3$) were mainly observed under magnetically quiet conditions. In the perturbed magnetosphere, the most frequently registered oscillations are those of the poloidal type, with a considerable portion of the compressible component [351–353]. Our calculations refer to the oscillations of the first type, with no considerable portion of the compressible component. Oscillations of another type are described within the framework of a theory taking into account the finite plasma pressure ($\beta \sim 1$) [274–276, 354]. Effects of resonant wave-particle interaction on the wave's spatial structure are considered in [262, 333].

It can be theorised that propagation of poloidal Alfvén oscillations across magnetic shells, related to their transverse dispersion, is universal in character, irrespective of their excitation mechanism. Thanks to this, the data of the above-cited observations can be interpreted as follows. Poloidal Alfvén waves excited by broadband sources can only be registered over the time interval $t \sim m/\Omega$ from the moment they are generated. As they travel across magnetic shells, they are transformed into toroidal oscillations due to their transverse dispersion. At the same time, toroidal oscillations generated by, e.g. Alfvén resonance (*field line resonance*) only increase their toroidal character as they travel across magnetic shells (see [224] and the results of Section 3.2). Such behaviour of broadband Alfvén waves predicted by the theory is also confirmed by satellite observations made during and after the interplanetary shock wave front passage through the magnetosphere [355–357].

Due to this, poloidal oscillations are considerably less likely to be observed than toroidal oscillations.

The theoretically predicted behaviour of hodographs were compared to the behaviour of hodographs plotted from STAR radar observations [351] in [251]. This experiment consisted in simultaneously plotting hodographs of Alfvén oscillations with $m \gg 1$ at the ionospheric level, using a dense enough network of observations points in the ionosphere. If poloidal Alfvén wave in the magnetosphere is assumed to cause these oscillations, the behaviour of these hodographs must be described by the theory in Section 3.7 of this monograph. The orientation of the observed hodographs has been found to coincide with the orientation of the hodographs of monochromatic Alfvén oscillations with $m \gg 1$, in the region between the poloidal and toroidal resonance surfaces. At the same time, the oscillation frequency varied during the observations in [351]. This can be explained as follows. Oscillation hodographs plotted for the time period $t \ll m/\Omega$ reflect the behaviour of a monochromatic wave slowly moving across magnetic shells, at the observation point. If the oscillation source is active a long enough time and varies little, we therefore must observe an oscillation smoothly changing from the poloidal to the toroidal type, at different points in space. Note that such behaviour is only possible if the wave dissipation in the ionosphere is weak enough. If the dissipation is strong, the oscillation will be observed to be poloidal at all observation points.

Note the characteristic features of the oscillations in question which can serve as their indicator in observations. The first feature is related to the behaviour of the hodographs of monochromatic oscillations. As follows from the results in Section 3.7, the behaviour of the hodograph strongly depends on the observation point, in the region between the resonance surfaces, where the oscillation is chiefly concentrated. For one and the same oscillation, the hodograph rotates in different directions, near the poloidal surface (where $B_1 \gg B_2$) and near the toroidal surface ($B_2 \gg B_1$). The direction of its rotation depends on the sign of k_2 (or, which is the same, on the sign of m). The oscillation polarisation is linear, in the region between the resonance surfaces, far enough from each (see Figure 3.42).

The second feature concerns the spectrum of broadband stochastic oscillations. When the oscillations decay moderately, the characteristic property of such a spectrum is the presence of peaks near the eigenfrequencies of each harmonic of standing Alfvén waves in the magnetosphere. Each of these peaks splits up into two closely located peaks, one

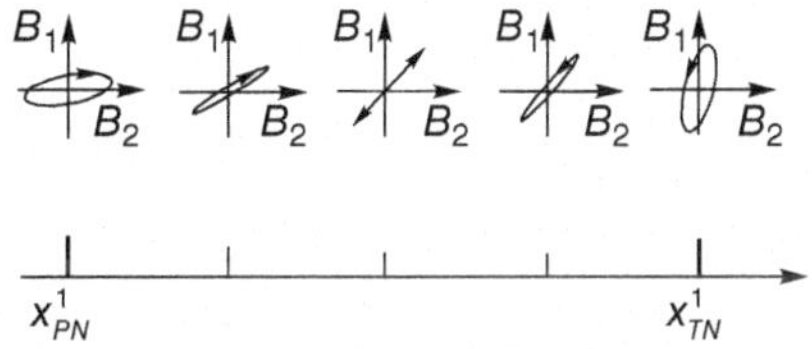

Figure 3.42 Hodographs of monochromatic Alfvén oscillations with $m \gg 1$, at different points in the transparent region between the resonance surfaces $x^1_{PN} < x^1 < x^1_{TN}$.

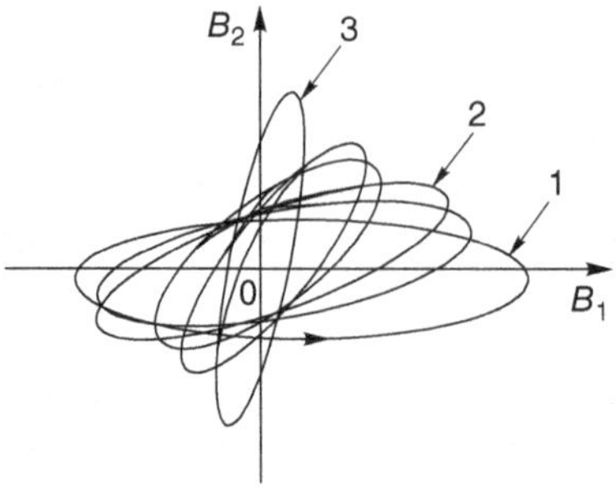

Figure 3.43 Hodographs of non-stationary Alfvén oscillations with $m \gg 1$ excited at the observation point by a source of the 'sudden pulse' type, at different moments of time: 1 – $t \ll m/\omega$ – oscillations of the poloidal type ($B_1 \gg B_2$); 2 – $t \sim m/\omega$ – oscillations of the intermediate type ($B_1 \sim B_2$); 3 – $t \gg m/\omega$ – oscillations of the toroidal type ($B_1 \ll B_2$).

corresponding to the poloidal, Ω_{PN}, and the other to the toroidal eigenfrequency Ω_{TN} (see Figures 3.40 and 3.41). The spectral splitting $\Delta\Omega_N = \Omega_{TN} - \Omega_{PN}$ is small compared to each eigenfrequency Ω_{PN}, Ω_{TN}. In a 'cold' plasma this splitting is largest for the main harmonic $N = 1$: $\Delta\Omega_1/\Omega_1 \sim 0.25$. Thus, the splitting of the Alfvén oscillation spectrum is easiest to observe near the frequency of the main harmonic $N = 1$. As N increases, the ratio $\Delta\Omega_N/\Omega_{TN}$ decreases dramatically.

A third feature is related to observations of Alfvén waves excited by sources of the 'sudden pulse' type. As follows from the findings in this work, oscillations of variable frequency must be observed when oscillations from such a source are registered at a fixed point in space. In the process of the observation (if the observation time $t > m/\Omega$), the frequency of the observed oscillations must increase from the poloidal Ω_{PN} to the toroidal eigenfrequency Ω_{TN} for each harmonic of standing Alfvén waves. Same as in the case of stochastic oscillations, the greatest effect must be observed for the main harmonic $N = 1$. The time variation of the characteristic form of the hodograph for such oscillations is shown in Figure 3.43.

3.13 Model Equation to Determine the Transverse Structure of Standing Alfvén Waves in the Magnetosphere

As we could see in Sections 3.7–3.12, Alfvén oscillations with $m \gg 1$ have a complex enough structure in the direction across magnetic shells. They are characterised by the presence of two prominent shells, poloidal ($x^1 = x^1_{PN}$) and toroidal ($x^1 = x^1_{TN}$), with the oscillation confined between them. The oscillations are generated on the poloidal surface by an external source (for example, external currents in the ionosphere) and slowly move across the magnetic shells towards the toroidal surface, where they are fully absorbed due to the Alfvén resonance.

The splitting of resonance surfaces into two and the phenomenon of transverse motion of standing Alfvén waves are determined by their specific transverse dispersion due to the magnetic field line curvature [338]. In realistic models of the magnetosphere, the distance between the poloidal and toroidal surfaces $\Delta = x^1_{TN} - x^1_{PN}$ (for the sake of convenience, we omit the subscript N in Δ_N in this section) is much smaller than their equatorial radius a (characteristic value $\Delta/a \sim 0.01 \div 0.2$). The theory developed in Sections 3.7–3.12 presumes that many wavelengths fit in between the resonance surfaces. This equality proves to be equivalent to the condition

$$m \gg a/\Delta. \tag{3.320}$$

Thanks to this inequality, the above-developed theory evidently refers to the $m \gg 1$ case. The main achievement of this theory is the separate investigation of the longitudinal (magnetic field-aligned) structure of the Alfvén wave, which makes the problem of describing the transverse (normal to the magnetic shells) structure a 1D problem.

We cannot obtain a universal equation for describing the transverse structure in the entire domain of existence of the wave. Equations describing the transverse structure of the oscillations, separately near the poloidal or the toroidal resonance surfaces were obtained in Sections 3.7.6 and 3.7.7. The WKB approximation in the transverse coordinate x^1 is applicable in between (but not too close to the surfaces), which made it possible to obtain the solution in this region (see Section 3.7.5). In this section we will obtain an equation reproducing this solution, in the WKB approximation. At the same time, it is a generalisation of the equations near the resonance surfaces and coincides with them in the domains of their applicability. It is impossible to obtain such a generalised equation from the initial MHD equations by means of a regular procedure, which is why it should be regarded as a model equation. At the same time, it can be derived semi-phenomenologically as shown below (see also [358]).

This model equation can also be solved in the case

$$m \leq a/\Delta, \tag{3.321}$$

when the results in Section 3.7 do not apply. To justify the applicability of this model equation for such m, Appendix I provides a rigorous derivation of the transverse equation for $m \ll a/\Delta$ in two regions, in the radial coordinate x^1: near the toroidal surface and in the asymptotically remote region (the restriction $m \gg 1$, excluding *field line resonance*, is retained). The resulting equations coincide qualitatively with the model equation. The above reasoning allows the model equation to be regarded as a reliable enough instrument for examining the spatial structure of the wave in a wide domain of its parameters.

3.13.1 Deriving the Homogeneous Model Equation (in the Absence of an Oscillation Source)

As was done in Sections 3.7–3.12, we will describe perturbation in Alfvén wave by means of potential $\tilde{\varphi}$, setting its dependence on time and coordinates as follows

$$\tilde{\varphi} = \varphi(x^1, x^3)\exp[i(k_2x^2 - \omega t)].$$

In Section 3.7, an equation was obtained for potential φ:

$$\hat{L}\varphi \equiv \nabla_1 \hat{L}_T(\omega)\nabla_1\varphi - k_2^2\hat{L}_P(\omega)\varphi = 0. \tag{3.322}$$

Here, $\hat{L}_T(\omega)$ and $\hat{L}_P(\omega)$ are the toroidal and the poloidal operators (see (3.109) and (3.110)) containing derivatives in the longitudinal coordinate l only, $p = (g_2/g_1)^{1/2}$ is the function playing a key role in the theory. Its dependence on coordinate l, determined by the magnetic field line curvature, leads to difference in operators $\hat{L}_T$ and $\hat{L}_P$, and, as a consequence, to transverse dispersion of Alfvén waves and the resonance surfaces splitting into a toroidal and a poloidal. Note that coordinate x^1 acts as a parameter in operators $\hat{L}_T$ and $\hat{L}_P$.

A large role in the theory propounded in Section 3.7 is played by the following boundary problems:

$$\hat{L}_T(\omega)R = 0, \qquad R|_{l_\pm} = 0, \tag{3.323}$$

$$\hat{L}_P(\omega)R = 0, \qquad R|_{l_\pm} = 0. \tag{3.324}$$

They define as their eigenvalues the toroidal $\omega = \Omega_{TN}(x^1)$ and the poloidal $\omega = \Omega_{PN}(x^1)$ frequencies, respectively, as well as the corresponding eigenfunctions $R = T_N(x^1, l)$ and $R = P_N(x^1, l)$. Here, $N = 1, 2, \ldots$ is the number of the longitudinal harmonic of Alfvén waves standing along the magnetic field lines. The eigenvalues and eigenfunctions depend on x^1 as a parameter. For given frequency ω, the equation $\omega = \Omega_{TN}(x^1)$ defines the coordinate of the toroidal surface, $x^1 = x^1_{TN}(\omega)$, and the equation $\omega = \Omega_{PN}(x^1)$ the coordinate of the poloidal surface, $x^1 = x^1_{PN}(\omega)$.

Let us introduce parameter $\overline{\kappa}$, which is equal, on order of magnitude, to $m\Delta/a$ (recall that a is the equatorial radius of the magnetic shell, see Figure 3.2). The ratio m/a coincides, on order of magnitude, with the azimuthal component of the wave vector, which we will denote k_y, keeping in mind the analogy with the model flat layer, where the curvilinear coordinates x^1, x^2, x^3 are replaced with the Cartesian x, y, z. Therefore, $\overline{\kappa} \sim k_y\Delta$.

We will start deriving the model equation with the $\overline{\kappa} \gg 1$ case, discussed in Section 3.7. The solution of (3.322) has the form

$$\varphi(x^1, l, \omega) = U_N(x^1, \omega)R_N(x^1, l, \omega), \tag{3.325}$$

where function $R_N(x^1, l, \omega)$ describes the longitudinal structure of the standing wave and relatively weakly depends on the x^1 coordinate (the characteristic scale of variation in this coordinate is Δ), and function $U_N(x^1, \omega)$ defines the transverse structure of the wave (its characteristic scale in x^1 is $1/k_y$).

In the space between the poloidal and the toroidal surfaces, where the WKB approximation applies,

$$U_N(x^1, \omega) = C\left(v^1_N\left\langle\frac{q_N}{v_A}\right\rangle\right)^{-1/2} \exp\left(i\int_{x^1_{PN}}^{x^1} k_{1N}(x^{1\prime}, \omega)dx^{1\prime}\right). \tag{3.326}$$

Here C is a constant, $k_{1N}(x^1, \omega)$ is quasiclassical wave vector,

$$v^1_N(x^1, \omega) = [\partial k_{1N}(x^1, \omega)\big/\partial\omega]^{-1}$$

is group velocity, $q_N = pk^2_{1N} + p^{-1}k^2_2$, and the angle brackets denote the field-line average:

$$\langle F\rangle = \left(\oint F\frac{dl}{v_A}\right)\bigg/\left(\oint\frac{dl}{v_A}\right),$$

where the curvilinear integral denotes forward-and-backward integration along the field line between the magnetoconjugated ionospheres.

Wave vector $k_{1N}(x^1, \omega)$ is determined along with longitudinal function $R_N(x^1, l, \omega)$ from the eigenvalue problem

$$\left[k^2_{1N}\hat{L}_T(\omega) + k^2_2\hat{L}_P(\omega)\right]R_N = 0, \quad R_N|_{l_\pm} = 0, \tag{3.327}$$

where k^2_{1N} plays the role of eigenvalues, and $R_N(x^1, l, \omega)$ are eigenfunctions. We will presume function R_N to be normalised by the condition

$$\left\langle\frac{q_N}{v_A}R^2_N\right\rangle = \left\langle\frac{q_N}{v_A}\right\rangle.$$

The normalising factor in this relation is chosen such that functions R_N are orthogonal

$$\left\langle \frac{q_N}{v_A} R_N R_{N'} \right\rangle = \delta_{NN'} \left\langle \frac{q_N}{v_A} \right\rangle,$$

where $\delta_{NN'}$ is the Kronecker symbol. We will also note here that

$$\left\langle \frac{q_N}{v_A} \right\rangle = \alpha_T k_{1N}^2 + \alpha_P k_2^2,$$

where

$$\alpha_T = \left\langle \frac{p}{v_A} \right\rangle, \quad \alpha_P = \left\langle \frac{1}{p v_A} \right\rangle.$$

Comparing the problem (3.327) to problems (3.323) and (3.324) makes it easy to see that, on the poloidal surface, for $x^1 = x^1_{PN}$,

$$k_{1N}^2 = 0, \ R_N = P_N,$$

and on the toroidal surface, for $x^1 = x^1_{TN}$,

$$k_{1N}^2 = \infty, \ R_N = T_N.$$

Expressions for k_{1N}^2 near the resonance surfaces were obtained in Section 3.7.2. Near the poloidal surface,

$$k_{1N}^2 = k_2^2 \alpha_P \left\langle \frac{\partial^2 p}{\partial l^2} v_A R_N^2 \right\rangle^{-1} \left(\omega^2 - \Omega_{PN}^2\right), \tag{3.328}$$

and near the toroidal surface,

$$k_{1N}^2 = k_2^2 \frac{1}{\alpha_T} \left\langle \frac{\partial^2}{\partial l^2} \left(-\frac{1}{p}\right) v_A R_N^2 \right\rangle \frac{1}{\Omega_{TN}^2 - \omega^2}. \tag{3.329}$$

We will take the following expression for $k_{1N}(x^1, \omega)$ as our first step to deriving the model equation:

$$k_{1N}^2 = k_y^2 \frac{\omega^2 - \Omega_{PN}^2}{\Omega_{TN}^2 - \omega^2}, \tag{3.330}$$

where

$$k_y^2 = \frac{\alpha_P}{\alpha_T} k_2^2. \tag{3.331}$$

Let us now define the parameter $\overline{\kappa}$ by

$$\overline{\kappa} = k_y \Delta.$$

On order of magnitude (see details in Section 3.7)

$$\frac{1}{\alpha_T} \left\langle \frac{\partial^2 p}{\partial l^2} v_A R_N^2 \right\rangle \sim \frac{1}{\alpha_P} \left\langle \frac{\partial^2}{\partial l^2} \left(-\frac{1}{p}\right) v_A R_N^2 \right\rangle \sim \left(\Omega_{TN}^2 - \Omega_{PN}^2\right),$$

therefore expression (3.330) correctly describes the limiting cases (3.328) and (3.329).

We will assume the model equation describing function $U_N(x^1, \omega)$ to be a second-order differential equation of the form

$$\frac{\partial}{\partial x^1} A(x^1, \omega) \frac{\partial U_N(x^1, \omega)}{\partial x^1} + D(x^1, \omega) U_N(x^1, \omega) = 0, \tag{3.332}$$

where A and D are the functions to be defined. It is shown in Appendix I that knowing the WKB solution of (3.332) (both the argument of the exponential function, and the pre-exponential factor) makes it possible to uniquely determine functions A and D. Assuming (3.326) to be the WKB solution of (3.332), we have, in accordance with the findings in Appendix I,

$$A = \frac{\alpha_T k_{1N}^2 + \alpha_P k_2^2}{\partial k_{1N}^2/\partial\omega}; \; D = \frac{k_{1N}^2(\alpha_T k_{1N}^2 + \alpha_P k_2^2)}{\partial k_{1N}^2/\partial\omega}.$$

Substituting the model equation (3.332) yields the following equation for function $U_N(x^1, \omega)$:

$$\frac{\partial}{\partial x^1}(\omega^2 - \Omega_{TN}^2)\frac{\partial U_N(x^1,\omega)}{\partial x^1} - k_y^2(\omega^2 - \Omega_{PN}^2)U_N(x^1,\omega) = 0. \tag{3.333}$$

The advantage of the above derivation of the model equation lies in the clarity of the assumptions. First, the assumption that it must be a second-order differential equation, and second, using model expression (3.330) for $k_{1N}^2(x^1,\omega)$. The drawback is the impossibility to obtain the right-hand side of the equation which plays the role of the oscillation source. This drawback is due to the WKB solution (3.326) used as a reference point, for which it was shown in Section 3.7 that the source can be neglected in its applicability domain. Because of this, we will give another derivation of the model equation which is not so transparent as the previous one for obtaining the left-hand side of the equation, but does allow its right-hand side to be obtained.

3.13.2 Inhomogeneous Model Equation

We will proceed from (3.322) and the boundary condition at the ionosphere taking into account the dissipation and the external current playing the role of the oscillation source (see Section 2.17.2):

$$\varphi|_{l_\pm} = \pm i\frac{v_{p\pm}}{\omega}\frac{\partial\varphi}{\partial l}\bigg|_{l_\pm} - \frac{J_\parallel^{(\pm)}}{V_{p\pm}}. \tag{3.334}$$

Here

$$v_{p\pm} = \frac{\bar{c}^2\cos\chi_\pm}{4\pi\Sigma_P^{(\pm)}}, \quad V_{p\pm} = \frac{\Sigma_P^{(\pm)}}{\cos\chi_\pm},$$

the $(\pm)$ signs refer to the conjugated ionospheres of the Northern and Southern hemispheres, $\chi_\pm$ is the angle between the vertical and the field line at the point where it intersects the ionosphere (see Figure 2.49), $\Sigma_P^{(\pm)}$ is the integral Pedersen conductivity of the ionosphere, function $J_\parallel^{(\pm)}$ is related to the density $j_\parallel^{(\pm)}$ of the external longitudinal current in the ionosphere by

$$\Delta_\perp^{(\pm)} J_\parallel^{(\pm)} = j_\parallel^{(\pm)}, \tag{3.335}$$

where $\Delta_\perp^{(\pm)} = \Delta_\perp|_{l_\pm} \equiv (1/g_1^{(\pm)})\nabla_1^2 - k_2^2/g_2^{(\pm)}$. We will assume the right-hand side of (3.334) to be small, which means that parameters $v_{p\pm}$ and $(1/V_{p\pm})$ are small.

Let us first address the $\overline{\kappa} \gg 1$ case. A perturbation theory was developed for this case in Section 3.7 using several small parameters at once: the small right-hand side of (3.334) and the small parameter $(1/\overline{\kappa})$, related to the inequality

$$\left|\frac{\nabla_1 U_N}{U_N}\right| \gg \left|\frac{\nabla_1 R_N}{R_N}\right|.$$

In the main order of perturbation theory the solution can be represented in the form (3.325), and we have an eigenvalue problem, (3.327), which we will now rewrite as

$$\left[-\frac{\nabla_1^2 U_N}{U_N}\hat{L}_T(\omega) + k_2^2 \hat{L}_P(\omega)\right] R_N = 0, \quad R_N\big|_{l_\pm} = 0. \tag{3.336}$$

This problem defines the eigenfunction R_N and the relation

$$-\frac{\nabla_1^2 U_N(x^1,\omega)}{U_N(x^1,\omega)} = k_{1N}^2(x^1,\omega), \tag{3.337}$$

playing the role of the eigenvalue. Knowing function $k_{1N}^2(x^1,\omega)$, the relation (3.337) can be regarded as an equation for function $U_N(x^1,\omega)$. The accuracy this relation was obtained with, however, is not sufficient for determining U_N, because small corrections arising in the next order of perturbation theory can change the solution significantly. In terms of the WKB approximation, the accuracy the relation (3.337) was obtained with allows the exponential function, but not the pre-exponential factor, of the solution to be determined correctly.

In the next order of perturbation theory, we assume

$$\varphi(x^1,l,\omega) = U_N(x^1,\omega)\left[R_N(x^1,l,\omega) + h_N(x^1,l,\omega)\right], \tag{3.338}$$

where h_N is a small correction. Our goal will not be to determine this correction, but to obtain the correct equation for the main approximation function U_N, the equation being the condition of solvability of the equation for the correction h_N. Substituting (3.338) into (3.322) produces

$$(\nabla_1^2)\hat{L}_T R_N + (\nabla_1 U_N)\hat{L}_T(\nabla_1 R_N) - k_2^2 U_N \hat{L}_P R_N = -\hat{L}U_N h_N. \tag{3.339}$$

Let us left-multiply this equality by R_N and integrate along the field line. We will represent the result as

$$\nabla_1 \varsigma_N(x^1,\omega)\nabla_1 U_N - k_2^2 \varpi_N(x^1,\omega)U_N = -\oint H_N \hat{L}U_N h_N dl, \tag{3.340}$$

where

$$\varsigma_N(x^1,\omega) = \oint R_N \hat{L}_T(\omega) R_N dl, \tag{3.341}$$

$$\varpi_N(x^1,\omega) = \oint R_N \hat{L}_P(\omega) R_N dl. \tag{3.342}$$

It is not difficult to derive a useful relation for functions ς_N and ϖ_N. Multiplying (3.336) by R_N and integrating with respect to l yields

$$\frac{\varpi_N(x^1,\omega)}{\varsigma_N(x^1,\omega)} = -\frac{k_{1N}^2(x^1,\omega)}{k_2^2}. \tag{3.343}$$

It appears impossible to obtain an explicit analytical expression for these functions. We will assume the following model expressions for them

$$\varpi_N(x^1,\omega) = \alpha_P(\omega^2 - \Omega_{PN}^2), \tag{3.344}$$
$$\varsigma_N(x^1,\omega) = \alpha_T(\omega^2 - \Omega_{TN}^2). \tag{3.345}$$

The advantage of this choice can be argued as follows. A perturbation theory is developed in Section 3.7 allowing various functions to be calculated explicitly near the poloidal and toroidal resonance surfaces. Its application results in formulas (3.344) and (3.345). Substituting these expressions into (3.343) results in the earlier-employed model equation (3.330) for $k_{1N}^2(x^1,\omega)$. Finally, these expressions produce the same left-hand side in (3.340) as in (3.333).

To calculate the right-hand side of (3.340) we will use the linearised boundary condition (3.334):

$$U_N h_N\big|_{l_\pm} = \mp i\frac{v_{p\pm}}{\omega} U_N \frac{\partial R_N}{\partial l}\bigg|_{l_\pm} - \frac{J_\parallel^{(\pm)}}{V_{p\pm}}.$$

By integrating by parts and given (3.335), we obtain

$$\oint R_N \hat{L} U_N h_N dl = 2i\alpha_T \omega \gamma_{TN} \nabla_1 U_N^2 - 2i\alpha_P k_2^2 \omega \gamma_{PN} U_N - I_N, \tag{3.346}$$

where

$$\gamma_{TN} = \frac{1}{\alpha_T}\frac{1}{\omega^2}\left[p_+ v_{p+}\left(\frac{\partial R_N}{\partial l}\right)^2_{l_+} + p_- v_{p-}\left(\frac{\partial R_N}{\partial l}\right)^2_{l_-}\right],$$

$$\gamma_{PN} = \frac{1}{\alpha_P}\frac{1}{\omega^2}\left[\frac{v_{p+}}{p_+}\left(\frac{\partial R_N}{\partial l}\right)^2_{l_+} + \frac{v_{p-}}{p_-}\left(\frac{\partial R_N}{\partial l}\right)^2_{l_-}\right],$$

$$I_N = 2\sqrt{g_1^{(+)} g_2^{(+)}}\left(\frac{\partial R_N}{\partial l}\right)_{l_+}\frac{j_\parallel^{(+)}}{V_{p+}} - 2\sqrt{g_1^{(-)} g_2^{(-)}}\left(\frac{\partial R_N}{\partial l}\right)_{l_-}\frac{j_\parallel^{(-)}}{V_{p-}}.$$

Parameters γ_{TN} and γ_{PN} can be treated as the local values of the decrement near, respectively, the toroidal and poloidal surfaces (for more detail, see Section 3.7.4). They are assumed small: $\gamma_{TN}, \gamma_{PN} \ll \omega$. The order of their magnitude is the same, and we will not distinguish between them in the framework of the model approach, setting $\gamma_{TN} = \gamma_{PN} = \gamma_N$.

Substituting (3.344)–(3.346) into (3.340) yields the final form of the model equation:

$$\alpha_T \nabla_1\left[(\omega + i\gamma_N)^2 - \Omega_{TN}^2\right]\nabla_1 U_N - \alpha_P k_2^2\left[(\omega + i\gamma_N)^2 - \Omega_{PN}^2\right] U_N = I_N. \tag{3.347}$$

This equation generalises (3.333), taking into account the dissipation in the ionosphere and the external current playing the role of the oscillation source. Let it be emphasised that, near resonance surfaces, (3.347) coincides with corresponding equations derived in Section 3.7 by means of rigorous perturbation theory.

Let us now proceed to the $\bar{\kappa} \ll 1$ case. In contrast to the opposite limiting case, $\bar{\kappa} \gg 1$, it does not appear possible to carry out a rigorous and complete mathematical analysis here that could serve as a foundation. A rigorous derivation is given in Appendix J of an equation for the transverse structure of the wave, in two limiting regions, in the x^1 coordinate:

$$|x^1 - x_{TN}^1| \ll 1/k_y, \quad |x^1 - x_{TN}^1| \gg 1/k_y. \tag{3.348}$$

The solution here can also be represented in the form (3.325), the longitudinal function R_N coinciding with the toroidal T_N in the former region, and with the poloidal P_N, in the latter. In both regions, equations for the transverse function U_N qualitatively coincide with (3.347). Based on this, we will assume the model equation (3.347) to be applicable to the $\overline{\kappa} \ll 1$ case too. It is then natural to assume that it is applicable to arbitrary values of $\overline{\kappa}$. Thus, it can be assumed that the model equation (3.347) is applicable in a wide range of parameters limited by the conditions

$$m \gg 1, \quad \gamma_N \ll \omega.$$

3.13.3 Analytical Solution of the Model Equation

As will be seen further, the mode localisation region in x^1 is much smaller than the characteristic variation scale of the equilibrium parameters. This means that, for functions $\Omega_{TN}(x^1)$ and $\Omega_{PN}(x^1)$, the linear expansions can be used in this region

$$\begin{aligned} \Omega_{TN} &= \omega\left(1 - \frac{x^1 - x^1_{TN}}{2a_N}\right), \\ \Omega_{PN} &= \omega\left(1 - \frac{x^1 - x^1_{PN}}{2a_N}\right). \end{aligned} \tag{3.349}$$

Here a_N is the characteristic variation scale of the functions near the resonance surfaces. We assume that $\nabla_1\Omega_{TN}, \nabla_1\Omega_{PN} < 0$, as is the case in most of the magnetosphere. The expansions (3.349) are inapplicable near the function $\Omega_{TN}(x^1)$, $\Omega_{PN}(x^1)$ extrema, requiring a special treatment.

We use decomposition (3.349), to reduce (3.347) to

$$\nabla_1(x^1 - x^1_{TN} + i\varepsilon)\nabla_1 U_N - k_y^2(x^1 - x^1_{PN} + i\varepsilon)U_N = b_N, \tag{3.350}$$

where

$$\varepsilon = 2a_N\frac{\gamma_N}{\omega}, \quad b_N = \frac{a_N I_N}{\alpha_T\omega^2}.$$

We will assume that the right-hand side of (3.350) has a much larger variation scale in x^1 than the mode localisation region, so that b_N can be assumed constant within this region.

The coefficients of (3.350) being linear functions of x^1, this equation can be solved be means of Fourier transformation. Let us denote

$$\tilde{U}_N(k) = \frac{1}{2\pi}\int_{-\infty}^{\infty} U_N(x^1)\exp(-ikx^1)dx^1. \tag{3.351}$$

Fourier-transforming the left- and right-hand sides of (3.350) yields

$$-i(k^2 + k_y^2)\frac{d\tilde{U}(k)}{dk} + \left[k^2x^1_{TN} + k_y^2x^1_{PN} - ik - i\varepsilon\left(k^2 + k_y^2\right)\right]\tilde{U}_N = b_N\delta(k). \tag{3.352}$$

The solution of this first-order equation can be found without difficulty. An inverse Fourier transformation produces

$$U_N(x^1) = i\frac{b_N}{k_y}\int_0^{\infty}\frac{dk}{(k^2 + k_y^2)^{1/2}}\exp\left[ik(x^1 - x^1_{TN}) - k\varepsilon + ik_y\Delta\arctan\frac{k}{k_y}\right]. \tag{3.353}$$

This integral representation can be used to analytically examine various limiting cases.

At large distances from the resonance surfaces, when $|x^1 - x^1_{TN}| \gg \Delta, 1/k_y$, the main contribution to the integral is made by the integration region $k \ll k_y$. Setting $k = 0$ in the pre-exponential factor, we have

$$U_N(x^1) = -\frac{b_N}{k_y^2(x^1 - x^1_{TN} + i\varepsilon)}. \tag{3.354}$$

The same expression is obtained for arbitrary values of x^1, for large decay, when $\varepsilon \gg 1/k_y, \Delta$.

In the opposite limiting case, $|x^1 - x^1_{TN}| \ll 1/k_y, \Delta$, the main contribution into the integral is made by the region $k \gg k_y$. In this case, it is more convenient to calculate $\nabla_1 U_N(x^1)$ first. Differentiating the integrand in (3.353) and later neglecting k_y, small compared to k, in the pre-exponential factor produces

$$\begin{aligned}\nabla_1 U_N(x^1) &= -\frac{b_N}{k_y} \exp\left(\frac{i\pi\overline{\kappa}}{2}\right) \int_0^\infty \exp\left[k(x^1 - x^1_{TN} + i\varepsilon)\right] dk \\ &= -i\frac{b_N \exp\left(i\pi\overline{\kappa}/2\right)}{k_y(x^1 - x^1_{TN} + i\varepsilon)}.\end{aligned} \tag{3.355}$$

Integrating this expression we obtain the main term in the asymptotic of function $U_N(x^1)$, near the toroidal surface:

$$U_N(x^1) = -\frac{b_N}{k_y} \exp\left(\frac{i\pi\overline{\kappa}}{2}\right) \ln\left(x^1 - x^1_{TN} + i\varepsilon\right). \tag{3.356}$$

Saddle-point technique can be used to calculate the integral (3.353) for the $\overline{\kappa} \gg 1$ case (see [337]). Let us denote the large phase of the integrand as

$$\Theta(x^1, k) = k(x^1 - x^1_{TN}) + k_y\Delta \arctan\frac{k}{k_y}. \tag{3.357}$$

The saddle-point is defined by $\partial\Theta/\partial k = 0$. This expression implies that

$$k = k_{1N}(x^1) \equiv k_y\left(\frac{x^1 - x^1_{PN}}{x^1_{TN} - x^1}\right)^{1/2}.$$

As was to be expected, this solution coincides (given the formulas (3.349)) with the model expression (3.330). The phase value at the saddle point is

$$\begin{aligned}\Theta(x^1, k_{1N}(x^1)) &\equiv \int_{x^1_{PN}}^{x^1} k_{1N}(x^{1\prime})dx^{1\prime} \\ &= k_y\sqrt{(x^1 - x^1_{PN})(x^1_{TN} - x^1)} + k_y\Delta \arctan\sqrt{\frac{x^1 - x^1_{PN}}{x^1_{TN} - x^1}}.\end{aligned}$$

The second derivative at the saddle point is

$$\left.\frac{\partial^2\Theta}{\partial k^2}\right|_{k=k_{1N}} = -\frac{2a_N}{\omega}v^1_N,$$

where the group velocity is given by

$$v^1_N = \frac{\omega}{k_y}\frac{\Delta}{a_N}\left(\frac{x^1 - x^1_{PN}}{\Delta}\right)^{1/2}\left(\frac{x^1_{TN} - x^1}{\Delta}\right)^{3/2}.$$

It is now not difficult to write down the result of integrating by means of the saddle-point technique:

$$U_N(x^1) = \frac{b_N}{k_y}\left[\frac{\omega}{a_N}\frac{1}{v_N^1}\frac{1}{(k_{1N}^2 + k_y^2)}\right]^{1/2} \times \exp\left[i\int_{x_{PN}^1}^{x^1} k_{1N}(x^{1\prime})dx^{1\prime} - k_{1N}(x^1)\varepsilon + i\frac{\pi}{4}\right]. \quad (3.358)$$

This expression coincides with the WKB formula (3.326), the difference of course being that the constant C is defined here and all the participating functions have simple analytical expressions. Approaching the resonance surfaces, v_N^1 tends to zero and the saddle-point technique ceases to be applicable. Other simplifications are however possible in (3.353) in these cases.

Near the poloidal surface, the main contribution is from integration with respect to $k \ll k_y$. This can be seen from the phase (3.357) featuring third-order zero at $k = 0$ for $x^1 = x_{PN}^1$. Expanding this phase with respect to small k results in

$$U_N(x^1) = i\frac{b_N}{k_y^2}\int_0^\infty dk \exp\left[ik(x^1 - x_{PN}^1 + i\varepsilon) - i\frac{\Delta}{3k_y^2}k^3\right].$$

Let us introduce the characteristic length $\lambda_P = (\Delta/k_y^2)^{1/3}$ and replace the variable $k = s/\lambda_P$. We will present the result as

$$U_N(x^1) = -\frac{b_N}{k_y^{4/3}\Delta^{1/3}} G\left(\frac{x^1 - x_{PN}^1 + i\varepsilon}{\lambda_P}\right), \quad (3.359)$$

where $G(z)$ is a function having an integral representation (3.260).

Near the toroidal surface, integration with respect to large k contributes most. This can be seen from the integral (3.353) diverging (in the absence of decay, $\varepsilon = 0$) at the upper limit, for $x^1 = x_{TN}^1$ (and in the absence of decay, $\varepsilon = 0$). Expanding the integrand with respect to large k, we have

$$U_N(x^1) = i\frac{b_N}{k_y}\exp\left(i\frac{\pi\overline{\kappa}}{2}\right)\int_0^\infty \frac{dk}{k}\exp\left[ik(x^1 - x_{TN}^1 + i\varepsilon) - i\frac{k_y^2\Delta}{k}\right].$$

Introducing the characteristic length $\lambda_T = 1/(k_y^2\Delta)^{1/3}$ and replacing the variable $k = s/\lambda_T$, we will represent the result as

$$U_N(x^1) = 2i\frac{b_N}{k_y}\exp\left(i\frac{\pi\overline{\kappa}}{2}\right) g\left(\frac{x^1 - x_{TN}^1 + i\varepsilon}{\lambda_T}\right), \quad (3.360)$$

where function $g(z) = K_0(2z^{1/2})$, expressed in terms of the modified Bessel function K_0, has an integral representation (3.261). Formulas (3.359)–(3.360) completely coincide with the corresponding formulas in Section 3.7. This means that, in the $\overline{\kappa} \gg 1$ case, the model equation reproduces all the significant results of the rigorous theory.

In the opposite limiting case, $\overline{\kappa} \ll 1$, the factor

$$\exp\left[ik_y\Delta \arctan\left(\frac{k}{k_y}\right)\right] \approx 1 + ik_y\Delta \arctan\left(\frac{k}{k_y}\right)$$

can be omitted, because it makes only a small correction to the solution. The integral (3.353) can then be represented as

$$U_N(x^1) = i\frac{b_N}{k_y} f\left[k_y(x^1 - x^1_{TN})\right], \tag{3.361}$$

where

$$f(\xi) = \int_0^\infty \frac{dq}{(1+q^2)^{1/2}} \exp(iq\xi - \overline{\varepsilon} q) \tag{3.362}$$

and the following notation is introduced:

$$\overline{\varepsilon} = k_y \varepsilon = 2k_y a_N (\gamma_N/\omega).$$

Parameter Δ is absent from the solution (3.361) and (3.362). This means that the poloidal surface fails to stand out for $\overline{\kappa} \ll 1$, and there are no reasons to call it 'the resonance'. The characteristic scale of the solution in this case is $1/k_y$. The behaviour of the solution for $|x^1 - x^1_{TN}| \gg 1/k_y$ and $|x^1 - x^1_{TN}| \ll 1/k_y$ is defined by formulas (3.354) and (3.356), respectively.

The expression for $U_N(x^1)$ can also be represented in a form analogous to (3.361) in the general case, for arbitrary values of $\overline{\kappa}$. Introducing the dimensionless variable

$$\xi = k_y(x^1 - x^1_{TN}) \tag{3.363}$$

and replacing $q = k/k_y$ in (3.353) yields

$$U_N(x^1) = i\frac{b_N}{k_y} f(\xi), \tag{3.364}$$

where

$$f(\xi) = \int_0^\infty \frac{dq}{(1+q^2)^{1/2}} \exp(iq\xi - \overline{\varepsilon} q + i\overline{\kappa} \arctan q). \tag{3.365}$$

The role of dimensionless parameters is conspicuous in the formula: $\overline{\varepsilon}$ characterises dissipation, and $\overline{\kappa}$ transverse dispersion of the wave.

Knowing function $U_N(x^1)$, it is possible to use formulas (E.1) and (E.2) (Appendix E) and (3.325) for writing the expressions for the components of the electromagnetic field of the oscillations:

$$E_1 = -ib_N f'(\xi) R_N, \qquad E_2 = (\alpha_T/\alpha_P)^{1/2} b_N f(\xi) R_N, \tag{3.366}$$

$$B_1 = \frac{c}{\omega}\frac{g_1}{\sqrt{g}} k_2 f(\xi) \frac{\partial R_N}{\partial l}, \qquad B_2 = i\frac{c}{\omega}\frac{g_2}{\sqrt{g}} k_2 f'(\xi) \frac{\partial R_N}{\partial l}. \tag{3.367}$$

An important characteristic of the oscillations is their polarisation. It is evident from the above formulas that the standing Alfvén waves with $m \gg 1$ we are discussing can be toroidal (i. e. $|B_2| \gg |B_1|$ and $|E_1| \gg |E_2|$) in one region, but poloidal ($|B_2| \ll |B_1|$ and $|E_1| \ll |E_2|$) in another, in the x^1 coordinate. We will introduce a parameter characterising the wave type integrally. It is natural to use for this purpose the oscillation energy. Hydromagnetic wave energy density is composed of the perturbed magnetic field and kinetic energy of the plasma:

$$w = \frac{\mathbf{B}^2}{8\pi} + \frac{\rho_0 \mathbf{v}_E^2}{2},$$

where ρ_0 is plasma density, $\mathbf{v}_E = \bar{c}[\mathbf{E}\mathbf{B}_0]/B_0^2$ perturbed plasma velocity equalling the electric drift velocity. Let us represent the oscillation energy as a sum, $w = w_T + w_P$, of the toroidal and poloidal component, where

$$w_T = \frac{B_2 B_2^*}{8\pi g_2} + \frac{\rho_0 \bar{c}^2}{2B_0^2}\frac{E_1 E_1^*}{g_1}, \qquad w_P = \frac{B_1 B_1^*}{8\pi g_1} + \frac{\rho_0 \bar{c}^2}{2B_0^2}\frac{E_2 E_2^*}{g_2}.$$

Substituting (3.366) and (3.367) produces

$$w_T = \frac{\bar{c}^2 |b_N|^2}{8\pi}\frac{1}{g_1}|f'(\xi)|^2 \left[\frac{1}{\omega^2}\left(\frac{\partial R_N}{\partial l}\right)^2 + \frac{1}{v_A^2}R_N^2\right],$$

$$w_P = \frac{\bar{c}^2 |b_N|^2}{8\pi}\frac{\alpha_T}{\alpha_P}\frac{1}{g_2}|f(\xi)|^2 \left[\frac{1}{\omega^2}\left(\frac{\partial R_N}{\partial l}\right)^2 + \frac{1}{v_A^2}R_N^2\right].$$

The energy contained in a thin field tube with unit dimensions in x^1 and x^2 is given by

$$W_T = \frac{1}{2}\oint \sqrt{g_1 g_2} w_T dl = \frac{\bar{c}^2 |b_N|^2}{16\pi}|f'(\xi)|^2 \oint p \left[\frac{1}{\omega^2}\left(\frac{\partial R_N}{\partial l}\right)^2 + \frac{1}{v_A^2}R_N^2\right] dl,$$

$$W_P = \frac{1}{2}\oint \sqrt{g_1 g_2} w_P dl = \frac{\bar{c}^2 |b_N|^2}{16\pi}\frac{\alpha_T}{\alpha_P}|f(\xi)|^2 \oint \frac{1}{p} \left[\frac{1}{\omega^2}\left(\frac{\partial R_N}{\partial l}\right)^2 + \frac{1}{v_A^2}R_N^2\right] dl.$$

We will assume the integrals along the field line in these formulas to equal $2\alpha_T$ and $2\alpha_P$, respectively. This is true on the order of magnitude, and such accuracy is sufficient for dealing with the model. Therefore,

$$\begin{aligned} W_T(x^1) &= \frac{\bar{c}^2 |b_N|^2}{8\pi}\alpha_T |f'(\xi)|^2, \\ W_P(x^1) &= \frac{\bar{c}^2 |b_N|^2}{8\pi}\alpha_T |f(\xi)|^2. \end{aligned} \tag{3.368}$$

Full oscillation energy integrated over x^1, within a unit interval in x^2 is given by

$$\overline{W}_T = \int_{-\infty}^{\infty} W_T(x^1) dx^1 = \frac{\alpha_T \bar{c}^2 |b_N|^2}{8\pi k_y}\int_{-\infty}^{\infty} |f'(\xi)|^2 d\xi,$$

$$\overline{W}_P = \int_{-\infty}^{\infty} W_P(x^1) dx^1 = \frac{\alpha_T \bar{c}^2 |b_N|^2}{8\pi}\int_{-\infty}^{\infty} |f(\xi)|^2 d\xi.$$

Substituting (3.365) yields, upon some simple transformations,

$$\overline{W}_T = \frac{\alpha_T \bar{c}^2 |b_N|^2}{4k_y}\int_0^{\infty} \frac{q^2 e^{-2\bar{\varepsilon}q}}{1+q^2} dq,$$

$$\overline{W}_P = \frac{\alpha_T \bar{c}^2 |b_N|^2}{4k_y}\int_0^{\infty} \frac{e^{-2\bar{\varepsilon}q}}{1+q^2} dq.$$

Notably, these values do not depend on parameter $\bar{\kappa}$. This means that the integral type of wave polarisation is only defined by decay (parameter $\bar{\varepsilon}$). Note also that there is a simple expression for full oscillation energy

$$\overline{W} = \overline{W}_T + \overline{W}_P = \frac{\alpha_T c^2 |b_N|^2}{8k_y \bar{\varepsilon}}.$$

We will accept

$$\overline{\eta} = \frac{\overline{W}_T}{\overline{W}_P} = \left(\int_0^\infty \frac{q^2 e^{-2\overline{\varepsilon}q}}{1+q^2} dq\right) \Big/ \left(\int_0^\infty \frac{e^{-2\overline{\varepsilon}q}}{1+q^2} dq\right).$$

as the parameter defining wave polarisation type. Asymptotic expressions for function $\overline{\eta} = \overline{\eta}(\overline{\varepsilon})$ have the form

$$\overline{\eta}(\overline{\varepsilon}) = \begin{cases} (\pi\overline{\varepsilon})^{-1}, & \text{for } \overline{\varepsilon} \ll 1, \\ \overline{\varepsilon}^{-2}/2, & \text{for } \overline{\varepsilon} \gg 1. \end{cases}$$

Thus, the oscillations are toroidal when the decay is small, and poloidal, when it is big, independent of $\overline{\kappa}$.

3.13.4 Numerical Investigation of the Model Equation Solutions

A full picture of the transverse structure of the wave and its polarisation depending on the values of $\overline{\kappa}$ and $\overline{\varepsilon}$, including their intermediate values, can be obtained by means of numerical investigation. This is done easiest by using an integral representation for the solution in the form (3.364) and (3.365). Characteristic results of the numerical calculations are in Figures 3.44–3.46.

Function $\text{Re} f(\xi)$ and $\text{Im} f(\xi)$ plots are shown in Figure 3.44 for three different values of parameter $\overline{\kappa} = 0.1; 3; 20$ and fixed value of parameter $\overline{\varepsilon} = 10^{-3}$. The latter was chosen to be very small so as to exclude the effect of dissipation and clearly demonstrate the role of

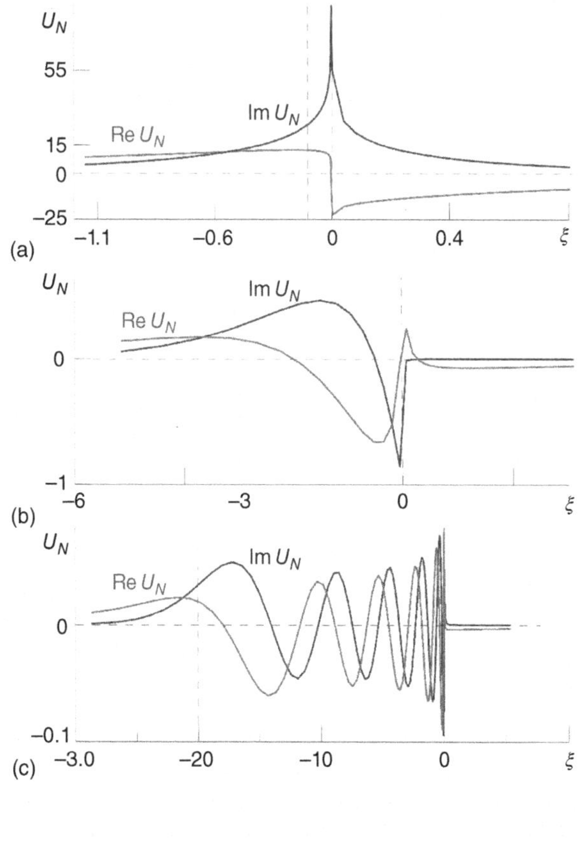

Figure 3.44 Field structure across magnetic shells of standing Alfvén waves with $m \gg 1$, for various values of parameter $\overline{\kappa}$: (a) $\overline{\kappa} = 0.1$ – a typical structure of resonant oscillations; (b) $\overline{\kappa} = 3$ – structure of intermediate type; (c) $\overline{\kappa} = 20$ – 'travelling wave'-type structure.

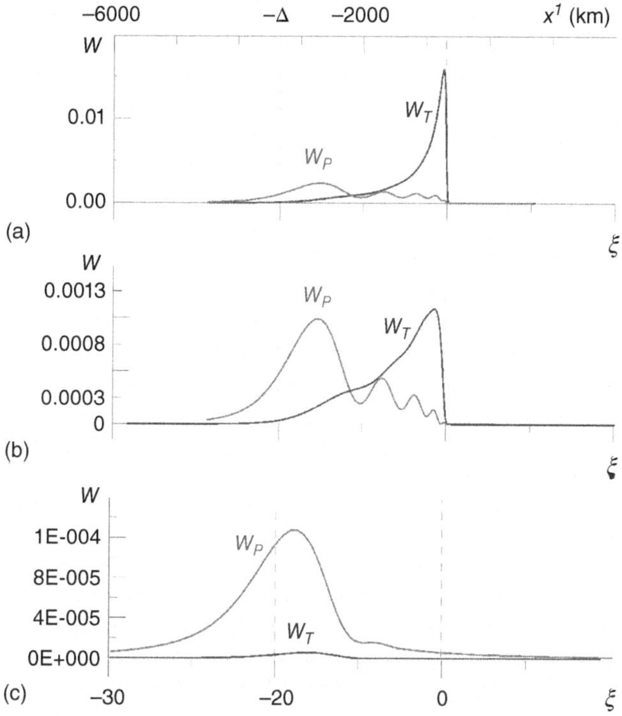

Figure 3.45 Spatial distribution over the transverse coordinate ξ for the poloidal (W_P) and toroidal (W_T) components of the full energy of standing Alfvén wave (in relative units) for 'travelling wave' type oscillations ($\overline{\kappa} \gg 1$, see Figure 3.44c) for various values of the dimensionless dissipation index $\overline{\epsilon} = 2k_y a_N(\gamma_N/\omega)$: (a) $\overline{\epsilon} = 0.4$; (b) $\overline{\epsilon} = 0.8$; (c) $\overline{\epsilon} = 4$.

dispersion. We have a structure typical of resonant oscillations, e.g. resonant Alfvén waves (*field line resonance*) in the first plot of Figure 3.44. Recall that normal Alfvén resonance is impossible in this case because $m \gg 1$ – external currents in the ionosphere are the field source in the neighbourhood of the resonance (i.e. the source is inside the magnetosphere), not fast magnetosonic wave from outside. However, the main features of the field behaviour are of the same character as those for resonant Alfvén waves. There is a singularity on the resonance (toroidal) surface – logarithmic for E_1 and B_2 and linear – for components E_2 and B_1 (see (3.355) and (3.356), regularised by weak dissipation. No noticeable details can be found in the field behaviour on the poloidal surface, for $\xi = -\overline{\kappa}$. Moving away from the resonance surface, the field monotonously tends to asymptotic values.

As $\overline{\kappa}$ increases, the structure changes into the 'travelling wave' type. It can be seen quite clearly in the third plot of Figure 3.44. The field here is confined between the poloidal and toroidal surfaces, where many wavelengths are contained. Wavelength decreases sharply when approaching the toroidal surface, where a singularity exists in the field. The dependence of the oscillation field structure on two parameters $\overline{\kappa}$ and $\overline{\epsilon}$ is better demonstrated by the plots for the toroidal and poloidal components $w_T(x^1)$ and $w_P(x^1)$ of the oscillation energy density. Figure 3.45 displays function $W_T(\xi)$ and $W_P(\xi)$ plots for the value $\overline{\kappa} = 20$ and three different values $\overline{\epsilon} \ll 1, \overline{\epsilon} \sim 1$ and $\overline{\epsilon} \gg 1$. Note that the numerical values of parameter $\overline{\epsilon}$ were chosen such that the best visual representation of the functions under study is achieved in Figures 3.45 and 3.46. The former case is clearly dominated

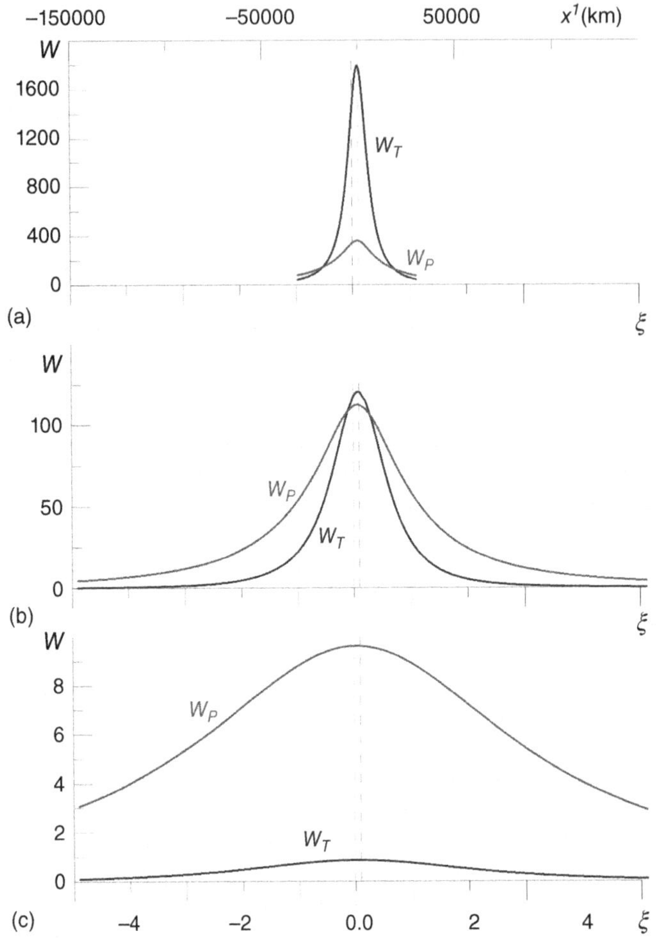

Figure 3.46 Same as Figure 3.45, for wave with resonant type transverse structure ($\overline{\kappa} \ll 1$, see Figure 3.44a): (a) $\overline{\epsilon} = 0.4$ – weak dissipation – toroidal type wave; (b) $\overline{\epsilon} = 1.2$ – moderate dissipation – intermediate type wave; (c) $\overline{\epsilon} = 6$ – strong dissipation – poloidal type wave.

by the toroidal component, while the latter by the poloidal component. An interesting feature easily visible in these plots is the oscillatory character of the energy density of the poloidal component. It is determined by the fact that, for $\overline{\kappa} \gg 1$, a wave is generated in the neighbourhood of the poloidal surface which is an almost standing (in x^1) wave, and the travelling wave structure only forms at a large enough distance from the poloidal surface.

Analogously, the $\overline{\kappa} = 0.1$ case is shown in Figure 3.46 for three values of parameter $\overline{\epsilon}$: $\overline{\epsilon} \ll 1, \overline{\epsilon} \sim 1$ and $\overline{\epsilon} \gg 1$. The picture for the ξ coordinate is more simple. Both functions, $W_T(\xi)$ and $W_P(\xi)$, are bell-shaped and exchange their roles when proceeding from the first to the last case. Note that, because $\overline{\epsilon} \gg \overline{\kappa}$ can be assumed in these cases, an asymptotic expression can be used for $f(\xi)$ for large $\overline{\epsilon}$. Hence, (cf. (3.354))

$$W_T(\xi) \sim \frac{1}{(\xi^2 + \overline{\epsilon}^{-2})^2}, \quad W_P(\xi) \sim \frac{1}{\xi^2 + \overline{\epsilon}^{-2}}.$$

The plots in Figure 3.46 are in good agreement with these formulas.

3.14 Spatial Structure of Alfvén Oscillations Excited in the Magnetosphere by Localised Monochromatic Source

This and the Sections 3.14 and 3.15 will examine Alfvén oscillations of the magnetosphere their source being localised perturbations in the ionosphere. They can be both monochromatic and broadband perturbations. Thanks to this, investigating the Alfvén wave field requires a switch from describing the structure of a separate Fourier harmonic of the oscillations, with fixed frequency ω and azimuthal wavenumber m, to the field of oscillations generated by broadband spatially distributed sources.

A model equation was proposed in Section 3.13 allowing the spatial structure of Alfvén waves to be studied practically within the entire range of azimuthal wavenumber m values, except for a few first harmonics with $m \sim 1$, describable by Alfvén resonance theory. When generated by a strongly localised source, however, these oscillation harmonics fail to play any significant role in the wave packet. When m are small, the MHD oscillation field consists of the coupled toroidal Alfvén and magnetosonic waves. A transparent region exists for magnetosonic wave with $m \sim 1$ where it can escape across magnetic shells towards infinity. Therefore, the field of Alfvén wave coupled with this magnetosonic wave will decay.

In this section, we will study the field of monochromatic Alfvén waves generated in the magnetosphere by a source localised in the ionosphere [359]. This means that the oscillation source can have an arbitrary structure across geomagnetic field line, i.e. it excites a full spectrum of harmonics over the entire range of azimuthal wavenumbers m. A specific attention will be paid to oscillations from a strongly localised source. A high frequency (HF) radar can be used to generate such oscillations periodically heating the ionosphere at frequency close to the frequency of Alfvén eigen oscillations in the magnetosphere, at the magnetic shell under study [196]. Characteristic features of the spatial distribution of the amplitude for such Alfvén oscillations can be used to measure polarisation splitting of their spectrum into toroidal and poloidal eigenmodes.

3.14.1 Structure of Monochromatic Alfvén Oscillations from a Source Localised Across Magnetic Field Lines

As was demonstrated in Sections 3.7 and 3.13, the spatial structure of standing Alfvén wave with fixed frequency ω and azimuthal wavenumber m can be represented as

$$\varphi_N = U_N(x^1, k_2, \omega) R_N(x^1, x^3, k^2, \omega) e^{-i\omega t + ik_2 x^2},$$

where $N = 1, 2, 3, \ldots$ is the number of the longitudinal eigenharmonic of standing waves, function R_N describes its structure along magnetic field lines (see Eq. (3.327)), and U_N describes its structure across magnetic shells (3.333). If azimuthal angle ϕ is used as the azimuthal coordinate x^2, $k_2 = m = 0, \pm 1, \pm 2, \ldots$ is the azimuthal wavenumber. We employ the general notations x^2 and k_2, because a large number of azimuthal harmonics ($m \sim 1$ to $m = \infty$) is studied and k_2 variation will be assumed as continuous.

A solution was found in Section 3.13 to (3.350), an equation describing Alfvén oscillations excited in the magnetosphere by a non-localised monochromatic source. Expression (3.353) was obtained for $U_N(x^1, k_2, \omega)$, a function describing the structure across magnetic shells of a separate harmonic of oscillations. The right-hand side of (3.350) was assumed constant.

Let us now obtain the solution to (3.350), assuming its right-hand side to be a function of the transverse coordinate: $b_N \equiv b_N(\xi, k_y)$, where the azimuthal wavenumber k_y is defined by

(3.331), and the dimensionless transverse coordinate ξ by (3.363). Function b_N also depends on frequency ω, but this fact plays no specific role in this case, and we will not describe the corresponding dependence in an explicit form. We will use the Fourier technique to find a solution to (3.350) (see Section 3.13), representing the desired solution in the form of (3.351). Substituting this expression into (3.350) yields a first-order equation for the Fourier transform $\tilde{U}_N(k)$, analogous to (3.352), the solution of which is easy to find.

Substituting the solution into (3.351) yields

$$U_N(\xi) = \frac{i}{2\pi} \int_{-\infty}^{\infty} b_N(\xi', k_y) d\xi' \int_{-\infty}^{\infty} \frac{e^{i\Theta_N(\xi,k,k_y)}}{\sqrt{k^2 + k_y^2}} dk \int_{-\infty}^{k} \frac{e^{-i\Theta_N(\xi',k',k_y)}}{\sqrt{k'^2 + k_y^2}} dk', \tag{3.369}$$

where

$$\Theta_N(\xi, k, k_y) = k(\xi + i\epsilon) + |k_y| \arctan \frac{k}{|k_y|}.$$

If b_N does not depend on ξ, the integral with respect to ξ' results in the $\delta(\xi')$-function and the solution (3.369) is transformed into solution (3.364).

Solution (3.369) describes a Fourier harmonic of oscillations with fixed value of the azimuthal wavenumber m (or k_y). If the source has an arbitrary structure in the azimuthal coordinate also, i.e. the source function can be represented as

$$\overline{b}_N(\xi, \eta) = \frac{1}{2\pi} \int_{-\infty}^{\infty} b_N(\xi, k_y) e^{ik_y\eta} dk_y,$$

where $\eta = \sqrt{\alpha_T/\alpha_P} x^2/\Delta_N$ is a dimensionless azimuthal coordinate, the full solution will then be

$$U_N(\xi, \eta) = \frac{1}{(2\pi)^2} \int_{-\infty}^{\infty} \int_{-\infty}^{\infty} \overline{b}_N(\xi', \eta') V_N(\xi, \xi', \eta, \eta') d\xi' d\eta', \tag{3.370}$$

where function

$$V_N = i \int_{-\infty}^{\infty} dk_y \int_{-\infty}^{\infty} \frac{\exp[i\Theta_N(\xi, k, k_y) + ik_y\eta]}{\sqrt{k^2 + k_y^2}} dk \times \int_{-\infty}^{k} \frac{\exp[-i\Theta_N(\xi', k', k_y) + ik_y\eta']}{\sqrt{k'^2 + k_y^2}} dk' \tag{3.371}$$

describes the transverse structure of the Nth harmonic of standing Alfvén waves excited by a source of the form $\delta(\xi - \xi_0)\delta(\eta - \eta_0)$, where (ξ_0, η_0) is the source localisation point.

3.14.2 Transverse Structure of Standing Alfvén Waves from a Source Strongly Localised in One of the Transverse Coordinates

To qualitatively understand the structure of the solution (3.370), let us consider two opposite limiting cases. Let $\overline{b}_N$ in (3.370) have the form $\overline{b}_N = \tilde{b}_N \delta(\xi - \xi_0) e^{i\overline{k}_y\eta}$, where $\tilde{b}_N$ is independent of ξ and η. Then

$$U_N = \overline{B}_N \int_{-\infty}^{\infty} \frac{e^{i\Theta_N(\xi,k,\overline{k}_y)}}{\sqrt{k^2 + \overline{k}_y^2}} dk \int_{-\infty}^{k} \frac{e^{-i\Theta_N(\xi_0,k',\overline{k}_y)}}{\sqrt{k'^2 + \overline{k}_y^2}} dk', \tag{3.372}$$

where $\overline{B}_N = i\tilde{b}_N e^{i\overline{k}_y\eta}/2\pi$. Let $\overline{k}_y\Delta_N \gg 1$, where $\Delta_N = x_{TN}^1 - x_{PN}^1$ is the distance between the poloidal and toroidal resonance surfaces. Then the saddle point technique can be used for approximate calculation both in the internal integral with respect to k', and in the external integral with respect to k. The saddle points $\overline{k}'$ and $\overline{k}$ in these integrals are defined from conditions $\partial\Theta_N(\xi_0, k', \overline{k}_y)/\partial k'|_{k'=\overline{k}'} = 0$, $\partial\Theta_N(\xi, k, \overline{k}_y)/\partial k|_{k=\overline{k}} = 0$ and have the form $\overline{k}' = \pm\overline{k}_y\kappa_N(\xi_0)$, $\overline{k} = \pm\overline{k}_y\kappa_N(\xi)$, where

$$\kappa_N = \sqrt{\frac{\xi+1}{-\xi}}. \tag{3.373}$$

Here, $-1 < \xi < 0$, and points $\xi = 0$ and $\xi = -1$ correspond to the toroidal and poloidal resonance surfaces. Taking these integrals sequentially yields

$$U_N = \frac{2\pi(\omega/a_N)\overline{B}_N}{\left[v_N v_N^0 (\overline{k}'^2 + \overline{k}_y^2)(\overline{k}^2 + \overline{k}_y^2)\right]^{1/2}}$$
$$\times \left\{ \theta(\xi - \xi_0)e^{i\Theta_N^0} + ie^{-i(\Theta_N + \Theta_N^0)} + \theta(\xi_0 - \xi)e^{i(\overline{\Theta}_N - \Theta_N^0)} \right\},$$

where

$$v_N(\xi, \omega) = \frac{\omega}{\overline{k}_y}\frac{\Delta_N}{a_N}(\xi+1)^{1/2}\xi^{3/2}$$

is the group velocity of the Nth harmonic of standing Alfvén waves in ξ; $v_N^0 = v_N(\xi_0, \omega)$, $\Theta_N = \Theta_N(\xi, \overline{k}, \overline{k}_y)$, $\Theta_N^0 = \Theta_N(\xi_0, \overline{k}', \overline{k}_y)$. In the transparent region, $-1 \leq \xi; \xi_0 \leq 0$, the first term in the braces describes the wave travelling from the source localisation point ($\xi = \xi_0$) towards the poloidal resonance surface ($\xi = -1$), the second term describes the wave reflected from the poloidal surface and travelling towards the toroidal resonance surface ($\xi = 0$), and the third term describes the wave travelling from the source localisation point towards the toroidal surface.

The other limiting case is a source strongly localised in azimuth at point $\eta = \eta_0$: $\overline{b}_N = \tilde{b}_N \delta(\eta - \eta_0)$. For this source,

$$\nabla_2 U_N = -i\overline{B}_{N1}\int_0^\infty \frac{d\kappa}{\sqrt{1+\kappa^2}}\left[\frac{1}{\Theta_N(\xi,\kappa)+\eta-\eta_0} + \frac{1}{\Theta_N(\xi,\kappa)-\eta+\eta_0}\right] \tag{3.374}$$

is a function defining the radial structure of the wave field components B_1 and E_2. Here, $\overline{B}_{N1} = i\tilde{b}_N\sqrt{\alpha_T/\alpha_P}/2\pi\Delta_N$ and

$$\Theta_N(\xi, \kappa) = \kappa(\xi + i\overline{\varepsilon}) + \arctan\kappa.$$

The integral (3.374) is obtained from (3.369) by successive integration with respect to η', ξ', k' and k_y and replacing the integration variable $\kappa = k/|k_y|$. Function $\nabla_1 U_N$, determining the radial structure of the field components B_2 and E_1, differs from (3.374) in having an additional multiplier κ in the integrand and in the amplitude $\overline{B}_{N2} = \sqrt{\alpha_P/\alpha_T}\overline{B}_{N1}$.

Based on the structure of the integrand (3.374), it is possible to assert that the main contribution into the integral is made at the beginning ($0 \leq \kappa \leq \overline{\kappa} \ll 1$) of the integration path and points $\kappa_\pm$, where singularities defined by the zeros in the denominators are located: $\Theta_N(\xi, \kappa_\pm) \pm (\eta - \eta_0) = 0$. If $\kappa < \overline{\kappa}$ the approximate expression $\Theta_N \approx (\xi + 1 + i\overline{\varepsilon})\kappa$ is valid for Θ_N and the integral is easily calculated for $\kappa \leq \overline{\kappa}$:

$$\nabla_2 U_N|_{\kappa \leq \overline{\kappa}} \approx -i\frac{\overline{B}_{N1}}{\xi+1+i\overline{\varepsilon}}\ln\frac{(\eta-\eta_0)^2 - \overline{\kappa}^2(\xi+1+i\overline{\varepsilon})^2}{(\eta-\eta_0)^2}.$$

There is a singularity in this expression for $\eta = \eta_0$, which cannot be regularised by the presence of dissipation $\overline{\epsilon}$. To regularise this singularity, it is enough to 'spread out' somewhat the source structure in the η coordinate by replacing the $\delta(\eta - \eta_0)$-function in the expression for $\overline{b}_N$, by, e.g.

$$\overline{b}_N(\eta) = \tilde{b}_N \frac{\Delta}{(\eta - \eta_0)^2 + \Delta^2},$$

which turns into $\overline{b}_N = \tilde{b}_N \delta(\eta - \eta_0)$ when $\Delta \to 0$. Using such a source model in all the previous calculations results in replacement $\eta - \eta_0 \to \eta - \eta_0 + i\Delta$.

For the two other singularities of the integrand, it is easy to demonstrate that they are poles at points $\kappa = \pm\kappa_N(\xi)$, where $\kappa_N(\xi)$ is defined by (3.373), and the coordinate ξ lies on the characteristics defined by

$$\frac{d\eta}{d\xi} = \pm\kappa_N(\xi). \tag{3.375}$$

The solution (3.374) will thus be determined by the contribution of the source localisation region $\eta = \eta_0$ and waves travelling from such a source along the characteristic lines (3.375).

Based on the two limiting cases, the following pattern can be expected for the field distribution of oscillations excited by a source localised in both transverse coordinates. A local maximum must be observed in the oscillation amplitude distribution, at source localisation point (ξ_0, η_0). If the source is located inside the localisation region of the standing wave $(-1 \leq \xi_0 \leq 0)$, local maxima in amplitude must be observed at the resonance surfaces, at the points where they are crossed by the characteristics passing through the source localisation point. Two characteristics corresponding to the different signs in (3.375) pass through this point (ξ_0, η_0), therefore local maxima at resonance surfaces will be located symmetrically about $\eta = \eta_0$.

3.14.3 Transverse Structure of Standing Alfvén Waves from a Source Localised in Two Transverse Coordinates

In Section 3.14.2, we analysed qualitatively the propagation across geomagnetic field lines of standing Alfvén waves excited by a source localised in one of the transverse coordinates, x^1 or x^2. Let us now integrate the solution (3.370) numerically for the case when the source is strongly localised in both transverse coordinates. To be definite, we will consider the transverse propagation of waves as mapped onto the ionosphere of the Northern hemisphere. The source is strongly localised, therefore integration with respect to ξ' and η' in (3.370) concerns the source function $b_N(\xi', \eta')$ only, and replacements $\xi' \to \xi_0$ and $\eta' \to \eta_0$ are made in the expression for V_N. As a result, the expressions for the electromagnetic field components of the Nth harmonic of standing Alfvén waves near the ionosphere have the form

$$E_{xN} = E_N \frac{I_1 \cos\chi}{\sqrt{g_1^{(i)}}}, \qquad E_{yN} = \sqrt{\frac{\alpha_T}{\alpha_P}} E_N \frac{I_2}{\sqrt{g_2^{(i)}}},$$

$$B_{xN} = \sqrt{\frac{\alpha_T}{\alpha_P}} B_N \frac{I_2 \cos\chi}{\sqrt{g_2^{(i)}}}, \qquad B_{yN} = -B_N \frac{I_1}{\sqrt{g_1^{(i)}}}, \tag{3.376}$$

where $E_N = A_N R_N^{(i)}$, $B_N = i\bar{c}A_N(\partial R_N/\partial l)^{(i)}/\omega$ are the characteristic amplitudes of the oscillation electric and magnetic fields,

$$A_N = 2\frac{a_N}{\omega^2}\left(\frac{\partial R_N}{\partial l}\right)^{(i)}\frac{\cos\chi}{\sqrt{\alpha_P\alpha_T\Delta_N^2}}\int_{-\infty}^{\infty}\int_{-\infty}^{\infty}\frac{j_\|(x,y)}{\Sigma_P}dxdy,$$

and the (i) index indicates that the values are selected at the upper ionospheric boundary.

The coordinates x and y, at the ionospheric level, are directed respectively, South-North and West-East (see Figure 2.49). They are related to dimensionless coordinates ξ and η by $x = \xi\Delta_N^{(i)}$, $y = \eta\sqrt{\alpha_P/\alpha_T}\Delta_N^{(i)}\cos\chi$, where $\Delta_N^{(i)} = \Delta_N\sqrt{g_1^{(i)}}/\cos\chi$ is the distance between resonance surfaces as mapped onto the ionosphere. Dimensionless functions $I_1(\xi,\eta)$ and $I_2(\xi,\eta)$ describe oscillation field distribution in the dimensionless coordinates ξ and η:

$$I_1 = \int_{-\infty}^{\infty}\frac{\kappa d\kappa}{\sqrt{1+\kappa^2}}\int_{-\infty}^{\kappa}\frac{d\kappa'}{\sqrt{1+\kappa'^2}}\left[\frac{1}{(\tilde{\eta}+\eta+i\Delta)^2}+\frac{1}{(\tilde{\eta}-\eta+i\Delta)^2}\right],$$

$$I_2 = \int_{-\infty}^{\infty}\frac{d\kappa}{\sqrt{1+\kappa^2}}\int_{-\infty}^{\kappa}\frac{d\kappa'}{\sqrt{1+\kappa'^2}}\left[\frac{1}{(\tilde{\eta}+\eta+i\Delta)^2}-\frac{1}{(\tilde{\eta}-\eta+i\Delta)^2}\right],$$

where

$$\tilde{\eta} = \kappa(\xi+i\bar{\varepsilon}) - \kappa'(\xi_0+i\bar{\varepsilon}) + \arctan\kappa - \arctan\kappa',$$

ξ_0 is the shell where the source is localised, Δ is the characteristic scale of its localisation in η, and $\eta_0 = 0$.

The numerical calculations below are based on a dipole model of the geomagnetic field, and the Alfvén speed distribution in the meridional plane is given by the model (3.152). The characteristic scale of functions Ω_{TN} and Ω_{PN} variation is defined as $a_N = (\partial\ln t_A/\partial a)^{-1} = a/\mu$. The small parameters regularising the solutions are chosen as follows: $\bar{\varepsilon} = 10^{-1}$, $\Delta = 10^{-1}$, meaning that the wave dissipation in the ionosphere is small and the scale of the source localisation is small compared to the transparent region size Δ_N.

The results of the calculations are illustrated in Figure 3.47, showing the distributions of the oscillation electric field amplitude $E_t = \sqrt{|E_{xN}|^2 + |E_{yN}|^2}$ over coordinates ξ and η, for unit characteristic amplitude $|E_N| = 1$ of the oscillations. An analogous distribution of the amplitude is also exhibited by the oscillation field magnetic components, which, conversely, have an antinode at the ionosphere. It should be noted that, in the zeroth order of the WKB approximation, the equality $R_N^{(i)} \equiv R_N(x, l_\pm, \omega) = 0$ is valid at the ionosphere, implying $|E_N| = 0$. This expression should, however, be regarded as approximate, being accurate to small parameters determined by the small dissipation of waves in the ionosphere. The exact value $|E_N| \neq 0$. Figure 3.47 shows E_t distributions for four different cases of the oscillation source location: a – in the opaque region behind the toroidal resonance surface ($\xi_0 = 1$), b – at the toroidal surface ($\xi_0 = 0$), c – at the poloidal resonance surface ($\xi_0 = -1$) and d – in the opaque region behind the poloidal surface ($\xi_0 = -2$).

The toroidal resonance surface ($\xi_0 = 0$) and source localisation point ($\xi = \xi_0, \eta = 0$) can be seen clearly in all four cases. The poloidal surface fails to manifest itself in these calculations. This can be understood from the following reasoning. When the dissipation of the excited waves is small, their amplitude at the toroidal resonance surface is much larger than their amplitude at the poloidal surface. Moreover, the poloidal surface fails to stand

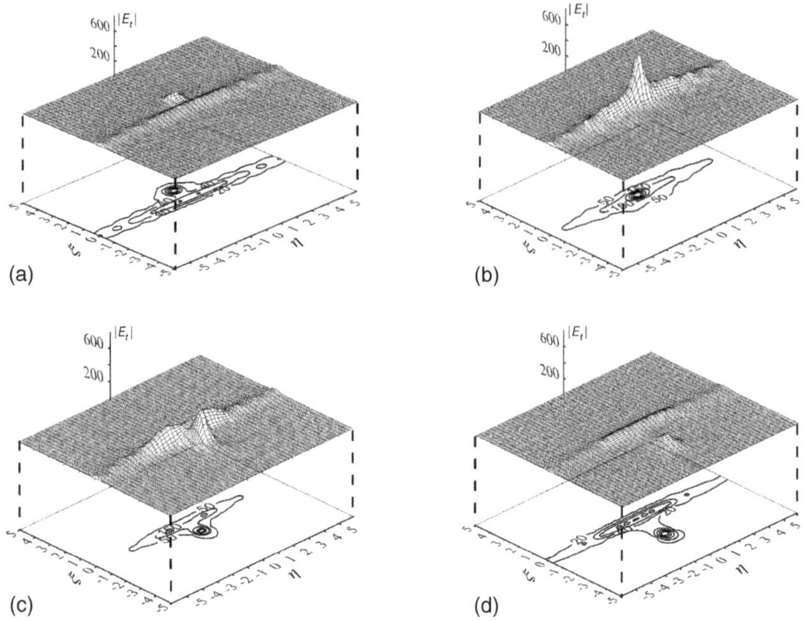

Figure 3.47 Distribution of the amplitude of standing Alfvén waves excited by strongly localised monochromatic sources over dimensionless transverse coordinates (ξ and η). Source location: (a) in the opaque region behind the toroidal surface, (b) at the toroidal surface, (c) at the poloidal surface, and (d) in the opaque region behind the poloidal surface.

out for harmonics with $m < a_N/\Delta_N$, whereas the oscillation amplitude has a singularity regularised by the small parameter $\overline{\epsilon}$ at the toroidal surface. At the source localisation point, the amplitude also has a singularity regularised by the small parameter Δ. Increasing these parameters can decrease the oscillation amplitude in these regions making it comparable to the amplitude at the poloidal surface. The amplitude of the oscillations, however, decreases so much in the process that they practically cease to be observable.

Another feature visible in Figure 3.47 is the fact that the oscillation amplitude is much larger when the source is localised in the transparent region of the wave ($-1 \le \xi_0 \le 0$) than when it is located in the opaque regions. This means that harmonics with $m \gg a_N/\Delta_N \gg 1$ play a decisive role in oscillations excited by a strongly localised source. This is also indicated by the presence of two local maxima at the toroidal surface when the source is localised at the poloidal surface. It follows from the findings in Section 3.14.3 that these maxima are related to waves propagating from the source along the characteristic lines described by (3.375). When the source is localised at the toroidal surface, the length of the characteristics becomes zero and all three maxima merge into one maximum.

3.14.4 On the Methods of Measuring the Polarisation Splitting of the Alfvén Oscillations

Certain experiments attempted to artificially generate geomagnetic pulsations by means of a periodic (at magnetospheric Alfvén eigen oscillation frequency) impact on the ionosphere [188, 360, 361]. Similar experiments could be used to measure the splitting $\Delta_N^{(i)}$ of magnetospheric resonance surfaces and related polarisation splitting $\Delta\Omega_N = \Omega_{TN} - \Omega_{PN}$

of the spectrum. This of course requires that a dense enough (at a step of $\Delta_N^{(i)}/2$; $\Delta_N^{(i)}/4$) observation network is organised in the neighbourhood of the oscillation source. This could be done if an HF radar is used for observation, e.g. in [351]. If the affected area of the ionosphere is much smaller than $\Delta_N^{(i)}$, the above proposed theory can be used for describing the field of the excited oscillations.

The source seems most likely to be localised in the transparent region between the resonance surfaces in such an experiment. If it is localised in the opaque region the amplitude of the excited oscillations will not be large enough for observation, and it is unlikely to be localised exactly at the resonance surfaces. If the source is localised in the transparent region three local maxima should be observed in the oscillation amplitude: one related to the oscillation source, and two located symmetrically about $y = 0$ – at the toroidal resonance surface. The distance between the maxima is easy to calculate by means of the equation of characteristic lines (3.375). Integration produces

$$\Delta\eta = 2\int_0^{\Delta\eta/2} d\eta = 2\int_{\xi_0}^{0} \kappa_N(\xi)d\xi = 2\left(\sqrt{-\xi_0(1+\xi_0)} + \arcsin\sqrt{-\xi_0}\right). \tag{3.377}$$

This equation can be used for determining the splitting value $\Delta_N^{(i)}$ between the poloidal and toroidal surfaces based on distances between local peaks at the toroidal surface, $\Delta_{yN} = \sqrt{\alpha_T/\alpha_P}\Delta\eta\Delta_N^{(i)}$, and from the source localisation point to the toroidal surface, $\Delta_0 = -\xi_0\Delta_N^{(i)}$, as measured in the experiment. We then have

$$\Delta_{yN} = 2\sqrt{\alpha_T/\alpha_P}\left(\sqrt{\Delta_0(\Delta_N^{(i)} - \Delta_0)} + \Delta_N^{(i)}\arcsin\sqrt{\frac{\Delta_0}{\Delta_N^{(i)}}}\right),$$

where $0 < \Delta_0 < \Delta_N^{(i)}$. This equation implicitly determines the splitting value $\Delta_N^{(i)}$ of resonance surfaces. If the source is localised at the toroidal surface, ($\Delta_0 = 0$) $\Delta_{yN} = 0$, i.e. all local maxima merge. If it is localised at the poloidal surface ($\Delta_0 = \Delta_N^{(i)}$), we have $\Delta_{yN} = \pi\Delta_N^{(i)}$ – the distance between the maxima at the toroidal surface is largest ($\sqrt{\alpha_T/\alpha_P} \approx 1$ at all geomagnetic latitudes). Defining $\Delta_N^{(i)}$ in this manner, it is also possible to define the polarisation splitting of the spectrum

$$\Delta\Omega_N = \Omega_{TN} - \Omega_{PN} = \frac{\Delta_N^{(i)}}{a_N^{(i)}}\omega,$$

where $a_N^{(i)}$ is the characteristic scale of magnetospheric plasma inhomogeneity, across magnetic shells as mapped onto the ionosphere. In our case, it is the characteristic scale of the function $\Omega_{TN}(x)$, $\Omega_{PN}(x)$ variation.

The characteristic calculated values of the parameters under study are given in Figures 3.48 and 3.49. The model (3.152) of Alfvén speed distribution in a meridional plane was used for these calculations. The dependencies of $\Delta_N^{(i)}$ and Ω_{TN} on magnetic shell parameter $L = a/R_E$ are shown for the first harmonics of magnetospheric Alfvén eigen oscillations. What attracts one's attention in Figure 3.48 is the anomalously large value $\Delta_1^{(i)} \approx 800$ km as compared to $\Delta_{2,3}^{(i)} \sim 10 - 50$ km. When studying the main harmonic of the oscillations, this makes it possible to rely on a less dense observation station network, or even on ground-based measurement devices. The amplitude of this harmonic must, however, be significantly smaller than that of the other harmonics, their source power being equal because $E_N \sim \Delta_N^{-2}$. It follows from Figure 3.49 that it is

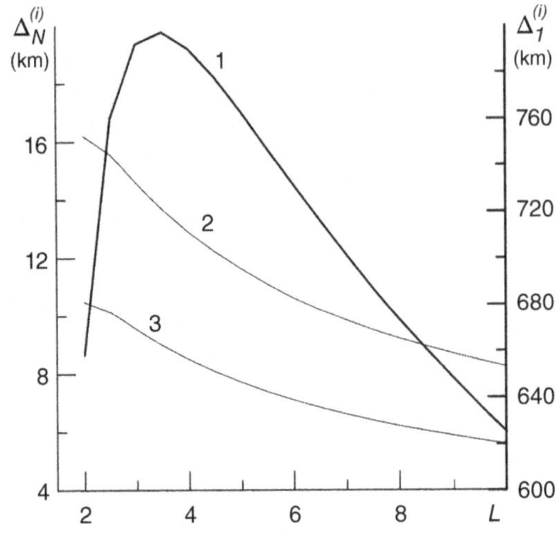

Figure 3.48 Calculated dependence of the splitting, $\Delta_N^{(i)}$ between the toroidal and poloidal resonance magnetic shells as mapped onto the ionosphere on the magnetic shell parameter $L = a/R_E$, for the first ($N = 1$, bold line, right-hand axis) and two next harmonics ($N = 2$, 3, left-hand axis) of standing Alfvén waves.

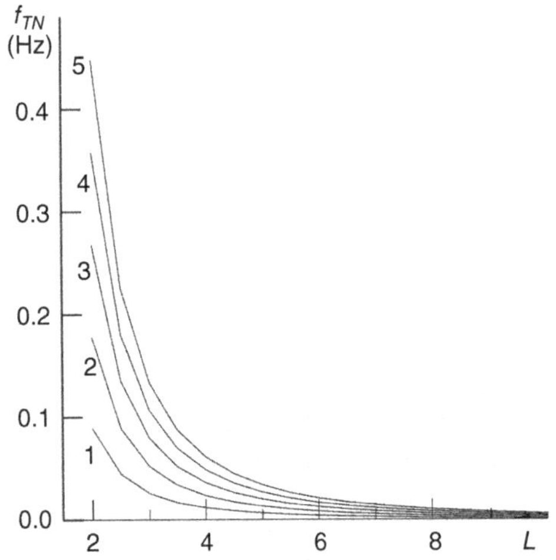

Figure 3.49 Calculated dependence of the eigenfrequencies $f_{TN} = \Omega_{TN}/2\pi$ of toroidal Alfvén oscillations on the magnetic shell parameter $L = a/R_E$, for the first five eigen harmonics ($N = 1, 2, 3, 4, 5$) of standing Alfvén waves.

easier to excite eigen oscillations at low latitudes due to their eigenfrequency increasing, because the monochromatic source must be active over a large enough number of oscillation periods.

3.15 High-m Alfvén Oscillations Generated in the Magnetosphere by Localised Pulse Sources

Section 3.14 addressed Alfvén oscillations generated by a localised monochromatic source. However, localised oscillations can also be excited in the magnetosphere by broadband sources. We will further examine the field dynamics of Alfvén oscillations triggered by pulse currents localised in the ionosphere (see [362]).

3.15.1 From Monochromatic to Broadband Oscillations

Expression (3.370) was obtained in Section 3.14.1, for function U_N, describing the structure of localised monochromatic Alfvén wave, the integrand containing the function $V_N(\xi, \xi', \eta, \eta')$ (3.371). If we proceed to integration variables $\kappa = k/|k_y|$ and $\kappa' = k'/|k_y|$ in (3.371), the outer integral over k_y can be calculated analytically. The expression for $U_N(\xi, \eta, \omega)$ can then be represented as

$$U_N(\xi, \eta, \omega) = \frac{\sqrt{\alpha_{TN}/\alpha_{PN}}}{4\pi^2} \overbrace{\underbrace{\iint}_{-\infty}}^{\infty} \tilde{b}_N(\xi', \eta', \omega)\tilde{V}_N(\xi, \xi', \eta, \eta', \omega) d\xi' d\eta', \tag{3.378}$$

where

$$\tilde{b}_N = \frac{a_N \Delta_N}{\alpha_{TN}\Omega^2_{TN}} I_N, \quad I_N = j_N^+ - j_N^-,$$

$$j_N = 2\sqrt{g_1 g_2} \frac{\partial R_N}{\partial l} \frac{\tilde{j}_\parallel(x^1, x^2, \omega)}{\Sigma_P} \cos\chi,$$

and the plus and minus signs mean that the parameters are taken at the upper boundary of the Northern or Southern ionosphere, respectively. Here, $\tilde{j}_\parallel$ is the Fourier harmonic of the density of the external longitudinal field-aligned current (Alfvén wave source) at the upper boundary of the ionosphere, Σ_P is the integral Pedersen conductivity of the ionosphere, χ is the magnetic inclination angle (see Figure 2.49). Dimensionless transverse coordinates

$$\xi = (x^1 - x^1_{TN})/\Delta_N, \qquad \eta = \sqrt{\frac{\alpha_{TN}}{\alpha_{PN}}} \frac{x^2}{\Delta_N}.$$

are used here. Function $\tilde{b}_N(\xi, \eta, \omega)$ in (3.378) describes the structure of Fourier harmonics of oscillations with frequency ω from a spatially distributed source. Function $\tilde{V}_N$ describes the transverse structure of the field of standing Alfvén oscillations excited by a source of the form $b_N = \delta(\xi - \xi')\delta(\eta - \eta')$, where $\delta(x)$ is the delta function. Dimensionless function $\tilde{V}_N$ has the form

$$\tilde{V}_N = \Delta\Omega_N \int_{-\infty}^{\infty} \frac{d\kappa}{\sqrt{1+\kappa^2}} \times \int_0^{\infty} \frac{dk}{k\sqrt{1+(\kappa - k^{-1})^2}} \left[\frac{1}{\omega - \omega_+(\kappa, k)} + \frac{1}{\omega - \omega_-(\kappa, k)} \right], \tag{3.379}$$

where

$$\omega_\pm = \Delta\Omega_N \Psi^\pm(\kappa, k) - i\gamma_N, \tag{3.380}$$
$$\Psi^\pm = k[\kappa(\xi - \xi') \pm (\eta - \eta') - \arctan\kappa + \arctan(\kappa - k^{-1})],$$

and

$$\gamma_N = (g_N^{(+)} + g_N^{(-)})/\Omega^2_N \alpha_{TN}, \qquad g_N = \sqrt{\frac{g_1}{g_2}} \frac{\bar{c}^2}{4\pi} \frac{\cos\chi}{\Sigma_P} \left(\frac{\partial R_N}{\partial l}\right)^2,$$

is decrement of standing Alfvén wave due to the Joule dissipation in the ionosphere. In contrast to (3.380), the '(±)' signs in the expression for γ_N denote variables corresponding to the Northern or Southern ionospheres.

In order to describe oscillations excited in the magnetosphere by broadband sources, we will carry out an inverse Fourier transformation (3.378) with respect to the source frequency spectrum. Expression (3.378) was obtained assuming $\omega > 0$. To take the spectrum region $\omega < 0$ into account, (3.378) should be integrated over ω from 0 to ∞ and a complex conjugate added to the resulting expression. It was shown in Section 3.14 that expression (3.378) has sharp peaks at points $\omega = \Omega_{TN}$, the bulk of the integral accumulated in their neighbourhood. Therefore it is possible in the integral over ω to switch to the integration variable $\omega' = \omega - \Omega_{TN}$ and extend the lower limit of integration over ω' down to $-\infty$. As can be seen from (3.379), there are poles in the integrand at points $\omega' = \omega_\pm - \Omega_{TN}$. This circumstance can be made use of by applying the residue theorem for calculating the resulting integrals. Closing the integration contour in the upper semiplane of the complex ω' produces the following expressions for the wave electromagnetic field components:

$$E_{1N} = Q_N^1 R_N, \qquad E_{2N} = Q_N^2 R_N,$$
$$B_{1N} = \frac{\bar{c}}{\Omega_{TN}} Q_N^3 \frac{\partial R_N}{\partial l}, \qquad B_{2N} = \frac{\bar{c}}{\Omega_{TN}} Q_N^4 \frac{\partial R_N}{\partial l},$$

where

$$Q_N^n = \frac{\Omega_{TN}}{\pi^2} \int_{-\infty}^{t} dt' \underbrace{\iint}_{-\infty}^{\infty} b_N(\xi', \eta', t') I_{Nn}(\xi - \xi', \eta - \eta', t - t') d\xi' d\eta', \tag{3.381}$$

$$I_{Nn}(\xi, \eta, t) = \tau_1 \int_{-\infty}^{\infty} d\kappa \int_0^{\infty} a_n(\kappa, k, t)[\cos(\Omega_N^+ t + \iota_n) + \delta_n \cos(\Omega_N^- t + \iota_n)] dk, \tag{3.382}$$

$n = 1, 2, 3, 4$, and the following notations are introduced

$$a_1 = -a_4 = \frac{\kappa e^{-\gamma_N t}}{\sqrt{(1+\kappa^2)(1+(\kappa - k^{-1})^2)}}, \qquad a_2 = -a_3 = a_1/\kappa,$$
$$\iota_1 = \iota_2 = 0, \quad \iota_3 = \iota_4 = -\pi/2, \quad \delta_1 = \delta_3 = -\delta_2 = -\delta_4 = 1,$$
$$\Omega_N^\pm = \Omega_{TN}(\xi') + \Delta\Omega_N \Psi^\pm, \quad \tau_1 = \Delta\Omega_1 t,$$

and $b_N(\xi, \eta, t)$ is the source function, its Fourier transform being $\tilde{b}_N(\xi, \eta, \omega)$. Note that, for the sake of convenience, the expression for I_{Nn} is multiplied by the dimensionless factor Δ_1/Δ_N, and the expression for b_N - by an inverse factor, Δ_N/Δ_1 (see (3.384)), in the calculations below. As can be seen from (3.381), the causality principle holds for the standing Alfvén wave excited in the magnetosphere: the oscillation field at moment t is determined by the source action over all the previous moments of time. Expression (3.381) describes the field structure of standing Alfvén wave excited in the magnetosphere by a source that is arbitrarily distributed in the ionosphere and exhibits an arbitrary behaviour in time.

If the source is extremely localised in transverse coordinates and operates over a small enough interval of time (more precise quantitative relations will be given below), expressions (3.378)–(3.382) can be simplified. Integration over ξ', η' and t' is fully applied to function b_N, describing the oscillation source, and functions I_{Nn} can be taken out from these integrals at source localisation point ($\xi' = \xi_0, \eta' = \eta_0$) at the moment it operated $t' = t_0$.

Switching to physical components of the standing wave field at the upper boundary of the ionosphere, to be definite, produces

$$\begin{aligned}
E_{xN} &= \theta(t-t_0)E_N I_{N1}\cos\chi/\sqrt{g_1},\\
E_{yN} &= \theta(t-t_0)\sqrt{\frac{\alpha_{TN}}{\alpha_{PN}}}E_N I_{N2}/\sqrt{g_2},\\
B_{xN} &= \theta(t-t_0)\sqrt{\frac{\alpha_{TN}}{\alpha_{PN}}}B_N I_{N3}\cos\chi/\sqrt{g_2},\\
B_{yN} &= -\theta(t-t_0)B_N I_{N4}/\sqrt{g_1},
\end{aligned} \tag{3.383}$$

where

$$E_N = A_N R_N,\ B_N = \frac{\bar{c}}{\Omega^0_{TN}}A_N\frac{\partial R_N}{\partial l}$$

are the characteristic amplitudes of the oscillations of the electric and magnetic field of standing Alfvén wave near the ionosphere,

$$A_N = \frac{\cos^2\chi}{\pi^2\sqrt{\alpha_{TN}\alpha_{PN}}}\frac{1}{\Omega^0_{TN}\Delta_1}\left(\frac{\partial R_N}{\partial l}\right)\overbrace{\underbrace{\iiint}_{-\infty}}^{\infty}\frac{j_{\parallel}(x,y,t)}{\Sigma_P(x,y,t)}dxdydt, \tag{3.384}$$

$\theta(t)$ is the stepwise Heaviside function. Hereafter, we will assume, without loss of generality, $\xi_0 = 0, \eta_0 = 0, t_0 = 0, \Omega^0_{TN} \equiv \Omega_{TN}(\xi_0 = 0)$. Note that, for the higher harmonics of standing Alfvén waves ($N \gg 1$), where the WKB approximation is applicable to the longitudinal coordinate, the characteristic amplitudes E_N and B_N are practically independent of the harmonic number N. Thus, E_N and B_N can be replaced with certain averaged values and taken from under the summation sign, for approximate calculations when summing up the fields of all the harmonics of standing waves.

Let us examine two different oscillation regimes of a standing Alfvén wave.

3.15.2 Initial Oscillation Regime ($\tau_N \ll 1$)

Let us inspect functions $I_{Nn}(\xi, \eta, t)$, describing the oscillations of the field components of standing Alfvén wave. Their behaviour significantly depends on parameter

$$\tau_N = \Delta\Omega_N t.$$

Let us examine the initial oscillation regime, when $\tau_N \ll 1$. The condition $\Delta\Omega_N \ll \Omega_{TN}$ being valid for all (except the very first) harmonics of standing waves in the magnetosphere, this regime can be maintained over many oscillation periods after pulse excitation. In the $\tau_N \ll 1$ limit, the bulk of the inner integral (3.382) comes from characteristic values $k \gg 1$. Therefore, it can be approximately assumed, in the integrand,

$$\begin{aligned}
&a_2 = -a_3 \approx -e^{-\gamma_N t}/(1+\kappa^2), \quad a_1 = -a_4 \approx -\kappa e^{-\gamma_N t}/(1+\kappa^2),\\
&\arctan(\kappa - k^{-1}) \approx \arctan\kappa.
\end{aligned}$$

After this, the inner integral over k is easily calculated, resulting in

$$I_{Nn} \simeq -\sin(\Omega^0_{TN}t + \iota_n)\int_{-\infty}^{\infty} a_n(\kappa)[(\kappa\xi+\eta)^{-1} + (\kappa\xi-\eta)^{-1}]d\kappa.$$

There are poles in the integrand at points $\kappa = \pm i$, $\kappa = \pm\eta/\xi$. Let us use the residue theorem to calculate these integrals. The result will be

$$I_{Nn} \simeq A_{Nn} \sin(\Omega^0_{TN} t + \iota_n), \tag{3.385}$$

where

$$\begin{aligned} A_{N1} &= A_{N4} \simeq -2\pi \tfrac{\Delta_1}{\Delta_N} \tfrac{|\xi| e^{-\gamma_N t}}{|\xi|^2+\eta^2}, \\ A_{N2} &= A_{N3} \simeq -2\pi \tfrac{\Delta_1}{\Delta_N} \tfrac{\eta e^{-\gamma_N t}}{|\xi|^2+\eta^2}. \end{aligned} \tag{3.386}$$

The coefficients A_{Nn} have a singularity at source localisation point ($\xi = 0, \eta = 0$). This is due to the fact that a singular function $b_N \sim \delta(\eta)$ is used as the oscillation source. If a function with no singularities is used,

$$b_N = \overline{b}_N \frac{\Delta^2_\eta}{\eta^2 + \Delta^2_\eta}, \tag{3.387}$$

where Δ_η is the characteristic scale of source localisation in coordinate η, the singularity in coefficients (3.386) vanishes. Moreover, the replacement $\pm\eta \to \pm\eta - i\Delta_\eta$ should be made in all the previous derivations, resulting in the replacement $|\xi| \to |\xi| + \Delta_\eta$ in the coefficients in (3.386).

As can be seen from (3.385), oscillations in the initial regime proceed at the same frequency, at all the points of the plane (ξ, η). The amplitude of these oscillations has a maximum at the source localisation point and decreases according to the power law $(\sim 1/\sqrt{\xi^2 + \eta^2})$ away from it. When using the source model (3.387) note also that integration with respect to y in the expression for the characteristic amplitude of oscillations (3.384) should be replaced with the term

$$\int dy \to \Delta_y = \sqrt{\frac{\alpha_{PN} g_2}{\alpha_{TN}}} \Delta_N \Delta_\eta,$$

equal to the characteristic scale of source localisation in the ionosphere, in the y coordinate.

3.15.3 Asymptotic Regime of Oscillations ($\tau_N \gg 1$)

Functions $I_{Nn}(\xi, \eta, t)$ consists of the sum of integrals of the form

$$I = \tau_1 e^{\pm i\Omega^0_{TN} t} \int_0^\infty dk \int_{-\infty}^\infty a_n(k, \kappa) \exp\left(\pm i\tau_N \Psi^\pm(k, \kappa)\right) d\kappa. \tag{3.388}$$

If $\tau_N \gg 1$, the saddle point method can be used for estimating these integrals. Equating the first derivative of the phase $\Psi^\pm$ with respect to κ to zero produces an equation defining the saddle point:

$$\nu_1 \equiv \left(\frac{\partial \Psi^\pm}{\partial \kappa}\right)_{\kappa=\tilde{\kappa}} = k\left[\xi - (1 + \tilde{\kappa}^2)^{-1} + (1 + (\tilde{\kappa} - k^{-1})^2)^{-1}\right] = 0. \tag{3.389}$$

The second derivative of the phase at the saddle point is

$$\nu_2 \equiv \left(\frac{\partial^2 \Psi^\pm}{\partial \kappa^2}\right)_{\kappa=\tilde{\kappa}} = 2k\left[\frac{\tilde{\kappa}}{(1 + \tilde{\kappa}^2)^2} - \frac{\tilde{\kappa} - k^{-1}}{(1 + (\tilde{\kappa} - k^{-1})^2)^2}\right].$$

We will use the standard formulas of the saddle point method (see [337]) to write down the asymptotic expression for the integral (3.388):

$$I \simeq \tau_1 \sqrt{\frac{2\pi}{\tau_N}} e^{\pm i\Omega_N^0 t} \int_0^\infty \frac{a_n(k,\tilde{\kappa})}{\sqrt{|\nu_2(k)|}} \exp\left[\pm i\tau_N \tilde{\Psi}^\pm \pm i\frac{\pi}{4}\mathrm{sign}(\nu_2)\right] dk, \tag{3.390}$$

where $\tilde{\Psi}^\pm \equiv \Psi^\pm(k,\tilde{\kappa})$. The large parameter $\tau_N \gg 1$ is conserved in the exponent of the integrand, therefore the saddle point method can also be applied to this integral. Equating the first derivative of the phase $\tilde{\Psi}^\pm$ with respect to k, we will find an equation defining the saddle point $\overline{k}$:

$$\overline{w}_1 \equiv \left(\frac{\partial \tilde{\Psi}^\pm}{\partial k}\right)_{k=\overline{k}} = \overline{\kappa}\xi \pm \eta - \arctan\overline{\kappa} + \arctan(\overline{\kappa} - \overline{k}^{-1})$$
$$+ \left[\overline{k}\left(1 + (\overline{\kappa} - \overline{k}^{-1})^2\right)\right]^{-1} = 0, \tag{3.391}$$

where $\overline{\kappa} \equiv \tilde{\kappa}(\overline{k})$. Let us also write down the expression for the second derivative of the phase $\tilde{\Psi}^\pm$ at the saddle point:

$$\overline{w}_2 \equiv \left(\frac{\partial^2 \tilde{\Psi}^\pm}{\partial k^2}\right)_{k=\overline{k}} = -\frac{4}{\overline{k}^{2}(1+\overline{\kappa}^2)} \frac{\overline{\kappa} - \overline{k}^{-1}}{\left(1 + (\overline{\kappa} - \overline{k}^{-1})^2\right)\overline{\nu}_2},$$

where $\overline{\nu}_2 \equiv \nu_2(\overline{k},\overline{\kappa})$. We will use the standard formulas of the saddle-point technique to write down the final asymptotic expression for the integral (3.388):

$$I \simeq A_{Nn} \exp(\pm i(\Omega_N^0 + \Delta\Omega_N \overline{\Psi}^\pm)t \pm i\alpha_n), \tag{3.392}$$

where

$$A_{Nn} = 2\pi \frac{\Delta_1}{\Delta_N} \frac{\bar{a}_n}{\sqrt{|\overline{\nu}_2 \overline{w}_2|}}, \tag{3.393}$$

$$\alpha_n = \frac{\pi}{4}(\mathrm{sign}(\overline{\nu}_2) + \mathrm{sign}(\overline{w}_2)), \tag{3.394}$$

$\overline{\Psi}^\pm \equiv \Psi^\pm(\overline{k},\overline{\kappa},\xi,\eta)$, $\bar{a}_n \equiv a_n(\overline{k},\overline{\kappa})$. To write down the expressions for the functions I_{Nn}, it is necessary to find the roots of (3.389) and (3.391) and define their domains of existence. We will seek such roots of (3.389) and (3.391), that leave the phase $\overline{\Psi}^\pm$ real. If $\overline{\Psi}^\pm$ is a complex phase, the integrals (3.392) yield the solutions that exponentially decay with time which we will ignore. The integration paths in (3.388) being along the real axes k and κ, we will restrict ourselves to seeking real roots $\overline{k}$ and $\overline{\kappa}$. If $\overline{k}$ and $\overline{\kappa}$ are real, the phase $\overline{\Psi}^\pm$ is also real. Expressing $\overline{k}^{-1}$ from (3.389) in terms of $\overline{\kappa}$ produces

$$\overline{k}^{-1} = \overline{\kappa} \pm \sqrt{\frac{\overline{\kappa}^2 + \xi(1+\overline{\kappa}^2)}{1 - \xi(1+\overline{\kappa}^2)}}. \tag{3.395}$$

It can be seen from this that $\overline{k}$ remains a positive real value if $\overline{\kappa}$ and ξ are in the following tolerance ranges

$$-\sqrt{(1-\xi)/\xi} \le \overline{\kappa} \le \sqrt{(1-\xi)/\xi}, \quad 0 \le \xi \le 1, \tag{3.396}$$

$$\overline{\kappa} \ge \sqrt{-\xi/(1+\xi)}, \quad \overline{\kappa} \le -\sqrt{-\xi/(1+\xi)}, -1 \le \xi \le 0. \tag{3.397}$$

Substituting (3.395) into (3.391) yields the equation for $\overline{\kappa}$

$$\frac{\overline{\kappa}}{(1+\overline{\kappa}^2)} - \arctan\overline{\kappa} \pm \left[\frac{\sqrt{(\overline{\kappa}^2+(1+\overline{\kappa}^2)\xi)(1-(1+\overline{\kappa}^2)\xi)}}{1+\overline{\kappa}^2} - \arctan\sqrt{\frac{\overline{\kappa}^2+\xi(1+\overline{\kappa}^2)}{1-\xi(1+\overline{\kappa}^2)}}\right] = \pm\eta. \tag{3.398}$$

Note that the '±' signs in the left and right sides of this equation are not interrelated, making four different equations possible. A detailed analysis demonstrates that the solutions of only three of these four equations satisfy the conditions (3.396) and (3.397). The first equation corresponding to the range of (3.396) is obtained when the plus sign is chosen in the left-hand side of (3.398) and such a sign is chosen in the right-hand side that would guarantee its negative value, i.e. $-|\eta|$.

The behaviour of the roots of (3.398) is easiest to analyse by plotting the left- and right-hand sides of those equations. The intersection points of these plots are the roots of (3.398). The graphical solution of the first equation in (3.398) is given in Figure 3.50a. For $1 \geq \xi \geq 0$, two roots can be seen to exist for this equation, $\overline{\kappa}_1$ and $\overline{\kappa}_2$, satisfying the condition (3.396). The $-|\eta|$ value can vary within $\eta_2 \leq -|\eta| \leq \eta_1$ for the root $\overline{\kappa}_1$ and within $\eta_3 \leq -|\eta| \leq \eta_1$ for the root $\overline{\kappa}_2$, where

$$\eta_1 = \sqrt{\xi(1-\xi)} - \arctan\sqrt{\xi/(1-\xi)},$$
$$\eta_2 = -\sqrt{\xi(1-\xi)} + \arctan\sqrt{(1-\xi)/\xi} - \pi/2,$$
$$\eta_3 = \sqrt{\xi(1-\xi)} - \arctan\sqrt{(1-\xi)/\xi} - \pi/2.$$

Obviously, both roots $\overline{\kappa}_1, \overline{\kappa}_2$ merge into one root, for $-|\eta| = \eta_1$. The other two equations in (3.398), that correspond to the range (3.397), result from selecting two different signs in the

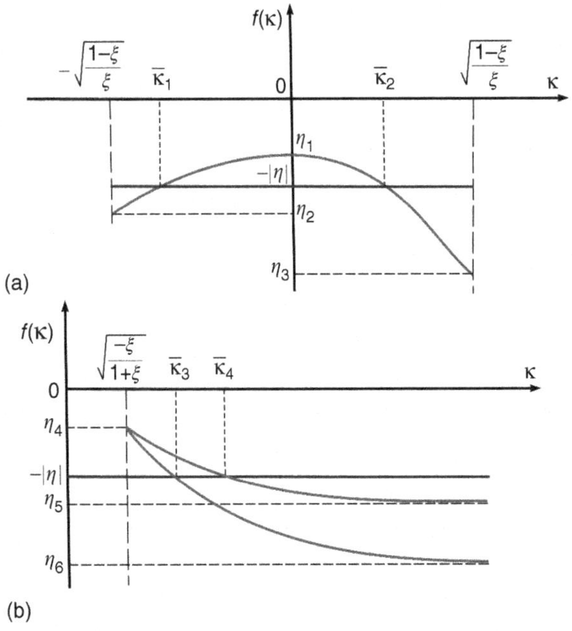

Figure 3.50 Graphical solution of (3.398): (a) in the tolerance range (3.396), (b) in the tolerance range (3.397). The dark gray curves are the left-hand sides (panel (b) the upper curve is for the '−' sign, the lower for '+'), and straight horizontal black lines are the right-hand sides of (3.398).

left-hand side of (3.398) and negative $-|\eta|$ in the right-hand side. The corresponding graphical solutions of these equations are shown in Figure 3.50b. For $0 \geq \xi \geq -1$, two roots $\overline{\kappa}_3$ and $\overline{\kappa}_4$ can be seen to exist satisfying the condition (3.397). The tolerance range of $-|\eta|$ for the root $\overline{\kappa}_3$ lies within $\eta_6 \leq -|\eta| \leq \eta_4$, and for the root $\overline{\kappa}_4$ – within $\eta_5 \leq -|\eta| \leq \eta_4$, where

$$\eta_4 = \sqrt{-\xi(1+\xi)} - \arctan\sqrt{-\xi/(1+\xi)},$$
$$\eta_5 = -\sqrt{-\xi(1+\xi)} + \arctan\sqrt{-(1+\xi)/\xi} - \pi/2,$$
$$\eta_6 = \sqrt{-\xi(1+\xi)} - \arctan\sqrt{-(1+\xi)/\xi} - \pi/2.$$

When $-|\eta| = \eta_4$, the roots $\overline{\kappa}_3$ and $\overline{\kappa}_4$ merge together. The domain of existence of the real roots of (3.395) and (3.398) in the plane (ξ, η) is thus limited by the characteristics η_j ($j = 1, 2, \ldots, 6$). Its form in the plane of dimensionless coordinates (ξ, η) is shown in Figure 3.51. Its shape resembles 'butterfly wings', therefore we will hereafter use this definition for brevity's sake to denote the domain of existence of oscillations in the asymptotic regime. While the roots $\overline{\kappa}_2$ and κ_3 exist throughout this range, the roots $\overline{\kappa}_1$ and $\overline{\kappa}_4$ only exist in the part of the 'butterfly wings' limited by the characteristics $\pm(\eta_1, \eta_2)$ and $\pm(\eta_3, \eta_4)$.

Let us write down the values of $\overline{v}_2^{(i)}, \overline{w}_2^{(i)}$ and $\overline{\Psi}^{\pm}_{(i)}$ that correspond to the roots $\overline{\kappa}_i$ on the characteristic lines η_j (Figure 3.51):

1. on the characteristic line η_1: $\overline{\kappa}_1 = \overline{\kappa}_2 = 0$, $\overline{k}_1 = \overline{k}_2 = \sqrt{\xi/(1-\xi)}$, $\overline{v}_2^{(1,2)} = 2(1-\xi)^2$, $\overline{w}_2^{(1,2)} = 0$, $\overline{\Psi}^{\pm}_{(1,2)} = -(1-\xi)$.
2. on the characteristic line η_2: $\overline{\kappa}_1 = -\sqrt{(1-\xi)/\xi}$, $\overline{k}_1 = 0$, $\overline{v}_2^{(1)} = 0$, $\overline{w}_2^{(1)} = 2$, $\overline{\Psi}^{\pm}_{(1)} = 0$.
3. on the characteristic line η_3: $\overline{\kappa}_2 = \sqrt{(1-\xi)/\xi}$, $\overline{k}_2 = 0$, $\overline{v}_2^{(2)} = 0$, $\overline{w}_2^{(2)} = 2$, $\overline{\Psi}^{\pm}_{(2)} = 0$.
4. on the characteristic line η_4: $\overline{\kappa}_3 = \overline{\kappa}_4 = \sqrt{-\xi/(1+\xi)}$, $\overline{k}_3 = \overline{k}_4 = \overline{\kappa}_3^{-1}$, $\overline{v}_2^{(3,4)} = 2(1+\xi)^2$, $\overline{w}_2^{(3,4)} = 0$, $\overline{\Psi}^{\pm}_{(3,4)} = -1$.
5. on the characteristic line η_5: $\overline{\kappa}_4 = \infty$, $\overline{k}_4 = 0$, $\overline{v}_2^{(4)} = 0$, $\overline{w}_2^{(4)} = 2$, $\overline{\Psi}^{\pm}_{(4)} = \xi$.
6. on the characteristic line η_6: $\overline{\kappa}_3 = \infty$, $\overline{k}_3 = 0$, $\overline{v}_2^{(3)} = 0$, $\overline{w}_2^{(3)} = 2$, $\overline{\Psi}^{\pm}_{(3)} = \xi$.

It can be seen that one of the second derivatives, $\overline{v}_2^{(i)}$ or $\overline{w}_2^{(i)}$, becomes zero on each of these characteristic lines. It can be shown that functions $\overline{v}_2^{(1)}(\xi, \eta)$ and $\overline{v}_2^{(4)}(\xi, \eta)$ reverse their sign when switching from η_1 to η_2 or from η_4 to η_5, respectively, passing through zero at certain lines η_0 located between these characteristic lines. Equations for η_0 are obtained

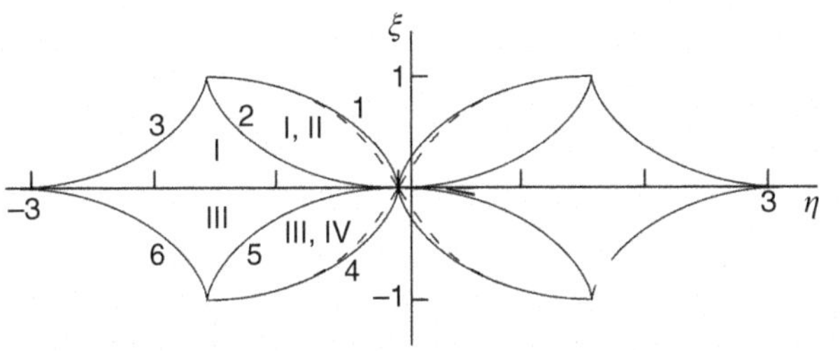

Figure 3.51 Region occupied by Alfvén wave oscillations in the asymptotic regime ('butterfly wings'). Solid lines (1–6) are characteristic lines $\eta_1, \eta_2, \ldots, \eta_6$, dashed lines are characteristic lines η_0. The Roman numerals I, II, III and IV indicate the domains of existence for the roots of (3.398) – $\overline{\kappa}_1, \overline{\kappa}_2, \overline{\kappa}_3$ and $\overline{\kappa}_4$.

in Appendix K, and their curves are shown in Figure 3.51 as dashed lines. The representation (3.392), obtained using the classical saddle point technique, cannot be applied at the characteristic lines η_j themselves. It is however possible to expand the applicability range of this technique by taking into account the higher-order derivatives in the phase $\Psi^{\pm}$ decomposition.

The third derivative of the phase $\overline{w}_3^{(i)} \equiv (\partial^3 \Psi^{\pm}_{(i)}/\partial k^3)$, where $i = 1, 2$ at the characteristic lines $\pm\eta_1$ and $i = 3, 4$ at $\pm\eta_4$, is

$$\overline{w}_3^{(i)} = -2\left(\frac{|\xi|}{1-|\xi|}\right)^{5/2}.$$

Taking it into account when using the saddle-point technique (see Appendix L) produces, for the integral (3.388), an asymptotic representation analogous to (3.392), where the following should be assumed

$$A_{Nn}^{(i)} = \bar{a}_n \frac{\tau_1}{\tau_N^{5/6}} \Gamma\left(\frac{1}{3}\right) \sqrt{\frac{2\pi}{3|\overline{v}_2^{(i)}|}} \left(\frac{6}{|\overline{w}_3^{(i)}|}\right)^{1/3}, \tag{3.399}$$

$$\alpha_n^{(i)} = \frac{\pi}{4} \operatorname{sign}(\overline{v}_2^{(i)}). \tag{3.400}$$

Here, $\Gamma(x)$ is the gamma function. These expressions should be used near the characteristic lines $\pm\eta_1$ and $\pm\eta_4$, if this condition is satisfied:

$$\frac{|\overline{w}_2^{(i)}|^3}{\overline{w}_3^{(i)2}} \tau_N \lesssim 1.$$

Analogously, taking into account the third derivative

$$\overline{v}_3^{(i)} \equiv \left(\frac{\partial^3 \Psi^{\pm}_{(i)}}{\partial \kappa^3}\right)_{\kappa=\overline{\kappa}_i} = 2\overline{k}_i \left[\frac{1-3\overline{\kappa}_i^2}{\left(1+\overline{\kappa}_i{}^2\right)^3} - \frac{1-3(\overline{\kappa}_i - \overline{k}_i^{-1})^2}{\left(1+(\overline{\kappa}_i - \overline{k}_i^{-1})^2\right)^3}\right]$$

on the characteristic lines η_0 produces, for the integral (3.388), an expression of the form (3.392), where

$$A_{Nn}^{(i)} = \bar{a}_n \frac{\tau_1}{\tau_N^{5/6}} \Gamma\left(\frac{1}{3}\right) \sqrt{\frac{2\pi}{3|\overline{w}_2^{(i)}|}} \left(\frac{6}{|\overline{v}_3^{(i)}|}\right)^{1/3}, \tag{3.401}$$

$$\alpha_n^{(i)} = \frac{\pi}{4} \operatorname{sign}(\overline{w}_2^{(i)}). \tag{3.402}$$

These expressions are applicable near the characteristic lines η_0 if this condition is satisfied:

$$\frac{|\overline{v}_2^{(i)}|^3}{\overline{v}_3^{(i)2}} \tau_N \lesssim 1.$$

Even such an extended application of the stationary phase technique, however, fails to produce a correct expression for the integral (3.388) at the characteristic lines $\pm\eta_2, \pm\eta_3, \pm\eta_5, \pm\eta_6$, where irrespective of their order all the derivatives of phase $\Psi^{\pm}_{(i)}$ with respect to κ become zero. In this case, a technique of approximate calculation of the integrals can be used (3.388) as introduced in Appendix M, which is based on the fact that

$\overline{k}_i \to 0$ when these characteristic lines are approached. Using this technique produces an expression for the integral (3.388) which is analogous to (3.392), where

$$A_{Nn}^{(i)} = \frac{\Delta_1}{\Delta_N}|\Lambda|e^{-\gamma_N t}, \tag{3.403}$$

$$\alpha_n^{(i)} = \arctan(\mathrm{Re}\overline{\Lambda}/\mathrm{Im}\overline{\Lambda}), \tag{3.404}$$

where

$$\overline{\Lambda} = -\int_0^1 \frac{dr}{\sqrt{1-r^2}}\left\{\left[\ln\sqrt{\frac{2-r}{1+r}} + i\left(\frac{3}{2}\pi - |\eta|\right)\right]^{-1} + \left[\ln\sqrt{\frac{2-r}{1+r}} + i(\frac{\pi}{2} - |\eta|)\right]^{-1}\right\}. \tag{3.405}$$

These expressions should be used near the above-mentioned characteristic lines if this condition is satisfied:

$$\overline{v}_2^{(i)}\tau_N \lesssim 1.$$

If $\tau_N \gg 1$ the following expressions can be written for functions $I_{Nn}(\xi, \eta, t)$, describing the oscillations of the electromagnetic field components for a standing Alfvén wave in the asymptotic regime:

$$I_{Nn} = -\sum_{i=1}^{4}\Theta^{(i)}(\xi,\eta)A_{Nn}^{(i)}\cos[(\Omega_N^0 + \Delta\Omega_N\overline{\Psi}_{(i)}^{\pm}(\xi,\eta))t + \overline{\iota}_n^{(i)}], \tag{3.406}$$

where $\Theta^{(i)}(\xi,\eta)$ is a function defining the existence domains for the roots $\overline{\kappa}_i, \overline{k}_i$:

$$\Theta^{(1)} = \theta(\xi)\theta(1-\xi)\theta(|\eta| - \eta_2)\theta(\eta_1 - |\eta|),$$

$$\Theta^{(2)} = \theta(\xi)\theta(1-\xi)\theta(|\eta| - \eta_3)\theta(\eta_1 - |\eta|),$$

$$\Theta^{(3)} = \theta(-\xi)\theta(1+\xi)\theta(|\eta| - \eta_6)\theta(\eta_4 - |\eta|),$$

$$\Theta^{(4)} = \theta(-\xi)\theta(1+\xi)\theta(|\eta| - \eta_5)\theta(\eta_4 - |\eta|).$$

The oscillation amplitudes $A_{Nn}^{(i)}$ are defined by (3.393), (3.399), (3.401) and (3.403) in the ranges of their applicability, the corrections to the oscillation frequency $\Delta\Omega_N\overline{\Psi}_{(i)}^{\pm}$ by (3.380), and the initial phases $\overline{\iota}_n^{(i)}$ can be written as

$$\overline{\iota}_n^{(i)} = \iota_n + \alpha_n^{(i)},$$

where $\alpha_n^{(i)}$ is defined by the corresponding expressions (3.394), (3.400), (3.402) and (3.404). If the model representation (3.380) is used for the oscillation source, these replacements $\gamma_N \to \gamma_N + \overline{k}_i\Delta_\eta\Delta\Omega_N$ and $|\eta| \to |\eta| + i\Delta_\eta$ should be made in the expressions for $\bar{a}_n$ and (3.403) and (3.405).

It can be seen from (3.406) that, in the asymptotic regime, the Alfvén oscillations have a complex character. Unlike oscillations in the initial regime, not only the amplitude but also the frequency and initial phase of the asymptotic oscillations depend on the observation point coordinate. In those areas of the 'butterfly wings' where two roots of (3.398) exist simultaneously, two oscillations of different frequency, phase and amplitude are superimposed.

3.15.4 Model Plasmasphere and Equations for the Field Components of Standing Alfvén Waves

For our numerical calculations, we will use the dipole model of the geomagnetic field and the Alfvén speed model (3.152). The electric field components of Alfvén oscillations excited in the magnetosphere by a pulse source are described by expressions (3.383). The source being strongly localised suggests that its characteristic scale across geomagnetic field lines is much smaller than the corresponding localisation scale $\Delta_N^{(i)}$ of standing Alfvén waves near the ionosphere. Our further calculations will use the values of $\Delta_N^{(i)}$ calculated for the magnetic shell $L = 1.5$ in order to compare them with what was found during observations in the course of an active experiment MASSA carried out at the latitude corresponding to this magnetic shell [269]. This experiment involved a powerful ground-level explosion, after which the Aureole 3 satellite travelling near the ionosphere, at a small distance from the explosion site, registered MHD oscillations of the Alfvén type in the magnetosphere [75, 269]. The values of $\Delta_N^{(i)}$ were calculated in Section 3.14.4 for the first three harmonics of standing Alfvén waves at magnetic shell $L = 6.6$. At the magnetic shell in question ($L = 1.5$) the values of $\Delta_N^{(i)}$ ($i = 2, 3, 4, 5$) near the ionosphere vary from $\Delta_2^{(i)} = 60$ km to $\Delta_5^{(i)} = 10$ km. For the first harmonic, $\Delta_1^{(i)} \approx 600$ km is much higher than the localisation sizes of all other harmonics. Thus, Eq. (3.383) are applicable if the size of the source of oscillations in the ionosphere is not larger than several kilometres, for harmonics $N = 2, 3, \ldots$, while being capable of reaching several tens of kilometres for the main harmonic.

The pulse character of the source implies that its characteristic duration Δ_t is much smaller than the periods of the oscillations it excites. In the WKB approximation, these periods can be defined approximately as $t_N = t_A/N$, where t_A is the characteristic 'there and back' transit time at Alfvén speed along a geomagnetic field line between the magnetoconjugated ionospheres (3.142). For the model of the Alfvén speed distribution in the magnetosphere at the magnetic shell under study, $L = 1.5$, we have $t_A \simeq 5$ seconds. However, this time is much smaller than the periods of oscillations of the main harmonic of standing Alfvén waves observable at this magnetic shell. This is explainable by the fact that a substantial contribution to the Alfvén oscillation periods here is from magnetic flux tubes filled with oxygen ions. It was shown in [358] that taking into account oxygen ions results in $t_A \simeq 15$–30 seconds at $L = 1.5$, which coincides well with the observed oscillation periods. In this work, we use the value $t_A \simeq 20$ seconds.

The full field of MHD oscillations excited in the magnetosphere by a broadband localised source is a superposition of all harmonics of the standing Alfvén waves. For the physical components of the electromagnetic field of these oscillations near the ionosphere, it can be written

$$E_{(x,y,z)} = \sum_{N=1}^{\infty} E_{(x,y,z)N}, \quad B_{(x,y,z)} = \sum_{N=1}^{\infty} B_{(x,y,z)N}, \tag{3.407}$$

where the field components of individual harmonics of the standing Alfvén waves, E_{xN}, E_{yN}, B_{xN}, B_{yN}, are defined by Eq. (3.383). As has been noted, the characteristic amplitudes of the electric and magnetic components of the field of these waves, (E_N and B_N), practically do not depend on the harmonic number N, in the WKB approximation ($N \gg 1$). At certain accuracy, this statement can be extended to include the first harmonics $N \sim 1$ as well. The amplitudes E_N and B_N can then be replaced with their mean values, $\bar{E}$ and $\overline{B}$, and taken out from under the sum symbol in the expressions (3.407). Restricting ourselves to the MHD oscillation field components that are tangential with respect to the ionosphere, we can write

$$E_x \simeq \bar{E}_x \sum_{N=1}^{\infty} I_{N1}(x,y,t), \qquad E_y \simeq \bar{E}_y \sum_{N=1}^{\infty} I_{N2}(x,y,t),$$
$$B_x \simeq \overline{B}_x \sum_{N=1}^{\infty} I_{N3}(x,y,t), \qquad B_y \simeq \overline{B}_y \sum_{N=1}^{\infty} I_{N4}(x,y,t), \tag{3.408}$$

where the functions $I_{Nn}(x,y,t)$, describing the behaviour of individual harmonics of the standing Alfvén wave are given by the expressions (3.382).

Note that using the WKB approximation for functions R_N to describe the longitudinal structure of standing waves gives amplitudes $\bar{E}_x = \bar{E}_y = 0$, because functions R_N have nodes at the ionosphere. However, taking into account the finite conductivity of the ionosphere produces $\bar{E}_x$, $\bar{E}_y \neq 0$, even if small enough. Conversely, functions $\overline{B}_x, \overline{B}_y$, proportionate to $\partial R_N/\partial l$, reach maximum values at the ionosphere. Our further calculations will only consider such magnetic components of the oscillation field, B_x, B_y.

As the harmonic number N increases, the characteristic localisation scale of the standing Alfvén wave near the ionosphere, $\Delta_N^{(i)}$, and the eigenfrequency splitting $\Delta\Omega_N$ decrease. Thus, starting from certain $\overline{N}$, the condition $\tau_N \equiv \Delta\Omega_N t < 1$ will be satisfied for all harmonics $N > \overline{N}$, at time t, and they will oscillate in the initial regime. Functions I_{Nn}, describing the oscillations of these harmonics in the initial regime, have the following form in the physical transverse coordinates x, y:

$$I_{Nn} = A_{Nn} \sin(\Omega_0 t + \iota_n), \tag{3.409}$$

where

$$A_{N1} = A_{N4} = -2\pi\Delta_1^{(i)} \frac{|x| e^{-\gamma_N t}}{|x|^2 + y^2 \cos^2 \chi},$$
$$A_{N2} = A_{N3} = -2\pi\Delta_1^{(i)} \frac{y e^{-\gamma_N t} \cos\chi}{|x|^2 + y^2 \cos^2 \chi}, \tag{3.410}$$

$\iota_1 = \iota_2 = 0$, $\iota_3 = \iota_4 = \pi/2$. It can be seen from this that coefficients A_{Nn} do not depend on the number N, and the series of (3.409) will therefore diverge when summed up. Moreover, these coefficients have a singularity when $|x| \to 0$ and $|y| \to 0$. As was demonstrated in Section 3.14, this singularity is related to using the source (field-aligned currents $j_\parallel$) model in the form of a singular function, $\delta(\eta)$. Employing the model function of the current distribution in the ionosphere,

$$\tilde{j}_\parallel = \bar{j}_\parallel \frac{\Delta_y^2}{y^2 + \Delta_y^2},$$

eliminates this singularity in the coefficients (3.410). Additionally, the replacement $|x| \to |x| + \Delta_y$ takes place.

Analogously, the divergence of the series with terms (3.409) is related to the fact that a singular function of the form $\delta(t)$ is used as the oscillation source. If a model function lacking any singularities is used,

$$j_\parallel = \tilde{j}_\parallel \frac{\Delta_t^2}{t^2 + \Delta_t^2} = \bar{j}_\parallel \frac{\Delta_y^2}{y^2 + \Delta_y^2} \frac{\Delta_t^2}{t^2 + \Delta_t^2}, \tag{3.411}$$

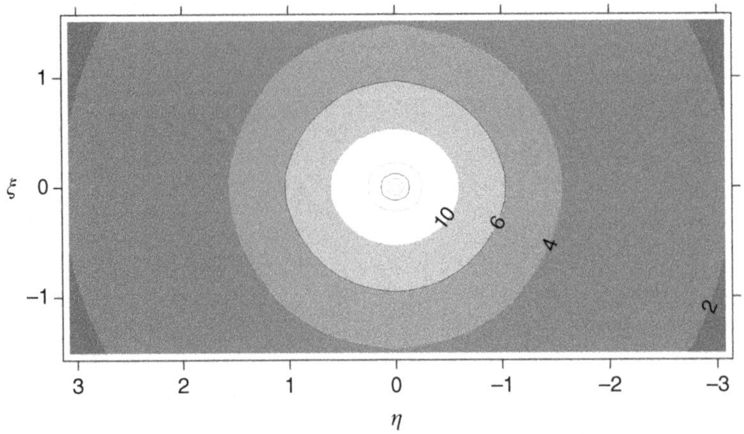

Figure 3.52 Amplitude distribution for a separate harmonic of standing Alfvén wave, in the initial regime of oscillations, in the plane of dimensionless transverse coordinates (ξ, η). Sectors with smallest amplitude are in shades of grey.

the series (3.408) become convergent. Each of the coefficients (3.409) is also multiplied by $e^{-\Omega_N^0 \Delta_t}$, to guarantee the convergence, because $\Omega_N^0 = 2\pi N/t_A$. Note that using the source model (3.411) removes limitations on the applicability of formulas (3.408) caused by the localisation scale and source duration. The characteristic oscillation amplitudes $\overline{B}_x$ and $\overline{B}_y$ have the form

$$\overline{B}_x = \overline{B}\sqrt{\alpha_T \alpha_P} \cos\chi / \sqrt{g_2^{(i)}}, \quad \overline{B}_y = -\overline{B}/\sqrt{g_1^{(i)}},$$

where

$$\overline{B} = \frac{\cos^2\chi}{\pi^2\sqrt{\alpha_T \alpha_P}} \frac{\Delta_y \Delta_t}{\Delta_1^{(i)}} \frac{\overline{c}}{v_{A(i)}^2} \int_{-\infty}^{\infty} \frac{\overline{j}_\parallel(x)}{\Sigma_P(x)} dx, \tag{3.412}$$

$$\alpha_P = 2\int_{x_-^3}^{x_+^3} \sqrt{\frac{g_1 g_3}{g_2}} \frac{dx^3}{v_A^2}, \quad \alpha_T = 2\int_{x_-^3}^{x_+^3} \sqrt{\frac{g_2 g_3}{g_1}} \frac{dx^3}{v_A^2}.$$

Here, $\overline{c}$ is the speed of light in vacuum, and the index (i) denotes the parameter values at the upper boundary of the ionosphere.

3.15.5 Calculating Alfvén Oscillation Field in the MASSA Experiment

Let us consider the field structure of an individual standing Alfvén wave of unit amplitude (with $\overline{B} = 1$). We will select $\Delta_y = 0.1\Delta_1^{(i)}$, $\Delta_t = 0.1/\Omega_1^0$ for our calculations. The Alfvén oscillation amplitude $(\sqrt{|B_x|^2 + |B_y|^2})$ distribution in the initial regime $(\tau_N \ll 1)$, in the plane of dimensionless transverse coordinates (ξ, η) is shown in Figure 3.52. It can be seen that the amplitude has a maximum at the source localisation point $(\xi = 0, \eta = 0)$ and decreases according to the power law $(1/\sqrt{\xi^2 + \eta^2})$ away from this point.

The oscillation amplitude distribution for standing Alfvén wave in the asymptotic regime $(\tau_N \gg 1)$ in the plane (ξ, η) is shown in Figure 3.53. The amplitude distribution for this

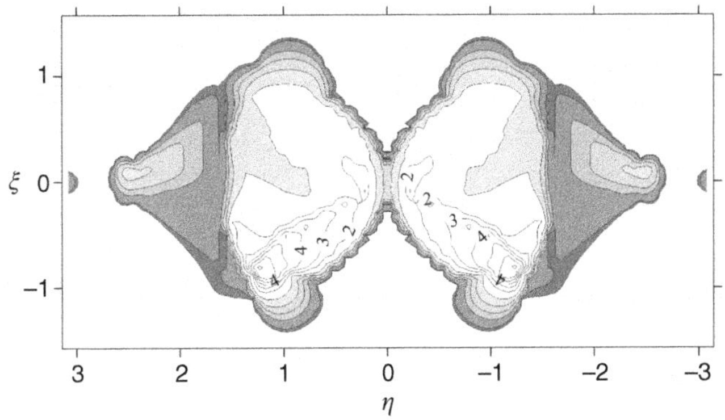

Figure 3.53 Amplitude distribution for a separate harmonic of standing Alfvén wave, in the asymptotic regime of oscillations, in the plane of dimensionless transverse coordinates (ξ, η). Sectors with smallest amplitude are in shades of grey.

regime clearly exhibits the 'butterfly wings'-type structure. Recall that the term 'butterfly wings' is used here to denote the region occupied, in the plane of transverse coordinates, by oscillations of standing Alfvén wave in the asymptotic regime. In contrast to oscillations in the initial regime, the amplitude maximum in the asymptotic regime is shifted away from the source localisation source into the equatorial part of the 'butterfly wings' (their lower part in Figure 3.53). These same sectors feature a sharp gradient in amplitude between the inner regions of the 'butterfly wings' and their outer regions, where the oscillations are practically absent. The width of the transition zone between these regions is zero in the model accepted here when the oscillations are absent outside the 'butterfly wings'. In reality, it does have a certain small width, determined by the intensity of the oscillation decay in the outer region. In the asymptotic region, the B_y component of the field dominates in most of the 'butterfly wings'. The further calculations will therefore concern this oscillation field component.

The full field of oscillations excited in the magnetosphere by broadband sources includes all possible harmonics of standing Alfvén waves. It is therefore necessary to determine which harmonics oscillate, by the chosen moment of time, in the initial regime, and which in the asymptotic regime. For each harmonic, this is determined by the magnitude of $\tau_N = \Delta\Omega_N t$. The largest relative value $\Delta\Omega_N/\Omega_N$ belongs to the main harmonic of the oscillations. For the magnetic shell in question $L = 1.5$, $\Delta\Omega_1/\Omega_1 \simeq 0.3$. For the other harmonics, this relation decreases sharply as N grows: $\Delta\Omega_2/\Omega_2 \simeq 0.01$, $\Delta\Omega_3/\Omega_3 \simeq 0.002$. Thus, after the very first oscillation period $t = t_A$, the first harmonic of standing waves switches to the asymptotic regime. All the other harmonics can oscillate in the initial regime over many periods of oscillations of the main harmonic. The further calculations concern the full field of oscillations, where the main harmonic oscillates in the asymptotic regime, and all the others in the initial regime.

When the source model (3.411) is used the series (3.408) converge, therefore it is possible to restrict ourselves to a finite number of summands when summing them up. The calculations below employ 10 first summands, even though the test calculations demonstrate that the general pattern of the oscillations changes little when more than three members of the series are taken into account.

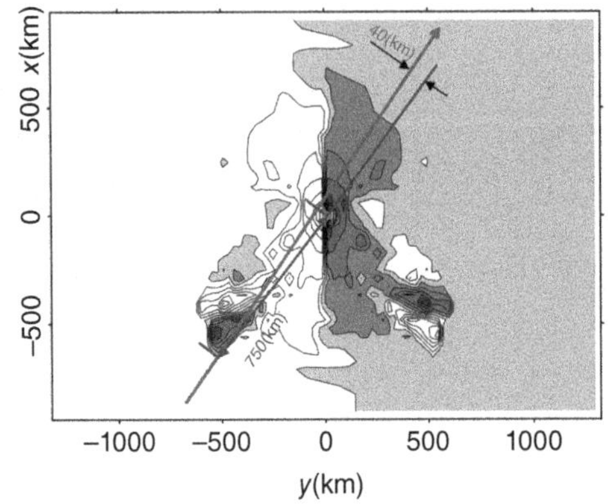

Figure 3.54 Amplitude distribution of the B_y-component of the full field of standing Alfvén waves excited by a pulsed source near the ionosphere, in the horizontal plane (x, y). Possible view of the Aureole-3 satellite trajectory (dark gray line) in the MASSA experiment relative to the region occupied by the Alfvén oscillations.

Let us compare the calculated pattern of the full field of MHD oscillations to data registered onboard the Aureole-3 satellite in the MASSA experiment. For this purpose we will calculate the B_y-component of the oscillation field along the satellite trajectory. The amplitude distribution for the full field including all the harmonics of standing Alfvén waves is shown in Figure 3.54. The main result of these calculations is the fact that the oscillations of the first harmonic practically entirely determine the wave field in the equatorial (lower part of Figure 3.54) areas of the 'butterfly wings' occupied by this harmonic. The oscillation amplitude of the first harmonic in these areas is many times higher than the oscillation amplitude of all the other harmonics combined. Moreover, a maximum appears in the amplitude in the source localisation region determined by the oscillations of the other harmonics in the initial region. The oscillation field in the regions removed from the source localisation point is also determined by the oscillations of higher harmonics (especially, of the second harmonic) of standing waves.

Data on the satellite movement in that experiment are provided in [269]. The minimum distance between the trajectory and the ground-level explosion point as mapped onto the satellite altitude (impact parameter) was 40 km. The satellite speed along the trajectory was 7.8 km/s. The Alfvén burst (aperiodic MHD oscillations lasting about 20 seconds, with polarisation typical of Alfvén waves) was registered onboard the satellite at a distance of 750 km from the explosion point as mapped onto the satellite altitude. If the pulse is assumed to be related to the satellite crossing the equatorial boundary of the left-hand wing of the 'butterfly', the satellite trajectory in the (x, y) plane must be as shown in Figure 3.54. The trajectory inclination to the coordinate axes depends essentially on the shape of the area occupied by the oscillations of the main harmonic in the real magnetosphere (i.e. on the degree to which it resembles the calculated 'butterfly wings'). Also essential is $\Delta_1^{(i)}$, its value determining the size of this area. The calculations employ the value $\Delta_1^{(i)} = 600$ km, which should however be treated as an order-of-magnitude estimate only. To make this value more accurate requires special dedicated experiments based on the principles presented in Section 3.14.

The exact timing of the initial pulse in the field-aligned currents in the ionosphere being unknown, we will assume that at least one oscillation period of the main harmonic of standing Alfvén waves occurred before the satellite crossed the boundary of the 'butterfly wings' (characteristic lines η_4). This time is needed for the structure of standing Alfvén waves to set in along magnetic field lines. It was shown in Section 3.15.2 that, for harmonic $N = 1$,

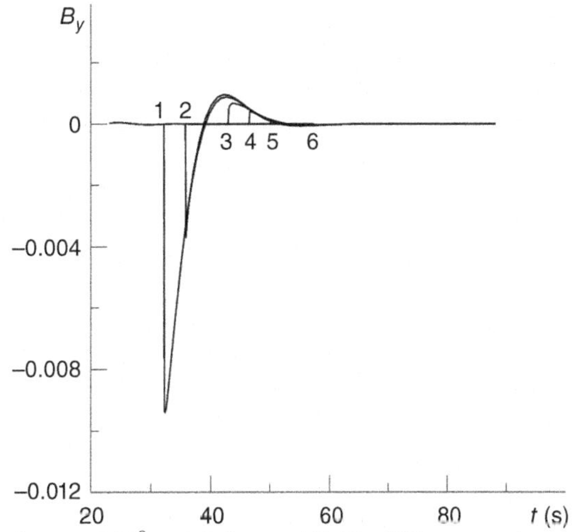

Figure 3.55 Behaviour of the B_y-component of the full field of Alfvén oscillations onboard the satellite crossing the equatorial boundary of the 'butterfly wing', in the strong decay case. Variants (1–6) correspond to the satellite crossing the boundary at time moments (3.413).

this boundary oscillates according to the law $\cos[(\Omega_1^0 - \Delta\Omega_1)t - \pi/4]$. Six different variants in Figure 3.55 correspond to the behaviour of the oscillation field component B_y when the satellite crosses the 'butterfly wing' boundary. The following moments were chosen for the satellite to cross the boundary as counted from the initial pulse:

$$\begin{aligned} t_1 &= (2\pi n - \pi/4)/(\Omega_1^0 - \Delta\Omega_1), & t_2 &= 2\pi n/(\Omega_1^0 - \Delta\Omega_1), \\ t_3 &= (2\pi n + \pi/2)/(\Omega_1^0 - \Delta\Omega_1), & t_4 &= 2\pi(n + \pi)/(\Omega_1^0 - \Delta\Omega_1), \\ t_5 &= (2\pi n + 3\pi/2)/(\Omega_1^0 - \Delta\Omega_1), & t_6 &= (2\pi n + 7\pi/4)/(\Omega_1^0 - \Delta\Omega_1). \end{aligned} \quad (3.413)$$

Here, n is the integer representing the number of periods of the main harmonic, from the moment when the pulse excited the standing Alfvén wave.

The calculation step was 1 seconds along practically the entire the considered satellite trajectory part. The exception was the time interval 1 seconds before and 1 seconds after the boundary was crossed, where 0.1 seconds was chosen as the time step. It can be seen from Figure 3.55 that all variants must exhibit a sharp (momentary, in our calculations) jump of the field component B_y, from an almost background level to a maximum value, at the moment the satellite crosses the 'butterfly wing' boundary. The sign of the value naturally depends on the oscillation phase that the boundary is crossed at. At time moments t_1, t_2, t_6 the value of B_y is negative, and at time moments t_3, t_4, t_5 – positive. If the decay is strong enough, it takes the component B_y one period t_A after the boundary is crossed to decay to practically background level. Such an oscillation pattern is very close to the 'Alfvén burst' registered by the Aureole-3 satellite in the MASSA experiment.

Accepting the above reasoning about the origin of the Alfvén impulse in the MASSA experiment could make it possible to explain some other features of this experiment. It was pointed out in [269] that such impulses had never been observed at higher or lower latitudes in other MASSA experiments. Presumably, the field of the Alfvén oscillations excited in experiments at higher latitudes had an amplitude which was not large enough to be registered by the satellite. As follows from (3.412), the oscillation amplitude is inversely proportionate to the volume of the magnetic flux tube the cross-section of which in the ionosphere coincides with the source localisation area (factor $\sqrt{a_T a_P}$ in the denominator of (3.412)). As latitude increases, the volume of such a flux tube increases. As latitude decreases, the amplitude of oscillations excited in the magnetosphere should increase.

However, decreased geomagnetic latitude is accompanied by increased decrement of standing Alfvén waves due to their dissipation in the ionosphere. This means that such oscillations can only be registered over a very short time interval after the initial pulse. It is possible that the satellite never crossed the oscillation localisation region during this interval in any of such experiments.

Another feature of the MASSA experiment is the sharp increase that was observed in the high-frequency electromagnetic noise registered by the satellite at the moment the Alfvén burst was registered. Sharp changes were also registered in the intensity of this noise at distances 140 and 280 km from the point where the leading front of the Alfvén burst was registered, as well as in the source localisation region. According to the hypothesis in [269], this noise is related to the plasma instabilities developing in the field of an intense Alfvén wave. This phenomenon can then be interpreted as follows. A sharp increase in noise when the frontal edge of the recorded impulse is registered is due to the satellite crossing the equatorial boundary of a 'butterfly wing'. Noise in the source localisation region is due to both the initial impulse and the oscillations of high harmonics of standing Alfvén waves. Sharp increases in intensity, 140 and 280 km from the 'butterfly wing' boundary, can be related to the satellite crossing the regions with maximum amplitude of Alfvén oscillations. As can be seen from Figure 3.53, the oscillation field of the main harmonic has a certain periodic structure, with spatial period ~150 km, along the satellite trajectory, inside the 'butterfly wings'. The satellite failed to register the oscillations of the Alfvén wave field itself due to the fast decay of the waves. The related electromagnetic noise can, however, feature a much smaller decrement.

Such a scenario of the Alfvén wave field generation in the magnetosphere makes it possible to draw the following conclusion, important for understanding this phenomenon. The field of the excited oscillations decaying strongly, the time interval between the impulse of the external currents in the ionosphere and the Alfvén oscillation pulse registration by the satellite cannot exceed one oscillation period of the main harmonic of standing waves at a given magnetic shell. At the magnetic shells in question, $L \sim 1.5$, this period is 15–25 seconds.

3.16 Ballooning Instability of Alfvén and SMS Oscillations on Field Lines Crossing the Current Sheet

It was demonstrated in Sections 2.15 and 2.16 that the solar wind plasma shear flow at the magnetopause is unstable to FMS oscillations (Kelvin–Helmholtz instability). According to the classification accepted in plasma physics, this instability belongs to the class of MHD instabilities characterised by the largest growth rates. They are followed by drift instabilities which are due to the drift of charged particles in inhomogeneous electric and magnetic fields. They are characterised by smaller growth rates than are the MHD instabilities. Finally, the smallest growth rates of driven MHD oscillations has micro-instabilities related to the presence in the background plasma of high-energy charged particle fluxes.

Earth's magnetosphere provides an opportunity for an MHD instability of another type to develop. It is a ballooning instability of Alfvén and SMS oscillations at magnetic field lines having segments with high curvature [363–365]. Such segments form, for example, on field lines crossing the current sheet of the geotail. For this instability to develop, high curvature of magnetic field lines must be combined with a plasma pressure gradient in the earthward direction [366, 367]. In geomagnetic storm periods, this instability is presumed to be capable of being responsible for magnetic reconnection onset in the near-Earth part of the geotail current sheet [368–370].

This section examines the spatial structure and spectrum of unstable Alfvén and SMS waves on geomagnetic field lines crossing the current sheet. An analytical model is developed of an axisymmetric magnetosphere describing the geometry of elongated magnetic field lines and equilibrium plasma distribution. The spectra of Alfvén and SMS oscillations are compared obtained in the local and WKB approximations, in the geotail models with thick or thin current sheets.

3.16.1 Equation for Ballooning Modes

It is well known that in a finite-β inhomogeneous plasma the Alfvén and slow magnetosonic modes are coupled due to the field line curvature [274–277, 287]. Let us consider the wave equations for these modes.

As throughout this section, to describe the MHD oscillation field, we will use an orthogonal curvilinear system of coordinates, (x^1, x^2, x^3), tied to the magnetic field lines, where the x^3 coordinate is along the field line, x^1 across magnetic shells, and x^2 completes the right-handed system of coordinates. We will simulate the magnetic field with closed field lines extended into the geotail as the vector sum of the dipole magnetic field and the field of the azimuthal current localised near the equatorial plane (see Figure 3.56).

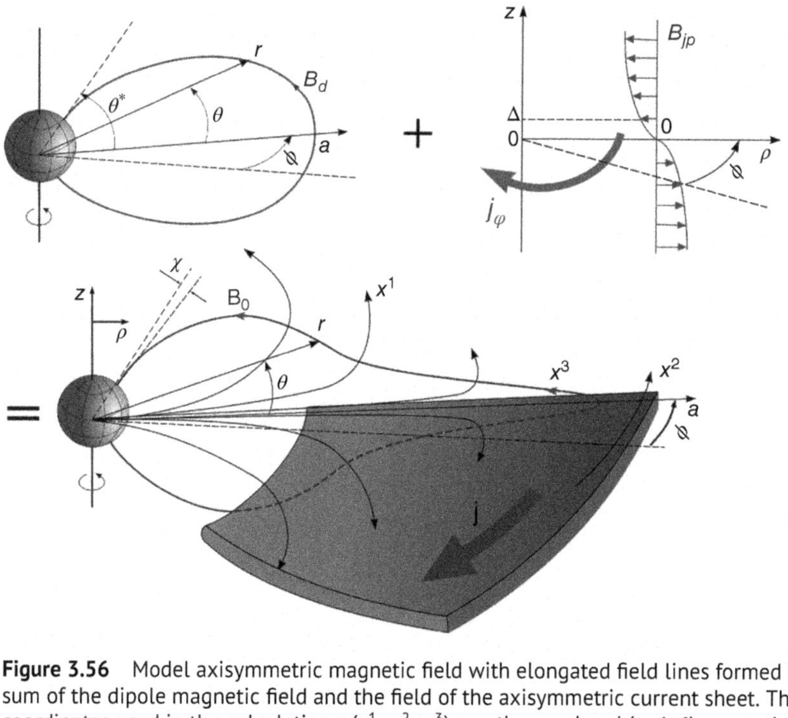

Figure 3.56 Model axisymmetric magnetic field with elongated field lines formed by the vector sum of the dipole magnetic field and the field of the axisymmetric current sheet. The systems of coordinates used in the calculations: (x^1, x^2, x^3) – orthogonal and (a, ϕ, θ) non-orthogonal curvilinear systems of coordinates, (ρ, ϕ, z) – cylindric system of coordinates.

Appendix E derives Eqs. (E.6) and (E.7) for the scalar φ and vector ψ potentials of the MHD oscillation field:

$$\nabla_1 B_0 \hat{L}_T \nabla_1 \varphi - k_2^2 B_0 \hat{L}_P \varphi$$
$$= ik_2 \left(\nabla_1 B_0 \hat{L}_T \frac{g_1}{\sqrt{g}} \psi - B_0 \hat{L}_P \frac{g_2}{\sqrt{g}} \nabla_1 \psi - B_0 \frac{\varkappa_{1g}}{\sqrt{g_3}} \tilde{\Delta}_\perp \psi \right), \tag{3.414}$$

$$\frac{B_0\sqrt{g_3}}{4\pi\rho_0} \hat{L}_0 \frac{B_0}{\sqrt{g_3}} \tilde{\Delta}\psi + v_s^2 \overline{\Delta}\psi - \hat{L}_2 \psi$$
$$= -i\frac{B_0\sqrt{g_3}}{4\pi k_2 \rho_0} \hat{L}_0 B_0 \hat{L}_T \nabla_1 \varphi + ik_2 v_s^2 \frac{g_3}{\sqrt{g}} \hat{L}_1 \varphi, \tag{3.415}$$

where g_i are diagonal components of the metric tensor ($i = 1, 2, 3$), $g = g_1 g_2 g_3$, $\nabla_i \equiv \partial/\partial x^i$, $v_s = \sqrt{\gamma P_0/\rho_0}$ is the speed of sound in plasma, $v_A = B_0/\sqrt{4\pi\rho_0}$ is the Alfvén speed, the longitudinal operators $\hat{L}_T$ and $\hat{L}_P$ are defined by (3.6) and (3.7),

$$\hat{L}_0 = \frac{v_s^2}{\omega^2} \frac{\rho_0}{P_0^\sigma \sqrt{g}} \nabla_3 \frac{\sqrt{g}}{g_3} \frac{P_0^\sigma}{\rho_0} \nabla_3 + 1,$$

is the longitudinal operator for magnetosonic waves ($\sigma = 1/\gamma$),

$$\overline{\Delta} = \overline{\Delta}_\perp + \frac{B_0}{P_0^\sigma} \frac{1}{\sqrt{g_1 g_2}} \nabla_3 \frac{\sqrt{g}}{g_3} \frac{P_0^\sigma}{\rho_0} \nabla_3 \frac{\rho_0}{B_0\sqrt{g_3}},$$

$$\overline{\Delta}_\perp = \frac{B_0}{P_0^\sigma} \frac{1}{\sqrt{g_1 g_2}} \left(\nabla_1 \frac{pP_0^\sigma}{B_0} \nabla_1 - \frac{k_2^2}{p} \frac{P_0^\sigma}{B_0} \right),$$

$$\tilde{\Delta} = \tilde{\Delta}_\perp + \nabla_3 \frac{g_2}{\sqrt{g}} \nabla_3 \frac{g_1}{\sqrt{g}},$$

$$\tilde{\Delta}_\perp = \frac{g_3}{\sqrt{g}} \nabla_1 \frac{g_2}{\sqrt{g}} \nabla_1 - \frac{k_2^2}{g_2}$$

are analogous to the Laplacian operator, $p = \sqrt{g_2/g_1}$,

$$\hat{L}_1 = \frac{B_0}{\omega^2 P_0^\sigma \sqrt{g_3}} \nabla_3 \frac{\varkappa_{1B} v_A^2 P_0^\sigma}{B_0\sqrt{g_3}} \nabla_3 - \varkappa_{1P},$$

$$\hat{L}_2 = \frac{B_0 v_s^2}{\omega^2 P_0^\sigma \sqrt{g_1 g_2}} \nabla_3 \frac{\varkappa_{1B} v_A^2 P_0^\sigma}{B_0\sqrt{g_3}} \nabla_3 \frac{g_2}{\sqrt{g}} \nabla_1 - \omega^2$$

are operators related to the transverse gradients of the background magnetic field and plasma parameters:

$$\varkappa_{1g} = \nabla_1(\ln g_3), \quad \varkappa_{1B} = \nabla_1(\ln \sqrt{g_3} B_0),$$
$$\varkappa_{1P} = \nabla_1(\ln \sqrt{g_3} P_0^\sigma / B_0). \tag{3.416}$$

Equations (3.414) and (3.415) form a system which is closed relative to the potentials φ and ψ. Note that, for a force-free magnetic field (curl $\mathbf{B}_0 = 0$ is in a plasma with no current), for example, such as a dipole field, $\varkappa_{1B} = 0$ and respectively, $P_0 = \text{const}$, $\varkappa_{1P} = \varkappa_{1g}$.

Our calculations will address azimuthally small-scale MHD oscillations with $|k_2\overline{L}/\sqrt{g_2}| \sim m \gg 1$, where $\overline{L}$ is the characteristic scale of magnetospheric plasma inhomogeneity. As we will see, the oscillations in question have the form of standing waves between the magneto-conjugated ionospheres, in the direction along field lines. The wavelength of the basic harmonics for such standing waves, $\sim \overline{L}$. Their characteristic wavelength across magnetic shells is rather small: $\overline{L}|\nabla_1 \ln(\varphi)/\sqrt{g_1}| \sim \overline{L}|\nabla_1 \ln(\psi)/\sqrt{g_1}| \gg 1$. In this case, operators $\Delta_\perp$ in (3.414) and (3.415) can be approximately represented as

$$\overline{\Delta}_\perp \approx \tilde{\Delta}_\perp \approx \Delta_\perp = g_1^{-1}\nabla_1^2 - k_2^2/g_2.$$

Leaving the largest ($\sim \Delta_\perp\psi$) term in the right-hand side of (3.414), we obtain

$$\Delta_\perp\psi = \frac{i\sqrt{g_3}}{k_2 B_0 \varkappa_{1g}}\left(\nabla_1 B_0 \hat{L}_T \nabla_1\varphi - k_2^2 B_0 \hat{L}_P\varphi\right). \tag{3.417}$$

We will substitute this expression into the left-hand side of (3.415) and, leaving the largest terms in it, will obtain

$$\hat{L}_s\nabla_1\hat{L}_T\nabla_1\varphi - k_2^2\left(\hat{L}_s\hat{L}_P + \hat{L}_C\right)\varphi = 0, \tag{3.418}$$

where

$$\hat{L}_C = \frac{\varkappa_{1g}\omega^2}{v_A^2\sqrt{g_1 g_2}}\hat{L}_1,$$

$$\hat{L}_s = \frac{\varkappa_{1g}\rho_0}{B_0 P_0^\sigma\sqrt{g}}\nabla_3\frac{\sqrt{g}\,P_0^\sigma}{g_3\ \rho_0}\nabla_3\frac{B_0}{\varkappa_{1g}} + \frac{\omega^2}{c_s^2}$$

is the longitudinal operator for SMS waves, $c_s = v_A v_s/\sqrt{v_A^2 + v_s^2}$ is SMS velocity in plasma. Equation (3.418) describes the spatial structure of azimuthally small-scale MHD oscillations, including 'ballooning modes'. An equation for them can be obtained in the limiting case $|k_2/\sqrt{g_2}| \gg |\nabla_1 \ln(\varphi)/\sqrt{g_1}| \gg \overline{L}^{-1}$, if the terms $\sim \nabla_1^2\varphi$, $\nabla_1\varphi$ are neglected in (3.418). The result will be an equation for describing the longitudinal structure of 'ballooning modes' for MHD oscillations:

$$\hat{L}_s\hat{L}_P\varphi + \hat{L}_C\varphi = 0. \tag{3.419}$$

It is easily verifiable that in a homogeneous plasma, (3.419) gives dispersion equations for Alfvén ($\omega^2 = k_\parallel^2 v_A^2$) and SMS ($\omega^2 = k_\parallel^2 c_s^2$) waves. These oscillations are called 'ballooning modes' when they switch to the 'ballooning instability' regime.

Equation (3.419) should be supplemented with boundary conditions at the ends of closed field lines crossing the ionosphere in the Northern ($x^3 = x_+^3$) and Southern ($x^3 = x_-^3$) hemispheres. In the zero approximation, we will assume the ionosphere to be ideally conducting. This means that the tangential components of the electric field of oscillations become zero (for $E_3 \equiv 0$ we have $E_1(x_\pm^3) = E_2(x_\pm^3) = 0$) at the points where the field line crosses the ionosphere. For the $k_2 \to \infty$ limit, this produces the following conditions for potential φ

$$\varphi(x_\pm^3) = 0, \quad \hat{L}_P\varphi|_{x^3=x_\pm^3} = 0. \tag{3.420}$$

The latter condition is obtained from (3.417) given $\psi(x_\pm^3) = 0$. Let us analyse the possible solutions of (3.419) for two approximations in the x^3 coordinate. One of them is the so-called 'local approximation', based on the assumption that the solution of (3.419) can be sought using the Fourier transformation with respect to harmonics of the form $\varphi \sim \exp(ik_3x^3)$. The solution in the eikonal form ($\varphi \sim \exp(i\int k_3 dx^3)$) is mentioned more frequently, but k_3 is fixed and selected at the top of the magnetic field line. However, this is the same as seeking a solution of the form $\varphi \sim \exp(ik_3x^3)$.

The coefficients of (3.419) must be assumed constant (see, e.g. [281, 287]). The result will be a local dispersion equation for the MHD oscillations in question. The values of background parameters of plasma are usually chosen corresponding to the field line top, where these parameters exhibit largest gradients in the transverse coordinate x^1. This is justified by the fact that the amplitude growth of the oscillations in question is determined by the near-equatorial region of the magnetosphere. The structure of such oscillations far away from the equatorial surface is not addressed.

Using the WKB approximation appears to be more suitable for solving (3.419). It well describes the structure and spectrum of longitudinal harmonics with large wavenumbers, but is also qualitatively applicable for analysing the main harmonics. The exact solution of (3.419) for the main harmonics can only be found by numerical integration. In this section, we will restrict ourselves to analysing solutions in the 'local' and WKB approximations and compare them.

3.16.2 Model of the Medium

Of course, we have now quite well-developed models of the geotail current sheet [165, 371, 372]. However, all of them are too complicated for our further studies of the stability problem for MHD oscillations at the magnetic shells crossing the current sheet. Therefore, we use a simpler equilibrium model of the geomagnetic field and plasma. Let us simulate the magnetic field in the following manner. We will use the radius-vector $\mathbf{r}(a, \theta)$, the coordinate origin being the centre of the Earth, to describe, in the meridional section, the shape of the magnetic field line (see Figure 3.56), where a is the equatorial radius of the field line, θ is the latitude of the point on the field line, as counted from the equator. Let us consider a model magnetosphere which is symmetrical about the equatorial plane. The field-line shape of a dipole magnetic field is described by an equation of the form $r \equiv r_d = a\cos^2\theta$, and its strength by

$$B_d = \overline{B}_d \left(\frac{R_E}{a} \right)^3 \frac{\sqrt{1 + 3\sin^2\theta}}{\cos^6\theta},$$

where $\overline{B}_d = 0.32$ G is geomagnetic field strength at Earth surface, at the equator (R_E is Earth radius). A cylindrical system of coordinates (ρ, ϕ, z) will also be used in further calculations (see Figure 3.56). In this system of coordinates, the dipole magnetic field components $\mathbf{B}_d = (B_{d\rho}, 0, B_{dz})$ have the form

$$B_{d\rho} = B_d \cos\tilde{\theta}, \quad B_{dz} = B_d \sin\tilde{\theta},$$

where $\tilde{\theta} = \arccos(3\sin\theta\cos\theta/\sqrt{1 + 3\sin^2\theta})$ is the angle between the tangent to the dipole field line and the ρ axis.

For the magnetic field $\mathbf{B}_j = (B_{j\rho}, 0, B_{jz})$ of the azimuthal current, we will adopt the following model of the component:

$$B_{j\rho} = \frac{B_{j\infty}}{2} \left[1 + \tanh \frac{\rho - \tilde{\rho}}{\Delta_\rho} \right] \tanh \frac{z}{\Delta},$$

where $\tilde{\rho} \approx 10\,R_E$ is the plasma sheet boundary closest to Earth, $\Delta_\rho \approx 2R_E$ is the characteristic thickness of the transition layer, Δ is the characteristic thickness of the current sheet, $B_{j\infty} \approx 20$ nT is the magnetic field strength in the geotail lobes far away from the current sheet. The current sheet thickness is known to vary within very wide limits [359, 373, 374], from $\Delta \sim 1\text{–}2\,R_E$ in magnetically quiet conditions (thick current sheet) to $\Delta \sim 0.1\text{–}0.4\,R_E$ in a perturbed magnetosphere (thin current sheet). Our further numerical calculations use the values $\Delta = 0.2\,R_E$ for the thin current sheet and $\Delta = 1.5\,R_E$ for the thick current sheet.

The current magnetic field must satisfy the equation

$$\mathrm{div}\mathbf{B}_j = \frac{1}{\rho}\frac{\partial \rho B_{j\rho}}{\partial \rho} + \frac{\partial B_{jz}}{\partial z} = 0,$$

which is why

$$B_{jz} = -\frac{B_{j\infty}\Delta}{2}\left[\frac{1}{\rho}\left(1 + \tanh\frac{\rho - \tilde{\rho}}{\Delta_\rho}\right) + \frac{1}{\Delta_\rho \cosh^2[(\rho - \tilde{\rho})/\Delta_\rho]}\right]\ln\cosh\frac{z}{\Delta}.$$

The B_{jz} component has a singularity for $\rho \to 0$ and $z \to \infty$, but the contribution of this component is negligibly small in the region of extended field lines $\Delta \ll \rho, z < \rho$ we are interested in. In the real magnetosphere, magnetic field has of course no singularities, and its distribution is not only determined by the current in the current sheet, but by currents at the magnetopause. The same distribution of the background magnetic field components can be achieved in the simulation area by introducing equivalent surface currents at a finite distance from the current sheet. Calculating the spatial distribution of such currents is, however, complicated enough and beyond the scope of the problem posed here.

The main component of the axi-symmetrical azimuthal current, $\mathbf{j} = (0, j_\phi, 0)$, corresponding to the above-set magnetic field components has the form

$$j_\phi = \frac{\overline{c}}{4\pi}(\mathrm{curl}\,\mathbf{B}_0)_\phi \approx \frac{\overline{c}B_{j\infty}}{8\pi}\frac{1}{\Delta\cosh^2(z/\Delta)}\left[1 + \tanh\frac{\rho - \tilde{\rho}}{\Delta_\rho}\right].$$

It describes the azimuthal current for $\rho > \tilde{\rho}$, localised near the equatorial surface at scale Δ. Thus, the full background magnetic field $\mathbf{B}_0 = (B_{0\rho}, 0, B_{0z})$ has components $B_{0\rho} = B_{d\rho} + B_{j\rho}$, $B_{0z} = B_{dz} + B_{jz}$. The shape of the magnetic field line for this field is determined by the following equation (see Appendix N)

$$r(a,\theta) = a\exp\left[\int_0^\theta \frac{\sin\theta'\cos\theta' - \sin\overline{\theta}\cos\overline{\theta}}{\sin^2\overline{\theta} - \sin^2\theta'}d\theta'\right], \tag{3.421}$$

where $\overline{\theta}$ is the angle between the tangent to the field line and the ρ axis ($\sin\overline{\theta} = B_{0z}/B_0$, $\cos\overline{\theta} = B_{0\rho}/B_0$). In Figure 3.57 we can see geomagnetic field lines in a model with a thin current sheet, calculated from (3.421).

Expressions for the metric tensor components, in the coordinate system (a, ϕ, θ) linked to magnetic field lines, have the form (see Appendix P)

$$g_1 = \frac{r^2(a,\theta)}{r^2(a,\theta) + (\partial r/\partial\theta)^2}\left(\frac{\partial r}{\partial a}\right)^2, \quad g_2 = r^2(a,\theta)\cos^2\theta. \tag{3.422}$$

Figure 3.57 Shape of magnetic field lines calculated from (3.421), in a geotail model with a thin current sheet.

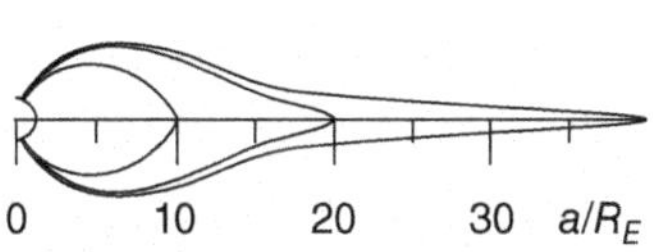

Determining the metric tensor component g_3 is more complicated. The further calculations will, however, not need a precise expression for it. They only employ a logarithmic derivative

$$\overline{\varkappa}_{1g} = \frac{\varkappa_{1g}}{\sqrt{g_1}} = \frac{\nabla_1(\ln g_3)}{\sqrt{g_1}} = 2/r_c, \tag{3.423}$$

where

$$r_c = \frac{[r^2 + (r')^2]^{3/2}}{r^2 + 2r'^2 - rr''}$$

is the field-line curvature radius, $r' \equiv \partial r/\partial\theta$. Note that the coordinates (a, ϕ, θ) are not orthogonal, but are uniquely related to the coordinates (x^1, x^2, x^3). In particular, the derivatives of arbitrary function $f(a, \theta)$ taken with respect to x^1, in the system of coordinates (a, θ), have the form

$$\nabla_1 f = \left(\frac{\partial f}{\partial a}\right)_\theta - \left(\frac{\partial f}{\partial \theta}\right)_a \frac{(\partial r/\partial\theta)_a(\partial r/\partial a)_\theta}{r^2(a,\theta) + (\partial r/\partial\theta)_a^2},$$

where the subscripts mean that the derivative is taken for constant a or θ.

We will set the model distribution of Alfvén speed as follows. In the region dominated by the dipole component of the magnetic field ($\rho < \tilde{\rho}$, where $\tilde{\rho} \approx 10\ R_E$ is the inner boundary of the plasma sheet), we will set $v_A(a, \theta) \equiv v_{Ad} = v_{Ai}/r_d(a, \theta)$, where $v_{Ai} \approx 5000$ km/s is Alfvén speed at the upper boundary of the ionosphere ($r_i = R_E + 1500$ km), r_d is the radius of the dipole field line. This simple model correctly simulates the averaged distribution of v_A along the field line from the ionosphere to the equatorial plane. Of course, it does not take into account the sharp change in v_A at the plasmapause, but we will not consider that region of the magnetosphere.

In the current sheet region ($\rho > \tilde{\rho}$), the main change in v_A is in the z coordinate. Velocity v_A changes from $v_{A0} \approx 100$ km/s in the current sheet ($z = 0$) to $v_{A\infty} \approx 6000$ km/s in the geotail lobes (where $z \gg \Delta$). We will assume the following model Alfvén speed in this region:

$$v_A(z) \equiv v_{Ac} = v_{A0}\left[1 - (1 - (v_{A\infty}/v_{A0}))\tanh\frac{z}{\Delta}\right].$$

The full model of the Alfvén speed can be represented as

$$v_A(\rho, z) = \frac{1}{2}\left[v_{Ad} + v_{Ac} - (v_{Ad} - v_{Ac})\tanh\frac{\rho - \tilde{\rho}}{\Delta_\rho}\right], \tag{3.424}$$

where $\Delta_\rho \approx 2R_E$ is the characteristic thickness of the transition layer of the near plasma sheet. The model distribution of the Alfvén speed in the meridional plane is shown in Figure 3.58a for the thick current sheet. For the given distribution of magnetic field, Eq. (3.424) defines the distribution of plasma density ρ_0.

To determine the sound speed, it is necessary to set a model distribution of plasma pressure. Similar to what was done above for Alfvén speed, we will set the pressure distribution separately for the $\rho < \tilde{\rho}$ region, dominated by a dipole magnetic field, and the $\rho > \tilde{\rho}$ region, dominated by the $B_{j\rho}(z)$ field component. In the former, we will neglect the azimuthal current field, and in the latter, the dipole magnetic field. Such a distribution of balanced plasma pressure can of course be regarded as approximate only.

We have the equilibrium condition of the plasma configuration from (1.1), in the stationary state ($\partial/\partial t = 0$)

$$\nabla P_0 = \frac{1}{4\pi}[\text{curl}\ \mathbf{B}_0 \times \mathbf{B}_0], \tag{3.425}$$

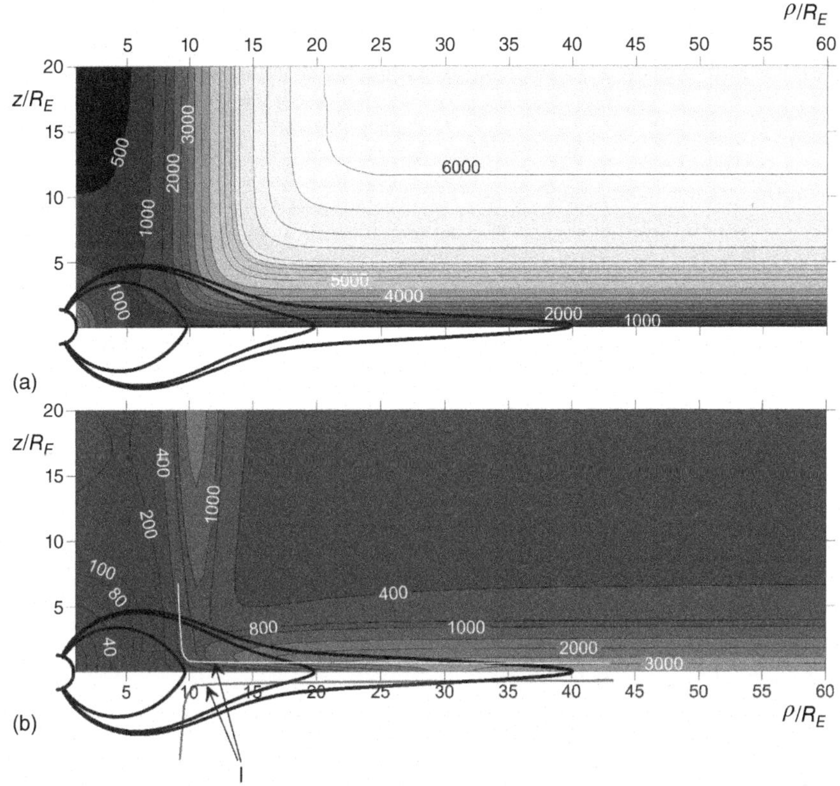

Figure 3.58 Distribution isolines for: (a) Alfvén speed v_A (km/s); (b) sound speed v_s (km/s), in the meridional plane, in the geotail model with a thick current sheet. Lines 1 in panel (b) indicate the meridional sections of the magnetic field-line inflexion surfaces.

which implies that pressure P_0 remains constant along magnetic field lines. If it were possible to set precise pressure distribution on some surface crossing all the field lines (e.g. near the ionosphere), it would be defined throughout the model magnetosphere. Unfortunately, it appears to be impossible to set such a distribution, with good accuracy, without taking into account the currents at the magnetospheric boundary. In this work, we will therefore use the above-mentioned technique of approximately describing the distribution of the balanced plasma pressure.

Under the above assumptions, $P_0 \equiv P_{0d} = \text{const}$ in the $\rho < \tilde{\rho}$ region, dominated by a force-free dipole magnetic field. In the $\rho > \tilde{\rho}$ region, following from the equilibrium condition of the plasma configuration (3.425),

$$P_0(z) \equiv P_{0j}(z) = P_{0j}(\infty) - \frac{B_{j\rho}^2(z)}{8\pi} + \frac{B_{j\rho}^2(\infty)}{8\pi},$$

where $P_{0j}(\infty)$ is plasma pressure in the geotail lobes away from the current sheet. We will choose it such that $\beta_\infty = 8\pi P_{0j}(\infty)/B_{j\rho}^2(\infty) = 0.005$. The resulting distribution $P_0(z)$ corresponds to the Harris layer model, regarded as a good enough approximation for describing the geotail current sheet. Given that all the effects defining the instability of MHD oscillations are determined by how the pressure behaves inside and near the current sheet, the results obtained below will hopefully not change too much when a more accurate model of the $P_0(\rho, z)$ distribution is used. For the sake of definiteness, we will also set $P_{0d} = B_{j\rho}^2(\infty)/8\pi$. Full pressure $P_0(z)$ distribution will be modelled by the formula

$$P_0(\rho, z) = \frac{1}{2}\left[P_{0d} + P_{0j} - (P_{0d} - P_{0j})\tanh\frac{\rho - \tilde{\rho}}{\Delta_\rho}\right]. \tag{3.426}$$

The ρ_0 distribution being determined by the Alfvén speed distribution, we can construct the sound speed distribution in plasma. For the model with a thick current sheet, the v_s distribution in the meridional plane is shown in Figure 3.58b.

3.16.3 The Ballooning Instability of MHD Oscillations as Studied in the Local Approximation

The local approximation assumes that the plasma parameters are set at the top of the field line and remain unchanged along it. The following equation is obtained from the background plasma equilibrium condition (3.425):

$$\nabla_1 P_0 + \frac{B_0^2}{4\pi}\nabla_1 \ln(B_0\sqrt{g_3}) = 0.$$

Using the earlier-introduced parameters $\varkappa_{1g}, \varkappa_{1B}$ and $\varkappa_{1P}$, it can be rewritten as

$$\varkappa_{1P} - \varkappa_{1g} + \left(1 + \frac{2\sigma}{\beta}\right)\varkappa_{1B} = 0, \tag{3.427}$$

where $\sigma = 1/\overline{\gamma} = 3/5$, $\beta = 8\pi P_0/B_0^2$. This relation can be used to determine the possible signs of $\varkappa_{1g}, \varkappa_{1B}$ and $\varkappa_{1P}$. In the model under study, $\varkappa_{1g}, \varkappa_{1B} < 0$, $\varkappa_{1P} > 0$ at the tops of the field lines crossing the current sheet.

We will seek the solution of (3.419) in the local approximation, in the form $\varphi \sim \exp(ik_3x^3)$. The result will be a dispersion equation of the form

$$\left(\frac{\omega^2}{k_\parallel^2} - v_A^2\right)\left(\frac{\omega^2}{k_\parallel^2} - c_s^2\right) = \frac{\overline{\varkappa}_{1g}c_s^2}{k_\parallel^2}\left(\frac{\omega^2}{k_\parallel^2}\overline{\varkappa}_{1P} + v_A^2\overline{\varkappa}_{1B}\right), \tag{3.428}$$

where $k_\parallel = k_3/\sqrt{g_3}$, $\overline{\varkappa}_1 = \varkappa_1/\sqrt{g_1}$ are physical components of the parameters. For convenient comparison with other authors' findings (see [279, 281, 287]), we will rewrite (3.428) using other notations:

$$\left(\frac{\omega^2}{k_\parallel^2} - v_A^2\right)\left(\frac{\omega^2}{k_\parallel^2} - c_s^2\right) = \frac{2\kappa_c c_s^2}{k_\parallel^2}\left[\frac{\omega^2}{k_\parallel^2}\left(2\kappa_c - (\frac{\beta}{2} + \frac{1}{\gamma})\kappa_p\right) + v_A^2\frac{\beta}{2}\kappa_P\right], \tag{3.429}$$

where $\kappa_c = -\overline{\varkappa}_g/2 = 1/r_c$, r_c is the field line curvature radius, $\kappa_P = g_1^{-1/2}\nabla_1 \ln(P_0)$, $\kappa_B = g_1^{-1/2}\nabla_1 \ln(B_0)$. We have $\overline{\varkappa}_B = \kappa_B - \kappa_c, \overline{\varkappa}_P = \kappa_P/\gamma - \kappa_B - \kappa_c$ and the plasma configuration equilibrium condition in the form

$$\frac{\beta}{2}\kappa_P + \kappa_B - \kappa_c = 0.$$

Let us analyse qualitatively the solutions of (3.428), assuming the wave vector component $k_\parallel$ to be set. For it, we will use values corresponding to the lengths of waves standing along the magnetic field line, when the solution of (3.419) can be written as

$$\varphi = C\sin(k_\parallel l),$$

where $k_{\|} = \pi N/L$, $N = 1, 2, 3, \ldots$ is the number of the standing wave harmonic, L is magnetic field line length, l is the coordinate of the point on the field line as counted from the ionosphere of the Southern Hemisphere ($dl = \sqrt{g_3}dx^3$), C is an arbitrary constant. The solution of (3.428) can be represented as

$$\left(\frac{\omega^2}{k_{\|}^2}\right)_{1,2} = \frac{1}{2}\left(v_A^2 + c_s^2 + \frac{\overline{\varkappa}_{1g}\overline{\varkappa}_{1P}}{k_{\|}^2}c_s^2\right) \pm \sqrt{\frac{1}{4}\left(v_A^2 + c_s^2 + \frac{\overline{\varkappa}_{1g}\overline{\varkappa}_{1P}}{k_{\|}^2}c_s^2\right)^2 + \left(\frac{\overline{\varkappa}_{1g}\overline{\varkappa}_{1B}}{k_{\|}^2}c_s^2 - 1\right)v_A^2c_s^2}. \tag{3.430}$$

Equation (3.428) describes two branches of MHD oscillations. The expression with '+' before the radical describes the Alfvén mode, and the one with '−' the SMS modes of MHD oscillations. To obtain a qualitative picture of the character of the resulting solutions, we will use a technique of defining them graphically. Figure 3.59 plots the left- and right-hand sides of the dispersion equation (3.428) as functions of the square of the phase velocity of the wave, $\omega^2/k_{\|}^2$. The left-hand side is described by a parabola, crossing the $\omega^2/k_{\|}^2$ axis at points c_s^2 and v_A^2, and the vertical axis, at point $c_S^2 v_A^2$. The right-hand side is represented by a straight line crossing the horizontal axis at point $-v_A^2\overline{\varkappa}_{1B}/\overline{\varkappa}_{1P}$, and the vertical axis at point $\overline{\varkappa}_{1g}\overline{\varkappa}_{1B}c_S^2 v_A^2/k_{\|}^2$.

It was shown in [281, 287] that solutions for the square of oscillation frequency, ω^2, described by (3.428) and (3.429), can only be real. Solutions corresponding to Alfvén waves always satisfy the condition $\omega^2 > 0$, and solutions for SMS waves can, under certain conditions, become aperiodically unstable, i.e. $\omega^2 < 0$. In the region of real ω^2, the solutions of (3.428) are determined by the points where the straight line crosses the parabola. If $\overline{\varkappa}_{1g}\overline{\varkappa}_{1B} < k_{\|}^2$, the solutions of (3.428) are an ordinary poloidal Alfvén wave and azimuthally small-scale slow magnetosonic wave (straight line *1* in Figure 3.59). If the curvature radius grows infinitely ($r_c \to \infty$, $\overline{\varkappa}_{1g} \to 0$), the solutions of (3.428) have the form $\omega^2 = k_{\|}^2 v_A^2$ and

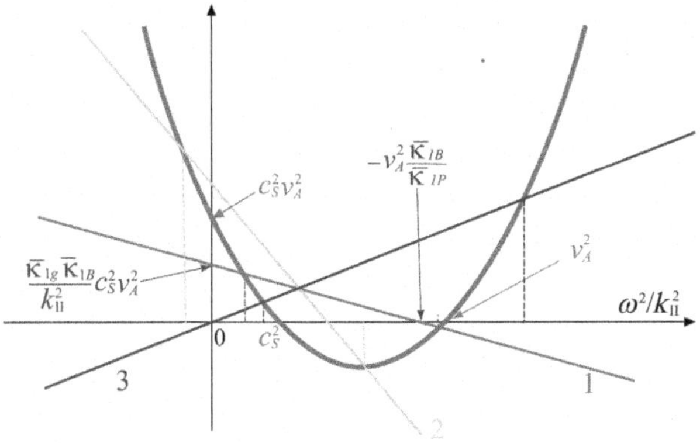

Figure 3.59 Graphical solution of dispersion equation (3.428). The solution is determined by the points where the parabola crosses the straight lines corresponding to the right-hand side of (3.428), for different ratios between the parameters. Straight line 1 corresponds to solutions for neutral poloidal Alfvén and azimuthally small-scale SMS waves; straight line 2 to the neutral Alfvén and aperiodically unstable SMS wave; straight line 3 to neutral Alfvén and SMS waves, in a force-free magnetic field.

$\omega^2 = k_\parallel^2 c_s^2$. Such oscillation modes are called 'neutral' (in contrast to 'unstable', increasing, and 'stable', decaying modes).

When $\overline{\varkappa}_{1g}\overline{\varkappa}_{1B} > k_\parallel^2$ the Alfvén wave remains in the region of positive ω^2, while SMS wave has $\omega^2 < 0$ and one of its branches becomes aperiodically unstable (straight line 2 in Figure 3.59). It is easily demonstrable that this instability threshold corresponds to the condition $\kappa_c\kappa_P > k_\parallel^2$, obtained in [281, 287]. Such an instability (and the MHD mode itself) is called 'ballooning': increasing field-line curvature results in increased growth rate of the perturbation leading to increased curvature etc.

What causes the unstable MHD oscillations is the current in the plasma. It would be more accurate to say: the rather intense and thin current sheet capable of imparting necessary curvature to the magnetic field lines. Strictly speaking, therefore, what we are dealing with here should be called MHD wave instability in a plasma with a current sheet. In a plasma with no current sheet ($\overline{\varkappa}_{1B} = 0$, $\overline{\varkappa}_{1P} = \overline{\varkappa}_{1g}$, e.g. in a dipole field), the solution of (3.428) are neutral poloidal Alfvén waves and azimuthally small-scale SMS waves with $\omega^2 > 0$ (straight line 3 in Figure 3.59).

Some investigations regard 'ballooning modes' as coupled Alfvén and SMS waves, and treat their instability as resulting from this coupling. As will be shown in Section 3.17 Alfvén and SMS waves can indeed form coupled oscillation modes, in the plasma configuration under study, and their amplitude can also increase under the influence of the ballooning instability. It is, however, incorrect to regard this coupling as causing such an instability. Both these waves can propagate along the same magnetic field lines. They do not always interact effectively, in the process.

Let us now find out which of the above scenarios are realised in the model we are using. Figure 3.60 shows calculated distributions of Eigen frequencies for standing Alfvén and SMS waves, over magnetic shells. Shown are the solutions of (3.428) for the 'thin' (Figure 3.60a) and 'thick' (Figure 3.60b) current sheet model. Three first odd harmonics ($N = 1, 3, 5$) are examined for standing waves of either type. The Alfvén wave frequencies can be seen to remain real throughout the range of the magnetic shells under study

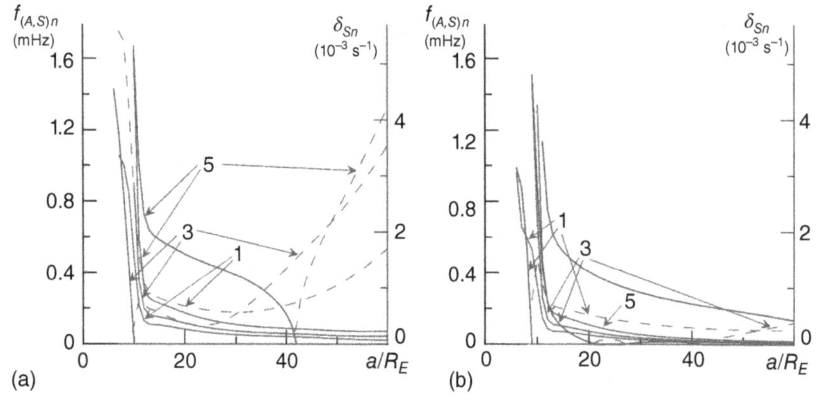

Figure 3.60 Eigen frequency distribution for first odd harmonics $N = 1, 3, 5$ of azimuthally small-scale standing Alfvén waves ($f_{AN} = \Omega_{PN}/2\pi$ – solid thick lines) and Eigen frequency ($f_{sN} = \Omega_{sN}/2\pi$ – solid thin lines) and growth rate (δ_{sN} – dashed lines) distribution for standing SMS waves, calculated in the local approximation from the dispersion equation (3.428): (a) for the geotail model with a thin current sheet and (b) for the model with a thick current sheet.

($5 < a/R_E < 60$). SMS waves on magnetic shells crossing the current sheet can switch to the regime of aperiodic instability. The higher the harmonic number and the thicker the current sheet, the farther toward the tail this switch takes place. Thus, the fifth harmonic of SMS waves did not switch to the instability regime, in the magnetic shell range in question, in the thick current sheet model. In the thin current sheet model, the first harmonic becomes aperiodically unstable immediately after crossing the near boundary of the sheet and remains unstable throughout the entire range of the magnetic shells.

Interestingly, Alfvén and SMS wave Eigen frequencies are close (but not identical), on order of magnitude, in the local approximation. This is related to $\beta \gg 1$ ($v_s \gg v_A$, $c_s \approx v_A$) in the current sheet, which is what produces close Eigen frequencies in the local approximation. Thus, a scenario is realised presented in Figure 3.59 by the intersection of the parabola and the straight line 2. This scenario implies that neutrally stable poloidal Alfvén and unstable SMS oscillations could be expected on the field lines passing through the current sheet. In Section 3.16.4 will examine the structure and Eigen frequency spectrum for standing Alfvén and SMS waves, in the WKB approximation, in the longitudinal x^3 coordinate.

3.16.4 Calculating the Structure and Spectrum of Standing Alfvén and SMS Waves on Elongated Field Lines, in the WKB Approximation

As was shown in Section 3.7, the solution of (3.419) in the WKB approximation for azimuthally small-scale oscillations can be sought in the form

$$\varphi = U(x^1)\exp(i\Theta(x^1, x^3)),$$

where the function $U(x^1)$ describes the small-scale structure of the solution, in the x^1 coordinate, and $\Theta(x^1, x^3)$ is the large quasiclassical phase describing the structure of the solution in the longitudinal coordinate. Its characteristic variation scale, in the x^1 coordinate, is of order the medium inhomogeneity scale. We will seek the solution as an expansion, $\Theta = \Theta_0 + \Theta_1 + \Theta_2 + \cdots$, where $|\Theta_0| \gg |\Theta_1| \gg |\Theta_2| + \cdots$. Substituting this solution into (3.419), we obtain, in the main order of perturbation theory, the solution for $k_\parallel = g_3^{-1/2}\partial\Theta_0(x^1, x^3)/\partial x^3$, analogous to (3.428). Solving it with respect to $k_\parallel$, we obtain

$$k_\parallel^2 = \frac{\omega^2}{v_A^2} + \frac{\omega^2}{2v_s^2} + \frac{\overline{\varkappa}_{1g}\overline{\varkappa}_{1B}}{2} \pm \sqrt{\frac{1}{4}\left(\frac{\omega^2}{v_s^2} + \overline{\varkappa}_{1g}\overline{\varkappa}_{1B}\right)^2 + \frac{\omega^2}{v_A^2}\overline{\varkappa}_{1g}(\overline{\varkappa}_{1P} + \overline{\varkappa}_{1B})}. \tag{3.431}$$

Here, the '+' sign before the radical corresponds to SMS wave ($k_\parallel \equiv k_{\parallel S}$), and the '−' sign to poloidal Alfvén wave ($k_\parallel \equiv k_{\parallel AP}$). In the first order of perturbation theory, we have this equation

$$4\Theta_0'^3\Theta_1' - i6\Theta_0'^2\Theta_0'' - i(2\kappa_{lB} + \kappa_{lP} - \kappa_{lg})\Theta_0'^3 = 0,$$

where the prime denotes the derivative $\nabla_l \equiv \partial/\partial l$ and the following parameters are introduced: $\kappa_{lB} = \nabla_l \ln(B_0/p\varkappa_{1g})$, $\kappa_{lP} = \nabla_l \ln(\sqrt{g_1 g_2}P_0^\sigma/\rho_0)$, $\kappa_{lg} = \nabla_l \ln p$. From this we obtain

$$\Theta_1 = \frac{i}{4}\ln(k_\parallel^6 v_A^2 P_0^\sigma g_1^2/g_2).$$

The full solution of (3.419) can be represented, in the WKB approximation, as

$$\varphi = \frac{U(x^1)}{(v_A^2 P_0^\sigma g_1^2/g_2)^{1/4}} \left[k_{\|AP}^{-3/2} \left(C_1 e^{i\int_{l_-}^{l} k_{\|AP} dl'} + C_2 e^{-i\int_{l_-}^{l} k_{\|AP} dl'} \right) \right.$$
$$\left. + k_{\|S}^{-3/2} \left(C_3 e^{i\int_{l_-}^{l} k_{\|S} dl'} + C_4 e^{-i\int_{l_-}^{l} k_{\|S} dl'} \right) \right], \tag{3.432}$$

where $C_{1,2,3,4}$ are arbitrary constants, defined using the boundary conditions (3.420). Applying these boundary conditions produces, in the main order of the WKB approximation $C_2 = -C_1$, $C_3 = -C_4$ and the following dispersion equation

$$(\varkappa_\|^2 - 1)^2 \sin \Psi_{AP} \sin \Psi_s = 0, \tag{3.433}$$

where $\varkappa_\| = (k_{\|S}/k_{\|AP})_{l=l_\pm}$, $l_\pm$ are the coordinates of the points where the field line crosses the ionosphere of the Northern ('+' sign) and Southern ('−' sign) hemispheres,

$$\Psi_{AP} = \int_{l_-}^{l_+} k_{\|AP}\, dl, \quad \Psi_s = \int_{l_-}^{l_+} k_{\|S}\, dl.$$

It follows from (3.433) that the solution of (3.419), in the WKB approximation, are standing poloidal Alfvén waves of the form

$$\varphi_{APN} = U_{PN}(x^1) P_N(x^1, x^3),$$
$$P_N(x^1, x^3) = C_{PN} \sin\left(\int_{l_-}^{l} k_{\|APN}\, dl \right) / k_{\|APN}^{3/2},$$

where C_{PN} is an arbitrary constant, $k_{\|APN} \equiv k_{\|AP}(\Omega_{PN})$, $N = 1, 2, 3, \ldots$ is the harmonic number, $\omega = \Omega_{PN}$ is the standing wave Eigen frequency determined from the dispersion equation

$$\Psi_{AP}(\Omega_{PN}) = N\pi, \tag{3.434}$$

or standing SMS waves

$$\varphi_{SN} = U_{SN}(x^1) \sin\left(\int_{l_-}^{l} k_{\|SN}\, dl \right) / k_{\|SN}^{3/2}, \tag{3.435}$$

the Eigen frequencies of which are determined by the dispersion equation

$$\Psi_s(\Omega_{SN}) = N\pi. \tag{3.436}$$

Let us determine provisionally the possible type of solutions to the dispersion equations. Obviously, the solutions of dispersion equations (3.434) and (3.436) can be sought in the domain of real values of ω^2 (in the form of neutral, $\omega^2 > 0$, or aperiodic, $\omega^2 < 0$, modes) if $k_\|^2 > 0$ on the entire field line, for any real ω^2. If, however, the expression under the radical in (3.431) passes through zero somewhere on the field line, for real ω^2, the solution may be periodic unstable modes with a complex value of ω^2. An analysis of (3.431) shows that, for the expression under the radical to remain positive for any real ω^2, the condition

$$\overline{\varkappa}_{1B}(\overline{\varkappa}_{1B} + \overline{\varkappa}_{1P}) < 0. \tag{3.437}$$

must hold on the entire field line. If this condition fails on the field line, a solution with a complex frequency will be more likely.

Figure 3.61 Distribution of parameter $\overline{\varkappa}_{1B}(\overline{\varkappa}_{1B} + \overline{\varkappa}_{1P})$ along the field line located on magnetic shell $L = 20$, in models with a thick (thick line) and a thin (thin line) current sheet. $\theta = 0$ in the equatorial plane.

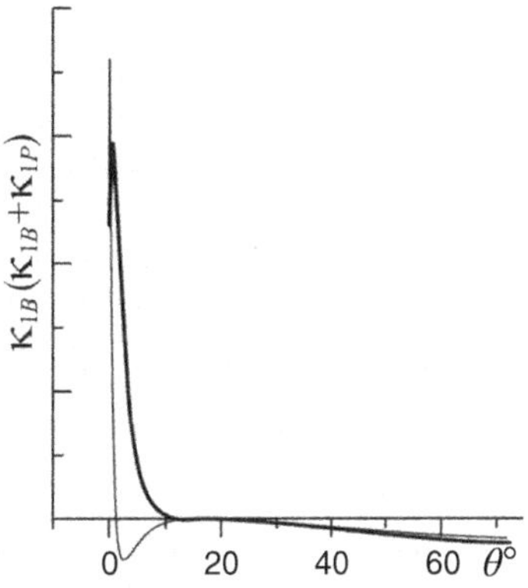

Figure 3.61 displays the distribution of parameter $\overline{\varkappa}_{1B}(\overline{\varkappa}_{1B} + \overline{\varkappa}_{1P})$ on the field line $a = 20R_E$ for models with a thick and a thin current sheet. Qualitatively, such a distribution is typical of all field lines crossing the current sheet. Even though the condition (3.437) is met on most of the field line, it is clear that $\overline{\varkappa}_{1B}(\overline{\varkappa}_{1B} + \overline{\varkappa}_{1P}) > 0$ in the current sheet region. The likely type of solution for dispersion equations (3.434) and (3.436), in the region of field lines crossing the current sheet, are therefore periodic unstable oscillation modes. The following procedure is chosen for solving the dispersion equations numerically. Solutions of (3.434) and (3.436) correspond to longitudinal harmonic (with eigen number N) in the near-Earth part of the magnetosphere, on a magnetic shell that does not cross the current sheet. Here, these solutions describe ordinary poloidal Alfvén waves and azimuthally small-scale SMS waves. Next, proceeding at a small step along across the magnetic shells, we find the solutions in the entire range of the shells.

Figure 3.62 shows the distribution of $k^2_{\|AP}$ on different magnetic shells, for the main harmonic of standing Alfvén waves ($N = 1$), in the thin current sheet model of the magnetosphere. On shells $a = 6R_E$ and $a = 8R_E$, the entire field line is a transparent region for the oscillations (here, $\mathrm{Re}\, k^2_{\|AP} > 0$). On field line, located in the near part of the current sheet (where $a = 10R_E$), the transparent region on the field line is split into three – a narrow near-equatorial region and two wide regions adjoining the ionosphere. Small opaque regions are located in between. In such cases, the eigen-oscillation spectrum is known to split: the wider and deeper the opaque regions the larger the splitting [375]. The opaque regions being small in our case, we will not examine effects related to the oscillation spectrum splitting.

Integration in (3.434) and (3.434) pertains to transparent regions only. In the dispersion equation (3.434), the full phase is the sum of integrals for all transparent regions (their number can be as high as 5 on a field line). On the field line with $a = 10R_E$, the oscillation remains an ordinary poloidal Alfvén wave of real frequency. On more remote field lines crossing the current sheet, however, the solution satisfying the boundary conditions (3.420) is a periodic unstable Alfvén wave with $\omega_N \equiv \Omega_{PN} = \overline{\Omega}_{PN} + i\delta_{PN}$, where $\overline{\Omega}_{PN}$ is the real part of the Eigen frequency of poloidal Alfvén wave, $\delta_{PN} > 0$ is its growth rate (only the unstable branch of the oscillations is examined here). This differs substantially from the local approximation result, where the poloidal Alfvén wave is neutrally stable. Figure 3.62 shows the distribution of $\mathrm{Re}\,(k^2_{\|AP})$ and $\mathrm{Im}(k^2_{\|AP})$ along the field line with $a = 13R_E$. The full integral of $\mathrm{Im}(k_{\|AP})$ along the entire field line turns zero.

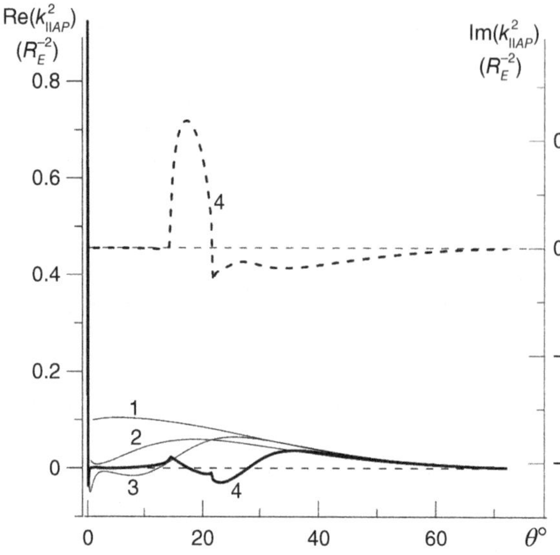

Figure 3.62 Distribution of the wavenumber squared, along different field lines ($\mathrm{Re}\,k_{\parallel}^2$ – solid lines, $\mathrm{Im}\,k_{\parallel}^2$ – dashed line), for the main harmonic ($N = 1$) of standing poloidal Alfvén waves, in the thin current sheet model. Neutrally stable oscillations on magnetic shells: (1) – $L = 6$, (2) – $L = 8$; (3) – $L = 10$; (4) – unstable ($\delta_{P1} > 0$) periodic oscillations on magnetic shell $L = 13$.

Figure 3.63 shows the distribution over magnetic shells of the Eigen frequencies of the first five harmonics of standing Alfvén waves, for the thick and thin current sheet models. Regions occupied by unstable oscillations are wider and the absolute values of the growth rate larger in the thin current sheet model than in the thick current sheet model. The largest growth rate is exhibited by harmonic $N = 2$ and reached at magnetic shells $a \approx 30R_E$. The higher the number of the harmonic, the smaller the growth rate and the farther into the geotail the neutral standing poloidal Alfvén waves transform into unstable oscillations.

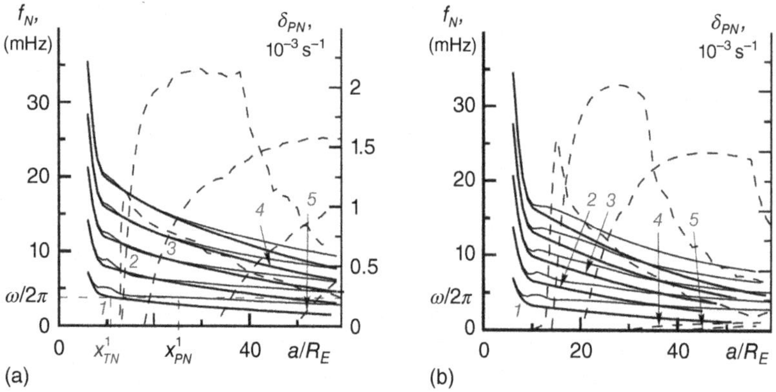

Figure 3.63 Eigen frequency distribution for the first five harmonics $N = 1, 2, 3, 4, 5$ of standing poloidal ($f_N = \mathrm{Re}\,(\Omega_{PN})/2\pi \equiv \overline{\Omega}_{PN}/2\pi$ – thin solid lines, $\delta_{Pn} = \mathrm{Im}\,\Omega_{Pn}$ – dashed lines) and toroidal ($f_N = \Omega_{TN}/2\pi$ – thick solid lines) Alfvén waves across magnetic shells: (a) for the geotail model with a thin current sheet and (b) for the model with a thick current sheet.

The same figure illustrates the Eigen frequency distribution for toroidal Alfvén oscillations described by

$$\hat{L}_T \varphi_{AT} = 0, \tag{3.438}$$

obtained from (3.418) in the $|\nabla_1 \ln \varphi_{AT}| \gg |k_2|$ limit. Its boundary conditions, determined, in the main order, from the condition that the ionosphere is ideally conducting, are: $\varphi_{AT}(l_\pm) = 0$. The solution of (3.438), describing standing toroidal Alfvén waves in the WKB approximation, has the form

$$\varphi_{ATN} = U_{TN}(x^1) T_N(x^1, x^3),$$
$$T_N(x^1, x^3) = C_{TN} \sin\left(\int_{l_-}^{l} k_{\|ATN} dl \right) / k_{\|ATN}^{1/2}, \tag{3.439}$$

where $k_{\|AT} = \omega / v_A$, and the Eigen frequencies are determined by the dispersion equation

$$\Psi_{AT}(\Omega_{TN}) = \int_{l_-}^{l_+} k_{\|AT} dl = N\pi. \tag{3.440}$$

It is interesting to note that the Eigen frequency plots for poloidal and toroidal Alfvén oscillations, differing little on magnetic shells in the near-Earth part of the geotail, diverge significantly farther into the geotail.

Figure 3.64 shows an analogous calculation for the Eigen frequencies of azimuthally small-scale SMS waves. All the oscillation harmonics can be seen to switch to the instability regime in practically one and the same region of the near part of the current sheet. It should be noted that, in contrast to the local approximation, the SMS Eigen frequencies and growth rates are much smaller than the frequencies and growth rates of the Alfvén waves. This is explained by the fact that, in the WKB approximation, the Eigen frequencies are determined by integrals along the field lines of Alfvén speed v_A and SMS speed c_s, while $v_A \gg c_s \approx v_s$ on most field lines. Another significant difference of the unstable standing SMS waves is that the solutions they are described by and satisfying the boundary conditions (3.420) only exist in a limited range of magnetic shells. Figure 3.64 shows the Eigen frequency distributions on those shells only to which it was possible to numerically trace

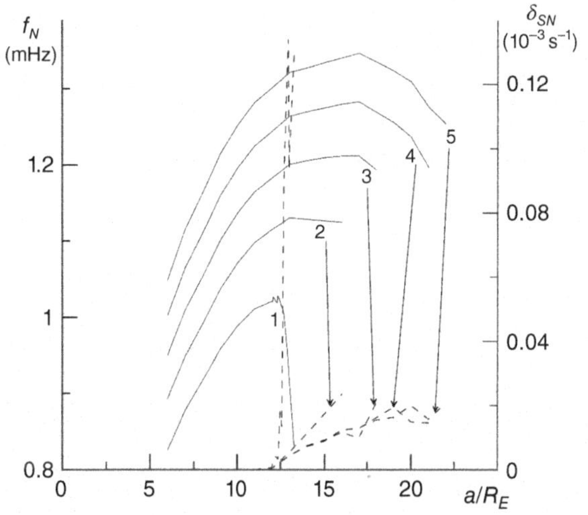

Figure 3.64 Eigen frequency distribution for the first five harmonics $N = 1, 2, 3, 4, 5$ of standing azimuthally small-scale SMS waves ($f_N = \mathrm{Re}\,(\Omega_{sN})/2\pi$ – solid lines, $\delta_{sN} = \mathrm{Im}\,\Omega_{sn}$ – dashed lines), across magnetic shells, in the model geotail with a thin current sheet.

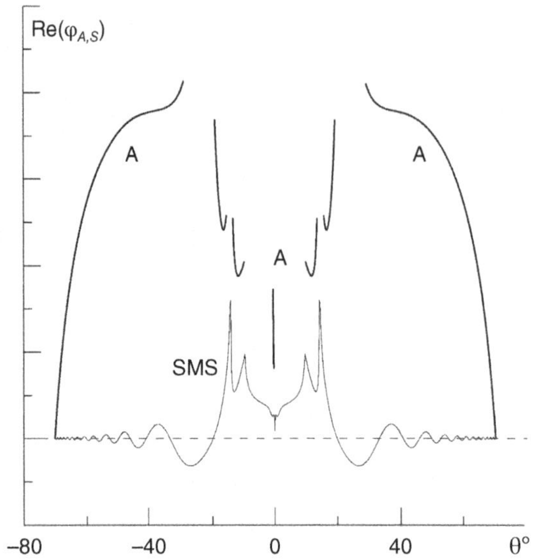

Figure 3.65 Longitudinal structure of the main harmonic ($N = 1$) of unstable poloidal Alfvén waves (A) on magnetic shell $L = 13$ and, for comparison, the structure of the SMS wave (SMS) of the same frequency (which fails to satisfy the boundary conditions on the ionosphere). Gaps in the Alfvén wave structure correspond to the neighbourhoods of turning points (where $\mathrm{Re}\,(k_{\parallel}^2) = 0$).

each harmonic of the SMS waves in question. The higher the number of the harmonic, the farther into the geotail its domain of existence reaches.

Another note regarding SMS waves is that in reality they are strongly decaying modes – an effect which is due to their phase velocity being close to the thermal velocity of plasma ions (see Chapter 1). Given the very small growth rates of standing SMS waves, therefore, it is hardly worth expecting them to be amplified substantially in the presence of even a very intense and thin current sheet.

The wave field structure along the field line $a = 13R_E$ is shown in Figure 3.65 for unstable main harmonic of standing poloidal Alfvén waves ($N = 1$). The wave function ϕ_{A1} distribution is shown along the entire field line, except the neighbourhoods of the turning points, where the WKB approximation does not apply. To accurately determine the wave function structure, it is necessary to numerically solve (3.436). The following conclusions can be made from the form of the wave function in Figure 3.65, obtained in the WKB approximation. On most of the field line, the oscillations under study have a structure analogous to the structure of an ordinary standing Alfvén wave. Near the equatorial surface, the structure of the oscillations in question along the field line should be expected to become substantially small-scale in the current sheet region. These conclusions are also valid for other harmonics of standing waves. For comparison, the same figure contains the wave function ϕ_s, describing the SMS wave structure at the same frequency $\omega \equiv \Omega_{P1}$. The ϕ_s distribution along the entire field line is shown, including the opaque region (at turning points, $k_{\parallel_s}^2$ is regularised by the imaginary part of the frequency). It of course fails to satisfy the boundary conditions (3.420). It can be seen that, at given frequency, the SMS wave has a significantly smaller wavelength, on the entire field line, than has the Alfvén wave. Note that ballooning instability developing in azimuthally small-scale MHD oscillations can result in the current sheet splitting into separate filaments, as shown in [362, 376].

This section addressed the structure of unstable standing Alfvén and SMS waves, along magnetic field lines. The Alfvén oscillation structure across magnetic shells is analogous to the structure of coupled Alfvén and SMS oscillations addressed in Section 3.17.3.

3.17 Coupled Alfvén and SMS Oscillation Modes in the Geotail

As was shown in Section 3.7, Alfvén oscillations with $m \gg 1$ can be excited near the poloidal resonance surface by external currents in the ionosphere. The source oscillation spectrum must then contain harmonics with frequencies corresponding to the eigenfrequencies of poloidal standing Alfvén waves. High-energy particle fluxes in the magnetosphere can also serve as the source of such oscillations [377]. If the structure of the oscillations is described in the form $\varphi \sim \exp(i \int k_1(x^1)dx^1)$, in the WKB approximation, in the x^1 coordinate, $|k_1/k_2| \to 0$ near the poloidal resonance surface. Oscillations generated near this surface escape across magnetic shells to the toroidal resonance surface (where $|k_1/k_2| \to \infty$), remaining standing along magnetic field lines. In the course of such propagation the oscillation polarisation changes from poloidal to toroidal. In the neighbourhood of the toroidal resonance surface (Alfvén resonance region) the oscillations are fully absorbed thanks to dissipation in the ionosphere. Hence, no wave is reflected from the toroidal surface.

Section 3.16 examined the longitudinal structure and spectra of azimuthally small-scale Alfvén and SMS waves standing between the magnetoconjugated ionospheres. Even though the WKB approximation used there is more suitable for their description than is the local approximation, it, however, rather roughly describes the basic harmonics of oscillations with wavelengths comparable to the magnetospheric plasma inhomogeneity scales. It is these harmonics that are of most interest, however, because they possess the largest growth rates.

This section will carry out a numerical integration of the equation describing the field of the basic harmonics of azimuthally small-scale MHD oscillations. The same axisymmetric model geotail is used as in Section 3.16. This makes it possible to not only verify the WKB approximation results but also to calculate the spatial distribution of the field of coupled Alfvén and SMS waves. The conditions are also analytically studied of the coupling between the Alfvén and SMS oscillations in the current sheet.

Frequencies and growth rates of SMS oscillations on magnetic field lines extended into the geotail are much smaller than the corresponding frequencies and growth rates of Alfvén waves. Thanks to this, unstable Alfvén oscillations must manifest themselves in the magnetosphere much more noticeably than SMS waves. As will be shown below, Alfvén waves are coupled to SMS oscillations in the current sheet region, on magnetic shells crossing the current sheet. We will therefore denote them as coupled modes of MHD oscillations, in the following calculations. In this section, we will restrict ourselves to calculating the fields of the main and first harmonics of the coupled modes, at frequencies close to the frequencies of standing poloidal Alfvén waves.

3.17.1 Coupled Mode Structure Along Magnetic Field Lines

As was shown in Section 3.16, the distance between the poloidal and toroidal resonance surfaces for Alfvén waves, in the current sheet region, is large enough, in contrast to the 'cold plasma' case. The poloidal oscillations are therefore highly likely to be absorbed near the poloidal resonance surface, where they are generated. To describe the structure of such oscillations, along the background magnetic field lines, near the poloidal resonance surface, let us use Eq. (3.418). As was shown in Section 3.7–3.15, the multiple-scale technique can be used for describing the spatial structure of such oscillations, by representing the scalar potential of MHD oscillations as

$$\varphi = U(x^1)H(x^1, x^3)e^{ik_2x^2 - i\omega t}, \quad (3.441)$$

where $U(x^1)$ describes the small-scale structure of oscillations across magnetic shells, and $H(x^1, x^3)$ their structure along magnetic field lines (x^1 dependence is determined by changing coefficients in (3.418), at the magnetospheric plasma inhomogeneity scales).

Substituting (3.441) into (3.418) and differentiating produces an equation for function $H(x^1, x^3)$:

$$[\nabla_l^4 + \kappa_3\nabla_l^3 + \kappa_2\nabla_l^2 + \kappa_1\nabla_l + \kappa_0]H = 0, \quad (3.442)$$

where $\nabla_l \equiv \partial/\partial l = (g_3)^{1/2}\nabla_3$ is the derivative over the longitudinal coordinate l, its length element having the form, in the system of coordinates (a, ϕ, θ) (see (N.1) in Appendix N)

$$dl = \sqrt{r^2(a,\theta) + (\partial r/\partial\theta)^2}d\theta.$$

The radius of the point on the field line, $r(a, \theta)$, is determined as (3.421), and the expressions for the coefficients κ_i are given in Appendix Q.

To numerically integrate (3.442), it should be complemented with boundary conditions on the ionosphere. In the same, main, order of perturbation theory as used for obtaining (3.442), we will assume the ionosphere to be ideally conducting. The boundary conditions are therefore reduced to the requirement that the tangential components of the oscillation electric field defined in terms of E_1 and E_2 become zero on the ionosphere. It follows from the boundary condition (2.309) that, in the main order, when $(|k_1/k_2| \to 0)$,

$$\varphi(l_\pm) = 0, \quad (3.443)$$

where $l_\pm$ are the coordinates of the points where the field lines cross the ionosphere in the Northern and Southern hemispheres, respectively. Based on the boundary condition $\psi(l_\pm) = 0$, we have $\hat{L}_P\varphi|_{l_\pm} = 0$ (see (3.420)), or, given (3.443)

$$\nabla_l^2\varphi|_{l_\pm} = \varkappa_{lp}\nabla_l\varphi|_{l_\pm}, \quad (3.444)$$

where $\varkappa_{lp} = \nabla_l(\ln p^{-1})$. When numerically integrating (3.442), the value of $\nabla_l\varphi|_{l_-}$ can be chosen arbitrarily because it sets the amplitude of the solution and is determined by the selected normalisation.

Integration of the fourth-order equation requires that the third derivative $\varphi_-''' \equiv \nabla_l^3\varphi|_{l_-}$ is also set on the ionosphere (assuming the integration to begin from the ionosphere of the Southern hemisphere). No other boundary conditions can help determine $\varphi'''(l_-)$. At the opposite end of the field line, the desired solution must also satisfy the two boundary conditions (3.443) and (3.444), therefore, it is determined by two eigenvalues of the parameters of the problem.

We will select the eigenfrequency (denoting it $\omega = \Omega_N$, where $N = 1, 2, 3$, is the longitudinal wave number of the large-scale standing coupled mode) of the waves standing between the ionospheres and the corresponding value of the third derivative of the desired function on the ionosphere $\varphi_{N-}''' \equiv \varphi_N'''(l_-)$ as such parameters. Note that the eigenvalue Ω_N should not be regarded as the frequency of eigen oscillations of the complete problem (3.418).

The eigenvalues Ω_N and φ'''_{N-} determine the spatial structure of the harmonic of standing waves on the magnetic shell in question. They are functions of the transverse coordinate, $\Omega_N \equiv \Omega_N(x^1)$, $\varphi'''_{N-} \equiv \varphi'''_{N-}(x^1)$, but can be assumed to be constant, when solving the longitudinal problem on an individual magnetic shell.

It will be shown in the following calculations that $\mathrm{Re}\,\Omega_N(x^1)$ can be regarded as the frequency that must be present in the spectrum of the source capable of exciting the Nth harmonic of standing waves on the resonance magnetic shell in question, while $\mathrm{Im}\,\Omega_N(x^1)$ determines the amplitude distribution for such oscillations, across magnetic shells. To simplify the numeric calculations, we will use the symmetry about the equatorial plane, in the model in question. The second pair of boundary conditions can then be formulated, not on the magnetoconjugated ionosphere, but in the equatorial plane, as follows. For even modes ($N = 1, 3, 5, \ldots$, in the notation used), this means that all odd derivatives of the desired function turn zero. For the problem under study, the requirements $\nabla_l\varphi|_{l_e} = \nabla_l^3\varphi|_{l_e} = 0$, are enough, where $l = l_e$ is the coordinate of the equatorial plane. For odd modes ($N = 2, 4, 6, \ldots$), the analogous boundary conditions are $\varphi(l_e) = \nabla_l^2\varphi|_{l_e} = 0$.

Figure 3.66 displays the structure of two first harmonics of waves standing along magnetic field lines as obtained using the numeric integration technique for (3.442) with boundary conditions (3.443) and (3.444) on the ionosphere. Standing wave structures are shown for two magnetic field lines, one located in the inner magnetosphere at magnetic shell $L = a/R_E = 6$, where a is the equatorial radius of the field line (Figure 3.66a,b), and the other crossing the current sheet at shell $L = 15$ (Figure 3.66c,d). The oscillation parameters in the current sheet (Figure 3.66c,d) were calculated by gradually transforming

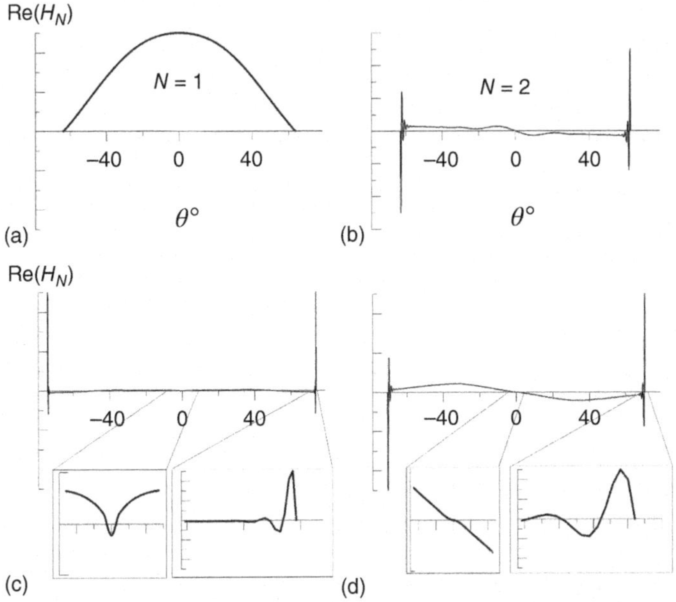

Figure 3.66 Distribution along the field line of the scalar potential of the electric field of the main ($N = 1$) and first ($N = 2$) harmonics of the coupled modes: (a, b) in the inner magnetosphere, at shell $L = 6$; (c, d) in the current sheet region, at magnetic shell $L = 15$.

the solutions found in the inner magnetosphere (Figure 3.66a,b), while proceeding, at a small step, across magnetic shells into the current sheet region.

Of most interest here is the presence of small-scale structure of the eigenfunctions of the oscillations near the ionosphere, which is a manifestation of small-scale SMS wave coupled with large-scale Alfvén wave. What is unusual here is that it is manifested, not only in the current sheet, as was expected in earlier papers (see [289, 378]), but near the ionosphere as well. A similar structure resulted from numeric calculations in [277, 278, 379], where it only manifested itself in individual components of the oscillation field. According to the above calculations, this feature must manifest itself in all wave field components, because they are all expressed in terms of the scalar potential φ.

The spatial structure of the oscillations under study is also refined in the current sheet. The amplitude of these oscillations is, however, smaller here than near the ionosphere. This is also related to the coupling of the large-scale Alfvén and small-scale SMS modes in the current sheet. Inside the current sheet, it becomes complicated enough to determine which of the standing-wave harmonics the oscillations belong to based on only the shape of their large-scale component because of the presence of small-scale structure of the oscillations.

Figure 3.67 shows the distributions along magnetic field lines of the electromagnetic field components E_y, B_x, B_z of the oscillations, for the main harmonic of standing waves

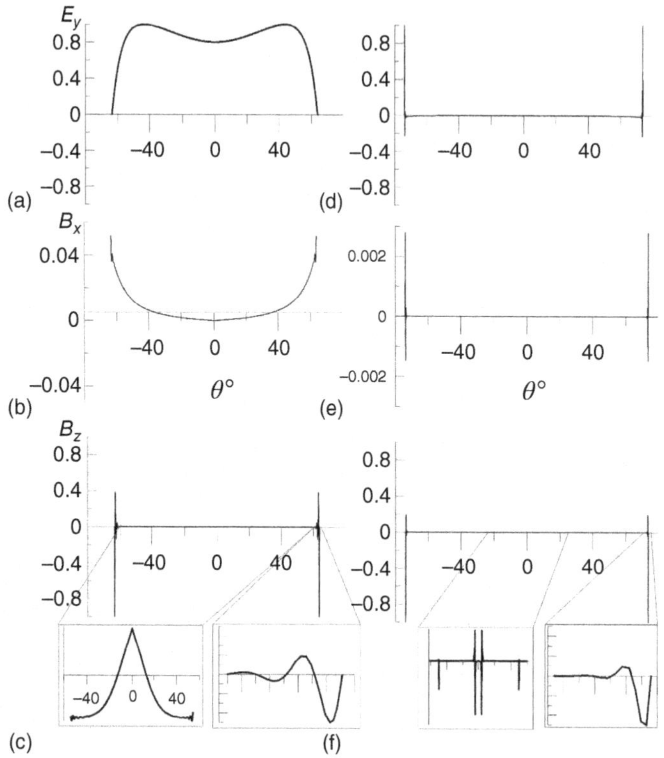

Figure 3.67 Distribution along the field line of the electromagnetic field components (E_y, B_x, B_z) of the coupled modes main harmonic ($N = 1$): (a–c) in the inner magnetosphere, at shell $L = 6$; (d–f) in the current sheet region, at magnetic shell $L = 15$.

($N = 1$), at the same magnetic shells, $L = 6$ and $L = 15$. Manifestations of the small-scale constituent in these components are even more noticeable near the ionosphere. This is due to the features of the plasma parameter distribution along the magnetic field lines converging towards the ionosphere (see [380]).

The spatial refinement of the oscillations is more complicated in character, in the current sheet. The bottom part of Figure 3.67 shows a higher-resolution distribution of the B_z-component of the field, for most part of the field line, except the region near the ionosphere. Dramatic changes in the oscillation field structure on field lines passing through the current sheet are due to the parameter $\varkappa_{1g}$ passing through zero in the region in question. This happens at field-line inflexion points. In sum, these points form inflexion surfaces, their cross-sections shown in Figure 3.58, in the meridional plane.

It can be seen from Figures 3.58 and 3.67 that there are four such points on the field line, two located near the current sheet, and the two others in the transition region. To regularise the singularities at these points, small, higher-order corrections must be taken into account in (3.417), an equation relating the potentials φ and ψ. This feature of the structure of the oscillation field components can be used for determining the inflexion points on geomagnetic field lines based on spacecraft observations of ULF oscillations – for example, in order to define the conditional boundaries of the current sheet.

It was demonstrated in Chapter 1 that SMS waves are strongly dissipative, which fact is not included in the framework of the ideal MHD used in the above calculations. In the real magnetosphere, therefore, the small-scale constituent of the oscillations is strongly dominant near the ionosphere but not to the same degree as in the structures calculated here. The oscillation structure will change more smoothly in the current sheet if the real dissipation of SMS waves is taken into account. To a certain degree, however, these features will still manifest themselves, and can be used for identifying coupled modes in spacecraft-observed ULF oscillations.

The calculated dependence of the first two frequencies Ω_1, Ω_2 of standing waves of the coupled modes, on the magnetic shell parameter $L = a/R_E$ is shown in Figure 3.68. Note how complex the numerical search for the eigen solutions of (3.442) is. Each root $\Omega_N(L)$ corresponding to large-scale standing Alfvén wave, in the inner magnetosphere, is split, in the current-sheet transition region, into multiple 'branches' differing in the

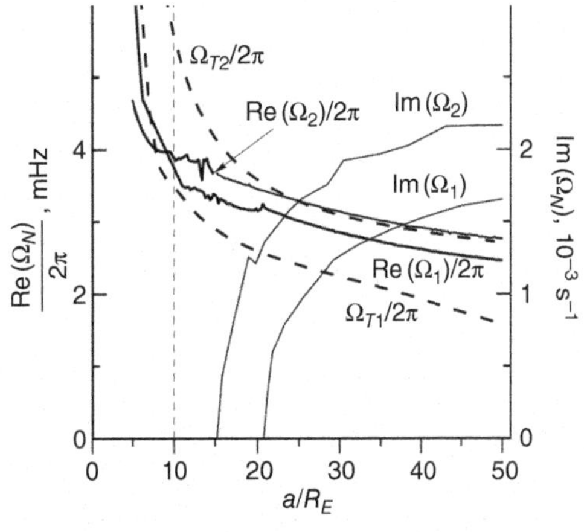

Figure 3.68 Distribution across magnetic shells of the eigenfrequencies of the main ($N = 1$) and first ($N = 2$) harmonics of toroidal Alfvén (thick dashed lines) and coupled (thick solid lines) modes of MHD oscillations, as well as their growth rates (thin solid lines) in the current sheet region. The transition layer region is shown in grey.

structure of their small-scale constituents. Newton's gradient technique used here to seek solutions for two eigen values is very sensitive to the initial parameters of the problem that are employed. Efforts to continuously trace the behaviour of each such solution in the entire calculation domain have therefore failed. By choosing corresponding initial parameters, however, one can minimise the number of jumps from one branch to another. The existence domains for all possible branches obtained by integrating the solutions for different large-scale harmonics of standing waves can overlap.

The solutions in Figure 3.68 are neutrally stable outside the current sheet ($\mathrm{Im}\,\Omega_{1,2} = 0$), and become unstable inside it ($\mathrm{Im}\,\Omega_{1,2} > 0$). This is a manifestation of the ballooning instability of poloidal Alfvén oscillations at magnetic shells crossing the current sheet. The instability in question is not related to coupling of Alfvén and SMS oscillations which was suggested in [279, 281]. Coupled Alfvén and SMS modes become unstable inside the current sheet only, where the conditions for ballooning instability of poloidal Alfvén waves are realised. An analogous problem is solved in Section 3.16 (see also [289]) in the WKB approximation. This approximation does not include coupling of Alfvén and SMS waves. In the geotail model in question, both Alfvén and SMS waves can become unstable in the current sheet region independent of each other.

We will demonstrate in Section 3.17.2 that coupling in the current sheet is in the form of linear transformation of the modes. SMS waves being highly dissipative, this coupling must enhance the decay of Alfvén waves competing with their ballooning instability. Note that this instability cannot be regarded as an instability of the oscillation eigenmodes of the plasma configuration under study. The presence of this instability can result in the amplitude of the oscillation increasing as they move away from the poloidal resonance surface, across magnetic shells.

For comparison, Figure 3.68 shows the distribution of the frequencies Ω_{T1}, Ω_{T2} of toroidal Alfvén waves that are the eigenvalues of the problem (3.438). Comparing the eigenfrequencies Ω_1, Ω_2 to Ω_{T1}, Ω_{T2} demonstrates that the resonance shells for toroidal and poloidal standing Alfvén waves with identical frequencies are widely spaced in the geotail model in question. Wave generated on the poloidal resonance surface cannot reach the toroidal resonance surface without decaying noticeably. They will be absorbed thanks to dissipation in the ionosphere near the poloidal resonance surface. When solving the problem of the structure of such oscillations across magnetic shells, we can therefore restrict ourselves to examining their structure in a certain neighbourhood of the poloidal resonance surface only.

3.17.2 Linear Transformation of Alfvén and SMS Waves in the Current Sheet

Let us do a qualitative investigation into the features of the spatial structure for the oscillations in question, in the current sheet. It was shown in Section 3.17.1 that this structure is determined by the large-scale, poloidal Alfvén wave-related constituent, on a most part of the field line. Outside the current sheet $c_s^2 \approx v_s^2 \ll v_A^2$ and $|\kappa_0 \overline{L}^4|, |\kappa_1 \overline{L}^3|, |\kappa_2 \overline{L}^2| \gg |\kappa_3 \overline{L}| \sim 1$, where $\overline{L}$ is the characteristic field-line length. Qualitative investigations of this structure can therefore neglect the contribution from the two first terms with higher derivatives in (3.442), for almost the entire field line, assuming

$$[\kappa_2 \nabla_l^2 + \kappa_1 \nabla_l + \kappa_0] H \approx 0, \tag{3.445}$$

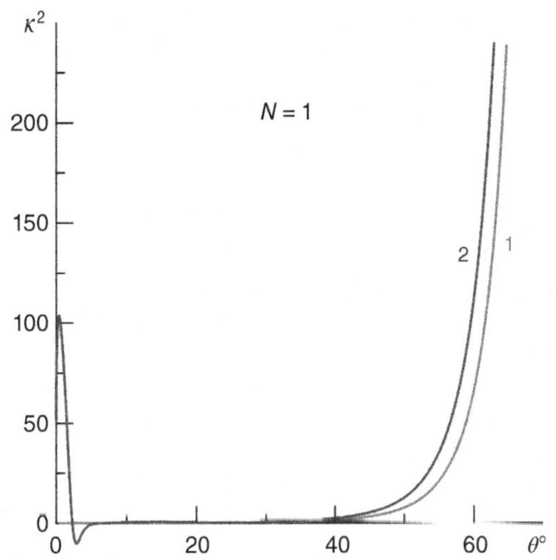

Figure 3.69 Parameter κ_2 distribution along the field line, for the main harmonic of coupled modes ($N = 1$), on magnetic shells $L = 6$ (line 1) and $L = 15$ (line 2).

and, in the main order, it is reduced to the well-known equation for poloidal Alfvén waves, $\hat{L}_P H \approx 0$. In the WKB approximation, Solution of (3.445) has the form

$$H \approx \frac{1}{(\kappa_0/\kappa_2)^{1/4}} \exp\left[\pm i \int \sqrt{\kappa_0/\kappa_2} dl - \frac{1}{2}\int \frac{\kappa_1}{\kappa_2} dl\right]. \tag{3.446}$$

Let us consider the distribution of parameter κ_2 along the entire field line. Such a distribution is shown in Figure 3.69 for the main harmonic of coupled modes ($N = 1$), on two above-discussed magnetic shells, $L = 6$ and $L = 15$. In the current sheet region $v_s^2 \gg c_s^2 \approx v_A^2$ and the characteristic feature of coefficient κ_2 is that it passes through zero, at certain points $l = l_0$. Obviously, the WKB approximation ceases to apply in the neighbourhood of l_0, and it is necessary to take into account higher derivatives in (3.442). Points l_0 are singular points of (3.446) and are singular turning points.

For a qualitative investigation of the solution in the neighbourhood of points $l = l_0$, the higher derivative can be left in (3.442), enabling the desired solution to be regularised. Including the third derivative results in these points shifting somewhat. Linearising the coefficients of such an equation near $l = l_0$, we will write it in this form:

$$[\nabla_z^4 + z\nabla_z^2 + \alpha_1 \nabla_z + \alpha_0] H_N \approx 0, \tag{3.447}$$

where $z = (l - l_0)/\lambda + i\lambda^2 \mathrm{Im}\,\kappa_2(l_0)$, $\lambda = (\partial \mathrm{Re}\,\kappa_2/\partial l)_{l_0}^{-1/3}$, $\alpha_1 = \kappa_1(l_0)\lambda^3$, $\alpha_0 = \kappa_0(l_0)\lambda^4$. Equation (3.447) has a standard form for seeking its solutions using the contour integral method (see [253, 339]). We will seek a solution to (3.447) in the form

$$H_N(x^1, z) = \int_{\tilde{C}} \tilde{H}_N(s) e^{sz} ds, \tag{3.448}$$

where $\tilde{C}$ is a certain integration path in the complex s plane. Substituting (3.448) into (3.447) and integrating the second term by parts produces this equation for $\tilde{H}_N$:

$$s^2 \tilde{H}_N - \frac{\partial s^2 \tilde{H}_N}{\partial s} + \alpha_1 s \tilde{H}_N + \alpha_0 \tilde{H}_N = 0 \tag{3.449}$$

provided the term outside the integral

$$s^2 \tilde{H}_N e^{sz}\Big|_{\tilde{C}} = 0, \tag{3.450}$$

which serves as the condition to select the integration contours $\tilde{C}$. It is clear from this that contours $\tilde{C}$ must be such that either expression (3.450) turns zero at their ends, or the contour start and end points coincide. The solution of (3.449) has the form

$$\tilde{H}_N = \frac{1}{s^{2-\alpha_1}} \exp\left(\frac{s^3}{3} - \frac{\alpha_0}{s}\right), \tag{3.451}$$

and the corresponding solution (3.448) has the form

$$H_N(x^1, z) = \int_{\tilde{C}} \exp\left(\frac{s^3}{3} - \frac{\alpha_0}{s} + sz\right) \frac{ds}{s^{2-\alpha_1}}, \tag{3.452}$$

and the contour $\tilde{C}$ selection condition has the form

$$s^{\alpha_1} \exp\left(\frac{s^3}{3} - \frac{\alpha_0}{s} + sz\right)\Big|_{\tilde{C}} = 0.$$

Let us represent the complex variable s as $s = r_s e^{i\psi_s}$ It follows from this that, if $s \to 0$, expression (3.450) turns zero in the $-\pi/2 < \psi_s < \pi/2$ sector for $\mathrm{Re}\,\alpha_0 > 0$ and in the $\pi/2 < \psi_s < 3\pi/2$ sector for $\mathrm{Re}\,\alpha_0 < 0$. If $s \to \infty$, they will be the $\pi/6 < \psi_s < \pi/2$, $5\pi/6 < \psi_s < 7\pi/6$ and $3\pi/2 < \psi_s < 11\pi/6$ sectors, left unshaded in Figure 3.70. We will not construct the full system of solutions of (3.447), as was done in [253, 339]. We will restrict ourselves to the contours $\tilde{C}$ the integrals (3.448) over which correspond to the solutions describing coupling of the Alfvén and SMS oscillations in question.

For this purpose, let us estimate integrals (3.448) for $|z| \to \infty$ by means of the saddle-point technique (see [337]). The saddle points in (3.448) are determined by the zero of the derivative of the exponent under the integral:

$$s^2 + \frac{\alpha_0}{s^2} + z = 0.$$

Hence,

$$s^2 = -\frac{z}{2} \pm \sqrt{\frac{z^2}{4} - \alpha_0}.$$

For $\mathrm{Re}\,z \to \infty$, we have four saddle points $S_{1,2} = \pm i\sqrt{z}$ and $S_{3,4} = \pm i\sqrt{\alpha_0/z}$ (for $\mathrm{Re}\,\alpha_0 > 0$). In the general case, z is a complex value, its imaginary part determined by the complex frequency ω. The bypass rule for singular point $z = 0$, from $\mathrm{Re}\,z \to \infty$ to $\mathrm{Re}\,z \to -\infty$, is

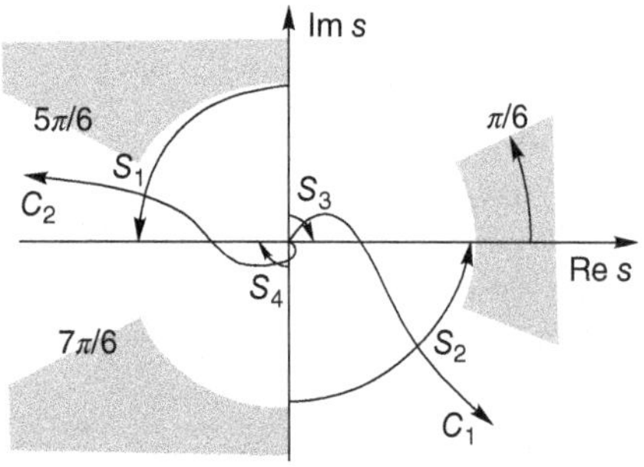

Figure 3.70 Azimuth variations for saddle points S_i ($i = 1, 2, 3, 4$) in (3.452), from $\mathrm{Re}\,z \to \infty$ to $\mathrm{Re}\,z \to -\infty$ and integration contours $\tilde{C}_{1,2}$ for the solutions of (3.447) for $\mathrm{Re}\,\alpha_0 > 0$.

determined such as is the case for $\text{Im}\omega > 0$, i.e. phase z varies 0 to π. For $\text{Re}\, z \to -\infty$ therefore, the saddle points move to $S_{1,2} = \pm\sqrt{-z}$, $S_{3,4} = \pm\sqrt{-\alpha_0/z}$ as shown in Figure 3.70.

Let us select integration contours $\tilde{C}_1$ and $\tilde{C}_2$ as shown in Figure 3.70. Either contour crosses two saddle points and describes coupled Alfvén and SMS modes. Let us denote $\alpha_0 = r_\alpha e^{i\nu}$. Using standard formulas for estimating the integrals by means of the saddle-point technique and taking into account the contribution of two such points in each solution produces the following expressions for solutions (3.448) on the asymptotics:

$$H_N^{(1)} = \begin{cases} \dfrac{z^{1/4-\alpha_1/2}}{\alpha_0^{3/4-\alpha_1/2}} \exp[2i\sqrt{\alpha_0 z} + i\varphi_{A1}] \\ \quad + \dfrac{\exp[-\frac{2}{3}iz^{3/2} + i\varphi_{s1}]}{z^{5/4-\alpha_1/2}}, & \text{Re}\, z \to \infty, \\ \dfrac{(-z)^{1/4-\alpha_1/2}}{\alpha_0^{3/4-\alpha_1/2}} \exp[-2\sqrt{-\alpha_0 z} + i\varphi_{A2}] \\ \quad + \dfrac{\exp[-\frac{2}{3}(-z)^{3/2} + i\varphi_{s2}]}{(-z)^{5/4-\alpha_1/2}}, & \text{Re}\, z \to -\infty, \end{cases}$$
$$H_N^{(2)} = \begin{cases} \dfrac{z^{1/4-\alpha_1/2}}{\alpha_0^{3/4-\alpha_1/2}} \exp[-2i\sqrt{\alpha_0 z} + i\varphi_{A3}] \\ \quad + \dfrac{\exp[\frac{2}{3}iz^{3/2} + i\varphi_{s3}]}{z^{5/4-\alpha_1/2}}, & \text{Re}\, z \to \infty, \\ \dfrac{(-z)^{1/4-\alpha_1/2}}{\alpha_0^{3/4-\alpha_1/2}} \exp[2\sqrt{-\alpha_0 z} + i\varphi_{A4}] \\ \quad + \dfrac{\exp[\frac{2}{3}(-z)^{3/2} + i\varphi_{s4}]}{(-z)^{5/4-\alpha_1/2}}, & \text{Re}\, z \to -\infty, \end{cases} \tag{3.453}$$

where $\varphi_{A1} = (\pi + \nu)/4$; $\varphi_{A2} = \nu/4 - \alpha_1\pi/2$; $\varphi_{A3} = \nu/4 + \alpha_1\pi + 3\pi/4$; $\varphi_{A4} = \nu/4 + \alpha_1\pi/2 + \pi/2$; $\varphi_{s1} = \alpha_1\pi + 3\pi/4$; $\varphi_{s2} = 3\alpha_1\pi/2$; $\varphi_{s3} = \pi/4$; $\varphi_{s4} = (\alpha_1 - 1)\pi/2$. The upper lines in (3.453) describe the Alfvén and SMS wave field in the transparent region ($\text{Re}\, z \to \infty$), while the lower lines describe them in the opaque region ($\text{Re}\, z \to -\infty$). By linearising coefficients $\kappa_{0,1,2}$ in (3.445) near $l = l_0$, the WKB solution for Alfvén waves can be represented for ($\text{Re}\, z \to 0$) as

$$H_N \sim z^{1/4-\alpha_1/2} \exp(\pm 2i\sqrt{\alpha_0 z}).$$

It can be seen that the inner asymptotic of the WKB solutions (3.446) matches the outer asymptotic of the first terms in (3.453). Thus the first terms in (3.453) correspond to the Alfvén wave field, and the second terms to the SMS field.

In the neighbourhood of points $l = l_0$, Alfvén waves are partially linearly transformed into SMS waves. There may be several such points in the current sheet region. Note that linear transformation differs from resonant interaction of modes found in Alfvén and magnetosonic resonances. In the current sheet, $c_s \approx v_A$ and the characteristic spatial structure of Alfvén and SMS waves become similar, but the oscillation amplitude fails to increase noticeably.

Linear transformation can occur even in the absence of transformation points. As can be seen from Figure 3.70, coefficient κ_2 passes through zero nowhere on magnetic shell $L = 6$. The small-scale constituent is, however, present in the solutions in Figures 3.66 and 3.67. This is due to the fact that linear transformation takes place even in the case when coefficient κ_2 becomes small enough somewhere on the field line. It can be seen from Figure 3.69 that this takes place in the near-equatorial region, on magnetic shell $L = 6$. It can be assumed therefore that the entire current sheet is where the Alfvén and SMS waves in question are linearly transformed.

The formalism propounded here cannot be used for accurately calculating wave fields because it is impossible to reach the real asymptotic of the functions under analysis, in the calculation region. It does, however, enable us to understand the coupling mechanism of Alfvén and SMS oscillations.

3.17.3 Coupled MHD Mode Structure Across Magnetic Shells

Let us determine the spatial structure of the oscillations, across magnetic shells. It is then necessary to find the solution of the full Eq. (3.418), with given boundary conditions, not only on the ionosphere, but on the asymptotics in the transverse coordinate x^1 as well. Let us formulate these boundary conditions. If a solution is found to (3.442), on some magnetic shell $x^1 = x^1_{PN}$, with the above-defined boundary conditions on the ionosphere, it is only natural to require that the wave field amplitude should be finite for the oscillations in question, away from this shell. Demonstrably, no solutions can be constructed for (3.418) satisfying the boundary conditions on two asymptotics $(x^1 - x^1_{PN}) \to \pm\infty$, simultaneously, in the regions where functions $\Omega_N(x^1)$ monotonously vary for monochromatic eigen oscillations (no outer source). The solution limited on one asymptotic must increase on the other.

If we have a broadband source of Alfvén oscillations with $m \gg 1$, the frequency $\omega = \mathrm{Re}\,(\Omega_N)$ present in the spectrum, this source excites poloidal standing Alfvén wave on magnetic shell $x^1 = x^1_{PN}$. In the current sheet region, Alfvén wave is partially transformed into SMS wave, resulting in coupled MHD mode field along the field line. Magnetic shell $x^1 = x^1_{PN}$ is a turning surface dividing the transparent and opaque regions of such a wave, in the x^1 coordinate. The wave transparent region is located between the poloidal $(x^1 = x^1_{PN})$ and toroidal $(x^1 = x^1_{TN})$ resonance surfaces of the Nth harmonic of standing Alfvén waves (see Figure 3.63). The distance between these surfaces on magnetic shells located in the current sheet region is large enough and most likely makes the excited coupled mode dissipate in the neighbourhood of the poloidal resonance surface (given also the high dissipativity of the constituent SMS wave). We can therefore restrict ourselves to exploring its structure in the neighbourhood of the poloidal resonance surface.

The wave excited on the poloidal resonance surface travels in the transparent region towards the toroidal resonance surface. The $x^1 = x^1_{TN}$ surface is a singular turning surface, the wave is fully absorbed in its neighbourhood. In the transparent region, such a wave travels across magnetic shells, remaining a standing wave, along magnetic field lines. Let us construct a solution describing the structure of (unstable, in the general case) azimuthally small-scale coupled Alfvén and SMS waves, in the neighbourhood of the poloidal resonance surface. In the main order of perturbation theory, the structure of the oscillations, along the magnetic field lines was defined by means of ideal boundary conditions on the ionosphere. Its structure across magnetic shells is described in the next, first, order of perturbation theory. We will take into account its finite conductivity and the presence of external currents (see (2.309)) in the boundary conditions on the ionosphere.

Let us seek the solution of (3.418) for the Nth harmonic of standing coupled modes in this form

$$\varphi_N = U_N(x^1)[H_N(x^1, x^3) + h_N(x^1, x^3)], \tag{3.454}$$

where function $H_N(x^1, x^3)$ describes the longitudinal structure of standing wave in the zeroth approximation, and $h_N(x^1, x^3)$ is the correction of the first approximation. For this function, the boundary conditions on the ionosphere are (see (3.183))

$$h_N(x^1, l_\pm) = \pm i \frac{v_{p\pm}}{\omega} \frac{\partial H_N}{\partial l}\bigg|_{l_\pm} - \frac{J_\parallel^\pm}{V_{p\pm}} U_N^{-1}(x^1). \tag{3.455}$$

In the main order of perturbation theory, the following equation was solved

$$\hat{L}_s(\Omega_N)\hat{L}_P(\Omega_N)H_N + \hat{L}_C(\Omega_N)H_N = 0.$$

Substituting the solution of the form (3.454) into (3.418), we have, in the first order of perturbation theory,

$$\nabla_1^2 U_N(x^1)\hat{L}_T(\Omega_N)H_N - k_2^2 U_N(x^1)\frac{(\omega^2 - \Omega_N^2)}{pv_A^2}H_N - k_2^2 U_N(x^1)\hat{L}_P(\Omega_N)h_N \approx 0.$$

It takes into account the fact that $c_s \ll v_A$ and $|\hat{L}_s(H_N + h_N)| \approx |\Omega_N^2(H_N + h_N)/c_s^2| \gg |\hat{L}_C(\Omega_N)(H_N + h_N)|$ on most of the field line. Left-multiplying this equation by H_N and integrating along the field line between the magneto-conjugated ionospheres produces

$$\beta_N \nabla_1^2 U_N + k_2^2[\alpha_N(\omega^2 - \Omega_N^2) + \overline{\delta}_N]U_N = 0, \tag{3.456}$$

where

$$\alpha_N = \int_{l_-}^{l_+} \frac{H_N^2}{pv_A^2} dl,$$

$$\beta_N = -\int_{l_-}^{l_+} H_N \hat{L}_T(\Omega_N)H_N dl,$$

$$\overline{\delta}_N = \int_{l_-}^{l_+} H_N \hat{L}_P(\Omega_N)h_N dl = -\frac{h_N}{p}\frac{\partial H_N}{\partial l}\bigg|_{l_-}^{l_+}.$$

Given the boundary conditions (3.455), we select the normalisation of the eigenfunctions H_N such that $\alpha_N = 1$ to obtain this equation

$$\nabla_1^2 U_N + \frac{k_y^2}{\omega^2}[(\omega + i\gamma_N)^2 - \Omega_N^2]U_N = I_{\parallel N}, \tag{3.457}$$

describing the structure of standing coupled modes, in the x^1 coordinate, near the poloidal resonance surface. Here, $k_y^2 = k_2^2\omega^2/\beta_N$,

$$\gamma_N = \frac{1}{2\omega^2}\left[\frac{v_{p+}}{p_+}\left(\frac{\partial H_N}{\partial l}\right)_+^2 + \frac{v_{p-}}{p_-}\left(\frac{\partial H_N}{\partial l}\right)_-^2\right]$$

is the decrement of Alfvén waves thanks to the finite conductivity of the ionosphere, near the poloidal surface,

$$I_{\parallel N} = \frac{k_y^2}{\omega^2}\left[\frac{J_\parallel^+}{p_+V_{p+}}\left(\frac{\partial H_N}{\partial l}\right)_+ - \frac{J_\parallel^-}{p_-V_{p-}}\left(\frac{\partial H_N}{\partial l}\right)_-\right]$$

is the function describing the Alfvén wave source related to the external currents in the ionosphere.

Let us write $\Omega_N = \overline{\Omega}_N + i\delta_N$, where $\overline{\Omega}_N \equiv \mathrm{Re}\,(\Omega_N)$, $\delta_N \equiv \mathrm{Im}\,(\Omega_N)$. We will use the following approximate linear expansion, near the resonance surface $x^1 = x^1_{PN}$:

$$\overline{\Omega}_N^2 \approx \omega^2 \left(1 - \frac{x^1 - x^1_{PN}}{a_N}\right),$$

where $a_N = (\nabla_1 \ln \overline{\Omega}_N^2)^{-1}$ is the characteristic $\overline{\Omega}_N$ variation scale at point $x^1 = x^1_{PN}$. Substituting this expansion into (3.457) and switching to the dimensionless transverse coordinate $\xi = (x^1 - x^1_{PN})/\Delta_N$ (where $\Delta_N = a_N^{1/3}/k_y^{2/3}$), it is possible to represent this equation as

$$\frac{\partial^2}{\partial \xi^2} U_N + (\xi + i\varepsilon_N) U_N = \frac{a_N^{2/3}}{k_y^{4/3}} I_{\|N}, \tag{3.458}$$

where $\varepsilon_N = 2(\gamma_N - \delta_N)(k_y a_N)^{2/3}/\omega$. We will assume the localisation scale of the wave source in the ionosphere is much larger than the characteristic transverse wavelength. On the localisation scale of the solution of (3.458), the righthand side can then be assumed practically constant here. The solution of (3.458) that would satisfy the given boundary conditions (finite oscillation amplitude on the asymptotics) is

$$U_N(x^1) = \frac{a_N^{2/3}}{k_y^{4/3}} I_{\|N} G(\xi + i\varepsilon_N), \tag{3.459}$$

where $G(\zeta)$ is the function solving the nonhomogeneous Airy equation (2.88) and having the integral representation (2.89) and asymptotics (2.90).

In the opaque region, the solution (3.459) amplitude decreases in accordance with the power law, while describing a wave travelling away from the resonance surface, in the transparent region. The structure of this solution, across magnetic shells, is shown in Figure 3.71. If the ballooning instability growth rate of the coupled modes is larger than their decrement due to dissipation in the ionosphere ($\varepsilon_N > 0$), the oscillation amplitude increases away from the resonance surface (Figure 3.71a). This only happens of course on those magnetic shells where the condition for the existence of unstable mode is satisfied

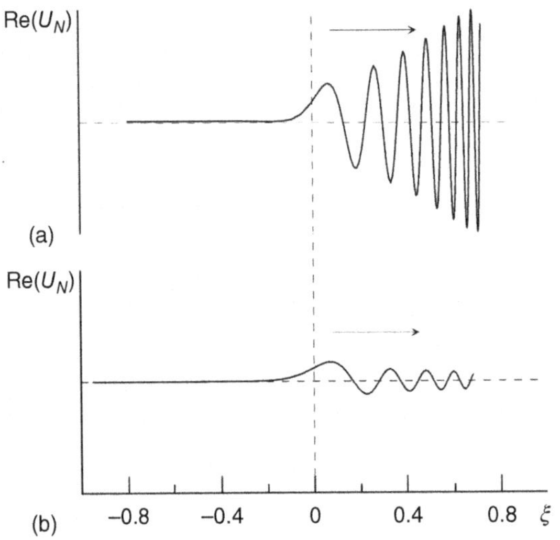

Figure 3.71 Coupled Alfvén and SMS mode structure across magnetic shells, near the resonance surface for poloidal Alfvén waves: (a) oscillations with a growth rate exceeding their decrement due to dissipation in the ionosphere ($\varepsilon_N > 0$); (b) oscillations with a lower instability growth rate than the decrement ($\varepsilon_N < 0$).

(i.e. as long as $|k_2| > |k_1|$). As the wave moves away from the resonance surface, k_1 increases and the wave structure switches somewhere to the $|k_2| < |k_1|$ state. Further away, the wave amplitude decreases until the toroidal resonance surface is reached. If $\varepsilon_N < 0$, the oscillation amplitude starts to decrease as early as the resonance surface (Figure 3.71b).

Let us now discuss the above-obtained results in terms of the energy principle. It was shown in [381] that, in an immobile plasma bounded by ideally conducting walls, the eigen oscillations can exist as either periodic neutral modes with $\omega^2 > 0$ (if the potential energy change is positive, $\Delta W > 0$, when the plasma is disturbed), or aperiodic unstable modes with $\omega^2 < 0$ (if $\Delta W < 0$). The magnetic field lines are assumed to be closed and not to cross the wall bounding the plasma, while the plasma itself is regarded as a dissipation-free medium.

These findings were generalised in [280, 382] for the case of a magnetic field similar to that found in planetary magnetospheres – when magnetic field lines cross the highly conducting ionosphere. It was also assumed, however, that ideally conducting walls are present bounding the plasma volume in question. This of course does not comply with the real situation in the magnetosphere, having a free outer boundary partially permeable to MHD waves. Where ballooning modes strongly localised across magnetic shells are considered, however, the presence of ideally conducting walls far from the oscillation localisation region is regarded as non-essential.

When solving the ballooning mode stability problem, researchers restrict themselves, as a rule, to solving the problem of the oscillation structure along magnetic field lines. The eigenfrequency spectrum of such oscillations is determined as a result of solving the corresponding problem with given boundary conditions on the ionosphere. These frequencies are regarded as the frequencies of eigen oscillations for the entire plasma system under study. The frequencies of Alfvén and SMS eigen oscillations as found in the local approximation in Section 3.16 and some earlier papers perfectly fit into the above-noted concepts. At the same time, the complex frequencies of azimuthally small-scale standing waves obtained in the WKB approximation in Section 3.16 and this section are not consistent with them.

Let us examine the causes of such a divergence using the example of azimuthally small-scale Alfvén waves. The main cause appears to be falsely identifying the harmonics of waves standing along field lines and satisfying the boundary conditions on the ionosphere with the eigen oscillations of the entire plasma system. The fact is that, in the curvilinear magnetic field, Alfvén waves acquire transverse dispersion related to magnetic field line curvature resulting in the splitting of resonance magnetic shells. In the presence of an external source, Alfvén oscillations with toroidal ($m = 0$) and poloidal ($m \to \infty$) polarisation are driven at different magnetic shells.

Our studies of the azimuthally small-scale ($m \gg 1$) Alfvén oscillation structure across magnetic shells have found that it is impossible to construct solutions for the eigenmodes of such oscillations that would satisfy the condition that their amplitudes should be finite simultaneously on two asymptotics, far from the resonance surfaces. This feature of azimuthally small-scale Alfvén waves is not related to the presence/absence of the current sheet or to finite plasma pressure, but is only determined by the magnetic field line curvature. Such solutions can only be successfully constructed for oscillations excited by an external source. External currents in the ionosphere can serve as such a source.

A spatially distributed monochromatic source excites azimuthally small-scale Alfvén wave on the poloidal resonance magnetic shell. This shell is a turning surface for the waves in question, in the transverse coordinate x^1. If a transverse component $k_1(x^1)$ of the wave vector is introduced in the WKB approximation, $k_1^2(x_{PN}^1) = 0$ on the poloidal magnetic shell. The transparent region for the waves is located on one side of the poloidal shell (where $k_1^2(x^1) > 0$), and the opaque region on the other (where $k_1^2(x^1) < 0$). Thanks to transverse dispersion related to magnetic field line curvature, Alfvén waves acquire a transverse (across magnetic shells) component of the group velocity. In the transparent region the wave travels away from the poloidal surface and reaches the toroidal resonance shell, where its energy is fully absorbed thanks to some dissipation mechanism. It could either be dissipation due to the Joule heating of the ionospheric plasma at magnetic field line ends, or refinement of the spatial structure of the oscillations due to other effects resulting in transverse dispersion of Alfvén waves [253], which will eventually result in the full dissipation of the wave.

Thus, azimuthally small-scale Alfvén wave has a complicated spatial structure across magnetic shells and is a strongly dissipative mode of oscillations. It therefore cannot be regarded as the eigenmode of MHD oscillations of the entire plasma system. Such oscillations can exist in the system in question as forced oscillations only, in the presence of an external source. If the ballooning instability emerging in the presence of a current sheet exceeds the local decrement of such oscillations due to their dissipation in the ionosphere, their amplitude increases away from the generation region on the poloidal resonance shell. In the neighbourhood of the toroidal resonance surface, however, the oscillations are fully absorbed.

This example demonstrates that it is not sufficient for ballooning instability investigations to find the solution of only the longitudinal problem describing the oscillation structure along magnetic field lines, and determine their eigenfrequency spectrum. It is also necessary to explore their complete spatial structure, including the wave field structure across magnetic shells. The eigenmodes of the oscillations in the longitudinal problem are not the eigenmodes of the entire plasma system.

Concluding this section, we remind that our theory of the ballooning instability was based on the single fluid MHD theory. Taking into account two fluid and kinetic effects can sufficiently alter this theory [383–389]. Related to the ballooning instability is the interchange (flute) instability [390, 391]. It also demands a sharp pressure gradients, but develops when field lines can freely slide on the ionosphere. So far, the theory of this instability has not been developed to sufficient detail [392–394].

4

MHD Oscillations in 3D-Inhomogeneous Models of the Magnetosphere

4.1 MHD Oscillation Properties in Non-homogeneous Models of the Magnetosphere of Different Dimension

Chapters 2 and 3 present MHD oscillation theory in 1D- and 2D-inhomogeneous models of the magnetosphere. However, the real magnetosphere, is a 3D-inhomogeneous plasma system, and all its features can only be identifiable in 3D-inhomogeneous models. Unfortunately, employing 3D-inhomogeneous models multiply problems in analytical research of MHD oscillations manifold. Models of smaller dimensions are therefore used most frequently, allowing individual features in magnetospheric MHD oscillations to be studied. Thus, 1D-inhomogeneous models allow effects related to the inhomogeneity of magnetospheric plasma to be studied in the radial (across magnetic shells) direction (see [34–37]). 2D-inhomogeneous models, e.g. axisymmetric models, allow the MHD oscillations to be examined in much better detail. Such models allow the oscillation features related to both the radial and longitudinal (along the magnetic field direction) plasma inhomogeneity to be studied, as well as effects of the magnetic field line curvature (see [222, 224, 225, 239, 251, 275, 354]).

Even though efforts to examine MHD oscillations in 3D-inhomogeneous models in as much detail as in models of smaller dimension have so far failed, a number of interesting findings have been obtained in this respect. Thus, it was shown in [395–397] that Alfvén resonance (Alfvén wave driven on the resonance surface by monochromatic fast magnetosonic [FMS] wave) is present not only in a 1D-inhomogeneous plasma but also in 2D- and 3D-inhomogeneous plasma mediums. Note also [398] calculating the spectra of standing Alfvén waves in a 3D model of the Earth's magnetosphere. Those investigations were done for large-scale waves in the azimuthal direction. In the case of an axisymmetric model, this corresponds to azimuthal harmonics of oscillations with $m \sim 1$. It is such oscillations that are often observed in the dayside magnetosphere [237, 348–351, 399].

Azimuthally, small-scale MHD oscillations are, however, also registered onboard spacecraft frequently enough [400–403]. In axisymmetric models, these oscillations correspond to harmonics with azimuthal wave numbers $m \gg 1$. It is therefore important to understand the laws according to which such oscillations propagate in a 3D-inhomogeneous

Magnetospheric MHD Oscillations: A Linear Theory, First Edition. Anatoly Leonovich, Vitalii Mazur, and Dmitri Klimushkin.

plasma too. The spectrum of azimuthally small-scale coupled modes of oscillations was calculated in [29] using a 3D-inhomogeneous model magnetosphere, but the oscillation structure across magnetic field lines was not addressed.

In this chapter, we will examine the features of the propagation of azimuthally small-scale Alfvén waves across magnetic field lines in a 3D-inhomogeneous model magnetosphere. The calculations below define the phase trajectories[1] of azimuthally small-scale standing Alfvén waves propagating in such a model (see [404]). Knowing the phase trajectories of transverse small-scale Alfvén waves, it is possible to find out where the oscillations generated at the poloidal resonance surface will find themselves and be absorbed at the toroidal resonance surface.

Recall that such Alfvén oscillations have the following features in an axisymmetric magnetosphere. A monochromatic source excites poloidal standing Alfvén wave in the magnetosphere, on the poloidal resonance surface, where its frequency coincides with the local frequency of one of the harmonics of standing waves. This wave travels across magnetic shells towards the toroidal resonance surface where it is fully absorbed. The dissipation mechanism is of no importance in this case. In the real magnetosphere, most of the Alfvén oscillation energy is absorbed at the ends of magnetic field lines crossing the ionospheric conducting layer. Between the two resonance surfaces lies the transparent region for transverse small-scale Alfvén waves, the waves not propagating outside of it. The research below is a generalisation of calculating the phase trajectories of transverse small-scale Alfvén waves in Section 3.7 for the case of a 3D-inhomogeneous model magnetosphere.

4.2 Coordinate System

To derive wave equations, it is necessary to set a system of coordinates to be used in further calculations. The natural requirement to such a system is that it must be convenient for describing Alfvén waves in a 3D-inhomogeneous magnetosphere. Such a coordinate system must satisfy two conditions:

(1) magnetic field $\mathbf{B}_0$ lines are lines of coordinates;
(2) magnetic field configuration providing, the system of coordinates must be orthogonal, i.e. the coordinate surfaces $x^1 = c_1$, $x^2 = c_2$ and $x^3 = c_3$ (where $c_{1,2,3} = \text{const}$) form a triorthogonal system.[2]

This choice is justified as follows. First, in the ideal MHD approximation, the electric field of hydromagnetic oscillations has only two transverse components (E_1 and E_2) in such a system. Second, non-diagonal components of the metric tensor vanish if the background magnetic field forms closed magnetic shells and has now shear (this means that longitudinal currents are absent, see [405]). This simplifies analytical calculations substantially.

Let us examine a model magnetosphere allowing a system of coordinates satisfying these conditions. Obviously, the axisymmetric models of the magnetosphere in Chapter 3 allow us

---

1 Phase trajectories are lines at each point of which the direction of the tangent coincides with the direction of the phase velocity of the oscillation propagation.
2 Three families of curved surfaces form a triorthogonal system if any two surfaces of different families intersect at right angles.

to do this. Defining such a coordinate system is based on Dupin's theorem, which is known from differential geometry (see [406, 407]). According to this theorem, lines along which the surfaces of one family of the triorthogonal system intersect with the surfaces of the two other families are the lines of maximum and minimum curvature of these surfaces (see Appendix O). The coordinate lines of the considered coordinate system are thus the curvature lines of the coordinate surfaces. We will select surfaces that are everywhere orthogonal to magnetic field lines as the coordinate surfaces x^3 = const (see Figure N.2 in Appendix O).

The lines of maximum curvature will be the coordinate x^1 lines and the lines of minimum curvature the coordinate x^2 lines.[3] Dupin's theorem implies that field lines are the curvature lines of the surfaces x^1 = const and x^2 = const because they are their intersection lines. The thus constructed system of coordinates satisfies the above-formulated conditions. Note that curvature lines are defined unambiguously, accurate to transformations $x^1 = f(x^1)$, $x^2 = f(x^2)$, $x^3 = f(x^3)$, that do not change the shape of the coordinate lines and surfaces, which is an important element of the following discussion. Simultaneously setting the coordinates x^1 and x^2 defines the field line, and x^3 defines a point on it. The length element ds in this system of coordinates is given by the quadratic form

$$ds^2 = g_1(dx^1)^2 + g_2(dx^2)^2 + g_3(dx^3)^2,$$

where $g_i = g_{ii}(x^1, x^2, x^3)$ are diagonal components of the metric tensor (all other components are zero due to orthogonality). In the particular case of an axisymmetric magnetosphere, the above-introduced coordinate x^1 is a radial coordinate (for example, McIlwain parameter L), and x^2 is an azimuthal coordinate (for example, azimuthal angle ϕ).

A system of coordinates constructed in [408] satisfies condition (1), formulated above. Constant pressure surfaces in it are the coordinate surfaces x^1 = const, and current and field lines are the coordinate lines x^2 and x^3. Obviously, this system also meets condition (2), $\mathbf{B} \cdot \mathbf{J} = 0$, because $\nabla P = [\mathbf{J} \times \mathbf{B}]$. It follows from the previous reasoning that the $x^i = c_i$ surfaces are defined unambiguously, hence the curvature lines of the surfaces x^3 = const must coincide with constant pressure lines and current lines if $\mathbf{J} \neq 0$.

4.3 Basic Equations

Let us now address directly to the theory of transverse small-scale MHD oscillations. We will use the ideal MHD approximation to represent the expression for Alfvén oscillation electric field as

$$\mathbf{E}_\perp = -\nabla_\perp \varphi, \tag{4.1}$$

where $\nabla_\perp = (\nabla_1, \nabla_2)$. Substituting these expressions into (2.69), where the dielectric permeability tensor components, in the ideal MHD approximation, have the form

$$\hat{\varepsilon}_{11} = \hat{\varepsilon}_{22} = \frac{\bar{c}^2}{v_A^2}, \qquad \hat{\varepsilon}_{33} = -\infty$$

(in this case, $\hat{\varepsilon}_{11} = \hat{\varepsilon}_{xx}$, $\hat{\varepsilon}_{22} = \hat{\varepsilon}_{yy}$, $\hat{\varepsilon}_{33} = \hat{\varepsilon}_{zz}$), yielding the following equation for the scalar potential of Alfvén oscillations:

$$\nabla_1 \hat{L}_T(\omega, l) \nabla_1 \varphi + \nabla_2 \hat{L}_P(\omega, l) \nabla_2 \varphi = 0, \tag{4.2}$$

3 The coordinate x^i lines are lines where the two other coordinates, except x^i, are constant.

where $\hat{L}_T(\omega, l)$ and $\hat{L}_P(\omega, l)$ are the operators describing the longitudinal structure of toroidal and poloidal Alfvén waves (see (3.109) and (3.110)). Unlike the equations obtained in Sections 3.1 and 3.7, no axial symmetry is assumed in (4.2) for the medium, and, correspondingly, the derivatives over the azimuthal coordinates, $\nabla_2 = \partial/\partial x^2$, are left. Here, l is the physical length of the field line whose element $dl = \sqrt{g_3}dx^3$, and $p = \sqrt{g_2(x^1, x^2, l)/g_1(x^1, x^2, l)}$ is the parameter playing a key role in the further calculations. We will also assume an ideally conducting ionosphere in the main order of perturbation theory:

$$\varphi|_{l_\pm} = 0, \tag{4.3}$$

where $l_\pm$ are the coordinates of the point where the field line crosses the ionospheres of the Northern and Southern hemispheres.

Of particular importance in the theory in question is the fact that the value of parameter p varies along magnetic field lines. If this were not the case, operators $\hat{L}_P$ and $\hat{L}_T$ would be proportionate to each other. It is the dependence of p on l (necessary condition for magnetic field line curvature, except circular field lines) that determines the propagation of the oscillations across magnetic shells (see more detail in [338]). Since p and v_A are functions of the coordinates, operators $\hat{L}_P$ and $\hat{L}_T$ are also functions of x^1, x^2, and l. The eigenfunctions P_N and T_N of these operators satisfy the boundary conditions on the ionosphere analogous to (4.3)

$$T_N, P_N\, |_{l_\pm} = 0,$$

where N is the number of half-waves fitting into a section of the field line between the magnetoconjugated ionospheres.

Let us also denote the eigenvalues of the frequency of the toroidal and poloidal operators as $\Omega_{TN}(x^1, x^2)$ and $\Omega_{PN}(x^1, x^2)$, respectively. We will call the surfaces defined by the equations

$$\omega = \Omega_{TN}(x^1, x^2), \quad \omega = \Omega_{PN}(x^1, x^2), \tag{4.4}$$

as toroidal and poloidal resonance surfaces, respectively. In the terrestrial magnetosphere, the distance between these surfaces near the equator varies from several hundred to several thousand kilometres. Note that these surfaces coincide (particularly, in a magnetic field with circular field lines) if p does not depend on the longitudinal coordinate.

Let us return to (4.2). We will represent Alfvén wave-related perturbation as

$$\varphi = \exp\left[iQ(x^1, x^2, l)\right].$$

These relations are valid for transverse small-scale waves:

$$\left|\frac{1}{\sqrt{g_1}}\frac{\partial\varphi}{\partial x^1}\right| \gg \left|\frac{\partial\varphi}{\partial l}\right|, \quad \left|\frac{1}{\sqrt{g_2}}\frac{\partial\varphi}{\partial x^2}\right| \gg \left|\frac{\partial\varphi}{\partial l}\right|,$$

which is why the solution to the problem (4.2) and (4.3) can be sought in the Wentzel–Kramers–Brillouin (WKB) approximation, in the coordinates x^1, x^2. In this case, the quasiclassical phase can be represented in the form of the following asymptotic decomposition:

$$Q(x^1, x^2, l) = Q_0(x^1, x^2) + Q_1(x^1, x^2, l) + \cdots,$$

where the main-order term Q_0 is a function of transverse coordinates only. Denoting $\exp(iQ_1) = H$, the perturbation can be represented as

$$\varphi = H(x^1, x^2, l) \exp\left[iQ_0(x^1, x^2)\right]. \tag{4.5}$$

The components of the quasiclassical wave vector have the form

$$k_1(x^1, x^2) = \frac{\partial Q_0}{\partial x^1}, \quad k_2(x^1, x^2) = \frac{\partial Q_0}{\partial x^2}. \tag{4.6}$$

Let us introduce

$$\kappa(x^1, x^2) = k_1/k_2.$$

Substituting (4.6) into (4.2) and (4.3), it is easy to obtain this equation:

$$\left[\kappa^2 \hat{L}_T(\omega) + \hat{L}_P(\omega)\right] H = 0 \tag{4.7}$$

and the corresponding boundary conditions

$$H\,|_{l_\pm} = 0. \tag{4.8}$$

For given x^1, x^2 and ω, these equations can be regarded as an eigenvalue problem for the parameter κ^2. Correspondingly, the dependence of the eigenfunctions $H = H_N(x^1, x^2, l, \omega)$ on l describes the longitudinal structure of standing Alfvén wave. Let us further denote the eigenvalues of κ^2 as $\kappa_N^2(x^1, x^2, \omega)$, i.e. $\kappa = \pm\kappa_N$ (we will select a positive κ_N). A first-order differential equation for phase Q_0 follows from (4.6):

$$\frac{\partial Q_0}{\partial x^1} \mp \kappa_N \frac{\partial Q_0}{\partial x^2} = 0. \tag{4.9}$$

The characteristics of this equation, i.e. lines in the manifold $\{x^1, x^2\}$, along which $Q_0 = \text{const}$, are determined by an ordinary differential equation (see [409])

$$\frac{dx^2}{dx^1} = \pm\kappa_N(x^1, x^2), \tag{4.10}$$

whence

$$k_1 dx^1 + k_2 dx^2 = 0, \tag{4.11}$$

where the differentials dx^1 and dx^2 are taken along the characteristics.

It is easy to understand the physical sense of the constant phase lines. For this we will find the components of the vector of the transverse group velocity for the oscillations in question. Let us introduce functions $\omega_N = \omega_N(x^1, x^2, \kappa)$, inverse to functions $\kappa_N = \kappa_N(x^1, x^2, \omega)$. Differentiating them yields, by definition,

$$v_N^1 = \frac{\partial}{\partial k_1}\omega_N(x^1, x^2, k_1/k_2) = \frac{1}{k_2}\frac{\partial \omega_N}{\partial \kappa},$$

$$v_N^2 = \frac{\partial}{\partial k_2}\omega_N(x^1, x^2, k_1/k_2) = -\frac{k_1}{k_2^2}\frac{\partial \omega_N}{\partial \kappa}.$$

Hence,

$$v_N^1 k_1 + v_N^2 k_2 = 0, \tag{4.12}$$

i.e. the characteristics are lines along which the group velocity vector of transverse small-scale Alfvén waves is directed. It was shown in Section 3.7 that the transverse contravariant components of the Poynting vector integrated over the volume of a flux tube of unit size, in the x^1 and x^2 coordinates, are proportionate to v^1 and v^2. This means that the characteristics are rays in the manifold $\{x^1, x^2\}$ along which the energy flux of the waves in question is transferred. The plus and minus signs in (4.9), (4.10) correspond to waves transporting energy in the opposite directions, in the x^2 coordinate, if their direction in the x^1 coordinate is defined. Let us emphasise again that the waves under study are standing waves along magnetic field lines (in the x^3 coordinate).

4.4 Qualitative Investigation of the Equation for Characteristics

To solve the equation for characteristics, (4.10), it is necessary to solve the problem (4.7) and (4.8), but, due to the complex dependence of the equilibrium parameters on the coordinates, it is better to solve this cumbersome problem numerically. It is, however, not obligatory when all we want is a qualitative picture of the behaviour of the characteristics. Let us make the following comment. Problems (4.7) and (4.8) can also be regarded as eigenvalue problems for ω, for given κ. Let us examine two special cases: $\kappa \to 0$ and $\kappa \to \infty$. In the latter case, (4.7) has the form

$$\hat{L}_P(\omega, l)H = 0.$$

This equation (given the boundary condition ((4.8)) is only valid for $\omega = \Omega_{PN}(x^1, x^2)$. The physical statement of the problem being such that a wave with a given frequency is under study, the conclusion is that κ is zero on the poloidal surface. Correspondingly, if $\kappa \to \infty$, (4.7) is reduced to

$$\hat{L}_T(\omega, l)H = 0.$$

This equality is valid on the toroidal surface.

It follows from (4.10) that, near the poloidal surface, the characteristics are directed along the coordinate surface $x^1 = \text{const}$, and along the surface $x^2 = \text{const}$ near the toroidal surface.[4] It should be emphasised again that the position of the coordinate surfaces is uniquely determined by the magnetic field geometry. Conversely, knowing the behaviour of the characteristics near the resonance surfaces, it is possible to determine the main features in their behaviour in the entire transparent region.

In the case of an axisymmetric magnetosphere, the toroidal and poloidal eigenfrequencies depend on the x^1 coordinate only, and the resonance surfaces are determined by the coordinate lines $x^1 = \text{const}$. This results in a simple enough pattern of energy fluxes: the wave energy is generated on the poloidal resonance surface, flows away from it at a right angle and flows further away towards the toroidal surface, changing its direction in such a manner that the wave fluxes enter the toroidal surface along a line that is tangential to it. The propagation of waves with $m > 0$ as x^2 increases is no different from the propagation of waves with $m < 0$ as x^2 decreases. The corresponding lines of the constant phase (characteristics) have identical shapes (see Figures 3.29 and 4.1). The wave field pattern does not depend on the azimuthal coordinate.

4 What we deal with hereafter are the projections of the coordinate and resonance surfaces onto a two-dimensional manifold of transverse coordinates $\{x^1, x^2\}$, for instance, onto the equatorial surface.

If there is no axial symmetry, however, the picture becomes more complicated. In this case, the poloidal and toroidal eigenfrequencies do not depend on the x^1 coordinate only, but on x^2 as well, and the resonance surfaces therefore do not coincide with the surfaces $x^1 = \text{const}$. The magnetosphere then consists of sectors differing in the sign of the inclination angle μ between the toroidal resonance surface and surface $x^1 = \text{const}$ (see Figure 4.1b).

The importance of the differences between the sectors becomes clear if one recalls that the characteristics near the toroidal resonance surface are tangential to the line $x^1 = \text{const}$. Let us examine the propagation of a wave with $\kappa > 0$ in the magnetospheric sector with $\mu < 0$. It follows from (4.11) that the corresponding characteristics pass in this sector between the resonance surfaces right-to-left so that the inequality $dx^1 > 0$ remains valid along the entire characteristic. These waves gradually change their direction and enter the toroidal surface along the line $x^1 = \text{const}$ (see characteristic *1* in Figure 4.1b).

For waves with $\kappa < 0$, the propagation pattern is completely different. These waves must propagate left to right, because $dx^2 > 0$ for $dx^1 > 0$. A simple analysis shows that, remaining in the $\mu < 0$ sector, waves with $\kappa < 0$ cannot reach the toroidal surface unless the sign of κ changes. This can happen in two cases: (i) at a certain point, κ passes through zero or (ii) a knee forms on the characteristic. It was shown above, however, the κ becomes zero on the characteristic on the poloidal surface only, its position defined uniquely by (4.4). Thus, the wave cannot change its direction because of κ changing its sign. Nor is the knee possible in the characteristics, because such a discontinuity corresponds to (4.10) being singular, which in fact it is not, because its right-hand side is regular.

We thus arrive at the only possible conclusion: a wave with $\kappa < 0$ cannot reach the toroidal surface in the $\mu < 0$ sector. Correspondingly, this wave travels further on into sectors with a different sign of μ, where it can enter the toroidal surface along the coordinate surface $x^1 = \text{const}$ without κ changing its sign in this case. On the other

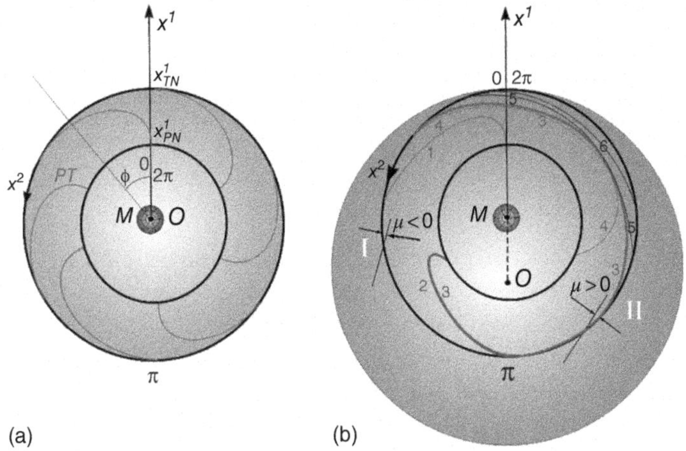

Figure 4.1 (a) Characteristics (energy flux lines) of transverse small-scale Alfvén waves for a fixed sign of κ, in an axisymmetrical model magnetosphere. PT-type characteristics only are present. (b) Characteristics in a 3D-inhomogeneous model magnetosphere. The types of characteristics, *1*– *6*, are described in the text. Asymmetry can be seen in the behaviour of the characteristics in sectors 0°–180° and 180°–360°. Both PT- (numbered *1*– *4*) and TT-type characteristics (numbered *5* and *6*) are present.

hand, characteristics with $\kappa > 0$ must also exist, entering the toroidal surface in the $\mu < 0$ sectors because they are the solutions of (4.10). There is only one possibility for such characteristics: leaving the toroidal surface along the coordinate line x^1 = const in the $\mu > 0$ sector, they can move further into the $\mu < 0$ sector, where they enter the toroidal surface again, tangentially to the coordinate line x^1 = const (see characteristics *5* and *6* in Figure 4.1).

Let us call the characteristics starting off on the poloidal surface and ending on the toroidal surface as 'PT-type' characteristics. It is the only type of characteristics possible in an axisymmetric model magnetosphere. We will call the characteristics starting off from and ending on the toroidal surface as 'TT-type' characteristics. They have no analogue in the axisymmetric case. Due to the above-mentioned reasons, PT- and TT-type characteristics neither overlap nor cross each other.

The same interpretation can also be given for sectors with positive sign of the angle between the toroidal and coordinate surface x^1 = const, $\mu > 0$. Obviously, PT characteristics correspond here to a $\kappa < 0$ wave, while PT characteristics starting off from the given sector and ending in an opposite-sign sector, $\mu < 0$, as well as TT characteristics, correspond to a $\kappa > 0$ wave.

To illustrate the above, Figure 4.1b demonstrates a set of characteristics with $\kappa > 0$ calculated for a model magnetosphere, its coordinate lines being circles, both resonance surfaces being concentric circles (again, the manifold $\{x^2, x^2\}$ is meant here) with a common centre M, which does not coincide with the coordinate origin O. The following model κ is chosen:

$$\kappa_N = \left(\frac{r - r_P}{r_T - r}\right)^{1/2},$$

where r is the coordinate counted from the centre of the resonance surfaces having corresponding values r_T and r_P on the toroidal and poloidal surfaces. Such a behaviour of resonance surfaces corresponds to (4.7), from which equation it is possible to obtain, by means of the perturbation technique:

$$\kappa_N^2 \sim \left(\omega^2 - \Omega_{PN}^2\right)$$

near the poloidal surface and

$$\kappa_N^2 \sim \left(\Omega_{TN}^2 - \omega^2\right)^{-1}$$

near the toroidal surface.

The above-described whimsical pattern of characteristics is shown in Figure 4.1b. Note the qualitative difference of this picture from the one in the left-hand panel of this figure displaying the characteristics in an axisymmetric magnetosphere. The sample lines of constant phase in the right-hand figure are numbered in ascending order with respect to the x^2 coordinate (labelled as angle ϕ in the figure). Characteristic *1*, starting with $\phi = 0$, lies completely in the $\mu < 0$ sector (sector I) and belongs to the PT type. Characteristic *2* (PT type) also starts in sector I and ends on the toroidal surface, at the boundary of sector I, where $\phi = \pi$. All characteristics between characteristics *1* and *2* belong to the PT type. Characteristic *3* (PT type) starts on the poloidal surface just after characteristic *2*. It does not end in sector II (where $\mu > 0$), but passes through the entire sector and ends in sector I.

All subsequent characteristics starting on the poloidal surface between characteristics *3* and *1* end in sector I and belong to the PT type. Characteristic *3* is a separatrix dividing the PT- and TT-type characteristics. Note that despite the fact that characteristics *4* and *5* visually merge with the separatrix in a certain region in the right-hand panel, they are actually located on different sides of it. Characteristic *6* (TT type) passes far from the separatrix. In the real magnetosphere, the number of sectors with different signs of μ can ofcourse be larger than, of course, exceed two and these can have different shapes. The above-discussed example, however, completely illustrates the main features in the behaviour of the characteristics. Numerical studies generally confirm results of analytical investigation presented in this chapter [309, 410–413].

4.5 Wave Singularity in the 3D-Inhomogeneous Magnetosphere

As we have already seen, both 1D- and 2D-inhomogeneous models show the singularity on the toroidal resonant surface: the logarithmic singularity in the azimuthal electric field and in the radial magnetic field and the pole singularity in the radial electric field and in the azimuthal magnetic field. Let us consider the case of the 3D-inhomogeneous model [414].

Our starting point can be the model equation for the azimuthally small-scale Alfvén waves' transverse structure (3.333). Let us rewrite it in the 2D-inhomogeneous model:

$$\frac{\partial}{\partial x^1}\left[\omega^2-\Omega_{TN}^2(x^1)\right]\frac{\partial U_N(x^1,\omega)}{\partial x^1}-k_y^2\left[\omega^2-\Omega_{PN}^2(x^1)\right]U_N(x^1,\omega)=0. \tag{4.13}$$

In the 3D-inhomogeneous model, the toroidal and poloidal eigenfrequencies Ω_{TN}, Ω_{PN} are functions not only the radial coordinate x^1, but the azimuthal coordinate x^2 as well. Moreover, the azimuthal wave number cannot be introduced in the 3D case. Instead, one should revert to the azimuthal derivative with respect to the azimuthal angle ϕ. As the radial coordinate, the L-parameter can be used. Thus, the model equation can be written as

$$\frac{\partial}{\partial L}\left[\omega^2-\Omega_{TN}^2(L,\phi)\right]\frac{\partial U_N(x^1,\phi,\omega)}{\partial L}+\frac{1}{L^2}\frac{\partial}{\partial \phi}\left[\omega^2-\Omega_{PN}^2(L,\phi)\right]\frac{\partial U_N(L,\phi,\omega)}{\partial \phi}=0 \tag{4.14}$$

(see Eq. (20) from [414]).

Due to the azimuthal inhomogeneity, the toroidal and poloidal surfaces $\Omega_{TN}(L,\phi)=\text{const}$, $\Omega_{PN}(L,\phi)=\text{const}$ are inclined with respect to the coordinate surfaces $x^1=\text{const}$ (see Figure 4.1b). Let us denote the corresponding angle α (Figure 4.2). The span between the toroidal and poloidal surfaces can be called the channel as it is the region where the wave energy propagates (see the discussion in Section 4.5). Unless in the vicinity of the extremes where the channel is tangent to the coordinate surfaces $x^1=\text{const}$, the α value is nearly constant. In general, in the 3D case, the channel has variable width Δ and inclination α. The only effect of the 3D inhomogeneity subject to investigation in this paper is a non-zero value of the inclination of the channel.

To solve Eq. (4.14) near some magnetic shell L_0, let us use the local approximation assuming $L=L_0+x_0$, $L_0\varphi=y_0$, where x_0 and y_0 are the new transverse coordinates, and $|x_0|\ll L_0$ (see Figure 4.2). As a result, we obtain

$$\frac{\partial}{\partial x_0}\left(\omega^2-\Omega_{TN}^2\right)\frac{\partial}{\partial x_0}U_N+\frac{\partial}{\partial y_0}\left(\omega^2-\Omega_{PN}^2\right)\frac{\partial}{\partial y_0}U_N=0. \tag{4.15}$$

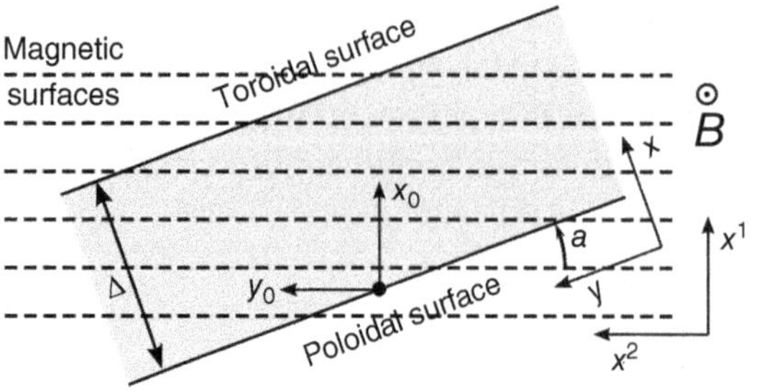

Figure 4.2 The inclination of the channel with respect to the coordinate surfaces.

Then, we rotate the coordinate system at angle α (see Figure 4.2): the new coordinates are $x = x_0 \cos\alpha + y_0 \sin\alpha$ and $y = -x_0 \sin\alpha + y_0 \cos\alpha$. The direction x is normal to the Ω_{TN} or Ω_{PN} gradient. As a result, Ω_{TN} and Ω_{PN} become functions depending on x only. It has the role analogous to the radial direction x^1 in an axisymmetric case, while the tangential (y) direction is analogous to the azimuthal direction x^2. Let us assume the wave to propagate along the y coordinate, then $U_N \propto \exp(-i\omega t + ik_c y)$, where k_c is the component of the wave vector along the channel, and we obtain an ordinary differential equation

$$\begin{aligned}&\frac{\partial}{\partial x}\left(\omega^2 - \Omega_{TN}^2 + \Delta\Omega^2 \sin^2\alpha\right)\frac{\partial}{\partial x}R_N + ik_c\Delta\Omega^2 \sin 2\alpha \frac{\partial}{\partial x}U_N \\ &- k_c^2\left(\omega^2 - \Omega_{PN}^2 - \Delta\Omega^2\sin^2\alpha\right)U_N = 0,\end{aligned} \tag{4.16}$$

where $\Delta\Omega^2 = \Omega_{TN}^2 - \Omega_{PN}^2$.

In the local approximation, the linear approximations $\Omega_{TN}^2 = \Omega_0^2(1 - x/a)$ and $\Omega_{PN}^2 = \Omega_0^2(1 - x/a - \Delta/a)$ can be used. Remind that Δ is the channel width (see Figure 4.2). Then Eq. (4.16) can be transformed into

$$\begin{aligned}&\frac{\partial}{\partial x}\left(x - x_T + \Delta\sin^2\alpha\right)\frac{\partial}{\partial x}R_N + ik_c\Delta \sin 2\alpha \frac{\partial}{\partial x}R_N \\ &- k_c^2\left(x - x_P - \Delta\sin^2\alpha\right)R_N = 0.\end{aligned} \tag{4.17}$$

It is the final equation to solve, where $x_T = a(\Omega_0^2 - \omega^2)/\Omega_0^2$ and $x_P = x_T - \Delta$ are coordinates of the toroidal and poloidal surfaces, respectively. The separatrix mentioned in Section 4.5 has the coordinate

$$x_S = x_T - \Delta\sin^2\alpha. \tag{4.18}$$

Let us solve Eq. (4.17) in the vicinity of the points x_P, x_T and x_S. Assuming the solution of Eq. (4.17) in the form

$$U_N(x) = \tilde{U}(x)\exp\{-ik_c\Delta \sin\alpha \cos\alpha \ln(x - x_S)\} \tag{4.19}$$

we obtain the equation

$$\frac{\partial}{\partial x}(x - x_S)\frac{\partial}{\partial x}\tilde{U} + k_c^2\frac{(x - x_P)(x_T - x)}{x - x_S}\tilde{U} = 0. \tag{4.20}$$

In the vicinity of x_S, Eq. (4.20) is reduced to the form

$$(x - x_S)^2\frac{\partial^2}{\partial x^2}\tilde{U} + (x - x_S)\frac{\partial}{\partial x}\tilde{U} + k_c^2\Delta\sin^2\alpha\cos^2\alpha\,\tilde{U} = 0 \tag{4.21}$$

corresponding to the Euler equation, which has a solution

$$\tilde{U} = C_{S1} \exp\{ik_c\Delta \sin\alpha \cos\alpha \ln(x - x_S)\} + C_{S2} \exp\{-ik_c\Delta \sin\alpha \cos\alpha \ln(x - x_S)\}. \quad (4.22)$$

Taking into account Eq. (4.19), we have the solution of Eq. (4.17) in the vicinity of x_S:

$$U_N = C_{S1} + C_{S2} \exp\{-ik_c\Delta \sin 2\alpha \ \ln(x - x_S)\}. \quad (4.23)$$

In the vicinity of the separatrix surface, there is a singularity at x_S due to the factor $\ln(x - x_S)$ in Eq. (4.23). Near this surface the radial wavelength becomes infinitely small. The wave amplitude remains finite. However, since the logarithm should be analytically continued according to the rule $\ln(x_S - x) = \ln(x - x_S) + i\pi$, the solution must experience jump on the value $\exp(\pi k_c\Delta \sin 2\alpha)$.

In order to calculate the components of the wave's electromagnetic field, one should take spatial derivatives of the U_N function. The E_y and B_x components of the wave behave as U_N,

$$E_y, B_x \propto k_c \exp\{-ik_c\Delta \sin 2\alpha \ln(x - x_S)\}, \quad (4.24)$$

which contrasts the simple logarithmic behaviour in the 1D and 2D cases. The E_x and B_y have the pole singularity $(x - x_S)^{-1}$,

$$E_x, B_y \propto \frac{k_c\Delta}{x - x_S} \exp\{-ik_c\Delta \sin 2\alpha \ln(x - x_S)\}. \quad (4.25)$$

This result is similar to the 1D and 2D cases, but in the 3D case, the singularity is accompanied by oscillations. That is, the separatrix surface is the resonant surface.

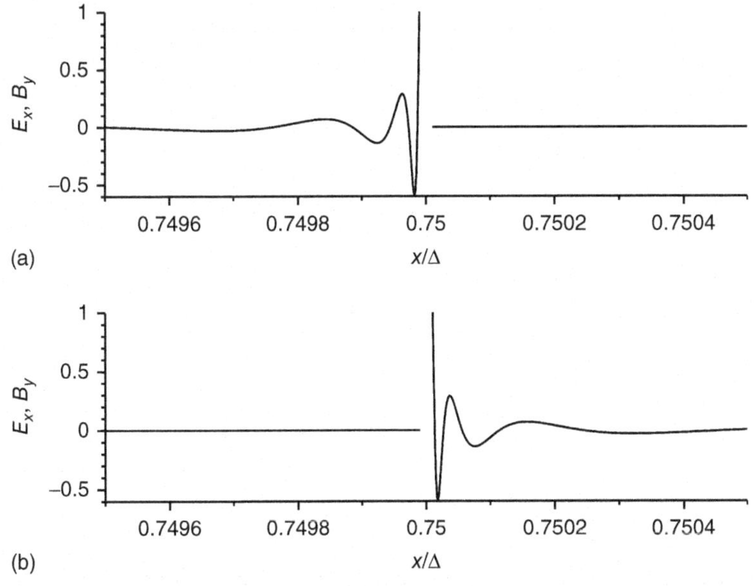

Figure 4.3 The structure transverses the channel of E_x and B_y electromagnetic components of the wave in vicinity of the resonant point x_S (corresponding $x/\Delta = 0.75$) for $k_c > 0$ (a) and $k_c < 0$ (b). The structure was calculated for value $|k_c\Delta| = 5$.

Due to the jump of the amplitude when passing through x_S, the energy of the waves with opposite k_c signs (that is, with different propagation directions along the channel) accumulates on opposite sides of the resonant point x_S (Figure 4.3). It is in accordance with the conclusion based on WKB analysis performed in Section 4.5: the wave propagating up to one side of the separatrix surface cannot leave it on the opposite one.

In vicinities of x_P and x_T, Eq. (4.20) is reduced to the form of Airy equation

$$\frac{\partial^2 \tilde{U}}{\partial z^2} - z\tilde{U} = 0, \tag{4.26}$$

where

$$z = (x_P - x)\left(\frac{k_c^2}{\Delta \cos^2\alpha}\right)^{1/3} \text{ at } x_P, \text{ and } z = (x - x_T)\left(\frac{k_c^2}{\Delta \sin^2\alpha}\right)^{1/3} \text{ at } x_T. \tag{4.27}$$

Airy equation is a differential equation with a turning point where the solution changes from oscillatory to exponential that confirms our assumption based on WKB analysis that the x_P and x_T are turning points. A solution of this equation is

$$\tilde{U} = C_1 Ai(z) + C_2 Bi(z), \tag{4.28}$$

where $Ai(z)$ and $Bi(z)$ are Airy functions. Here, solution with $Ai(z)$ function is suitable only, since $Ai(z)$ provides exponential decrease outside the transparent region. It should be emphasised that the Airy function describes the wave structure near both poloidal and toroidal surfaces. It constitutes yet another difference from the 2D model where the transverse structure was described by a Bessel function with the logarithmic singularity.

5

Conclusion

Let us briefly summarise the basic findings of this monograph. In an inhomogeneous magnetospheric plasma, different branches of MHD oscillations interact generating a complicated pattern of wave fields. The launch of multi-satellite missions over the recent years has enabled a detailed exploration of MHD oscillations generated and propagating in the terrestrial magnetosphere (see [135, 136, 352, 415]). This in turn demands a deeper theoretical investigation into them. A detailed comparison of the theory propounded here to observation data for ultra-low frequency (ULF) waves in the magnetosphere is beyond the scope of this monograph and should be a subject of a special investigation. We can find out, however, which types of MHD oscillations are typical of different parts of the magnetosphere and what their characteristic features are.

It was suggested at the early stages of the theoretical investigations into magnetospheric MHD oscillations that the entire magnetospheric cavity is one gigantic MHD resonator. The resonator properties of the magnetosphere are most conspicuous for Alfvén waves. These waves propagate along magnetic field lines that cross the high-conducting ionosphere twice (in the Northern and Southern hemispheres), the latter being a reflecting surface for them. According to theoretical perceptions, therefore, Alfvén oscillations form standing waves on closed magnetic field lines bounded by the points where they cross the ionosphere [19, 216, 416].

This is true of the lowest-frequency Alfvén oscillations, their wavelength being comparable to the size of the magnetosphere itself. High-frequency Alfvén waves can propagate along geomagnetic field lines as wave packets (which can be considered as a sum of a large number of standing wave harmonics). Geomagnetic pulsations of the Pc1 frequency range are often registered in the form of such wave packets in magneto-conjugated regions of Earth. These oscillations are found to have the highest amplitude near the plasmapause. Their unusual form as recorded on magnetograms (resembling a thread of pearls) has caused them to be called 'pearls' [417].

But what are the generation mechanisms for magnetospheric Alfvén waves? For high-frequency Pc1 oscillations, cyclotron instability of high-energy protons of the radiation belt in the near-equatorial region of the magnetosphere is regarded as such a mechanism (see [418, 419]). This instability is peculiar in that it amplifies quasilongitudinal Alfvén waves only (for which $|k_\parallel| \gg \sqrt{u}|k_\perp|$, where $u = \omega/\omega_i \ll 1$). It was shown in the Introduction section, however, that the transverse vector of Alfvén waves increases

Magnetospheric MHD Oscillations: A Linear Theory, First Edition. Anatoly Leonovich, Vitalii Mazur, and Dmitri Klimushkin.

rapidly in a transverse-inhomogeneous plasma (which is especially important near the plasmapause), and they must quickly exit from the quasilongitudinal propagation regime. The only possibility for such waves to be amplified effectively is to retain the quasilongitudinal propagation regime for many periods of bounce oscillations of wave packets between the ionospheres. This is only possible if Alfvén waves are captured in a waveguide resulting from their transverse dispersion and preventing the transverse component of the wave vector from increasing (see Sections 2.9 and 2.12). Such a waveguide is indeed formed on the outer boundary of the plasmapause, as well as in the magnetospheric plasma filaments detaching themselves from the dusk bulge of the plasmasphere [72, 92, 256].

What is regarded as a generation mechanism for low-frequency Alfvén waves is their resonant interaction (field line resonance [FLR]) with fast magnetosonic waves [34–37]. Note that we are dealing with waves that are large scale in the azimuthal directions (in axisymmetric models, this corresponds to azimuthal harmonics with $m \sim 1$). Transverse inhomogeneity of the magnetospheric plasma is also important for this interaction. Monochromatic fast magnetosonic (FMS) wave propagating in a transverse-inhomogeneous plasma drives Alfvén wave on the magnetic shell where its frequency coincides with the local frequency of Alfvén oscillations.

For a fairly long time, the Alfvén waves examined in 1D inhomogeneous models were regarded as completely different types of magnetospheric oscillations. It was only when 2D inhomogeneous models began to be used for exploring magnetospheric MHD oscillations theoretically that theories developed in 1D inhomogeneous models were found to describe one and the same phenomenon – Alfvén resonance in an inhomogeneous magnetospheric plasma. The structure of resonant Alfvén oscillations along magnetic field lines are standing waves and their structure across magnetic shell as is described by equations analogous to those describing resonant Alfvén oscillations in a transverse-inhomogeneous plasma [57, 223–225].

1D inhomogeneous models successfully applied to describing resonant oscillations in a 2D inhomogeneous plasma are explained by a strong difference between the characteristic scales of these oscillations along magnetic field lines and across magnetic shells. Thanks to this difference, the analytical expression describing the MHD oscillation structure is factorised. Meanwhile, the partial differential equation for Alfvén oscillations in a 2D inhomogeneous plasma can be subdivided into two ordinary differential equations describing their structure along the magnetic field and across magnetic shells (see Section 3.1). Applying the results of Alfvén resonance theory has enabled us to explain the distribution of the mean amplitude and the spectrum of geomagnetic pulsations Pc3 constantly observed in the magnetospheric dayside sector. In particular, the harmonic structure and dependence of the recorded oscillations on magnetic shell location have been successfully explained – the closer to the magnetopause, the lower the frequency of each harmonic.

The situation is analogous for SMS waves. They can also be driven at resonance magnetic shells by FMS oscillations propagating in the magnetosphere [51, 297], or generated by currents in the ionosphere as the solar terminator passes through it [28, 301]. The main difference between resonant SMS oscillations and Alfvén oscillations is the strong dissipativity of the former. SMS waves decay because of interaction with background plasma ions (see Chapter 1).

This feature of SMS waves can serve as a mechanism for transferring the momentum from the solar wind into the magnetosphere. FMS waves penetrating into the magnetosphere from the solar wind excite resonant SMS waves in the region adjoining the magnetopause. Thanks to their high dissipation rate due to Landau damping, the energy and momentum of SMS waves is efficiently transferred to background plasma ions on resonance magnetic shells. This can result in cells with inverse convection of plasma in magnetospheric regions adjoining the magnetopause [61]. That is what can indeed be observed in magnetospheric regions adjoining the magnetopause (and high-latitude ionospheric regions that are magneto-conjugated to them) during periods when the interplanetary magnetic field has a 'north' component and no magnetic reconnection takes place at the magnetopause [420].

The question arises what are the sources of FMS oscillations in the magnetosphere. The solar wind flow shear instability (Kelvin–Helmholtz instability) at the magnetopause is regarded as one such source [97, 117]. This instability results in surface FMS wave driven at the magnetopause. Decreasing in amplitude, the field of this wave penetrates into the magnetosphere, where it can drive Alfvén waves on resonance magnetic shells. Until recently, however, no magnetospheric MHD oscillations had been directly observed that could be identified with such unstable modes. It was only recently that the THEMIS and Double Star satellites registered oscillations that could be regarded as Kelvin–Helmholtz instability-driven surface FMS waves [421]. The amplitude of the oscillations decreases away from the magnetopause and increases in the direction they propagate into the geotail.

Analogous observations are cited in [317], where the propagation velocity of unstable oscillations (~200 km/s) coincides with the expected propagation velocity of surface FMS wave. Moreover, one of the satellites (THEMIS 5), located inside the magnetosphere, registered a narrow-localised amplitude jump in the oscillations identified as resonant Alfvén waves driven by the FLR mechanism. This can be regarded as the first direct evidence of magnetospheric Alfvén resonance (FLR) related to the Kelvin–Helmholtz instability at the magnetopause.

Most theoretical studies of shear flow instabilities at the magnetopause rely on models involving a tangential discontinuity of the parameters. In such models, the parameters change jump-wise when passing from one homogeneous halfspace with an immobile plasma to another, where the plasma is mobile. In the magnetosphere, however, the plasma is strongly inhomogeneous. Thanks to this, studies appeared using models where the plasma parameters change in the direction that is transverse to the shear layer, in the halfspace simulating the magnetosphere. This inhomogeneity is simplest to simulate by a reflecting wall present at a distance from the shear layer [118, 120, 149].

An important feature of such models is a waveguide for fast magnetosonic waves travelling between the reflecting wall and the magnetopause [146]. The magnetopause being a boundary that is only partially reflecting, the main harmonic in such a waveguide is a quarter-wave, i.e. its frequency is lower than in a waveguide with two ideally reflecting walls. Its frequency (~1 mHz) is located in the range of 'magic frequency' spectrum (stable oscillations with a discrete spectrum, in the ~1 mHz range, observable in the terrestrial magnetosphere). FMS waves have a lower threshold with respect to the value of the plasma velocity jump at the magnetopause, which when exceeded causes a Kelvin–Helmholtz instability to develop. Under certain conditions, however, this threshold disappears (see Sections 2.15, 2.16, and [144].

A more complex variety are models with monotonous variation of parameters in the halfspaces describing the magnetosphere [422]. Such models are much better at describing the plasma parameter distribution in the real magnetosphere. In this case, the role of the reflecting wall for FMS waves is played by the reflecting surfaces dividing their transparent and opaque regions inside the magnetosphere. In the opaque region located behind the turning point, a resonance surface appears for Alfvén wave, where the energy of FMS wave propagating in the magnetospheric waveguide is partially absorbed [156].

Such a waveguide can be excited by FMS wave falling from the solar wind onto the magnetopause. The wave reflection coefficient in this case has ostensible peaks at frequencies corresponding to the eigenmode frequencies of the magnetospheric waveguide [79]. The maximum values of the reflection coefficient can exceed unity. In this case, the wave reflected from the shear layer has a larger amplitude than the incident wave. The amplitude is increased thanks to the energy transferred by the plasma shear flow to the wave, while the wave itself is referred to as a wave with negative energy [146, 152, 154, 422].

The waveguide limits the propagation of FMS waves in the direction across magnetic shells but does not hinder their propagation in other directions. The magnetospheric plasma being inhomogeneous in all directions; however, the transparent region for FMS waves, under certain conditions, is bounded on all sides by reflection surfaces. An FMS resonator forms in the process with a discrete eigenfrequency spectrum. Earlier studies regarded the entire magnetospheric cavity as a global FMS resonator bounded by a sharp boundary, the magnetopause (see [238, 303]). The eigenfrequencies of such a resonator were called 'global' modes of the magnetosphere. The characteristic frequencies of such oscillations were calculated in a dipole model magnetosphere [226, 234]. The frequency of the basic tone of oscillations in such a resonator is approximately 10 mHz. The region occupied by such a resonator was found to contain, not the entire magnetospheric cavity, but only its near-equatorial part which adjoins the magnetopause. The conditions for it to exist are only realised in the dayside magnetosphere, which is why its eigenmodes can hardly be regarded as global magnetospheric oscillations. That there exists another FMS resonator, with the magnetopause as its boundary, was first proved in [235, 236]. Its eigen oscillation frequencies are 30 mHz and higher.

Attempts have been made over a fairly long period of time to discover magnetospheric oscillations with a discrete frequency spectrum in the above-mentioned ranges. Unfortunately, attempts to reliably prove the existence of such oscillations experimentally have been unsuccessful. However, magnetospheric oscillations with a discrete spectrum have been discovered that are very similar to them, but exhibit lower frequencies, first recorded by ground-based low-frequency radar and magnetometer networks [304, 305]. The reproducibility of the frequency spectrum (0.8, 1.3, 1.9, 2.6 … mHz) during different observation sessions and the stability of the frequencies for each observation made them to be called 'magic' frequencies. Such oscillations are as a rule recorded in the midnight-morning sector of the magnetosphere, at latitudes 60°–80°. Oscillations with analogous spectral characteristics have recently been found by satellites in the neighbourhood of the dayside magnetopause [423, 424] and even in the solar wind region [425].

Several theoretical concepts have been proposed to explain the nature of these oscillations. It was suggested in [305] that the observable oscillations with a discrete frequency spectrum are the eigenmodes of an FMS waveguide in the geotail lobes. Certain difficulties are encountered, however, when using this approach. The main is the fact that the recorded oscillations have a polarisation that is characteristic of standing waves, not waves propagating in a waveguide [76].

A mechanism is proposed in [318, 425] of oscillations with a 'magic' frequency spectrum directly penetrating from the solar wind into the magnetosphere. These oscillations are regarded in [423, 424] as the eigenmodes of Alfvén oscillations excited on the magnetopause by solar wind pressure pulses related to the solar wind inhomogeneities. Note the intrinsic difficulties of the two concepts. The dayside magnetosphere being an opaque region for the oscillations in question, the amplitude of such oscillations decreases exponentially, from the magnetopause into the magnetosphere (see Section 3.5 and [233]). It is therefore difficult to explain why such oscillations are present in the midnight-morning sector of the internal magnetosphere, where they were first discovered.

A model magnetospheric resonator was proposed in [159, 160] (see also Section 3.6.3 of this monograph) that explained most features of the registered oscillations of the 'magic' frequency spectrum. This resonator for FMS waves forms in the near-Earth part of the plasma sheet, having a global minimum in the Alfvén speed distribution in the magnetosphere. The calculated frequency spectrum of such a resonator coincides closely enough with the spectrum of the observed 'magic' frequencies, which, it should be specially noted, are not equidistant. The resonator localisation region as mapped (along the geomagnetic field lines) onto the ionosphere coincides with the region where the oscillations are registered (the magnetospheric midnight-morning sector, latitudes 60°–80°). The resonator is only formed under conditions of a weakly perturbed magnetosphere, when the plasma sheet parameters are approximately identical. This explains why the spectrum of the observed frequencies is stable.

The following can be concluded based on the above. Several isolated resonators for FMS waves are formed in the terrestrial magnetosphere. The largest-scale and lowest-frequency resonator is in the near-Earth part of the plasma sheet. Even it occupies a relatively small portion of the magnetosphere however. This means that it would be better to completely reject the concept of global modes of magnetospheric cavity oscillations and stop using it any further.

Of the MHD oscillations of the magnetosphere, the azimuthally small-scale Alfvén waves (in the axisymmetric models of the magnetosphere they are waves with large azimuthal wave numbers $m \gg 1$) have the most complex structure. These oscillations also form standing waves between the magnetoconjugated ionospheres of the Northern and Southern hemispheres, in the direction along the magnetic field, on closed field lines. The structure of the monochromatic Alfvén wave with $m \gg 1$ strongly depends on the Alfvén speed profile across magnetic shells. The Alfvén wave structure in the region with a monotonous distribution of Alfvén speed is as follows. The monochromatic source (for example, external currents in the ionosphere) generates a standing Alfvén wave with poloidal polarisation, on the poloidal resonance surface. This wave travels away from the poloidal surface across magnetic shells towards the other, toroidal, resonance surface, where it is fully absorbed due to dissipation in the ionospheric conducting layer. In the process of this displacement, wave polarisation changes from poloidal to toroidal (see Section 3.7 and [251]). Waves with such a spatial-temporal structure were recently registered when one of the Van Allen Probe spacecrafts crossed the plasmapause [426, 427].

In the region where the Alfvén speed profile has a minimum, a resonator forms for poloidal standing Alfvén waves [252]. It is only Alfvén waves confined in such a resonator that are capable of keeping their poloidal polarisation for many oscillation periods. Non-monotonous distribution of plasma parameters is typical of the inner boundary of the plasmapause. It is there that satellites most frequently register poloidal Alfvén waves (see [428]).

ULF oscillations have some peculiarities in the geotail. The current and plasma sheets that are present there leave their imprint on the structure and spectra of Alfvén waves (see [15, 429, 430]). During geomagnetic substorms, the current sheet suffers a disruption [431], resulting in an FMS wave pulse which is transformed, on resonance magnetic shells, into Alfvén waves [432]. The Alfvén wave generated in such a process looks like a pulse of field-aligned currents, which is registered in the magnetosphere and on Earth as substorm geomagnetic Pi2 pulsations [433, 434]. Alfvén waves generated in this and other similar magnetospheric processes can result in accelerating charged particles into the auroral ionosphere and structured auroras emerging there [46, 435].

Similar to Alfvén waves, slow magnetosonic (SMS) waves travel nearly along the geomagnetic field lines. On magnetic shells crossing the geotail current sheet, the Alfvén and SMS waves can interact forming a special type of oscillations – coupled modes (see [274, 275, 279] and Section 3.17 of this monograph). If magnetic field lines are curved and a background plasma pressure gradient is directed outside, such coupled oscillations can become unstable [277, 280, 281]. Such a (ballooning) instability develops under large enough magnetic field line curvature and plasma pressure gradient that are reached in the geotail current sheet, under nearly substorm conditions (see Sections 3.16 and 3.17). The ballooning instability is supposed to result in reconnection of magnetic field lines and disruption in the geotail current sheet, at the initial stage of magnetospheric substorms [385, 436–438].

Until recently, researching coupled modes has chiefly been a subject of theoretical studies because they are rather hard to distinguish among other MHD-oscillation modes, in observations. Multi-satellite missions such as THEMIS, Cluster, Double Star coming into operation have enabled oscillations registered at different, closely located points in space to be analysed. This made it possible to identify various modes of MHD oscillations and analyse their mutual relations.

Low-frequency MHD oscillations observed near the equator by the Geotail satellite, immediately after a substorm onset, were analysed in detail in [437]. It was demonstrated that the FMS mode of oscillations can be identified confidently enough in observations, but the Alfvén and SMS waves are hard to separate. SMS wave with a period of 30 seconds related to periodic reconnections in the geotail was observed in [439] by the Cluster satellites. Those observations can be regarded as the first evidence of unstable coupled MHD modes.

In [440], the time history of events and macroscale interactions during substorms (THEMIS) satellites first discovered simultaneous Alfvén and SMS oscillations of the same frequency range (period ~100 seconds) near the current sheet (distance from Earth, $\sim 11R_E$). No substantial plasma pressure gradient was observed, meaning the oscillations in question were unlikely to be unstable. Phenomena related to such oscillations are also observed on Earth. They frequently manifest themselves as structured aurorae [441]. Wave-like structures with azimuthal wave number $m \approx 76$ and period $T \sim 120$ seconds, characteristic of coupled MHD modes, were observed, at the substorm growth stage, on images recorded by an all-sky camera located in Dawson City (magnetic latitude 65,7°), in [442].

Their spatial structure features can be used to identify coupled Alfvén and SMS oscillations in observations. The full spatial structure of such oscillations was examined in [289, 443] on magnetic shells with closed magnetic field lines extending into the tail. External currents in the ionosphere were discussed as the oscillation source. Thanks to the high conductivity of the ionosphere, these oscillations are waves standing along the geomagnetic field. Sharp, narrow-localised peaks should appear in the oscillation amplitude distribution, on magnetic field lines extending into the tail, at their inflexion points. Each such field line has four inflexion points: two near the current sheet, on both sides, and two others in the transition region between the dipole magnetic field and the field with magnetic field lines elongated by the tail current.

In the direction across magnetic shells, these coupled modes can propagate in the transparent region located between the poloidal and toroidal resonance surfaces for Alfvén waves. Opaque regions are located outside this region, where the oscillation amplitude decreases. In the transparent region, coupled modes, same as azimuthally small-scale Alfvén waves, travel away from the poloidal resonance surface towards the toroidal resonance surface. Near the toroidal surface, they are fully absorbed thanks to the finite ionospheric conductivity.

The real magnetosphere is a 3D inhomogeneous medium. Such a medium has its own peculiarities as to Alfvén wave propagation. There are sectors in a 3D inhomogeneous magnetosphere where transverse-small-scale Alfvén waves with different signs of the azimuthal wave number propagate differently. Wave propagation in sectors with the same sign of the transverse phase speed and the Alfvén speed inclination angle to magnetic shells is, qualitatively, similar to that in an axisymmetric magnetosphere. Waves are generated on the poloidal surface, travel towards the toroidal surface, where they are absorbed in the same sector of the magnetosphere. Speaking of waves with the opposite sign of the phase speed, they travel a significantly large distance before being absorbed near the toroidal surface, in a sector where its inclination angle to the coordinate surface, determined by the magnetic field geometry, also has the opposite sign. In contrast to an axisymmetric magnetosphere, transverse-small-scale Alfvén waves in a 3D inhomogeneous magnetosphere can be generated, not only on the poloidal resonance surface but also on the toroidal resonance surface.

Note the main achievements of and problems yet to be resolved in the analytical theory of magnetospheric MHD oscillations. All the main features of MHD oscillations appear to have been explored in 1D inhomogeneous models. The main achievement of these models is that they make it simple enough to study the resonant interaction and transformation of various modes of MHD oscillations in a homogeneous plasma. The spatial structure of resonant Alfvén and SMS waves has been examined. It has so far been possible to examine analytically the instability of solar wind plasma shear flow at the magnetopause in such models only. Developing the theory using simple 1D inhomogeneous models appears to come to an end.

What can be listed as one of the main achievements of MHD oscillation theory in 2D inhomogeneous models of the magnetosphere is synthesised investigations of the spatial structure of guided modes of the MHD oscillations (Alfvén and SMS waves) along magnetic field lines and across magnetic shells, obtained earlier by means of simple, 1D inhomogeneous models. 2D inhomogeneous, axisymmetric models have been successfully used to construct a detailed picture of FMS resonators in the terrestrial magnetosphere and explore their eigen oscillations. The magnetosphere has been found not to be a global MHD resonator, but to consist of several large-scale resonators for FMS waves [160, 233, 234].

It is only in 2D inhomogeneous models of the geotail that coupling is observed between Alfvén and SMS oscillations, which become unstable under certain conditions. These unstable modes of MHD oscillations can play a significant role in magnetospheric substorm evolution mechanisms. We believe that the main gap in the analytical investigations using axisymmetric models concerns the absence of Kelvin–Helmholtz instability theory for the magnetopause. Recent works [444, 445], which investigated the Kelvin–Helmholtz instability at the dipole magnetosphere boundary wrapped by an azimuthal solar wind flow, can be considered successful partially only. Self-consistent analytical models of an equilibrium dipole magnetosphere with a magnetopause in the form of both a tangential discontinuity [444] and a smooth transition layer [445] are developed in them. However, in the problem on magnetopause stability, within these models framework, oversimplified methods were used for solving of MHD equations describing the magnetopause oscillations. The surface wave growth rates calculated in these models were found too small to take them into account in studying the magnetopause dynamics. This problem is imminent and awaits a solution. Considerable advances are also possible in all other research directions.

Finally, what remains effectively a blank spot is investigating MHD oscillations by means of 3D inhomogeneous models of the magnetosphere. All the main achievements in this area are at present related to numerical simulations of oscillations using such models. As has been noted above, however, numerical simulation cannot replace analytical investigations capable of providing some understanding of the full picture of wave fields in the magnetosphere. It is only analytical research that can reveal all relations between various oscillation modes in a non-homogeneous plasma. From this standpoint, theory has just half-opened the door into a vast area of investigations into MHD oscillations of a 3D inhomogeneous plasma. The main theoretical achievements are yet to be made here.

Appendixes

A Transverse Dispersion of MHD Waves in a 'Cold' and 'Hot' Plasma

The solution of the dispersion Eq. (1.21) can be written as

$$q^2 = \frac{1}{2\nu}\left[(1+\nu)(\alpha-1) \pm \sqrt{(1-\nu)^2(\alpha-1)^2 + 4u^2\nu\alpha}\right].$$

It describes two branches of MHD oscillations, Alfvén and fast magnetosonic (FMS) waves, Figure A.1 providing a general idea of their behaviour. Points $\alpha_{01} = 1 - u$ and $\alpha_{02} = 1 + u$, where $q^2(\alpha) = 0$, determine the transition from the state when the waves freely propagate in the plasma across magnetic field lines ($q^2 > 0$) to oscillations for which the plasma is opaque ($q^2 < 0$). Figure A.1(a) illustrates the behaviour of the oscillations in the case of a 'cold' dispersion of Alfvén waves ($\nu = -\mu^4 < 0$). In this case, function $q^2(\alpha)$ has branchpoints on the real axis $\alpha_{1,2} = 1 \mp 2u\mu^2$ between which $q^2(\alpha)$ is complex. Analytical continuation through these points sees the FMS branch transforming into a large-scale Alfvén branch, A1, its transverse dispersion determined by (1.23). The small-scale branch of Alfvén oscillations, A2, its dispersion determined by (1.24) for $\beta \ll m_e/m_i$, transforms into branch A3, for which the plasma is opaque. If 'hot' dispersion of Alfvén waves is dominant ($\nu = \varkappa^4 > 0$), the branchpoints of function $q^2(\alpha)$ lie on a complex plane, the function having no singularities on the real axis (see Figure A.1(a)).

B Deriving an Equation for MHD Oscillations in a 1D-Inhomogeneous Moving Plasma

Let us write out (1.5) componentwise, assuming the plasma to move along the z axis (directed along $\mathbf{B}_0$) at velocity $v_0(x)$ ($\mathbf{v}_0 = (0,0,v_0(x))$). In a plasma inhomogeneous in the x coordinate, for oscillations of the form $\Phi = \widetilde{\Phi}(x)e^{(ik_y y + ik_z z - i\omega t)}$ (where Φ is any of the wave field components) we have the following:

$$-i\bar{\omega}\rho_0 v_x = -\nabla_x P + \frac{B_0}{4\pi}(ik_z B_x - \nabla_x B_z) - \frac{B_z}{4\pi}\nabla_x B_0, \tag{B.1}$$

$$-i\bar{\omega}\rho_0 v_y = -ik_y P - i\frac{B_0}{4\pi}(k_y B_z - k_z B_y), \tag{B.2}$$

Magnetospheric MHD Oscillations: A Linear Theory, First Edition. Anatoly Leonovich, Vitalii Mazur, and Dmitri Klimushkin.

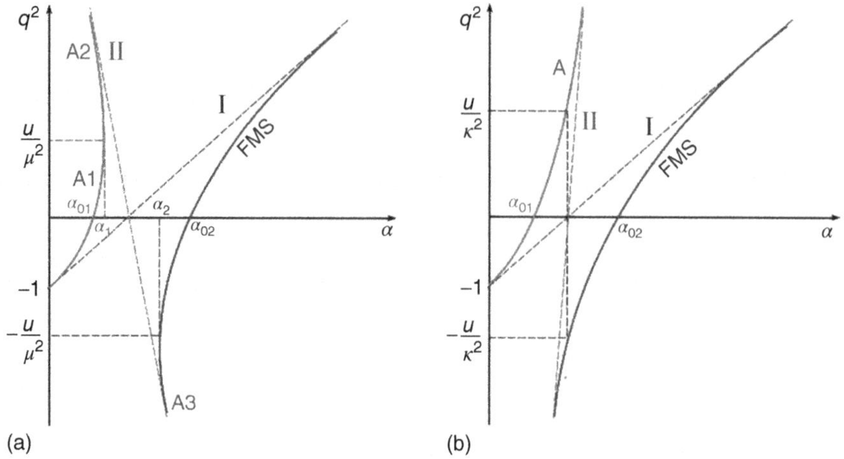

Figure A.1 Dependence $q^2(\alpha)$ for: (*a*) the 'cold' dispersion of Alfvén waves ($\beta \ll m_e/m_i$). Asymptote equations: (I) $q^2 = \alpha - 1$ and (II) $q^2 = -(\alpha - 1)/\mu^4$; (*b*) the 'hot' dispersion of Alfvén waves ($\beta \ll m_e/m_i$), asymptote equations: (I) $q^2 = \alpha - 1$ and (II) $q^2 = (\alpha - 1)/x^4$.

$$-i\bar{\omega}\rho_0 v_z = -\rho_0 v_x \nabla_x v_0 - ik_z P + \frac{B_x}{4\pi}\nabla_x B_0, \tag{B.3}$$

$$-i\bar{\omega}B_x = ik_z B_0 v_x, \tag{B.4}$$

$$-i\bar{\omega}B_y = ik_z B_0 v_y, \tag{B.5}$$

$$-i\bar{\omega}B_z = \nabla_x(v_0 B_x - v_x B_0) + ik_y(v_0 B_y - v_y B_0), \tag{B.6}$$

$$-i\frac{\bar{\omega}}{\rho_0}P = -v_s^2\left(\nabla_x v_x + ik_y v_y + ik_z v_z\right) - v_x \frac{1}{\rho_0}\frac{dP_0}{dx}, \tag{B.7}$$

where k_y and k_z are the wave vector components in y and z, ω is the oscillation frequency, $v_s = \sqrt{\gamma P_0/\rho_0}$ is the sound speed in plasma. We have, from (B.4) and (B.5):

$$B_x = -\frac{k_z B_0}{\bar{\omega}} v_x, \tag{B.8}$$

$$B_y = -\frac{k_z B_0}{\bar{\omega}} v_y. \tag{B.9}$$

Substituting (B.8) and (B.9) into (B.2) and (B.6) yields

$$v_y K_A^2 = \frac{k_y}{\rho_0\bar{\omega}}P + \frac{k_y B_0}{4\pi\rho_0\bar{\omega}}B_z,$$

$$B_z = -i\nabla_x\left(\frac{B_0}{\bar{\omega}}v_x\right) + \frac{k_y B_0}{\bar{\omega}}v_y,$$

where $K_A^2 = 1 - k_z^2 v_A^2/\bar{\omega}^2$, $v_A = B_0/\sqrt{4\pi\rho_0}$ is the Alfvén speed. Thus,

$$v_y = k_y K_s^{-2}\left[-i\frac{v_A^2}{\bar{\omega}}\nabla_x(\frac{1}{\bar{\omega}}v_x) - i\frac{B_0\nabla_x(B_0)}{4\pi\rho_0\bar{\omega}^2}v_x + \frac{1}{\rho_0\bar{\omega}}P\right], \tag{B.10}$$

$$B_z = -i\frac{K_A^2}{K_s^2}\nabla_x(\frac{B_0}{\bar{\omega}}v_x) + \frac{k_y^2 B_0}{\bar{\omega}^2\rho_0 K_s^2}P, \tag{B.11}$$

where $K_s^2 = K_A^2 - k_y^2 v_A^2/\bar{\omega}^2$. Expressing v_z by means of (B.3) and substituting the resulting expression and (B.10) into (B.7) produces

$$P = -i\frac{K_A^2 \rho_0 v_s^2}{\chi_s^2 \bar{\omega}}\left[\nabla_x v_x + \left(\frac{k_z \nabla_x (v_0)}{\bar{\omega}} + (1 - K_s^2(1 + \frac{v_A^2}{v_s^2 K_A^2}))\nabla_x (\ln B_0)\right) v_x\right], \tag{B.12}$$

where $\chi_s^2 = 1 - (k_z^2 + k_y^2)(v_A^2 + v_s^2 - k_z^2 v_A^2 v_s^2/\bar{\omega}^2)/\bar{\omega}^2$. Expressing now the speed component v_x in terms of the plasma shift ζ ($v_x = \mathrm{d}\zeta/\mathrm{d}t = \partial\zeta/\partial t + (\mathbf{v}_0\nabla)\zeta = -i\bar{\omega}\zeta$) and substituting the former into the above-obtained expressions for the wave field components yields

$$v_x = -i\bar{\omega}\zeta, \quad v_y = -\frac{1}{K_s^2}\left(v_A^2 + \frac{K_A^2 v_s^2}{\chi_s^2}\right)\frac{k_y}{\bar{\omega}}\frac{\partial\zeta}{\partial x}, \tag{B.13}$$

$$v_z = -\frac{k_z K_A^2 v_s^2}{\bar{\omega}\chi_s^2}\frac{\partial\zeta}{\partial x} - \zeta\frac{dv_0}{dx},$$

$$B_x = ik_z B_0 \zeta, \quad B_y = -\frac{k_z B_0}{\bar{\omega}} v_y, \tag{B.14}$$

$$B_z = -\frac{K_A^2 B_0}{\chi_s^2}\left(1 - \frac{k_z^2 v_s^2}{\bar{\omega}^2}\right)\frac{\partial\zeta}{\partial x} - \zeta\frac{dB_0}{dx},$$

$$P = -\gamma P_0 \frac{K_A^2}{\chi_s^2}\frac{\partial\zeta}{\partial x} + \zeta\frac{d}{dx}\left(\frac{B_0^2}{8\pi}\right), \tag{B.15}$$

Employing also the frozen-in condition (1.8) for the components of the perturbed electric field of the oscillations, we have

$$E_x = \frac{k_y \omega\, B_0}{\bar{c}\bar{\omega}^2\, K_s^2}\left(v_A^2 + \frac{K_A^2 v_s^2}{\chi_s^2}\right)\frac{\partial\zeta}{\partial x}, \quad E_y = -i\frac{\omega B_0}{\bar{c}}\zeta, \quad E_z = 0. \tag{B.16}$$

Given (B.8) and (B.11), Eq. (B.1) can be written as

$$\frac{1}{\bar{\omega}^2 \rho_0}\frac{\partial}{\partial x}\left(P + \frac{B_0 B_z}{4\pi}\right) - K_A^2 \zeta = 0, \tag{B.17}$$

where the bracketed expression $(P + B_0 B_z/4\pi)$ is full perturbed pressure. Substituting P and B_z from (B.14) and (B.15) into this expression will produce the basic Eq. (2.4).

C A Model of the Spectral Function of the Solar Wind FMS Oscillations

There are currently many spacecraft-based observations of oscillations in the solar wind parameters. It was demonstrated in [103, 104, 446] that solar wind oscillations have the character of a stochastic 'white noise', their correlation functions being a δ-function of frequencies and wave vectors.

The literature cites two types of statistical observations of the spectra of such oscillations. A number of papers present the oscillation energy density W spectra as functions of their frequency ω. According to these data (see [103]), in most of the frequency range, $W(\omega)$ can be approximated by an exponential function of the form $W \sim \omega^{-\alpha}$, where the index lies within $1 \lesssim \alpha \lesssim 3$. The mean value of α is close to $\alpha = 5/3$, which corresponds to the Kolmogorov spectrum for fully-developed turbulent oscillations. Below the frequency $\omega = \omega_i$, where ω_i is the solar wind ion gyrofrequency, the spectrum $W(\omega)$ is abruptly cut off thanks to the resonant ion-cyclotron absorption of waves.

Papers by other authors (see [446]) provide data on the spectrum of the solar wind plasma concentration oscillations depending on wavelength, i.e. function $\bar{\rho}(k_t)$ effectively. The overbar means a certain averaged value, because turbulent oscillations are concerned. Denoting a separate Fourier harmonic of the oscillation density as $\rho(\mathbf{k}_t, \omega)$ we have

$$\bar{\rho}(k_t) = \frac{2}{\pi}\left[\int_0^\infty \rho^2(k_t, \theta, \omega)d\omega\right]^{1/2}.$$

Observations [446] imply that $\bar{\rho}(k_t)$ could be approximated by a function of the form $\bar{\rho} \sim k_t^{-\beta}$, where the value of β is close to the Kolmogorov spectral index, $\beta \approx 5/3$.

Oscillation energy density is a quadratic function of the amplitude, making it possible to suggest the following model expression for a Fourier harmonic of the mean square of the plasma oscillation density:

$$\rho^2 = C\omega^{-\alpha}k_t^{-2\beta}, \tag{C.1}$$

where C is a certain amplitude which can be found by inverse Fourier transformation over the entire spectrum of frequencies and wave vectors. The spectrum (C.1) should be cut off for $k_t \to 0$. It was demonstrated in [103] that a maximum correlation scale $\hat{l}$ exists related to the inhomogeneous structure of the solar wind. It is $\hat{l} = 1{,}12 \cdot 10^6$ km for magnetic field oscillations. Presumably, it is also a maximum scale for magnetosonic oscillation as well, making it possible to limit the oscillation spectrum (C.1) from below by the minimum value $\hat{k}_t = 2\pi/\hat{l}$. Correspondingly, a minimum frequency can also be introduced, $\hat{\omega} = \hat{k}_t v_0$, limiting the oscillation frequency spectrum from below. 'Transparency zones' should also be taken note of in the oscillation spectrum, as described in Section 2.13. It can be demonstrated that the amplitude C satisfying these requirements has the following form in the spectral function (C.1)

$$C = \widetilde{C}\bar{\rho}^2\Phi(\mathbf{k}_t, \omega), \tag{C.2}$$

where $\bar{\rho}^2$ is the mean square of the amplitude of density oscillations incident to the wave transition layer, $\Phi(\mathbf{k}_t, \omega)$ is the filter function taking care of the characteristic scales of the spectrum cut off and the 'transparent zones':

$$\begin{aligned}\Phi(\mathbf{k}_t, \omega) = \Theta(\omega - \hat{\omega})\Theta(\omega_i^I - \omega)\Big[&\Theta(\theta - \pi + \theta^*)\Theta(k_t - \hat{k}_t)\\ &+ \Theta(\theta^* - \theta)\Big(\Theta(k_t - \hat{k}_t)\Theta(k_t^{(1)} - k_t) + \Theta(k_t - k_t^{(2)})\Big)\\ &+ \Theta(\pi - \theta^* - \theta)\Theta(\theta - \theta^*)\Theta(k_t - \hat{k}_t)\Theta(k_t^{(1)} - k_t)\Big],\end{aligned}$$

$\Theta(x)$ is the Heaviside step function ($\Theta(x) = 0$ for $x < 0$, $\Theta(x) = 1$ for $x \geq 0$), and θ^* and $k_t^{(1,2)}$ are defined in Section 2.13. The normalising factor $\widetilde{C}$ in (C.2) has been chosen such that

$$\frac{2\widetilde{C}}{(2\pi)^{3/2}}\int_0^\infty \omega^{-\alpha}d\omega\int_0^\infty \Phi(\mathbf{k}_t, \omega)k_t^{-2\beta+1}dk_t = 1.$$

D Stability of MHD Oscillations with $\mathbf{k}_t \parallel \mathbf{B}_0$, in the Shear Layer, for $\beta^* < 1$ in a Boundless Medium

Reversing the sign in one of the radicals in the dispersion Eq. (2.184) would not change the form of the solution (2.186). Hence, the solution (2.186) contains roots related to two different signs of the radicals in (2.184). To find out which sign in (2.184) corresponds to the chosen solution (2.186) which is unstable for $\beta^* < 1$, let us scrutinise it near the neutral point $M = M_0$, where $c_i = 0$. In this case, $c^2 = 1 + M_0^{-2}$. Substituting this expression into (2.179) we find that $k_x^2 > 0$ for $M_0 < M_{01} = \sqrt{\sqrt{5} - 2/2}$, while $k_x^2 < 0$ for $M_0 > M_{01}$.

In the former case, the medium is transparent to the waves in question and enjoys a 'well-defined' concept of 'group velocity' the wave energy is transferred at. Retreating somewhat in M into the region where the imaginary constituent appears, $c_i > 0$ $(M = M_0(1 - \varepsilon)$, where $0 < \varepsilon \ll 1)$, produces a complex-valued expression for k_x. The sign of the radical in (2.179) should be chosen such that $\mathrm{Im}(k_x) > 0$ for $x > 0$, and $\mathrm{Im}(k_x) < 0$ for $x < 0$, which corresponds to a solution that exponentially decreases in amplitude away from the shear layer. The signs of the group velocity correspond to oscillations carrying the energy away from the layer. Such a choice of signs of k_x, however, results in a dispersion equation differing from (2.184) in that the radicals in it will have the same sign. Hence, this solution is not a solution of the dispersion Eq. (2.184).

In the latter case, $(M_0 > M_{01})$, the medium is opaque to the oscillations and the choice of k_x signs at the neutral point $M = M_0$ is determined by the requirement that their amplitude should decrease exponentially away from the shear layer. The group velocity is not defined for such oscillations. Let us retreat into the $M < M_0$ region, where a positive imaginary constituent, $c_i > 0$, and the related small constituent of the group velocity, $\mathrm{Re}\,(v_{gx})$, appear. It is easily verifiable that, same as in the previous case, the group velocity signs correspond to oscillations carrying the energy away from the layer, but the signs of the radicals, in the thus obtained dispersion equation, do not coincide with the signs in (2.184). These signs can only be made to agree by choosing an exponentially increasing solution one side of the shear layer, which contradicts the physical statement of the problem.

Hence, the group velocity should be regarded as 'ill-defined' in both these cases and cannot be used for calculating wave energy flux.

E Deriving Equations for Potentials φ and ψ for MHD Waves in a 'Warm' Plasma, in a Curvilinear Orthogonal System of Coordinates (x^1, x^2, x^3)

Decomposition (2.72) yields the following equations for the components of the perturbed electric field, in the orthogonal system of curvilinear coordinates:

$$\begin{aligned} E_1 &= -\nabla_1 \varphi + i k_2 \frac{g_1}{\sqrt{g}} \psi, \\ E_2 &= -i k_2 \varphi - \frac{g_2}{\sqrt{g}} \nabla_1 \psi, \\ E_3 &= 0. \end{aligned} \tag{E.1}$$

These equations, the second equation in (1.5) and the frozen-in condition (1.8) are used to obtain the following, in stationary plasma ($\mathbf{v}_0 = 0$), for the components of the perturbed magnetic field and the velocity field

$$\begin{aligned} B_1 &= \frac{c}{\omega}\frac{g_1}{\sqrt{g}}\nabla_3\left(k_2\varphi - i\frac{g_2}{\sqrt{g}}\nabla_1\psi\right), \\ B_2 &= \frac{c}{\omega}\frac{g_2}{\sqrt{g}}\nabla_3\left(i\nabla_1\varphi + k_2\frac{g_1}{\sqrt{g}}\psi\right), \\ B_3 &= i\frac{c}{\omega}\frac{g_3}{\sqrt{g}}\left(\nabla_1\frac{g_2}{\sqrt{g}}\nabla_1\psi - k_2^2\frac{g_1}{\sqrt{g}}\psi\right), \end{aligned} \quad \text{(E.2)}$$

$$\begin{aligned} v_1 &= -\frac{cp^{-1}}{B_0}\left(ik_2\varphi + \frac{g_2}{\sqrt{g}}\nabla_1\psi\right), \\ v_2 &= \frac{cp}{B_0}\left(\nabla_1\varphi - ik_2\frac{g_1}{\sqrt{g}}\psi\right). \end{aligned} \quad \text{(E.3)}$$

The third component of the first vector Eq. (1.5) helps us obtain

$$v_3 = -\frac{i}{\omega\rho_0}\nabla_3 P + \frac{icB_0\varkappa_{1B}}{4\pi\rho_0\omega^2\sqrt{g_1g_2}}\nabla_3\left(k_2\varphi - i\frac{g_2}{\sqrt{g}}\nabla_1\psi\right),$$

where $\varkappa_{1B} = \nabla_1(\ln\sqrt{g_3}B_0)$. The third equation in (1.5) enables us to obtain the following equation for the perturbed pressure:

$$P = -i\frac{v_s^2\rho_0}{\omega P_0^\sigma\sqrt{g}}\left[\nabla_1\left(\frac{\sqrt{g}}{g_1}P_0^\sigma v_1\right) + ik_2\frac{\sqrt{g}}{g_2}P_0^\sigma v_2 + \nabla_3\left(\frac{\sqrt{g}}{g_3}P_0^\sigma v_3\right)\right], \quad \text{(E.4)}$$

where $\sigma = 1/\gamma$. Substituting the above-obtained expressions for the velocity field components produces

$$\begin{aligned} \hat{L}_0 P = &-i\frac{cB_0}{4\pi\omega\sqrt{g_3}}\frac{v_s^2}{v_A^2} \\ &\times\left[i\frac{k_2g_3}{\sqrt{g}}\hat{L}_1\varphi - \bar{\Delta}_\perp\psi + \frac{B_0}{\omega^2P_0^\sigma\sqrt{g_1g_2}}\nabla_3\frac{\varkappa_{1B}v_A^2P_0^\sigma}{B_0\sqrt{g_3}}\nabla_3\frac{g_2}{\sqrt{g}}\nabla_1\psi\right], \end{aligned} \quad \text{(E.5)}$$

an equation determining the perturbed pressure via the scalar and vector potentials. The following operators are introduced here:

$$\hat{L}_0 = \frac{v_s^2}{\omega^2}\frac{\rho_0}{P_0^\sigma\sqrt{g}}\nabla_3\frac{\sqrt{g}}{g_3}\frac{P_0^\sigma}{\rho_0}\nabla_3 + 1,$$

– longitudinal operator for magnetosonic waves,

$$\hat{L}_1 = \frac{B_0}{\omega^2P_0^\sigma\sqrt{g_3}}\nabla_3\frac{\varkappa_{1B}v_A^2P_0^\sigma}{B_0\sqrt{g_3}}\nabla_3 - \varkappa_{1P},$$

where $\varkappa_{1P} = \nabla_1(\ln\sqrt{g_3}P_0^\sigma/B_0)$, and

$$\bar{\Delta}_\perp = \frac{B_0}{P_0^\sigma}\frac{1}{\sqrt{g_1g_2}}\left(\nabla_1\frac{pP_0^\sigma}{B_0}\nabla_1 - \frac{k_2^2}{p}\frac{P_0^\sigma}{B_0}\right).$$

Left-multiplying the first component of the first vector Eq. (1.5) by $-ik_2$, taking derivative ∇_1 of the second component of this equation, adding the resulting equations and expressing the remaining terms in terms of potentials φ and ψ produces

$$\nabla_1 B_0 \hat{L}_T \nabla_1 \varphi - k_2^2 B_0 \hat{L}_P \varphi$$
$$= ik_2 \left(\nabla_1 B_0 \hat{L}_T \frac{g_1}{\sqrt{g}} \psi - B_0 \hat{L}_P \frac{g_2}{\sqrt{g}} \nabla_1 \psi - B_0 \frac{\varkappa_{1g}}{\sqrt{g_3}} \tilde{\Delta}_\perp \psi \right), \quad \text{(E.6)}$$

where $\varkappa_{1g} = \nabla_1(\ln g_3)$, the longitudinal toroidal and poloidal operators $\hat{L}_T, \hat{L}_P$ are defined by expressions (3.6) and (3.7), and the operator $\tilde{\Delta}_\perp$ by (3.9). In a homogeneous plasma, the right-hand side of this equation becomes zero, and the left-hand side determines a dispersion equation for Alfvén waves $\omega^2 = k_\parallel^2 v_A^2$. Thus, in an inhomogeneous plasma, (E.25) can be assumed to describe Alfvén waves possibly triggered by magnetosonic waves the field of which determines the right-hand side of this equation.

Let us left-operate on the second component of the first vector Eq. (1.5) by $(-i/k_2)\hat{L}_0$ and expressing the terms of the resulting equation in terms of potentials φ and ψ yields

$$\frac{B_0\sqrt{g_3}}{4\pi\rho_0} \hat{L}_0 \frac{B_0}{\sqrt{g_3}} \tilde{\Delta}\psi + v_s^2 \Delta\psi - \hat{L}_2\psi = -i\frac{B_0\sqrt{g_3}}{4\pi k_2 \rho_0} \hat{L}_0 B_0 \hat{L}_T \nabla_1 \varphi + ik_2 v_s^2 \frac{g_3}{\sqrt{g}} \hat{L}_1 \varphi, \quad \text{(E.7)}$$

where operator $\tilde{\Delta}$ is defined as (3.8) the following operators are introduced:

$$\bar{\Delta} = \bar{\Delta}_\perp + \frac{B_0}{P_0^\sigma} \frac{1}{\sqrt{g_1 g_2}} \nabla_3 \frac{\sqrt{g}\, P_0^\sigma}{g_3\ \rho_0} \nabla_3 \frac{\rho_0}{B_0\sqrt{g_3}},$$
$$\hat{L}_2 = \frac{B_0 v_s^2}{\omega^2 P_0^\sigma \sqrt{g_1 g_2}} \nabla_3 \frac{\varkappa_{1B} v_A^2 P_0^\sigma}{B_0\sqrt{g_3}} \nabla_3 \frac{g_2}{\sqrt{g}} \nabla_1 - \omega^2.$$

In a homogeneous plasma, a dispersion equation for Alfvén waves is obtained in the right-hand side, and an equation for magnetosonic waves (1.10) in the left-hand side of (E.7). In an inhomogeneous plasma, (E.7) can be regarded as an equation describing the impact of Alfvén wave on the field of magnetosonic oscillations.

F WKB Solution of the Longitudinal Problem for FMS Waves Having Two Turning Points on the Field Line

Let us address the type I solutions in Figure 3.18. In the $x_-^3 \le x^3 < x_1^3$ region, a general solution of (3.85) consists of the sum of the growing and falling exponents. The boundary condition at the ionosphere ($H_n(x_-^3) = 0$) implies that the solution in this region has the form

$$H_n = Cf \sinh\left(\int_{x_-^3}^{x^3} |k_3| dx^{3'} \right). \quad \text{(F.1)}$$

Near the turning point $x^3 = x_3^3$ (see Figure 3.18), the potential can be linearised in (3.85) and represented as

$$\frac{\partial^2 H}{\partial \xi^2} + \xi H = 0, \quad \text{(F.2)}$$

where $\xi = \bar{f}(x^3 - x_1^3)$ is dimensionless longitudinal coordinate, $\bar{f} = |\nabla_3 k_3^2|^{1/3}_{x^3 = x_1^3}$. (F.2) is an Airy equation its general solution being

$$H = A_1 Ai(-\xi) + B_1 Bi(-\xi), \quad \text{(F.3)}$$

where $Ai(-\xi)$, $Bi(-\xi)$ are Airy functions. Using the asymptotic expressions for them as $\xi \to -\infty$ in this solution and matching it to solution (F.1) yields

$$A_1 - 2B_1 e^{2\bar{\psi}} = \sqrt{\pi} f_1 C e^{\bar{\psi}},$$

where $f_1 = (g_3\sqrt{g}/\bar{f}g_1)^{3/2}_{x^3=x_1^3}$,

$$\bar{\psi} = \int_{x_-^3}^{x_1^3} |k_3| dx^3.$$

In the $x_1^3 < x^3 < x_2^3$ region, the Wentzel–Kramers–Brillouin (WKB) solution (3.87) can be matched to the solution near the turning point x_1^3, which, for coefficients A and B in (3.87), results in

$$\begin{aligned} A &= \frac{C}{2}(e^{\bar{\psi}} - \frac{i}{2}e^{-\bar{\psi}})e^{i\pi/4}, \\ B &= i\frac{A}{2}(e^{\bar{\psi}} + \frac{i}{2}e^{-\bar{\psi}})e^{-i\pi/4}. \end{aligned} \tag{F.4}$$

Analogously, near the second turning point $x^3 = x_2^3$, the solution of (3.85) can be represented as

$$H = A_2 Ai(\xi) + B_2 Bi(\xi), \tag{F.5}$$

where $\xi = \bar{f}(x^3 - x_2^3)$. Matching to the solution (3.87), where coefficients A, B are expressed as (F.4), yields

$$A_2 = -\sqrt{\pi} C f_2 \left(e^{\bar{\psi}} \cos\left(\psi + \frac{\pi}{2}\right) + \frac{e^{-\bar{\psi}}}{3} \sin\left(\psi + \frac{\pi}{2}\right)\right),$$

$$B_2 = \sqrt{\pi} C f_2 \left(e^{\bar{\psi}} \sin\left(\psi + \frac{\pi}{2}\right) - \frac{e^{-\bar{\psi}}}{2} \cos\left(\psi + \frac{\pi}{2}\right)\right),$$

where $f_2 = (g_3\sqrt{g}/\bar{f}g_1)^{1/2}_{x^3=x_2^3}$,

$$\psi(x^3) = \int_{x_3^3}^{x^3} k_3 dx^{3'}.$$

In the $x_2^3 < x^3 \leq x_+^3$ region, the solution satisfying the boundary condition at the ionosphere ($H(x_+^3) = 0$) is

$$H_n = C_1 f \sinh\left(\int_{x_2^3}^{x^3} |k_3| dx^{3'}\right).$$

Matching this solution to the solution (F.5) is only possible when the quantisation condition is satisfied

$$\tan\left(\psi(x_2^3) + \frac{\pi}{2}\right) = 0,$$

or

$$\int_{x_1^3}^{x_2^3} k_3 dx^3 = \pi(n + \frac{1}{2}),$$

where $n = 0, 1, 2, 3, \ldots$ Also, $C_1 = \pm C$ ('±' for even or odd n, respectively), yielding a general solution of the form (3.89).

G Integrals of Functions $G(z)$ and $g(z)$ Describing the Transverse Structure of Standing Alfvén Waves

To calculate the integral

$$I_1 = \int_{-\infty}^{\infty} |G(\eta + i\varepsilon)|^2 d\eta$$

we will use an integral representation of the function $G(z)$ (3.260). We obtain

$$\begin{aligned} I_1 &= \frac{1}{\pi} \int_{-\infty}^{\infty} d\eta \int_0^{\infty} ds \int_0^{\infty} ds' \exp\left[i(s - s')\eta - \varepsilon(s + s') - i\frac{s^3}{3} + i\frac{s'^3}{3}\right] \\ &= \frac{1}{\pi} \int_0^{\infty} ds \int_0^{\infty} ds' \exp\left[-i\frac{s^3}{3} + i\frac{s'^3}{3} - \varepsilon(s + s')\right] \int_{-\infty}^{\infty} d\eta \exp\left[i(s - s')\eta\right] \\ &= 2 \int_0^{\infty} ds \int_0^{\infty} ds' \exp\left[-i\frac{s^3}{3} + i\frac{s'^3}{3} - \varepsilon(s + s')\right] \delta(s - s') = 2 \int_0^{\infty} dse^{-2\varepsilon s} = \frac{1}{\varepsilon}. \end{aligned}$$

Analogously, in calculating

$$I_2 = \int_{-\infty}^{\infty} |g'(\eta + i\varepsilon)|^2 d\eta$$

we will use an integral representation $g(z)$ (3.261). From this, we have

$$g'(z) = \frac{i}{\sqrt{\pi}} \int_0^{\infty} \exp\left(isz + \frac{i}{s}\right) ds.$$

Hence

$$I_2 = 2 \int_0^{\infty} ds \int_0^{\infty} ds' \exp\left[\frac{i}{s} - \frac{i}{s'} - \varepsilon(s + s')\right] \delta(s - s') = 2 \int_0^{\infty} dse^{-2\varepsilon s} = \frac{1}{\varepsilon}.$$

H Parameters of the Polarisation Ellipse of Stochastic Oscillations

Let $\mathbf{r} = (x, y)$ be a two-dimensionless vector its components $x = x(t), y = y(t)$ being random functions of time. Let us assume correlators $\langle x^2 \rangle$, $\langle y^2 \rangle$, $\langle xy \rangle$. We will examine the projection of vector $\mathbf{r}$ onto a straight line inclined α to the x axis:

$$r_\alpha = x \cos\alpha + y \sin\alpha.$$

The mean value of its square can be represented as

$$\left\langle r_\alpha^2 \right\rangle = r_{\min}^2 + \left(r_{\max}^2 - r_{\min}^2\right) \cos^2(\alpha - \alpha_0), \tag{H.1}$$

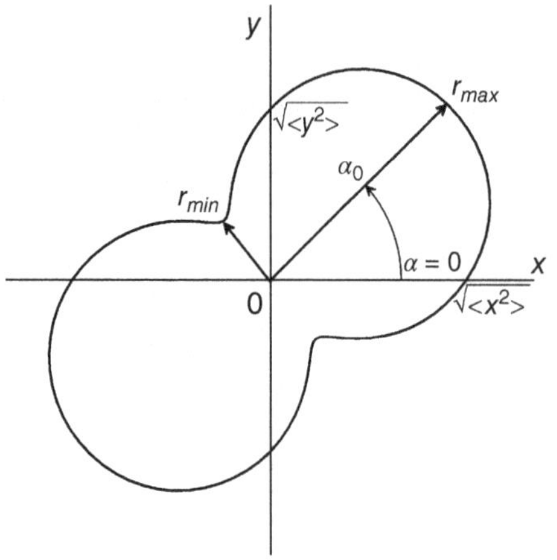

Figure H.1 Distribution, in the (x, y) plane, of the mean square of the amplitude of a 2D random vector $\mathbf{r} = (x, y)$ with given quadratic correlators $\langle x^2\rangle$, $\langle y^2\rangle$, $\langle xy\rangle$.

where

$$r_{\min}^2 = \frac{1}{2}\left[\langle x^2\rangle + \langle y^2\rangle - \sqrt{\left(\langle x^2\rangle - \langle y^2\rangle\right)^2 + 4\langle xy\rangle^2}\right],$$

$$r_{\max}^2 = \frac{1}{2}\left[\langle x^2\rangle + \langle y^2\rangle + \sqrt{\left(\langle x^2\rangle - \langle y^2\rangle\right)^2 + 4\langle xy\rangle^2}\right],$$

$$\tan 2\alpha_0 = \frac{2\langle xy\rangle}{\langle x^2\rangle - \langle y^2\rangle}.$$

The configuration described by (H.1) in polar coordinates (α is polar angle, $r = \sqrt{\langle r_\alpha^2\rangle}$ is radius) is not an ellipse. Its characteristic form is shown in Figure H.1. The maximum radius $r_{\max}$ of this configuration is in the direction $\alpha = \alpha_0$, and the minimum radius $r_{\min}$ in the perpendicular direction. However, if the end of the vector $\mathbf{r}$ is harmonically rotated along an ellipse with semiaxes a and b inclined ϕ to the x axis, i.e.

$$x = a\cos\phi\cos\omega t - b\sin\phi\sin\omega t,$$

$$y = a\sin\phi\cos\omega t + b\cos\phi\sin\omega t,$$

we can calculate the time-averaged $\langle x^2\rangle$, $\langle y^2\rangle$, $\langle xy\rangle$ and substitute them into the above formulas to obtain $\alpha_0 = \phi$, $r_{\max} = a$, $r_{\min} = b$. In the general case, this justifies calling $r_{\min}$ and $r_{\max}$ the polarisation ellipse semiaxes, and α_0 the inclination as well.

I Deriving Coefficients of the Differential Equation Based on the Given WKB Solution

Let us find the solution of (3.332) using the WKB approximation. We will assume

$$U_N(x^1) = e^{iS(x^1)}.$$

For the phase $S(x^1)$, we obtain the equation

$$-AS'^2 + i(AS')' + D = 0.$$

Using the WKB approximation, this equation is solved in the form of an asymptotic series

$$S = S_0 + S_1 + \cdots.$$

In the main order,

$$-AS_0'^2 + D = 0.$$

Hence,

$$S_0 = \int \left(\frac{D}{A}\right)^{1/2} dx.$$

In the next order,

$$-2AS_0'S_1' + i(AS_0')' = 0,$$

yielding

$$S_1 = \frac{i}{4}\ln(AD) - i\ln C,$$

where C is an arbitrary constant. The resulting solution can be written as

$$U_N = e^{i(S_0+S_1)} = Cr^{-1/2}\exp\left(i\int kdx\right),$$

where

$$r = (AD)^{1/2}, \quad k = (D/A)^{1/2}. \tag{I.1}$$

Presuming the solution of (3.332) is given, in the WKB approximation, i.e. functions r and k are set, and the equation coefficients must be restored. From (I.1), we immediately obtain

$$A = r/k, \qquad D = rk.$$

J Strictly Deriving a Transverse Model Equation for Standing Alfvén Waves, for the $\kappa \ll 1$ Case

One of the regions in question is located in the neighbourhood of the toroidal surface. We will assume that the mode in this region is nearly toroidal. This means that

$$\left|\nabla_1^2\varphi\right| \gg k_y^2|\varphi|. \tag{J.1}$$

In this case, its longitudinal structure must be close to toroidal harmonic T_N and, hence, its solution can be sought in the form

$$\varphi(x^1, l, \omega) = U_N(x^1, \omega)T_N(x^1, l, \omega) + h_N(x^1, l, \omega), \tag{J.2}$$

where the first term in the right-hand side represents the first order of approximation, and h_N is a first-order correction. Let us also assume the difference $\omega^2 - \Omega_{TN}^2$ to be a small first-order value. Substituting (J.2) into (3.322) and retaining the main- and first-order terms only yields

$$\nabla_1\left(\frac{\partial}{\partial l}p\frac{\partial T_N}{\partial l} + p\frac{\omega^2}{v_A^2}T_N\right)\nabla_1 U_N$$
$$- k_2^2 U_N\left(\frac{\partial}{\partial l}\frac{1}{p}\frac{\partial T_N}{\partial l} + \frac{1}{p}\frac{\omega^2}{v_A^2}T_N\right) = -\hat{L}(\Omega_{TN})\varphi_N. \tag{J.3}$$

Given the fact that function T_N satisfies the equation

$$\frac{\partial}{\partial l}p\frac{\partial T_N}{\partial l} + p\frac{\Omega_{TN}^2}{v_A^2}T_N = 0,$$

we have

$$\frac{\partial}{\partial l}p\frac{\partial T_N}{\partial l} + p\frac{\omega^2}{v_A^2}T_N = p\frac{\omega^2 - \Omega_{TN}^2}{v_A^2}T_N,$$

$$\frac{\partial}{\partial l}\frac{1}{p}\frac{\partial T_N}{\partial l} + \frac{1}{p}\frac{\omega^2}{v_A^2}T_N = \frac{1}{p}\frac{\omega^2 - \Omega_{TN}^2}{v_A^2}T_N + 2\left(\frac{\partial}{\partial l}\frac{1}{p}\right)\frac{\partial T_N}{\partial l}.$$

We will use these relations to rewrite (J.3) as

$$\nabla_1\left(p\frac{\omega^2 - \Omega_{TN}^2}{v_A^2}T_N\right)\nabla_1 U_N$$
$$-k_2^2\left[\frac{1}{p}\frac{\omega^2 - \Omega_{TN}^2}{v_A^2}T_N 2\left(\frac{\partial}{\partial l}\frac{1}{p}\right)\frac{\partial T_N}{\partial l}\right]U_N = -\hat{L}(\Omega_{TN})\varphi_N. \quad (J.4)$$

Multiply this equation by T_N and integrate along the field line. The result will be

$$\alpha_T\nabla_1(\omega^2 - \Omega_{TN}^2)\nabla_1 U_N - \alpha_P k_2^2(\omega^2 - \widetilde{\Omega}_{PN}^2)U_N = -\oint T_N\hat{L}(\Omega_{TN})\varphi_N dl. \quad (J.5)$$

The notation is introduced here:

$$\widetilde{\Omega}_{PN}^2 = \Omega_{TN}^2 + \frac{1}{\alpha_P}\oint\left(\frac{\partial^2}{\partial l^2}\frac{1}{p}\right)T_N^2 dl. \quad (J.6)$$

The right-hand side in (J.5) is calculated in the same manner as we did in the main text when deriving (3.347). Upon which, (J.4) takes the form

$$\alpha_T\nabla_1[(\omega + \gamma_{TN})^2 - \Omega_{TN}^2]\nabla_1 U_N - \alpha_P k_2^2[(\omega + \gamma_{PN})^2 - \widetilde{\Omega}_{PN}^2]U_N = I_N. \quad (J.7)$$

This equation differs from the model Eq. (3.347) in Ω_{PN}^2 replaced with $\widetilde{\Omega}_{PN}^2$. Expression (J.6) and estimates in Section 3.7, imply that the differences $\Omega_{TN}^2 - \Omega_{PN}^2$ and $\Omega_{TN}^2 - \widetilde{\Omega}_{PN}^2$ are of the same sign and order of magnitude. The above difference is therefore insignificant for a qualitative inspection. A wider affirmation can be made, however, this difference is insignificant (i.e. small) in quantitative terms as well. It was shown in Section 3.13.3 that the poloidal surface loses its significance as a resonance surface for $\kappa \ll 1$, and the solution of the equation is practically independent of its location (provided, of course, $\Delta \ll 1/k_y$). In particular, we can set $\Omega_{PN}^2 = \Omega_{TN}^2$ in this case, which is the same as neglecting the term $\kappa \arctan q$ in the exponent of the integral representation (3.365). Thanks to the above-said the model Eq. (3.347) can be used for inspecting the spatial structure to the same degree as the strictly derived (J.7).

Let us estimate the size of the domain where (J.7) is applicable. It was derived assuming that the toroidality condition (J.1) was satisfied. Near the toroidal surface $U_N \sim \ln(x^1 - x_{TN}^1)$, so this condition means

$$|x^1 - x_{TN}^1| \ll 1/k_y. \quad (J.8)$$

Let us emphasise that the domain (J.8) includes the entire interval between the toroidal and poloidal surfaces, when $k_y\Delta \ll 1$ (irrespective of whether the poloidal surface is determined by frequency Ω_{PN} or $\widetilde{\Omega}_{PN}$). Therefore, the mode is toroidal on the poloidal surface as well, in this case.

Another domain in the x^1 coordinate, where a transverse equation can be strictly derived is the mode poloidality region, where

$$|\nabla_1^2\varphi| \ll k_y^2|\varphi|. \tag{J.9}$$

It is logical to presume that this condition is met in the region defined by an inverse inequality to (J.8):

$$|x^1 - x^1_{TN}| \gg 1/k_y. \tag{J.10}$$

This presumption can be supported by the following argument. If the model Eq. (3.347) is applicable to the (J.10) region, its solution (3.354) satisfying the condition (J.9), leads to inequality (J.10).

In the mode poloidality region, the solution should be sought in the form

$$\varphi = U_N(x^1,\omega)P_N(x^1,l,\omega) + h_N(x^1,l,\omega),$$

where U_NP_N is the main order of the approximation, and h_N is a small correction. Assuming the difference $\omega^2 - \Omega^2_{PN}$ to be small as well, we obtain this equation instead of (J.37)

$$\nabla_1\left(p\frac{\omega^2-\Omega^2_{PN}}{v_A^2} + 2\frac{\partial p}{\partial l}\frac{\partial P_N}{\partial l}\right)\nabla_1 U_N - k_2^2\frac{\omega^2-\Omega^2_{PN}}{pv_A^2}U_N = -\hat{L}(\Omega_{PN})\varphi_N.$$

Multiplying it by P_N, integrating along the field line and calculating the integral in the right-hand side in the same manner as before produces

$$\alpha_T\nabla_1[(\omega+\gamma_{TN})^2 - \widetilde{\Omega}^2_{TN}]\nabla_1 U_N - \alpha_P k_2^2[(\omega+\gamma_{PN})^2 - \Omega^2_{PN}]U_N = I_N. \tag{J.11}$$

Here,

$$\widetilde{\Omega}^2_{TN} = \Omega^2_{PN} + \frac{1}{\alpha_T}\oint\left(\frac{\partial^2}{\partial l^2}\frac{1}{p}\right)P_N^2\,dl.$$

The same reasoning as above can be used to demonstrate that the differences $\Omega^2_{PN} - \Omega^2_{TN}$ and $\Omega^2_{PN} - \widetilde{\Omega}^2_{TN}$ are of the same sign and order of magnitude, and hence (J.11) can very well be replaced with the model Eq. (3.347).

Note that, in terms of the dimensionless variable $\xi = k_y(x^1 - x^1_{TN})$, the (J.8) and (J.10) regions, where the transverse equation can be strictly derived, are defined by inequalities $|\xi| \ll 1$ and $|\xi| \gg 1$.

Let another important circumstance be emphasised. In the (J.8) region, where (J.11) is valid, the position of the singular (toroidal) surface is determined by the 'actual' toroidal frequency Ω_{TN}. In (J.11), however, which describes the mode in an asymptotically remote region (J.10), the role of the toroidal frequency is played by $\widetilde{\Omega}_{TN}$ which differs somewhat from Ω_{TN}. But the precise position of the toroidal surface is not important any longer for a solution in this region.

K Calculating Characteristics η_0

Substituting (3.395) into the expression for v_2 and equating the resulting equation to zero produces the following equation for calculating $\kappa_0 \equiv \bar{\kappa}(\xi, \eta_0(\xi))$:

$$\tilde{z}^3(1+\xi^{-1}) - \tilde{z}^2(4+3\xi^{-1}) + 3(2+\xi^{-1})\tilde{z} - 4 = 0,$$

where $\tilde{z} = (1+\kappa_0^2)\xi$. Denoting

$$r = -\frac{4\xi+3}{\xi+1}, \quad s = 3\frac{2\xi+1}{\xi+1}, \quad t = -\frac{4\xi}{\xi+1}$$

and replacing the variable $z = \tilde{z} + r/3$ reduces the equation to the canonical form:

$$z^3 + pz + q = 0,$$

where

$$p = \frac{3s - r^2}{3}, \quad q = \frac{2r^3}{27} - \frac{rs}{3} + t.$$

The form of the roots of this equation depends on the discriminant

$$D = \frac{p^3}{27} + \frac{q^2}{4}.$$

It is easily verifiable that, for $-1 \leq \xi \leq 1$, the discriminant $D > 0$. According to the general theory of cubic equations, this means that the equation has only one real root,

$$z_1 = u + v - r/3,$$

where

$$u = (-(q/2) + \sqrt{D})^{1/3}, \quad v = (-(q/2) - \sqrt{D})^{1/3}.$$

Thus,

$$\kappa_0 = \pm\sqrt{(\tilde{z}_1 - \xi)/\xi},$$

where

$$\tilde{z}_1 = z_1 + r/3.$$

Substituting $\kappa = \kappa_0$ into (3.398) and setting $\pm\eta = \eta_0$ in the right-hand side, we obtain an equation determining characteristics η_0. If $\xi > 0$, the '+' sign is chosen in the left-hand side of (3.398), and '−' in the definition of κ_0. For $\xi < 0$, the opposite is true: the '−' sign is chosen in (3.398), and '+' in the definition of κ_0.

L Calculating the Integral (3.390) Near the Characteristics $\pm\eta_1, \pm\eta_4$ and η_0

Taking an inner integral over κ in (3.388) produces an integral of the form (3.390). If the second derivative of phase $\bar{w}_2$ is zero near the saddle point $\bar{k}$, the phase $\tilde{\Psi}_\pm$ can be decomposed to an accuracy of third-order of smallness, near $k = \bar{k}$:

$$\tilde{\Psi}_\pm(k, \tilde{\kappa}(k)) \approx \bar{\Psi}_\pm(\bar{k}, \bar{\kappa}) + \frac{\bar{w}_3}{6}(k - \bar{k})^3,$$

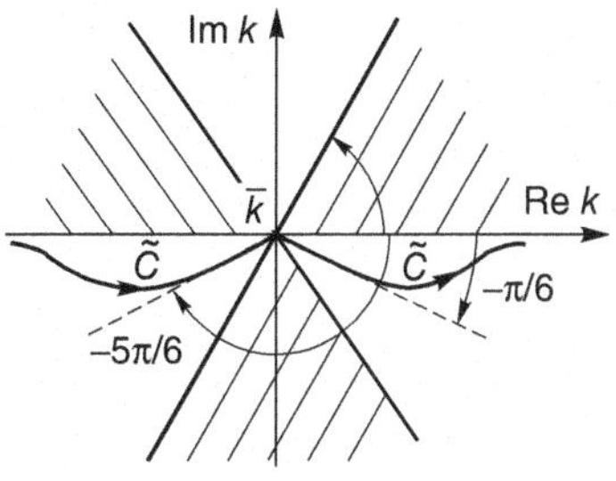

Figure L.1 Integration contour $\tilde{C}$ and the location of sectors of exponentially growing (hatched) and decreasing integrand in (3.390), near the characteristics $\pm\eta_1, \pm\eta_4$ and η_0.

where $\bar{w}_3 \equiv (\partial^3\widetilde{\Psi}_\pm/\partial k^3)_{k=\bar{k}}$. Figure L.1 shows sectors where the integration function either decreases or grows exponentially, away from the saddle point $\bar{k}$. Choosing the integration path $\tilde{C}$ as shown in Figure L.1, we reduce the integral (3.390) to the following form

$$I \simeq \tau_1\sqrt{\frac{2\pi}{\tau_N|\bar{v}_2|}}\bar{a}_n \exp[i(\Omega_N^0 t + \bar{\Psi}_\pm\tau_N + \pi\mathrm{sign}(\bar{v}_2)/4)] \times$$
$$\left(e^{-i\pi/6} - e^{-i5\pi/6}\right)\int_0^\infty e^{-i|\bar{w}_3|\tau_N r^3/6}dr$$

in the neighbourhood of the saddle point. Making the resulting integral dimensionless and given

$$\int_0^\infty e^{-z^3}dz = \Gamma(1/3)/3,$$

we obtain

$$I \approx \frac{\tau_1}{\tau_N^{5/6}}\bar{a}_n\Gamma\left(\frac{1}{3}\right)\sqrt{\frac{2\pi}{3|\bar{v}_2|}}\left(\frac{6}{|\bar{w}_3|}\right)^{1/3} \exp[i(\Omega_N^0 t + \bar{\Psi}_\pm\tau_N + \frac{\pi}{4}\mathrm{sign}(\bar{v}_2))].$$

The expression for integral (3.390) is also obtained analogously on the characteristics η_0, where $\bar{v}_2 = 0$.

M Calculating the Integral (3.388) Near the Characteristics $\pm\eta_2, \pm\eta_3, \pm\eta_5$ and $\pm\eta_6$

The integration expression in (3.388) has four branchpoints: $\kappa = \pm i$ and $\kappa = k^{-1} \pm i$. Let us inspect the region $-1 \le \xi \le 0$. We will draw branch cuts in the complex κ plane and construct a closed integration contour $\tilde{C}$ as shown in Figure L.2. The integration expression has no singularities inside this contour, so the inner integral (3.388) over κ is zero along it. The integral $\tilde{I}$ along the contour $\tilde{C}$ is the sum of these integrals:

$$\tilde{I} = \tilde{I}_0 + \tilde{I}_1 + \tilde{I}_2 + I = 0,$$

where $\tilde{I}_0$ is the integral over the remote contour $\tilde{C}_0$, $\tilde{I}_1, \tilde{I}_2$ are the integrals along the branch cuts $\tilde{C}_1$ and $\tilde{C}_2$, and I is the initial integral along the real axis. As the contour $\tilde{C}_0$ expands to infinity, integration expression for it exponentially tends to zero, whence $\tilde{I}_0 = 0$. We examine the characteristics where $\bar{k} \to 0$, therefore we will set, in the integrand $\tilde{I}_1$ (where $\mathrm{Re}\,\kappa = 0$):

$$\arctan(\kappa - k^{-1}) \approx -\pi/2, \qquad 1 + (\kappa - k^{-1})^2 \approx k^{-2}.$$

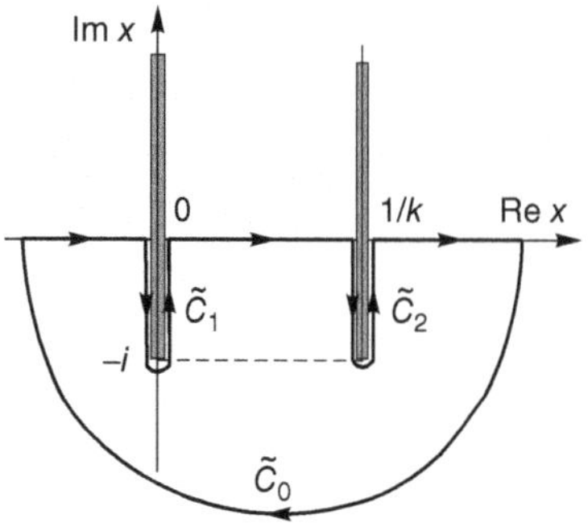

Figure L.2 Integration contour $\tilde{C}$ for the integral (3.388) near the characteristics $\pm\eta_2, \pm\eta_3, \pm\eta_5$ and $\pm\eta_6$, consisting of the real axis of complex κ, paths $\tilde{C}_1$ and $\tilde{C}_2$ along banks of cut and the remote contour $\tilde{C}_0$.

As a result, the integral $\tilde{I}_1$ along the banks of the branch cut $\tilde{C}_1$ is reduced to

$$\tilde{I}_1 \simeq -2\tau_1 e^{-i\Omega_N^0 t} \int_0^\infty k e^{-ik(\pm\eta+\pi/2)\tau_N} \cos(\pi k \tau_N) dk \int_0^1 \frac{r e^{-k\xi\tau_N r}}{\sqrt{1-r^2}} \left(\frac{1+r}{2-r}\right)^{k\tau_N/2} dr.$$

Calculating the integral over k, we obtain

$$\tilde{I}_1 \simeq -\frac{\tau_1}{\tau_N^2} e^{-i\Omega_N^0 t} \int_0^1 \frac{dr}{\sqrt{1-r^2}} \left\{ \left[r\xi - \ln\sqrt{\frac{2-r}{1+r}} + i\left(\frac{3}{2}\pi - |\eta|\right) \right]^{-1} \right.$$
$$\left. + \left[r\xi - \ln\sqrt{\frac{2-r}{1+r}} + i(\frac{\pi}{2} - |\eta|) \right]^{-1} \right\}.$$

Analogous calculations of the integral $\tilde{I}_2$ over the contour $\tilde{C}_2$ (where Re $\kappa = k^{-1} \to \infty$) yield

$$\tilde{I}_2 \simeq -\frac{\tau_1}{\tau_N} e^{-i(\Omega_N^0 t + \xi\tau_N)} \int_0^1 \frac{dr}{\sqrt{1-r^2}} \left\{ \left[\ln\sqrt{\frac{2-r}{1+r}} + i(\frac{3}{2}\pi - |\eta|) \right]^{-1} \right.$$
$$\left. + \left[\ln\sqrt{\frac{2-r}{1+r}} + i(\frac{\pi}{2} - |\eta|) \right]^{-1} \right\}.$$

Since $\tau_N \gg 1$, the contribution of $\tilde{I}_1$ can be neglected and it may be assumed that $I \approx -\tilde{I}_2$. The integral (3.388) can be calculated analogously for $0 \le \xi \le 1$.

N Determining the Shape of a Field Line from Given Components of Background Magnetic Field

Let the field line shape be described by its radius $r(a, \theta)$, which is $r(a, 0) = a$ in the equatorial plane. We will define the length element along field line, dl, as angle θ changes by an infinitely small value $d\theta$. It can be seen from Figure N.1 that $dl = \sqrt{\Delta_1^2 + \Delta_2^2}$, where $\Delta_1 \approx r(a, \theta) - r(a, \theta + d\theta) = -(\partial r/\partial\theta)d\theta$, $\Delta_2 \approx r(a, \theta)d\theta$. Therefore

$$dl = \sqrt{r^2(a, \theta) + (\partial r/\partial\theta)^2} d\theta. \tag{N.1}$$

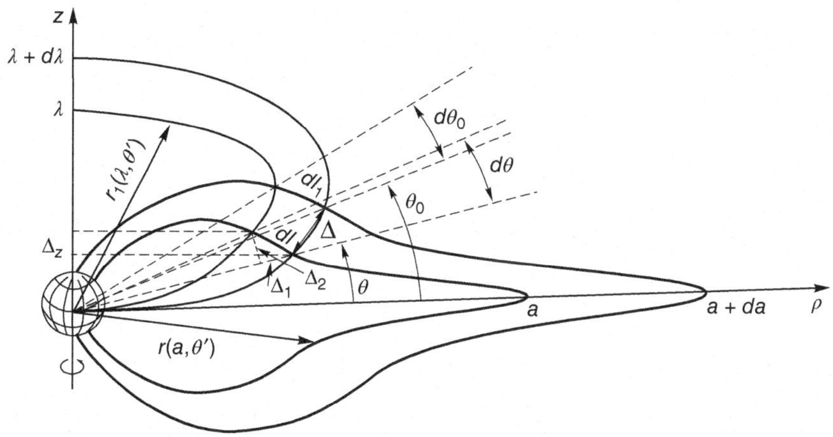

Figure N.1 Length elements necessary for calculating the components of metric tensor in a non-orthogonal curvilinear system of coordinates (a, ϕ, θ). Two infinitely proximate magnetic field lines $r(a, \theta')$, $r(a + da, \theta')$ are shown as well as orthogonal coordinate lines $r(\lambda, \theta')$, $r(\lambda + d\lambda, \theta')$ in the plane of magnetic meridian $\phi = \text{const}$.

Figure N.1 also implies that $\sin\bar{\theta} = \Delta_z/dl$, where $\bar{\theta}$ is the angle between the field line tangent and the ρ axis, $\Delta_z = r(a, \theta + d\theta)\sin(\theta + d\theta) - r(a, \theta)\sin\theta \approx (r(a, \theta)\cos\theta + (\partial r/\partial\theta)\sin\theta)d\theta$. Therefore,

$$\sin\bar{\theta} = \frac{r(a, \theta)\cos\theta + (\partial r/\partial\theta)\sin\theta}{\sqrt{r^2(a, \theta) + (\partial r/\partial\theta)^2}}.$$

Squaring the right- and left-hand sides of the equation produces

$$\frac{\partial \ln r}{\partial\theta} = \frac{\sin\theta\cos\theta \pm \sin\bar{\theta}\cos\bar{\theta}}{\sin^2\bar{\theta} - \cos^2\bar{\theta}}.$$

The correct result is obtained if the minus sign is chosen in the enumerator. Integrating this expression produces (3.421), an equation describing the field-line shape for given values of the background magnetic-field components.

O Defining Tri-orthogonal System of Coordinates Related to Magnetic-Field Lines

To construct a system of coordinates satisfying the two conditions in Section 4.2, let us recall the definition of curvature lines and some of their properties (see [406]). We will examine a small area of an arbitrary curved surface near some point M (see Figure N.2). Let us draw the normal O to this surface at point M and draw an intersecting plane containing the normal O. The intersection line between the plane and the surface has a certain curvature radius r. Varying r by rotating the transecting plane around the O axis, we will identify the directions of transecting planes having the maximum and minimum curvature radii. These directions are normal to each other. The lines along which the curved surface has extreme values of the curvature radius are curvature lines. Except for special cases (sphere and plane), the surface curvature lines are uniquely definable.

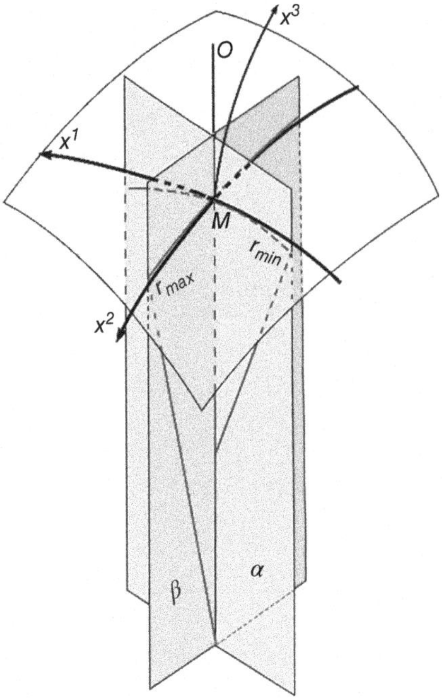

Figure N.2 Defining tri-orthogonal system of coordinates. The x^3 coordinate coincides with magnetic field line. The $x^3 = \text{const}$ surface is shown, where M is an arbitrary point, and O is the normal to this surface, at point M. The α and β planes contain the normal OM. Their intersection lines on surface $x^3 = \text{const}$ have maximum r_{max} and minimum r_{min} curvature radii at point M. Maximum and minimum curvature lines tangent to the intersection lines, but not coinciding with them, are the coordinate lines x^1 and x^2.

Even if the surface is not completely spherical or plane, it can have points where the maximum and minimum curvature radii are equal, namely umbilical and flattening points (in the latter case, the curvature radius is infinitely large). The curvature directions are undefined at these points. But these directions are uniquely defined in the infinitely small neighbourhoods of the umbilical and flattening points. Consequently, the curvature lines can also be drawn through the umbilical and flattening points, uniquely defining them in this manner at these points as well.

To avoid misunderstanding, let us emphasise that, in the general case, the curvature lines do not coincide with the transection lines between the surface and the plane in question, even if they have a common tangent line.

P Determining Metric Tensor Components in a Curvilinear Orthogonal System of Coordinates

Let us define a family of coordinate surfaces $r_1(\lambda, \theta)$ (see Figure N.1) that are orthogonal to magnetic shells $r(a, \theta)$ satisfying the orthogonality condition

$$\frac{\partial \ln r(a, \theta)}{\partial \theta} \frac{\partial \ln r_1(\lambda, \theta)}{\partial \theta} = -1. \tag{P.1}$$

We will examine two infinitely close field lines $r(a, \theta')$ and $r(a + da, \theta')$. According to the definition

$$\Delta = \sqrt{g_1} da \tag{P.2}$$

is coordinate line segment $r_1(\lambda + d\lambda, \theta')$ cut off by these field lines at latitude θ. It can be seen in Figure N.1 that

$$dl^2 = (r(a, \theta + d\theta) - r(a + da, \theta + d\theta))^2 - \Delta^2 \approx \left[\left(\frac{\partial r}{\partial a}\right)^2 - g_1\right] da^2.$$

Comparing this to (N.1), we obtain

$$d\theta^2 = \frac{(\partial r/\partial a)^2 - g_1}{r^2(a, \theta) + (\partial r/\partial \theta)^2} da^2. \tag{P.3}$$

Let us write the coordinates of vectors $\mathbf{r}(a, \theta) = (\rho_1, z_1)$ and $\mathbf{r}(a + da, \theta + d\theta) = (\rho_2, z_2)$ in a cylindrical system of coordinates (ρ, ϕ, z), in the $\phi = $ const coordinate plane:

$$\rho_1 = r(a, \theta) \cos\theta, \quad z_1 = r(a, \theta) \sin\theta,$$

$$\rho_2 = r(a, \theta + d\theta) \cos(\theta + d\theta) \approx \rho_1 + \left[\frac{\partial r}{\partial a} da + \frac{\partial r}{\partial \theta} d\theta\right] \sin\theta - r(a, \theta) \sin\theta d\theta,$$

$$z_2 = r(a, \theta + d\theta) \sin(\theta + d\theta) \approx z_1 + \left[\frac{\partial r}{\partial a} da + \frac{\partial r}{\partial \theta} d\theta\right] \sin\theta + r(a, \theta) \cos\theta d\theta.$$

We have

$$\Delta^2 = (\rho_1 - \rho_2)^2 + (z_1 - z_2)^2 = g_1 da^2.$$

We use expressions for ρ_1, ρ_2, z_1, z_2, as well as (P.3) to obtain expression (3.422) for g_1.

To determine g_2, we have, by definition, $\Delta_\phi = \sqrt{g_2} d\phi$ a segment of the azimuthal coordinate line lying on magnetic shell $a = $ const, cut off by two infinitely close meridional planes ϕ and $\phi + d\phi$, at latitude θ. On the other hand, $\Delta_\phi = r(a, \theta) \cos\theta d\phi$. Equating these two expressions, we obtain expression (3.422) for g_2.

Q Coefficients of the Equation for the Coupled Modes of MHD Oscillations

Let us write the coefficients of (3.442) in the following form:

$$\kappa_3 = 3\varkappa_{lp} + \varkappa_{lb},$$

$$\kappa_2 = \frac{\varkappa_{1g}\varkappa_{1B}}{g_1} + \varkappa_a^{(2)} + \varkappa_{ls}\varkappa_{la} + 2\varkappa_{lb}\varkappa_{lp} + 3p\nabla_l^2 p^{-1} + \omega^2(v_A^{-2} + c_s^{-2}),$$

$$\kappa_1 = \frac{\varkappa_{1g}\varkappa_{1B}}{g_1}\varkappa_{lc} + \varkappa_{lp}\varkappa_{ls}\varkappa_{la} + \varkappa_{lp}\varkappa_a^{(2)}$$
$$+ \varkappa_{lb} p\nabla_l^2 p^{-1} + p\nabla_l^3 p^{-1} + \omega^2 \left(\frac{\varkappa_{lb}}{v_A^2} + \frac{\varkappa_{lp}}{c_s^2} + 2p\nabla_l \frac{p^{-1}}{v_A^2}\right),$$

$$\kappa_0 = \frac{\omega^4}{v_A^2 c_s^2} + \omega^2 \left(\frac{\varkappa_a^{(2)} + \varkappa_{ls}\varkappa_{la} - \varkappa_{1g}\varkappa_{1P}/g_1}{v_A^2} + \varkappa_{lb} p\nabla_l \frac{p^{-1}}{v_A^2} + p\nabla_l^2 \frac{p^{-1}}{v_A^2}\right),$$

where

$$p = \sqrt{g_2/g_1}, \quad \varkappa_{lp} = \nabla_l(\ln p^{-1}), \quad \varkappa_{la} = \nabla_l\left(\ln \frac{B_0}{\varkappa_{1g}}\right), \quad \varkappa_{ls} = \nabla_l\left(\ln \frac{\sqrt{g_1 g_2} P_0^\sigma}{\rho_0}\right),$$

$$\varkappa_{lb} = \varkappa_{ls} + 2\varkappa_{la}, \qquad \varkappa_{lc} = \nabla_l\left(\ln \frac{\varkappa_{1B} B_0 P_0^\sigma}{\rho_0}\right), \qquad \varkappa_a^{(2)} = \frac{\varkappa_{1g}}{B_0}\nabla_l^2 \frac{B_0}{\varkappa_{1g}},$$

and $\varkappa_{1g}$, $\varkappa_{1B}$ and $\varkappa_{1P}$ are defined in (3.416).

R Equation for MHD Oscillations in a Cylindrical Coordinate System

Putting the vectors of the background medium parameters $\mathbf{B}_0 = (0, B_0, 0)$ and $\mathbf{v}_0 = (0, \rho\Omega, v_{0z})$ in a cylindrical coordinate system (ρ, ϕ, z), let us write the following results of the action of the vector operators in (2.229), (2.232):

$$(\mathbf{v}_0\nabla)\mathbf{v} = \left(i(\omega-\bar{\omega})v_\rho - \Omega v_\phi,\ i(\omega-\bar{\omega})v_\phi + \Omega v_\rho,\ i(\omega-\bar{\omega})v_z\right),$$
$$(\mathbf{v}\nabla)\mathbf{v}_0 = \left(-\Omega v_\phi,\ v_\rho\frac{d\rho\Omega}{d\rho},\ v_\rho v'_{0z}\right),$$
$$(\mathbf{v}_0\nabla)P = i(\omega-\bar{\omega})P,$$
$$(\mathbf{v}\nabla)P_0 = v_\rho P'_0,$$

where the prime denotes a derivative with respect to ρ ($P'_0 = dP_0/d\rho$, $v'_{0z} = dv_{0z}/d\rho$). Let us rewrite the Eqs. (2.229)–(2.230) component-wise:

$$-i\bar{\omega}\rho_0 v_\rho - \rho\Omega^2\tilde{\rho} - 2\Omega\rho_0 v_\phi = -\frac{dP}{d\rho} - \frac{B_0}{4\pi}\left(\frac{1}{\rho}\frac{d\rho B_\phi}{d\rho} - i\frac{m}{\rho}B_\rho\right) - \frac{B_\phi}{4\pi}\frac{1}{\rho}\frac{d\rho B_0}{d\rho}, \tag{R.1}$$

$$-i\bar{\omega}\rho_0 v_\phi + 2\Omega\rho_0 v_\rho + \rho_0 v_\rho \rho\Omega' = -i\frac{m}{\rho}P + \frac{B_\rho}{4\pi}\frac{1}{\rho}\frac{d\rho B_0}{d\rho}, \tag{R.2}$$

$$-i\bar{\omega}\rho_0 v_z + \rho_0 v'_{0z} v_\rho = -ik_z P + i\frac{B_0}{4\pi}\left(\frac{m}{\rho}B_z - k_z B_\phi\right), \tag{R.3}$$

$$-i\bar{\omega}B_\rho = i\frac{m}{\rho}B_0 v_\rho, \tag{R.4}$$

$$-i(\bar{\omega} + m\Omega)B_\phi = ik_z(\rho\Omega B_z - B_0 v_z) + \frac{d}{d\rho}(\rho\Omega B_r - B_0 v_\rho), \tag{R.5}$$

$$-i(\bar{\omega} + k_z v_{0z})B_z = i\frac{m}{\rho}v_{0z}B_\phi + \frac{1}{\rho}\frac{d}{d\rho}(\rho v_{0z}B_\rho) + i\frac{m}{\rho}B_0 v_z. \tag{R.6}$$

(2.235) follows immediately from (R.4). Moreover, using the condition

$$\operatorname{div}\mathbf{B} = \frac{1}{\rho}\frac{d\rho B_\rho}{d\rho} + i\frac{m}{\rho}B_\phi + ik_z B_z = 0,$$

we obtain from (R.5) and (R.6):

$$B_\phi = \frac{k_z B_0}{\bar{\omega}}v_z - i\frac{m\Omega' B_0}{\bar{\omega}^2}v_\rho - \frac{i}{\bar{\omega}}\frac{dB_0 v_\rho}{d\rho}, \tag{R.7}$$

$$B_z = -i\frac{m B_0}{\rho\bar{\omega}}\left(\frac{v'_{0z}}{\bar{\omega}}v_\rho - iv_z\right). \tag{R.8}$$

From (2.232) we have

$$P = -\frac{i}{\bar{\omega}}\left(v_\rho P'_0 + \gamma P_0 \operatorname{div}\mathbf{v}\right). \tag{R.9}$$

Substituting P from (R.9) and B_ρ from (2.235) in (R.2) produces

$$\varkappa_s^2 v_\phi = -i\frac{m v_s^2}{\rho\bar{\omega}^2}\left(\frac{1}{\rho}\frac{d\rho v_\rho}{d\rho} + ik_z v_z\right)$$
$$-iv_\rho\left[\frac{m v_s^2}{\rho\bar{\omega}^2}\frac{(\ln P_0)'}{\gamma} + \frac{(\rho\Omega)' + \Omega}{\bar{\omega}} + \frac{m v_A^2}{\rho\bar{\omega}^2}(\ln \rho B_0)'\right],$$

where $\varkappa_s^2$ is defined by (2.244). Using also the balance condition for the background plasma (2.225), we have

$$v_\phi = -\frac{i}{\varkappa_s}\left[\frac{m v_s^2}{\rho\bar{\omega}^2}\left(\frac{1}{\rho}\frac{d\rho v_\rho}{d\rho} + ik_z v_z\right) + \left(\frac{m\Omega^2}{\bar{\omega}^2} + \frac{(\rho\Omega)' + \Omega}{\bar{\omega}}\right)v_\rho\right]. \tag{R.10}$$

Similarly, substituting P from (R.9), B_z from (R.56), and v_ϕ from (R.10) in (R.51), we obtain expression (2.234) for v_z. Substituting it in (R.10), we obtain expression (2.233) for v_ϕ. Substituting also v_z from (2.234) in (R.7) and (R.8), we obtain expressions (2.236) and (2.237) for the field oscillation components B_ϕ and B_z. Substitution of v_ϕ and v_z from (2.233) and (2.234) in (R.9) yields the expression (2.238) for perturbed pressure.

Expressing div $\mathbf{v}$ from (2.232) and substituting in (2.231), we obtain

$$\tilde{\rho} = v_s^{-2}\left(P + i\frac{P_0'}{\bar{\omega}}v_\rho\right).$$

Finally, substituting expressions for the field oscillation components expressed in terms of v_ρ in (R.1) and using the variable $\zeta = iv_\rho/\bar{\omega}$, we obtain, after simple but rather cumbersome transformations, Eq. (2.245) for MHD oscillations in a plasma that moves at velocity v_{0z} along the z axis and rotates around it with angular velocity Ω.

S Equality of the Alfvén Oscillation Specific Power Absorbed Near the Resonance Surface and the Density of Energy Carried Away by KAWs

Let us demonstrate that the specific power the resonant Alfvén oscillations lose as they decay near the resonance surface is equal to the density of the energy flux carried away by kinetic Alfvens (KAWs). To do this, we will take into account the small dissipation of the Alfvén waves by making a formal replacement $\omega \to \omega - i\gamma$ in (2.77), where $\gamma \ll |\omega|$ is a local decrement of the Alfvén oscillations. It may be associated, for example, with the decay of standing Alfvén waves in the ionosphere due to its finite conductivity (see [205]). Near the resonance surface, where the decomposition (2.81) is valid, we represent (2.77) in the form

$$\Lambda^2\frac{d^2\varphi'}{dx^2} + \frac{x - x_A + i\varepsilon}{a_A}\varphi' \approx \tilde{C}, \tag{S.1}$$

where x_A is coordinate of the resonance surface, $\varepsilon = 2\gamma a_A/\omega$ is the characteristic scale of the resonant Alfvén oscillation, $\tilde{C} \approx$ const. Most of the energy of Alfvén oscillations is contained in their magnetic field, or, when near the resonance surface, in the component $B_y = -(k_z\bar{c}/\omega)\varphi'$, exhibiting the highest amplitude. In further calculations, we will only use this main component of the field of resonant Alfvén oscillations.

Away from the resonance surface, where the WKB approximation is applicable, the KAW field described by (S.1) can be represented in the form

$$\varphi' = \frac{C}{\sqrt{k_x}} \exp\left(i \int_{x_A}^{x} k_x(x')dx' \right), \tag{S.2}$$

where $C = const$. If the condition $|\gamma/\omega| \ll |\Lambda/a_A|^{2/3}$ is satisfied near the resonance surface, the resonant Alfvén oscillations have a KAW structure in the entire region of their existence. If the inverse inequality $|\gamma/\omega| \gg |\Lambda/a_A|^{2/3}$ holds, the first term in (S.1) is small near the resonance surface and can be neglected. In this case, we have an Alfvén oscillation field having a classical resonant structure of the form

$$\varphi' \approx \frac{\tilde{C} a_A}{x - x_A + i\varepsilon}, \tag{S.3}$$

which is also obtained in the ideal MHD approximation. Let us define the specific (per unit area of the resonance surface) absorbed power of Alfvén oscillations as

$$W \approx -\int_{-\Delta}^{\Delta} \frac{\partial}{\partial t} \frac{|B|^2}{8\pi} dx \approx 2\gamma \int_{-\Delta}^{\Delta} \frac{|B_y|^2}{8\pi} dx,$$

where $\Delta \gg \varepsilon$. For extremely small values of ε, we have

$$\begin{aligned} W &\approx \lim_{\varepsilon \to 0} |\tilde{C}|^2 \frac{\gamma a_A^2}{4\pi} \left(\frac{k_z \bar{c}}{\omega} \right)^2 \int_{-\Delta}^{\Delta} \frac{d(x - x_A)}{(x - x_A)^2 + \varepsilon^2} \\ &= |\tilde{C}|^2 \frac{a_A (k_z \bar{c})^2}{8\pi\omega} \int_{-\infty}^{\infty} \frac{dy}{y^2 + 1} = |\tilde{C}|^2 \frac{a_A (k_z \bar{c})^2}{8\omega}. \end{aligned} \tag{S.4}$$

It can be seen that the absorbed specific power does not depend on the value of decrement γ. This is explained by the fact that a decrement causes the characteristic size of the localisation of the field of resonant Alfvén oscillations to decrease and their amplitude to increase. The integral (S.4) meanwhile remains unchanged.

In the $|\gamma/\omega| \ll |\Lambda/a_A|^{2/3}$ case, the energy is carried away from the resonance surface by KAWs. The KAW structure near the resonance surface is described by an expression of the form (2.87), where $\tilde{\varphi}' \equiv \tilde{C}(a_A/\Lambda)^{2/3}$. Away from the resonance surface, where the WKB approximation is applicable, we can represent the solution of (S.1) using an expression of the form (S.2), where $k_x = \sqrt{(x - x_A)/\tilde{a}^3}$, ($\tilde{a} = |a_A|^{1/3}|\Lambda|^{2/3}$). Substituting this expression for k_x in (S.2) and comparing the resulting expression with (2.87), employing an asymptotic expression for the function $G(z)$ for $z \to \infty$, we obtain the relationship

$$C = \tilde{C} \frac{\sqrt{\pi a_A}}{\Lambda} e^{-i3\pi/4}$$

between the constants determining the amplitudes in these expressions. The density of the flux of energy carried by KAWs along the x axis is determined as

$$S_x \approx \frac{|B_y|^2}{8\pi} v_{gx},$$

where $v_{gx} = \partial\omega/\partial k_x \approx \omega k_x \Lambda^2$ is the KAW group velocity component along the x axis. Substituting here the expression for B_y with φ' of the form (S.2) and using the above relationship between the constants C and $\widetilde{C}$, we obtain

$$S_x \approx \frac{|C|^2}{8\pi k_x}\left(\frac{k_z\bar{c}}{\omega}\right)^2 \omega k_x \Lambda^2 = |\widetilde{C}|^2 \frac{a_A (k_z\bar{c})^2}{8\omega},$$

which is exactly the same as in (S.4). Hence, the specific power of the Alfvén oscillations which is absorbed near the resonance surface and the flux density of the energy carried away by KAWs from the resonance surface are identical and independent of the magnitude of the small parameters determining these processes.

References

1 Gul'elmi, A. (1974). Diagnostics of the magnetosphere and interplanetary medium by means of pulsations. *Space Science Reviews* 16: 331–345. https://doi.org/10.1007/BF00171562.

2 Nishida, A. (1978). *Geomagnetic Diagnosis of the Magnetosphere, Physics and Chemistry in Space*. New York: Springer.

3 Stewart, B. (1861). On the great magnetic disturbance which extended from August 28 to September 7, 1859, as recorded by photography at the Kew Observatory. *Royal Society of London Philosophical Transactions Series I* 151: 423–430.

4 Rolf, B. (1931). Giant micropulsations at Abisko. *Terrestrial Magnetism and Atmospheric Electricity (Journal of Geophysical Research)* 36: 9–14. https://doi.org/10.1029/TE036i001p00009.

5 Harang, L. (1932). Observations of micropulsations in the magnetic records at Tromsø. *Terrestrial Magnetism and Atmospheric Electricity (Journal of Geophysical Research)* 37: 57–61. https://doi.org/10.1029/TE037i001p00057.

6 Alfvén, H. (1950). *Cosmical Electrodynamics*. Oxford: Clarendon Press.

7 Vernov, S.N., Savenko, I.A., Shavris, P.I. et al. (1963). The Earth's radiation belts at altitudes of 180-250 km. *Planetary and Space Science* 11: 567–574. https://doi.org/10.1016/0032-0633(63)90080-1.

8 Ness, N.F., Scearce, C.S., and Seek, J.B. (1964). Initial results of the Imp 1 magnetic field experiment. *Journal of Geophysical Research* 69: 3531–3569. https://doi.org/10.1029/JZ069i017p03531.

9 Ness, N.F. (1965). The Earth's magnetic tail. *Journal of Geophysical Research* 70: 2989–3005. https://doi.org/10.1029/JZ070i013p02989.

10 Hones, E.W. Jr., Asbridge, J.R., and Bame, S.J. (1971). Time variations of the magnetotail plasma sheet at 18 R_E determined from concurrent observations by a pair of Vela satellites. *Journal of Geophysical Research* 76: 4402–4419. https://doi.org/10.1029/JA076i019p04402.

11 Guglielmi, A.V. and Troitskaya, V.A. (1973). *Geomagnetic Pulsations and Diagnostics of the Magnetosphere*. Moskow: Nauka (in Russian).

12 Pudovkin, M.I., Raspopov, O.M., and Kleimenova, N.G. (1976). *Perturbations of the Earth's Electromagnetic Field, Part 1*. Leningrad: Leningrad University Publisher (in Russian).

13 Guglielmi, A.V. (1979). *MHD-Waves in the Near-Earth Plasma*. Moskow: Nauka (in Russian).

Magnetospheric MHD Oscillations: A Linear Theory, First Edition. Anatoly Leonovich, Vitalii Mazur, and Dmitri Klimushkin.

14 Menk, F.W. and Waters, C.L. (2013). *Magnetoseismology: Ground-Based Remote Sensing of Earth's Magnetosphere*, 1e. Weinheim, Germany: Wiley-VCH.

15 Pilipenko, V.A. (1990). ULF waves on the ground and in space. *Journal of Atmospheric and Terrestrial Physics* 52: 1193–1209. https://doi.org/10.1016/0021-9169(90)90087-4.

16 Guglielmi, A. and Pokhotelov, O. (1993). Nonlinear problems of physics of the geomagnetic pulsations. *Space Science Reviews* 65: 5–57. https://doi.org/10.1007/BF00749761.

17 Kangas, J., Guglielmi, A., and Pokhotelov, O. (1998). Morphology and physics of short-period magnetic pulsations. *Space Science Reviews* 83: 435–512.

18 Glassmeier, K.H., Othmer, C., Cramm, R. et al. (1999). Magnetospheric field line resonances: a comparative planetology approach. *Surveys in Geophysics* 20: 61–109. https://doi.org/10.1023/A:1006659717963.

19 Dungey, J.W. (1954). Electrodynamics of the Outer Atmospheres. Pensylvania State Univ. Ionos. Res. Lab., Sci. Rep., Cambridge, PA, p. 69.

20 Dungey, J.W. (1961). Interplanetary magnetic field and the auroral zones. *Physical Review Letters* 6: 47–48. https://doi.org/10.1103/PhysRevLett.6.47.

21 Johnson, J.R., Wing, S., and Delamere, P.A. (2014). Kelvin Helmholtz instability in planetary magnetospheres. *Space Science Reviews* 184: 1–31. https://doi.org/10.1007/s11214-014-0085-z.

22 Hasegawa, H., Fujimoto, M., Takagi, K. et al. (2006). Single-spacecraft detection of rolled-up Kelvin-Helmholtz vortices at the flank magnetopause. *Journal of Geophysical Research: Space Physics* 111: A09203 https://doi.org/10.1029/2006JA011728.

23 Krall, N.A. and Trivelpiece, A.W. (1973). *Principles of Plasma Physics*, International Student Edition –International Series in Pure and Applied Physics. Tokyo: McGraw-Hill Kogakusha.

24 Akhiezer, A.I., Akhiezer, I.A., Polovin, R.V. et al. (1975). *Plasma Electrodynamics. Linear Theory*, International Series on Natural Philosophy, vol. 1. Oxford: Pergamon Press.

25 Kadomtsev, B.B. (1976). *Collective Phenomena in Plasma*. Moscow: Nauka (in Russian).

26 Mikhailovskii, A.B. (1975). *Plasma Instability Theory. Instabilities in a Homogeneous Plasma /2nd Revised and Enlarged Edition*, vol. 1. Moscow: Atomizdat.

27 Abramowitz, M. and Stegun, I.A. (1972). *Handbook of Mathematical Functions*. New York: Dover.

28 Leonovich, A.S., Kozlov, D.A., and Edemskiy, I.K. (2010). Standing slow magnetosonic waves in a dipole-like plasmasphere. *Planetary and Space Science* 58: 1425–1433. https://doi.org/10.1016/j.pss.2010.06.007.

29 Cheng, C.Z. and Zaharia, S. (2003). Field line resonances in quiet and disturbed time three-dimensional magnetospheres. *Journal of Geophysical Research: Space Physics* 108: 1001. https://doi.org/10.1029/2002JA009471.

30 Osterbrock, D.E. (1961). The heating of the solar chromosphere, plages, and corona by magnetohydrodynamic waves. *Astrophysical Journal* 134: 347–388. https://doi.org/10.1086/147165.

31 Ofman, L. (2005). MHD waves and heating in coronal holes. *Space Science Reviews* 120: 67–94. https://doi.org/10.1007/s11214-005-5098-1.

32 Tataronis, J.A. and Grossmann, W. (1976). On Alfven wave heating and transit time magnetic pumping in the guiding-centre model of a plasma. *Nuclear Fusion* 16: 667–678.

33 Karney, C.F.F., Perkins, F.W., and Sun, Y.C. (1979). Alfven resonance effects on magnetosonic modes in large tokamaks. *Physical Review Letters* 42: 1621–1624. https://doi.org/10.1103/PhysRevLett.42.1621.

34 Tamao, T. (1965). Transmission and coupling resonance of hydromagnetic disturbances in the non-uniform Earth's magnetosphere. *Science Reports Tohoku University* 17: 43–54.

35 Chen, L. and Hasegawa, A. (1974). A theory of long-period magnetic pulsations: 1. Steady state excitation of field line resonance. *Journal of Geophysical Research* 79: 1024–1032. https://doi.org/10.1029/JA079i007p01024.

36 Radoski, H.R. (1974). A theory of latitude dependent geomagnetic micropulsations: the asymptotic fields. *Journal of Geophysical Research* 79: 595–603. https://doi.org/10.1029/JA079i004p00595.

37 Southwood, D.J. (1974). Some features of field line resonances in the magnetosphere. *Planetary and Space Science* 22: 483–491. https://doi.org/10.1016/0032-0633(74)90078-6.

38 Green, C.A. (1978). Meridional characteristics of a Pc 4 micropulsation event in the plasmasphere. *Planetary and Space Science* 26: 955–967. https://doi.org/10.1016/0032-0633(78)90078-8.

39 Lessard, M.R., Hudson, M.K., Samson, J.C., and Wygant, J.R. (1999). Simultaneous satellite and ground-based observations of a discretely driven field line resonance. *Journal of Geophysical Research: Space Physics* 104: 12361–12378. https://doi.org/10.1029/1998JA900117.

40 Guglielmi, A.V. (1989). Diagnostics of the plasma in the magnetosphere by means of measurement of the spectrum of Alfven oscillations. *Planetary and Space Science* 37: 1011–1012. https://doi.org/10.1016/0032-0633(89)90055-X.

41 Kawano, H., Yumoto, K., Pilipenko, V.A. et al. (2002). Using two ground stations to identify magnetospheric field line eigenfrequency as a continuous function of ground latitude. *Journal of Geophysical Research: Space Physics* 107: 1202. https://doi.org/10.1029/2001JA000274.

42 Menk, F.W., Orr, D., Clilverd, M.A. et al. (1999). Monitoring spatial and temporal variations in the dayside plasmasphere using geomagnetic field line resonances. *Journal of Geophysical Research: Space Physics* 104: 19955–19970. https://doi.org/10.1029/1999JA900205.

43 Denton, R.E., Lessard, M.R., Anderson, R. et al. (2001). Determining the mass density along magnetic field lines from toroidal eigenfrequencies: polynomial expansion applied to CRRES data. *Journal of Geophysical Research: Space Physics* 106: 29915–29924. https://doi.org/10.1029/2001JA000204.

44 Menk, F.W., Mann, I.R., Smith, A.J. et al. (2004). Monitoring the plasmapause using geomagnetic field line resonances. *Journal of Geophysical Research: Space Physics* 109: A04216 https://doi.org/10.1029/2003JA010097.

45 Samson, J.C., Cogger, L.L., and Pao, Q. (1996). Observations of field line resonances, auroral arcs, and auroral vortex structures. *Journal of Geophysical Research: Space Physics* 101: 17373–17384. https://doi.org/10.1029/96JA01086.

46 Stasiewicz, K., Bellan, P., Chaston, C. et al. (2000). Small scale Alfvénic structure in the aurora. *Space Science Reviews* 92: 423–533.

47 Samson, J.C., Rankin, R., and Tikhonchuk, V.T. (2003). Optical signatures of auroral arcs produced by field line resonances: comparison with satellite observations and modeling. *Annales Geophysicae* 21: 933–945. https://doi.org/10.5194/angeo-21-933-2003.

48 Hasegawa, A. (1976). Particle acceleration by MHD surface wave and formation of aurora. *Journal of Geophysical Research* 81: 5083–5090. https://doi.org/10.1029/JA081i028p05083.

49 Goertz, C.K. (1984). Kinetic Alfven waves on auroral field lines. *Planetary and Space Science* 32: 1387–1392. https://doi.org/10.1016/0032-0633(84)90081-3.

50 Saka, O., Hayashi, K., and Leonovich, A.S. (2015). Ionospheric loop currents and associated ULF oscillations at geosynchronous altitudes during preonset intervals of substorm aurora. *Journal of Geophysical Research: Space Physics* 120: 2460–2468. https://doi.org/10.1002/2014JA020842.

51 Yumoto, K. (1985). Characteristics of localized resonance coupling oscillations of the slow magnetosonic wave in a non-uniform plasma. *Planetary and Space Science* 33: 1029–1036. https://doi.org/10.1016/0032-0633(85)90021-2.

52 Cadez, V.M., Csik, A., Erdelyi, R., and Goossens, M. (1997). Absorption of magnetosonic waves in presence of resonant slow waves in the solar atmosphere. *Astronomy and Astrophysics* 326: 1241–1251.

53 Leonovich, A.S., Kozlov, D.A., and Pilipenko, V.A. (2006). Magnetosonic resonance in a dipole-like magnetosphere. *Annales Geophysicae* 24: 2277–2289. https://doi.org/10.5194/angeo-24-2277-2006.

54 Leonovich, A.S. and Kozlov, D.A. (2009). Alfvenic and magnetosonic resonances in a nonisothermal plasma. *Plasma Physics and Controlled Fusion* 51 (8): 085007 https://doi.org/10.1088/0741-3335/51/8/085007.

55 Vdovin, V.L., Zinov'ev, O.A., Ivanov, A.A. et al. (1971). Excitation of magnetosonic resonance in the Tokamak plasma. *ZhETF Pisma Redaktsiiu* 14: 228.

56 Arregui, I., Oliver, R., and Ballester, J.L. (2003). Coupling of fast and Alfvén waves in a straight bounded magnetic field with density stratification. *Astronomy and Astrophysics* 402: 1129–1143. https://doi.org/10.1051/0004-6361:20030312.

57 Kivelson, M.G. and Southwood, D.J. (1985). Resonant ULF waves – a new interpretation. *Geophysical Research Letters* 12: 49–52. https://doi.org/10.1029/GL012i001p00049.

58 Leonovich, A.S., Kozlov, D.A., and Vlasov, A.A. (2021). Kinetic Alfven waves near a dissipative layer. *Journal of Geophysical Research: Space Physics* 126 (10): e29580 https://doi.org/10.1029/2021JA029580.

59 Landau, L.D. (1946). On the vibration of the electronic plasma. *The Journal of Physics, USSR* 16: 574–586.

60 Landau, L.D. and Lifshitz, E.M. (1960). *Electrodynamics of Continuous Media*. Oxford: Pergamon Press.

61 Leonovich, A.S. and Kozlov, D.A. (2013). Magnetosonic resonances in the magnetospheric plasma. *Earth, Planets and Space* 65: 369–384. https://doi.org/10.5047/eps.2012.07.002.

62 Korn, G.A. and Korn, T.M. (1968). *Mathematical Handbook for Scientists and Engineers. Definitions, Theorems, and Formulas for Reference and Review*, 2nd enlarged and revised edition. New York: McGraw-Hill.

63 Landau, L.D. and Lifshitz, E.M. (1969). *Mechanics*, vol. 1. Oxford: Pergamon Press.

64 Tamao, T. (1986). Direct contribution of oblique field-aligned currents to ground magnetic fields. *Journal of Geophysical Research: Space Physics* 91: 183–189. https://doi.org/10.1029/JA091iA01p00183.

65 Klimushkin, D.Y. (1994). Method of description of the Alfvén and magnetosonic branches of inhomogeneous plasma oscillations. *Plasma Physics Reports* 20: 280–286.

66 Leonovich, A.S., Mazur, V.A., and Senatorov, V.N. (1983). Dispersion effects of magnetohydrodynamic waves in an inhomogeneous plasma. *Issledovaniia po Geomagnetizmu Aeronomii i Fizike Solntsa* 66: 3–17.

67 Leonovich, A.S., Mazur, V.A., and Senatorov, V.N. (1985). MHD waveguides in a nonuniform plasma. *Soviet Journal of Plasma Physics* 11: 1106–1115.

68 Gulelmi, A.V. (1985). Ray theory of MHD-wave propagation – review. *Geomagnetism and Aeronomy* 25: 356–370.

69 Landau, L.D. and Lifshitz, E.M. (1977). *Quantum Mechanics: Non-Relativistic Theory*, vol. 3. Oxford: Pergamon Press. ISBN: 978-0-08-020940-1

70 Leonovich, A.S., Mazur, V.A., and Senatorov, V.N. (1983). An Alfven waveguide. *Zhurnal Eksperimental'noi i Teoreticheskoi Fiziki* 85: 141–145.

71 Gulelmi, A.V. and Poliakov, A.R. (1983). On the discreteness of the spectrum of Alfven oscillations. *Geomagnetism and Aeronomy* 23: 341–343.

72 Dmitrienko, I.S. and Mazur, V.A. (1985). On waveguide propagation of Alfven waves at the plasmapause. *Planetary and Space Science* 33: 471–477. https://doi.org/10.1016/0032-0633(85)90092-3.

73 Mikhailova, O.S. (2014). The spatial structure of ULF-waves in the equatorial resonator localized at the plasmapause with the admixture of the heavy ions. *Journal of Atmospheric and Solar-Terrestrial Physics* 108: 10–16. https://doi.org/10.1016/j.jastp.2013.12.007.

74 Mikhailova, O.S., Mager, P.N., and Klimushkin, D.Y. (2020). Transverse resonator for ion-ion hybrid waves in dipole magnetospheric plasma. *Plasma Physics and Controlled Fusion* 62 (9): 095008 https://doi.org/10.1088/1361-6587/ab9be9.

75 Galperin, I.I., Leonovich, A.S., Mazur, V.A., and Tashchilin, A.V. (1986). Motion of a packet of small-scale Alfven waves in the midlatitude plasmasphere. *Cosmic Research (Engl. Transl.); (United States)* 24: 598–609.

76 Samson, J.C. and Rankin, R. (1994). The coupling of solar wind energy to MHD cavity modes, waveguide modes, and field line resonances in the Earth's magnetosphere. In: *Solar Wind Sources of Magnetospheric Ultra-Low-Frequency Waves*, Geophysical Monograph Series, vol. 81 (ed. M.J. Engebreston, K. Takahashi, and M. Scholer), 253–264. Washington, DC: American Geophysical Union. https://doi.org/10.1029/GM081p0253.

77 Walker, A.D.M. (1998). Excitation of magnetohydrodynamic cavities in the magnetosphere. *Journal of Atmospheric and Solar-Terrestrial Physics* 60: 1279–1293. https://doi.org/10.1016/S1364-6826(98)00077-7.

78 Wright, A.N. and Mann, I.R. (2006). Global MHD eigenmodes of the outer magnetosphere. In: *Magnetospheric ULF Waves: Synthesis and New Directions*, Geophysical Monograph Series, vol. 169 (ed. K. Takahashi, P.J. Chi, R.E. Denton, and R.L. Lysak), 51. Washington, DC: American Geophysical Union.

79 Mazur, V.A. (2010). Resonance excitation of the magnetosphere by hydromagnetic waves incident from solar wind. *Plasma Physics Reports* 36: 953–963. https://doi.org/10.1134/S1063780X10110048.

80 Mazur, V.A. and Chuiko, D.A. (2013). Kelvin-Helmholtz instability on the magnetopause, magnetohydrodynamic waveguide in the outer magnetosphere, and Alfvén

resonance deep in the magnetosphere. *Plasma Physics Reports* 39: 488–503. https://doi.org/10.1134/S1063780X13060068.
81 Roberts, B. and Webb, A.R. (1978). Vertical motions in an intense magnetic flux tube. *Solar Physics* 56: 5–35. https://doi.org/10.1007/BF00152630.
82 Farahani, S.V., Nakariakov, V.M., and van Doorsselaere, T. (2010). Long-wavelength torsional modes of solar coronal plasma structures. *Astronomy and Astrophysics* 517: A29. https://doi.org/10.1051/0004-6361/201014502.
83 Chappell, C.R. (1974). Detached plasma regions in the magnetosphere. *Journal of Geophysical Research* 79: 1861–1870. https://doi.org/10.1029/JA079i013p01861.
84 Posch, J.L., Engebretson, M.J., Murphy, M.T. et al. (2010). Probing the relationship between electromagnetic ion cyclotron waves and plasmaspheric plumes near geosynchronous orbit. *Journal of Geophysical Research: Space Physics* 115: A11205 https://doi.org/10.1029/2010JA015446.
85 Usanova, M.E., Darrouzet, F., Mann, I.R., and Bortnik, J. (2013). Statistical analysis of EMIC waves in plasmaspheric plumes from Cluster observations. *Journal of Geophysical Research: Space Physics* 118: 4946–4951. https://doi.org/10.1002/jgra.50464.
86 Kaye, S.M. and Kivelson, M.G. (1979). Observations of Pc 1–2 waves in the outer magnetosphere. *Journal of Geophysical Research: Space Physics* 84: 4267–4276. https://doi.org/10.1029/JA084iA08p04267.
87 Sato, N., Hirasawa, T., and Shirokura, Y. (1987). Fingerprint structure Pc 1 geomagnetic pulsations. *Geophysical Research Letters* 14: 664–667. https://doi.org/10.1029/GL014i006p00664.
88 Roux, A., Gendrin, R., Wehrlin, N. et al. (1973). Fine structure of Pc 1 pulsations: 2. Theoretical interpretation. *Journal of Geophysical Research* 78: 3176–3181. https://doi.org/10.1029/JA078i016p03176.
89 Liemohn, H.B. (1967). Cyclotron-resonance amplification of VLF and ULF whistlers. *Journal of Geophysical Research* 72: 39–55. https://doi.org/10.1029/JZ072i001p00039.
90 Tverskoy, B.A. (1969). Main mechanisms in the formation of the Earth's radiation belts. *Reviews of Geophysics* 7: 219–231. https://doi.org/10.1029/RG007i001p00219.
91 Leonovich, A.S. (1984). Propagation of geomagnetic pulsations in magnetospheric ducts. *Geomagnetism and Aeronomy/Geomagnetizm i Aeronomiia* 24: 94–98.
92 Buzevich, A.V., Leonovich, A.S., and Parkhomov, V.A. (1987). Geomagnetic pulsations associated with MHD-waveguide in magnetospheric ducts. *Planetary and Space Science* 35 (9): 1093–1100. https://doi.org/10.1016/0032-0633(87)90014-6.
93 Rycroft, M.J. (1975). A review of in situ observations of the plasmapause. *Annales de Geophysique* 31: 1–16.
94 Krinberg, I.A. and Taschilin, A.V. (1984). *Ionosphere and Plasmasphere*. Moskow: Nauka.
95 Mazur, V.A. and Potapov, A.S. (1983). The evolution of pearls in the Earth's magnetosphere. *Planetary and Space Science* 31: 859–863. https://doi.org/10.1016/0032-0633(83)90139-3.
96 Axford, W.I. (1964). Viscous interaction between the solar wind and the Earth's magnetosphere. *Planetary and Space Science* 12: 45–53. https://doi.org/10.1016/0032-0633(64)90067-4.
97 McKenzie, J.F. (1970). Hydromagnetic wave interaction with the magnetopause and the bow shock. *Planetary and Space Science* 18: 1–23. https://doi.org/10.1016/0032-0633(70)90063-2.

98 Verzariu, P. (1973). Reflection and refraction of hydromagnetic waves at the magnetopause. *Planetary and Space Science* 21: 2213–2225. https://doi.org/10.1016/0032-0633(73)90194-3.

99 Wolfe, A. and Kaufmann, R.L. (1975). MHD wave transmission and production near the magnetopause. *Journal of Geophysical Research* 80: 1764–1775. https://doi.org/10.1029/JA080i013p01764.

100 Leonovich, A.S., Mishin, V.V., and Cao, J.B. (2003). Penetration of magnetosonic waves into the magnetosphere: influence of a transition layer. *Annales Geophysicae* 21: 1083–1093. https://doi.org/10.5194/angeo-21-1083-2003.

101 Brekhovskikh, L.M., Lieberman, D., Beyer, R.T. et al. (1962). Waves in layered media. *Physics Today* 15: 70. https://doi.org/10.1063/1.3058151.

102 Anderson, J.E. (1963). *Magnetohydrodynamic Shock Waves*. Cambridge: MIT Press.

103 Matthaeus, W.H. and Goldstein, M.L. (1982). Measurement of the rugged invariants of magnetohydrodynamic turbulence in the solar wind. *Journal of Geophysical Research: Space Physics* 87: 6011–6028. https://doi.org/10.1029/JA087iA08p06011.

104 Goldstein, M.L., Roberts, D.A., and Matthaeus, W.H. (1995). Magnetohydrodynamic turbulence in the solar wind. *Annual Review of Astronomy and Astrophysics* 33: 283–326. https://doi.org/10.1146/annurev.aa.33.090195.001435.

105 Mishin, V.M., Russell, C.T., Saifudinova, T.I., and Bazarzhapov, A.D. (2000). Study of weak substorms observed during December 8, 1990, Geospace Environment Modeling campaign: timing of different types of substorm onsets. *Journal of Geophysical Research: Space Physics* 105: 23263–23276. https://doi.org/10.1029/1999JA900495.

106 Dungey, J.W. (1955). Electrodynamics of the outer atmosphere. In: *Physics of the Ionosphere, Report of the Conference Held at the Cavendish Laboratory*, Cambridge (September 1954), 229–236. London: The Physical Society.

107 Parker, E.N. (1964). Dynamical properties of stellar coronas and stellar winds. I. Integration of the momentum equation. *Astrophysical Journal* 139: 72–91. https://doi.org/10.1086/147740.

108 Landau, L.D. (1944). Stability of tangential discontinuities in compressible fluid. *Doklady Akademii Nauk SSSR* 44: 139–142.

109 Syrovatsky, S.I. (1954). Instability of tangential discontinuities in a compressible medium. *Journal of Experimental and Theoretical Physics* 27: 121–123.

110 Michalke, A. (1964). On the inviscid instability of the hyperbolictangent velocity profile. *Journal of Fluid Mechanics* 19: 543–556. https://doi.org/10.1017/S0022112064000908.

111 Mishin, V.V. and Morozov, A.G. (1983). On the effect of oblique disturbances on Kelvin-Helmholtz instability at magnetospheric boundary layers and in solar wind. *Planetary and Space Science* 31: 821–828. https://doi.org/10.1016/0032-0633(83)90135-6.

112 Blumen, W. (1970). Shear layer instability of an inviscid compressible fluid. *Journal of Fluid Mechanics* 40: 769–781. https://doi.org/10.1017/S0022112070000435.

113 Blumen, W., Drazin, P.G., and Billings, D.F. (1975). Shear layer instability of an inviscid compressible fluid. Part 2. *Journal of Fluid Mechanics* 71: 305–316. https://doi.org/10.1017/S0022112075002595.

114 Drazin, P.G. and Davey, A. (1977). Shear layer instability of an inviscid compressible fluid. Part 3. *Journal of Fluid Mechanics* 82: 255–260. https://doi.org/10.1017/S0022112077000640.

115 Ray, T.P. and Ershkovich, A.I. (1983). Kelvin–Holmholtz instabilities in a sheared compressible plasma. *Monthly Notices of the Royal Astronomical Society* 204 (3): 821–831.
116 Rankin, R., Frycz, P., Samson, J.C., and Tikhonchuk, V.T. (1997). Shear flow vortices in magnetospheric plasmas. *Physics of Plasmas* 4: 829–840. https://doi.org/10.1063/1.872149.
117 Miura, A. and Pritchett, P.L. (1982). Nonlocal stability analysis of the MHD Kelvin-Helmholtz instability in a compressible plasma. *Journal of Geophysical Research: Space Physics* 87: 7431–7444. https://doi.org/10.1029/JA087iA09p07431.
118 Miura, A. (1992). Kelvin-Helmholtz instability at the magnetospheric boundary: dependence on the magnetosheath sonic Mach number. *Journal of Geophysical Research: Space Physics* 97: 10655–10675. https://doi.org/10.1029/92JA00791.
119 Kolykhalov, P.I. (1984). Instability of a shear discontinuity in a gas moving near a wall. *Fluid Dynamics* 19: 465–469. https://doi.org/10.1007/BF01093913.
120 Leonovich, A.S. and Mishin, V.V. (2005). Stability of magnetohydrodynamic shear flows with and without bounding walls. *Journal of Plasma Physics* 71: 645–664. https://doi.org/10.1017/S002237780400337X.
121 Sen, A.K. (1964). Effect of compressibility on Kelvin-Helmholtz instability in a plasma. *The Physics of Fluids* 7: 1293–1298. https://doi.org/10.1063/1.1711374.
122 Chandrasekhar, S. and Gillis, J. (1962). Hydrodynamic and hydromagnetic stability. *Physics Today* 15: 58. https://doi.org/10.1063/1.3058072.
123 Miles, J.W. (1957). On the reflection of sound at an interface of relative motion. *The Journal of the Acoustical Society of America* 29: 226–228. https://doi.org/10.1121/1.1908836.
124 Ribner, H.S. (1957). Reflection, transmission, and amplification of sound by a moving medium. *The Journal of the Acoustical Society of America* 29: 435–441. https://doi.org/10.1121/1.1908918.
125 Mann, I.R. and Wright, A.N. (1999). Diagnosing the excitation mechanisms of Pc5 magnetospheric flank waveguide modes and FLRs. *Geophysical Research Letters* 26: 2609–2612. https://doi.org/10.1029/1999GL900573.
126 Mann, I.R., Voronkov, I., Dunlop, M. et al. (2002). Coordinated ground-based and Cluster observations of large amplitude global magnetospheric oscillations during a fast solar wind speed interval. *Annales Geophysicae* 20: 405–426. https://doi.org/10.5194/angeo-20-405-2002.
127 Hartinger, M.D., Plaschke, F., Archer, M.O. et al. (2015). The global structure and time evolution of dayside magnetopause surface eigenmodes. *Geophysical Research Letters* 42 (8): 2594–2602. https://doi.org/10.1002/2015GL063623.
128 Kozyreva, O., Pilipenko, V., Lorentzen, D. et al. (2019). Transient oscillations near the dayside open-closed boundary: evidence of magnetopause surface mode? *Journal of Geophysical Research: Space Physics* 124 (11): 9058–9074. https://doi.org/10.1029/2018JA025684.
129 Walker, A.D.M. (1981). The Kelvin-Helmholtz instability in the low-latitude boundary layer. *Planetary and Space Science* 29: 1119–1133. https://doi.org/10.1016/0032-0633(81)90011-8.
130 Pu, Z.Y. and Kivelson, M.G. (1983). Kelvin-Helmholtz instability at the magnetopause: energy flux into the magnetosphere. *Journal of Geophysical Research: Space Physics* 88: 853–862. https://doi.org/10.1029/JA088iA02p00853.

131 Kivelson, M.G. and Pu, Z.Y. (1984). The Kelvin-Helmholtz instability on the magnetopause. *Planetary and Space Science* 32 (11): 1335–1341. https://doi.org/10.1016/0032-0633(84)90077-1.

132 Mishin, V.V. (1993). Accelerated motions of the magnetopause as a trigger of the Kelvin-Helmholtz instability. *Journal of Geophysical Research: Space Physics* 98 (A12): 21365–21371. https://doi.org/10.1029/93JA00417.

133 Mills, K.J., Wright, A.N., and Mann, I.R. (1999). Kelvin-Helmholtz driven modes of the magnetosphere. *Physics of Plasmas* 6: 4070–4087. https://doi.org/10.1063/1.873669.

134 Guglielmi, A.V., Potapov, A.S., and Klain, B.I. (2010). Rayleigh-Taylor-Kelvin-Helmholtz combined instability at the magnetopause. *Geomagnetism and Aeronomy* 50: 958–962. https://doi.org/10.1134/S0016793210080050.

135 Foullon, C., Farrugia, C.J., Fazakerley, A.N. et al. (2008). Evolution of Kelvin-Helmholtz activity on the dusk flank magnetopause. *Journal of Geophysical Research: Space Physics* 113: A11203 https://doi.org/10.1029/2008JA013175.

136 Agapitov, O.V. and Cheremnykh, O.K. (2013). Magnetospheric ULF waves driven by external sources. *Advances in Astronomy and Space Physics* 3: 12–19.

137 Horvath, I. and Lovell, B.C. (2021). Subauroral flow channel structures and auroral undulations triggered by Kelvin-Helmholtz Waves. *Journal of Geophysical Research: Space Physics* 126 (6): e29144 https://doi.org/10.1029/2021JA029144.

138 Engebretson, M., Glassmeier, K.H., Stellmacher, M. et al. (1998). The dependence of high-latitude PcS wave power on solar wind velocity and on the phase of high-speed solar wind streams. *Journal of Geophysical Research: Space Physics* 103: 26271–26384. https://doi.org/10.1029/97JA03143.

139 Nakariakov, V.M., Pilipenko, V., Heilig, B. et al. (2016). Magnetohydrodynamic oscillations in the solar corona and Earth's magnetosphere: towards consolidated understanding. *Space Science Reviews* 200 (1–4): 75–203. https://doi.org/10.1007/s11214-015-0233-0.

140 Southwood, D.J. (1968). The hydromagnetic stability of the magnetospheric boundary. *Planetary and Space Science* 16: 587–605. https://doi.org/10.1016/0032-0633(68)90100-1.

141 McKenzie, J.F. (1970). Hydromagnetic oscillations of the geomagnetic tail and plasma sheet. *Journal of Geophysical Research* 75: 5331–5339. https://doi.org/10.1029/JA075i028p05331.

142 Yumoto, K. and Saito, T. (1980). Hydromagnetic waves driven by velocity shear instability in the magnetospheric boundary layer. *Planetary and Space Science* 28: 789–798. https://doi.org/10.1016/0032-0633(80)90076-8.

143 Wright, A.N., Mills, K.J., Ruderman, M.S., and Brevdo, L. (2000). The absolute and convective instability of the magnetospheric flanks. *Journal of Geophysical Research: Space Physics* 105: 385–394. https://doi.org/10.1029/1999JA900417.

144 Turkakin, H., Rankin, R., and Mann, I.R. (2013). Primary and secondary compressible Kelvin-Helmholtz surface wave instabilities on the Earth's magnetopause. *Journal of Geophysical Research: Space Physics* 118: 4161–4175. https://doi.org/10.1002/jgra.50394.

145 Pu, Z.Y. and Kivelson, M.G. (1983). Kelvin-Helmholtz instability at the magnetopause: solution for compressible plasmas. *Journal of Geophysical Research: Space Physics* 88: 841–852. https://doi.org/10.1029/JA088iA02p00841.

146 Mann, I.R., Wright, A.N., Mills, K.J., and Nakariakov, V.M. (1999). Excitation of magnetospheric waveguide modes by magnetosheath flows. *Journal of*

Geophysical Research: Space Physics 104: 333–354. https://doi.org/10.1029/1998JA900026.

147 Korzhov, N.P., Mishin, V.V., and Tomozov, V.M. (1984). On the role of plasma parameters and the Kelvin-Helmholtz instability in a viscous interaction of solar wind streams. *Planetary and Space Science* 32: 1169–1178. https://doi.org/10.1016/0032-0633(84)90142-9.

148 Potapov, A.S. and Guglielmi, A.V. (2011). Upward acceleration of magnetospheric ions by oscillatory centrifugal force. *Geomagnetism and Aeronomy* 51: 843–847. https://doi.org/10.1134/S001679321107019X.

149 Turkakin, H., Mann, I.R., and Rankin, R. (2014). Kelvin-Helmholtz unstable magnetotail flow channels: deceleration and radiation of MHD waves. *Geophysical Research Letters* 41: 3691–3697. https://doi.org/10.1002/2014GL060450.

150 Leonovich, A.S. and Mazur, V.A. (1989). Resonance excitation of standing Alfvén waves in an axisymmetric magnetosphere (nonstationary oscillations). *Planetary and Space Science* 37: 1109–1116. https://doi.org/10.1016/0032-0633(89)90082-2.

151 Fujita, S., Glassmeier, K.H., and Kamide, K. (1996). MHD waves generated by the Kelvin-Helmholtz instability in a nonuniform magnetosphere. *Journal of Geophysical Research: Space Physics* 101: 27317–27326. https://doi.org/10.1029/96JA02676.

152 Mazur, V.A. and Chuiko, D.A. (2015). Azimuthal inhomogeneity in the MHD waveguide in the outer magnetosphere. *Journal of Geophysical Research: Space Physics* 120: 4641–4655. https://doi.org/10.1002/2014JA020819.

153 Leonovich, A.S. (2011). A theory of MHD instability of an inhomogeneous plasma jet. *Journal of Plasma Physics* 77: 315–337. https://doi.org/10.1017/S0022377810000346.

154 Mazur, V.A. and Chuiko, D.A. (2013). Influence of the outer-magnetospheric magnetohydrodynamic waveguide on the reflection of hydromagnetic waves from a shear flow at the magnetopause. *Plasma Physics Reports* 39: 959–975. https://doi.org/10.1134/S1063780X13120064.

155 Leonovich, A.S. (2011). MHD-instability of the magnetotail: global modes. *Planetary and Space Science* 59: 402–411. https://doi.org/10.1016/j.pss.2011.01.006.

156 Mazur, V.A. and Chuiko, D.A. (2011). Excitation of a magnetospheric MHD cavity by Kelvin-Helmholtz instability. *Plasma Physics Reports* 37: 913–934. https://doi.org/10.1134/S1063780X11090121.

157 Leonovich, A.S., Kozlov, D.A., and Zong, Q. (2018). Stability of plasma cylinder with current in a helical plasma flow. *Journal of Plasma Physics* 84 (2): 905840203 https://doi.org/10.1017/S0022377818000235.

158 Leonovich, A.S. and Kozlov, D.A. (2018). Kelvin-Helmholtz instability in the geotail low-latitude boundary layer. *Journal of Geophysical Research: Space Physics* 123: 6548–6561. https://doi.org/10.1029/2018JA025552.

159 Leonovich, A.S. and Mazur, V.A. (2005). *Letter to the Editor* Why do ultra-low-frequency MHD oscillations with a discrete spectrum exist in the magnetosphere? *Annales Geophysicae* 23: 1075–1079. https://doi.org/10.5194/angeo-23-1075-2005.

160 Mazur, V.A. and Leonovich, A.S. (2006). ULF hydromagnetic oscillations with the discrete spectrum as eigenmodes of MHD-resonator in the near-Earth part of the plasma sheet. *Annales Geophysicae* 24: 1639–1648. https://doi.org/10.5194/angeo-24-1639-2006.

161 Sergeev, V.A. and Tsyganenko, N.A. (1980). *The Earth's Magnetosphere*. Moscow: Nauka (in Russian).

162 Borovsky, J.E., Thomsen, M.F., Elphic, R.C. et al. (1998). The transport of plasma sheet material from the distant tail to geosynchronous orbit. *Journal of Geophysical Research: Space Physics* 103: 20297–20332. https://doi.org/10.1029/97JA03144.

163 Eastman, T.E. and Hones, E.W. Jr. (1979). Characteristics of the magnetospheric boundary layer and magnetopause layer as observed by Imp 6. *Journal of Geophysical Research: Space Physics* 84: 2019–2028. https://doi.org/10.1029/JA084iA05p02019.

164 Wang, C.P., Lyons, L.R., Weygand, J.M. et al. (2006). Equatorial distributions of the plasma sheet ions, their electric and magnetic drifts, and magnetic fields under different interplanetary magnetic field B_z conditions. *Journal of Geophysical Research: Space Physics* 111: A04215 https://doi.org/10.1029/2005JA011545.

165 Tsyganenko, N.A. and Mukai, T. (2003). Tail plasma sheet models derived from Geotail particle data. *Journal of Geophysical Research: Space Physics* 108: 1136. https://doi.org/10.1029/2002JA009707.

166 Kivelson, M.G. and Chen, S.H. (1995). The magnetopause: surface waves and instabilities and their possible dynamical consequences. In: *Physics of the Magnetopause*, Geophysical Monograph Series, vol. 90 (ed. P. Song, B.U. Sonnerup, and M.F. Thomsen), 257–268. Washington, DC: American Geophysical Union. https://doi.org/10.1029/GM090p0257.

167 Owen, C., Taylor, M., Krauklis, I. et al. (2004). Cluster observations of surface waves on the dawn flank magnetopause. *Annales Geophysicae* 22: 971–983. https://doi.org/10.5194/angeo-22-971-2004.

168 Gnavi, G., Gratton, F.T., Farrugia, C.J., and Bilbao, L.E. (2009). Supersonic mixing layers: stability of magnetospheric flanks models. *Journal of Physics Conference Series* 166: 012022 https://doi.org/10.1088/1742-6596/166/1/012022.

169 Hwang, K.J., Kuznetsova, M.M., Sahraoui, F. et al. (2011). Kelvin-Helmholtz waves under southward interplanetary magnetic field. *Journal of Geophysical Research: Space Physics* 116: A08210 https://doi.org/10.1029/2011JA016596.

170 Marin, J., Pilipenko, V., Kozyreva, O. et al. (2014). Global Pc5 pulsations during strong magnetic storms: excitation mechanisms and equatorward expansion. *Annales Geophysicae* 32: 319–331. https://doi.org/10.5194/angeo-32-319-2014.

171 Baumjohann, W., Junginger, H., Haerendel, G., and Bauer, O.H. (1984). Resonant Alfven waves excited by a sudden impulse. *Journal of Geophysical Research: Space Physics* 89: 2765–2769. https://doi.org/10.1029/JA089iA05p02765.

172 Southwood, D.J. and Kivelson, M.G. (1990). The magnetohydrodynamic response of the magnetospheric cavity to changes in solar wind pressure. *Journal of Geophysical Research: Space Physics* 95: 2301–2309. https://doi.org/10.1029/JA095iA03p02301.

173 Korotova, G.I. and Sibeck, D.G. (1994). Generation of ULF magnetic pulsations in response to sudden variations in solar wind dynamic pressure. In: *Solar Wind Sources of Magnetospheric Ultra-Low-Frequency Waves*, Geophysical Monograph Series (ed. M.J. Engebreston, K. Takahashi, and M. Scholer), 265–271. Washington, DC: American Geophysical Union https://doi.org/10.1029/GM081p0265.

174 Archer, M.O., Horbury, T.S., Eastwood, J.P. et al. (2013). Magnetospheric response to magnetosheath pressure pulses: a low-pass filter effect. *Journal of Geophysical Research: Space Physics* 118 (9): 5454–5466. https://doi.org/10.1002/jgra.50519.

175 Klibanova, Y.Y., Mishin, V.V., and Tsegmed, B. (2014). Specific features of daytime long-period pulsations observed during the solar wind impulse against a background of the substorm of August 1, 1998. *Cosmic Research* 52 (6): 421–429. https://doi.org/10.1134/S0010952514040054.

176 Zhu, X. and Kivelson, M.G. (1988). Analytic formulation and quantitative solutions of the coupled ULF wave problem. *Journal of Geophysical Research: Space Physics* 93 (A8): 8602–8612. https://doi.org/10.1029/JA093iA08p08602.

177 Lee, D.H. and Lysak, R.L. (1989). Magnetospheric ULF wave coupling in the dipole model – the impulsive excitation. *Journal of Geophysical Research: Space Physics* 94: 17097–17103. https://doi.org/10.1029/JA094iA12p17097.

178 Dmitrienko, I.S. (2013). Evolution of FMS and Alfvén waves produced by the initial disturbance in the FMS waveguide. *Journal of Plasma Physics* 79: 7–17. https://doi.org/10.1017/S0022377812000608.

179 Hartinger, M.D., Welling, D., Viall, N.M. et al. (2014). The effect of magnetopause motion on fast mode resonance. *Journal of Geophysical Research: Space Physics* 119 (10): 8212–8227. https://doi.org/10.1002/2014JA020401.

180 Mishin, V.V., Klibanova, Y.Y., and Tsegmed, B. (2013). Solar wind inhomogeneity front inclination effect on properties of front-caused long-period geomagnetic pulsations. *Cosmic Research* 51 (2): 96–107. https://doi.org/10.1134/S0010952513020020.

181 Morozov, A.I. (1961). Cerenkov generation of magneto-acoustic waves. In: *Plasma Physics and the Problem of Controlled Thermonuclear Reactions*, vol. 4 (ed. M.A. Leontovich), 392. New York: Pergamon Press.

182 Klimushkin, D.Y. and Mager, P.N. (2021). Cherenkov radiation of the fast magnetoacoustic waves in the non-uniform magnetospheric plasma. *Physics of Plasmas* 28 (2): 022901 https://doi.org/10.1063/5.0035904.

183 Glassmeier, K.H. (1992). Traveling magnetospheric convection twin-vortices – observations and theory. *Annales Geophysicae* 10: 547–565.

184 Hughes, W.J. and Southwood, D.J. (1976). The screening of micropulsation signals by the atmosphere and ionosphere. *Journal of Geophysical Research* 81: 3234–3240. https://doi.org/10.1029/JA081i019p03234.

185 Hughes, W.J. and Southwood, D.J. (1976). An illustration of modification of geomagnetic pulsation structure by the ionosphere. *Journal of Geophysical Research* 81: 3241–3247. https://doi.org/10.1029/JA081i019p03241.

186 Newton, R.S., Southwood, D.J., and Hughes, W.J. (1978). Damping of geomagnetic pulsations by the ionosphere. *Planetary and Space Science* 26: 201–209. https://doi.org/10.1016/0032-0633(78)90085-5.

187 Greifinger, C. and Greifinger, P.S. (1968). Theory of hydromagnetic propagation in the ionospheric waveguide. *Journal of Geophysical Research* 73: 7473–7490. https://doi.org/10.1029/JA073i023p07473.

188 Greifinger, C. (1972). Feasibility of ground-based generation of artificial micropulsations. *Journal of Geophysical Research* 77: 6761–6773. https://doi.org/10.1029/JA077i034p06761.

189 Rudenko, G.V., Churilov, S.M., and Shukhman, I.G. (1985). Excitation of the ionospheric waveguide by a localized packet of Alfvén waves. *Planetary and Space Science* 33: 1103–1108. https://doi.org/10.1016/0032-0633(85)90068-6.

190 Fujita, S. and Tamao, T. (1988). Duct propagation of hydromagnetic waves in the upper ionosphere. I – Electromagnetic field disturbances in high latitudes associated with

localized incidence of a shear Alfven wave. *Journal of Geophysical Research: Space Physics* 93: 14665–14673. https://doi.org/10.1029/JA093iA12p14665.

191 Fujita, S. (1988). Duct propagation of hydromagnetic waves in the upper ionosphere. II – Dispersion characteristics and loss mechanism. *Journal of Geophysical Research: Space Physics* 93: 14674–14682. https://doi.org/10.1029/JA093iA12p14674.

192 Sorokin, V.M. and Fedorovich, G.V. (1982). Propagation of short-period waves in the ionosphere. *Radiofizika* 25: 495–501.

193 Mazur, V.A. (1987). A waveguide for low-frequency whistlers in the lower ionosphere and its relevance to the magnetospheric resonator. *Geomagnetism and Aeronomy* 27 (1): 147–148.

194 Poliakov, S.V. and Rapoport, V.O. (1981). The ionospheric Alfven resonator. *Geomagnetism and Aeronomy* 21: 816–822.

195 Beliaev, P.P., Poliakov, S.V., Rapoport, V.O., and Trakhtengerts, V.I. (1990). The ionospheric Alfven resonator. *Geomagnetism and Aeronomy* 52: 781–788.

196 Alperovich, L.S. and Fedorov, E.N. (1981). Generation of low-frequency electromagnetic oscillations by the field of a high-power radio wave in the ionosphere. *Radiofizika* 24: 276–292.

197 Hughes, W.J. (1974). The effect of the atmosphere and ionosphere on long period magnetospheric micropulsations. *Planetary and Space Science* 22: 1157–1172. https://doi.org/10.1016/0032-0633(74)90001-4.

198 Alperovich, L.S. and Fedorov, E.N. (1984). Effect of the ionosphere on the propagation of MHD wave beams. *Radiofizika* 27: 1238–1247.

199 Maltsev, I.P., Leontev, S.V., and Liatskii, V.B. (1974). Pi-2 pulsations as a result of evolution of an Alfven impulse originating in the ionosphere during a brightening of aurora. *Planetary and Space Science* 22: 1519–1533. https://doi.org/10.1016/0032-0633(74)90017-8.

200 Ellis, P. and Southwood, D.J. (1983). Reflection of Alfven waves by nonuniform ionospheres. *Planetary and Space Science* 31: 107–117. https://doi.org/10.1016/0032-0633(83)90035-1.

201 Poliakov, S.V. (1988). Reflection of Alfven waves from the horizontally irregular ionosphere. *Geomagnetism and Aeronomy* 28: 592–597.

202 Yarker, J.M. and Southwood, D.J. (1986). The effect of non-uniform ionospheric conductivity on standing magnetospheric Alfven waves. *Planetary and Space Science* 34: 1213–1221. https://doi.org/10.1016/0032-0633(86)90058-9.

203 Glassmeier, K.H. (1983). Reflection of MHD-waves in the PC4-5 period range at ionospheres with non-uniform conductivity distributions. *Geophysical Research Letters* 10: 678–681. https://doi.org/10.1029/GL010i008p00678.

204 Glassmeier, K.H. (1984). On the influence of ionospheres with non-uniform conductivity distribution on hydromagnetic waves. *Journal of Geophysics* 54: 125–137.

205 Leonovich, A.S. and Mazur, V.A. (1991). An electromagnetic field, induced in the ionosphere and atmosphere and on the Earth's surface by low-frequency Alfven oscillations of the magnetosphere: general theory. *Planetary and Space Science* 39: 529–546. https://doi.org/10.1016/0032-0633(91)90048-F.

206 Leonovich, A.S. and Mazur, V.A. (1991). An electromagnetic field, induced on the Earth's surface by standing Alfven waves in the magnetosphere: the particular kinds of oscillations. *Planetary and Space Science* 39: 547–556. https://doi.org/10.1016/0032-0633(91)90049-G.

207 Crowley, G., Hughes, W.J., and Jones, T.B. (1987). Observational evidence of cavity modes in the Earth's magnetosphere. *Journal of Geophysical Research: Space Physics* 92: 12233–12240. https://doi.org/10.1029/JA092iA11p12233.

208 Allan, W. and Poulter, E.M. (1989). Damping of magnetospheric cavity modes. *Journal of Geophysical Research: Space Physics* 94: 11843–11853. https://doi.org/10.1029/JA094iA09p11843.

209 Yumoto, K. (1988). External and internal sources of low-frequency MHD waves in the magnetosphere – a review. *Journal of Geomagnetism and Geoelectricity* 40: 293–311.

210 Pilipenko, V.A., Parrot, M., Fedorov, E.N., and Mazur, N.G. (2019). Electromagnetic field in the upper ionosphere from ELF ground-based transmitter. *Journal of Geophysical Research: Space Physics* 124 (10): 8066–8080. https://doi.org/10.1029/2019JA026929.

211 Fedorov, E.N., Mazur, N.G., and Pilipenko, V.A. (2021). Electromagnetic response of the mid-latitude ionosphere to power transmission lines. *Journal of Geophysical Research: Space Physics* 126 (10): e29659 https://doi.org/10.1029/2021JA029659.

212 Liatskii, V.B. and Maltsev, I.P. (1983). *The Magnetosphere-Ionosphere Interaction*. Moscow: Nauka (in Russian).

213 Alperovich, L.S. and Fedorov, E.N. (2007). *Hydromagnetic Waves in the Magnetosphere and the Ionosphere*, Astrophysics and Space Science Library is a Series, Springer Science+Busines Media B.V. New York: Springer.

214 Mazur, V.A. (1988). The propagation of a low-frequency whistler in the ionosphere. *Radiophysics and Quantum Electronics* 31: 1423–1430. https://doi.org/10.1007/BF01044811.

215 Dungey, J.W. (1967). Hydromagnetic waves. In: *Physics of Geomagnetic Phenomena* (ed. S. Matsushita and W.H. Campbell), 913. New York: Academic Press.

216 Radoski, H.R. (1967). Highly asymmetric MHD resonances: the guided poloidal mode. *Journal of Geophysical Research* 72: 4026–4027. https://doi.org/10.1029/JZ072i015p04026.

217 Krylov, A.L., Lifshits, A.E., and Fedorov, E.N. (1981). Resonance properties of the magnetosphere. *Akademiia Nauk SSSR, Izvestiia, Fizika Zemli* 49–58.

218 Krylov, A.L. and Lifshits, A.E. (1984). Quasi-Alfven oscillations of magnetic surfaces. *Planetary and Space Science* 32: 481–492. https://doi.org/10.1016/0032-0633(84)90127-2.

219 Allan, W., White, S.P., and Poulter, E.M. (1986). Impulse-excited hydromagnetic cavity and field-line resonances in the magnetosphere. *Planetary and Space Science* 34: 371–385. https://doi.org/10.1016/0032-0633(86)90144-3.

220 Allan, W., Poulter, E.M., and White, S.P. (1987). Structure of magnetospheric MHD resonances for moderate 'azimuthal' asymmetry. *Planetary and Space Science* 35: 1193–1198. https://doi.org/10.1016/0032-0633(87)90025-0.

221 Leonovich, A.S. and Mazur, V.A. (1989). Alfven resonance in an axially symmetric magnetosphere. *Fizika Plazmy* 15: 660–673.

222 Wright, A.N. (1992). Coupling of fast and Alfven modes in realistic magnetospheric geometries. *Journal of Geophysical Research: Space Physics* 97: 6429–6438. https://doi.org/10.1029/91JA02655.

223 Lifshits, A.E. and Fedorov, E.N. (1986). Hydromagnetic oscillations of the magnetospheric-ionospheric resonator. *Akademiia Nauk SSSR Doklady* 287: 90–94.

224 Leonovich, A.S. and Mazur, V.A. (1989). Resonance excitation of standing Alfvén waves in an axisymmetric magnetosphere (monochromatic oscillations). *Planetary and Space Science* 37: 1095–1116. https://doi.org/10.1016/0032-0633(89)90081-0.

225 Chen, L. and Cowley, S.C. (1989). On field line resonances of hydromagnetic Alfven waves in dipole magnetic field. *Geophysical Research Letters* 16: 895–897. https://doi.org/10.1029/GL016i008p00895.

226 Lee, D.H. and Lysak, R.L. (1991). Monochromatic ULF wave excitation in the dipole magnetosphere. *Journal of Geophysical Research: Space Physics* 96: 5811–5817. https://doi.org/10.1029/90JA01592.

227 Lee, D.H. (1996). Dynamics of MHD wave propagation in the low-latitude magnetosphere. *Journal of Geophysical Research: Space Physics* 101: 15371–15386. https://doi.org/10.1029/96JA00608.

228 Yumoto, K., Saito, T., Akasofu, S.I. et al. (1985). Propagation mechanism of daytime Pc 3–4 pulsations observed at synchronous orbit and multiple ground-based stations. *Journal of Geophysical Research: Space Physics* 90: 6439–6450. https://doi.org/10.1029/JA090iA07p06439.

229 Cheng, C.Z. and Lin, C.S. (1987). Eigenmode analysis of compressional waves in the magnetosphere. *Geophysical Research Letters* 14: 884–887. https://doi.org/10.1029/GL014i008p00884.

230 Engebretson, M.J., Zanetti, L.J., Potemra, T.A. et al. (1987). Simultaneous observation of Pc 3–4 pulsations in the solar wind and in the Earth's magnetosphere. *Journal of Geophysical Research: Space Physics* 92: 10053–10062. https://doi.org/10.1029/JA092iA09p10053.

231 Fedorov, E., Mazur, N., Pilipenko, V., and Yumoto, K. (1998). MHD wave conversion in plasma waveguides. *Journal of Geophysical Research: Space Physics* 103: 26595–26606. https://doi.org/10.1029/98JA02578.

232 Pilipenko, V., Fedorov, E., Mazur, N. et al. (1999). Magnetohydrodynamic waveguide/resonator for Pc3 ULF pulsations at cusp latitudes. *Earth, Planets and Space* 51: 441–448.

233 Leonovich, A.S. and Mazur, V.A. (2000). Structure of magnetosonic eigenoscillations of an axisymmetric magnetosphere. *Journal of Geophysical Research: Space Physics* 105: 27707–27716. https://doi.org/10.1029/2000JA900108.

234 Leonovich, A.S. and Mazur, V.A. (2001). On the spectrum of magnetosonic eigenoscillations of an axisymmetric magnetosphere. *Journal of Geophysical Research: Space Physics* 106: 3919–3928. https://doi.org/10.1029/2000JA000228.

235 Gul'El'Mi, A.V. (1970). Annular trap for low-frequency wave in the Earth's magnetosphere. *JETP Letters (USSR) (Engl. Transl.)* 12: 25–28.

236 Gul'Yel'Mi, A.V. (1972). Magnetoacoustic channel under the dome of the plasmasphere. *Geomagnetism and Aeronomy* 12: 147–150.

237 Takahashi, K. and Anderson, B.J. (1992). Distribution of ULF energy ($f < 80$ mHz) in the inner magnetosphere – a statistical analysis of AMPTE CCE magnetic field data. *Journal of Geophysical Research: Space Physics* 97: 10751–10773. https://doi.org/10.1029/92JA00328.

238 Kivelson, M.G. and Southwood, D.J. (1986). Coupling of global magnetospheric MHD eigenmodes to field line resonances. *Journal of Geophysical Research: Space Physics* 91: 4345–4351. https://doi.org/10.1029/JA091iA04p04345.

239 Southwood, D.J. and Kivelson, M.G. (1986). The effect of parallel inhomogeneity on magnetospheric hydromagnetic wave coupling. *Journal of Geophysical Research: Space Physics* 91: 6871–6876. https://doi.org/10.1029/JA091iA06p06871.

240 Harrold, B.G. and Samson, J.C. (1992). Standing ULF modes of the magnetosphere – a theory. *Geophysical Research Letters* 19: 1811–1814. https://doi.org/10.1029/92GL01802.

241 Mazur, V.A. (2011). Resonant excitation of the magnetosphere by stochastic and unsteady hydromagnetic waves. *Plasma Physics Reports* 37: 422–432. https://doi.org/10.1134/S1063780X11040076.

242 Timofeev, A.V. (1979). The theory of Alfven oscillations in an inhomogeneous plasma. *Moscow Atomizdat* 205–232.

243 Mazur, V.A., Mikhailovskii, A.B., Frenkel, A.L., and Shukhman, I.G. (1979). Extrinsic mode instabilities. *Moscow Atomizdat* 233–255.

244 Mann, I.R., Wright, A.N., and Hood, A.W. (1997). Phase-mixing poloidal Alfvén wave polarisations. *Advances In Space Research* 20: 489–492. https://doi.org/10.1016/S0273-1177(97)00716-3.

245 Southwood, D.J. (1980). Low frequency pulsation generation by energetic particles. *Journal of Geomagnetism and Geoelectricity* 32 (Suppl. II): 75–88.

246 Chen, L. and Hasegawa, A. (1991). Kinetic theory of geomagnetic pulsations: 1. Internal excitations by energetic particles. *Journal of Geophysical Research: Space Physics* 96 (A2): 1503–1512. https://doi.org/10.1029/90JA02346.

247 Baddeley, L.J., Yeoman, T.K., Wright, D.M. et al. (2004). Statistical study of unstable particle populations in the global ring current and their relation to the generation of high *m* ULF waves. *Annales Geophysicae* 22: 4229–4241. https://doi.org/10.5194/angeo-22-4229-2004.

248 Mager, P.N. and Klimushkin, D.Y. (2008). Alfvén ship waves: high-m ULF pulsations in the magnetosphere, generated by a moving plasma inhomogeneity. *Annales Geophysicae* 26: 1653–1663. https://doi.org/10.5194/angeo-26-1653-2008.

249 Zolotukhina, N.A., Mager, P.N., and Klimushkin, D.Y. (2008). Pc5 waves generated by substorm injection: a case study. *Annales Geophysicae* 26: 2053–2059. https://doi.org/10.5194/angeo-26-2053-2008.

250 Yamakawa, T., Seki, K., Amano, T. et al. (2020). Excitation of internally driven ULF waves by the drift-bounce resonance with ring current ions based on the drift-kinetic simulation. *Journal of Geophysical Research: Space Physics* 125 (11): e28231 https://doi.org/10.1029/2020JA028231.

251 Leonovich, A.S. and Mazur, V.A. (1993). A theory of transverse small-scale standing Alfven waves in an axially symmetric magnetosphere. *Planetary and Space Science* 41: 697–717. https://doi.org/10.1016/0032-0633(93)90055-7.

252 Leonovich, A.S. and Mazur, V.A. (1995). Magnetospheric resonator for transverse-small-scale standing Alfven waves. *Planetary and Space Science* 43: 881–883. https://doi.org/10.1016/0032-0633(94)00206-7.

253 Leonovich, A.S. and Mazur, V.A. (1995). Linear transformation of the standing Alfven wave in an axisymmetric magnetosphere. *Planetary and Space Science* 43: 885–893. https://doi.org/10.1016/0032-0633(94)00207-8.

254 Takahashi, K. and McPherron, R.L. (1982). Harmonic structure of Pc 3–4 pulsations. *Journal of Geophysical Research: Space Physics* 87: 1504–1516. https://doi.org/10.1029/JA087iA03p01504.

255 Takahashi, K. and McPherron, R.L. (1984). Standing hydromagnetic oscillations in the magnetosphere. *Planetary and Space Science* 32: 1343–1359.

256 Engebretson, M.J., Zanetti, L.J., Potemra, T.A., and Acuna, M.H. (1986). Harmonically structured ULF pulsations observed by the AMPTE CCE magnetic field experiment. *Geophysical Research Letters* 13: 905–908. https://doi.org/10.1029/GL013i009p00905.

257 Mitchell, D.G., Williams, D.J., Engebretson, M.J. et al. (1990). Pc 5 pulsations in the outer dawn magnetosphere seen by ISEE 1 and 2. *Journal of Geophysical Research: Space Physics* 95: 967–975. https://doi.org/10.1029/JA095iA02p00967.

258 Wolfe, A., Venkatesan, D., Slawinski, R., and Maclennan, C.G. (1990). A conjugate area study of Pc 3 pulsations near cusp latitudes. *Journal of Geophysical Research: Space Physics* 95: 10695–10698. https://doi.org/10.1029/JA095iA07p10695.

259 Waters, C.L., Menk, F.W., Fraser, B.J., and Ostwald, P.M. (1991). Phase structure of low-latitude Pc3–4 pulsations. *Planetary and Space Science* 39: 569–577. https://doi.org/10.1016/0032-0633(91)90052-C.

260 Potapov, A.S. and Mazur, V.A. (1994). Pc3 pulsations: from the source in the upstream region to Alfven resonances in the magnetosphere. Theory and observations. In: *Solar Wind Sources of Magnetospheric Ultra-Low-Frequency Waves* (ed. M.J. Engebretson, K. Takahashi, and M. Scholer), 135–145. Washington, DC: American Geophysical Union.

261 Hasegawa, A., Tsui, K.H., and Assis, A.S. (1983). A theory of long period magnetic pulsations. III – Local field line oscillations. *Geophysical Research Letters* 10: 765–767. https://doi.org/10.1029/GL010i008p00765.

262 Klimushkin, D.Y. and Mager, P.N. (2004). The spatio-temporal structure of impulse-generated azimuthal small-scale Alfvén waves interacting with high-energy charged particles in the magnetosphere. *Annales Geophysicae* 22: 1053–1060. https://doi.org/10.5194/angeo-22-1053-2004.

263 Yumoto, K., Takahashi, K., Sakurai, T. et al. (1990). Multiple ground-based and satellite observations of global Pi 2 magnetic pulsations. *Journal of Geophysical Research: Space Physics* 95: 15175–15184. https://doi.org/10.1029/JA095iA09p15175.

264 Borisov, N.D. and Moiseev, B.S. (1989). Excitation of MHD disturbances in the ionosphere by a Rayleigh wave. *Geomagnetism and Aeronomy* 29: 614–620.

265 Liperovsky, V.A., Pokhotelov, O.A., Liperovskaya, E.V. et al. (2000). Modification of sporadic e-layers caused by seismic activity. *Surveys in Geophysics* 21 (5): 449–486. https://doi.org/10.1023/A:1006711603561.

266 Roussel-Dupré, R. and Gurevich, A.V. (1996). On runaway breakdown and upward propagating discharges. *Journal of Geophysical Research: Space Physics* 101: 2297–2312. https://doi.org/10.1029/95JA03278.

267 Karlov, V.D., Kozlov, S.I., and Tkachev, G.N. (1980). Large-scale disturbance in the ionosphere occurring during the flight of a rocket with an operating engine /Survey/. *Cosmic Research* 18: 214–224.

268 Koons, H.C. and Pongratz, M.B. (1981). Electric fields and plasma waves resulting from a barium injection experiment. *Journal of Geophysical Research: Space Physics* 86: 1437–1446. https://doi.org/10.1029/JA086iA03p01437.

269 Galperin, Y. and Hayakawa, M. (1998). On a possibility of parametric amplifier in the stratosphere-mesosphere suggested by active MASSA experiments with the AUREOL-3 satellite. *Earth, Planets and Space* 50: 827–832.

270 Pokhotelov, O.A., Pokhotelov, D.O., Gokhberg, M.B. et al. (1996). Alfvén solitons in the Earth's ionosphere and magnetosphere. *Journal of Geophysical Research: Space Physics* 101: 7913–7916. https://doi.org/10.1029/95JA03803.

271 Leonovich, A.S. (2001). Magnetospheric MHD response to a localized disturbance in the ionosphere. *Journal of Geophysical Research: Space Physics* 105: 2507–2520. https://doi.org/10.1029/1999JA900312.

272 Blanc, M., Kallenbach, R., and Erkaev, N.V. (2005). Solar system magnetospheres. *Space Science Reviews* 116: 227–298. https://doi.org/10.1007/s11214-005-1958-y.

273 Kivelson, M.G. (2007). Planetary magnetospheres. In: *Handbook of the Solar-Terrestrial Environment* (ed. Y. Kamide and A.C.L. Chian), 470. Netherlands: Springer.

274 Southwood, D.J. and Saunders, M.A. (1985). Curvature coupling of slow and Alfven MHD waves in a magnetotail field configuration. *Planetary and Space Science* 33: 127–134. https://doi.org/10.1016/0032-0633(85)90149-7.

275 Walker, A.D.M. (1987). Theory of magnetospheric standing hydromagnetic waves with large azimuthal wave number. I – Coupled magnetosonic and Alfven waves. *Journal of Geophysical Research: Space Physics* 92: 10039–10045. https://doi.org/10.1029/JA092iA09p10039.

276 Klimushkin, D.Y. (1998). Theory of azimuthally small-scale hydromagnetic waves in the axisymmetric magnetosphere with finite plasma pressure. *Annales Geophysicae* 16: 303–321. https://doi.org/10.1007/s00585-998-0303-7.

277 Cheremnykh, O.K. and Parnowski, A.S. (2006). Influence of ionospheric conductivity on the ballooning modes in the inner magnetosphere of the Earth. *Advances in Space Research* 37: 599–603. https://doi.org/10.1016/j.asr.2005.01.073.

278 Mazur, N.G., Fedorov, E.N., and Pilipenko, V.A. (2014). Longitudinal structure of ballooning MHD disturbances in a model magnetosphere. *Cosmic Research* 52: 175–184. https://doi.org/10.1134/S0010952514030071.

279 Ohtani, S., Miura, A., and Tamao, T. (1989). Coupling between Alfven and slow magnetosonic waves in an inhomogeneous finite-beta plasma. I – Coupled equations and physical mechanism. II – Eigenmode analysis of localized ballooning-interchange instability. *Planetary and Space Science* 37: 567–577. https://doi.org/10.1016/0032-0633(89)90097-4.

280 Hameiri, E., Laurence, P., and Mond, M. (1991). The ballooning instability in space plasmas. *Journal of Geophysical Research: Space Physics* 96: 1513–1526. https://doi.org/10.1029/90JA02100.

281 Liu, W.W. (1997). Physics of the explosive growth phase: ballooning instability revisited. *Journal of Geophysical Research: Space Physics* 102: 4927–4931. https://doi.org/10.1029/96JA03561.

282 Rubtsov, A.V., Mager, P.N., and Klimushkin, D.Y. (2020). Ballooning instability in the magnetospheric plasma: two-dimensional eigenmode analysis. *Journal of Geophysical Research: Space Physics* 125 (1): e27024 https://doi.org/10.1029/2019JA027024.

283 Zeleny, L.M. and Taktakishvili, A.L. (1987). Spontaneous magnetic reconnection mechanisms in plasma. *Astrophysics and Space Science* 134: 185–196. https://doi.org/10.1007/BF00636466.

284 Zelenyi, L.M., Malova, H.V., Artemyev, A.V. et al. (2011). Thin current sheets in collisionless plasma: equilibrium structure, plasma instabilities, and particle acceleration. *Plasma Physics Reports* 37: 118–160. https://doi.org/10.1134/S1063780X1102005X.

285 Golovchanskaya, I.V. and Maltsev, Y.P. (2005). On the identification of plasma sheet flapping waves observed by Cluster. *Geophysical Research Letters* 32: L02102 https://doi.org/10.1029/2004GL021552.

286 Zelenyi, L.M., Artemyev, A.V., Petrukovich, A.A. et al. (2009). Low frequency eigenmodes of thin anisotropic current sheets and Cluster observations. *Annales Geophysicae* 27: 861–868. https://doi.org/10.5194/angeo-27-861-2009.

287 Mazur, N.G., Fedorov, E.N., and Pilipenko, V.A. (2012). Dispersion relation for ballooning modes and condition of their stability in the near Earth plasma. *Geomagnetism and Aeronomy* 52: 603–612. https://doi.org/10.1134/S0016793212050118.

288 Kozlov, D.A., Mazur, N.G., Pilipenko, V.A., and Fedorov, E.N. (2014). Dispersion equation for ballooning modes in two-component plasma. *Journal of Plasma Physics* 80 (3): 379–393. https://doi.org/10.1017/S0022377813001347.

289 Leonovich, A.S. and Kozlov, D.A. (2013). On ballooning instability in current sheets. *Plasma Physics and Controlled Fusion* 55 (8): 085013 https://doi.org/10.1088/0741-3335/55/8/085013.

290 Titchmarsh, E.C. and Teichmann, T. (1958). Eigenfunction expansions associated with second-order differential equations, Part 2. *Physics Today* 11: 34. https://doi.org/10.1063/1.3062231.

291 Slater, D.W., Kleckner, E.W., Gurgiolo, C. et al. (1987). A possible energy source to power stable auroral red arcs – precipitating electrons. *Journal of Geophysical Research: Space Physics* 92: 4543–4552. https://doi.org/10.1029/JA092iA05p04543.

292 Rankin, R., Gillies, D.M., and Degeling, A.W. (2021). On the relationship between shear Alfvén waves, auroral electron acceleration, and field line resonances. *Space Science Reviews* 217 (4): 60. https://doi.org/10.1007/s11214-021-00830-x.

293 Cornwall, J.M., Coroniti, F.V., and Thorne, R.M. (1971). Unified theory of SAR arc formation at the plasmapause. *Journal of Geophysical Research* 76: 4428–4445. https://doi.org/10.1029/JA076i019p04428.

294 Archer, M.O., Southwood, D.J., Hartinger, M.D. et al. (2022). How a realistic magnetosphere alters the polarizations of surface, fast magnetosonic, and Alfvén waves. *Journal of Geophysical Research: Space Physics* 127 (2): e30032 https://doi.org/10.1029/2021JA030032.

295 Cummings, W.D., O'Sullivan, R.J., and Coleman, P.J. Jr. (1969). Standing Alfvén waves in the magnetosphere. *Journal of Geophysical Research* 74: 778–793. https://doi.org/10.1029/JA074i003p00778.

296 Vero, J., Hollo, L., Potapov, A.S., and Poliushkina, T.N. (1985). Analysis of the connections between solar wind parameters and day-side geomagnetic pulsations based on data from the observatories Nagycenk (Hungary) and Uzur (U.S.S.R.). *Journal of Atmospheric and Terrestrial Physics* 47: 557–565.

297 Southwood, D.J. (1977). Localised compressional hydromagnetic waves in the magnetospheric ring current. *Planetary and Space Science* 25: 549–554. https://doi.org/10.1016/0032-0633(77)90061-7.

298 Leonovich, A.S. (2012). Wave mechanism of the magnetospheric convection. *Planetary and Space Science* 65: 67–75. https://doi.org/10.1016/j.pss.2012.01.009.

299 Carpenter, D.L. and Anderson, R.R. (1992). An ISEE/Whistler model of equatorial electron density in the magnetosphere. *Journal of Geophysical Research: Space Physics* 97: 1097–1108. https://doi.org/10.1029/91JA01548.

300 Kozlov, D.A. and Leonovich, A.S. (2008). The structure of field line resonances in a dipole magnetosphere with moving plasma. *Annales Geophysicae* 26: 689–698. https://doi.org/10.5194/angeo-26-689-2008.

301 Afraimovich, E.L., Edemskiy, I.K., Leonovich, A.S. et al. (2009). MHD nature of night-time MSTIDs excited by the solar terminator. *Geophysical Research Letters* 36 (15): 106. https://doi.org/10.1029/2009GL039803.

302 Barnes, A. (1970). Theory of generation of bow-shock-associated hydromagnetic waves in the upstream interplanetary medium. *Cosmic Electrodynamics* 1: 90–114.

303 McClay, J.F. (1970). On the resonant modes of a cavity and the dynamical properties of micropulsations. *Planetary and Space Science* 18: 1673–1690. https://doi.org/10.1016/0032-0633(70)90002-4.

304 Ruohoniemi, J.M., Greenwald, R.A., Baker, K.B., and Samson, J.C. (1991). HF radar observations of Pc 5 field line resonances in the midnight/early morning MLT sector. *Journal of Geophysical Research: Space Physics* 96: 15697–15710. https://doi.org/10.1029/91JA00795.

305 Samson, J.C., Harrold, B.G., Ruohoniemi, J.M. et al. (1992). Field line resonances associated with MHD waveguides in the magnetosphere. *Geophysical Research Letters* 19: 441–444. https://doi.org/10.1029/92GL00116.

306 Mathie, R.A., Menk, F.W., Mann, I.R., and Orr, D. (1999). Discrete field line resonances and the Alfvén continuum in the outer magnetosphere. *Geophysical Research Letters* 26: 659–662. https://doi.org/10.1029/1999GL900104.

307 Wanliss, J.A., Rankin, R., Samson, J.C., and Tikhonchuk, V.T. (2002). Field line resonances in a stretched magnetotail: CANOPUS optical and magnetometer observations. *Journal of Geophysical Research: Space Physics* 107: 1100. https://doi.org/10.1029/2001JA000257.

308 Hartinger, M., Angelopoulos, V., Moldwin, M.B. et al. (2012). Observations of a Pc5 global (cavity/waveguide) mode outside the plasmasphere by THEMIS. *Journal of Geophysical Research: Space Physics* 117 (A6): A06202 https://doi.org/10.1029/2011JA017266.

309 Wright, A.N. and Elsden, T. (2020). Simulations of MHD wave propagation and coupling in a 3-D magnetosphere. *Journal of Geophysical Research: Space Physics* 125 (2): e27589 https://doi.org/10.1029/2019JA027589.

310 Zhu, X. and Kivelson, M.G. (1989). Global mode ULF pulsations in a magnetosphere with a nonmonotonic Alfven velocity profile. *Journal of Geophysical Research: Space Physics* 94: 1479–1485. https://doi.org/10.1029/JA094iA02p01479.

311 Siscoe, G.L. (1969). Resonant compressional waves in the geomagnetic tail. *Journal of Geophysical Research* 74: 6482–6486. https://doi.org/10.1029/JA074i026p06482.

312 Sutcliffe, P.R. and Yumoto, K. (1989). Dayside Pi 2 pulsations at low latitudes. *Geophysical Research Letters* 16: 887–890. https://doi.org/10.1029/GL016i008p00887.

313 Yeoman, T.K. and Orr, D. (1989). Phase and spectral power of mid-latitude Pi2 pulsations – evidence for a plasmaspheric cavity resonance. *Planetary and Space Science* 37: 1367–1383. https://doi.org/10.1016/0032-0633(89)90107-4.

314 Lee, D.H. (1998). On the generation mechanism of Pi 2 pulsations in the magnetosphere. *Geophysical Research Letters* 25: 583–586. https://doi.org/10.1029/98GL50239.

315 Madelung, E. (1964). *Die mathematischen Hilfsmittel des Physikers*, Die Grundlehren der mathematischen Wissenschaften, 7e. Berlin: Springer-Verlag.

316 Sibeck, D.G., Slavin, J.A., and Smith, E.J. (1987). *ISEE 3 Magnetopause Crossings – Evidence for the Kelvin-Helmholtz Instability. Magnetotail Physics*, 73–76. Baltimore, MD: Johns Hopkins University Press.

317 Agapitov, O., Glassmeier, K.H., Plaschke, F. et al. (2009). Surface waves and field line resonances: a THEMIS case study. *Journal of Geophysical Research: Space Physics* 114: A00C27 https://doi.org/10.1029/2008JA013553.

318 Kepko, L., Spence, H.E., and Singer, H.J. (2002). ULF waves in the solar wind as direct drivers of magnetospheric pulsations. *Geophysical Research Letters* 29: 1197. https://doi.org/10.1029/2001GL014405.

319 Potapov, A.S., Polyushkina, T.N., and Pulyaev, V.A. (2013). Observations of ULF waves in the solar corona and in the solar wind at the Earth's orbit. *Journal of Atmospheric and Solar-Terrestrial Physics* 102: 235–242. https://doi.org/10.1016/j.jastp.2013.06.001.

320 Rankin, R., Kabin, K., and Marchand, R. (2006). Alfvénic field line resonances in arbitrary magnetic field topology. *Advances in Space Research* 38: 1720–1729. https://doi.org/10.1016/j.asr.2005.09.034.

321 Kozlov, D.A., Leonovich, A.S., and Cao, J.B. (2006). The structure of standing Alfvén waves in a dipole magnetosphere with moving plasma. *Annales Geophysicae* 24: 263–274. https://doi.org/10.5194/angeo-24-263-2006.

322 Sarris, T.E., Liu, W., Kabin, K. et al. (2009). Characterization of ULF pulsations by THEMIS. *Geophysical Research Letters* 36: L04104 https://doi.org/10.1029/2008GL036732.

323 Leonovich, A.S. and Mazur, V.A. (2008). Eigen ultra-low-frequency magnetosonic oscillations of the near plasma sheet. *Cosmic Research* 46: 327–334. https://doi.org/10.1134/S0010952508040072.

324 Klimushkin, D.Y., Mager, P.N., and Glassmeier, K.H. (2004). Toroidal and poloidal Alfvén waves with arbitrary azimuthal wave numbers in a finite pressure plasma in the Earth's magnetosphere. *Annales Geophysicae* 22: 267–288. https://doi.org/10.5194/angeo-22-267-2004.

325 Klimushkin, D.Y. and Mager, P.N. (2004). The structure of low-frequency standing Alfvén waves in the box model of the magnetosphere with magnetic field shear. *Journal of Plasma Physics* 70: 379–395. https://doi.org/10.1017/S0022377803002563.

326 Elsden, T. and Wright, A. (2020). Evolution of high-*m* poloidal Alfvén waves in a dipole magnetic field. *Journal of Geophysical Research: Space Physics* 125 (8): e28187 https://doi.org/10.1029/2020JA028187.

327 Poliakov, A.R., Potapov, A.S., and Tsegmid, B. (1992). Experimental evaluation of the resonant frequency and Q of midlatitude Alfven resonators. *Geomagnetism and Aeronomy/Geomagnetizm i Aeronomiia* 32: 156–159.

328 Mikhailovskii, A.B. and Pokhotelov, O.A. (1975). New mechanism for generation of geomagnetic pulsations by fast particles. *Soviet Technical Physics Letters (Engl. Transl.); (United States)* 1: 430–433.

329 Hasegawa, A. (1969). Drift mirror instability of the magnetosphere. *Physics of Fluids* 12: 2642–2650. https://doi.org/10.1063/1.1692407.

330 Southwood, D.J., Dungey, J.W., and Etherington, R.J. (1969). Bounce resonant interaction between pulsations and trapped particles. *Planetary and Space Science* 17: 349–361. https://doi.org/10.1016/0032-0633(69)90068-3.

331 Pokhotelov, O.A. and Pilipenko, V.A. (1976). Contribution to the theory of the drift-mirror instability of the magnetospheric plasma. *Geomagnetism and Aeronomy* 16: 504–510.

332 Karpman, V.I., Meerson, B.I., Mikhailovskii, A.B., and Pokhotelov, O.A. (1977). The effects of bounce resonances on wave growth rates in the magnetosphere. *Planetary and Space Science* 25: 573–585. https://doi.org/10.1016/0032-0633(77)90064-2.

333 Klimushkin, D.Y. (2000). The propagation of high-*m* Alfvén waves in the Earth's magnetosphere and their interaction with high-energy particles. *Journal of Geophysical Research: Space Physics* 105: 23303–23310. https://doi.org/10.1029/1999JA000396.

334 Crabtree, C., Horton, W., Wong, H.V., and van Dam, J.W. (2003). Bounce-averaged stability of compressional modes in geotail flux tubes. *Journal of Geophysical Research* 108: 1084. https://doi.org/10.1029/2002JA009555.

335 Mager, P.N., Klimushkin, D.Y., and Kostarev, D.V. (2013). Drift-compressional modes generated by inverted plasma distributions in the magnetosphere. *Journal of Geophysical Research: Space Physics* 118: 4915–4923. https://doi.org/10.1002/jgra.50471.

336 Klimushkin, D.Y., Mager, P.N., Chelpanov, M.A., and Kostarev, D.V. (2021). Interaction of the long-period ULF waves and charged particle in the magnetosphere: theory and observations (overview). *Solar-Terrestrial Physics* 7 (4): 33–66. https://doi.org/10.12737/stp-74202105.

337 Budden, K.G. (1985). *The Propagation of Radio Waves: The Theory of Radio Waves of Low Power in the Ionosphere and Magnetosphere*. Cambridge and New York: Cambridge University Press.

338 Leonovich, A.S. and Mazur, V.A. (1990). The spatial structure of poloidal Alfven oscillations of an axisymmetric magnetosphere. *Planetary and Space Science* 38: 1231–1241. https://doi.org/10.1016/0032-0633(90)90128-D.

339 Erokhin, N.S. and Moiseev, S.S. (1966). Some aspects of problems in magnetohydrodynamic instability theory which can be reduced to a differential equation with an arbitrary parameter associated with the leading derivative. *Journal of Applied Mechanics and Technical Physics* 7: 17–19. https://doi.org/10.1007/BF00916966.

340 Vetoulis, G. and Chen, L. (1994). Global structures of Alfvén-ballooning modes in magnetospheric plasmas. *Geophysical Research Letters* 21: 2091–2094. https://doi.org/10.1029/94GL01703.

341 Klimushkin, D.Y. (1998). Resonators for hydromagnetic waves in the magnetosphere. *Journal of Geophysical Research: Space Physics* 103: 2369–2375. https://doi.org/10.1029/97JA02193.

342 Denton, R.E. and Vetoulis, G. (1998). Global poloidal mode. *Journal of Geophysical Research: Space Physics* 103: 6729–6739. https://doi.org/10.1029/97JA03594.

343 Denton, R.E., Lessard, M.R., and Kistler, L.M. (2003). Radial localization of magnetospheric guided poloidal Pc4-5 waves. *Journal of Geophysical Research: Space Physics* 108: 1105. https://doi.org/10.1029/2002JA009679.

344 Yeoman, T.K., James, M., Mager, P.N., and Klimushkin, D.Y. (2012). SuperDARN observations of high-*m* ULF waves with curved phase fronts and their interpretation in terms of transverse resonator theory. *Journal of Geophysical Research: Space Physics* 117: A06231 https://doi.org/10.1029/2012JA017668.

345 Mager, P.N. and Klimushkin, D.Y. (2013). Giant pulsations as modes of a transverse Alfvénic resonator on the plasmapause. *Earth, Planets and Space* 65: 397–409. https://doi.org/10.5047/eps.2012.10.002.

346 Rytov, S.M., Kravtsov, Y.A., and Tatarskii, V.I. (1988). *Principles of Statistical Radiophysics, 2. Correlation Theory of Random Processes*. Berlin, Heidelberg: Springer-Verlag.

347 Leonovich, A.S. and Mazur, V.A. (1998). Standing Alfvén waves with $m \gg 1$ in an axisymmetric magnetosphere excited by a stochastic source. *Annales Geophysicae* 16: 900–913. https://doi.org/10.1007/s00585-998-0900-5.

348 Anderson, B.J., Engebretson, M.J., Rounds, S.P. et al. (1990). A statistical study of Pc 3-5 pulsations observed by the AMPTE/CCE magnetic fields experiment. I – Occurrence distributions. *Journal of Geophysical Research: Space Physics* 95: 10495–10523. https://doi.org/10.1029/JA095iA07p10495.

349 Engebretson, M.J., Zanetti, L.J., Potemra, T.A. et al. (1988). Observations of intense ULF pulsation activity near the geomagnetic equator during quiet times. *Journal of Geophysical Research: Space Physics* 93: 12795–12816. https://doi.org/10.1029/JA093iA11p12795.

350 Takahashi, K., McEntire, R.W., Lui, A.T.Y., and Potemra, T.A. (1990). Ion flux oscillations associated with a radially polarized transverse Pc 5 magnetic pulsation. *Journal of Geophysical Research: Space Physics* 95: 3717–3731. https://doi.org/10.1029/JA095iA04p03717.

351 Walker, A.D.M., Greenwald, R.A., Korth, A., and Kremser, G. (1982). Stare and GEOS 2 observations of a storm time Pc 5 ULF pulsation. *Journal of Geophysical Research: Space Physics* 87: 9135–9146. https://doi.org/10.1029/JA087iA11p09135.

352 Takahashi, K. (1988). Multisatellite studies of ULF waves. *Advances in Space Research* 8: 427–436. https://doi.org/10.1016/0273-1177(88)90156-1.

353 Odera, T.J., van Swol, D., Russell, C.T., and Green, C.A. (1991). PC 3,4 magnetic pulsations observed simultaneously in the magnetosphere and at multiple ground stations. *Geophysical Research Letters* 18: 1671–1674. https://doi.org/10.1029/91GL01297.

354 Taylor, J.P.H. and Walker, A.D.M. (1987). Theory of magnetospheric standing hydromagnetic waves with large azimuthal wave number. II – Eigenmodes of the magnetosonic and Alfven oscillations. *Journal of Geophysical Research: Space Physics* 92: 10046–10052. https://doi.org/10.1029/JA092iA09p10046.

355 Zong, Q.G., Zhou, X.Z., Li, X. et al. (2007). Ultralow frequency modulation of energetic particles in the dayside magnetosphere. *Geophysical Research Letters* 34 (12): L12105 https://doi.org/10.1029/2007GL029915.

356 Zong, Q.G., Leonovich, A.S., and Kozlov, D.A. (2018). Resonant Alfven waves excited by plasma tube/shock front interaction. *Physics of Plasmas* 25 (12): 122904 https://doi.org/10.1063/1.5063508.

357 Leonovich, A.S., Zong, Q.G., Kozlov, D.A., and Kotovschikov, I.P. (2021). The field of shock-generated Alfven oscillations near the plasmapause. *Journal of Geophysical Research: Space Physics* 126 (10): e2021JA029488 https://doi.org/10.1029/2021JA029488.

358 Hattingh, S.K.F. and Sutcliffe, P.R. (1987). Pc 3 pulsation eigenperiod determination at low latitudes. *Journal of Geophysical Research: Space Physics* 92: 12433–12436. https://doi.org/10.1029/JA092iA11p12433.

359 Zelenyi, L.M., Milovanov, A.V., and Zimbardo, G. (1998). Multiscale magnetic structure of the distant tail: self-consistent fractal approach. In: *New Perspectives on the Earth's Magnetotail*, Geophysical Monograph Series, vol. 105 (ed. A. Nishida, D.N. Baker, and S.W.H. Cowley), 321–339. Washington, DC: American Geophysical Union.

360 Willis, J.W. and Davis, J.R. (1976). VLF stimulation of geomagnetic micropulsations. *Journal of Geophysical Research* 81: 1420–1432. https://doi.org/10.1029/JA081i007p01420.

361 Getmantsev, G.G., Gul'El'Mi, A.V., Klain, B.I. et al. (1978). Excitation of magnetic pulsations under the action on the ionosphere of radiation of a powerful short-wave transmitter. *Radiophysics and Quantum Electronics* 20 (7): 703–705. https://doi.org/10.1007/BF01040635.

362 Buechner, J., Kuznetsova, M., and Zelenyi, L.M. (1991). Sheared field tearing mode instability and creation of flux ropes in the Earth magnetotail. *Geophysical Research Letters* 18: 385–388. https://doi.org/10.1029/91GL00235.

363 Keiling, A. (2012). Pi2 pulsations driven by ballooning instability. *Journal of Geophysical Research: Space Physics* 117: A03228 https://doi.org/10.1029/2011JA017223.

364 Mazur, N.G., Fedorov, E.N., and Pilipenko, V.A. (2013). Global stability of the ballooning mode in a cylindrical model. *Geomagnetism and Acronomy* 53: 448–456. https://doi.org/10.1134/S0016793213030110.

365 Rubtsov, A.V., Mager, P.N., and Klimushkin, D.Y. (2018). Ballooning instability of azimuthally small scale coupled Alfvén and slow magnetoacoustic modes in two-dimensionally inhomogeneous magnetospheric plasma. *Physics of Plasmas* 25 (10): 102903 https://doi.org/10.1063/1.5051474.

366 Coppi, B. (1977). Topology of ballooning modes. *Physical Review Letters* 39: 939–942. https://doi.org/10.1103/PhysRevLett.39.939.

367 Cheremnykh, O.K., Parnowski, A.S., and Burdo, O.S. (2004). Ballooning modes in the inner magnetosphere of the Earth. *Planetary and Space Science* 52: 1217–1229. https://doi.org/10.1016/j.pss.2004.07.014.

368 Roux, A., Perraut, S., Robert, P. et al. (1991). Plasma sheet instability related to the westward traveling surge. *Journal of Geophysical Research: Space Physics* 96: 17697–17714. https://doi.org/10.1029/91JA01106.

369 Kalmoni, N.M.E., Rae, I.J., Watt, C.E.J. et al. (2015). Statistical characterization of the growth and spatial scales of the substorm onset arc. *Journal of Geophysical Research: Space Physics* 120 (10): 8503–8516. https://doi.org/10.1002/2015JA021470.

370 Rae, I.J. and Watt, C.E.J. (2016). ULF waves above the nightside auroral oval during substorm onset. In: *Low-Frequency Waves in Space Plasmas*, Geophysical Monograph Series, vol. 216 (ed. A. Keiling, D.-H. Lee, and V. Nakariakov), 99–120. Washington, DC: American Geophysical Union. https://doi.org/10.1002/9781119055006.ch7.

371 Kan, J.R. (1973). On the structure of the magnetotail current sheet. *Journal of Geophysical Research* 78 (19): 3773. https://doi.org/10.1029/JA078i019p03773.

372 Erkaev, N.V., Semenov, V.S., Kubyshkin, I.V. et al. (2009). MHD model of the flapping motions in the magnetotail current sheet. *Journal of Geophysical Research: Space Physics* 114: A03206 https://doi.org/10.1029/2008JA013728.

373 Zelenyi, L.M., Sitnov, M.I., Malova, H.V., and Sharma, A.S. (2000). Thin and superthin ion current sheets. Quasi-adiabatic and nonadiabatic models. *Nonlinear Processes in Geophysics* 7: 127–139.

374 Birn, J. (2011). Magnetotail dynamics: survey of recent progress. In: *The Dynamic Magnetosphere*, IAGA special Sopron Book Series 3 (ed. W. Liu and M. Fujimoto), 49–54. New York: Springer Science+Busines Media B.V. https://doi.org/10.1007/978.94.007.0501.2.4.

375 Mager, P.N., Klimushkin, D.Y., Pilipenko, V.A., and Schäfer, S. (2009). Field-aligned structure of poloidal Alfvén waves in a finite pressure plasma. *Annales Geophysicae* 27: 3875–3882. https://doi.org/10.5194/angeo-27-3875-2009.

376 Zeleny, L.M., Malova, H.V., and Popov, V.Y. (2003). Splitting of thin current sheets in the Earth's magnetosphere. *Soviet Journal of Experimental and Theoretical Physics Letters* 78: 296–299. https://doi.org/10.1134/1.1625728.

377 Mager, P.N. and Klimushkin, D.Y. (2005). Spatial localization and azimuthal wave numbers of Alfvén waves generated by drift-bounce resonance in the magnetosphere. *Annales Geophysicae* 23: 3775–3784. https://doi.org/10.5194/angeo-23-3775-2005.

378 Walker, A.D.M. and Pekrides, H. (1996). Theory of magnetospheric standing hydromagnetic waves with large azimuthal wave number 4. Standing waves in the ring current region. *Journal of Geophysical Research: Space Physics* 101: 27133–27148. https://doi.org/10.1029/96JA02701.

379 Parnowski, A.S. (2007). Eigenmode analysis of ballooning perturbations in the inner magnetosphere of the Earth. *Annales Geophysicae* 25: 1391–1403. https://doi.org/10.5194/angeo-25-1391-2007.

380 Erkaev, N.V., Shaidurov, V.A., Semenov, V.S. et al. (2005). Peculiarities of Alfvén wave propagation along a nonuniform magnetic flux tube. *Physics of Plasmas* 12 (1): 012905 https://doi.org/10.1063/1.1833392.

381 Frieman, E. and Rotenberg, M. (1960). On hydromagnetic stability of stationary Equilibria. *Reviews of Modern Physics* 32: 898–902. https://doi.org/10.1103/RevModPhys.32.898.

382 Miura, A. (2007). A magnetospheric energy principle for hydromagnetic stability problems. *Journal of Geophysical Research: Space Physics* 112: A06234 https://doi.org/10.1029/2006JA011992.

383 Cheremnykh, O.K. (1989). Dispersion equation and stability limit for ballooning flute modes in a tokamak with circular magnetic surfaces and an arbitrary pressure profile. *Nuclear Fusion* 29 (11): 1899–1905. https://doi.org/10.1088/0029-5515/29/11/005.

384 Miura, A., Ohtani, S., and Tamao, T. (1989). Ballooning instability and structure of diamagnetic hydromagnetic waves in a model magnetosphere. *Journal of Geophysical Research: Space Physics* 94: 15231–15242. https://doi.org/10.1029/JA094iA11p15231.

385 Cheng, C.Z. and Lui, A.T.Y. (1998). Kinetic ballooning instability for substorm onset and current disruption observed by AMPTE/CCE. *Geophysical Research Letters* 25: 4091–4094. https://doi.org/10.1029/1998GL900093.

386 Pritchett, P.L. and Coroniti, F.V. (2010). A kinetic ballooning/interchange instability in the magnetotail. *Journal of Geophysical Research: Space Physics* 115: A06301 https://doi.org/10.1029/2009JA014752.

387 Klimushkin, D.Y., Mager, P.N., and Pilipenko, V.A. (2012). On the ballooning instability of the coupled Alfvén and drift compressional modes. *Earth, Planets and Space* 64: 777–781. https://doi.org/10.5047/eps.2012.04.002.

388 Mager, P.N., Berngardt, O.I., Klimushkin, D.Y. et al. (2015). First results of the high-resolution multibeam ULF wave experiment at the Ekaterinburg SuperDARN radar: ionospheric signatures of coupled poloidal Alfvén and drift-compressional modes. *Journal of Atmospheric and Solar-Terrestrial Physics* 130–131: 112–126. https://doi.org/10.1016/j.jastp.2015.05.017.

389 Mager, P.N. and Klimushkin, D.Y. (2017). Non-resonant instability of coupled Alfvén and drift compressional modes in magnetospheric plasma. *Plasma Physics and Controlled Fusion* 59 (9): 095005 https://doi.org/10.1088/1361-6587/aa790c.

390 Gold, T. (1959). Motions in the magnetosphere of the Earth. *Journal of Geophysical Research* 64 (9): 1219–1224. https://doi.org/10.1029/JZ064i009p01219.

391 Sonnerup, B.U.Ö. and Laird, M.J. (1963). On magnetospheric interchange instability. *Journal of Geophysical Research* 68 (1): 131–139. https://doi.org/10.1029/JZ068i001p00131.

392 Southwood, D.J. and Kivelson, M.G. (1987). Magnetospheric interchange instability. *Journal of Geophysical Research: Space Physics* 92: 109–116. https://doi.org/10.1029/JA092iA01p00109.

393 Mingalev, O.V., Golovchanskaya, I.V., and Maltsev, Y.P. (2006). Simulation of the interchange instability in a magnetospheric substorm site. *Annales Geophysicae* 24 (6): 1685–1693. https://doi.org/10.5194/angeo-24-1685-2006.

394 Toffoletto, F.R., Wolf, R.A., and Derr, J. (2022). Thin filaments in an average magnetosphere: pure interchange versus ballooning oscillations. *Journal of Geophysical Research: Space Physics* 127 (10): e2022JA030556 https://doi.org/10.1029/2022JA030556.

395 Inhester, B. (1986). Resonance absorption of Alfven oscillations in a nonaxisymmetric magnetosphere. *Journal of Geophysical Research: Space Physics* 91: 1509–1518. https://doi.org/10.1029/JA091iA02p01509.

396 Schulze-Berge, S., Gowley, S., and Chen, L. (1992). Theory of field line resonances of standing shear Alfven waves in three-dimensional inhomogeneous plasmas. *Frontiers in Astronomy and Space Sciences* 97: 3219–3222. https://doi.org/10.1029/91JA02804.

397 Fedorov, E.N., Mazur, N.G., Pilipenko, V.A., and Yumoto, K. (1995). On the theory of field line resonances in plasma configurations. *Physics of Plasmas* 2: 527–532. https://doi.org/10.1063/1.870977.

398 Cheng, C.Z. (2003). MHD field line resonances and global modes in three-dimensional magnetic fields. *Journal of Geophysical Research: Space Physics* 108: 1002. https://doi.org/10.1029/2002JA009470.

399 Ostwald, P.M., Menk, F.W., Fraser, B.J. et al. (1993). Spatial and temporal characteristics of 15-100 mHz ULF waves recorded across a low-latitude azimuthal array. *Annales Geophysicae* 11: 742–752.

400 Glassmeier, K.H., Buchert, S., Motschmann, U. et al. (1999). Concerning the generation of geomagnetic giant pulsations by drift-bounce resonance ring current instabilities. *Annales Geophysicae* 17: 338–350. https://doi.org/10.1007/s00585-999-0338-4.

401 Agapitov, A.V. and Cheremnykh, O.K. (2011). Polarization of ULF waves in the Earth's magnetosphere. *Kinematics and Physics of Celestial Bodies* 27: 117–123.

402 Claudepierre, S.G., Mann, I.R., Takahashi, K. et al. (2013). Van Allen Probes observation of localized drift resonance between poloidal mode ultra-low frequency waves and 60 keV electrons. *Geophysical Research Letters* 40: 4491–4497. https://doi.org/10.1002/grl.50901.

403 Schäfer, S., Glassmeier, K.H., Eriksson, P.T.I. et al. (2008). Spatio-temporal structure of a poloidal Alfvén wave detected by Cluster adjacent to the dayside plasmapause. *Annales Geophysicae* 26: 1805–1817. https://doi.org/10.5194/angeo-26-1805-2008.

404 Klimushkin, D.Y., Leonovich, A.S., and Mazur, V.A. (1995). On the propagation of transversally small-scale standing Alfven waves in a three-dimensionally inhomogeneous magnetosphere. *Journal of Geophysical Research: Space Physics* 100: 9527–9534. https://doi.org/10.1029/94JA03233.

405 Salat, A. and Tataronis, J.A. (2000). Conditions for existence of orthogonal coordinate systems oriented by magnetic field lines. *Journal of Geophysical Research: Space Physics* 105: 13055–13062. https://doi.org/10.1029/1999JA000221.

406 Arfken, G.B. (1970). *Mathematical Methods for Physicists*. New York and London: Academic Press.

407 Sternberg, S. (1999). *Lectures on Differential Geometry*. Providence, RI: American Mathematical Society.

408 Hamada, S. (1962). Hydromagnetic equilibria and their proper coordinates. *Nuclear Fusion* 2: 23–37.

409 Farlow, S.J. (1982). *Partial Differential Equations for Scientists and Engineers*. New York: Wiley.

410 Elsden, T. and Wright, A.N. (2017). The theoretical foundation of 3-D Alfvén resonances: time-dependent solutions. *Journal of Geophysical Research: Space Physics* 122 (3): 3247–3261. https://doi.org/10.1002/2016JA023811.

411 Elsden, T. and Wright, A.N. (2018). The broadband excitation of 3-D Alfvén resonances in a MHD waveguide. *Journal of Geophysical Research: Space Physics* 123 (1): 530–547. https://doi.org/10.1002/2017JA025018.

412 Wright, A., Degeling, A.W., and Elsden, T. (2022). Resonance maps for 3D Alfvén waves in a compressed dipole field. *Journal of Geophysical Research: Space Physics* 127 (4): e30294 https://doi.org/10.1029/2022JA030294.

413 Elsden, T., Wright, A., and Degeling, A. (2022). A review of the theory of 3-D Alfvén (field line) resonances. *Frontiers in Astronomy and Space Sciences* 9: 917817 https://doi.org/10.3389/fspas.2022.917817.

414 Mager, P. and Klimushkin, D. (2021). The field line resonance in the three dimensionally inhomogeneous magnetosphere: principal features. *Journal of Geophysical Research: Space Physics* 126 (1): e28455 https://doi.org/10.1029/2020JA028455.

415 Hao, Y.X., Zong, Q.G., Wang, Y.F. et al. (2014). Interactions of energetic electrons with ULF waves triggered by interplanetary shock: Van Allen Probes observations in the magnetotail. *Journal of Geophysical Research: Space Physics* 119: 8262–8273. https://doi.org/10.1002/2014JA020023.

416 Radoski, H.R. (1966). Magnetic toroidal resonances and vibrating field lines. *Journal of Geophysical Research* 71: 1891–1893. https://doi.org/10.1029/JZ071i007p01891.

417 Troitskaya, V.A. and Gul'elmi, A.V. (1967). Geomagnetic micropulsations and diagnostics of the magnetosphere. *Space Science Reviews* 7: 689–768. https://doi.org/10.1007/BF00542894.

418 Cornwall, J.M. (1965). Cyclotron instabilities and electromagnetic emission in the ultra low frequency and very low frequency ranges. *Journal of Geophysical Research* 70 (1): 61–69. https://doi.org/10.1029/JZ070i001p00061.

419 Guglielmi, A., Kangas, J., and Potapov, A. (2001). Quasiperiodic modulation of the Pc1 geomagnetic pulsations: an unsettled problem. *Journal of Geophysical Research: Space Physics* 106: 25847–25856. https://doi.org/10.1029/2001JA000136.

420 Sergeev, V.A., Pellinen, R.J., and Pulkkinen, T.I. (1996). Steady magnetospheric convection: a review of recent results. *Space Science Reviews* 75: 551–604. https://doi.org/10.1007/BF00833344.

421 Volwerk, M., Glassmeier, K.H., Nakamura, R. et al. (2007). Flow burst-induced Kelvin-Helmholtz waves in the terrestrial magnetotail. *Geophysical Research Letters* 34: L10102 https://doi.org/10.1029/2007GL029459.

422 Walker, A.D.M. (2005). Excitation of field line resonances by sources outside the magnetosphere. *Annales Geophysicae* 23: 3375–3388. https://doi.org/10.5194/angeo-23-3375-2005.

423 Plaschke, F., Glassmeier, K.H., Auster, H.U. et al. (2009). Standing Alfvén waves at the magnetopause. *Geophysical Research Letters* 36: L02104 https://doi.org/10.1029/2008GL036411.

424 Archer, M.O., Hartinger, M.D., and Horbury, T.S. (2013). Magnetospheric "magic" frequencies as magnetopause surface eigenmodes. *Geophysical Research Letters* 40: 5003–5008. https://doi.org/10.1002/grl.50979.

425 Viall, N.M., Kepko, L., and Spence, H.E. (2009). Relative occurrence rates and connection of discrete frequency oscillations in the solar wind density and dayside magnetosphere. *Journal of Geophysical Research: Space Physics* 114: A01201 https://doi.org/10.1029/2008JA013334.

426 Dai, L., Takahashi, K., Wygant, J.R. et al. (2013). Excitation of poloidal standing Alfvén waves through drift resonance wave-particle interaction. *Geophysical Research Letters* 40: 4127–4132. https://doi.org/10.1002/grl.50800.

427 Leonovich, A.S., Klimushkin, D.Y., and Mager, P.N. (2015). Experimental evidence for the existence of monochromatic transverse small-scale standing Alfvén waves with spatially dependent polarization. *Journal of Geophysical Research: Space Physics* 120: 5443–5454. https://doi.org/10.1002/2015JA021044.

428 Liu, W., Cao, J.B., Li, X. et al. (2013). Poloidal ULF wave observed in the plasmasphere boundary layer. *Journal of Geophysical Research: Space Physics* 118 (7): 4298–4307. https://doi.org/10.1002/jgra.50427.

429 Rankin, R., Fenrich, F., and Tikhonchuk, V.T. (2000). Shear Alfvén waves on stretched magnetic field lines near midnight in Earth's magnetosphere. *Geophysical Research Letters* 27: 3265–3268. https://doi.org/10.1029/2000GL000029.

430 Keiling, A. (2009). Alfvén waves and their roles in the dynamics of the Earth's magnetotail: a review. *Space Science Reviews* 142: 73–156. https://doi.org/10.1007/s11214-008-9463-8.

431 Zelenyi, L.M., Lipatov, A.S., Lominadze, D.G., and Taktakishvili, A.L. (1984). The dynamics of the energetic proton bursts in the course of the magnetic field topology reconstruction in the Earth's magnetotail. *Planetary and Space Science* 32: 313–324. https://doi.org/10.1016/0032-0633(84)90167-3.

432 Allan, W. and Wright, A.N. (1998). Hydromagnetic wave propagation and coupling in a magnetotail waveguide. *Journal of Geophysical Research: Space Physics* 103: 2359–2368. https://doi.org/10.1029/97JA02874.

433 Lee, D.H. and Lysak, R.L. (1999). MHD waves in a three-dimensional dipolar magnetic field: a search for Pi2 pulsations. *Journal of Geophysical Research: Space Physics* 104: 28691–28700. https://doi.org/10.1029/1999JA900377.

434 Lysak, R.L., Song, Y., and Jones, T.W. (2009). Propagation of Alfvén waves in the magnetotail during substorms. *Annales Geophysicae* 27: 2237–2246. https://doi.org/10.5194/angeo-27-2237-2009.

435 Leonovich, A.S., Kozlov, D.A., and Cao, J.B. (2008). Standing Alfvén waves with $m \gg 1$ in a dipole magnetosphere with moving plasma and aurorae. *Advances in Space Research* 42: 970–978. https://doi.org/10.1016/j.asr.2007.05.019.

436 Cheng, C.Z. (2004). Physics of substorm growth phase, onset, and dipolarization. *Space Science Reviews* 113: 207–270. https://doi.org/10.1023/B:SPAC.0000042943.59976.0e.

437 Saito, M.H., Miyashita, Y., Fujimoto, M. et al. (2008). Modes and characteristics of low-frequency MHD waves in the near-Earth magnetotail prior to dipolarization: fitting method. *Journal of Geophysical Research: Space Physics* 113: A06201 https://doi.org/10.1029/2007JA012778.

438 Zhu, P. and Raeder, J. (2014). Ballooning instability-induced plasmoid formation in near-Earth plasma sheet. *Journal of Geophysical Research: Space Physics* 119: 131–141. https://doi.org/10.1002/2013JA019511.

439 Cao, J.B., Wei, X.H., Duan, A.Y. et al. (2013). Slow magnetosonic waves detected in reconnection diffusion region in the Earth's magnetotail. *Journal of Geophysical Research: Space Physics* 118: 1659–1666. https://doi.org/10.1002/jgra.50246.

440 Du, J., Zhang, T.L., Nakamura, R. et al. (2011). Mode conversion between Alfvén and slow waves observed in the magnetotail by THEMIS. *Geophysical Research Letters* 38: L07101 https://doi.org/10.1029/2011GL046989.

441 Galperin, I.I., Zelenyi, L.M., and Kuznetsova, M.M. (1986). Pinching of field-aligned currents as a possible mechanism for the formation of raylike auroral forms. *Kosmicheskie Issledovaniia* 24: 865–874.

442 Saka, O., Hayashi, K., and Thomsen, M. (2014). Pre-onset auroral signatures and subsequent development of substorm auroras: a development of ionospheric loop currents at the onset latitudes. *Annales Geophysicae* 32: 1011–1023. https://doi.org/10.5194/angeo-32-1011-2014.

443 Leonovich, A.S. and Kozlov, D.A. (2014). Coupled guided modes in the magnetotails: spatial structure and ballooning instability. *Astrophysics and Space Science* 353: 9–23. https://doi.org/10.1007/s10509-014-1999-3.

444 Leonovich, A.S. and Kozlov, D.A. (2019). Kelvin-Helmholtz instability in a dipole magnetosphere: the magnetopause as a tangential discontinuity. *Journal of Geophysical Research: Space Physics* 124 (10): 7936–7953. https://doi.org/10.1029/2019JA026842.

445 Leonovich, A.S. and Kozlov, D.A. (2020). Focusing of fast magnetosonic waves in the dayside magnetosphere. *Journal of Geophysical Research: Space Physics* 125 (7): e27925 https://doi.org/10.1029/2020JA027925.

446 Marsch, E. and Tu, C.Y. (1990). Spectral and spatial evolution of compressible turbulence in the inner solar wind. *Journal of Geophysical Research: Space Physics* 95: 11945–11956. https://doi.org/10.1029/JA095iA08p11945.

Index

a

Airy equation 18, 34, 41
Alfvén resonance 10, 12, 17, 41, 133, 137
  back influence of Alfvén wave to FMS field 146, 153
  in a dipole magnetosphere 149
  energy dissipation in the resonance region 19
  resonance point 13, 16, 87, 89
Alfvén waves
  absorption in the ionosphere 213
  boundary conditions at the ionosphere 120, 128
  kinetic 41
  m≫1
    excitation by a localised monochromatic source 275
    excitation by a localised pulse source 282
    excitation by a stochastic source 241
    excitation by HF radar 280
    pulse excitation 253
    transverse spatial structure 215, 222
  m∼1
    excitation by broadband sources 158
    excitation by FMS wave packet 161
    excitation by pulse sources 21, 160
    excitation by stochastic sources 162
  oscillation modes
    poloidal 203
    toroidal, m≫1 203
    toroidal, m∼1 141
  oscillations field on the ground 130, 226
  polarisation spectral splitting 204
  transverse dispersion 7, 37
    due to kinetic effects 349
    due to the field line curvature 234
approximation
  of deep potential well 47
  of geometrical optics xlii
  of ideal plasma 16
  local 301
  of shallow potential well 47
  of tangential discontinuity 63
  of thin layer 231

b

ballooning instability 316
  growth rate calculation
    in a local approximation 306
    in WKB approximation 309
Bohr-Sommerfeld quantisation condition 239

c

causality principle 74
characteristics (constant phase lines), for Alfvén waves with m≫1 on the coordinate surfaces x^3=const 333
coordinate surfaces 331
coordinate system
  Cartesian
    for box model 12
    for tangential discontinuity near the magnetopause 63
  curvilinear, field line oriented
    dipole-like 138
    for near-Earth's geotail 299

Magnetospheric MHD Oscillations: A Linear Theory, First Edition. Anatoly Leonovich, Vitalii Mazur, and Dmitri Klimushkin.

coordinate system (*contd.*)
for 3D-inhomogeneous magnetosphere 330, 365
cylindrical, for geotail 99
coupled MHD oscillations 315
slow magnetosonic and Alfvén waves 299
ballooning instability 316
linear transformation in the geotail current sheet 320

d

damping
collisional 4
collisionless 5
dispersion equation
for Alfvén waves 2, 7, 8
for FMS waves 8
in a shear flow 91
for magnetosonic waves 2

e

equations
for Alfvén waves with m≫1 331
for ballooning modes 301
of ideal MHD 1
linearized 1
for MHD-oscillations
in a dipole magnetosphere with a cold plasma 148
in a non-ideal plasma 39
in a 1D-inhomogeneous plasma 15, 16

f

fast magnetosonic mode 2
Cherenkov emission 111, 115
decay in the Alfvén resonance region 46
energy flux 67
reflection coefficient from magnetopause in a dipole magnetosphere model 191
reflection from inhomogeneity 13
spatial structure in a dipole magnetosphere 151, 178
transparent regions in a dipole magnetosphere model 182
turning point 13, 17, 153
wave packet 21
field line resonance (Alfvén resonance) 10
Friedrichs (polar) diagrams 2

g

geomagnetic pulsations xxxix
classification xxxix
magic frequencies 185
Pc1, high-latitude 62
Pc3, dayside 162
Pc3, high-latitude 62
Pi2 and SSC oscillations 251
Pi2, substorm 161, 193
group velocity xlii
of Alfvén waves 2, 8
of FMS waves 2
of SMS waves 2
transverse, of Alfvén waves with m≫1 205

h

Hamilton equations xlii
hodograph
for oscillations in the Alfvén resonance region 18, 35
for oscillations in the magnetosonic resonance region 35

i

instability
of global modes, in a cylindrical geotail model 95
growth rate xlii
of MHD-oscillations xlii

k

Kelvin-Helmholtz instability 63, 70, 83

l

linear oscillation theory xlii
linear transformation
of Alfvén and slow magnetosonic waves in the geotail current sheet 320
of large-scale Alfvén wave to kinetic Alfvén wave 232
low-latitude boundary layer (LLBL) 98

m

Mach cone 111, 115
magic frequencies 185
magnetohydrodynamics (MHD) xxxix
magnetosonic resonance 10, 87
  in a dipole magnetosphere 164
  resonance point 31, 32, 87, 89
MASSA experiment 292, 294
method
  contour integral 321
  Fourier transform 267
  Green's function 50
  Laplace transform 235
  multiple-scale 316
  saddle point 228–230, 237, 254, 276, 286, 322
metric tensor 165
MHD modes
  guided 2
  isotropic 2
  phase shift between their field components 3
MHD waves xxxix
  decomposition into potential and vortical components 38
model
  of Alfvén speed distribution 188
  of Alfvén wave generation in MASSA experiment 294
  of geotail 194
    with a current sheet 302
    cylindrical 69, 84, 88, 98, 99
  of ionosphere
    with inclined geomagnetic field 120
    as a thin layer 117
  of magnetopause in form of tangential discontinuity 68
  of magnetosphere xl
    box model 12
    dipole 138, 147
    parabolic 196
    with rotating plasma 164
    3D-inhomogeneous 330
  of plasma filament, cylindrical 56
  of shear flow 71
    in form of transition layer 94
  of transition layer
    in a cold plasma 15
    in a warm plasma 31
  of white noise source for stochastic oscillations 245
model equation, for transverse structure of Alfvén waves with $m \gg 1$ 260

n

near-Earth plasma sheet (NEPS) 189

o

opaque region 13

p

plasma equilibrium condition 30, 304
plasma parameter β 2, xl
Poynting vector 26

r

residue theorem 286
resonance surface
  poloidal 332
  toroidal 332
resonator
  for FMS waves
    in the 'box model' 185
    dipole model, outer magnetosphere 191
    dipole model, plasmasphere 192
    in the near-Earth plasma sheet 193, 196
  for poloidal Alfvén waves near the plasmapause 239

s

Schrödinger equation 44, 57
shear flow 70
  stability 70, 83
    smooth transition layer 94
    tangential discontinuity 88
singular point, bypassing rule 15
slow magnetosonic mode 2
  boundary conditions at the ionosphere 120
stochastic oscillations
  pair correlators 245
  polarisation ellipse 357

stochastic oscillations (*contd.*)
parameters, for Alfvén oscillations with $m \gg 1$ 248
spectral function 351
white noise source 245

t

tangential discontinuity, matching conditions 66, 90
tensor
of dielectric permeability 7, 25, 38
of plasma conductivity 26, 118
transparent region 13
for Alfvén waves with $m \gg 1$ 205
for fast magnetosonic mode in a dipole magnetosphere model 182
turning point
ordinary 13, 89, 205, 239
singular 205, 222, 239, 321

w

wave frequency 2
Doppler-shifted 16
local xlii
waveguide 43
for FMS waves 44, 52
for quasilongitudinal Alfvén waves 47
cylindrical 56
for quasi-transverse Alfvén waves 49, 51
cylindrical 58
WKB approximation 17, 21, 142, 152, 178, 179, 201, 208